Also from Wiley. . .

FLUIDS SOLUTIONS POWERED BY JUSTASK!

A website with answers

FLUIDS Solutions innvites you to be a part of the solution as it walks you step-by-step through over 100 end-of-chapter problems and 15 interactive applications from the text. This powerful online problem-solving tool takes select end-of-chapter problems and provides you with more than just the answers. FLUIDS Solutions is your passport to greater understanding of key concepts. Save time in your studies while building valuable analysis skills.

Wherever you see FLUIDS/SOLUTIONS in front of an end-of-chapter problem, you know that the FLUIDS Solutions will provide you with your choice of:

- Step-by-Step, detailed solutions and answers
- Detailed hints if you choose to solve the problems yourseld
- Convenient pop up windows that highlight relevant concept, background theory methods, and laws that should be applied when solving each problem
- Solution guidelines that illustrate the steps you need to take in order to solve the problem, while allowing you to solve the problem on your own
- A complete glossary of key terms and definitions.

Find out more about FLUIDS Solutions at

www.wiley.com/college/crowe

Fluid Mechanics like you have never seen it!

Engineering Fluid Mechanics

Eighth Edition

Clayton T. Crowe
Washington State University, Pullman

Donald F. Elger
University of Idaho, Moscow

John A. Roberson
Washington State University, Pullman

John Wiley & Sons, Inc.

ACQUISITIONS EDITOR Jennifer Welter
MARKETING MANAGER Frank Lyman
PRODUCTION SERVICES MANAGER Jeanine Furino
SENIOR PRODUCTION EDITOR Sandra Dumas
SENIOR DESIGNER Karin Kincheloe
COVER DESIGNER Carole C. Grobe
SENIOR PHOTO EDITOR Lisa Gee
PRODUCTION MANAGEMENT SERVICES Publication Services
COVER PHOTOGRAPH Photo Disc, Inc. / Getty Images

This book was set in Times Roman by Publication Services and printed and bound by Donnelley/Willard. The cover was printed by Phoenix Color.

This book is printed on acid-free paper. ♾

Crowe, C. T. (Clayton T.)
Engineering fluid mechanics /Clayton T. Crowe, Don Elger, John A. Roberson.--8th ed.

ISBN 0-471-48737-6

ISBN 0-471-66161-9 (Wiley International Edition)

Printed in the United States of America

10 9 8 7 6 5 4 3 2 1

Dedicated to our gracious wives,
Linda L., Linda M., and Amy

and to our loving children.

Contents

Preface

Audience

This book is written for engineering students taking a sophomore- or junior-level course in fluid mechanics. Students should have background knowledge in statics and calculus.

This text is best suited to a professor who has an instructional philosophy aligned with helping students develop skills and knowledge for engineering practice.

Approach

We want students to develop an awareness of the powerful appeal of fluid mechanics. To this end, the text uses photos, line drawings, and many problems that idealize real-world situations.

We want students to learn with meaning. Meaningful learning occurs when people link new knowledge to their everyday world and to concepts that they have previously learned. Information in this text is linked with content from mechanics courses, and we introduce new knowledge using simple verbal descriptions, while keeping the level of mathematics matched to the level of students.

Organization

Chapters 1 to 11 plus 13 are devoted to foundational concepts of fluid mechanics that support applications in many contexts. Relevant content includes fluid properties; forces and pressure variations in static fluids; qualitative descriptions of flow and pressure variations; the Bernoulli equation; the control volume concept; control volume equations for mass, momentum, and energy; dimensional analysis; head loss in conduits; measurements; drag force; and lift force.

Chapters 12, 14, and 15 are devoted to compressible flow, turbomachinery, and open-channel flow, respectively. These topics are optional for a first course in fluid mechanics.

New Features

Significant improvements have been made to this 8th edition. The most important changes are described next. As noted, a number of these changes represent trends that will be further developed in future editions of this text.

- **Control volume approach.** Chapters 4 and 5 of the 7th edition were rewritten so that the control volume approach is now presented in a sequence of three chapters. In particular, Chapter 5 in this edition presents the control volume approach plus the continuity principle. Chapter 6 presents the momentum principle, and Chapter 7 presents the energy principle.

- **Pressure variation**. Chapters 4 and 5 of the 7th edition were rewritten. Information about flow patterns, acceleration, Euler's equation, the Bernoulli equation, and rotation is now presented in Chapter 4 of this edition.
- **Turbomachinery**. Chapter 12, on turbomachinery, was rewritten to improve the organization and presentation of the material.
- **Solutions manual**. The formatting of the instructor's solutions manual was improved so that the problem solutions are organized in a way that aligns with engineering practice. Each solution includes a description of the situation; a statement of the problem goals, assumptions, sources, and values of fluid properties; a summary of the strategy; problem analysis; and review comments. In addition, each problem analysis is organized using text labels, such as "momentum equation (x direction)," so that the labels themselves provide a summary of the solution approach. Improved formatting of the solutions manual will continue with future editions of the text. This resource is available only to instructors. To request access to the password-protected solutions manual, visit the Instructor Companion Site portion of the web site located at `www.wiley.com/college/crowe`, and register for a password.
- **PowerPoint slides**. The figures from the text are available in PowerPoint format, for easy inclusion in lecture presentations. This resource is available only to instructors. To request access to this password-protected resource, visit the Instructor Companion Site portion of the web site located at `www.wiley.com/college/crowe`, and register for a password.
- **Interdisciplinary approach**. Historically, this text was written for the civil engineer. In the 8th edition, we are retaining this approach while adding material so that the text is also appropriate for other engineering disciplines. For example, the text presents the Bernoulli equation using both head terms (civil engineering approach) and terms with units of pressure (the approach used by chemical and mechanical engineers). We are also adding problems that are relevant to product development as practiced by mechanical and electrical engineers. Some problems feature other disciplines such as exercise physiology. The reason for this interdisciplinary approach is that the world of today's engineer is becoming more and more interdisciplinary.

FLUIDS SOLUTIONS

Logo used to indicate FLUIDS Solution Problem

- **FLUIDS Solutions.** Students may also wish to enhance their understanding by using FLUIDS Solutions, an innovative student-oriented website that provides the problem-solving approach and step-by-step solution to marked problems from the text. Included in the website are interactive visualizations of fluids theory that are impossible to easily convey on the printed page. Users rave as to the simplicity and usefulness of the website. Students can access FLUIDS Solutions on `www.wiley.com/college/crowe`.

Distinguishing Features

Some of the notable features of this text are

- **Visual approach.** The text makes generous use of photographs, line drawings, and animations to describe fluid flow phenomena and to illustrate problems. This approach helps students learn to visualize concepts and understand how engineers represent real-world systems with engineering sketches.
- **Derivations**. The text includes most steps in the derivation of equations, along with complete word descriptions to assist students in following the derivation. The text is thorough in the derivation of the Bernoulli equation, the Reynolds transport theorem, and the control volume equations for conservation of mass, momentum, and energy, respectively.
- **Example problems.** Each chapter contains numerous example problems, helping students grasp the meaning of concepts and equations by seeing a practical application.
- **Problems.** The text includes several types of end-of-chapter problems, including

  - *Analysis problems.* These problems are traditional problems that require a systematic, or step-by-step, approach. They may involve multiple concepts and generally cannot be solved by using a memorized solution. These problems help students learn engineering analysis.
  - *Concept problems.* These problems, which do not involve calculations, help students learn to reason with concepts.
  - *Computer problems.* These problems involve use of a computer program. Regarding the choice of software, we have left this open so that instructors may select a program, or may allow their students to select a program.
  - *Design problems.* These problems have multiple possible solutions and require assumptions and decision making. These problems help students learn to manage the messy, ill-structured problems that typify professional practice.

- **Chapter summaries.** Each chapter ends with a summary of the most important concepts and equations. This provides a way for students to quickly review the most important material in each chapter.

Acknowledgments

Special recognition is given to our colleagues and mentors. Ronald Adams, Dean at Oregon State University, mentored Donald Elger during his Ph.D. research and introduced him to a new way of thinking about fluid mechanics. Ralph Budwig, a fluid mechanics researcher and educator at Idaho, has provided many hours of delightful discussion. Charles L. Barker introduced John Roberson to the field of fluid mechanics and motivated him to write a textbook.

The reviewers of the 8th edition made valuable suggestions for improving the text. We thank Samuel Coates, Michigan Technological University; M. Mukaddes Darwish, Texas Tech University; Thomas Fontaine, South Dakota School of Mines and Technology; Richard McCuen, University of Maryland; Julia Muccino, Arizona State University; Bart Nijssen, University of Arizona; Philip Parker, University of Wisconsin–Platteville; Michael Robinson, Rose-Hulman Institute of Technology; David Werth, Clemson

University; Anthony Wheeler, San Francisco State University; and Clinton Willson, Louisiana State University.

Finally, we owe a debt of gratitude to Andrew DuBuisson and Anna Henson, who helped with the solutions manual. We are also grateful to our families for their encouragement and understanding during the writing and editing of the text.

We welcome feedback and ideas for interesting end-of-chapter problems Please contact us at the e-mail addresses given below.

Clayton T. Crowe (crowe@mail.wsu.edu)
Donald F. Elger (delger@uidaho.edu)
John A. Roberson (emeritus)

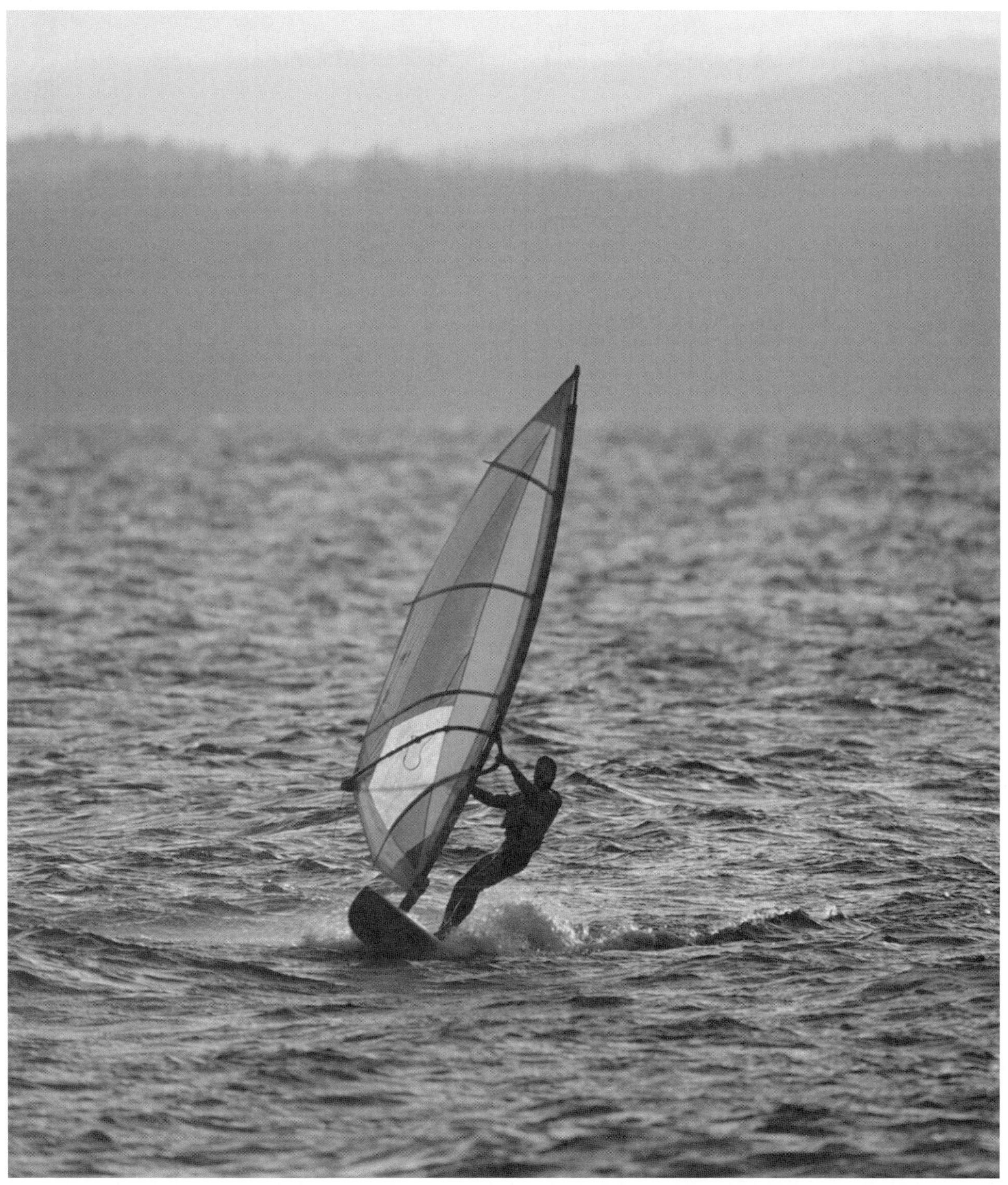

Windsurfing represents many phenomena associated with fluid mechanics. The force on the sail is due to the pressure distribution created by the wind. There is a drag force on the board due to the relative velocity between the board and the water. The choppy surface of the water is due to surface-wind interaction. (PhotoDisc, Inc. / Getty Images.)

C H A P T E R 1

Introduction

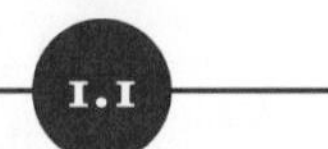

Fluids

Physical Characteristics of Fluids

Fluid mechanics is the science that deals with the action of forces on fluids. In contrast to a solid, a fluid is a substance the particles of which easily move and change their relative position. More specifically, a fluid is defined as a substance that will continuously deform—that is, flow under the action of a shear stress—no matter how small that shear stress may be. A solid, on the other hand, can resist a shear stress, assuming, of course, that the shear stress does not exceed the elastic limit of the material. The rate of deformation of the fluid is related to the applied shear stress by *viscosity,* a property of the fluid. Thus very viscous fluids, such as honey and cold oils, flow very slowly for a given shear stress. An example of this effect is observed when one pours a cup of oil or honey on an inclined surface. It takes a considerable length of time for these very viscous fluids to flow down the incline, whereas a cup of water would flow down and off very rapidly.

Distinction Between Solids, Liquids, and Gases

A fluid can be either a gas or a liquid. The molecular structure of liquids is such that the spacing between molecules is essentially constant (the spacing changes only slightly with temperature and pressure), so that a given mass of liquid occupies a definite volume of space. Therefore, when one pours a liquid into a container, it assumes the shape of the container for the volume it occupies. The molecules of solids also have definite spacing. However, the solid's molecules are arranged in a specific lattice formation and their movement is restricted, whereas liquid molecules can move with respect to each other when a shearing force is applied. The spacing of the molecules of gases is much wider than that of either solids or liquids, and it is also variable. Thus a gas completely fills the container in which it is placed. The gas molecules travel in straight lines through space until they either bounce off the walls of the container or are deflected by interaction with other gas molecules.

Fluid as a Continuum

In considering the action of forces on fluids, one can either account for the behavior of each and every molecule of fluid in a given field of flow or simplify the problem by considering the average effects of the molecules in a given volume. In most problems in fluid dynamics the latter approach is possible, which means that the fluid can be regarded as a *continuum*—that is, a hypothetically continuous substance.

The justification for treating a fluid as a continuum depends on the physical dimensions of the body immersed in the fluid and on the number of molecules in a given volume. Let us say that we are studying the flow of air past a sphere with a diameter of 1 cm. A continuum is said to prevail if the number of molecules in a volume much smaller than the sphere's is sufficiently great so that the average effects (pressure, density, and so on) within the volume either are constant or change smoothly with time. The number of molecules in a cubic meter of air at room temperature and sea-level pressure is about 10^{25}. Thus the number of molecules in a volume of 10^{-19} m^3 (about the size of a dust particle, which is very much smaller than the sphere) would be 10^6. This number of molecules is so large that the average effects within the microvolume are indeed virtually constant. On the other hand, if the 1-cm sphere were at an altitude of 305 km, there would be only one chance in 10^8 of finding a molecule in the microvolume, and the concept of an average condition would be meaningless. In this case, the continuum assumption would not be valid. It may thus be concluded that the assumption of a continuum is valid for fluid flow except in the rarest conditions, such as those encountered in outer space.

1.2 Flow Classification

The subject of fluid mechanics can be subdivided into two broad categories: hydrodynamics and gas dynamics. *Hydrodynamics* deals primarily with the flow of fluids for which there is virtually no density change, such as the flow of liquid or the flow of gas at low speeds. Hydraulics, for example—the study of liquid flows in pipes or open channels—falls within this category. The study of fluid forces on bodies immersed in flowing liquids or in low-speed gas flows can also be classified as hydrodynamics.

Gas dynamics, on the other hand, deals with fluids that undergo significant density changes. High-speed flows of gas through a nozzle or over a body, the flow of chemically reacting gases, and the movement of a body through the low-density air of the upper atmosphere fall within the general category of gas dynamics.

An area of fluid mechanics not classified as either hydrodynamics or gas dynamics is *aerodynamics,* which deals with the flow of air past aircraft or rockets, whether low-speed incompressible flow or high-speed compressible flow.

1.3 Historical Note

The science of fluid mechanics began with the need to control water for irrigation purposes in ancient Egypt, Mesopotamia, and India. Although these civilizations understood the nature of channel flow, there is no evidence that any quantitative relationships had been developed to guide them in their work. It was not until 250 B.C. that Archimedes discovered and recorded the principles of hydrostatics and flotation. Although the empirical understanding of hydrodynamics continued to improve with the development of fluid machinery, better sailing vessels, and more intricate canal systems, the fundamental principles of classical hydrodynamics were not set forth until the seventeenth and eighteenth centuries. Newton (6), Daniel Bernoulli (3), and Euler (11) made the greatest contributions to establishing these principles.

Classical hydrodynamics, though a fascinating subject that appealed to mathematicians, was not applicable to many practical problems because the theory was based on inviscid fluids. The practicing engineers at that time needed design procedures that involved the flow of viscous fluids. Consequently, they developed empirical equations that were usable but narrow in scope. Thus, on the one hand, the mathematicians and physicists developed theories that in many cases could not be used by the engineers, and on the other, the engineers used empirical equations that could not be used outside the limited range of application from which they were derived.

FIGURE 1.1

Leonhard Euler (1707–1783) (left), Professor of Physics and Mathematics. Euler had a greater interest in mathematics than Bernoulli and, in fact, Euler formalized the equation we now call Bernoulli's. Euler developed the basic equations of fluid motion.

Daniel Bernoulli (1700–1782) (right), Professor of Mathematics. Bernoulli published works on the statics and dynamics of fluids and made the first observations and notes relating to the equation that bears his name. (Culver Pictures)

Near the beginning of the twentieth century, however, it was necessary to merge the general approach of physicists and mathematicians with the experimental approach of engineers to bring about significant advances in the understanding of flow processes. Osborne Reynolds' (9) paper in 1883 on turbulence and later papers on the basic equations of motion contributed immeasurably to the development of fluid mechanics. After the turn of the century, Ludwig Prandtl (8) proposed the concept of the boundary layer. This concept not only paved the way to sophisticated analyses needed in the development of the airplane, but also resolved many of the paradoxes involved with the flow of a low-viscosity fluid.

Gas dynamics is a relatively new field, in that the earliest works on it did not appear until the nineteenth century. Riemann (10) published his paper on compression (shock) waves in 1876, and 20 years later Mach (5) observed such waves on supersonic projectiles. Once again it was Prandtl who organized the systematic study of gas dynamics around the turn of the century (7). Interest in gas dynamics increased tremendously after World War I, which led to the development of supersonic wind tunnels before World War II and supersonic flight soon after. The advent of space flight led to still another area of gas dynamics, called rarefied flow, in which the density of the air is so low that the motion and impact of individual molecules must be considered. This is in contrast to the treatment of fluid as a continuum, which is what we do in most of our earthbound problems.

Significance of Fluid Mechanics

The significance of fluid mechanics becomes apparent when we consider the vital role it plays in our everyday lives. When we turn on our kitchen faucets, we activate flow in a complex hydraulic network of pipes, valves, and pumps. When we flick on a light switch, we are drawing energy either from a hydroelectric source that operates by the flow of water through turbines or from a thermal power source derived from the flow of steam past turbine blades. When we drive our cars, pneumatic tires provide suspension, hydraulic shock absorbers reduce road shocks, gasoline is pumped through tubes and later atomized, and air resistance creates a drag on the auto as a whole; and when we stop, we are confident in the operation of the hydraulic brakes. Very complex fluid processes are also involved in the manufacture of the paper on which this book is printed. And our very lives depend on a very important fluid mechanic process—the flow of blood through our veins and arteries.

Some of the most significant environmental problems facing society today involve fluid mechanics. For example, coastal cities often discharge their wastewater (usually treated) into the sea, near the sea bed, far enough from shore so that the wastes become sufficiently diluted with the ambient sea water to render the resulting mixture harmless. The process involves mixing the wastewater with the ambient liquid, a complex turbulence phenomenon. The degree of mixing is a function of the characteristics of the wastewater and the ambient liquid (such as density) as well as the discharge velocity of the wastewater. Also involved in this process are the velocity and pattern of coastal currents. In addition to the fluid mechanics of such a problem, the contaminants in the mixture may change both chemically and biologically in the process. Thus sophisticated models linking the basic flow model with other aspects of the problem are required to design a satisfactory waste disposal system. Such models are generally developed and used by multidisciplinary teams that may include engineers, mathematicians, chemists, and bioscientists. There is an increasing need for engineers who have the ability and mathematical skills to assist in the generation of, and to use, sophisticated computational models of this type. Other problems, similar in nature, that involve fluid mechanics include air pollution and underground hazardous waste problems. By mastering the subject matter in this text, you will acquire the foundation necessary for study of the more involved processes noted here.

1.5

Trends in Fluid Mechanics

Modern developments in fluid mechanics, as in all fields, involve the use of high-speed computers in the solution of problems. Remarkable progress is being made in this area, and the use of the computer in fluid dynamic design is increasing. In the design of aircraft, computers are used to predict the flow over engine nacelles and appendages in order to select configurations that minimize aerodynamic drag. The NASA publication on wind tunnels (2) explains the role of computers in aircraft design. Currently it is possible to develop computational models to predict the flow properties over the entire aircraft. This capability represents a significant technological advance in design and development of flight vehicles (1). Computational solutions for wind forces on buildings and structures are used to complement measurements on wind tunnel models to ensure the safety and structural integrity of the full-scale structures.

The ever-increasing speed and memory capacity of modern computers are leading to even more exciting applications of computers in fluid mechanics. Computer solutions for the motion of terrestrial winds and weather fronts are leading to more accurate forecasting of local weather conditions. The coupling of fluid mechanics with heat transfer and chemical kinetics in computational solutions will lead to improved designs for industrial power and propulsion systems. As space stations and space travel become more feasible, computers will play a vital role in the design of flow systems in microgravity environments that are difficult to examine through terrestrial experimentation. The application of computers to the analysis of flows in biological systems is only beginning, but it will continue to grow as the mechanics of flows in these systems becomes better understood.

The science and engineering of fluid mechanics are also finding application in the sports and recreation industries. The design of sailboats is a particularly challenging fluid mechanics application because of the need to maximize the aerodynamic forces on the sails and minimize the hydrodynamic drag on the hull. Sailboats for the highly competitive America's Cup are designed with the help of sophisticated computer modeling. The helmets for professional bicyclists are designed with the application of aerodynamic principles to reduce drag. The design of modern parachutes is based on aerodynamic concepts to enhance maneuverability.

The capability of micromachining, developed over the past decade, has led to the study of fluid flows in channels on the order of 100 microns (one-tenth of a millimeter) or less. Micromotors the size of a human hair have been fabricated. The traditional friction laws break down in these situations. This is an exciting field because of its far-reaching applications in medicine, such as drug delivery systems, and in aerodynamics, such as the control of turbulent skin friction (4).

The science of fluid mechanics is developing at a rapid rate. Armed with more detailed measurements and numerical models, fluid mechanicians have developed higher levels of understanding that have led to sophisticated designs and applications of fluid systems. Still, there are many areas in which only rudimentary information is available. Turbulence is a prime example. Even though we presently have high-speed computers at our disposal, the solutions are only as valid as the equations we use to describe the basic flow phenomena. And there is currently no general analytic model that completely describes the nature of turbulence. We have good data on turbulence in straight pipes, so reliable empirical formulas have been developed to describe the turbulence in such a simple case. But turbulence in high-shear flows, buoyant flows, and compressible flows is still the subject of extensive study. Analyses of the flow of multiphase mixtures such as solids in a liquid (slurries) and bubbles in a liquid still rely heavily on empiricism. In oil recovery operations, the engineer is confronted with the problem of the flow of immiscible liquids, such as oil in water, which is not well understood. These are areas which represent exciting challenges to current and future practitioners of fluid mechanics.

References

1. Agarwal, R. "Computational Fluid Dynamics of Whole-Body Aircraft." *Annu. Rev. Fluid Mech.* 31 (1999), 125–169.

2. Baals, D. B., and W. R. Corliss. *Wind Tunnels of NASA,* Sup. of Doc., U.S. Govt. Printing Office, Washington, D.C., 1981.

3. Bernoulli, Daniel, *Hydrodynamics,* and Bernoulli, Johann, *Hydraulics.* Trans. Thomas Carmody and Helmut Kobus. Dover Publications, New York, 1968.

4. Ho, C.-H., and Y.-C. Tai. "Micro-Electro-Mechanical Systems (MEMS) and Fluid Flows." *Annu. Rev. Fluid Mech.* 30 (1998), 579–612.

5. Mach, E., and P. Salcher. "Photographische Fixierung der durch Projektile in der Luft eingeleiteten Vorgänge." *S. B. Akad. Wiss. Wien.*, 95 (1887), 764–780.

6. Newton, Isaac S. *Philosophie Naturalis Principia Mathematica.* Trans. Florian Cajori. University of California Press, Berkeley, 1934.

7. Oswatitsch, K. *Gas Dynamics.* Trans. G. Kuerti. Academic Press, New York, 1956.

8. Prandtl, L. "Über Flussigkeitsbewegung bei sehr kleiner Reibung." In *Verhandlungen des III. Internationalen Mathematiker Kongresses.* Leipzig, 1905.

9. Reynolds, O. "An Experimental Investigation of the Circumstances Which Determine Whether the Motion of Water Shall Be Direct or Sinuous, and of the Law of Resistance in Parallel Channels." *Phil. Trans. Roy. Soc. London,* 174 (1883).

10. Riemann, B. "Über die Fortplanzung ebener Luftwellen von endlicher Schwing-ungsweite." In *Gesammelte Werke.* Leipzig, 1876.

11. Rouse, H., and S. Ince. *History of Hydraulics.* Iowa Institute of Hydraulic Research, State University of Iowa, 1957.

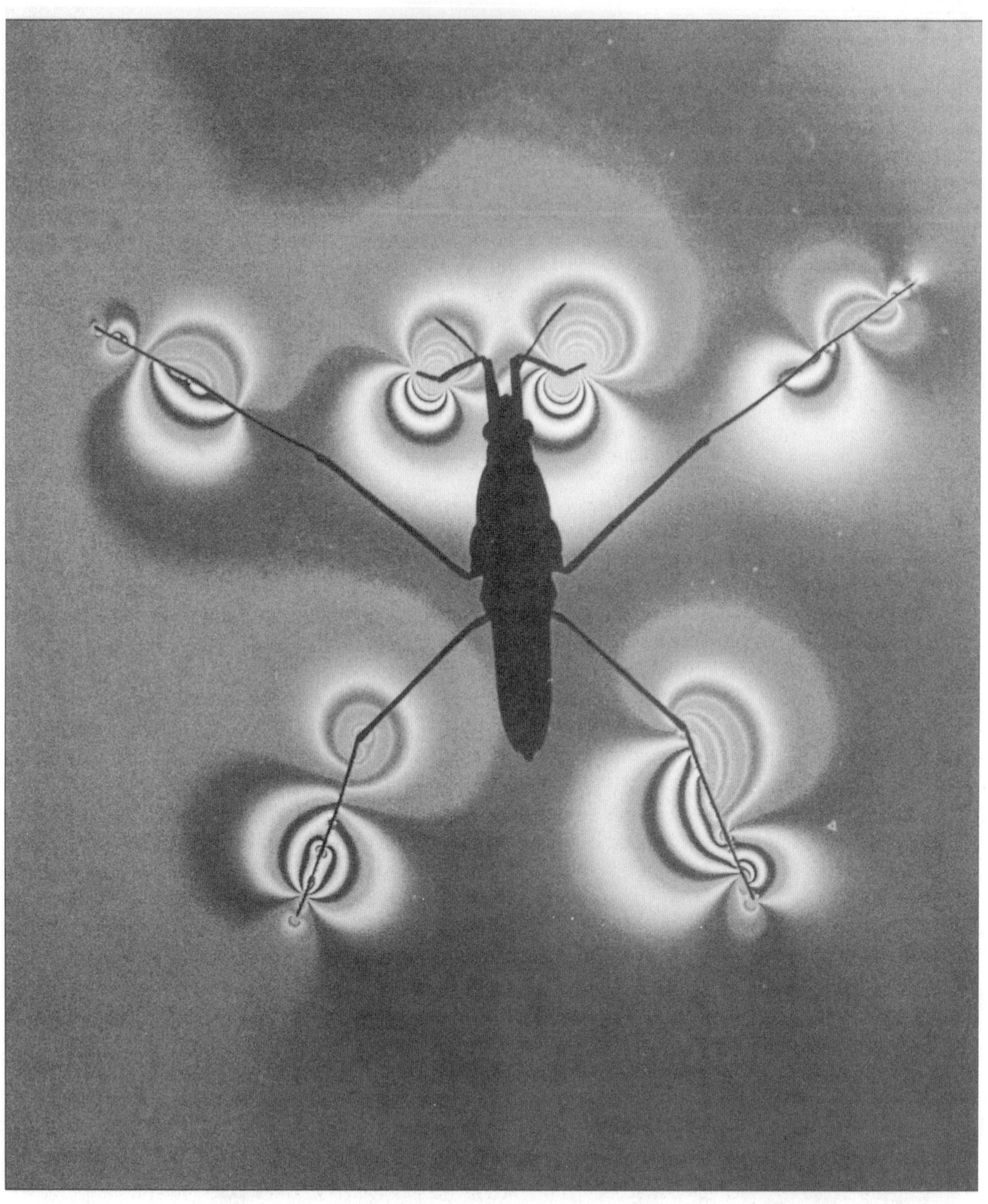

Surface tension of liquids is important in many engineering applications, but it is also paramount in much of the natural world. The non-wetting properties of this water strider's legs and the surface tension of the water allow it to move on the water surface environment with the greatest of ease. (Nuridsany et Perennou/Photo Researchers.)

CHAPTER 2

Fluid Properties

2.1

Basic Units

Every fluid has certain characteristics by which its physical condition may be described. We call such characteristics *properties* of the fluid. These properties are expressed in terms of a limited number of basic dimensions (length, mass or force, time, and temperature), which in turn are quantified by basic units. The traditional system of units in the United States has been the foot–pound–second system. However, because all the engineering societies are urging the adoption of the SI (Système International) system, this text uses both SI and traditional units. Approximately half of the problems and the majority of the examples are given in the SI system; the remainder are given in the foot–pound–second system.

SI System of Units

The basic units of mass, length, and time in the SI system are the kilogram (kg), meter (m), and second (s). The corresponding unit of force is derived from Newton's second law: the force required to accelerate a kilogram at one meter per second per second is defined as the *newton* (N). The acceleration due to gravity varies by 0.5% over the earth's surface, and the standard accepted value is 9.80655 m/s^2 or 9.81 m/s^2 for engineering calculations. Thus the weight of a kilogram at the earth's surface is

$$\begin{aligned} W &= Mg \\ &= (1)(9.81)\ \text{kg} \cdot \text{m/s}^2 = 9.81\ \text{N} \end{aligned} \tag{2.1}$$

From Eq. (2.1) we can also determine the units for kilograms in terms of the meter–newton–second units. From Eq. (2.1),

$$M = \frac{W}{g}\frac{\text{N}}{\text{m/s}^2}$$

Thus the mass M of a body in kilograms is given by its weight in newtons at the earth's surface divided by 9.81 m/s^2 (the acceleration due to gravity at the earth's surface). As shown, the units of mass may also be expressed in equivalent meter–newton–second units as $N \cdot s^2/m$.

The basic unit of temperature in the SI system is the kelvin (K). The Kelvin scale is 0 K at absolute zero and 273.16 K at the freezing point of water. The Celsius scale (°C) is 0°C at the freezing point of water. Therefore the conversion formula is

$$K = 273 + °C \tag{2.2}$$

The unit of work and energy in the SI system is the *joule* (J), which is a newton meter $(N \cdot m)$. The unit of power is the *watt* (W), which is a joule per second.

The prefixes used in the SI system to indicate multiplication of units by powers of 10 are

$$
\begin{aligned}
&\text{G (giga)} = 10^9 && \text{c (centi)} = 10^{-2} \\
&\text{M (mega)} = 10^6 && \text{m (milli)} = 10^{-3} \\
&\text{k (kilo)} = 10^3 && \mu \text{ (micro)} = 10^{-6}
\end{aligned}
$$

For example, km stands for kilometer, or 1000 meters, and mm signifies millimeter, or 0.001 meter.

Traditional Units

The system of units that preceded SI units in several countries is the so-called English system. The fundamental units of length and mass are the foot (ft), equal to 30.48 cm, and the slug, equal to 14.59 kg. The time unit of second is common to both systems. The force required to accelerate a mass of one slug at one foot per second per second is one pound force (lbf). The mass unit more common to mechanical engineers in the traditional system is the pound mass (lbm). The conversion factor for changing pounds mass to slugs is g_c, which is equal to 32.2 lbm/slug. In other words,

$$1 \text{ slug} = 32.2 \text{ lbm}$$

example 2.1

What is the weight of a pound mass on the earth's surface, where the acceleration due to gravity is 32.2 ft/s^2, and on the moon's surface, where the acceleration is 5.31 ft/s^2?

Solution First find the mass in slugs:

$$1 \text{ lbm} \times \frac{1 \text{ slug}}{32.2 \text{ lbm}} = \frac{1}{32.2} \text{ slugs}$$

By the law of universal gravitation,

$$W = Mg$$

where the unit for W is lbf, for M slugs, and for g ft/s^2. Therefore, the weight on the earth's surface is

$$W = \frac{1 \text{ slug}}{32.2} \times 32.2 \frac{\text{ft}}{\text{s}^2} = 1 \text{ lbf}$$

and on the moon's surface is

$$W = \frac{1 \text{ slug}}{32.2} \times 5.31 \frac{\text{ft}}{\text{s}^2} = 0.165 \text{ lbf}$$

Thus the weight of one pound mass on the earth's surface is one pound force. However, a pound force does *not* equal a pound mass because they have different units.

In the traditional system it is good practice to use the units lbm and lbf and not simply lb.

The fundamental unit of work in the traditional system is the foot-pound (ft-lbf), and the fundamental unit of power is the ft-lbf/s. A common unit of power is the horsepower, which is 550 ft-lbf/s. One horsepower is 746 W.

The traditional unit for temperature is the degree Fahrenheit (°F), which is $\frac{5}{9}$ of the Celsius degree. The corresponding absolute temperature scale is in degrees Rankine (°R). The Fahrenheit temperature at the freezing point of water is 32°F, and the formula for conversion between °F and °R is

$$°\text{R} = 460 + °\text{F} \tag{2.3}$$

2.2 System; Extensive and Intensive Properties

From Chapter 4 through the remainder of the text, wide use is made of the control-volume approach. In its application we use the concept of a system of particles and the intensive and extensive properties related to this concept. A *system,* in fluid mechanics and thermodynamics, is defined as a given quantity of matter. To illustrate, if at an instant a quantity of matter were designated as a system and dyed red to distinguish it from the remaining matter, this red matter would always constitute the system, even though it moved through space and might change in shape and volume. Because a system always consists of the same matter, the mass of a given system is constant. Properties related to the total mass of the system are called *extensive properties* and are usually represented by uppercase letters—for example, mass M and weight W. Properties that are independent of the amount of fluid are called *intensive properties* and are often designated by lowercase letters, such as pressure p (force per unit area) and *mass density* ρ (mass per unit volume). The distinction between extensive and intensive properties is very important in the derivation and application of the control-volume approach introduced in Chapter 5.

2.3 Properties Involving the Mass or Weight of the Fluid

Mass Density, ρ

The mass per unit volume is *mass density* and has units of kilograms per cubic meter or pounds mass per cubic foot. The mass density of water at 4°C is 1000 kg/m^3 (62.4 lbm/ft^3) and decreases slightly with increasing temperature, as shown in

Table A.5. The mass density of air at 20°C and standard atmospheric pressure is 1.2 kg/m^3 (0.075 lbm/ft^3). Mass density, often simply called density, is represented by the Greek symbol ρ (rho). The densities of common fluids are given in Tables A.2 to A.5.

Specific Weight, γ

The gravitational force per unit volume of fluid, or simply the weight per unit volume, is defined as *specific weight.* It is given the symbol γ (gamma). Water at 20°C has a specific weight of 9.79 kN/m^3. In contrast, the specific weight of air at the same temperature and at standard atmospheric pressure is 11.8 N/m^3. Specific weight and density are related by

$$\gamma = \rho g$$

Specific weights of common liquids are given in Table A.4 in the Appendix.

Variation in Density

The density of some fluids is more easily changed than that of others. For example, air can be compressed at moderate pressure with a consequent density change, whereas a very large pressure is needed to effect a relatively small density change in water. For most applications, water can be considered incompressible and, in turn, can be assumed to have constant density. Air, on the other hand, is a relatively compressible fluid with variable density. However, at velocities much less than the speed of sound, the air density changes only slightly and the air can also be treated as incompressible.

Incompressibility does not always imply constant density. For example, a mixture of salt in water changes the density of the water without changing its volume. Therefore there are some flows, such as in estuaries, in which density variations may occur within the flow field even though the fluid is essentially incompressible. Such a fluid is termed *nonhomogeneous.* This text emphasizes the flow of *homogeneous* fluids, so the term *incompressible,* used throughout the text, implies constant density.

Specific Gravity, S

The ratio of the specific weight of a given fluid to the specific weight of water at a standard reference temperature is defined as *specific gravity,* S:

$$S = \frac{\gamma_{\text{fluid}}}{\gamma_{\text{water}}} = \frac{\rho_{\text{fluid}}}{\rho_{\text{water}}}$$

The standard reference temperature for water is often taken as 4°C, where the specific weight of water at atmospheric pressure is 9810 N/m^3. With this reference, the specific gravity of mercury at 20°C is

$$S_{\text{Hg}} = \frac{133 \text{ kN/m}^3}{9.81 \text{ kN/m}^3} = 13.6$$

Because specific gravity is a ratio of specific weights, it is dimensionless and, of course, independent of the system of units used.

Ideal Gas Law

The fundamental equation of state for an ideal gas is

$$p\mathcal{V} = nR_uT$$

where p is the absolute pressure,* $\mathcal{V}$ is the volume, n is the number of moles, R_u is the universal gas constant (the same for all gases), and T is absolute temperature. The universal gas constant is 8.314 kJ/kmol-K in the SI system and 1545 ft-lbf/lbmol-R in the traditional system. The equation of state can be rewritten as

$$p = \frac{n\mathcal{M}}{\mathcal{V}}\frac{R_u}{\mathcal{M}}T$$

where $\mathcal{M}$ is the molecular weight of the gas. The product of the number of moles and the molecular weight is the mass of the gas. Thus $n\mathcal{M}/\mathcal{V}$ is the mass per unit volume, or density. The quotient $R_u/\mathcal{M}$ is the gas constant, R. Thus the equation of state, or *ideal gas law*, can be expressed as

$$p = \rho RT \tag{2.4}$$

The ideal gas law is a form of the general equation of state, which relates pressure, specific volume, and temperature for a pure substance. Actually, no gas is ideal. However, a gas far removed from the liquid phase, which is generally what we encounter in gas-flow problems, behaves like an ideal gas. Values of R for a number of gases are given in Table A.2. To determine the mass density of a gas, we simply solve Eq. (2.4) for ρ:

$$\rho = \frac{p}{RT} \tag{2.5}$$

example 2.2

Air at standard sea-level pressure (p = 101 kN/m^2) has a temperature of 4°C. What is the density of the air?

Solution We apply the ideal gas law to solve for ρ:

$$\rho = \frac{p}{RT}$$

$$= \frac{101 \times 10^3\ \text{N/m}^2}{287\ \text{J/kg K} \times (273 + 4)\ \text{K}} = 1.27\ \text{kg/m}^3 \quad \triangleleft$$

* We discuss pressure in detail in Chapter 3.

Properties Involving the Flow of Heat

Specific Heat, c

The property that describes the capacity of a substance to store thermal energy is called *specific heat.* By definition, it is the amount of thermal energy that must be transferred to a unit mass of substance to raise its temperature by one degree. The specific heat of a gas depends on the process accompanying the change in temperature. If the specific volume v of the gas ($v = 1/\rho$) remains constant while the temperature changes, then the specific heat is identified as c_v. However, if the pressure is held constant during the change in state, then the specific heat is identified as c_p. The ratio c_p/c_v is given the symbol k. Values for c_p and k for various gases are given in Table A.2.

Specific Internal Energy, u

The energy that a substance possesses because of the state of the molecular activity in the substance is termed *internal energy.* Internal energy is usually expressed as a specific quantity—that is, internal energy per unit mass. In the SI system, the *specific internal energy, u,* is given in joules per kilogram. The internal energy is generally a function of temperature and pressure. However, for an ideal gas, it is a function of temperature alone.

Specific Enthalpy, h

The combination $u + p/\rho$ is encountered frequently in the equations for thermodynamics and compressible flow; it has been given the name *specific enthalpy.* For an ideal gas, u and p/ρ are functions of temperature alone. Consequently their sum, specific enthalpy, is also a function solely of temperature.

Viscosity

The most important distinction between a solid such as steel and a viscous fluid such as water or air is that shear stress in a solid material is proportional to shear strain, and the material ceases to deform when equilibrium is reached, whereas the shear stress in a viscous fluid is proportional to the *time rate* of strain. The proportionality factor for the solid is the shear modulus. The proportionality factor for the viscous fluid is the *dynamic,* or *absolute, viscosity.* For example, near a wall, as shown in Fig. 2.1, the rate of strain is dV/dy, where V is the fluid velocity and y is the distance measured from the wall, so the shear stress is

$$\tau = \mu \frac{dV}{dy} \tag{2.6}$$

where τ (tau) is the shear stress, μ (mu) is the dynamic viscosity, and dV/dy is the time rate of strain, which is also the velocity gradient normal to the wall. Thus the definition of the viscosity, μ, is the ratio of the shear stress to the velocity gradient, $\mu = \tau/(dV/dy)$.

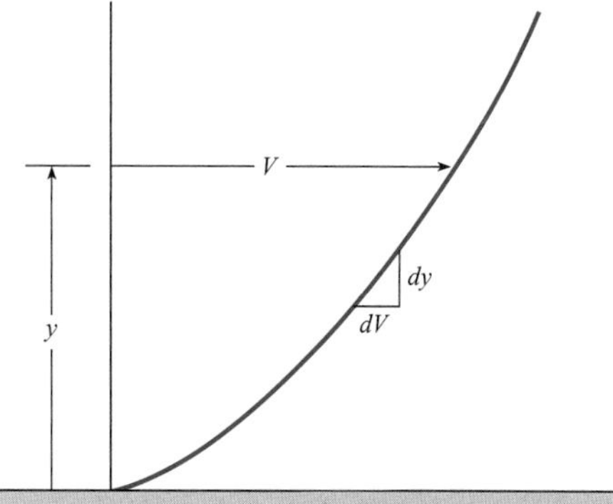

FIGURE 2.1

Velocity distribution next to a boundary.

Consider the flow shown in Fig. 2.1. This velocity distribution is typical of that for laminar (nonturbulent) flow next to a solid boundary. Several observations relating to this figure will help you appreciate the interaction between viscosity and velocity distribution. First, the velocity gradient at the boundary is finite. The curve of velocity variation cannot be tangent to the boundary because this would imply an infinite velocity gradient and, in turn, an infinite shear stress, which is impossible. Second, a velocity gradient that becomes less steep (dV/dy becomes smaller) with distance from the boundary has a maximum shear stress at the boundary, and the shear stress decreases with distance from the boundary. Also note that the velocity of the fluid is zero at the stationary boundary. This is characteristic of all flows dealt with in this basic text. That is, at the boundary surface the fluid has the velocity of the boundary—no slip occurs.

From Eq. (2.6) it can be seen that the units of μ are $\mathrm{N \cdot s/m^2}$.

$$\mu = \frac{\tau}{dV/dy} = \frac{\mathrm{N/m^2}}{\mathrm{(m/s)/m}} = \mathrm{N \cdot s/m^2}$$

A common unit of viscosity is the *poise,* which is 1 dyne-s/cm^2 or 0.1 $\mathrm{N \cdot s/m^2}$. The viscosity of water at 20°C is one centipoise (10^{-2} poise) or 10^{-3} $\mathrm{N \cdot s/m^2}$. The unit of viscosity in the traditional system is $\mathrm{lbf \cdot s/ft^2}$.

Many of the equations of fluid mechanics include the combination μ/ρ. Because it occurs so frequently, this combination has been given the special name *kinematic viscosity* (so called because the force dimension cancels out in the combination μ/ρ). The symbol used to identify kinematic viscosity is ν (nu). The units of kinematic viscosity ν are $\mathrm{m^2/s}$.

$$n = \frac{m}{r} = \frac{\mathrm{N \cdot s/m^2}}{\mathrm{kg/m^3}} = \mathrm{m^2/s}$$

The units for kinematic viscosity in the traditional system are $\mathrm{ft^2/s}$.

Whenever shear stress is applied to a fluid, motion occurs. This is the basic difference between fluids and solids. Solids can resist shear stress in a static condition, but fluids deform continuously under the action of a shear stress. Another important characteristic of fluids is that the viscous resistance is independent of the normal force (pressure) acting within the fluid. In contrast, for two solids sliding relative to each other, the shearing resistance is totally dependent on the normal force between the two.

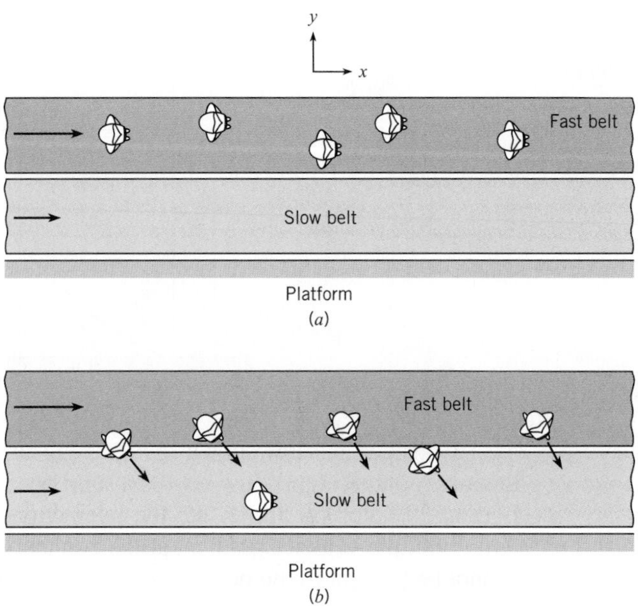

FIGURE 2.2
Conveyor-belt transportation system.

The manner in which viscous forces are produced can be seen in the conveyor-belt analogy. Consider a type of transit system in which people are carried from one part of a city to another on conveyor belts (Fig. 2.2*a*). People ride the fast-moving belt from left to right—an equilibrium condition exists. Next visualize the action when the people step off the fast belt onto a slower-moving belt (Fig. 2.2*b*). Before they step off the fast-moving belt, each possesses a certain amount of momentum in the x direction. But as soon as they acquire the speed of the slower belt, their momentum in the x direction is reduced by a significant amount. It is known from basic mechanics that a change in momentum of a body results from an external force acting on that body. In our example, it is the slower belt that exerts a force in the negative x direction as each person steps on the belt. Conversely, as each person steps on the slower belt, a force is exerted on the belt in the positive x direction. Now, if the people step from the faster belt to the slower belt at a rather steady rate, then a rather continuous force is exerted on the slower belt. In effect, by the action of the people moving in the negative y direction, they produce a force on the slow belt in the positive x direction. One may think of this as a shear force in the x direction. In a similar manner, it can be visualized that if people stepped from the slow-moving belt to the faster one, a "shear force" in the negative x direction would be imposed on the faster belt.

If the people were continuously going both ways (back and forth) from one belt to the other, there would be, in effect, a continuous augmenting force (force in the direction of motion) on the slow belt and a like retarding force on the fast belt. Furthermore, as the relative speed between the belts changes (analogous to a change in velocity gradient), the shear force is increased or decreased in direct proportion to the increase or decrease in relative velocity. Thus if both belts were made to have the same speed, the shear force would be zero.

In fluid flow we can think of streams of fluid traveling in a given general direction, such as in a pipe, with the fluid nearer the pipe center traveling faster (analogous to the

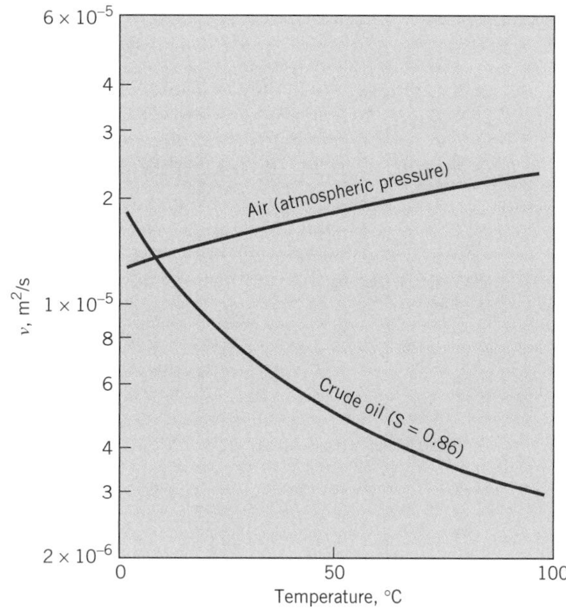

FIGURE 2.3

Kinematic viscosity for air and crude oil.

faster belt) while the fluid nearer the wall is traveling more slowly. The interaction between streams, in the case of gas flow, occurs when the molecules of gas travel back and forth between adjacent streams, thus creating a shear stress in the fluid. Because the rate of activity (back-and-forth motion) of the gas molecules increases with an increase in temperature, it follows that the viscosity of a gas should increase with the temperature of the gas. Such is indeed the case, as can be seen in Fig. 2.3.

A reasonable estimate for the variation of gas viscosity with absolute temperature is *Sutherland's equation,*

$$\frac{\mu}{\mu_0} = \left(\frac{T}{T_0}\right)^{3/2} \frac{T_0 + S}{T + S} \tag{2.7}$$

where μ_0 is the viscosity at temperature T_0 and S is Sutherland's constant. Sutherland's constant for air is 111 K; values for other gases are given in Table A.2. Using Sutherland's equation for air yields viscosities with an accuracy of ±2% for temperatures between 170 K and 1900 K. In general the effect of pressure on the dynamic viscosity of common gases is minimal for pressures less than 10 atmospheres.

For liquids, the shear stress is involved with the cohesive forces between molecules. These forces decrease with temperature, which results in a decrease in viscosity with an increase in temperature (see Fig. 2.3). An equation for the variation of liquid viscosity with temperature is (1)

$$\mu = Ce^{b/T} \tag{2.8}$$

where C and b are empirical constants that require viscosity data at two temperatures for evaluation. This equation should be used primarily for data interpolation. The variation of viscosity (dynamic and kinematic) for other fluids is given in Figs. A.2 and A.3 in the Appendix.

example 2.3

The dynamic viscosity of water at 20°C is $1.00 \times 10^{-3}\ \text{N} \cdot \text{s/m}^2$, and the viscosity at 40°C is $6.53 \times 10^{-4}\ \text{N} \cdot \text{s/m}^2$. Using Eq. (2.8), estimate the viscosity at 30°C.

Solution Taking the logarithm of Eq. (2.8) gives

$$\ln \mu = \ln C + b/T$$

Substituting in the data for μ and T at the two data points, we get

$$-6.908 = \ln C + 0.00341b$$
$$-7.334 = \ln C + 0.00319b$$

Solving for $\ln C$ and b gives

$$\ln C = -13.51 \qquad b = 1936\ (\text{K})$$

Substituting back into Eq. (2.8) results in

$$\mu = 1.357 \times 10^{-6} e^{1936/T}$$

Evaluating for the viscosity at 30°C gives

$$\mu = 8.08 \times 10^{-4}\ \text{N} \cdot \text{s/m}^2 \qquad \triangleleft$$

This value differs by 1% from the reported value but provides a much better estimate than would be obtained using a linear interpolation.

example 2.4

A board 1 m by 1 m that weighs 25 N slides down an inclined ramp (slope = 20°) with a velocity of 2.0 cm/s. The board is separated from the ramp by a thin film of oil with a viscosity of $0.05\ \text{N} \cdot \text{s/m}^2$. Neglecting edge effects, calculate the spacing between the board and the ramp.

Solution The board and ramp (left) and a free body of the board (right) are shown below. For a constant sliding velocity, the resisting shear force is equal to the component of weight parallel to the inclined ramp. Therefore,

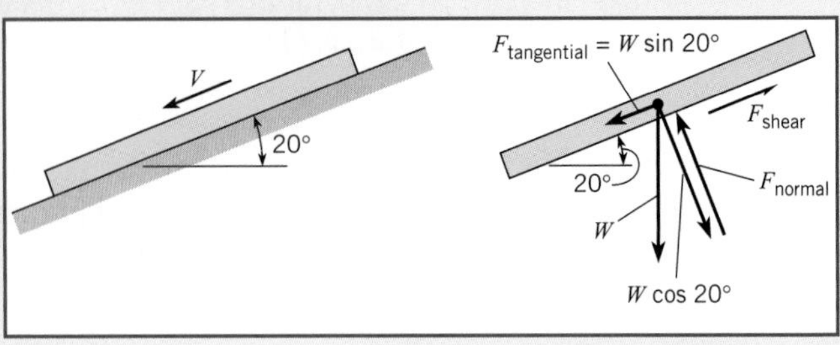

$$F_{\text{tangential}} = F_{\text{shear}}$$
$$W \sin 20^\circ = \tau A$$
$$W \sin 20^\circ = \mu \frac{dV}{dy} A$$

In this case we can assume a linear velocity distribution in the oil, so dV/dy can be expressed as $\Delta V/\Delta y$, where ΔV is the velocity of the board and Δy is the spacing between the board and the ramp. We then have

$$W \sin 20^\circ = \mu \frac{\Delta V}{\Delta y} A$$

or

$$\begin{aligned} \Delta y &= \frac{\mu \Delta V A}{W \sin 20^\circ} \\ &= \frac{0.05 \text{ N} \cdot \text{s/m}^2 \times 0.020 \text{ m/s} \times 1 \text{ m}^2}{25 \text{ N} \times \sin 20^\circ} \\ &= 0.000117 \text{ m} \\ &= 0.117 \text{ mm} \end{aligned}$$ ◁

Newtonian versus Non-Newtonian Fluids

Fluids for which the shear stress is directly proportional to the rate of strain are called *Newtonian fluids.* Because shear stress is directly proportional to the shear strain, dV/dy, a plot relating these variables (see Fig. 2.4) results in a straight line passing through the origin. The slope of this line is the value of the dynamic viscosity. For some fluids the shear stress may not be directly proportional to the rate of strain; these are called *non-Newtonian fluids.* One class of non-Newtonian fluids, shear-thinning fluids,

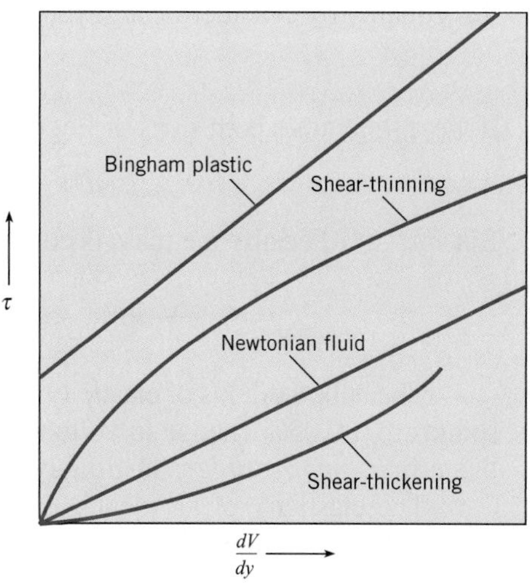

FIGURE 2.4

Shear stress relations for different types of fluids.

has the interesting property that the ratio of shear stress to shear strain decreases as the shear strain increases (see Fig. 2.4). Some common shear-thinning fluids are toothpaste, catsup, paints, and printer's ink. Fluids for which the viscosity increases with shear rate are *shear-thickening fluids*. Some examples of these fluids are mixtures of glass particles in water and gypsum-water mixtures. Another type of non-Newtonian fluid, called a *Bingham plastic,* acts like a solid for small values of shear stress and then behaves as a fluid at higher shear stress. The shear stress versus shear strain rate for a Bingham plastic is also shown in Fig. 2.4. This book will focus on the theory and applications involving Newtonian fluids. For more information on the theory of flow of non-Newtonian fluids, please see references (2) and (3).

2.6 Elasticity

When the pressure acting on a mass of fluid increases, the fluid contracts; when the pressure decreases, it expands. The elasticity of a fluid is related to the amount of deformation (expansion or contraction) for a given pressure change. The elasticity is often called the compressibility of the fluid. Quantitatively, the degree of elasticity is given by E_v, the definition of which is

$$dp = -E_v \frac{d\maltese}{\maltese} \tag{2.9}$$

or

$$E_v = -\frac{dp}{d\maltese/\maltese} \tag{2.10}$$

where E_v is the bulk modulus of elasticity, dp is the incremental pressure change, $d\maltese$ is the incremental volume change, and $\maltese$ is the volume of fluid. Because $d\maltese/\maltese$ is negative for a positive dp, a negative sign is used in the definition to yield a positive E_v. An alternative form of Eq. (2.10) is

$$E_v = \frac{dp}{d\rho/\rho} \tag{2.11}$$

By comparing Eqs. (2.10) and (2.11), one can see that $d\rho/\rho = -d\maltese/\maltese$. We can verify this equality by considering a given mass M of fluid, where

$$M = \rho\maltese \tag{2.12}$$

If we differentiate both sides of Eq. (2.12), we have

$$dM = \rho\, d\maltese + \maltese\, d\rho \tag{2.13}$$

But $dM = 0$ because the mass is constant. Hence we find that

$$\maltese\, d\rho = -\rho\, d\maltese \qquad \text{or} \qquad \frac{d\rho}{\rho} = -\frac{d\maltese}{\maltese}$$

The bulk modulus of elasticity of water is approximately 2.2 GN/m^2, which corresponds to a 0.05% change in volume for a change of 1 MN/m^2 in pressure. Obviously, the term *incompressible* is justifiably applied to water.

The elasticity of an ideal gas is proportional to the pressure. For an isothermal (constant-temperature) process,

$$\frac{dp}{d\rho} = RT$$

so

$$E_v = \rho \frac{dp}{d\rho} = \rho RT = p$$

For an adiabatic process, $E_v = kp$, where k is the ratio of specific heats, c_p/c_v.

The elasticity or compressibility of a gas is important in high-speed gas flows in which pressure variations can cause significant density changes. As will be shown in Chapter 12, the elasticity of a gas is related to the speed of sound. The ratio of the flow velocity to the speed of sound is the Mach number, which relates to the importance of elasticity effects.

2.7 Surface Tension

According to the theory of molecular attraction, molecules of liquid considerably below the surface act on each other by forces that are equal in all directions. However, molecules near the surface have a greater attraction for each other than they do for molecules below the surface. This produces a surface on the liquid that acts like a stretched membrane. Because of this membrane effect, each portion of the liquid surface exerts "tension" on adjacent portions of the surface or on objects that are in contact with the liquid surface. This tension acts in the plane of the surface, and its magnitude per unit length is defined as *surface tension*, σ (sigma). Surface tension for a water–air surface is 0.073 N/m at room temperature. The effect of surface tension is illustrated for the case of *capillary action* in a small tube (Fig. 2.5). Here the end of a small-diameter tube is put into a reservoir of water, and the characteristic curved water surface occurs within the tube. The relatively great attraction of the water molecules for the glass causes the water surface to curve upward in the region of the glass wall. Then the surface-tension force acts around the circumference of the tube, in the direction indicated. It may be assumed that θ (theta) is equal to 0° for water against glass. This produces a net upward force on the water that causes the water in the tube to rise above the water surface in the reservoir.

The *surface-tension force* is given by

$$F_s = \rho L \tag{2.14}$$

where L is the length over which the surface tension acts. A quantitative illustration of the surface-tension force acting to raise the water in a small-diameter tube is given in the following example.

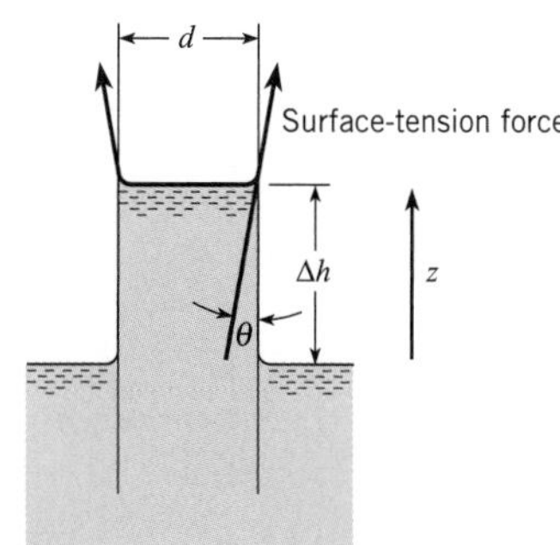

FIGURE 2.5

Capillary action in a small tube.

example 2.5

To what height above the reservoir level will water (at 20°C) rise in a glass tube, such as that shown in Fig. 2.5, if the inside diameter of the tube is 1.6 mm?

Solution By taking the summation of forces in the vertical direction on the water in the tube that has risen above the reservoir level, we have

$$F_{\sigma,z} - W = 0$$

$$\sigma \pi d \cos\theta - \gamma(\Delta h)(\pi d^2/4) = 0$$

However, θ for water against glass is so small it can be assumed to be 0°; therefore $\cos\theta \approx 1$. Then

$$\sigma \pi d - \gamma(\Delta h)\left(\frac{\pi d^2}{4}\right) = 0$$

or $$\Delta h = \frac{4\sigma}{\gamma d} = \frac{4 \times 0.073 \text{ N/m}}{9790 \text{ N/m}^3 \times 1.6 \times 10^{-3} \text{ m}} = 18.6 \text{ mm}$$ ◁

Other manifestations of surface tension include the excess pressure (over and above atmospheric pressure) created inside droplets and bubbles, the transformation of a liquid jet into droplets, and the binding together of wetted granular material, such as fine, sandy soil.

Surface-tension forces for several different cases are shown in Fig. 2.6. Case (*a*) is a spherical droplet of radius *r*. The surface-tension force is balanced by the internal pressure.

$$F_\sigma = \sigma L = pA$$

$$2\pi r\sigma = p\pi r^2$$

FIGURE 2.6

Surface-tension forces for several different cases.

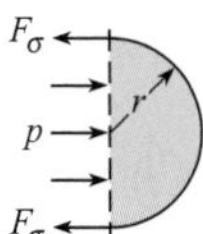

(*a*) Spherical droplet

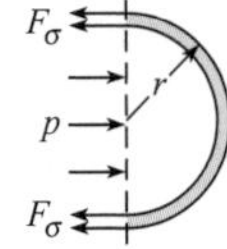

(*b*) Spherical bubble

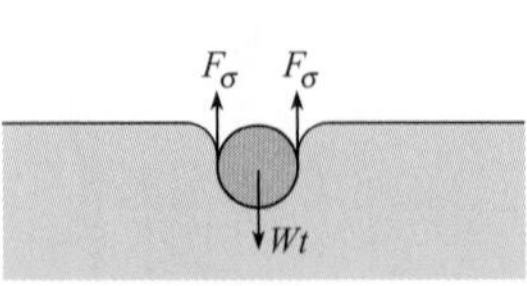

(*c*) Cylinder supported by surface tension (liquid does not wet cylinder)

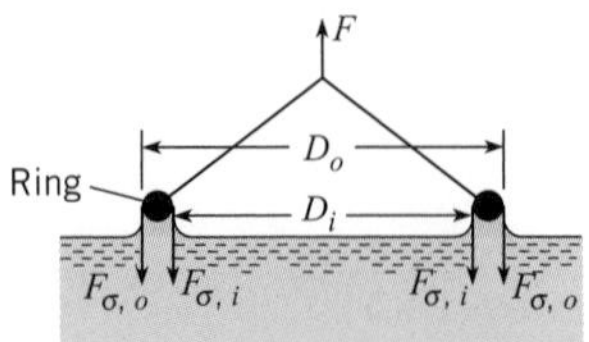

(*d*) Ring pulled out of liquid (liquid wets the ring)

or

$$p = \frac{2\sigma}{r}$$

Case (*b*) is a bubble of radius *r* that has internal and external surfaces and the surface-tension force acts on both surfaces, so

$$p = \frac{4r}{\rho}$$

Case (*c*) is a cylinder supported by surface-tension forces. The liquid does not wet the cylinder surface. The maximum weight the surface tension can support is

$$Wt = 2F_s = 2\sigma L$$

where L is the length of the cylinder.

Case (*d*) is a ring being pulled out of a liquid. This is a technique to measure surface tension. The force due to surface tension on the ring is

$$\begin{aligned} F_\sigma &= F_{\sigma,i} + F_{\sigma,o} \\ &= \pi\sigma(D_i + D_o) \end{aligned}$$

In addition to the preceding cases, surface tension is an important force in the shattering of liquid droplets, the shape and motion of bubbles, and the structure of foams.

Vapor Pressure

The pressure at which a liquid will boil is called its *vapor pressure*. This pressure is a function of temperature (vapor pressure increases with temperature). In this context we usually think about the temperature at which boiling occurs. For example, water boils at 212°F at sea-level atmospheric pressure (14.7 psia). However, in terms of vapor pressure, we can say that by increasing the temperature of water at sea level to 212°F, we increase the vapor pressure to the point at which it is equal to the atmospheric pressure (14.7 psia), so that boiling occurs. When we think of incipient boiling in terms of vapor pressure, it is easy to visualize that boiling can also occur in water at temperatures much below 212°F if the pressure in the water is reduced to its vapor pressure. For example, the vapor pressure of water at 50°F (10°C) is 0.178 psia (approximately 1% of standard atmospheric pressure). Therefore, if the pressure within water at that temperature is reduced to that value, the water boils.* Such boiling often occurs in flowing liquids, such as on the suction side of a pump. When such boiling does occur in flowing liquids, vapor bubbles start growing in local regions of very low pressure

* Actually, boiling can occur at this vapor pressure only if there is a gas–liquid surface present to allow the process to start. Boiling at the bottom of a pot of water is usually initiated in crevices in the material of the pot, in which minute bubbles of air are entrapped even when the pot is filled with water.

and then collapse in regions of higher pressure downstream. This phenomenon, which is called *cavitation,* is discussed in Chapter 5.

Table A.5 in the Appendix gives values of vapor pressure for water.

2.9 Summary

The two systems of units used for fluid mechanics are the traditional (foot–pound–second) system and the Système International (kilogram–meter–second), known as the SI system. The fundamental mass unit in the traditional system is the slug, which is related to the pound mass (lbm) by

$$1 \text{ slug} = 32.2 \text{ lbm}$$

The commonly used properties are

Mass density (ρ): mass per unit volume

Specific weight (γ): weight per unit volume

Specific gravity (S): ratio of specific weight to specific weight of water at reference conditions.

The relationship between pressure, density, and temperature for an ideal gas is

$$p = \rho R T$$

where R is the gas constant, and pressure and temperature must be expressed in absolute values.

In a fluid the shear stress is proportional to the rate of strain, and the constant of proportionality is the viscosity. The shear stress at a wall is given by

$$\tau = \mu \frac{dV}{dy}$$

where dV/dy is the velocity gradient of the fluid evaluated at the wall. In a Newtonian fluid, the viscosity is independent of the rate of strain. A fluid for which the effective viscosity decreases with increasing strain rate is a shear-thinning fluid.

Surface tension is the result of molecular attraction near a free surface, causing the surface to act like a stretched membrane.

The bulk modulus relates to the pressure required to change the density of a fluid.

The vapor pressure at a given temperature is the pressure at which a liquid boils.

References

1. Bird, R. B., Stewart, W. E., and Lightfoot, E. N. *Transport Phenomena*. John Wiley & Sons, New York, 1960.

2. Harris, J. *Rheology and Non-Newtonian Flow.* Longman, New York, 1977.

3. Schowalter, W. R. *Mechanics of Non-Newtonian Fluids.* Pergamon Press, New York, 1978.

Problems

2.1 An engineer living at an elevation of 2500 ft is conducting experiments to verify predictions of glider performance. To process data, density of ambient air is needed. The engineer measures temperature (74.3°F) and atmospheric pressure (27.3 inches of mercury). Calculate density in units of kg/m^3. Compare the calculated value with data from Table A.3 and make a recommendation about the effects of elevation on density; that is, are the effects of elevation significant?

2.2 Calculate the density and specific weight of carbon dioxide at 300 kN/m^2 absolute and 60°C.

2.3 Calculate the density and specific weight of methane at 500 kN/m^2 absolute and 60°C.

2.4 Natural gas is stored in a spherical tank at a temperature of 10°C. At a given initial time, the pressure in the tank is 100 kPa gage, and the atmospheric pressure is 100 kPa absolute. Some time later, after considerably more gas is pumped into the tank, the pressure in the tank is 200 kPa gage, and the temperature is still 10°C. What will be the ratio of the mass of natural gas in the tank when p = 200 kPa gage to that when the pressure was 100 kPa gage?

2.5 At a temperature of 100°C and an absolute pressure of 5 atmospheres, what is the ratio of the density of water to the density of air, ρ_w/ρ_a?

2.6 What is the weight of a 10-ft^3 tank of oxygen if the oxygen is pressurized to 400 psia, the tank itself weighs 100 lbf, and the temperature is 70°F?

2.7 What are the specific weight and density of air at an absolute pressure of 600 kPa and a temperature of 50°C?

2.8 Meteorologists often refer to air masses in forecasting the weather. Estimate the mass of 1 mi^3 of air in slugs and kilograms. Make your own reasonable assumptions with respect to the conditions of the atmosphere.

2.9 A bicycle rider has several reasons to be interested in the effects of temperature on air density. The aerodynamic drag force decreases linearly with density. Also, a change in temperature will affect the tire pressure.

a. To visualize the effects of temperature on air density, write a computer program that calculates the air density at atmospheric pressure for temperatures from –10°C to 50°C.

b. Also assume that a bicycle tire was inflated to an absolute pressure of 450 kPa at 20°C. Assume the volume of the tire does not change with temperature. Write a program to show how the tire pressure changes with temperature in the same temperature range, –10°C to 50°C.

Prepare a table or graph of your results for both problems. What engineering insights do you gain from these calculations?

2.10 A design team is developing a prototype CO_2 cartridge for a manufacturer of rubber rafts.This cartridge will allow a user to quickly inflate a raft. A typical raft is shown in the sketch. Assume a raft inflation pressure of 3 psi (this means that the absolute pressure is 3 psi greater than local atmospheric pressure). Estimate the volume of the raft and the mass of CO_2 in grams in the prototype cartridge.

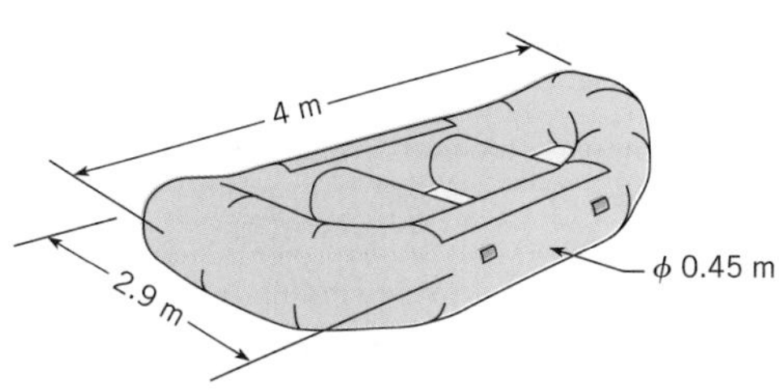

PROBLEM 2.10

2.11 A team is designing a helium-filled balloon that will fly to an altitude of 80,000 ft. As the balloon ascends, the upward force (buoyant force) will need to exceed the total weight. Thus, weight is critical. Estimate the weight (in newtons) of the helium inside the balloon. The balloon is inflated at a site where the atmospheric pressure is 0.89 bar and the temperature is 22°C. When inflated prior to launch, the balloon is spherical (radius 1.3 m) and the inflation pressure equals the local atmospheric pressure.

2.12 Hydrometers are used in the wine and beer industries to measure the alcohol content of the product. This is accomplished by measuring the specific gravity of the liquid before fermentation, during fermentation, or after fermentation is complete. During fermentation, glucose ($C_6H_{12}O_6$) is converted to ethyl alcohol (CH_3CH_2OH) and carbon dioxide gas, which escapes from the vat.

$$C_6H_{12}O \longrightarrow 2(CH_3CH_2OH) + 2(CO_2)$$

Brewer's yeast tolerates alcohol contents to approximately 5% before fermentation stops, whereas wine yeast tolerates alcohol contents up to 21% depending on the yeast strain. The specific gravity of alcohol is 0.80, and the maximum specific gravity of sugar in solution is 1.59. If a wine has a specific gravity of 1.08 before fermentation, and all the sugar is converted to alcohol, what will be the final specific gravity of the wine and the percent alcohol content by volume? Assume that the initial liquid (the unfermented wine is called must) contains only sugar and water.

2.13 What is the change in the viscosity and density of water between 10°C and 70°C? What is the change in the viscosity and density of air between 10°C and 70°C? Assume standard atmospheric pressure ($p = 101$ kN/m^2 absolute).

2.14 What is the change in the kinematic viscosity of air between 10°C and 60°C? Assume standard atmospheric pressure.

2.15 Find the dynamic and kinematic viscosities of kerosene, SAE 10W-30 motor oil, and water at a temperature of 38°C (100°F).

2.16 What is the ratio of the dynamic viscosity of air to that of water at standard pressure and a temperature of 20°C? What is the ratio of the kinematic viscosity of air to that of water for the same conditions?

2.17 Write a computer program to calculate the dynamic and kinematic viscosities of air using Sutherland's equation. Using Sutherland's constants from Table A.2, compare several calculated values with the data provided in Fig. A.3.

2.18 Using Sutherland's equation and the ideal gas law, develop an expression for the kinematic viscosity ratio ν/ν_0 in terms of pressures p and p_0 and temperatures T and T_0, where the subscript 0 refers to a reference condition.

2.19 The dynamic viscosity of air at 15°C is 1.78×10^{-5} N · s/m^2. Using Sutherland's equation, find the viscosity at 200°C.

2.20 The kinematic viscosity of methane at 15°C and atmospheric pressure is 1.59×10^{-5} m^2/s. Using Sutherland's equation and the ideal gas law, find the kinematic viscosity at 200°C and 2 atmospheres.

2.21 The dynamic viscosity of nitrogen at 59°F is 3.59×10^{-7} lbf · s/ft^2. Using Sutherland's equation, find the dynamic viscosity at 200°F.

2.22 The kinematic viscosity of helium at 59°F and 1 atmosphere is 1.22×10^{-3} ft^2/s. Using Sutherland's equation and the ideal gas law, find the kinematic viscosity at 30°F and a pressure of 1.5 atmospheres.

2.23 The absolute viscosity of propane at 100°C is 1.00×10^{-5} N · s/m^2 and at 400°C is 1.72×10^{-5} N · s/m^2. Find Sutherland's constant for propane.

2.24 The absolute viscosity of ammonia at 68°F is 2.07×10^{-7} lbf · s/ft^2 and at 392°F is 3.46×10^{-7} lbf · s/ft^2. Using these two data points, find Sutherland's constant for ammonia.

2.25 The viscosity of SAE 10W-30 motor oil at 38°C is 0.067 N · s/m^2 and at 99°C is 0.011 N · s/m^2. Using Eq. (2.8) for interpolation, find the viscosity at 60°C. Compare this value with that obtained by linear interpolation.

2.26 The viscosity of grade 100 aviation oil at 100°F is 4.43×10^{-3} lbf · s/ft^2 and at 210°F is 3.9×10^{-4} lbf · s/ft^2. Using Eq. (2.8), find the viscosity at 150°F.

2.27 Develop a program that gives the Sutherland constant in terms of μ/μ_0 and T/T_0, where the subscript represents the reference value. Enter the following data for carbon monoxide, find the Sutherland constant for each data point, and take the average to find the best value. Using this value for Sutherland's constant, calculate the viscosity ratio using Sutherland's equation. Find the percentage error at each data point and make a statement about the adequacy of Sutherland's equation in this application. The reference temperature is 273 K.

μ/μ_0	0.9605	0.9906	1.020
T (K)	260	270	280
μ/μ_0	1.049	1.078	1.213
T (K)	290	300	350
μ/μ_0	1.574	2.519	3.285
T (K)	500	1000	1500

2.28 Two plates are separated by a 1/8-in. space. The lower plate is stationary; the upper plate moves at a velocity of 25 ft/s. Oil (SAE 10W-30, 150°F), which fills the space between the plates, has the same velocity as the plates at the surface of contact. The variation in velocity of the oil is linear. What is the shear stress in the oil?

2.29 Find the kinematic and dynamic viscosities of air and water at a temperature of 40°C (104°F) and an absolute pressure of 170 kPa (25 psia).

2.30 The velocity distribution for water (20°C) near a wall is given by $u = a(y/b)^{1/6}$, where $a = 10$ m/s, $b = 2$ mm, and y is the distance from the wall in mm. Determine the shear stress in the water at $y = 1$ mm.

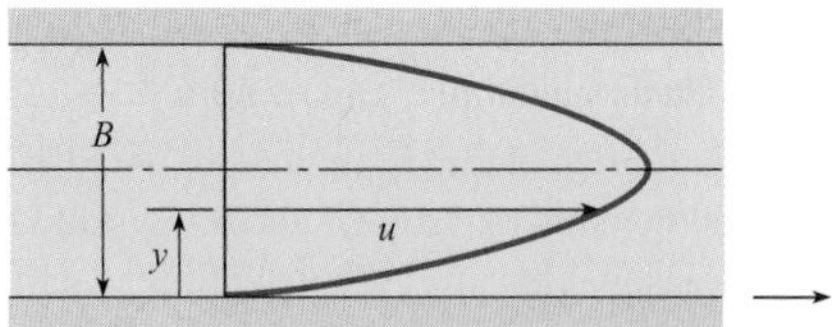

PROBLEMS 2.31, 2.32, 2.33

2.31 The velocity distribution for the flow of crude oil at 100°F ($\mu = 8 \times 10^{-5}$ lbf · s/ft^2) between two walls is given by $u = 100y(0.1 - y)$ ft/s, where y is measured in feet and the space between the walls is 0.1 ft. Plot the velocity distribution and determine the shear stress at the walls.

2.32 A liquid flows between parallel boundaries as shown above. The velocity distribution near the lower wall is given in the following table.

y in mm	V in m/s
1.0	1.00
2.0	1.99
3.0	2.98

a. If the viscosity of the liquid is 10^{-3} N · s/m^2, what is the maximum shear stress in the liquid?

b. Where will the minimum shear stress occur?

2.33 TSuppose that glycerin is flowing ($T = 20°$C) and that the pressure gradient dp/dx is –1.6 kN/m^3. What are the velocity and shear stress at a distance of 12 mm from the wall if the space B between the walls is 5.0 cm? What are the shear stress and velocity at the wall? The velocity distribution for viscous flow between stationary plates is

$$u = -\frac{1}{2\mu}\frac{dp}{dx}(By - y^2)$$

2.34 A laminar flow occurs between two horizontal parallel plates under a pressure gradient dp/ds (p decreases in the positive s direction). The upper plate moves left (negative) at velocity u_t. The expression for local velocity u is given as

$$u = -\frac{1}{2\mu}\frac{dp}{ds}(Hy - y^2) + u_t\frac{y}{H}$$

Is the magnitude of the shear stress greater at the moving plate ($y = H$) or at the stationary plate ($y = 0$)?

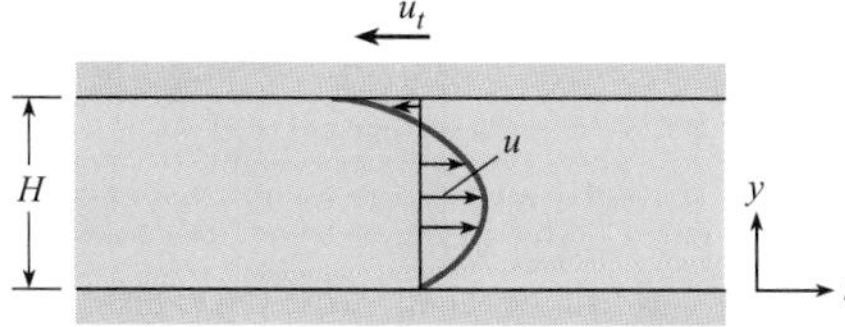

PROBLEM 2.34

2.35 TFor the conditions of Prob. 2.34, derive an expression for the y position of zero shear stress.

2.36 For the conditions of Prob. 2.34, derive an expression for the plate speed u_t required to make the shear stress zero at $y = 0$.

2.37 TA special damping device consists of a sphere, as shown in the figure. The sphere is enclosed in a spherical cavity with the distance between the sphere surface and the interior wall of the cavity being 1 mm. The space between the sphere and the wall is filled with oil (SAE 10W at 38°C). The diameter of the sphere is 100 mm. The sphere is turned by a shaft that has a diameter much less than the diameter of the sphere. Neglect viscous effects on the shaft. Determine the torque on the shaft for a rotation rate of 10 rpm.

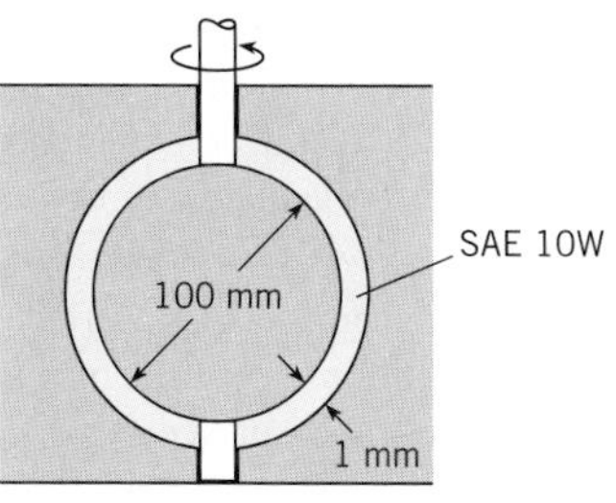

PROBLEM 2.37

2.38 Consider the ratio μ_{100}/μ_{50}, where μ is the viscosity of oxygen and the subscripts 100 and 50 are the temperatures of the oxygen in degrees Fahrenheit. Does this ratio have a value (a) less than 1, (b) equal to 1, or (c) greater than 1?

2.39 A solid circular cylinder of diameter d and length ℓ slides inside a vertical smooth pipe that has an inside diameter D. The small space between the cylinder and the pipe is lubricated with an oil film that has a viscosity μ. Derive a formula for the steady rate of descent of the cylinder in the vertical pipe. Assume that the cylinder has a weight W and is concentric with the pipe as it falls. Use the formula to find the rate of descent of a cylinder 100 mm in diameter that slides inside a 100.5-mm pipe. The cylinder is 200 mm long and weighs 20 N. The lubricant is SAE 20W oil at 10°C.

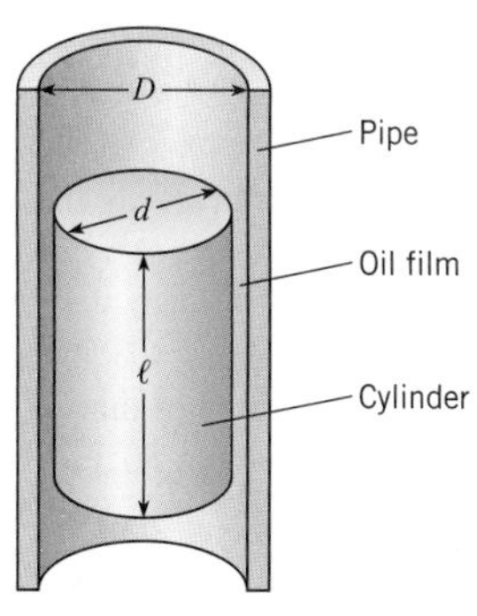

PROBLEMS 2.39, 2.40

2.40 Consider the pipe, cylinder, and oil described in Prob. 2.39. Suppose that the cylinder has a downward velocity of 0.5 m/s and is observed to be decelerating at a rate of 14 m/s^2. What is its weight?

2.41 The device shown consists of a disk that is rotated by a shaft. The disk is positioned very close to a solid boundary. Between the disk and the boundary is viscous oil.

a. If the disk is rotated at a rate of 1 rad/s, what will be the ratio of the shear stress in the oil at $r = 2$ cm to the shear stress at $r = 3$ cm?

b. If the rate of rotation is 2 rad/s, what is the speed of the oil in contact with the disk at $r = 3$ cm?

c. If the oil viscosity is 0.01 N · s/m^2 and the spacing y is 2 mm, what is the shear stress for the conditions noted in part (b)?

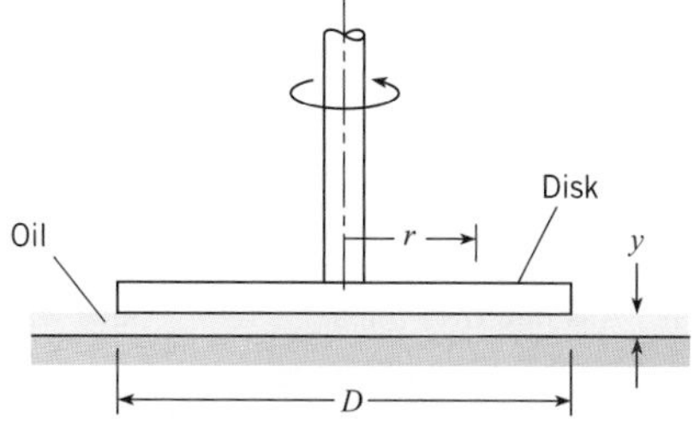

PROBLEMS 2.41, 2.42

2.42 What torque is required to rotate the disk of Prob. 2.41 at a rate of 5 rad/s, with $D = 10$ cm and with the same viscosity and spacing as in part (c)?

FLUID SOLUTIONS **2.43** Some instruments having angular motion are damped by means of a disk connected to the shaft. The disk, in turn, is immersed in a container of oil, as shown. Derive a formula for the damping torque as a function of the disk diameter D, spacing S, rate of rotation ω, and oil viscosity μ.

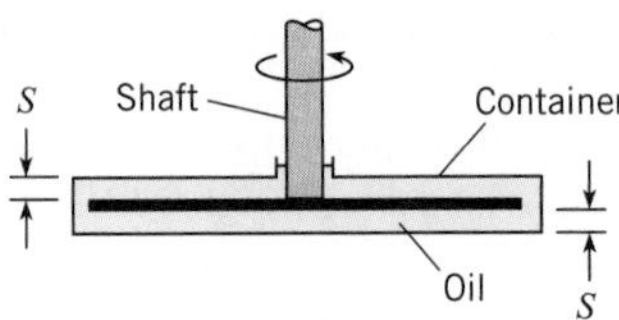

PROBLEM 2.43

2.44 One type of viscometer involves the use of a rotating cylinder inside a fixed cylinder The gap between the cylinders must be very small to achieve a linear velocity distribution in the liquid. (Assume the maximum spacing for proper operation is 0.05 in.). Design a viscometer that will be used to measure the viscosity of motor oil from 50°F to 200°F.

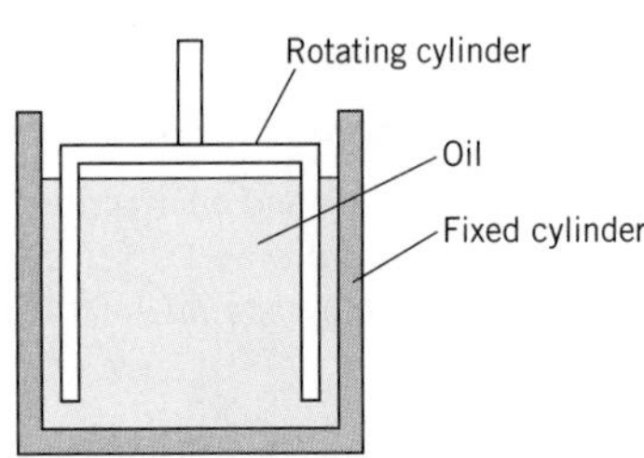

PROBLEM 2.44

2.45 A pressure of 2×10^6 N/m^2 is applied to a mass of water that initially filled a 1000-cm volume. Estimate its volume after the pressure is applied.

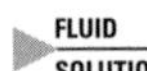

2.46 What pressure increase must be applied to water to reduce its volume by 1%?

2.47 Which of the following is the formula for the gage pressure within a very small spherical droplet of water?
(a) $p = \sigma/d$, (b) $p = 4\sigma/d$, (c) $p = 8\sigma/d$.

2.48 A spherical soap bubble has an inside radius R, a film thickness t, and a surface tension σ. Derive a formula for the pressure within the bubble relative to the outside atmospheric pressure. What is the pressure difference for a bubble with a 4-mm radius? Assume σ is the same as for pure water.

2.49 A water bug is suspended on the surface of a pond by surface tension (water does not wet the legs). The bug has six legs, and each leg is in contact with the water over a length of 5 mm. What is

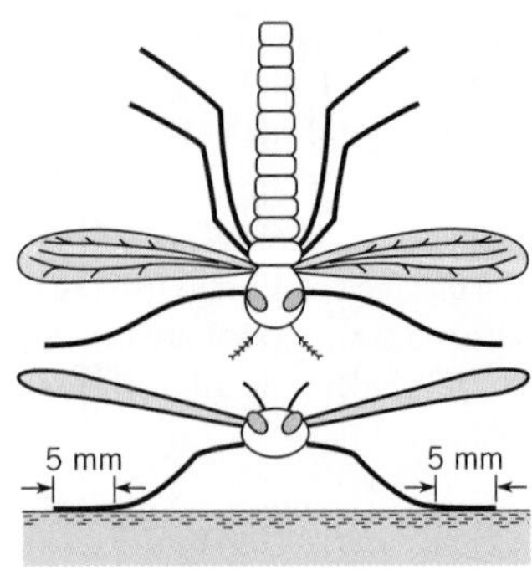

PROBLEM 2.49

2.50 A water column in a glass tube is used to measure the pressure in a pipe. The tube is 1/4 in. (6.35 mm) in diameter. How much of the water column is due to surface-tension effects? What would be the surface-tension effects if the tube were 1/8 in. (3.2 mm) or 1/32 in. (0.8 mm) in diameter?

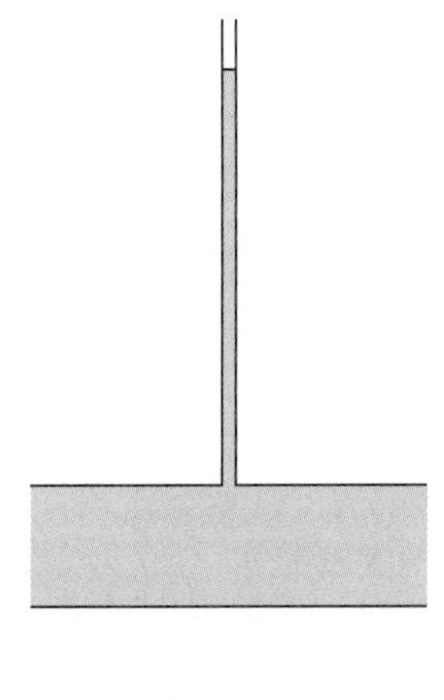

PROBLEM 2.50

2.51 Calculate the maximum capillary rise of water between two vertical glass plates spaced 1 mm apart.

2.52 What is the pressure within a 1-mm spherical droplet of water relative to the atmospheric pressure outside?

2.53 By measuring the capillary rise in a tube, one can calculate the surface tension. The surface tension of water varies linearly with temperature from 0.0756 N/m at 0°C to 0.0589 N/m at 100°C. Size a tube (specify diameter and length) that uses capillary rise of water to measure temperature in the range from 0°C to 100°C. Is this concept a good idea?

2.54 Mercury does not adhere to a glass surface, so when a glass tube is immersed in a pool of mercury, the meniscus is depressed, as shown in the figure. The surface tension of mercury is 0.514 N/m and the angle of contact is 40°. Find the depression distance in a 1-mm glass tube.

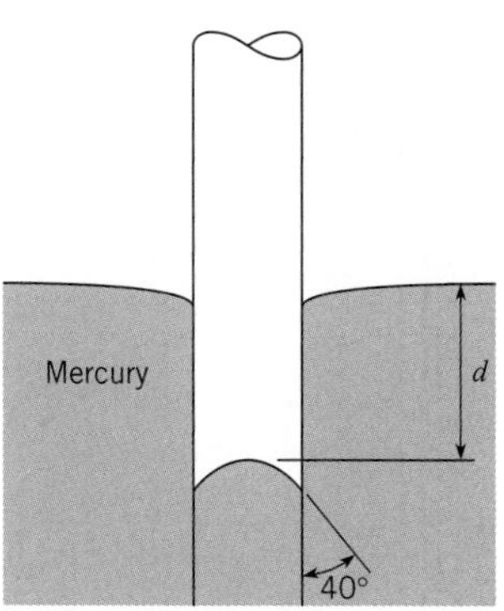

PROBLEM 2.54

2.55 Consider a soap bubble 2 mm in diameter and a droplet of water, also 2 mm in diameter, that are falling in air. If the value of the surface tension for the film of the soap bubble is assumed to be the same as that for water, which has the greater pressure inside it? (a) the bubble, (b) the droplet, (c) neither—the pressure is the same for both.

2.56 A drop of water at 20°C is forming under a surface. The configuration just before separating and falling as a drop is shown in the figure. Assume the forming drop has the volume of a hemisphere. What is the diameter of the hemisphere just before separating?

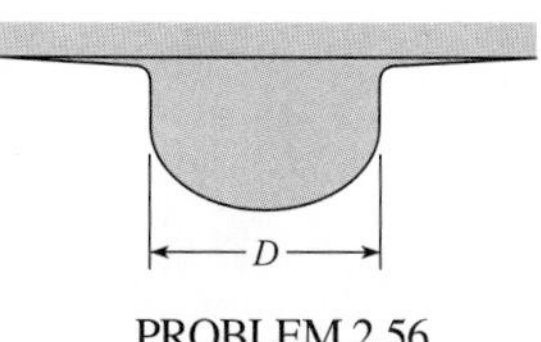

PROBLEM 2.56

2.57 The surface tension of a liquid is being measured with a ring as shown in Fig. 2.6d. The ring has an outside diameter of 10 cm and an inside diameter of 9.5 cm. The mass of the ring is 10 g. The force required to pull the ring from the liquid is the weight corresponding to a mass of 14 g. What is the surface tension of the liquid (in N/m)?

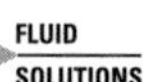

2.58 The vapor pressure of water at 100°C is 101 kN/m^2, because water boils under these conditions. The vapor pressure of water decreases approximately linearly with decreasing temperature at a rate of 3.1 $\text{kN/m}^2/°\text{C}$. Calculate the boiling temperature of water at an altitude of 3000 m, where the atmospheric pressure is 69 kN/m^2 absolute.

Tornados are intense, rotating columns of air that extend from thunderclouds to ground level. They become visible when condensation occurs in the funnel due to very low pressures or when dust, dirt, and debris are transported upward. They can vary from the order of 100 ft to a mile in diameter, with peak wind speeds as high as 300 mph. (Corbis Digital Stock.)

C H A P T E R 3

Fluid Statics

In general, fluids exert both normal and shearing forces on surfaces that are in contact with them. However, only fluids with velocity gradients produce shearing forces. For fluids at rest, only normal forces exist. These normal forces in fluids are called *pressure forces.*

3.1 Pressure

Definition of Pressure

At every point in a static fluid a certain pressure intensity exists. Specifically, this pressure intensity, usually simply called pressure, is defined as follows:

$$p = \lim_{\Delta A \to 0} \frac{\Delta F}{\Delta A} = \frac{dF}{dA} \tag{3.1}$$

where F is the normal force acting over the area A. Pressure intensity is a scalar quantity; that is, it has magnitude only and acts equally in all directions. This is easily demonstrated by considering the wedge-shaped element of fluid in equilibrium in Fig. 3.1. The forces that act on the element are the surface forces and the weight force.

By writing the equation of equilibrium for the x direction, we obtain

$$(p_n \Delta y \Delta \ell) \sin\alpha - p_x(\Delta y \Delta \ell \ \sin\alpha) = 0$$

or

$$p_n = p_x \tag{3.2}$$

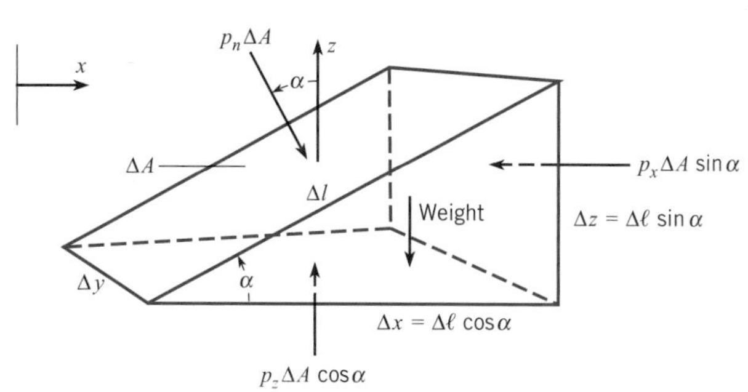

FIGURE 3.1

Pressure forces on a fluid element in equilibrium.

For the z direction, we obtain

$$-(p_n \Delta y \Delta \ell)\cos\alpha + p_z(\Delta y \Delta \ell \cos\alpha) - \tfrac{1}{2}\gamma \Delta \ell \cos\alpha \Delta \ell \sin\alpha \Delta y = 0$$

Now, when we divide this equation by the product $\Delta\ell\Delta y\cos\alpha$ and shrink the element to a point $(\Delta\ell \longrightarrow 0)$, the last term disappears. Thus we have

$$p_n = p_z \tag{3.3}$$

Combining Eqs. (3.2) and (3.3), we finally arrive at the result

$$p_n = p_x = p_z \tag{3.4}$$

Since the angle α (alpha) is arbitrary and p_n is independent of α, we conclude that the pressure at a point in a static fluid acts with the same magnitude in all directions:

$$p_n = p_x = p_y = p_z$$

Pressure Transmission

In a closed system, a pressure change produced at one point in the system will be transmitted throughout the entire system. The principle is known as Pascal's law after Blaise Pascal, the French scientist who first stated it in 1653. This phenomenon of pressure transmission, along with the ease with which fluids can be moved, has led to the widespread development of hydraulic controls for operating equipment such as aircraft-control surfaces, heavy earthmoving equipment, and hydraulic presses. Figure 3.2 is an illustration of the application of this principle in the form of a hydraulic lift used in service stations. Here air pressure from a compressor establishes the pressure in the oil system, which in turn acts against the piston in the lift. It can be seen that if a pressure of 600 kN/m^2, for example, acts on the 25-cm-diameter piston, then a force equal to pA, or 29.5 kN, will be exerted on the piston. To handle larger or smaller loads it is necessary only to increase or decrease the pressure.

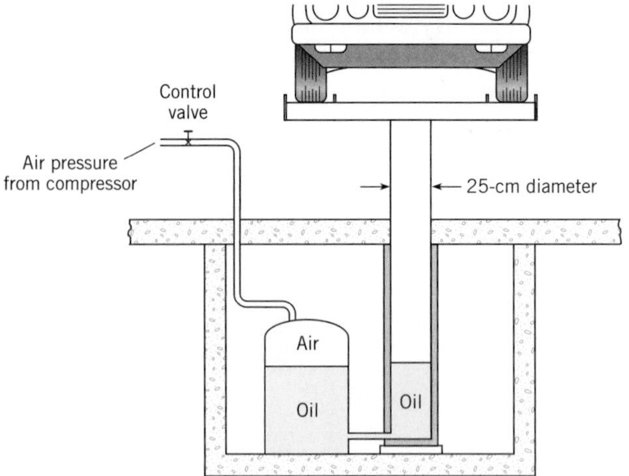

FIGURE 3.2

Hydraulic lift.

example 3.1

A hydraulic jack has the dimensions shown. If one exerts a force F of 100 N on the handle of the jack, what load, F_2, can the jack support? Neglect lifter weight.

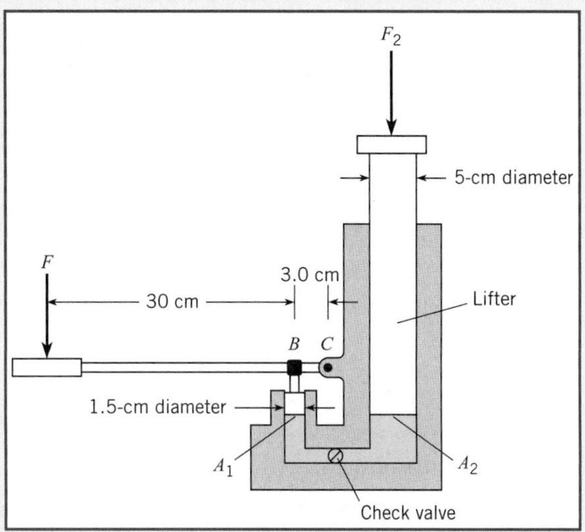

Solution The force F_1 exerted on the small piston is obtained by taking moments about C. Therefore,

$$(0.33 \text{ m}) \times (100 \text{ N}) - (0.03 \text{ m})F_1 = 0$$

$$F_1 = \frac{0.33 \text{ m} \times 100 \text{ N}}{0.03 \text{ m}} = 1100 \text{ N}$$

Because the small piston is in equilibrium, this force is equal to the pressure force on the piston, or

$$p_1 A_1 = 1100 \text{ N}$$

Hence
$$p_1 = \frac{1100}{A_1} = \frac{1100}{\pi d^2/4} = 6.22 \times 10^6 \text{ N/m}^2$$

Now we know the pressure in the liquid. Therefore, we can solve for the force on the large piston. Since surfaces 1 and 2 are at the same elevation, $p_1 = p_2$, so

$$F_2 = p_1 A_2$$

where A_2 is the area of the large piston. Finally,

$$F_2 = 6.22 \times 10^6 \frac{\text{N}}{\text{m}^2} \times \frac{\pi}{4} \times (0.05 \text{ m})^2 = 12.22 \text{ kN}$$ ◁

Absolute Pressure, Gage Pressure, and Vacuum

In a region such as outer space, which is virtually void of gases, the pressure is essentially zero. Such a condition can be approached very nearly in the laboratory when a vacuum pump is used to evacuate a bottle. The pressure in a vacuum is called *absolute zero,* and all pressures referenced with respect to this zero pressure are termed *absolute pressures.* Therefore, atmospheric pressure at sea level on a particular day might be given as 101 kN/m^2, which is equivalent to 760 mm of deflection on a mercury barometer.

Many pressure-measuring devices measure not absolute pressure but only differences in pressure. For example, a common Bourdon-tube gage (see Sec. 3.3) indicates only the difference between the pressure in the fluid to which it is tapped and the pressure in the atmosphere. In this case, then, the reference pressure is actually the atmospheric pressure at the gage. This type of pressure reading is called *gage pressure.*

The fundamental unit of pressure in the SI system is the pascal (Pa), which is one newton per square meter (N/m^2). Gage and absolute pressures are usually identified after the unit.* For example, if a pressure of 50 kPa is measured with a gage referenced to the atmosphere and the absolute atmospheric pressure is 100 kPa, then the pressure can be expressed as either

$$p = 50 \text{ kPa gage} \quad \text{or} \quad p = 150 \text{ kPa absolute}$$

Whenever atmospheric pressure is used as a reference (or, in other words, when gage pressure is being measured), the possibility exists that the pressure thus measured can be either positive or negative. Negative gage pressures are also termed *vacuum pressures.* Hence, if a gage tapped into a tank indicates a vacuum pressure of 31.0 kPa, this can also be stated as 70.0 kPa absolute, or –31.0 kPa gage, assuming that the atmospheric pressure is 101 kPa absolute. An example of this reference system is depicted in Fig. 3.3 for arbitrary pressures of p_A = 200 kPa gage and p_B = 51 kPa absolute with an atmospheric pressure of 101 kPa absolute.

FIGURE 3.3

Example of pressure relations.

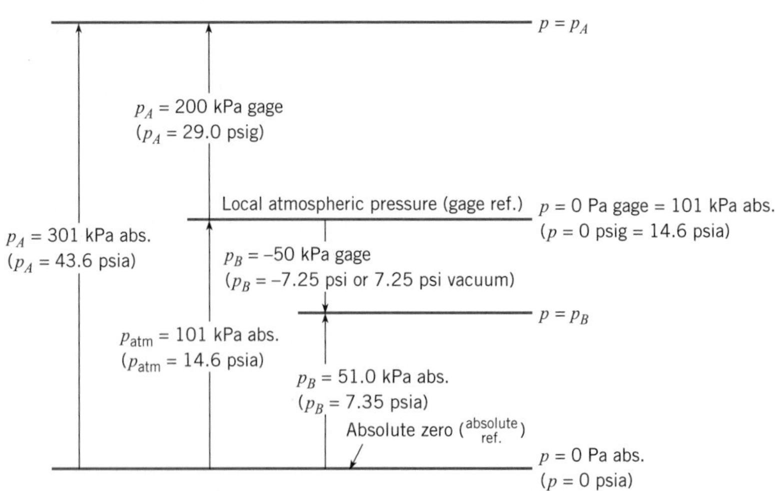

* In the traditional foot–pound–second system of units, the gage or absolute designations are usually included as part of the abbreviated unit. For example, a gage pressure of 10 pounds per square foot is designated as psfg. Other combinations are psfa, psig, psia. The latter two designations are for pounds per square inch gage and pounds per square inch absolute.

Pressure Variation with Elevation

Basic Differential Equation

For a static fluid, pressure varies only with the elevation within the fluid. This may be shown by isolating a cylindrical element of fluid and applying the equation of equilibrium to the element. Consider the element shown in Fig. 3.4. Here the element is oriented so that its longitudinal axis is parallel to an arbitrary ℓ direction. The element is $\Delta\ell$ long, ΔA in cross-sectional area, and inclined at an angle α with the horizontal. The equation of equilibrium for the ℓ direction, considering the pressure forces and gravitational force acting on the element in this direction, is

$$\Sigma F_\ell = 0$$

$$p\Delta A - (p + \Delta p)\Delta A - \gamma\Delta A\Delta\ell \sin\alpha = 0$$

Upon simplifying, and dividing by the volume of the element, $\Delta\ell\Delta A$, this reduces to

$$\frac{\Delta p}{\Delta\ell} = -\gamma\sin\alpha$$

However, if we let the length of the element approach zero, then in the limit $\Delta p/\Delta\ell = dp/d\ell$. Also one notes that $\sin\alpha = dz/d\ell$. Therefore,

$$\frac{dp}{d\ell} = -\gamma\frac{dz}{d\ell} \tag{3.5}$$

This can also be written as

$$\frac{dp}{dz} = -\gamma \tag{3.6}$$

which is the basic equation for hydrostatic pressure variation with elevation. Equation (3.5) states that for static fluids a change of pressure in the ℓ direction, $dp/d\ell$, occurs only when there is a change of elevation in the ℓ direction, $dz/d\ell$. In other words, if one considers a path through the fluid that lies in a horizontal plane, the pressure everywhere along this path is constant. On the other hand, the greatest possible change in hydrostatic pressure occurs along a vertical path through the fluid. Furthermore, Eqs. (3.5) and (3.6)

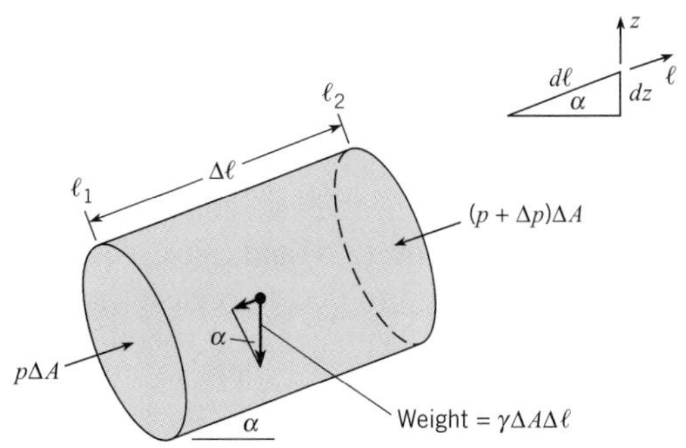

FIGURE 3.4

Variation in pressure with elevation.

state that the pressure changes inversely with elevation. If one travels upward in the fluid (positive z direction), the pressure decreases; and if one goes downward (negative z), the pressure increases. Of course, a pressure increase is exactly what a diver experiences when descending in a lake or pool.

example 3.2

Compare the rate of change of pressure with elevation for air at sea level, 101.3 kPa absolute, at a temperature of 15.5°C, and for fresh water at the same pressure and temperature. Assuming constant specific weights for air and water, determine also the total pressure change that occurs for both with a 4-m decrease in elevation.

Solution First, determine specific weights of water and air:

$$\rho_{\text{air}} = \frac{p}{RT} = \frac{101.3 \times 10^3 \text{ N/m}^2}{287 \text{ J/kg K} \times (15.5 + 273) \text{ K}}$$

Then

$$\rho_{\text{air}} = 1.22 \text{ kg/m}^3$$

$$\gamma_{\text{air}} = \rho g = 1.22 \text{ kg/m}^3 \times 9.81 \text{ m/s}^2$$

$$= 11.97 \text{ kg/m}^2\text{s}^2 = 11.97 \text{ N/m}^3$$

and

$$\gamma_{\text{water}} = 9799 \text{ N/m}^3 \quad \text{(interpolated from Table A.5)}$$

$$\frac{dp}{dz} = -\gamma$$

Then

$$\left(\frac{dp}{dz}\right)_{\text{air}} = -11.97 \text{ N/m}^3 \qquad \left(\frac{dp}{dz}\right)_{\text{water}} = -9799 \text{ N/m}^3$$

$$\text{Total pressure change for air} = (-11.97 \text{ N/m}^3) \times (-4 \text{ m})$$

$$= 47.9 \text{ N/m}^2 = 47.9 \text{ Pa} \quad \triangleleft$$

$$\text{Total pressure change for water} = (-9799 \text{ N/m}^3) \times (-4 \text{ m})$$

$$= 39.2 \text{ kN/m}^2 = 39.2 \text{ kPa} \quad \triangleleft$$

Pressure Variation for a Uniform-Density Fluid

Equations (3.5) and (3.6) are completely general in the sense that they describe the rate of change of pressure for all fluids in static equilibrium. However, much simplification accrues in practical applications of the equations if it can be assumed that the density, and thus the specific weight, of the fluid are uniform throughout the fluid. Then γ is a constant in Eqs. (3.5) and (3.6). The reason for the simplification is that the integration of Eq. (3.5) or (3.6) becomes easier and the resulting equation is simpler than if γ were a function of z. With constant specific weight, the following equation results from integration of Eq. (3.6):

$$p + \gamma z = \text{constant} \tag{3.7}$$

This sum of pressure and γz is known as the *piezometric pressure, p_z*. Dividing Eq. (3.7) by γ gives

$$\left(\frac{p}{\gamma} + z\right) = \text{constant} \tag{3.8}$$

The sum of the terms p/γ and z on the left-hand side of Eq. (3.8) is called the *piezometric head.* As shown by the equation, this is constant throughout an incompressible static fluid. Therefore, one can relate the pressure and elevation at one point to the pressure and elevation at another point in the fluid in the following manner:

$$\frac{p_1}{\gamma} + z_1 = \frac{p_2}{\gamma} + z_2 \tag{3.9}$$

or

$$\Delta p = -\gamma \Delta z \tag{3.10}$$

Equations (3.7) through (3.10) represent the principle that pressure increases with depth in a fluid. Since the equations are basically the same, they are identified as the *hydrostatic equation*. The hydrostatic equation applies only in a fluid with a constant specific weight. Thus Eqs. (3.9) and (3.10) apply to two points in the same fluid but not across an interface of two fluids having different specific weights.

example 3.3

What is the water pressure at a depth of 35 ft in the tank shown?

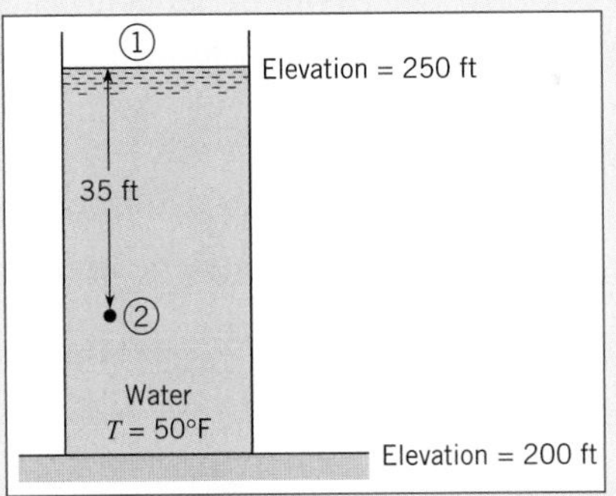

Solution At an elevation of 250 ft the gage pressure is zero, so

$$\frac{p_1}{\gamma} + z_1 = \frac{p_2}{\gamma} + z_2$$

$$0 + 250 = \frac{p_2}{\gamma} + 215$$

$$\frac{p_2}{\gamma} = 35 \text{ ft}$$

$$p_2 = 35 \times 62.4 = 2180 \text{ psfg} = 15.2 \text{ psig}$$ ◁

example 3.4

Oil with a specific gravity of 0.80 forms a layer 0.90 m deep in an open tank that is otherwise filled with water. The total depth of water and oil is 3 m. What is the gage pressure at the bottom of the tank?

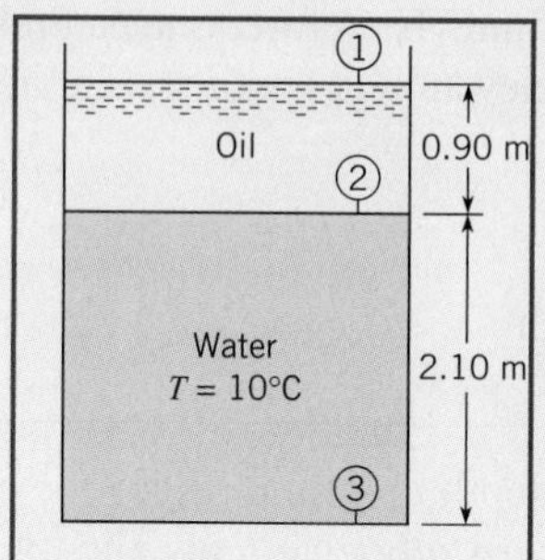

Solution First determine the pressure at the oil–water interface, staying within the oil, and then calculate the pressure at the bottom.

$$\frac{p_1}{\gamma} + z_1 = \frac{p_2}{\gamma} + z_2$$

where p_1 is the pressure at free surface of oil, z_1 is the elevation of free surface of oil, p_2 is the pressure at interface between oil and water, and z_2 is the elevation at interface between oil and water. For this example, $p_1 = 0$, $\gamma = 0.80 \times 9810 \text{ N/m}^3$, $z_1 = 3$ m, and $z_2 = 2.10$ m. Therefore,

$$p_2 = 0.90 \text{ m} \times 0.80 \times 9810 \text{ N/m}^3 = 7.06 \text{ kPa gage}$$

Now obtain p_3 from

$$\frac{p_2}{\gamma} + z_2 = \frac{p_3}{\gamma} + z_3$$

where p_2 has already been calculated and $\gamma = 9810 \text{ N/m}^3$.

$$p_3 = 9810\left(\frac{7060}{9810} + 2.10\right) = 27.7 \text{ kPa gage} \quad \triangleleft$$

Pressure Variation for Compressible Fluids

The preceding section dealt with pressure variation in fluids for which the specific weight is constant. However, when the specific weight varies significantly throughout the fluid, it must be expressed in such a form that Eq. (3.6) can be integrated. For the case of an ideal gas, this is accomplished through the equation of state, which relates the density of the gas to pressure and temperature:

$$\frac{p}{\rho} = RT$$

or

$$\rho = \frac{p}{RT} \tag{3.11}$$

This can be expressed as follows when both sides of Eq. (3.11) are multiplied by g:

$$\gamma = \frac{pg}{RT} \tag{3.12}$$

where R is the gas constant, 287 J/kg · K, for dry air, T is the absolute temperature in K, and p is the absolute pressure in Pa.

Equation (3.12) introduces another variable, temperature, so it becomes necessary to have additional data relating temperature and elevation. If one is interested in the pressure variation in the atmosphere, and if temperature-versus-elevation data for a local area at a given time are available, then one can quite accurately compute pressure versus elevation. Lacking such data, one can resort to the so-called *U.S. standard atmosphere* (1), which is shown in Fig. 3.5. This set of data compiled by the U.S. National Weather Service represents average conditions over the United States at 45° N latitude in July. At sea level the standard atmospheric pressure is 101.3 kPa and the temperature is 296 K. The atmosphere is divided into two layers, the *troposphere* and the *stratosphere.* In the troposphere, defined as the layer between sea level and 13.7 km (45,000 ft), the temperature decreases nearly linearly with increasing elevation at a *lapse rate* of 5.87 K/km. The stratosphere begins at the top of the troposphere, where the temperature is constant at –57.5°C, to an altitude of 16.8 km (55,000 ft), and then the temperature increases monotonically to –38.5°C at 30.5 km (100,000 ft).

We now have sufficient information to calculate the pressure and density at any elevation. Let us first consider the troposphere.

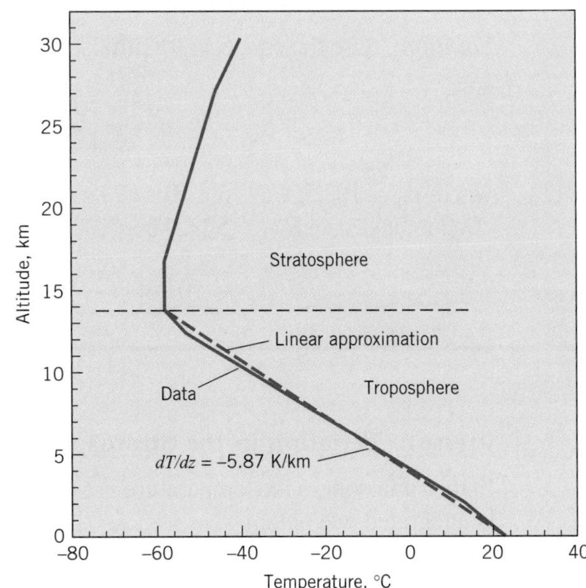

FIGURE 3.5

Temperature variation with altitude for the U.S. standard atmosphere in July (2).

Pressure Variation in the Troposphere

Let the temperature T be given by

$$T = T_0 - \alpha(z - z_0) \tag{3.13}$$

In this equation T_0 is the temperature at a reference level where the pressure is known and α is the lapse rate. If we use the specific weight of a gas from Eq. (3.12) in the basic hydrostatic equation, we obtain

$$\frac{dp}{dz} = -\frac{pg}{RT} \tag{3.14}$$

Substituting Eq. (3.13) for T, we get

$$\frac{dp}{dz} = -\frac{pg}{R[T_0 - \alpha(z - z_0)]}$$

Now we must separate the variables and integrate to obtain

$$\frac{p}{p_0} = \left[\frac{T_0 - \alpha(z - z_0)}{T_0}\right]^{g/\alpha R}$$

Thus the *atmospheric pressure variation* in the troposphere is given by

$$p = p_0\left[\frac{T_0 - \alpha(z - z_0)}{T_0}\right]^{g/\alpha R} \tag{3.15}$$

example 3.5

If at sea level the absolute pressure and temperature are 101.3 kPa and 23°C, what is the pressure at an elevation of 2000 m, assuming that standard atmospheric conditions prevail?

Solution Use the equation for atmospheric pressure variation:

$$p = p_0\left[\frac{T_0 - \alpha(z - z_0)}{T_0}\right]^{g/\alpha R}$$

where $p_0 = 101{,}300\ \text{N/m}^2$, $T_0 = 273 + 23 = 296\ \text{K}$, $\alpha = 5.87 \times 10^{-3}\ \text{K/m}$, $z - z_0 = 2000\ \text{m}$, and $g/\alpha R = 5.823$. Then

$$p = 101.3\left(\frac{296 - 5.87 \times 10^{-3} \times 2000}{296}\right)^{5.823} = 80.0\ \text{kPa} \quad \triangleleft$$

Pressure Variation in the Stratosphere

In the stratosphere the temperature is assumed to be constant. Therefore, when Eq. (3.14) is integrated, we obtain

$$\ln p = -\frac{zg}{RT} + C$$

At $z = z_0$, $p = p_0$, so the foregoing equation reduces to

$$\frac{p}{p_0} = e^{-(z-z_0)g/RT}$$

so the *atmospheric pressure variation* in the stratosphere takes the form

$$p = p_0 e^{-(z-z_0)g/RT} \tag{3.16}$$

example 3.6

If the pressure and temperature are 2.31 psia (p = 15.9 kPa absolute) and –71.5°F (–57.5°C) at an elevation of 45,000 ft (13.72 km), what is the pressure at 55,000 ft (16.77 km), assuming isothermal conditions over this range of elevation?

Solution For isothermal conditions,

$$T = -71.5 + 460 = 388.5°\text{R}$$

$$p = p_0 e^{-(z-z_0)g/RT} = 2.31e^{-(10{,}000)(32.2)/(1716 \times 388.5)} = 2.31e^{-0.483}$$

Therefore the pressure at 55,000 ft is

$$p = 1.43 \text{ psia} \quad \triangleleft$$

SI units $$p = 9.82 \text{ kPa absolute} \quad \triangleleft$$

3·3 Pressure Measurements

Numerous instruments have been devised to indicate the magnitude of pressure intensity, and most of these operate on either the principle of manometry or the principle of flexing of an elastic member whose deflection is directly proportional to the applied pressure. These principles and representative pressure gages are described in the following sections.

Manometry

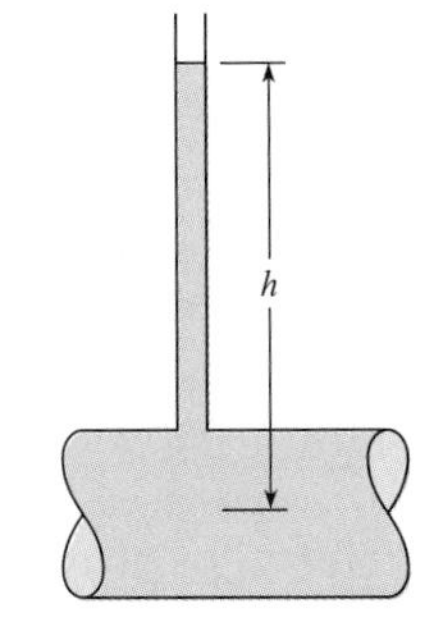

FIGURE 3.6

Piezometer attached to a pipe.

Basically, this method utilizes the change in pressure with elevation to evaluate pressure. Consider the *piezometer,* or simple manometer, attached to a pipe as shown in Fig. 3.6. It is easy to compute the gage pressure at the center of the pipe; here the pressure is simply $p = \gamma h$, which follows directly from Eq. (3.10). This type of pressure-indicating device is accurate and simple. However, the student may visualize how impractical it might become for measuring high pressures, and of course it is useless in its present form for pressure measurement in gases. For both of these cases, a U-tube (a *differential manometer*) such as that shown in Fig. 3.7 can be employed. In this case, a knowledge of the specific weights of the fluids involved and of the linear measurements ℓ and Δh is needed to calculate the pressure in the pipe. Here the procedure is to calculate the pressure changes, step by step, from one level to the next in each fluid and to apply these changes finally to evaluate the unknown pressure. The following example illustrates the procedure for the case shown in Fig. 3.7.

FIGURE 3.7

U-tube manometer.

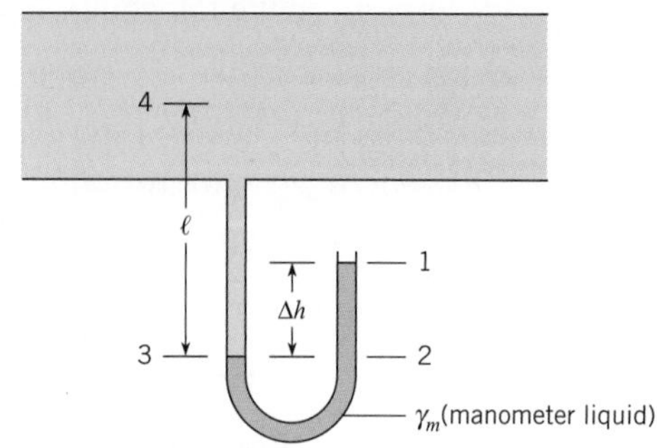

example 3.7

Water is the liquid in the pipe of Fig. 3.7 and mercury is the manometer fluid. If the deflection Δh is 60 cm and ℓ is 180 cm, what is the gage pressure at the center of the pipe?

Solution Since the manometer is open to the atmosphere, we know that the gage pressure at point 1, the mercury surface, is zero. Then the pressure at 2 will be

$$
\begin{aligned}
p_2 &= p_1 + \text{change in pressure between 1 and 2} = 0 - \gamma_m \Delta h \\
&= 0 - \gamma_m(-0.60) \qquad \text{where } \gamma_m = 133 \text{ kN/m}^3 \\
&= 79.8 \text{ kPa}
\end{aligned}
$$

Point 3 is at the same elevation as point 2 and in the same fluid; therefore, $p_3 = p_2$. The next step is to evaluate Δp from 3 to 4 and to apply this to the pressure at 3:

$$
\begin{aligned}
\Delta p_{3\to 4} &= -\gamma \times 1.80 \qquad \text{where } \gamma = 9810 \text{ N/m}^3 \\
&= -17.66 \text{ kPa}
\end{aligned}
$$

Then $$p_4 = p_3 - 17.66 \text{ kPa} = 62.1 \text{ kPa gage}$$ ◁

Once one is familiar with the basic principle of manometry, it should be easy to write a single equation rather than separate equations for each step in Example 3.7. The single equation for evaluation of the pressure in the pipe of Fig 3.7 is

$$0 + \gamma_m \Delta h - \gamma \ell = p_p$$

One can read the equation in this way: zero pressure at the open end, plus the change in pressure from 1 to 2, minus the change in pressure from 3 to 4, equals the pressure in the pipe. The main point that the student must remember in this process is that when one travels downward in the fluid, the pressure increases, and when one travels upward, the pressure decreases.

It is possible to write a general equation for the pressure difference measured by the manometer. The general *manometer equation* is

$$p_2 = p_1 + \sum_{\text{down}} \gamma_i h_i - \sum_{\text{up}} \gamma_i h_i \qquad (3.17)$$

where γ_i and h_i are the specific weight and deflection in each leg of the manometer. It does not matter where you start, that is, where you define the initial point 1 and final point 2. In Example 3.7, one can take point 1 as the initial point and point 4, the pressure in the pipe, as the final point. Then

$$p_p = 0 + \gamma_m \Delta h - \gamma \ell$$

Or, using the pressure in the pipe as the initial point and point 1 as the final point,

$$p_1 = 0 = p_p + \gamma \ell - \gamma_m \Delta h$$

$$p_p = \gamma_m \Delta h - \gamma \ell$$

which gives the same answer.

Up to this point we have considered only liquid-filled manometers. Let us consider Fig. 3.7 again with a gas-filled pipe. This is illustrated in the next example.

example 3.8

Air at 20°C is the fluid in the pipe of Fig. 3.7, and water is the manometer fluid. If the deflection Δh is 70 cm and ℓ is 140 cm, what is the gage pressure in the pipe? Also compute this pressure by neglecting the pressure change due to the 140-cm column of air. Assume standard atmospheric pressure.

Solution The specific weight of air is found from Eq. (3.12), which requires that the air pressure be known. Therefore, the air pressure at the bottom of the 70-cm column is first calculated:

$$p_{\text{air}} = 9790 \text{ N/m}^3 \times 0.70 \text{ m} = 6853 \text{ Pa gage}$$

Then the absolute air pressure is given as

$$p_{\text{air}} = 6853 \text{ Pa} + 101{,}300 \text{ Pa} = 108.15 \text{ kPa}$$

Then $$\rho_{\text{air}} = \frac{p}{RT} = \frac{108{,}150 \text{ N/m}^2}{287 \text{ J/kg K} \times (20+273) \text{ K}} = 1.286 \text{ kg/m}^3$$

or $$\gamma_{\text{air}} = 1.286 \text{ kg/m}^3 \times 9.81 \text{ m/s}^2 = 12.62 \text{ N/m}^3$$

Now compute the gage pressure in the pipe:

$$p_{\text{pipe}} = 6853 \text{ Pa} - 1.4 \text{ m} \times 12.62 \text{ N/m}^3 = 6835 \text{ Pa} \quad \triangleleft$$

If the effect of the air column is neglected, the gage pressure in the pipe is

$$p_{\text{pipe}} = 9790 \text{ N/m}^3 \times 0.70 \text{ m} = 6853 \text{Pa} \quad \triangleleft$$

Results of the foregoing example show that when liquids and gases are both involved in a manometer problem, it is well within engineering accuracy to neglect the pressure changes due to the columns of gas.

example 3.9

What is the pressure of the air in the tank shown in the accompanying figure if $\ell_1 = 40$ cm (1.31 ft), $\ell_2 = 100$ cm (3.28 ft), and $\ell_3 = 80$ cm (2.62 ft)?

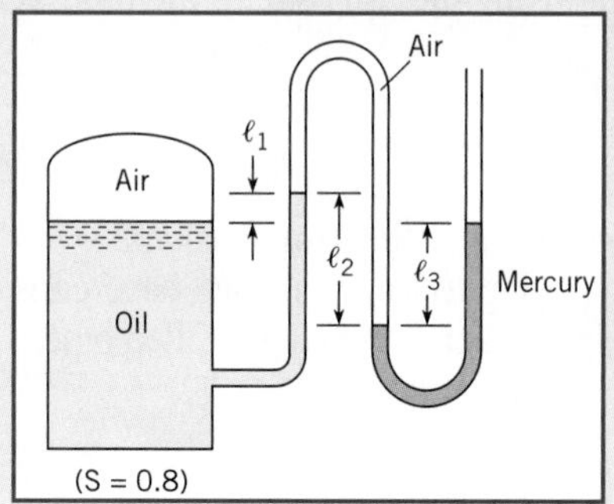

Solution

SI units

$$0 + 0.80\text{ m} \times 133{,}000\text{ N/m}^3 + 0.4\text{ m} \times 9810\text{ N/m}^3 \times 0.8 = p_{\text{air}}$$

$$p_{\text{air}} = 109.5\text{ kPa gage} \qquad \triangleleft$$

Traditional units

$$0 + 2.62\text{ ft} \times 846\text{ lbf/ft}^3 + 1.31\text{ ft} \times 62.4\text{ lbf/ft}^3 \times 0.8 = p_{\text{air}}$$

$$p_{\text{air}} = 2282\text{ psfg} = 15.85\text{ psig} \qquad \triangleleft$$

Differential Manometer

It is often desirable to measure the difference in pressure between two points in a pipe. For this application a manometer is connected to the two points between which the pressure difference is to be measured. Such a setup is shown in Fig. 3.8. The pressure difference between points 1 and 2 is given by $\Delta p = (\gamma_m - \gamma_f)\Delta h$, where γ_m is the specific weight of the manometer liquid, γ_f the specific weight of the fluid, and Δh is the deflection of this liquid.

FIGURE 3.8

Differential manometer.

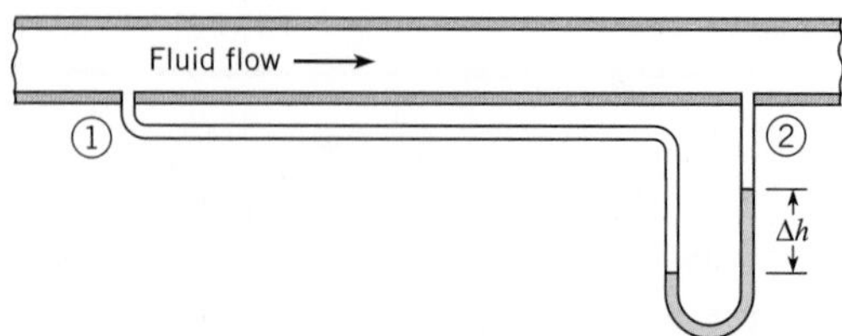

example 3.10

A differential mercury manometer is connected to two pressure taps in an inclined pipe as shown. Water at 50°F is flowing through the pipe. The deflection of mercury in the manometer is 1 inch. Find the change in piezometric pressure and piezometric head between the two points.

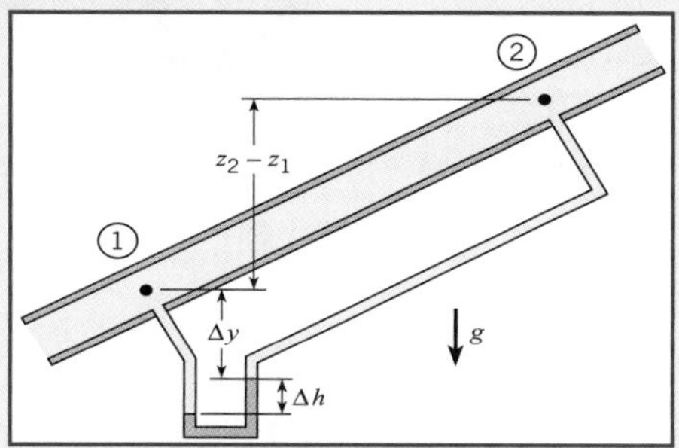

Solution The elevation difference between points 1 and 2 is $z_2 - z_1$. Define Δy as the elevation distance between point 1 and the surface of the mercury in the manometer leg. The deflection of the manometer is Δh.

Apply the *manometer equation* between points 1 and 2:

$$p_2 = p_1 + \gamma_w(\Delta y + \Delta h) - \gamma_m \Delta h - \gamma_w(\Delta y + z_2 - z_1)$$

The $\gamma_w \Delta y$'s cancel out and the equation can be written as

$$p_2 + \gamma_w z_2 - (p_1 + \gamma_w z_1) = \Delta h(\gamma_w - \gamma_m)$$
$$p_{z_2} - p_{z_1} = \Delta h(\gamma_w - \gamma_m)$$

The change in piezometric pressure is

$$p_{z_2} - p_{z_1} = \frac{1}{12}(\text{ft})(62.4 - 847)(\text{lbf/ft}^3)$$
$$= -65.4 \text{ psf}$$

The change in piezometric head is

$$h_2 - h_1 = \frac{p_{z_2} - p_{z_1}}{\gamma_w}$$
$$= \Delta h\left(1 - \frac{\gamma_m}{\gamma_w}\right)$$
$$= \frac{1}{12}(\text{ft})\left(1 - \frac{847}{62.4}\right)$$
$$= -1.05 \text{ ft}$$

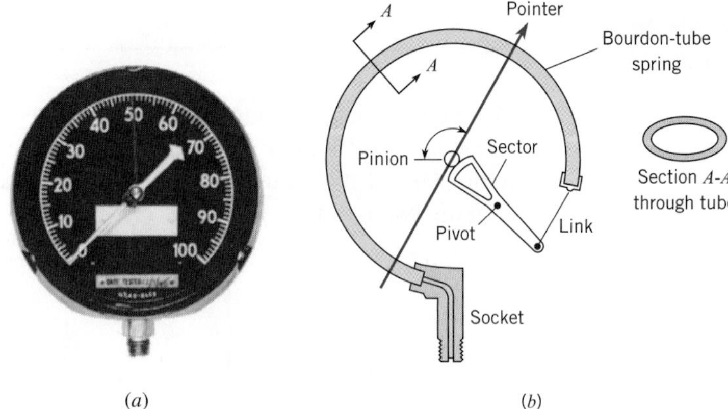

FIGURE 3.9

Bourdon-tube gage.
(a) View of typical gage.
(b) Internal mechanism (schematic).

Bourdon-Tube Gage

A Bourdon-tube gage consists of a tube having an elliptical cross section and bent into a circular arc, as shown in Fig. 3.9*b*. When atmospheric pressure (zero gage pressure) prevails in the gage, the tube is undeflected, and for this condition the gage pointer is calibrated to read zero pressure. When pressure is applied to the gage, the curved tube tends to straighten (much like the party favors that straighten out when one blows into them), thereby actuating the pointer to read correspondingly higher pressure. The Bourdon-tube gage is a very common type that is reliable if not subjected to excessive pressure pulsations or undue external shock. However, because both these conditions sometimes prevail in engineering applications, pulsation dampers should be installed in the line leading to such gages and the gages should be periodically calibrated to check their accuracy.

Pressure Transducers

Modern factories and systems that involve flow processes are controlled automatically, and much of their operation involves sensing of pressure at critical points of the system. Therefore, pressure-sensing devices, such as pressure transducers, are designed to produce electronic signals that can be transmitted to oscillographs or digital devices for record-keeping and/or to control other devices for process operation. Basically, most transducers are tapped into the system with one side of a small diaphragm exposed to the active pressure of the system. When the pressure changes, the diaphragm flexes and a sensing element connected to the other side of the diaphragm produces a signal that is usually linear with the change in pressure in the system. There are many types of sensing elements; one common type is the resistance-wire strain gage attached to a flexible diaphragm as shown in Fig. 3.10. As the diaphragm flexes, the wires of the strain gage change length, thereby changing the resistance of the wire. This change in resistance is utilized electronically to produce a voltage change that can then be used in various ways.

Another type of pressure transducer used for measuring rapidly changing high pressures, such as the pressure in the cylinder head of an internal combustion engine, is the piezoelectric transducer (2). These transducers operate with a quartz crystal that gen-

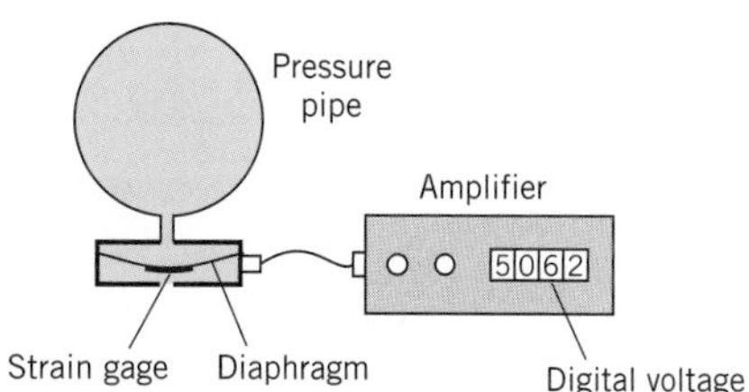

FIGURE 3.10

Schematic diagram of strain-gage pressure transducer.

erates a charge when subjected to a pressure. Sensitive electronic circuitry is required to convert the charge to a measurable voltage signal.

Computer data acquisition systems are used widely with pressure transducers. The analog signal from the transducer is converted (through an A/D converter) to a digital signal that can be processed by a computer. This expedites the data acquisition process and facilitates storing data on magnetic tapes or floppy disks.

3.4 Hydrostatic Forces on Plane Surfaces

Surfaces that are horizontal or are subjected to gas pressure have essentially uniform pressure over their entire surface. Therefore, the total force resulting from the pressure is equal to the product of the pressure and the area of the surface. For this case the resultant force acts at the centroid of the area, and its line of action is normal to the area.

If a plane surface is not horizontal and if it is acted on by a hydrostatic force such as that produced by static liquids, then the pressure is linearly distributed over the surface, and a more general type of analysis must be made to evaluate the magnitude of the resultant force and the location of its line of action. The following derivations assume atmospheric pressure at the liquid surface.

Magnitude of Resultant Hydrostatic Force

Consider the force on the top side of the plane surface *AB* in Fig. 3.11. Line *AB* is the edge view of a surface entirely submerged in the liquid. The plane of this surface intersects the horizontal liquid surface at axis 0-0 with an angle α. The distance from the axis 0-0 to the horizontal axis through the centroid of the area is given by $\bar{y}$. The distance from 0-0 to the differential area dA is y. The pressure on the differential area can be computed if the y distance to the point is known; that is, $p = \gamma y \sin\alpha$. Then it follows that the differential force on the differential area is

$$dF = p\,dA \quad \text{or} \quad dF = \gamma y \sin\alpha\, dA$$

The total force on the area is obtained by integrating the differential force over the entire area:

$$F = \int_A p\,dA$$

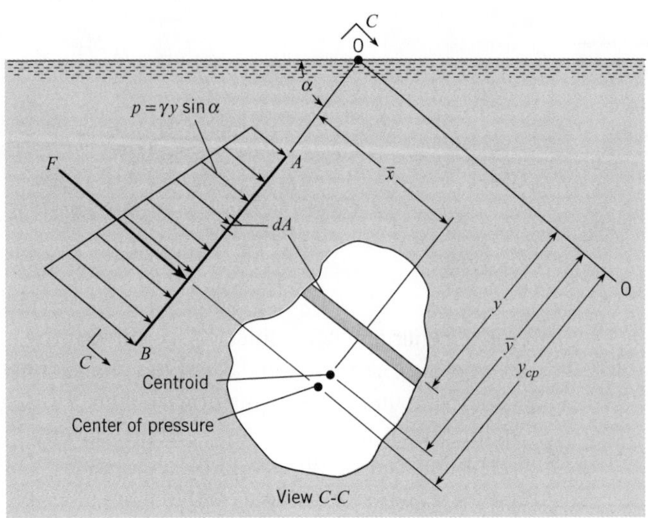

FIGURE 3.11

Distribution of hydrostatic pressure on a plane surface.

or

$$F = \int_A \gamma y \sin\alpha \, dA \tag{3.18}$$

In Eq. (3.18), γ and $\sin\alpha$ are constants. Therefore, we obtain

$$F = \gamma \sin\alpha \int_A y \, dA \tag{3.19}$$

Now, the integral in Eq. (3.19) is the first moment of the area. Consequently, this is replaced by its equivalent, $\overline{y}A$. Therefore, we obtain

$$F = \gamma \overline{y} A \sin\alpha$$

which can be rewritten in the following form:

$$F = (\gamma \overline{y} \sin\alpha) A \tag{3.20}$$

Reference to Fig. 3.11 will show that the product of the variables within the parentheses of Eq. (3.20) is the pressure at the centroid of the area. Consequently, we arrive at the conclusion that the magnitude of the resultant hydrostatic force on a plane surface is the product of the pressure at the centroid of the surface and the area of the surface:

$$F = \overline{p} A \tag{3.21}$$

which is identified as the *hydrostatic force* equation.

For most hydrostatic problems we are interested only in the forces created in excess of the ambient atmospheric pressures, because atmospheric pressure usually acts on the opposite side of the area in question. Therefore, unless otherwise specified, the pressures used in the following section will be gage pressures.

example 3.11

Assuming that freshly poured concrete exerts a hydrostatic force similar to that exerted by a liquid of equal specific weight, determine the force acting on one side of a concrete form 2.44 m high and 1.22 m wide (8 ft by 4 ft) that is used for pouring a basement wall. *Note:* The specific weight of concrete may be taken as 23.6 kN/m^3 (150 lbf/ft^3).

Solution

$$F = \overline{p}A$$

$$\overline{p} = 1.22 \text{ m} \times 23.6 \times 10^3 \text{ N/m}^3 = 28.79 \text{ kPa}$$

$$A = 1.22 \times 2.44 = 2.98 \text{ m}^2$$

Then

$$F = 28.79 \times 10^3 \text{ N/m}^2 \times 2.98 \text{ m}^2 = 85.8 \text{ kN}$$ ◁

Vertical Location of Line of Action of Resultant Hydrostatic Force

In general, the location of the line of action of the resultant hydrostatic force lies below the centroid because pressure increases with depth. We can derive an equation for this location by taking moments of the pressure forces about the horizontal axis 0-0. We call the point where the resultant force intersects the surface the *center of pressure* and identify the slant distance from 0-0 to this point by y_{cp} (Fig. 3.11). Then, by definition of the location of a resultant force, the following moment equation can be written:

$$y_{cp}F = \int y\,dF$$

But dF is given by $dF = p\,dA$; therefore,

$$y_{cp}F = \int_A yp\,dA$$

Also,

$$p = \gamma y \sin\alpha$$

so

$$y_{cp}F = \int_A \gamma y^2 \sin\alpha\,dA \tag{3.22}$$

Again, as in Eq. (3.18), γ and $\sin\alpha$ are constants, so we obtain

$$y_{cp}F = \gamma \sin\alpha \int_A y^2 dA \tag{3.23}$$

The integral on the right-hand side of Eq. (3.23) is the second moment of the area (often called the area moment of inertia). This shall be identified as I_0. However, for engineering applications it is convenient to express the second moment with respect to the horizontal centroidal axis of the area. Hence by the parallel-axis theorem we have

$$I_0 = \overline{I} + \overline{y}^2 A \tag{3.24}$$

When this is substituted into Eq. (3.23), we obtain

$$y_{cp}F = \gamma \sin\alpha(\bar{I} + \bar{y}^2 A)$$

However, from Eq. (3.20), $F = \gamma \bar{y} \sin\alpha A$. Therefore,

$$y_{cp}(\gamma \bar{y} \sin\alpha A) = \gamma \sin\alpha(\bar{I} + \bar{y}^2 A)$$

$$y_{cp} = \bar{y} + \frac{\bar{I}}{\bar{y}A} \tag{3.25}$$

or

$$y_{cp} - \bar{y} = \frac{\bar{I}}{\bar{y}A} \tag{3.26}$$

It can be seen from Eq. (3.26) that for a given area the center of pressure comes closer to the centroid as the area is lowered deeper into the liquid. Equation (3.26) is valid only when one liquid is involved. In addition, it is restricted to the case where $p = 0$ gage at the liquid surface. If the pressure is not zero at the surface, then an equivalent problem can be found that satisfies the restriction. That is, y must be measured from an equivalent free surface located above the centroid of the area a distance $\bar{p}/\gamma$.

The equations for hydrostatic force and pressure are valid only for a fluid with constant density. They would not be valid for layers of fluid with different densities.

example 3.12

An elliptical gate covers the end of a pipe 4 m in diameter. If the gate is hinged at the top, what normal force F is required to open the gate when water is 8 m deep above the top of the pipe and the pipe is open to the atmosphere on the other side? Neglect the weight of the gate.

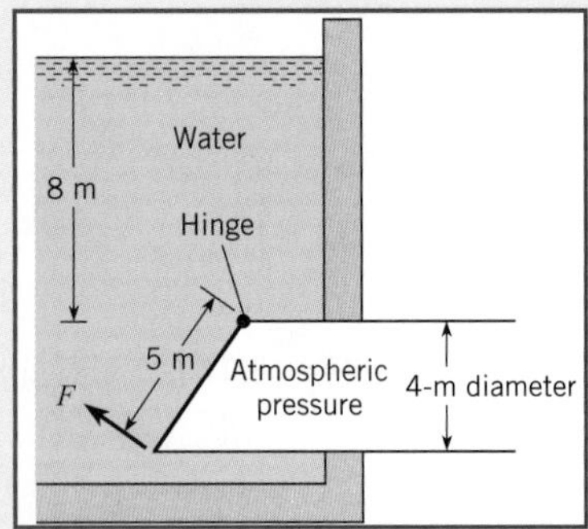

Solution First evaluate the magnitude of the hydrostatic force:

$$F = \bar{p}A$$

The area in question is an ellipse with major and minor axes of 5 m and 4 m. The area is given by the formula $A = \pi ab$ (from Fig. A.1 in the Appendix). Then

$$F = 10 \text{ m} \times 9810 \text{ N/m}^3 \times \pi \times 2 \text{ m} \times 2.5 \text{ m} = 1.541 \text{ MN}$$

Now calculate the slant distance between the centroid of the elliptical area and the center of pressure:

$$y_{cp} - \bar{y} = \frac{\bar{I}}{\bar{y}A} = \frac{\frac{1}{4}\pi a^3 b}{\bar{y}\pi ab} = \frac{\frac{1}{4}a^2}{\bar{y}}$$

Here $\bar{y}$ = 12.5 m (slant distance from the water surface to the centroid). Thus

$$y_{cp} - \bar{y} = \frac{1}{4} \times \frac{6.25 \text{ m}^2}{12.5 \text{ m}} = 0.125 \text{ m}$$

Now take moments about the hinge at the top of the gate to obtain F:

$$\sum M_{\text{hinge}} = 0$$

$$1.541 \times 10^6 \text{ N} \times 2.625 \text{ m} - F \times 5 \text{ m} = 0$$

$$F = 809 \text{ kN}$$ ◁

Note: Students are sometimes uncertain which axis to take the moment of inertia about when computing the distance to the center of pressure. A check of the derivation will reveal that the area moment of inertia as used in Eqs. (3.25) and (3.26) is always taken about the *horizontal-centroidal axis.* (Formulas for moments of inertia of selected areas are given in Fig. A.1 in the Appendix.)

Lateral Location of Line of Action of Resultant Hydrostatic Force

The same principles used for the vertical location of the line of action may be used for the lateral location—that is, by taking moments about a line normal to line 0-0 in Fig. 3.11. Areas that are symmetrical about an axis normal to 0-0 always yield a position for the center of pressure that is along the axis of symmetry and below the centroid. However, for asymmetrical areas it is necessary to carry out the analysis to evaluate the location.

example 3.13

Determine the magnitude of the hydrostatic force acting on one side of the submerged vertical plate shown in the figure and determine the location of the center of pressure.

Solution The centroid of the plate is at a depth of 4 m. Therefore $F = 4 \text{ m} \times 9810 \text{ N/m}^3 \times \frac{1}{2} \times 60 \text{ m}^2 = 1.177$ MN. The vertical location of the center of pressure is obtained from the center-of-pressure equation:

$$y_{cp} - \bar{y} = \frac{\bar{I}}{\bar{y}A} = \frac{bh^3/36}{\bar{y}\frac{1}{2}bh} = \frac{h^2}{18\bar{y}} = \frac{36}{72}$$

$$y_{cp} = 4 + \tfrac{1}{2} = 4.50 \text{ m}$$

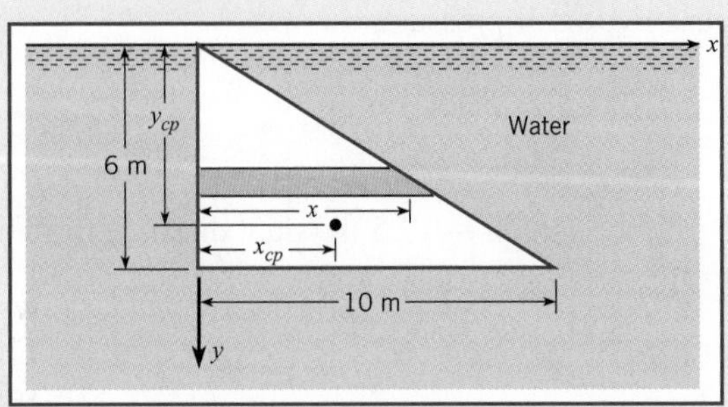

Obtain the lateral location of the center of pressure by summing moments of forces acting on the elemental strips and then dividing by F. Moments are taken about the vertical edge:

$$dM = \tfrac{1}{2}x\,dF = \tfrac{1}{2}x\gamma y x\,dy$$

But $$x = \frac{10}{6}y$$

so $$M = \frac{50}{36}\gamma\int_0^6 y^3\,dy$$

Then $$M = \frac{50}{36}(9810\text{ N/m}^3)\frac{y^4}{4}\bigg|_0^6 = 4.414\text{ MN}\cdot\text{m}$$

But $$Fx_{cp} = M$$

so $$x_{cp} = \frac{M}{F} = \frac{4.414\text{ N}\cdot\text{m}}{1.177\text{ N}} = 3.75\text{ m}$$ ◁

3·5 Hydrostatic Forces on Curved Surfaces

Consider the curved surface AB in Fig. 3.12a with the indicated pressure distribution. The goal of the engineer is to model the force due to pressure with an equivalent force vector acting through the center of pressure. One approach is to integrate the pressure force along the curved surface and find the equivalent force. However, it is easier to sum forces for a free body as defined in the upper part of Fig. 3.12b. The lower sketch in Fig. 3.12b shows how the force acting on the curved surface relates to the force F acting on the free body. Using the free-body diagram and summing forces in the horizontal direction shows that

$$F_x = F_{AC} \tag{3.27}$$

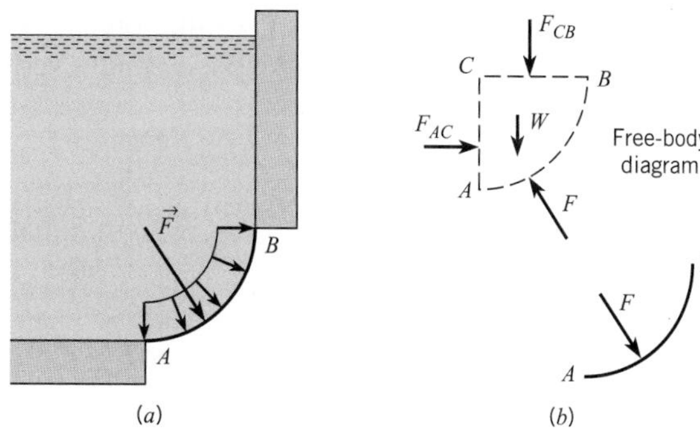

FIGURE 3.12

(a) Pressure distribution and equivalent force. (b) Free-body diagram and action-reaction force pair.

The line of action for the force F_{AC} is through the center of pressure for side AC as discussed in the previous section and designated as y_{cp}.

The vertical component of the equivalent force is

$$F_y = W + F_{CB} \tag{3.28}$$

where W is the weight of the fluid in the free body and F_{CB} is the force on the side CB.

The force F_{CB} acts through the centroid of surface CB and the weight acts through the center of gravity of the free body. The line of action for the vertical force may be found by summing the moments about any convenient axis. As illustrated in this section, the concepts embodied in Eqs. (3.27) and (3.28) are used to solve many curved-surface problems.

example 3.14

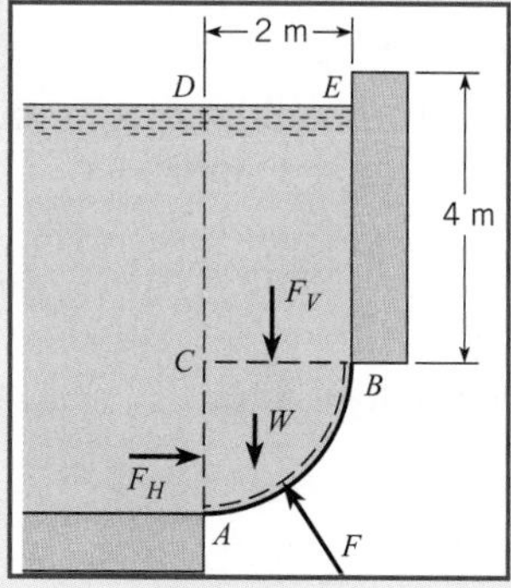

In the figure, surface AB is a circular arc with a radius of 2 m and a depth of 1 m into the paper. The distance EB is 4 m. The fluid above surface AB is water, and atmospheric pressure prevails on the free surface of the water and on the bottom side of surface AB. Find the magnitude and line of action of the hydrostatic force acting on surface AB.

Solution The sketch shows the forces acting on the water in volume *ABC*. The force F has a horizontal component F_x and a vertical component F_y. Balancing forces in the horizontal direction gives

$$F_x = F_H = \overline{p}A = (5 \text{ m})(9810 \text{ N/m}^3)(2 \times 1 \text{ m}^2)$$
$$= 98.1 \text{ kN}$$

The component of force in the vertical direction is

$$F_y = W + F_V$$

The vertical force on side *CB* is due to the weight of the water above:

$$F_V = \overline{p}_0 A = 9.81 \text{ kN/m}^3 \times 4 \text{ m} \times 2 \text{ m} \times 1 \text{ m} = 78.5 \text{ kN}$$

The weight of the water in volume *ABC* is

$$W = \gamma \maltese_{ABC} = (\gamma)(\tfrac{1}{4}\pi r^2)(1 \text{ m})$$
$$= (9.81 \text{ kN/m}^3) \times (0.25 \times \pi \times 4 \text{ m}^2)(1 \text{ m}) = 30.8 \text{ kN}$$

Therefore, the vertical force component is

$$F_y = W + F_V = 109.3 \text{ kN}$$

The line of action (y_{cp}) for the horizontal force is

$$y_{cp} = \overline{y} + \frac{\overline{I}}{\overline{y}A} = (5 \text{ m}) + \left(\frac{1 \times 2^3/12}{5 \times 2 \times 1} \text{ m}\right)$$
$$y_{cp} = 5.067 \text{ m}$$

The line of action (x_{cp}) for the vertical force is found by summing moments about point *C*:

$$x_{cp}F_y = F_V \times 1 \text{ m} + W \times \overline{x}_W$$

The horizontal distance from point *C* to the centroid of the area *ABC* is found using Fig. A.1 in the Appendix: $\overline{x}_W = 4r/3\pi = 0.849$ m. Thus,

$$x_{cp} = \frac{78.5 \text{ kN} \times 1 \text{ m} + 30.8 \text{ kN} \times 0.849 \text{ m}}{109.3 \text{ kN}} = 0.957 \text{ m}$$

The resultant force that acts on the curved surface is shown in the following figure.

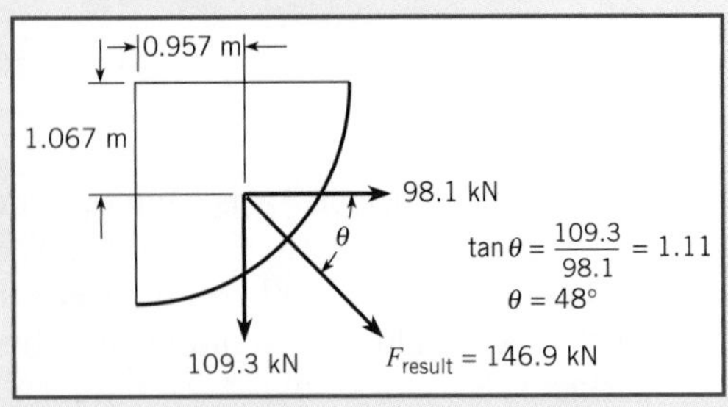

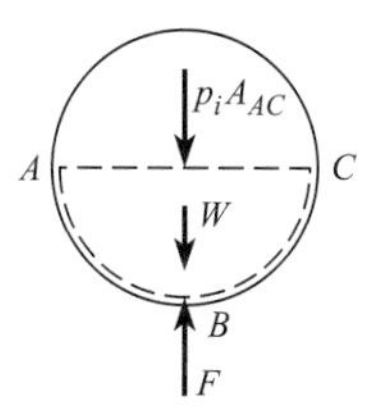

FIGURE 3.13

Pressurized spherical tank showing forces that act on the fluid inside the marked region.

The central idea of this section is that forces on curved surfaces may be found by applying equilibrium concepts to systems comprised of the fluid in contact with the curved surface. Notice how equilibrium concepts are used in each of the situations described in the next few paragraphs.

Consider a sphere holding a gas pressurized to a gage pressure p_i as shown in Fig. 3.13. The indicated forces act on the fluid in volume ABC. Applying equilibrium in the vertical direction gives

$$F = p_i A_{AC} + W$$

Because the specific weight for a gas is quite small, engineers usually neglect the weight of the gas:

$$F = p_i A_{AB} \tag{3.29}$$

Another example is finding the force on a curved surface submerged in a reservoir of liquid as shown in Fig. 3.14*a*. If atmospheric pressure prevails above the free surface and on the outside of surface AB, then force caused by atmospheric pressure cancels out and equilibrium gives

$$F = \gamma \forall_{ABCD} = W\downarrow \tag{3.30}$$

Hence the force on surface AB equals the weight of liquid above the surface, and the arrow indicates that the force acts downward.

Now consider the situation where the pressure distribution on a thin-curved surface comes from the liquid underneath, as shown in Fig. 3.14*b*. If the region above the surface, volume $ABCD$, were filled with the same liquid, the pressure acting at each point on the upper surface of AB would equal the pressure acting at each point on the lower surface. In other words, there would be no net force on the surface. Thus, the equivalent force on surface AB is given by

$$F = \gamma \forall_{abcd} = W\uparrow \tag{3.31}$$

where W is the weight of liquid needed to fill a volume that extends from the curved surface to the free surface of the liquid.

FIGURE 3.14

Curved surface with (a) liquid above and (b) liquid below. In (a), arrows represent forces acting on the liquid. In (b), arrows represent the pressure distribution on surface AB.

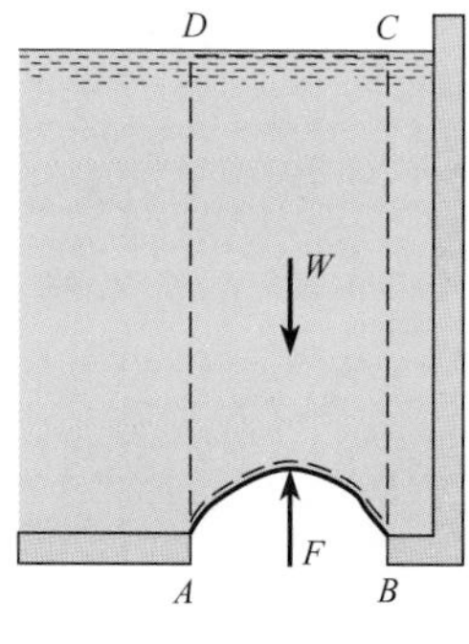

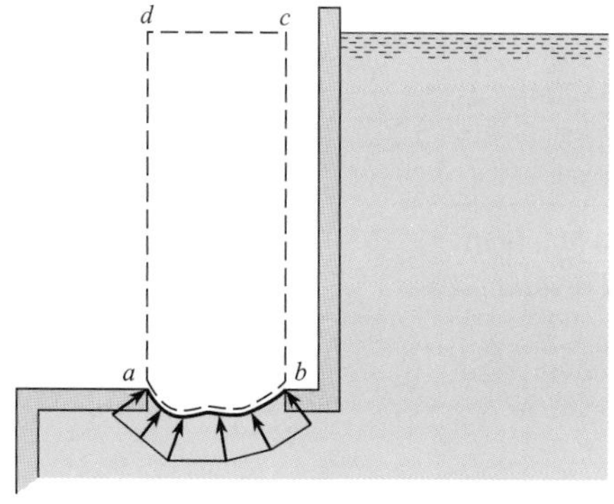

Buoyancy

Principle of Buoyancy

The principles of buoyancy can be developed using equilibrium concepts from the previous section. Consider a body *ABCD* submerged in a liquid of specific weight γ as shown in Fig. 3.15. The sketch on the left shows the pressure distribution acting on the body. As shown by Eq. (3.31), pressures acting on the lower portion of the body create an upward force equal to the weight of liquid needed to fill the volume above surface *ADC*. The upward force is

$$F_{\text{up}} = \gamma(\forall_b + \forall_a)$$

where $\forall_b$ is the volume of the body (i.e., volume *ABCD*) and $\forall_a$ is the volume of liquid above the body (i.e., volume *ABCFE*). As shown by Eq. (3.30), pressures acting on the top surface of the body create a downward force equal to the weight of the liquid above the body:

$$F_{\text{down}} = \gamma \forall_a$$

Subtracting the downward force from the upward force gives the net or buoyant force F_B acting on the body:

$$F_B = F_{\text{up}} - F_{\text{down}} = \gamma \forall_b \tag{3.32}$$

Hence, the net force or buoyant force (F_B) equals the weight of liquid that would be needed to occupy the volume of the body.

Consider a body that is floating as shown in Fig. 3.16. The marked portion of the object has a volume $\forall_D$. Pressure acts on curved surface *ADC* causing an upward force equal to the weight of liquid that would be needed to fill volume $\forall_D$. The buoyant force is given by

$$F_B = F_{\text{up}} = \gamma \forall_D \tag{3.33}$$

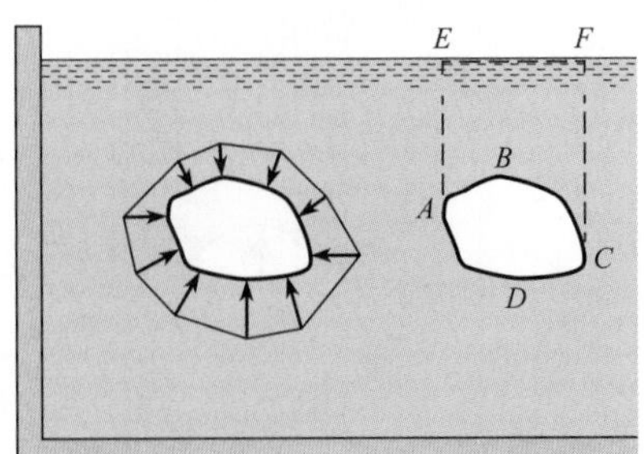

FIGURE 3.15

Two views of a body immersed in a liquid.

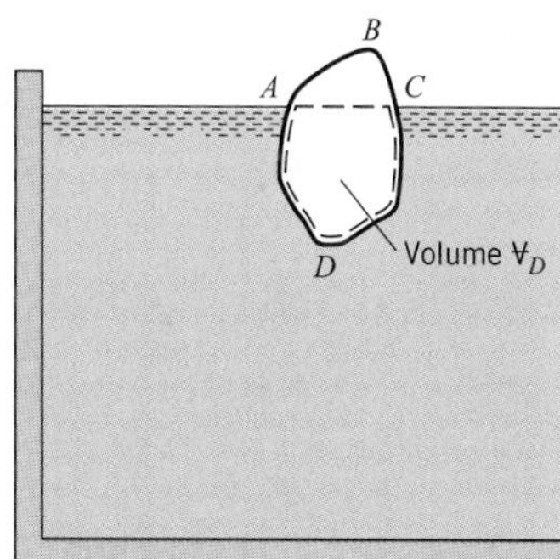

FIGURE 3.16

A body partially submerged in a liquid.

Hence, the buoyant force equals the weight of liquid that would be needed to occupy the volume $\forall_D$. This volume is called the displaced volume. Comparison of Eqs. (3.32) and (3.33) shows that we can write a single equation for the *buoyant force* as

$$F_B = \gamma \forall_D \tag{3.34}$$

Although Eq. (3.34) was derived for a liquid, it is equally valid for a gas. If the body is totally submerged, the displaced volume is the volume of the body. If a body is partially submerged, the displaced volume is the portion of the volume that is submerged. For a fluid of uniform density, the line of action of the buoyant force passes through the centroid of the displaced volume.

The general principle of buoyancy embodied in Eq. (3.34) is called *Archimedes' principle:* for an object partially or completely submerged in a fluid, there is a net upward force (buoyant force) equal to the weight of the displaced fluid.

example 3.15

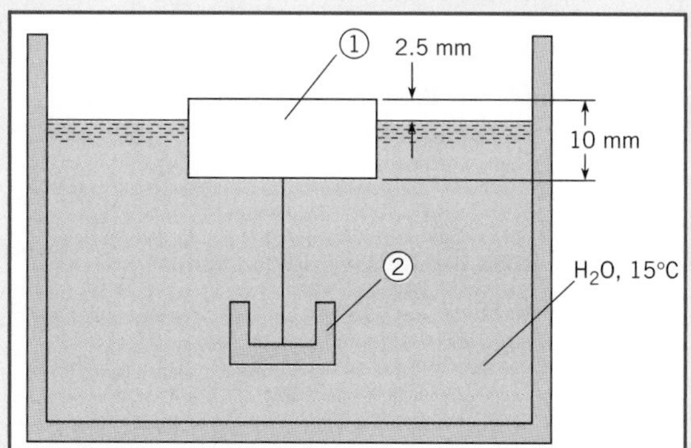

The figure shows a metal part (object 2) hanging by a thin cord from a floating wood block (object 1). The wood block has a specific gravity $S_1 = 0.3$ and dimensions of $50 \times 50 \times 10$ mm. The metal part has a volume of 6600 mm^3. Find the mass m_2 of the metal part and the tension T in the cord.

Solution First draw the free-body diagrams.

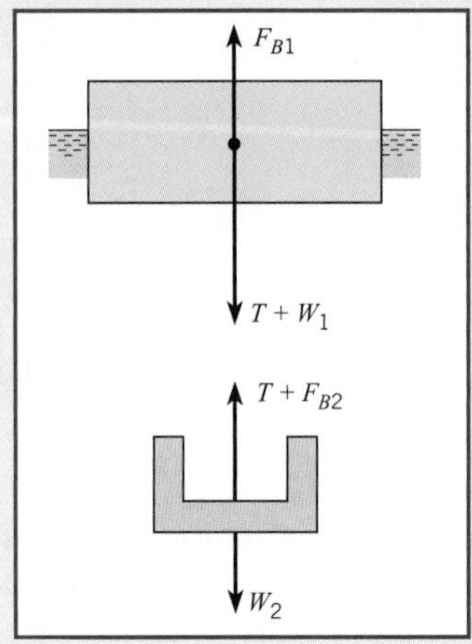

Sum forces on the block:

$$T = F_{B1} - W_1$$

The buoyant force on the floating block is $F_{B1} = \gamma \text{\textyen}_{D1}$, where $\text{\textyen}_{D1}$ is the submerged volume:

$$\begin{aligned} F_{B1} = \gamma \text{\textyen}_{D1} &= (9800 \text{ N/m}^3)(50 \times 50 \times 7.5 \text{ mm}^3)(10^{-9} \text{ m}^3/\text{mm}^3) \\ &= 0.184 \text{ N} \end{aligned}$$

The weight of the block is

$$\begin{aligned} W_1 = \gamma S_1 \text{\textyen}_1 &= (9800 \text{ N/m}^3)(0.3)(50 \times 50 \times 10 \text{ mm}^3)(10^{-9} \text{ m}^3/\text{mm}^3) \\ &= 0.0735 \text{ N} \end{aligned}$$

Hence the tension on the cord is

$$T = (0.184 - 0.0735) = 0.110 \text{ N}$$

Apply force equilibrium to the metal part:

$$W_2 = T + F_{B2}$$

Because the metal part is submerged, use the volume of the part to calculate the buoyant force:

$$F_{B2} = \gamma \text{\textyen}_2 = (9800 \text{ N/m}^3)(6600 \text{ mm}^3)(10^{-9}) = 0.0647 \text{ N}$$

Hence, the weight is given by $W_2 = (0.110 + 0.0647) = 0.175$ N, and the mass is found from

$$m_2 = W_2/g = 17.8 \text{ g}$$

◁

Notice that the tension in the cord (0.11 N) is less than the weight of the metal part (0.18 N). This result is consistent with the common observation that an object will "weigh less in water than in air." As the free-body diagram of the metal part shows, the perceived weight of an object in water (i.e., the tension in the cord) is less than the weight of the object because of the buoyant force.

Hydrometry

Precise measurement of the specific weight of a liquid is done by utilizing the principle of buoyancy. The device used for this, the *hydrometer,* is a glass bulb that is weighted on one end to make the hydrometer float in a vertical position and has a stem of constant diameter extending from the other end (Fig. 3.17). The hydrometer is so designed that only the stem end extends above the liquid surface. Therefore, appreciable vertical movement of the hydrometer is required to change the buoyant force or displaced volume of the device. Because the buoyant force (equal to the weight of the hydrometer) must be constant, the hydrometer will float deeper or shallower depending on the specific weight of the liquid. Consequently, graduations on the stem, corresponding to different depths of submergence of the hydrometer, can be made to indicate directly the specific weight or specific gravity of the liquid being measured.

3·7 Stability of Immersed and Floating Bodies

Immersed Bodies

The stability of an immersed body depends on the relative positions of the *center of gravity* of the body and the centroid of the displaced volume of fluid, which is called the *center of buoyancy.* If the center of buoyancy is above the center of gravity, such as in Fig. 3.18*a*, any tipping of the body produces a righting couple, and consequently, the body is stable. However, if the center of gravity is above the center of buoyancy, any tipping produces an increasing overturning moment, thus causing the body to turn through 180°. This is the condition shown in Fig. 3.18*c*. Finally, if the center of buoyancy and center of gravity are coincident, the body is neutrally stable—that is, it has the tendency for neither righting nor overturning.

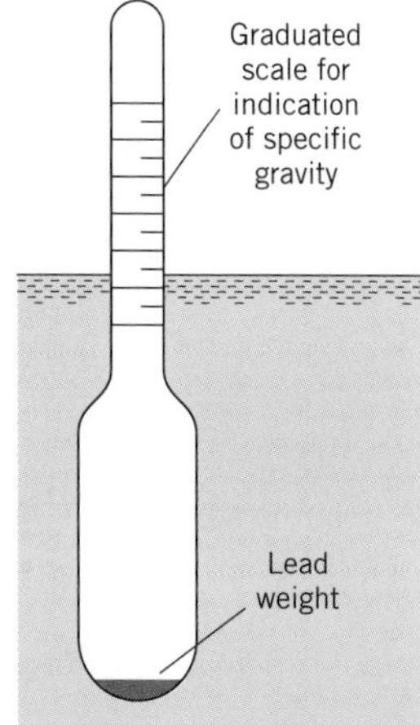

FIGURE 3.17

Hydrometer.

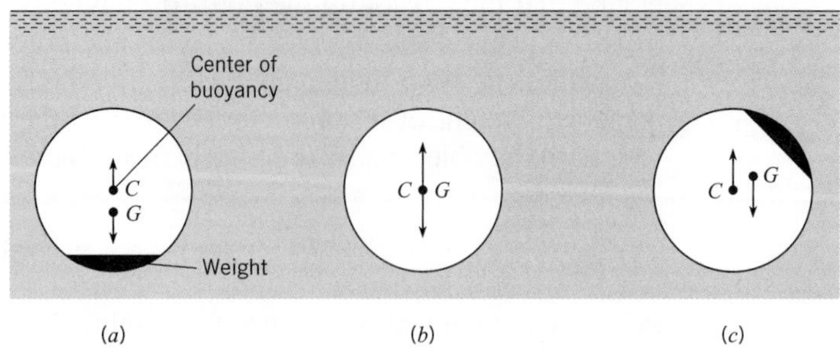

FIGURE 3.18

Conditions of stability for immersed bodies. (a) Stable. (b) Neutral. (c) Unstable.

Floating Bodies

The question of stability is more involved for floating bodies than for immersed bodies because the center of buoyancy may take different positions with respect to the center of gravity, depending on the shape of the body and the position in which it is floating. For example, consider the cross section of a ship shown in Fig. 3.19*a*. Here the center of gravity G is above the center of buoyancy C. Therefore, at first glance it would appear that the ship is unstable and could flip over. However, if we observe the position of C and G after the ship has taken a small angle of heel, as shown in Fig. 3.19*b*, we see that the center of gravity is in the same position but the center of buoyancy has moved outward of the center of gravity, thus producing a righting moment. A ship having such characteristics is stable.

The reason for the change in the center of buoyancy for the ship is that part of the original buoyant volume, as shown by the wedge shape AOB, is transferred to a new buoyant volume EOD. Because the buoyant center is at the centroid of the displaced volume, it follows that for this case the buoyant center must move laterally to the right. The point of intersection of the lines of action of the buoyant force before and after heel is called the *metacenter M,* and the distance GM is called the *metacentric height.* If GM is positive—that is, if M is above G—the ship is stable; however, if GM is negative, the ship is unstable. Quantitative relations involving these basic principles of stability are presented in the next paragraph.

Consider the ship shown in Fig. 3.20, which has taken a small angle of heel α. First we evaluate the lateral displacement of the center of buoyancy, CC'; then it will be easy by simple trigonometry to solve for the metacentric height GM or to evaluate the righting moment. Recall that the center of buoyancy is at the centroid of the displaced volume. Therefore, we must resort to the basic fundamentals of centroids to evaluate the displacement CC'. From the basic definition of the centroid of a volume, we can write the following equation:

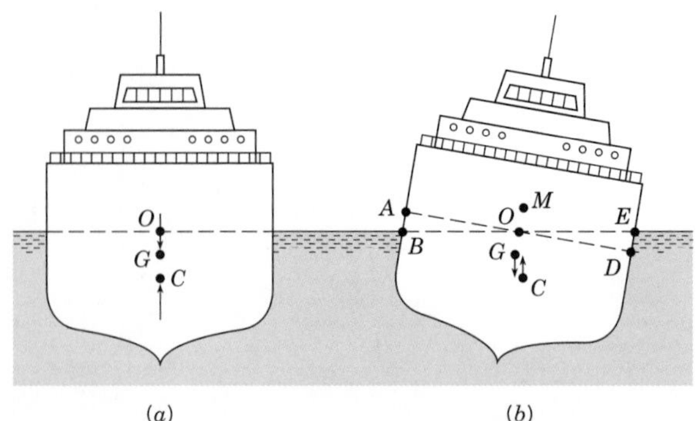

FIGURE 3.19

Ship stability relations.

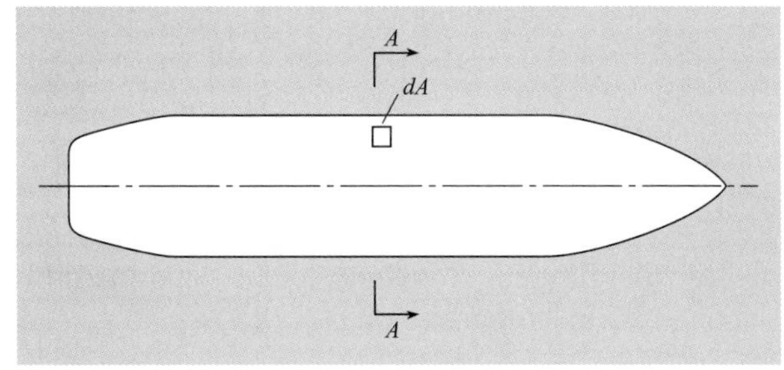

(a)

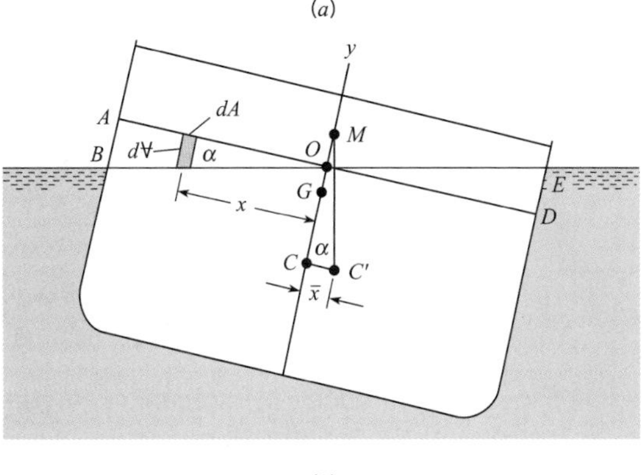

(b)

FIGURE 3.20

(a) Plan view of ship at waterline. (b) Section A-A of ship.

$$\bar{x}\forall = \Sigma x_i \Delta\forall_i \tag{3.35}$$

where $\bar{x} = CC'$, which is the distance from the plane about which moments are taken to the centroid of $\forall$; $\forall$ is the total volume displaced; $\Delta\forall_i$ is the volume increment; and x_i is the moment arm of the increment of volume.

Here we take moments about the plane of symmetry of the ship. Recall from mechanics that when we apply this equation, volumes to the left produce negative moments and volumes to the right produce positive moments. For the right side of Eq. (3.35) we write terms for the moment of the submerged volume about the plane of symmetry. A convenient way to do this is to consider the moment of the volume before heel, subtract the moment of the volume represented by the wedge *AOB*, and then add the moment represented by the wedge *EOD*. In a general way this is given by the following equation:

$$\bar{x}\forall = \text{moment of } \forall \text{ before heel} - \text{moment of } \forall_{AOB} + \text{moment of } \forall_{EOD} \tag{3.36}$$

Because the original buoyant volume is symmetrical with *y-y*, the moment for the first term on the right is zero. Also, the sign of the moment of $\forall_{AOB}$ is negative; therefore, when this negative moment is subtracted from the right-hand side of Eq. (3.36), we arrive at the following equation:

$$\bar{x}\forall = \Sigma x_i \Delta\forall_{i_{AOB}} + \Sigma x_i \Delta\forall_{i_{EOD}} \tag{3.37}$$

Now, expressing Eq. (3.37) in integral form yields

$$\bar{x}\forall = \int_{AOB} x\, d\forall + \int_{EOD} x\, d\forall \tag{3.38}$$

But it may be seen from Fig. 3.20*b* that $d\forall$ can be given as the product of the length of the differential volume, $x\tan\alpha$, and the differential area, dA. Consequently, Eq. (3.38) can be written as

$$\bar{x}\forall = \int_{AOB} x^2 \tan\alpha\, dA + \int_{EOD} x^2 \tan\alpha\, dA$$

Here $\tan\alpha$ is a constant with respect to the integration. Also, since the two terms on the right-hand side are identical except for the area over which integration is to be performed, we combine them as follows:

$$\bar{x}\forall = \tan\alpha \int_{A_{\text{waterline}}} x^2 dA \tag{3.39}$$

The second moment, or moment of inertia of the area defined by the waterline, is given the symbol I_{00}, and the following is obtained:

$$\bar{x}\forall = I_{00}\tan\alpha$$

Next, replacing $\bar{x}$ by CC' and solving for CC', we get

$$CC' = \frac{I_{00}\tan\alpha}{\forall}$$

From Fig. 3.20,

$$CC' = CM\tan\alpha$$

Thus eliminating CC' and $\tan\alpha$ yields

$$CM = \frac{I_{00}}{\forall}$$

However,

$$GM = CM - CG$$

Therefore the *metacentric height* is

$$GM = \frac{I_{00}}{\forall} - CG \tag{3.40}$$

Equation (3.40) is used to determine the stability of floating bodies. As already noted, if GM is positive, the body is stable, and if GM is negative, it is unstable.

example 3.16

A block of wood 30 cm square in cross section and 60 cm long weighs 318 N. Will the block float with sides vertical as shown?

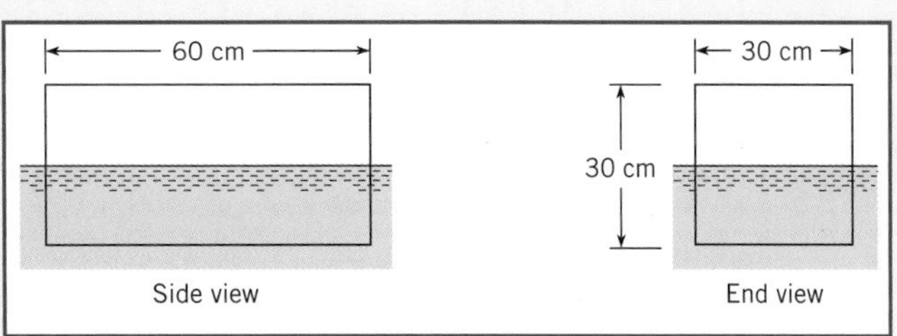

Solution First determine the depth of submergence of the block. This is calculated by applying the equation of equilibrium in the vertical direction.

$$\Sigma F_y = 0$$

$$-\text{weight} + \text{buoyant force} = 0$$

$$-318 \text{ N} + 9810 \text{ N/m}^3 \times 0.30 \text{ m} \times 0.60 \text{ m} \times d = 0$$

$$d = 0.18 \text{ m} = 18 \text{ cm}$$

Determine whether the block is stable about the longitudinal axis:

$$GM = \frac{I_{00}}{\text{V}\!\!\!\!-} - CG = \frac{\frac{1}{12} \times 60 \times 30^3}{18 \times 60 \times 30} - (15 - 9)$$

$$= 4.167 - 6 = -1.833 \text{ cm}$$

Because the metacentric height is negative, the block is not stable about the longitudinal axis. Thus a slight disturbance will make it tip. Next, check to see if the block is stable about the transverse axis:

$$GM = \frac{\frac{1}{12} \times 30 \times 60^3}{18 \times 30 \times 60} - 6 = 10.67 \text{ cm}$$ ◁

The block is stable about the transverse axis and will float with the short sides vertical.

Note that for small angles of heel α the righting moment or overturning moment is given as follows:

$$\text{R.M.} = \gamma \text{V}\!\!\!\!- GM\alpha \quad (3.41)$$

However, for large angles of heel, direct methods of calculation based on these same principles would have to be employed to evaluate the righting or overturning moment.

3.8 Summary

Pressure p expresses the magnitude of normal force per unit area. Pressure is a scalar quantity, meaning that the pressure at a specific point is the same in all directions. Absolute pressure p_{abs} is measured relative to absolute zero while gage pressure p_{gage} is measured relative to atmospheric pressure p_{atm}. Gage and absolute pressure are related by

$$p_{abs} = p_{gage} + p_{atm}$$

The weight of a fluid causes pressure to increase with increasing depth, giving the equation

$$\frac{dp}{dz} = -\gamma = -\rho g$$

where z is the elevation, γ is fluid weight per volume, ρ is fluid density, and g is gravitational acceleration. The pressure variation for a uniform-density fluid is given by

$$\frac{p}{\gamma} + z = \text{constant}$$

Hence, a fluid with uniform density has a constant value of piezometric head $(p/\gamma + z)$ or piezometric pressure $(p + \gamma z)$ throughout.

A fluid in contact with a surface exerts a pressure distribution on it. The pressure distribution can be modeled as a statically equivalent force F acting at the center of pressure. For a plane surface, the equivalent force is

$$F = \overline{p}A$$

where $\overline{p}$ is pressure at the centroid of the area A. For a horizontal surface, the center of pressure is at the centroid. Otherwise, the slant distance between the centroid $\overline{y}$ and the center of pressure y_{cp} is given by

$$y_{cp} - \overline{y} = \frac{\overline{I}}{\overline{y}A}$$

where $\overline{I}$ is the moment of area with respect to a horizontal axis through the centroid.

When a surface is curved, one can find the equivalent force by applying force equilibrium to a free body comprised of the fluid in contact with the surface.

When an object is either partially or totally submerged in a fluid, a buoyant force F_B acts. The magnitude is equal to the weight of the displaced volume of fluid:

$$F_B = \gamma \forall_D$$

where $\forall_D$ is the volume of displaced fluid. For a fluid with a uniform density, the center of buoyancy is the centroid of the displaced volume of fluid.

When an object is floating, it may be unstable or stable. Stable means that if the object is tipped, the buoyant force causes a moment that rotates the object back to its equilibrium position. An object is stable if the metacentric height is positive. In this case tipping the object causes the center of buoyancy to move such that the buoyant force produces a righting moment.

References

1. Bolz, R. E., and G. L. Tuve. eds. *Handbook of Tables for Applied Engineering Science.* Chemical Rubber Co., Cleveland, OH, 1970.

2. Holman, J. P., and W. J. Gajda, Jr. *Experimental Methods for Engineers.* McGraw-Hill, New York, 1984.

Problems

3.1 The Crosby gage tester shown in the figure is used to calibrate or to test pressure gages. When the weights and the piston together weigh 140 N, the gage being tested indicates 200 kPa. If the piston diameter is 30 mm, what percent error exists in the gage?

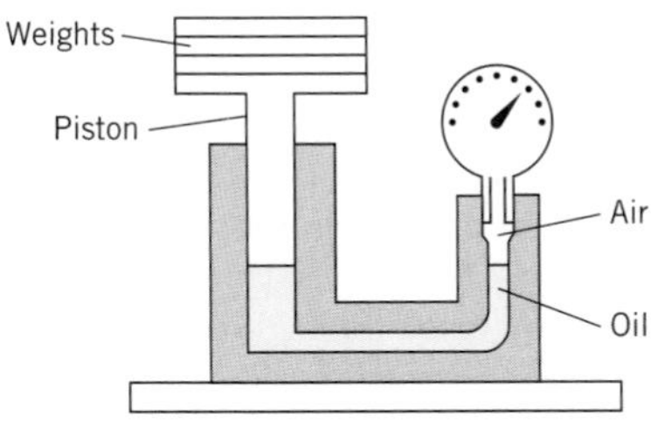

PROBLEM 3.1

3.2 Two hemispheric shells are perfectly sealed together and the internal pressure is reduced to 10% of atmospheric pressure. The inner radius is 6 in., and the outer radius is 6.25 in. The seal is located halfway between the inner and outer radius. If the atmospheric pressure is 14.5 psia, what force is required to pull the shells apart?

3.3 Find a parked automobile for which you have information on tire pressure and weight. Make your best estimate of the area of tire contact with the pavement. Next, using the weight information and tire pressure, use engineering principles to calculate the contact area. Compare your estimate with your calculation and discuss.

3.4 If exactly 20 bolts of 2.5 cm diameter are needed to hold the air chamber together at *A-A* as a result of the high pressure within, how many bolts will be needed at *B-B*? Here $D = 40$ cm and $d = 20$ cm.

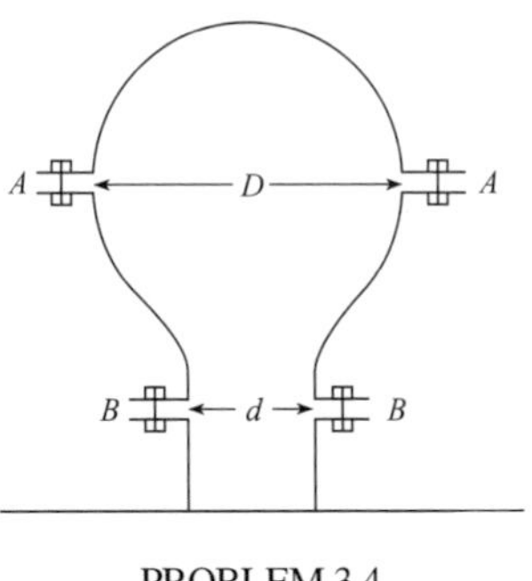

PROBLEM 3.4

3.5 A glass tube 10 cm long and of 0.5 mm internal diameter has one end closed. The tube is inserted into water to a depth of 2 cm, as shown. In the process of inserting the tube, the air is trapped inside and undergoes a constant temperature compression. The atmospheric pressure is 100 kPa, and the water density is 1000 kg/m^3. Find the location of the water level in the tube including the effects of surface tension.

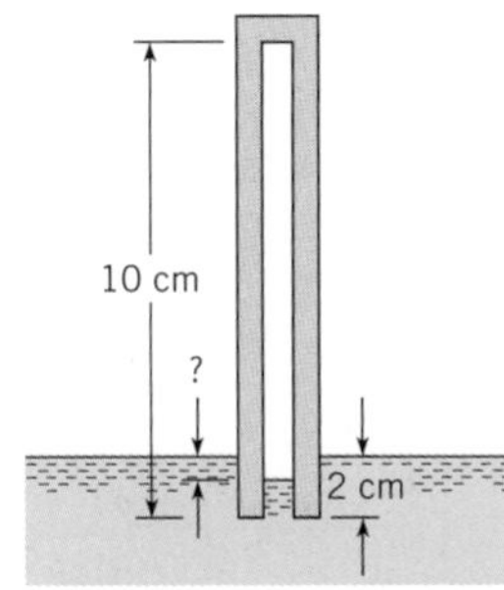

PROBLEM 3.5

3.6 The reservoir shown in the figure contains two immiscible liquids of specific weights γ_A and γ_B, respectively, one above the other. $\gamma_A > \gamma_B$. Which graph depicts the correct distribution of gage pressure along a vertical line through the liquids? Explain your answer.

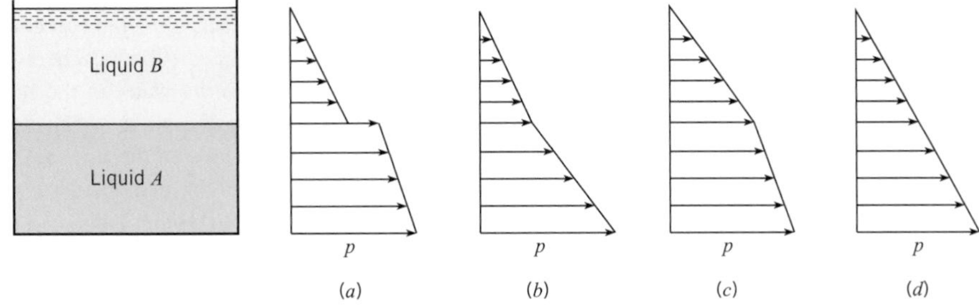

PROBLEM 3.6

FLUID SOLUTIONS **3.7** For the closed tank with Bourdon-tube gages tapped into it, what is the specific gravity of the oil and the pressure reading on gage *C*?

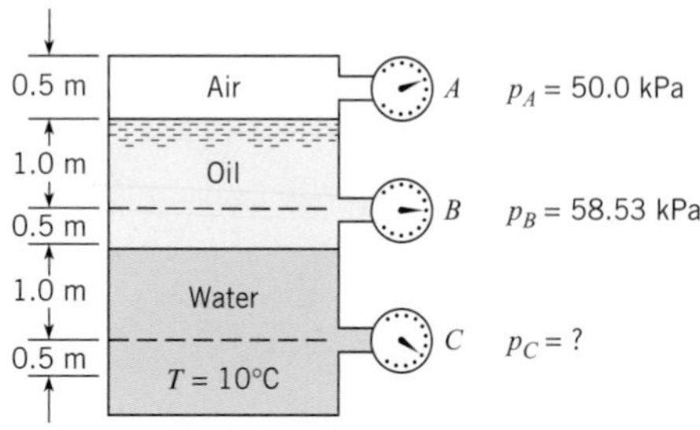

PROBLEM 3.7

3.8 This manometer contains water at room temperature. The glass tube on the left has an inside diameter of 1 mm (d = 1.0 mm).The glass tube on the right is three times as large. For these conditions, the water surface level in the left tube will be (a) higher than the water surface level in the right tube, (b) equal to the water surface level in the right tube, (c) less than the water surface level in the right tube. State your main reason or assumption for making your choice.

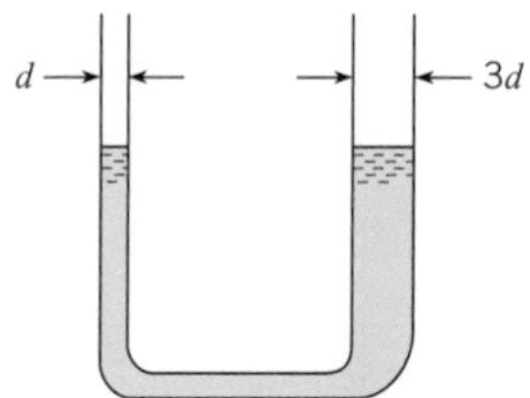

PROBLEM 3.8

FLUID SOLUTIONS **3.9** If a 200-N force F_1 is applied to the piston with the 4-cm diameter, what is the magnitude of the force F_2 that can be resisted by the piston with the 10-cm diameter? Neglect the weights of the pistons.

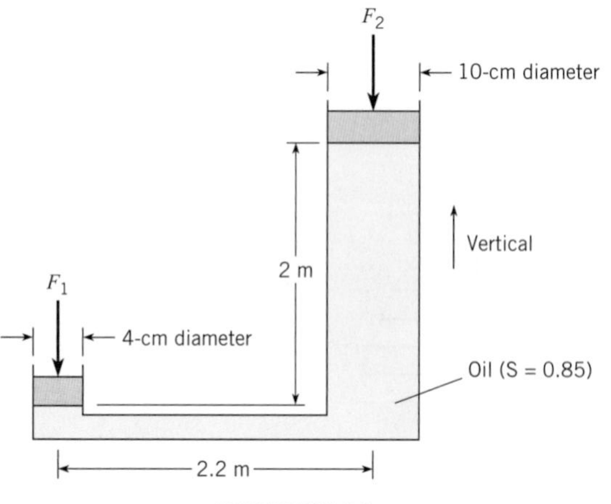

PROBLEM 3.9

3.10 Some skin divers go as deep as 50 m. What is the gage pressure at this depth in fresh water, and what is the ratio of the absolute pressure at this depth to normal atmospheric pressure? Assume T = 20°C.

3.11 Water occupies the bottom 1.0 m of a cylindrical tank. On top of the water is 0.5 m of kerosene, which is open to the atmosphere. If the temperature is 20°C, what is the gage pressure at the bottom of the tank?

3.12 An engineer is designing a hydraulic lift with a capacity of 10 tons. The moving parts of this lift weigh 1000 lbf. The lift should raise the load to a height of 6 ft in 20 seconds. This will be accomplished with a hydraulic pump that delivers fluid to a cylinder. Hydraulic cylinders with a stroke of 72 inches are available with bore sizes from 2 to 8 inches. Hydraulic piston pumps with an operating pressure range from 200 to 3000 psig are available with pumping capacities of 5, 10, and 15 gallons per minute. Select a hydraulic pump size and a hydraulic cylinder size that can be used for this application.

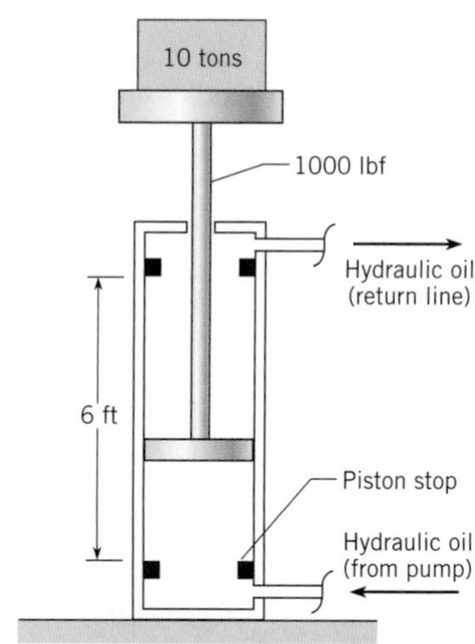

PROBLEM 3.12

3.13 The gage pressure at a depth of 5 m in an open tank of liquid is 75 kPa. What are the specific weight and the specific gravity of the liquid? FLUID SOLUTIONS

3.14 A tank with an attached manometer contains water at 20°C. The atmospheric pressure is 100 kPa. There is a stopcock located 1 m from the surface of the water in the manometer. The stopcock is closed, trapping the air in the manometer, and water is added to the tank to the level of the stopcock. Find the increase in elevation of the water in the manometer assuming the air in the manometer is compressed isothermally.

3.15 A tank is fitted with a manometer on the side, as shown. The liquid in the bottom of the tank and in the manometer has a specific gravity (S) of 3.0. The depth of this bottom liquid is 20 cm. A 15-cm layer of water lies on top of the bottom liquid. Find the position of the liquid surface in the manometer.

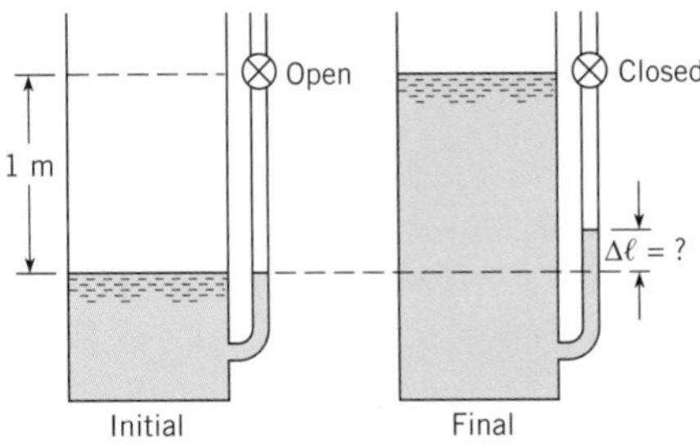

PROBLEM 3.14

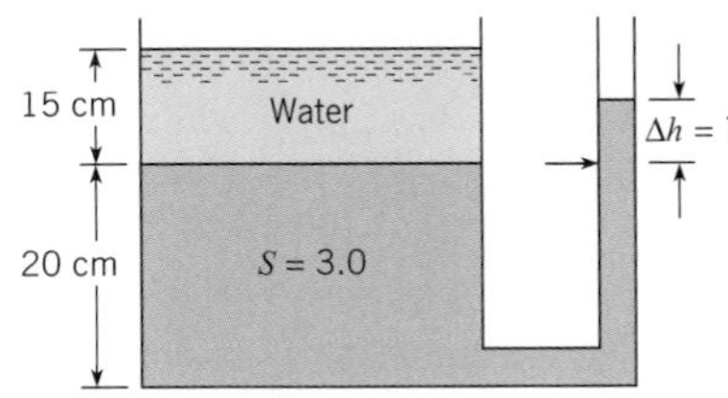

PROBLEM 3.15

3.16 What is the maximum gage pressure in the odd tank shown in the figure? Where will the maximum pressure occur? What is the hydrostatic force acting on the top (*CD*) of the last chamber on the right-hand side of the tank? Assume $T = 10°C$.

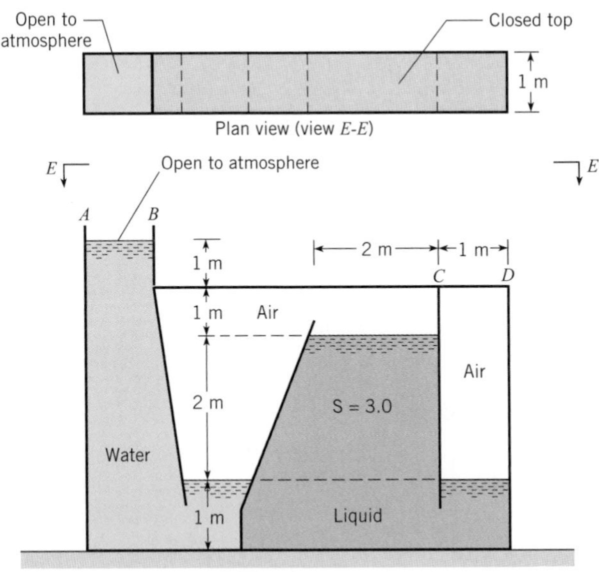

PROBLEM 3.16

3.17 Usually water is assumed to be incompressible for hydrostatic computations; however, extreme pressures may cause significant changes in density. Estimate the percentage difference in density of sea water between the surface and a point 6 km deep. Assume a constant temperature of 10°C and a bulk modulus of elasticity of 2.2 GPa.

3.18 The steel pipe and steel chamber together weigh 600 lbf. What force will have to be exerted on the chamber by all the bolts to hold it in place? The dimension ℓ is equal to 2.5 ft. *Note:* There is no bottom on the chamber—only a flange bolted to the floor.

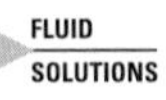

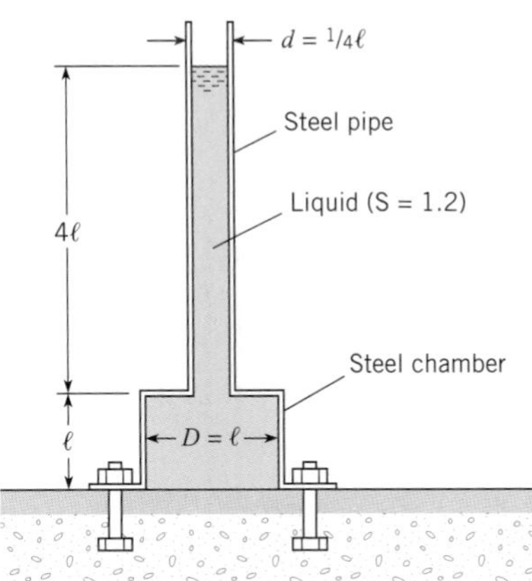

PROBLEM 3.18

3.19 What force must be exerted through the bolts to hold the dome in place? The metal dome and pipe weigh 1300 lbf. The dome has no bottom. Here $\ell = 3.0$ ft.

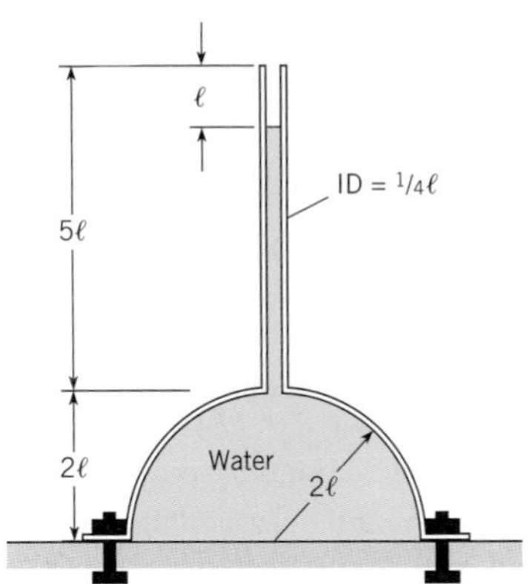

PROBLEMS 3.19, 3.20

3.20 What force must be exerted through the bolts to hold the dome in place? The metal dome and pipe weigh 6 kN. The dome has no bottom. Here $\ell = 80$ cm.

3.21 Find the vertical component of force in the metal at the base of the spherical dome shown when gage *A* reads 5 psig. Indicate whether the metal is in compression or tension. The specific gravity of the enclosed fluid is 1.5. The dimension *L* is 2 ft. Assume the dome weighs 1000 lbf.

3.22 The piston shown weighs 10 lbf. In its initial position, the piston is restrained from moving to the bottom of the cylinder by means of the metal stop. Assuming there is neither friction nor leakage between piston and cylinder, what volume of oil (S = 0.85) would have to be added to the 1-in. tube to cause the piston to rise 1 in. from its initial position?

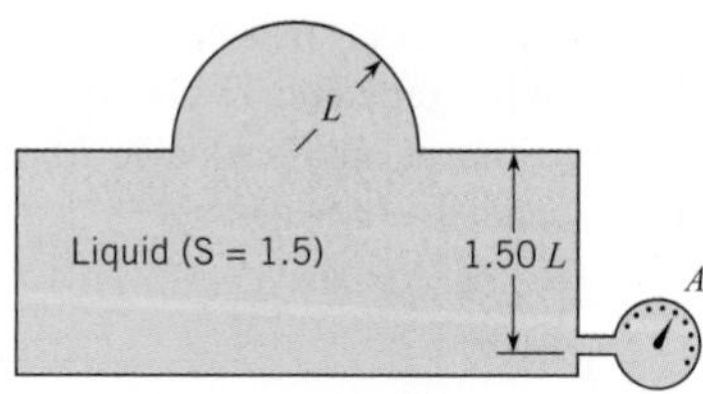

PROBLEM 3.21

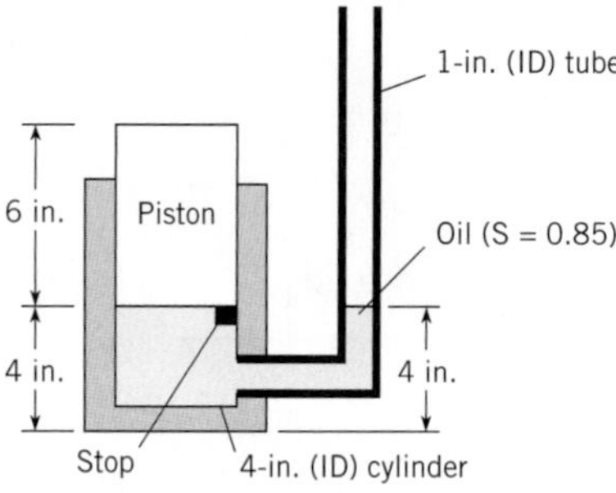

PROBLEM 3.22

FLUID SOLUTIONS **3.23** Consider an air bubble rising from the bottom of a lake. Neglecting surface tension, determine approximately what the ratio of the density of the air in the bubble will be at a depth of 34 ft to its density at a depth of 8 ft.

3.24 A liquid has the peculiar property that its mass density increases linearly with depth according to the expression $\rho = \rho_{water}(1 + 0.01d)$, where d is the depth below the liquid surface in meters. At a depth of 10 m, what is the gage pressure?

3.25 Consider the conditions given for Problem 3.24. The gage pressure at a particular point in the liquid is found by measurement to be 60 kPa. At what depth is that point?

3.26 The specific weight of a liquid increases with depth as follows: $\gamma = 50 + 0.1d$, where d = depth in feet below the surface and γ is given in lbf/ft^3. What is the pressure at a depth of 20 ft?

3.27 The gage pressure at the center of the pipe has a value that is (a) negative, (b) zero, (c) positive. (Neglect surface tension effects.)

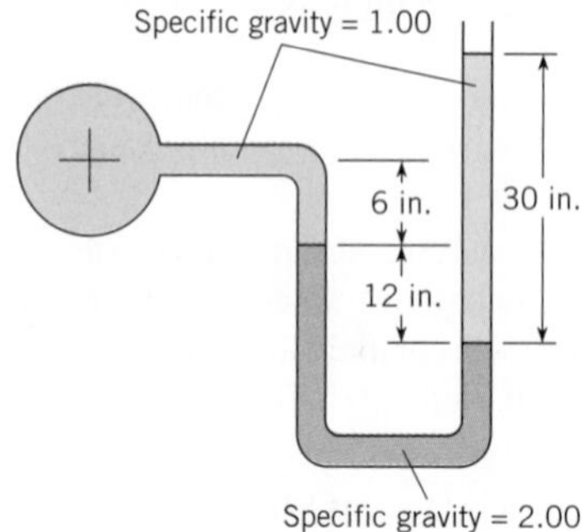

PROBLEM 3.27

3.28 Determine the gage pressure at the center of pipe A in pounds per square inch when the temperature is 70°F.

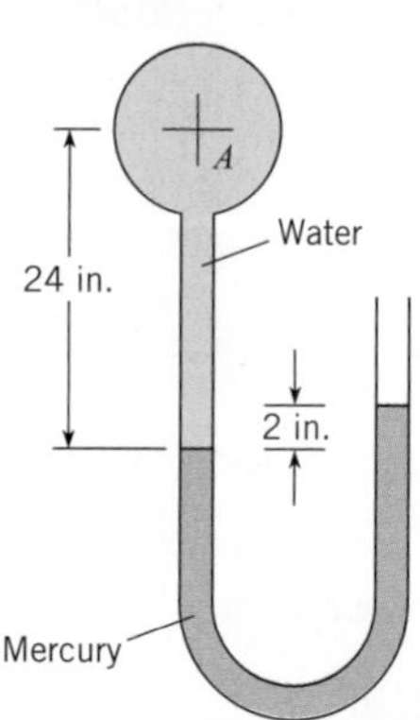

PROBLEM 3.28

3.29 Considering the effects of surface tension, estimate the gage pressure at the center of pipe A. T = 20°C.

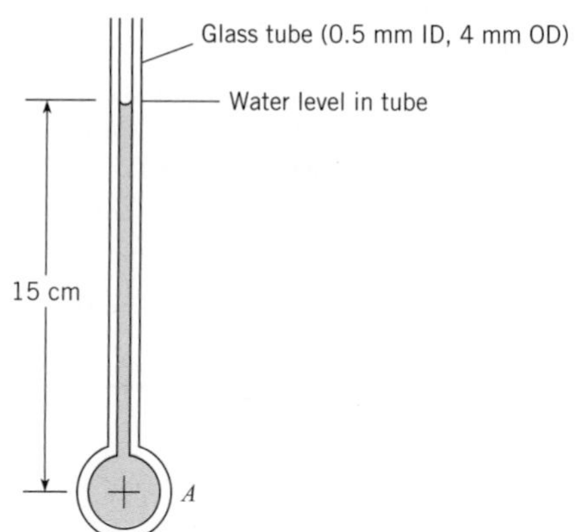

PROBLEM 3.29

3.30 What is the pressure at the center of pipe B?

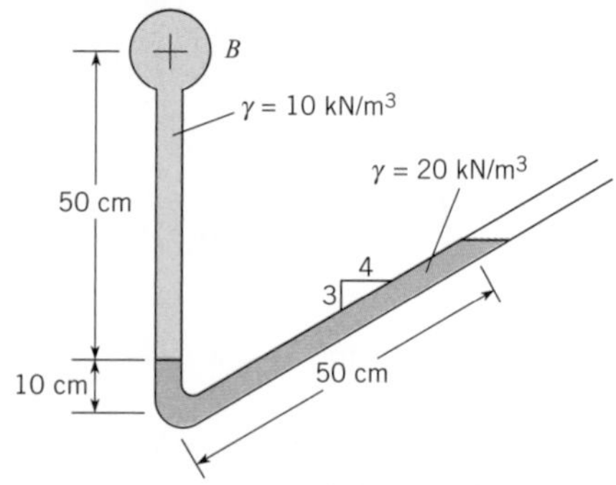

PROBLEM 3.30

3.31 The ratio of container diameter to tube diameter is 8. When air in the container is at atmospheric pressure, the free surface in the tube is at position 1. When the container is pressurized, the liquid in the tube moves 40 cm up the tube from position 1 to position 2. What is the container pressure that causes this deflection? The liquid density is 800 kg/m^3.

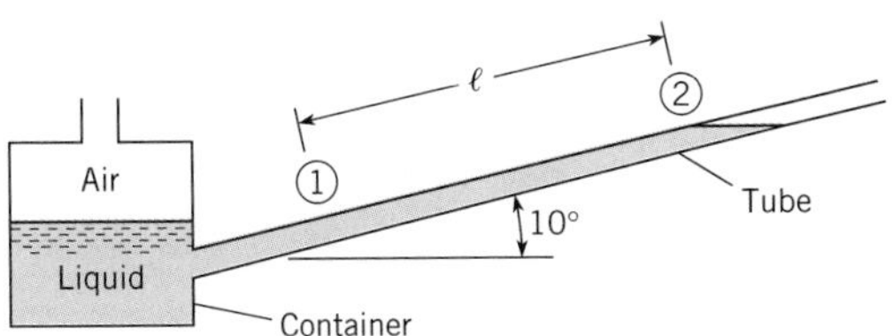

PROBLEMS 3.31, 3.32

3.32 The ratio of container diameter to tube diameter is 10. When air in the container is at atmospheric pressure, the free surface in the tube is at position 1. When the container is pressurized, the liquid in the tube moves 3 ft up the tube from position 1 to position 2. What is the container pressure that causes this deflection? The specific weight of the liquid is 50 lbf/ft^3.

3.33 A novelty scale for measuring a person's weight by having the person stand on a piston connected to a water reservoir and stand pipe is shown in the diagram. The level of the water in the stand pipe is to be calibrated to yield the person's weight in pounds force. When the person stands on the scale, the height of the water in the stand pipe should be near eye level so the person can read it. There is a seal around the piston that prevents leaks but does not cause a significant frictional force. The scale should function for people who weigh between 60 and 250 lbf and are between 4 and 6 feet tall. Choose the piston size and standpipe diameter. Clearly state the design features you considered. Indicate how you would calibrate the scale on the standpipe. Would the scale be linear?

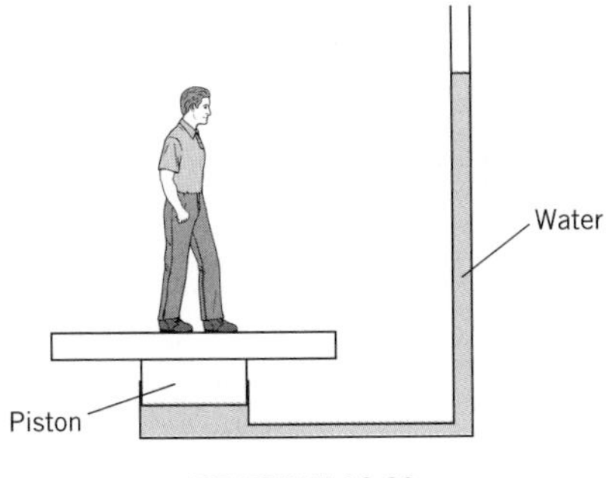

PROBLEM 3.33

3.34 Determine the gage pressure at the center of pipe A in pounds per square inch and in kilopascals.

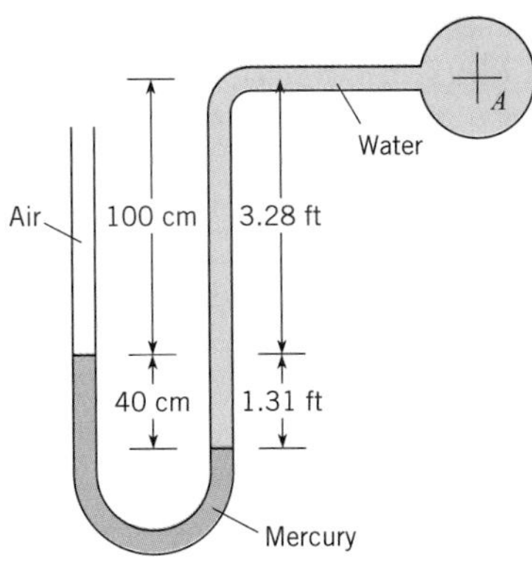

PROBLEM 3.34

3.35 A device for measuring the specific weight of a liquid consists of a U-tube manometer as shown. The manometer tube has an internal diameter of 0.5 cm and originally has water in it. Exactly 2 cm^3 of unknown liquid is then poured into one leg of the manometer, and a displacement of 5 cm is measured between the surfaces as shown. What is the specific weight of the unknown liquid?

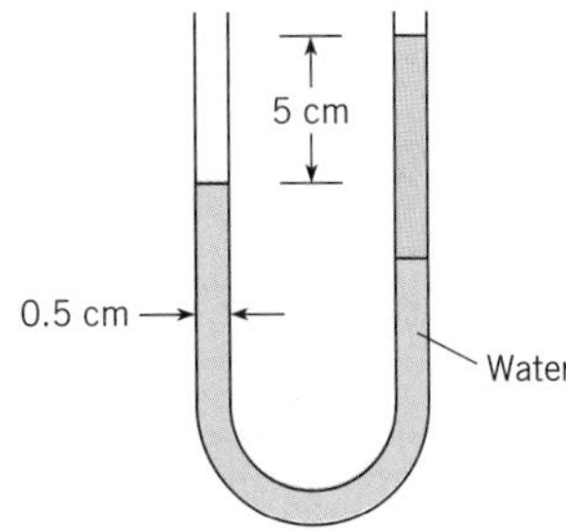

PROBLEM 3.35

3.36 Mercury is poured into the tube in the figure until the mercury occupies 1.0 ft of the tube's length. An equal volume of water is then poured into the left leg. Locate the water and mercury surfaces. Also determine the maximum pressure in the tube. FLUID SOLUTIONS

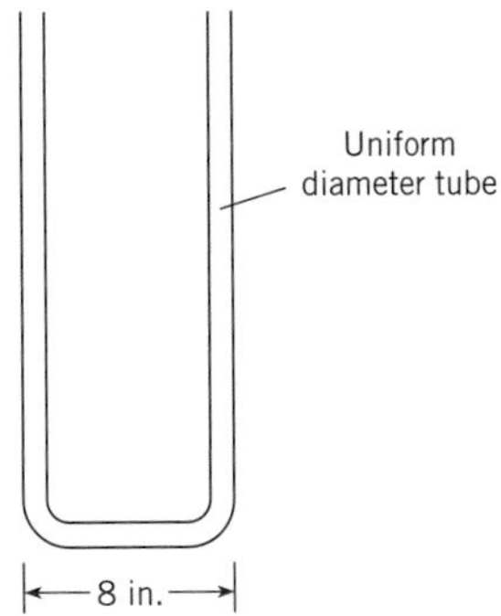

PROBLEM 3.36

3.37 A U-tube manometer is needed that will measure the difference in pressure between two points 100 m apart in a horizontal 6 cm pipe. The pipe carries water, and the maximum pressure difference is expected to be 60 kPa. Design the manometer and predict the probable degree of accuracy of measurement of Δp for your design.

3.38 It is necessary to measure the specific weight of several different liquids. The specific weights of these liquids may range from 50 lbf/ft^3 to 100 lbf/ft^3. Design an apparatus using the manometer principle to measure the specific weights of these liquids. Also estimate the expected degree of error of measurement using your equipment.

3.39 Find the pressure at the center of pipe A. Assume $T = 10°C$.

3.40 Determine (a) the difference in pressure and (b) the difference in piezometric head between points A and B. The elevations z_A and z_B are 10 m and 11 m, respectively, $\ell_1 = 1$ m, and the manometer deflection ℓ_2 is 50 cm.

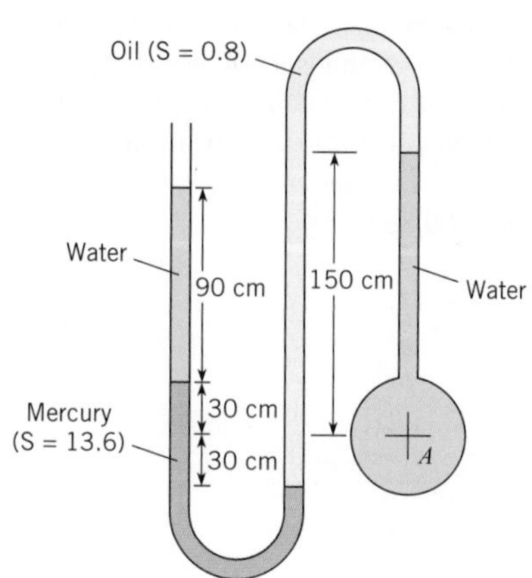

PROBLEMS 3.39

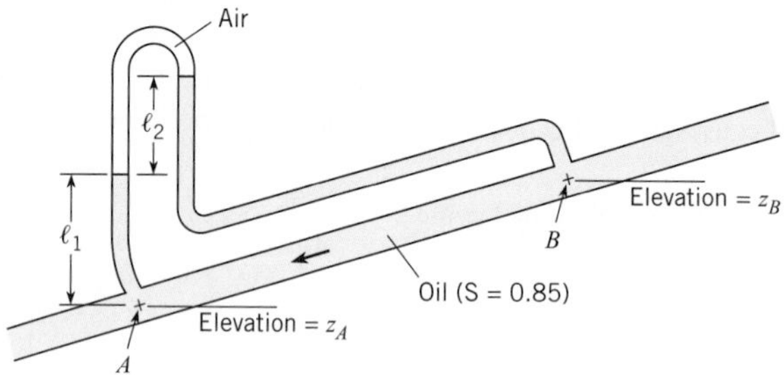

PROBLEM 3.40

3.41 The deflection on the manometer is h meters when the pressure in the tank is 150 kPa absolute. If the absolute pressure in the tank is doubled, what will the deflection on the manometer be?

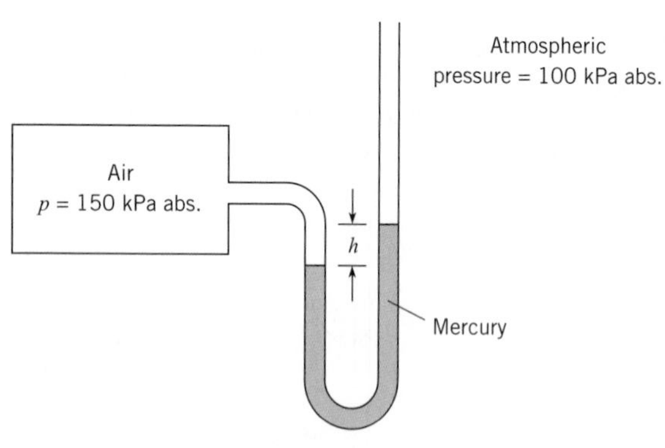

PROBLEM 3.41

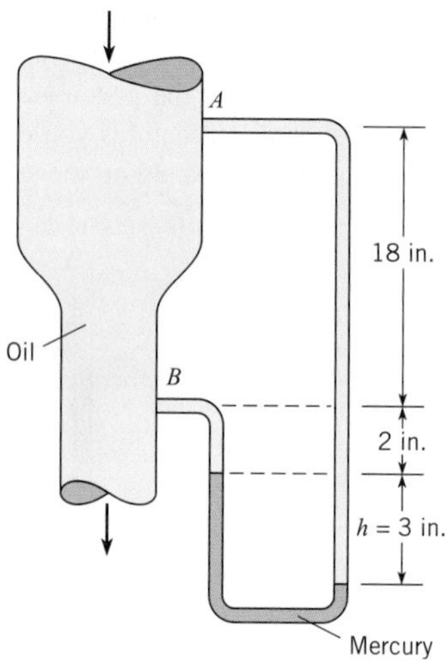

PROBLEM 3.42

3.42 A vertical conduit is carrying oil (S $= 0.95$). A differential mercury manometer is tapped into the conduit at points A and B. Determine the difference in pressure between A and B when $h = 3$ in. What is the difference in piezometric head between A and B?

3.43 Two water manometers are connected to a tank of air. One leg of the manometer is open to 100 kPa pressure (absolute) while the other leg is subjected to 90 kPa. Find the difference in deflection between both manometers, $\Delta h_a - \Delta h_b$.

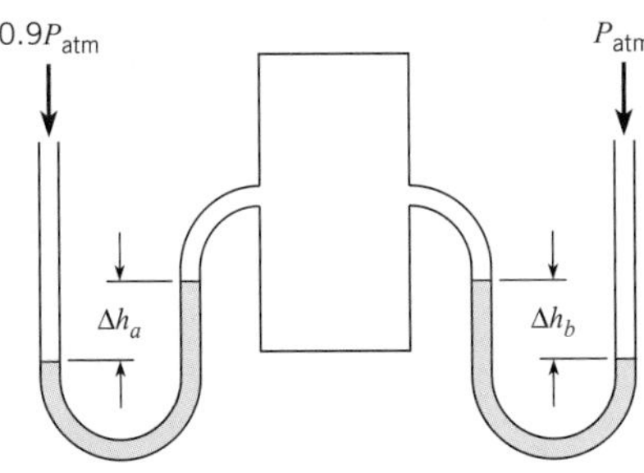

PROBLEM 3.43

3.44 A manometer is used to measure the pressure difference between points A and B in a pipe as shown. Water flows in the pipe and the specific gravity of the manometer fluid is 3.0. The distances and manometer deflection are indicated on the figure. Find (a) the pressure differences $p_A - p_B$, and (b) the difference in piezometric pressure, $p_{z,A} - p_{z,B}$. Express both answers in kPa.

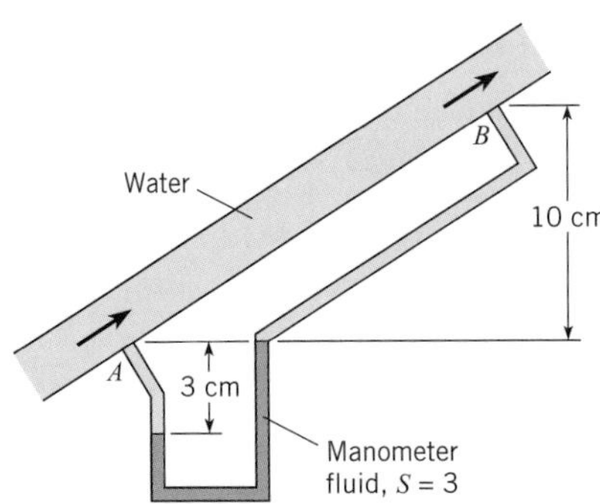

PROBLEM 3.44

3.45 One means of determining the surface level of liquid in a tank is by discharging a small amount of air through a small tube, the end of which is submerged in the tank, and reading the pressure on the gage that is tapped into the tube. Then the level of the liquid surface in the tank can be calculated. If the pressure on the gage is 20 kPa, what is the depth d of liquid in the tank?

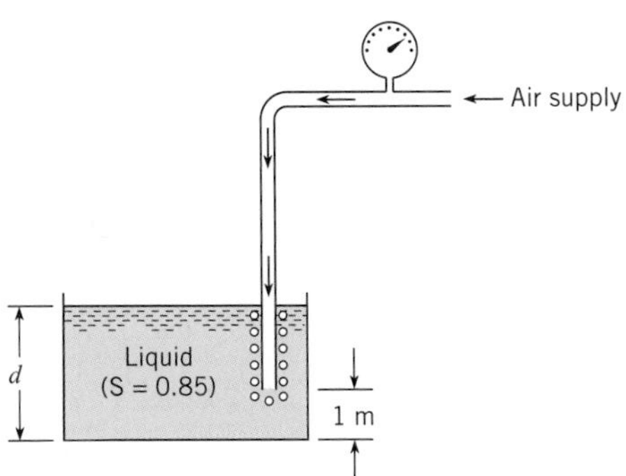

PROBLEM 3.45

3.46 Consider the ratio $(dp/dz)_0/(dp/dz)_{2000}$, where dp/dz is the pressure gradient in the atmosphere (z is positive up) and the subscripts 0 and 2000 refer to elevation in feet above sea level. Choose the correct statement: (a) The ratio has a value less than 1. (b) The ratio has a value equal to 1. (c) The ratio has a value greater than 1.

3.47 The boiling point of water decreases with elevation because of the pressure change. What is the boiling point of water at an elevation of 1500 m and at an elevation of 3000 m for standard atmospheric conditions?

3.48 From a depth of 10 m in a lake to an elevation of 4000 m in the atmosphere, plot the variation of absolute pressure. Assume that the lake water surface elevation is at mean sea level and assume standard atmospheric conditions.

3.49 Assume that a woman must breathe a constant mass rate of air to maintain her metabolic processes. If she inhales and exhales 16 times per minute at sea level where the temperature is 59°F (15°C) and the pressure is 14.7 psia (101 kPa), what would you expect her rate of breathing at 18,000 ft (5486 m) to be? Use standard atmospheric conditions. FLUID SOLUTIONS

3.50 A pressure gage in an airplane indicates a pressure of 95 kPa at takeoff, where the airport elevation is 1 km and the temperature is 10°C. If the standard lapse rate of 5.87°C/km is assumed, at what elevation is the plane when a pressure of 75 kPa is read? What is the temperature for that condition?

3.51 A pressure gage in an airplane indicates a pressure of 13.6 psia at takeoff, where the airport elevation is 2000 ft and the temperature is 70°F. If the standard lapse rate of 0.003221°F/ft is assumed, at what elevation is the plane when a pressure of 10 psia is read?

3.52 Denver, Colorado, is called the "mile-high" city. What are the pressure, temperature, and density of the air when standard atmospheric conditions prevail? Give your answer in traditional and SI units.

3.53 The mean atmospheric pressure on the surface of Mars is 7 m̄ and the mean surface temperature is –63°C. The atmosphere consists primarily of CO_2 (95.3%) with small amounts of nitrogen and argon. The acceleration due to gravity on the surface is 3.72 m/s^2. Data from probes entering the Martian atmosphere show that the temperature variation with altitude can be approximated as constant at –63°C from the Martian surface to 14 km, and then a linear decrease with a lapse rate of 1.5°C/km up to 34 km. Find the pressure at 8 km and 30 km altitude. Assume the atmosphere is pure carbon dioxide. Note that the temperature distribution in the atmosphere of Mars differs from that of the earth because the region of constant temperature is adjacent to the surface and the region of decreasing temperature starts at an altitude of 14 km.

3.54 An airplane is flying at 10 km altitude in a U.S. standard atmosphere. The dimensions of a typical window on the airplane are 30 cm by 25 cm. If the internal pressure of the aircraft interior is 100 kPa, what is the outward force on the window?

3.55 Design a computer program that calculates the pressure and density for the U.S. standard atmosphere from 0 to 30 km altitude. Assume the temperature profiles are linear and are approximated by the following ranges, where z is the altitude in kilometers.

0–13.72 km	$T = 23.1 - 5.87z$ (°C)
13.7–16.8 km	$T = -57.5$°C
16.8–30 km	$T = -57.5 + 1.387(z - 16.8)$°C

3.56 A submerged square gate (pivoted about its vertical centroidal axis) is set between two reservoirs of equal depth as shown. What is the net hydrostatic force on the gate? What moment about the pivot axis is required to keep the gate closed?

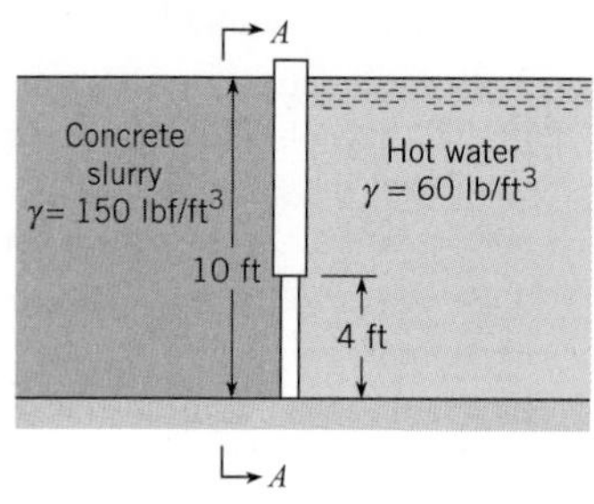

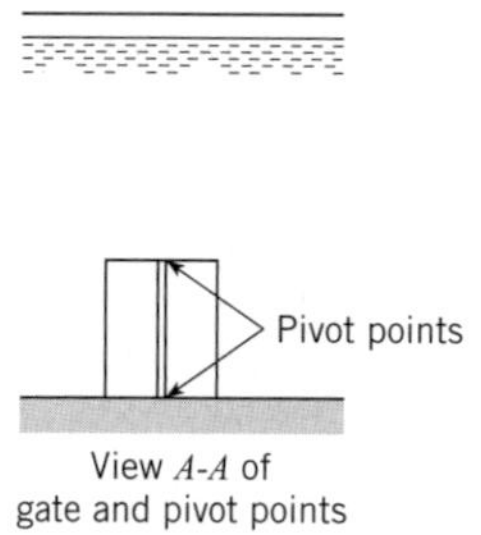

PROBLEM 3.56

FLUID SOLUTIONS **3.57** Consider the two rectangular gates shown in the figure. They are both the same size, but one (Gate A) is held in place by a horizontal shaft through its midpoint and the other (Gate B) is cantilevered to a shaft at its top. Now consider the torque T required to hold the gates in place as H is increased. Choose the valid statement(s): (a) T_A increases with H. (b) T_B increases with H. (c) T_A does not change with H. (d) T_B does not change with H.

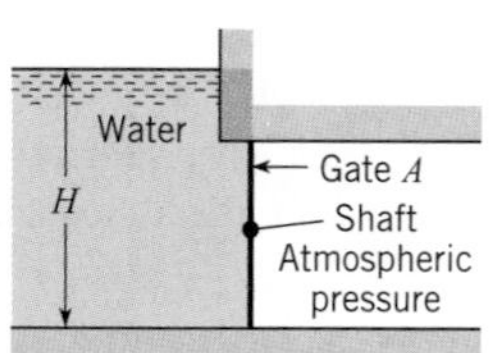

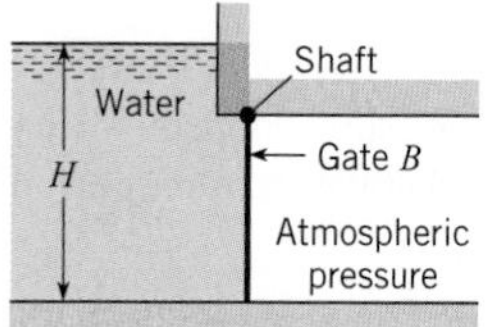

PROBLEMS 3.57, 3.58

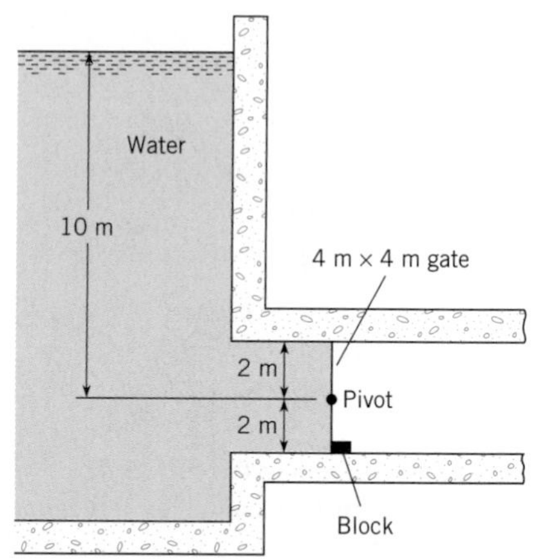

PROBLEM 3.59

3.58 For gate A, choose the statements that are valid: (a) The hydrostatic force acting on the gate increases as H increases. (b) The distance between the center of pressure on the gate and the centroid of the gate decreases as H increases. (c) The distance between the center of pressure on the gate and the centroid of the gate remains constant as H increases. (d) The torque applied to the shaft to prevent the gate from turning must be increased as H increases. (e) The torque applied to the shaft to prevent the gate from turning remains constant as H increases.

3.59 Find the force of the gate on the block.

3.60 Assume that wet concrete ($\gamma = 150$ lbf/ft^3) behaves as a liquid. Determine the force per unit foot of length exerted on the forms. If the forms are held in place as shown, with ties between vertical braces spaced every 2 ft, what force is exerted on the bottom tie?

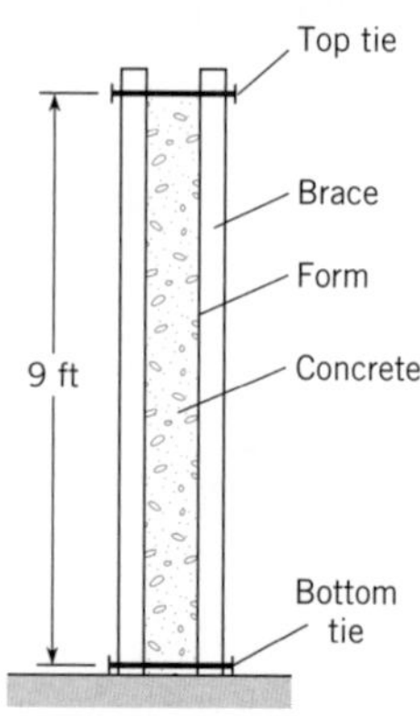

PROBLEM 3.60

3.61 A rectangular gate is hinged at the water line, as shown. The gate is 4 ft high and 12 ft wide. The specific weight of water is 62.4 lbf/ft^3. Find the necessary force (in lbf) applied at the bottom of the gate to keep it closed.

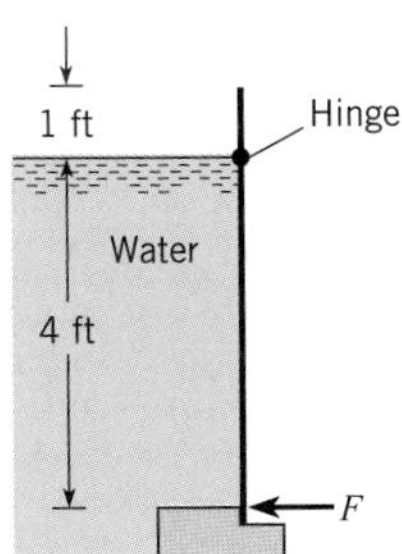

PROBLEM 3.61

3.62 The gate shown is rectangular and has dimensions 6 m × 6 m. What is the reaction at point A? Neglect the weight of the gate.

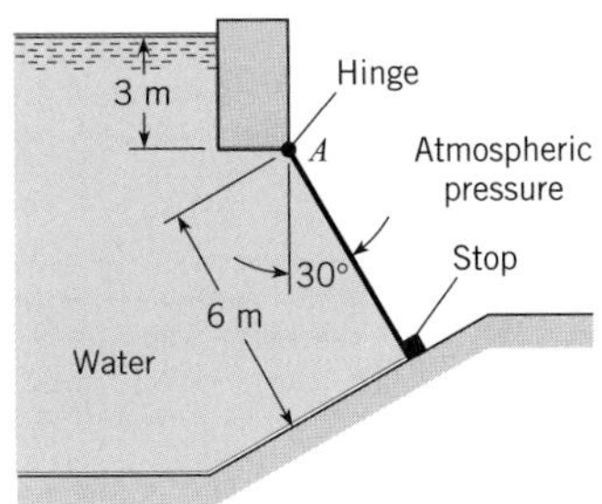

PROBLEM 3.62

3.63 Determine P necessary to just start opening the 2-m-wide gate.

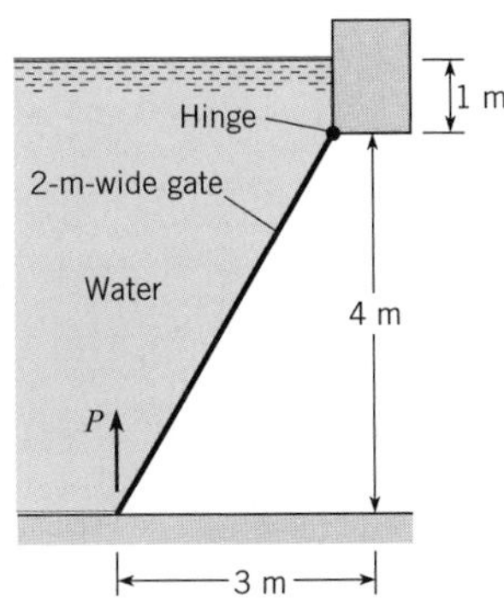

PROBLEM 3.63

3.64 The square gate shown is eccentrically pivoted so that it automatically opens at a certain value of h. What is that value in terms of ℓ?

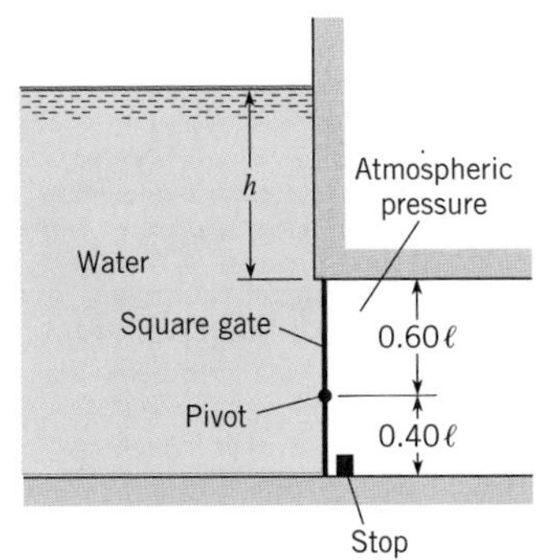

PROBLEM 3.64

3.65 This 10-ft-diameter butterfly valve is used to control the flow in a 10-ft-diameter outlet pipe in a dam. In the position shown, it is closed. The valve is supported by a horizontal shaft through its center. What torque would have to be applied to the shaft to hold the valve in the position shown?

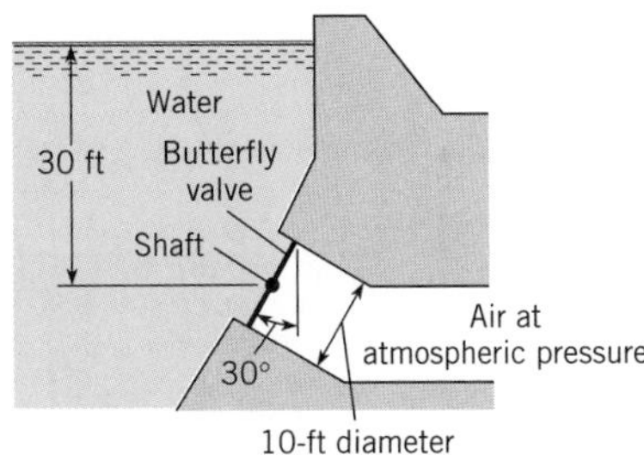

PROBLEM 3.65

3.66 For this gate, $\alpha = 45°$, $y_1 = 1$ m, and $y_2 = 3$ m. Will the gate fall or stay in position under the action of the hydrostatic and gravity forces if the gate itself weighs 90 kN and is 1.0 m wide? Assume $T = 10°C$.

3.67 For this gate, $\alpha = 45°$, $y_1 = 4$ ft, and $y_2 = 7.07$ ft. Will the gate fall or stay in position under the action of the hydrostatic and gravity forces if the gate itself weighs 18,000 lb and is 3 ft wide? Assume $T = 50°F$.

3.68 Determine the hydrostatic force F on the triangular gate, which is hinged at the bottom edge and held by the reaction R_T at the upper corner. Express F in terms of γ, h, and W. Also determine the ratio R_T/F. Neglect the weight of the gate.

3.69 For the plane rectangular gate ($\ell \times w$ in size), Figure (*a*), what is the magnitude of the reaction at A in terms of γ_w and the dimensions ℓ and w? For the cylindrical gate, Figure (*b*), will the magnitude of the reaction at A be greater than, less than, or the same as that for the plane gate? Neglect the weight of the gates.

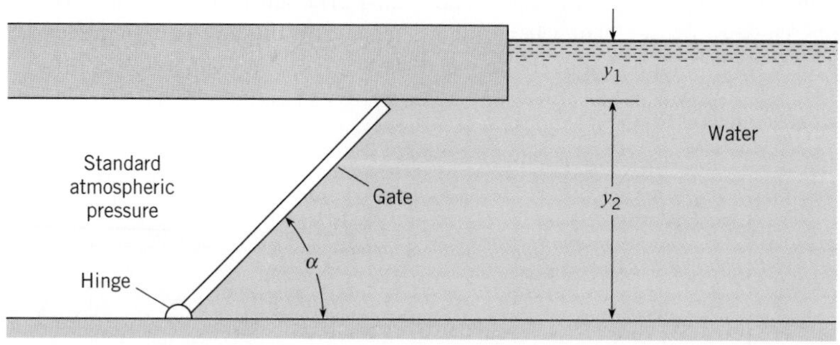

PROBLEMS 3.66, 3.67

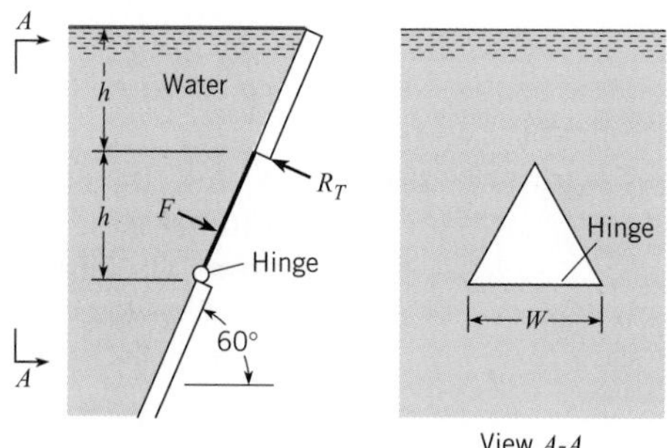

PROBLEM 3.68

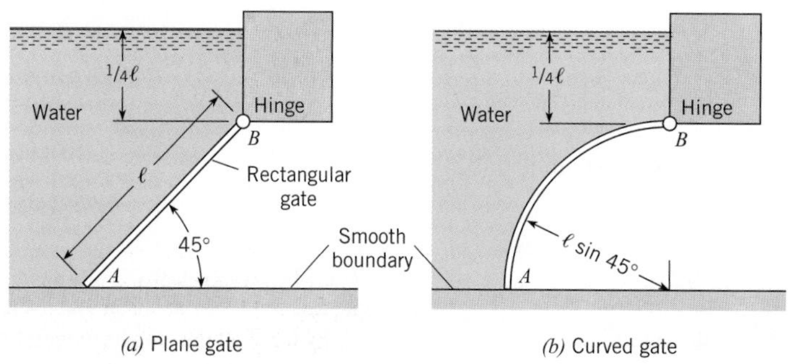

(a) Plane gate (b) Curved gate

PROBLEM 3.69

FLUID SOLUTIONS **3.70** The air above the liquid is under a pressure of 5 psig, and the specific gravity of the liquid in the tank is 0.80. If the rectangular gate is 6 ft wide and if $y_1 = 1$ ft and $y_2 = 10$ ft, what force P is required to hold the gate in place?

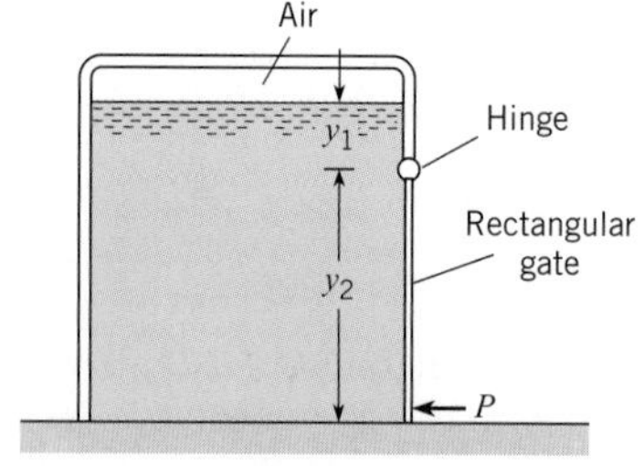

PROBLEM 3.70

3.71 In constructing dams, the concrete is poured in lifts of approximately 1.5 m ($y_1 = 1.5$ m). The forms for the face of the dam are reused from one lift to the next. The figure shows one such form, which is bolted to the already cured concrete. For the new pour, what moment will occur at the base of the form per meter of length (normal to the page)? Assume that concrete acts as a liquid when it is first poured and has a specific weight of 24 kN/m^3.

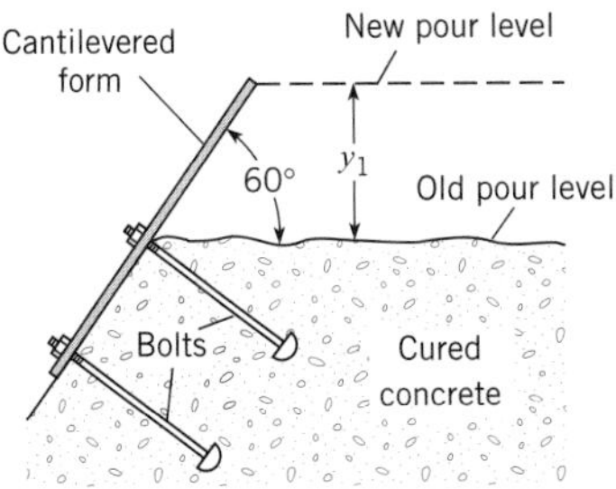

PROBLEM 3.71

3.72 The plane rectangular gate can pivot about the support at B. For the conditions given, is it stable or unstable? Neglect the weight of the gate.

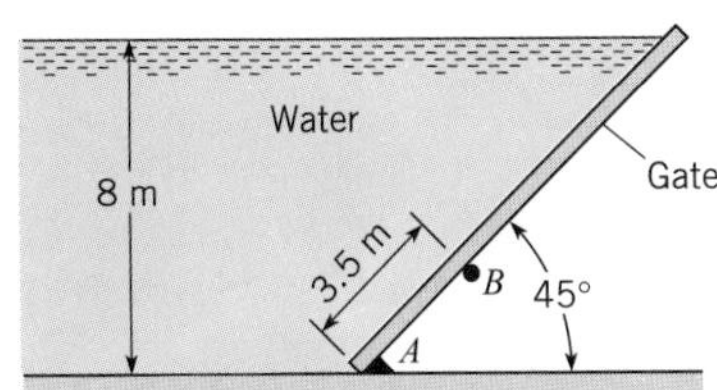

PROBLEM 3.72

3.73 Determine the minimum volume of concrete ($\gamma = 23.6$ kN/m^3) needed to keep the gate (1 m wide) in a closed position; $\ell = 2$ m. Note the hinge at the bottom of the gate.

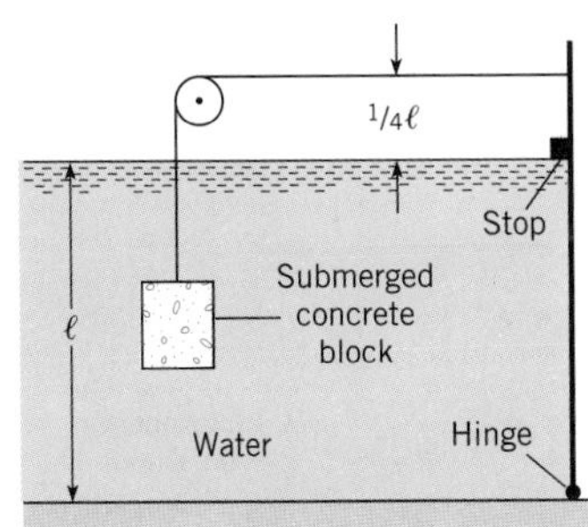

PROBLEMS 3.73, 3.74

3.74 Determine the minimum volume of concrete ($\gamma = 150$ lbf/ft^3) needed to keep the gate (2 ft wide) in a closed position; $\ell = 4$ ft.

FLUID SOLUTIONS **3.75** A gate with a circular cross section is held closed by a lever 1 m long attached to a buoyant cylinder. The cylinder is 25 cm in diameter and weighs 200 N. The gate is attached to a horizontal shaft so it can pivot about its center. The liquid is water. The chain and lever attached to the gate have negligible weight. Find the length of the chain such that the gate is just on the verge of opening when the water depth above the gate hinge is 10 m.

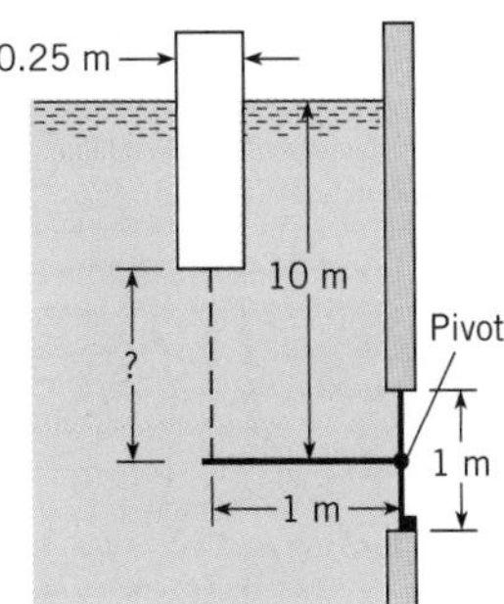

PROBLEM 3.75

3.76 The three walls are holding back water as shown. Per unit of width (normal to the page), which wall requires the greatest resisting moment (at A', B', or C') to resist the hydrostatic force, or are the moments equal? Explain. Neglect the weights of the walls.

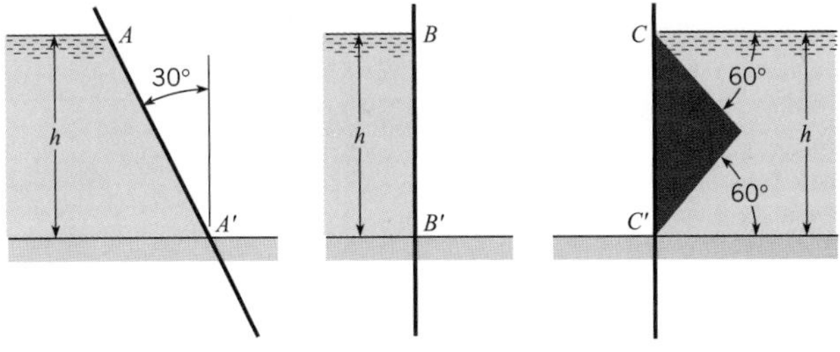

PROBLEM 3.76

3.77 Water is held back by this radial gate. Does the resultant of the pressure forces acting on the gate pass above the pin, through the pin, or below the pin?

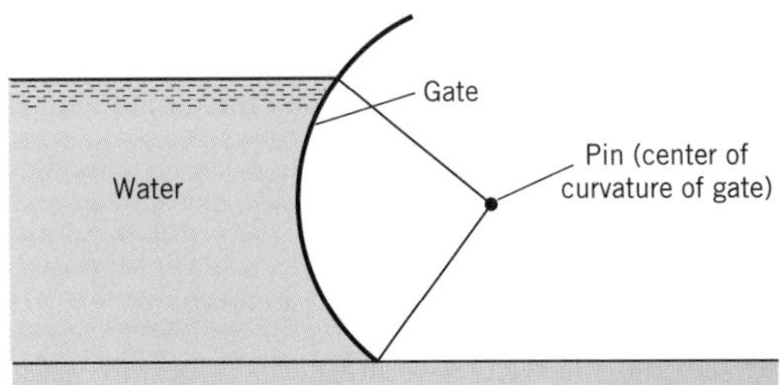

PROBLEM 3.77

3.78 For the curved surface AB:

a. Determine the magnitude, direction, and line of action of the vertical component of hydrostatic force acting on the surface. Here $\ell = 1$ m.

b. Determine the magnitude, direction, and line of action of the horizontal component of hydrostatic force acting on the surface.

c. Determine the resultant hydrostatic force acting on the surface.

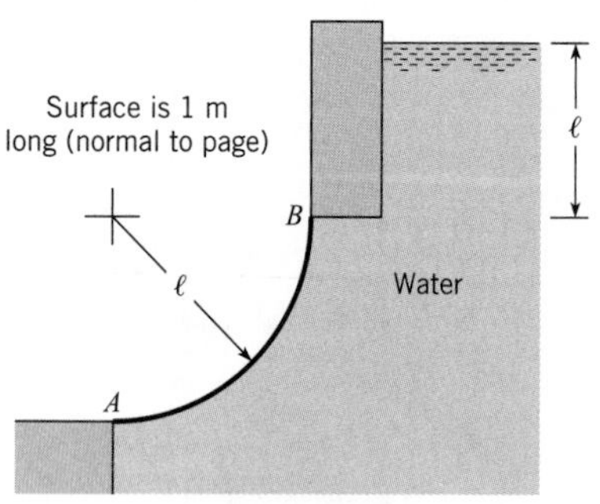

PROBLEM 3.78

3.79 Determine the hydrostatic force acting on the radial gate if the gate is 40 ft long (normal to the page). Show the line of action of the hydrostatic force acting on the gate.

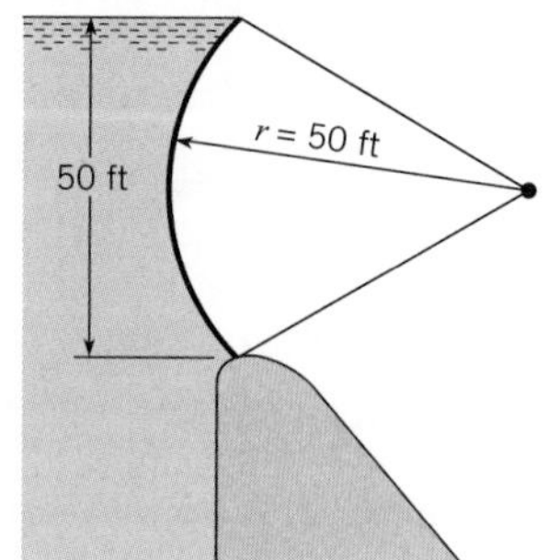

PROBLEM 3.79

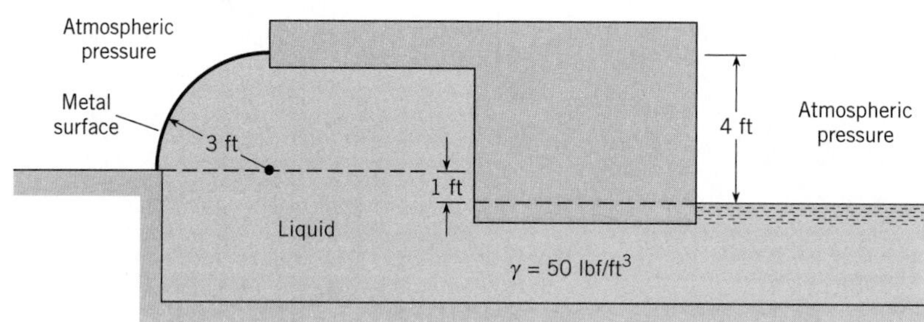

PROBLEM 3.80

3.80 Determine the magnitude and direction of the horizontal and vertical components of the hydrostatic force acting on the two-dimensional curved metal surface per foot of width. Determine the location of the horizontal component of hydrostatic force.

3.81 A plug in the shape of a hemisphere is inserted in a hole in the side of a tank as shown in the figure. The plug is sealed by an O-ring with a radius of 0.2 m. The radius of the hemispherical plug is 0.25 m. The depth of the center of the plug is 2 m in fresh water. Find the horizontal and vertical forces on the plug due to hydrostatic pressure.

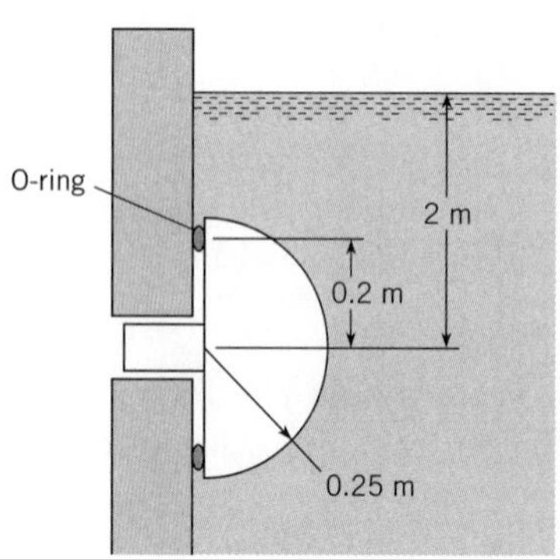

PROBLEM 3.81

3.82 This dome (hemisphere) is located below the water surface as shown. Determine the magnitude and sign of the force components needed to hold the dome in place and the line of action of the horizontal component of force. Here $y_1 = 1$ m and $y_2 = 2$ m. Assume $T = 10°C$.

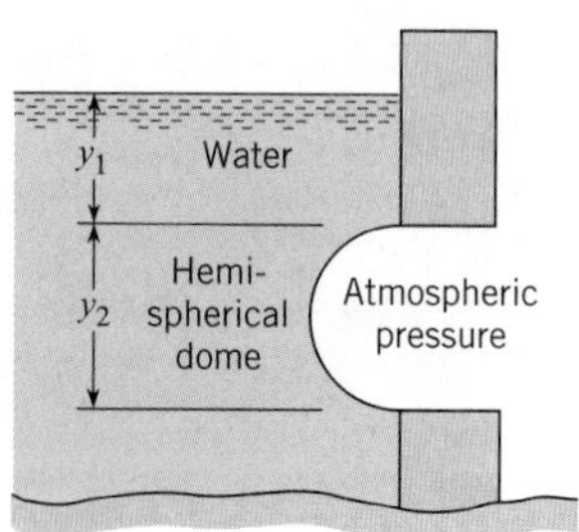

PROBLEMS 3.82, 3.83

3.83 Consider the dome of Prob. 3.82. This dome is 10 ft in diameter, but now the dome is not submerged. The water surface is at the level of the center of curvature of the dome. For these conditions, determine the magnitude and direction of the resultant hydrostatic force acting on the dome.

3.84 A block of material of unknown volume is submerged in water and found to weigh 400 N (in water). The same block weighs 700 N in air. Determine the specific weight and volume of the material.

3.85 A weather balloon is constructed of a flexible material such that the internal pressure of the balloon is always 10 kPa higher than the local atmospheric pressure. At sea level the diameter of the balloon is 1 m, and it is filled with helium. The balloon material, structure, and instruments have a mass of 100 g. This does not include the mass of the helium. As the balloon rises, it will expand. The temperature of the helium is always equal to the local atmospheric temperature, so it decreases as the balloon gains altitude. Calculate the maximum altitude of the balloon in a standard atmosphere.

3.86 A rock weighs 918 N in air and 609 N in water. Find its volume.

3.87 This uniform-diameter rod is weighted at one end and is floating in the liquid as shown. The liquid (a) is lighter than water, (b) must be water, (c) is heavier than water.

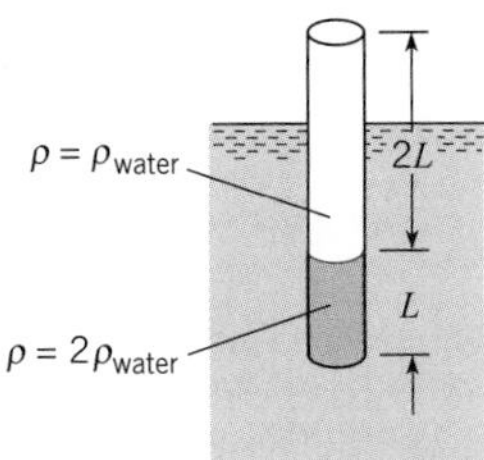

PROBLEM 3.87

3.88 A person is floating in a boat with an aluminum anchor. The anchor has a specific gravity of 2.2 and a volume of 0.5 ft^3. The surface area of the water in the pond is 500 ft^2. The person throws the anchor out of the boat into the water. By how much (if any) will the water level in the pond change?

3.89 A 90° inverted cone contains water as shown. The volume of the water in the cone is given by $\mathcal{V} = (\pi/3)h^3$. The original depth of the water is 10 cm. A block with a volume of 200 cm^3 and a specific gravity of 0.6 is floated in the water. What will be the change (in cm) in water surface height in the cone?

3.90 Large concrete cylindrical shells sealed at each end are to be used in the construction of an offshore oil drilling rig. These cylinders are floated out to the site and then uprighted to form part of the structure. The cylinders are 20 m in diameter and, when floating in the horizontal position (axis horizontal), sink to a depth of 15 m. What will be their height above the water when erected? The cylinder length is 40 m.

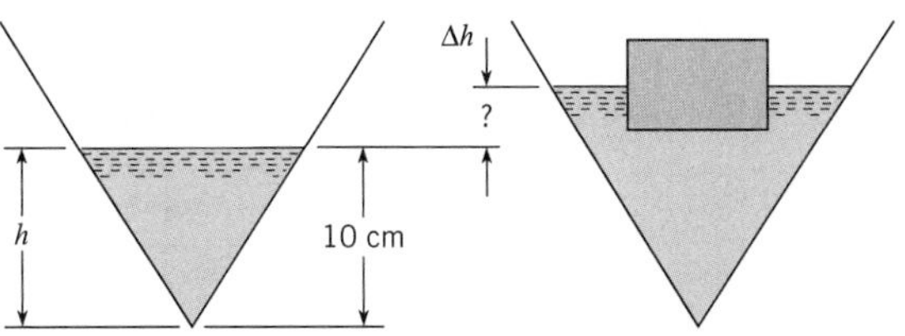

PROBLEM 3.89

3.91 A 1-ft-diameter cylindrical tank is filled with water to a depth of 2 ft. A cylinder of wood 6 in. in diameter and 3 in. long is set afloat on the water. The weight of the wood cylinder is 2 lbf. Determine the change (if any) in the depth of the water in the tank.

3.92 The floating platform shown is supported at each corner by a hollow sealed cylinder 1 m in diameter. The platform itself weighs 30 kN in air, and each cylinder weighs 1.0 kN per meter of length. What total cylinder length L is required for the platform to float 1 m above the water surface? Assume that the specific weight of the water (brackish) is 10,000 N/m^3. The platform is square in plan view.

3.93 To what depth d will this square block (it has a density 0.8 times that of water) float in the two-liquid reservoir?

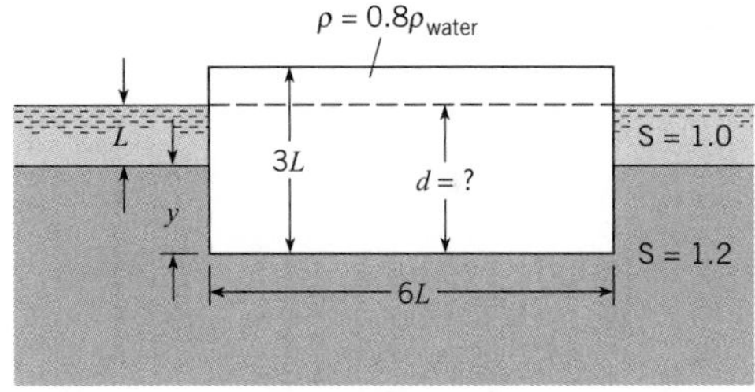

PROBLEM 3.93

3.94 A cylindrical container 4 ft high and 2 ft in diameter holds water to a depth of 2 ft. How much does the level of the water in the tank change when a 5-lb block of ice is placed in the container? Is there any change in the water level in the tank when the block of ice melts? Does it depend on the specific gravity of the ice? Explain all the processes.

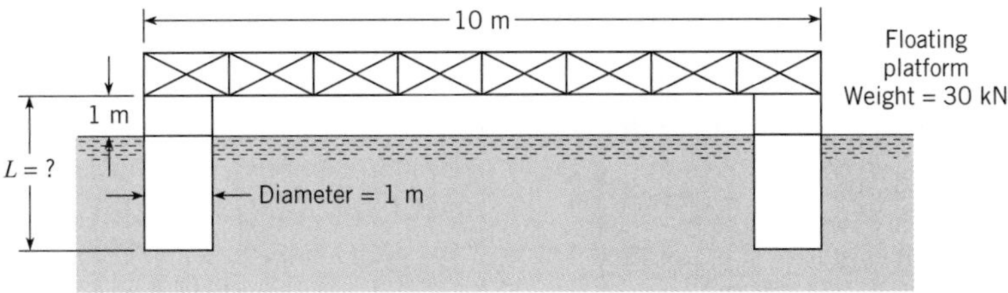

PROBLEM 3.92

3.95 The partially submerged wood pole is attached to the wall by a hinge as shown. The pole is in equilibrium under the action of the weight and buoyant forces. Determine the density of the wood.

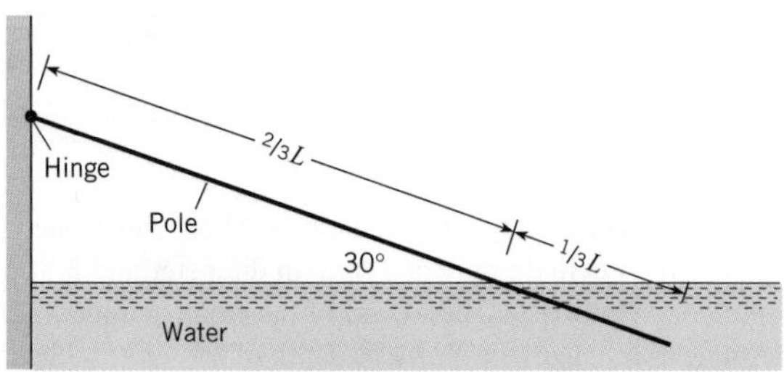

PROBLEM 3.95

3.96 This pole is pinned (hinged) at *A* and *B*, and the pole itself has a specific weight of 160 lb/ft^3. The liquid has a specific weight of 200 lb/ft^3. Upon release of the pin at *B*, will the end *B* rise, fall, or remain in its initial position?

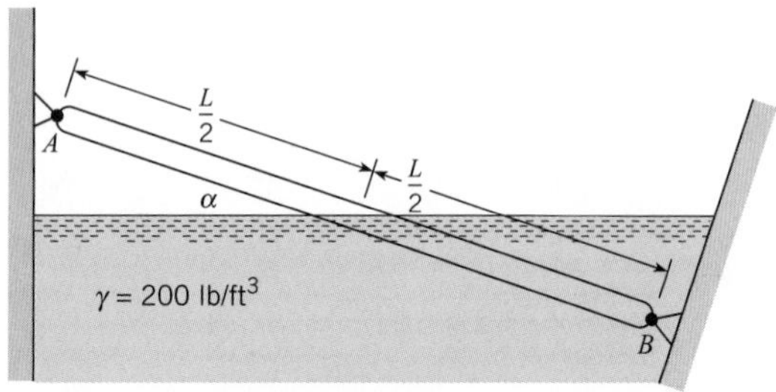

PROBLEM 3.96

3.97 An 800-ft ship has a displacement of 40,000 tons, and the area defined by the waterline is 38,000 ft^2. Will the ship take more or less draft when steaming from salt water to fresh water? How much will it settle or rise?

3.98 A submerged spherical steel buoy that is 1.2 m in diameter and weighs 1600 N is to be anchored in salt water 20 m below the surface. Find the weight of scrap iron that should be sealed inside the buoy in order that the force on its anchor chain will not exceed 4.5 kN.

FLUID SOLUTIONS **3.99** A balloon is to be used to carry meteorological instruments to an elevation of 15,000 ft where the air pressure is 8.3 psia. The balloon is to be filled with helium, and the material from which it is to be fabricated weighs 0.01 lbf/ft^2. If the instruments weigh 10 lbf, what diameter should the spherical balloon have?

3.100 A buoy is designed with a hemispherical bottom and conical top as shown in the figure. The diameter of the hemisphere is 1 m and the half angle of the cone is 30°. The buoy has a mass of 460 kg. Find the location of the water level on the buoy floating in sea water ($\rho = 1010$ kg/m^3).

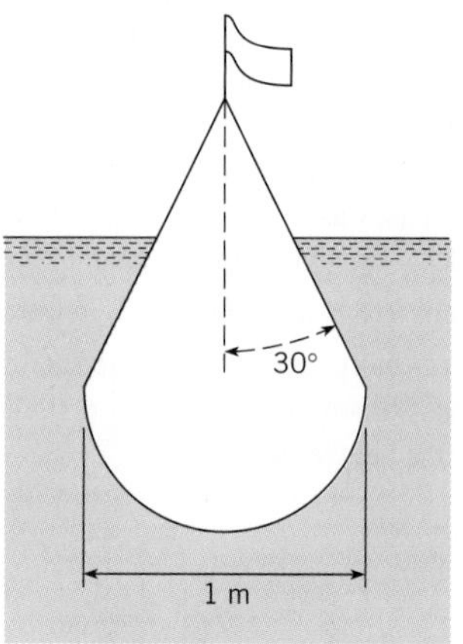

PROBLEM 3.100

3.101 The hydrometer shown sinks 5.3 cm in water (15°C). The bulb displaces 1.0 cm^3, and the stem area is 0.1 cm^2. Find the weight of the hydrometer.

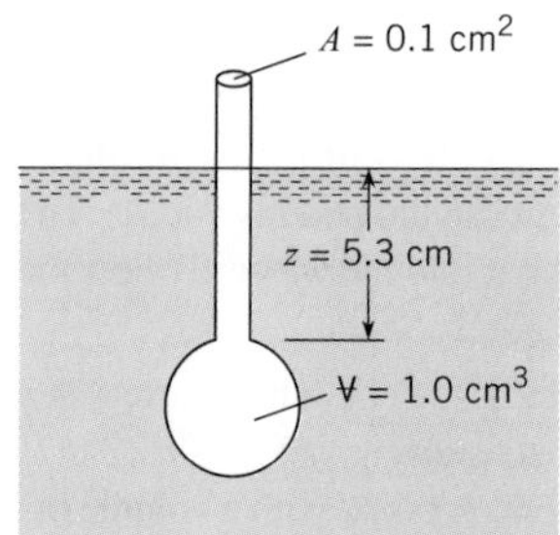

PROBLEMS 3.101, 3.102

3.102 The hydrometer of Prob. 3.101 weighs 0.015 N. If the stem sinks 6.3 cm in oil ($z = 6.3$ cm), what is the specific gravity of the oil?

3.103 A common commercial hydrometer for measuring the amount of antifreeze in the coolant system of an automobile engine consists of a chamber with differently colored balls. The system is calibrated to give the range of specific gravity by distinguishing between the balls that sink and those that float. The specific gravity of an ethylene glycol–water mixture varies from 1.012 to 1.065 for 10% to 50% by weight of ethylene glycol. Assume there are six balls, 1 cm in diameter each, in the chamber. What should the weight of each ball be to provide a range of specific gravities between 1.01 and 1.06 with 0.01 intervals?

3.104 A hydrometer with the configuration shown has a bulb diameter of 2 cm, a bulb length of 8 cm, a stem diameter of 1 cm, a length of 8 cm, and a mass of 35 g. What is the range of specific gravities that can be measured with this hydrometer? (*Hint:* Liquid levels range between bottom and top of stem.)

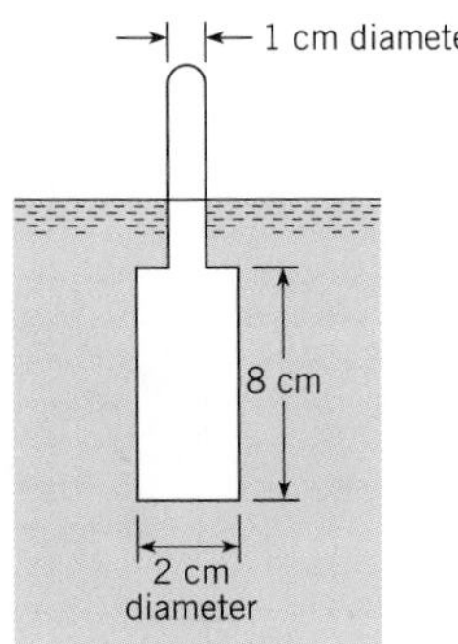

PROBLEM 3.104

3.105 Design a hydrometer to measure the specific weight of liquids that may have a range of specific gravities from 60 lbf/ft^3 to 70 lbf/ft^3. Also estimate the degree of accuracy of your hydrometer.

3.106 A barge 20 ft wide and 50 ft long is loaded with rocks as shown. Assume that the center of gravity of the rocks and barge is located along the centerline at the top surface of the barge. If the rocks and the barge weigh 400,000 lbf, will the barge float upright or tip over?

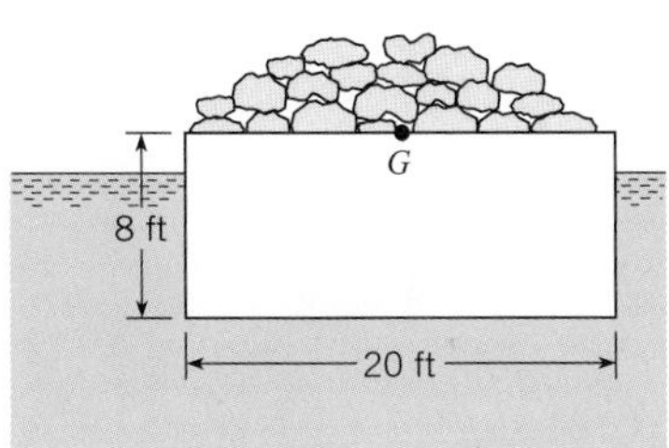

PROBLEM 3.106

3.107 A floating body has a square cross section with side w as shown in the figure. The center of gravity is at the centroid of the cross section. Find the location of the water line, ℓ/w, where the body would be neutrally stable ($GM = 0$). If the body is floating in water, what would be the specific gravity of the body material?

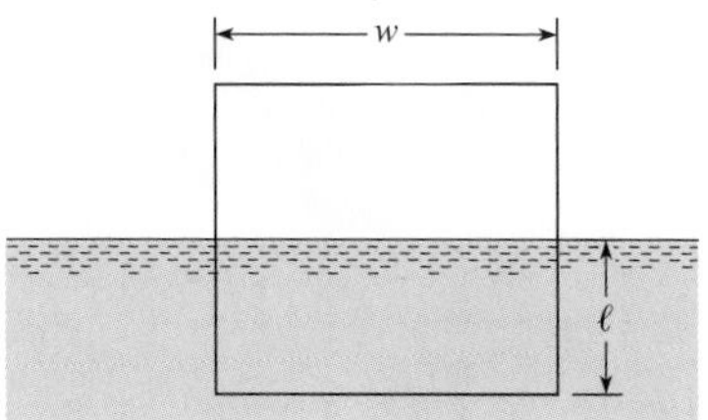

PROBLEM 3.107

3.108 A cylindrical block of wood 1 m in diameter and 1 m long has a specific weight of 7500 N/m^3. Will it float in water with its axis vertical?

3.109 A cylindrical block of wood 1 m in diameter and 1 m long has a specific weight of 5000 N/m^3. Will it float in water with the ends horizontal?

3.110 Is the block in this figure stable floating in the position shown? Show your calculations.

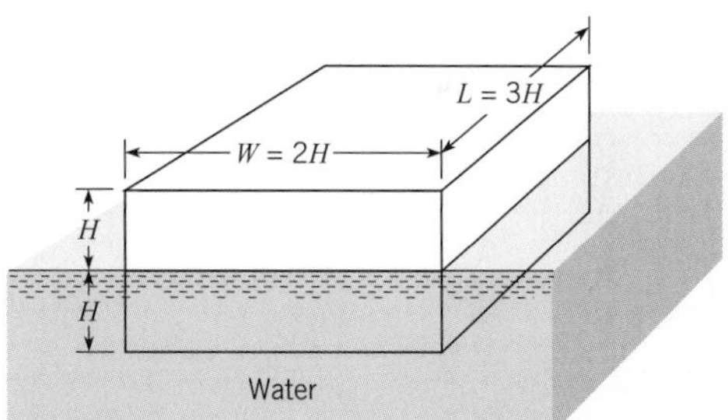

PROBLEM 3.110

This computer image of hurricane Fran (September 1996) reveals a very large-scale vortex. Hurricanes and cyclonic storms are the largest atmospheric vortices; however, tornadoes produce the greatest wind speeds. (Courtesy NASA)

CHAPTER 4

Flowing Fluids and Pressure Variation

Pressure variation in flowing fluids is important to the engineer for several reasons. In the design of tall structures, the pressure forces from the wind may dictate the design of individual elements, such as windows, as well as the basic structure to withstand wind loads. In aircraft design, the pressure distribution is primarily responsible for lift and contributes to the drag of the aircraft. In the design of flow systems, such as heating and air conditioning, the pressure distribution is responsible for flow in the ducts. The variation of pressure plays a key role in the many areas of engineering design and analysis.

Many phenomena that affect us in our daily lives are related to pressure in flowing fluids. For example, one indicator of our health, blood pressure, is related to the flow of blood through veins and arteries. The atmospheric pressure readings reported in weather forecasts control atmospheric flow patterns related to local weather conditions. Even the rotary motion generated when we stir a cup of coffee gives rise to pressure variations and flow patterns that enhance mixing.

This chapter introduces the basic concepts underlying fluid motion and pressure variation. The ideas of pathlines and streamlines help us visualize and understand fluid motion. The definition of fluid velocity and acceleration leads to an application of Newton's second law relating forces on a fluid element to the product of mass and acceleration. These relationships lead to the Bernoulli equation, which relates local pressure and elevation to fluid velocity and is fundamental to many fluid mechanic applications. This chapter also introduces the idea of fluid rotation and the concept of irrotationality.

4.1 Velocity and Descriptions of Flow

The velocity of flow is of primary concern for many engineering problems. For flow past structures, information on the local velocity allows the engineer to estimate pressures and forces that may exceed design loads and lead to failure. In canal or bridge pier design, the engineer may be interested in the velocity from the point of view of the scouring action on the channel bottom. In the design of air-cooling systems, the magnitude of the velocity establishes the cooling rate. In any case, the engineer involved with flow problems must be able to determine the velocity through experimental or analytical means.

Lagrangian and Eulerian Descriptions of Fluid Motion

There are two ways to describe fluid motion. One way is to identify a small mass of fluid in a flow, called a "fluid particle," and describe the fluid particle's motion with time. This is the Lagrangian approach. The path of a fluid particle is given by the vector $\mathbf{r}(t)$, shown in Fig. 4.1, and can be expressed in terms Cartesian coordinates as

$$\mathbf{r}(t) = x(t)\mathbf{i} + y(t)\mathbf{j} + z(t)\mathbf{k} \tag{4.1}$$

where $\mathbf{i}$, $\mathbf{j}$, and $\mathbf{k}$ are the unit vectors in the x, y, and z directions. The corresponding fluid velocity would be obtained by taking the time derivative

$$\mathbf{V}(t) = \frac{d\mathbf{r}(t)}{dt} = \frac{dx}{dt}\mathbf{i} + \frac{dy}{dt}\mathbf{j} + \frac{dz}{dt}\mathbf{k} \tag{4.2}$$

or

$$\mathbf{V}(t) = u\mathbf{i} + v\mathbf{j} + w\mathbf{k} \tag{4.3}$$

where u, v, and w are the component velocities in the respective coordinate directions.

The fluid particle path shown in Fig. 4.1 represents only one particle. To have a complete and general description of the fluid motion over some field, as shown in Fig. 4.2*a*, the paths of many, many fluid particles would have to be available. Then the velocity at a specific point in the field would be obtained by locating the fluid path that passed through that point and calculating the velocity using Eq. (4.2). Obviously, it is an enormous task to keep track of all the fluid particle paths required to find the flow velocity at an arbitrary point in a flow field.

The other way to describe fluid motion is to imagine an array of "windows" in the flow field, as shown in Fig. 4.2*b*, and have information for the velocity of fluid particles that pass each window for all time. This is the Eulerian approach. In this case, the velocity is a function of the window position (x, y, and z) and time, so

$$u = f_1(x, y, z, t) \qquad v = f_2(x, y, z, t) \qquad w = f_3(x, y, z, t) \tag{4.4}$$

The level of detail would depend on the number of windows available. In the limit, there would be an infinite number of windows of infinitesimal size (points), and the velocity would be available at every point in the field. The Eulerian approach is generally favored over the Lagrangian approach because the Eulerian approach is more suited to the analysis of a continuum. In some cases, such as rarefied gas dynamics, where we are con-

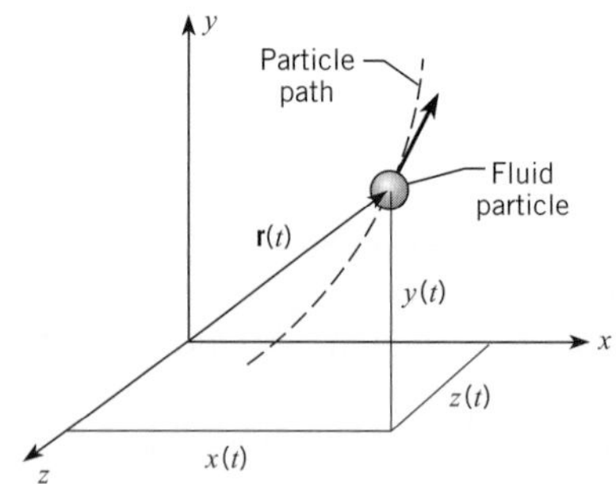

FIGURE 4.1

Fluid particle path and velocity.

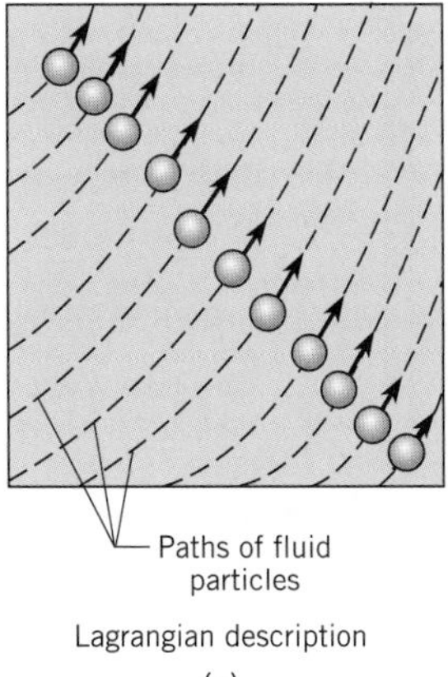

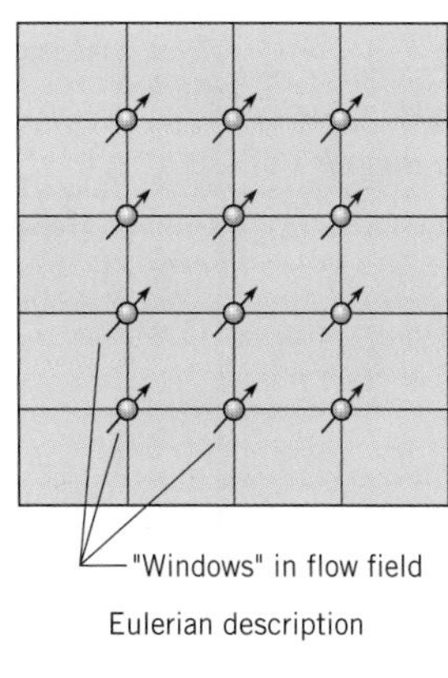

FIGURE 4.2

The Lagrangian and Eulerian descriptions of a flow field.

cerned with the motion of individual molecules, the Lagrangian approach is used. Also, in the analysis of gas-particle mixtures the particle field is often described using the Lagrangian approach (1). However, the vast majority of fluid dynamic analyses are based on the Eulerian approach.

Another useful way to express the velocity is in terms of the position along a streamline and time. This is given as

$$\mathbf{V} = \mathbf{V}(s, t)$$

The streamline will be defined in the next section.

Streamlines and Flow Patterns

To visualize the flow field it is desirable to construct lines that show the flow direction. Such a construction is called a *flow pattern* and the lines are called *streamlines*. The streamline is defined as a line drawn through the flow field in such a manner that the local velocity vector is tangent to the streamline at every point along the line at that instant. Thus the tangent of the streamline at a given time gives the direction of the velocity vector. A streamline, however, does not indicate the magnitude of the velocity. The flow pattern provided by the streamlines is an instantaneous visualization of the flow field.

An example of streamlines and a flow pattern is shown in Fig. 4.3*a* for water flowing through a slot in the side of a tank. The velocity vectors have been sketched at three different locations: *a*, *b*, and *c*. The streamlines, according to their definition, are tangent to the velocity vectors at these points. Also, the velocities are parallel to the wall in the wall region, so the streamlines adjacent to the wall follow the contour of the wall. One sees that the generation a flow pattern is a very effective way of illustrating the geometry of the flow.

Whenever flow occurs around a body, part of it will go to one side and part to the other as shown in Fig. 4.3*b* for flow over an airfoil section. The streamline that follows the flow division (that divides on the upstream side and joins again on the downstream side) is called the *dividing streamline.* At the location where the dividing streamline intersects the body, the velocity will be zero. This is the *stagnation point.*

FIGURE 4.3

Flow through an opening in a tank and over an airfoil section.

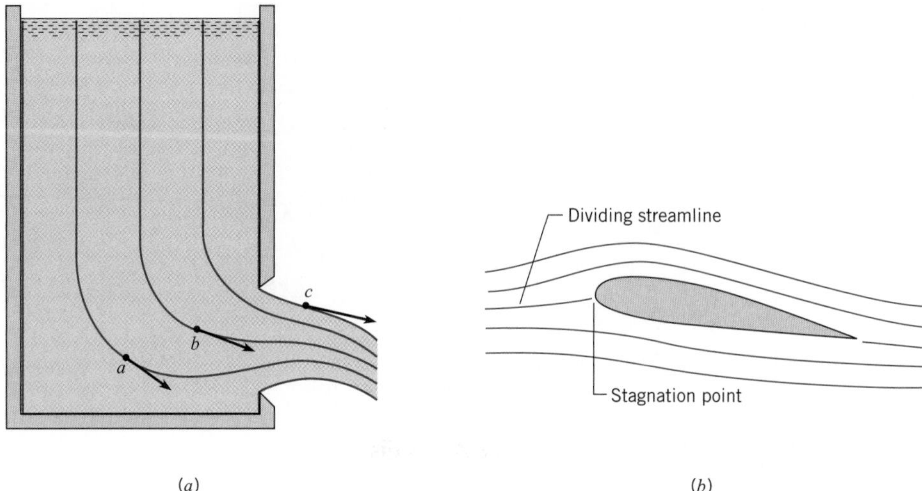

Having introduced the general concepts of flow patterns, it is convenient to make distinctions between different types of flow. Flows can be either uniform or nonuniform. In a uniform flow, the velocity does not change along a streamline; that is,

$$\frac{\partial \mathbf{V}}{\partial s} = 0$$

where s is the distance along the streamline. It follows that in uniform flows the streamlines are straight and parallel as shown in Fig. 4.4*a* and *b*. In nonuniform flows, the velocity changes along a streamline, so

$$\frac{\partial \mathbf{V}}{\partial s} \neq 0$$

For the converging duct in Fig. 4.5*a*, the magnitude of the velocity increases as the duct converges, so the flow is nonuniform. For the vortex flow shown in Fig. 4.5*b*, the magnitude of the velocity does not change along the streamline but the direction does, so the flow is nonuniform.

Flows can be either steady or unsteady. In a steady flow the velocity at a given point on a streamline does not change with time:

$$\frac{\partial \mathbf{V}}{\partial t} = 0$$

FIGURE 4.4

Uniform flow patterns. (a) Open-channel flow. (b) Flow in a pipe.

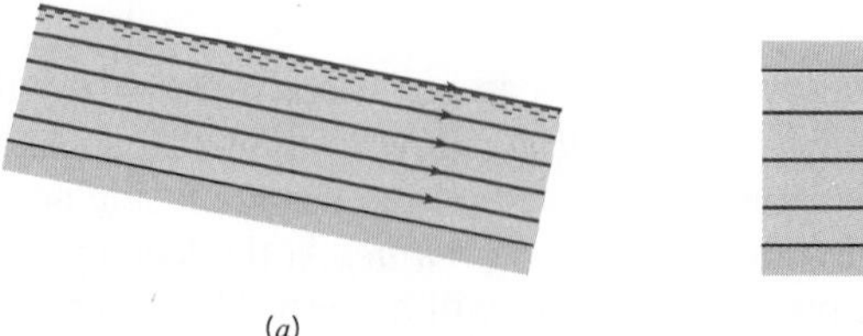

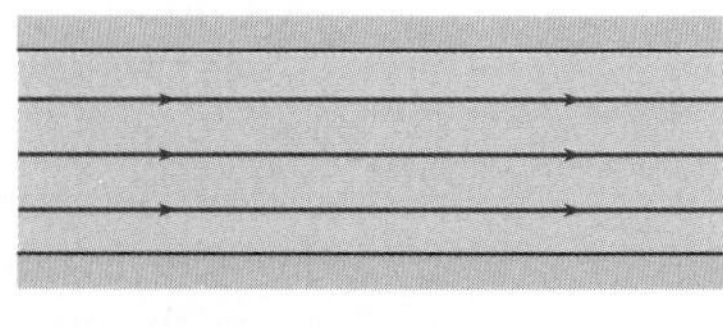

The flow in a pipe, shown in Fig. 4.4*b*, would be an example of steady flow if there was no change in velocity with time. An unsteady flow exists if

$$\frac{\partial \mathbf{V}}{\partial t} \neq 0$$

If the flow in the pipe changed with time due to a valve opening or closing, the flow would be unsteady; that is, the velocity at any point selected on a streamline would be increasing or decreasing with time. Though unsteady, the flow would still be uniform.

By studying the flow pattern one can generally decide if the flow is uniform or nonuniform. The flow pattern gives no indication of the steadiness or unsteadiness of the flow because the streamlines are only an instantaneous representation of the flow velocity.

Laminar and Turbulent Flow

Laminar flow is associated with low fluid velocities and characterized by a smooth appearance. A typical laminar flow would be the flow of honey or thick syrup from a pitcher. Laminar flow in a pipe has a smooth, parabolic velocity distribution as shown in Fig. 4.6*a*.

Turbulent flow is characterized by intense mixing and unsteady flow. For example, the flow in the wake of a ship is turbulent. The eddies observed in the wake cause intense mixing. The transport of smoke from a smoke stack on a windy day exemplifies a turbulent flow. The mixing is apparent as the plume widens and disperses.

Turbulent flow is nearly uniformly distributed across the pipe as shown in Fig. 4.6*b*. This occurs because the high-velocity fluid at the pipe center is transported by turbulent eddies across the pipe to the low-velocity region near the wall, and the low velocity at the wall is transported to the pipe center, effecting a nearly uniform velocity distribution. The velocity at any point in the pipe fluctuates with time, so the flow is unsteady. The standard approach to treating turbulent flow is to represent the velocity as an average value plus a fluctuating quantity. The average value is designated by $\bar{u}$ in Fig. 4.6*b*. We often define the flow as "steady" if the average value is unchanging with time.

FIGURE 4.5

Flow patterns for nonuniform flow. (a) Converging flow. (b) Vortex flow.

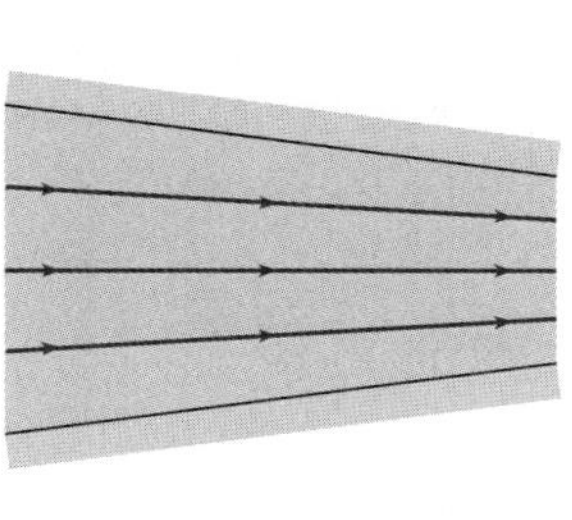

(*a*)

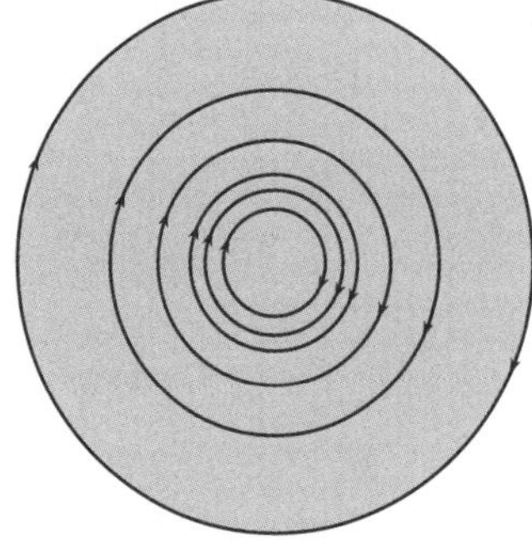

(*b*)

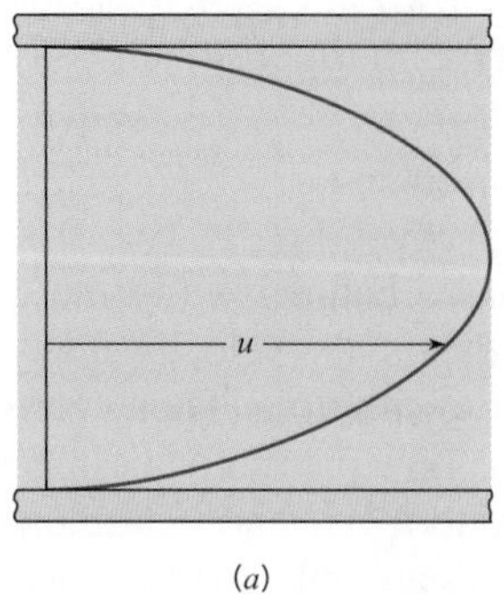

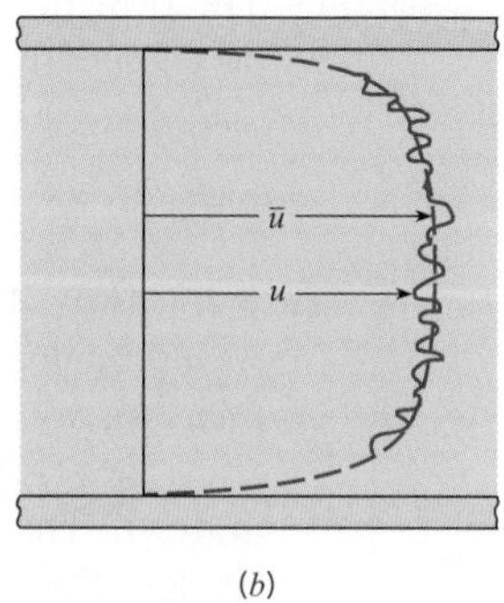

FIGURE 4.6

Laminar and turbulent flow in a straight pipe. (a) Laminar flow. (b) Turbulent flow.

In general, laminar pipe flows are associated with low velocities and turbulent flows with high velocities. A more accurate index is the Reynolds number, defined as $\mathrm{Re} = \rho VD/\mu$, where D is the pipe diameter. If the Reynolds number is large (Re > 2000), the flow will generally be turbulent, and if the Reynolds number is small (Re < 2000), the flow is probably laminar. Turbulent flows are most common. The significance of the Reynolds number will be considered in more detail in later chapters.

4.2 Methods for Developing Flow Patterns

The basic methods used to develop flow patterns are analytical, computational, and experimental.

Analytical Methods

Analytical methods use mathematical solutions of the equations describing fluid flow. Because the general equations are nonlinear and solution techniques are limited, there are very few closed-form mathematical solutions.

The oldest and best known of the analytical methods derive from ideal-flow theory—that is, theory concerning the flow of incompressible, nonviscous fluids. The differential equation fundamental to ideal-flow theory is Laplace's equation, which is encountered frequently in science and engineering. Solutions of Laplace's equation can be applied to many flow problems, such as low-speed flows past airfoils and free-surface flows such as wave motion.

Ideal-flow theory is not applicable to flows where viscous and/or compressibility effects are important. In these cases, few closed-form analytical solutions exist, and these few are restricted to simple flow configurations.

Computational Methods

Computational methods relate to the application of numerical techniques and procedures to solve the fluid flow equations using a computer. This area of expertise, called computational fluid mechanics (CFD), has rapidly developed in conjunction with enhanced computer capabilities.

The applications of CFD cover a wide range of problems, ranging from blood flow to the drag on vehicles reentering the earth's atmosphere. It is becoming increasingly common to use CFD predictions to complement the design of industrial systems. There are several software companies that market general-purpose fluid dynamics codes for use by government laboratories, industry, and universities. Fluid dynamists also continue to develop special-purpose computer codes to address problems beyond the capability of general-purpose codes.

An example of an application of CFD to automobile aerodynamics is the prediction of streamlines over a conceptual automobile designed by Volvo, as shown in Fig. 4.7. By analyzing such streamline patterns the experienced engineer can better understand how contour modifications might improve aerodynamic performance parameters such as drag and download.

Even with the phenomenal computational capability that has been developed over the past few years, the results from the CFD are only as valid as the fundamental physics built into them. The reliable modeling of turbulence remains one of the most difficult issues and may lead one to question the validity of certain numerical predictions. Also, the numerical schemes may lead to predictions that correspond to fluids with artificially modified properties. Thus it is important, whenever feasible, to verify numerical predictions with experimental results.

Experimental Method

For many flows, analytical or computational methods have not been developed to the point where they can reliably predict the flow pattern; therefore, it is necessary to resort to physical models. In other cases, such as basic research studies, experimental methods are also employed to define the pattern of flow. In these experimental setups, dye streams and floating or immersed particles are often used to define the flow pattern. When a photograph is taken of floating or neutrally buoyant particles in a flowing fluid (as in Fig. 4.8, which shows the approach channel to a spillway of a model dam), the particles produce light marks, which indicate the paths of the particles for the period of time of exposure. Each one of these light marks is a segment of a pathline for a given particle. By definition, then, a *pathline* is a line drawn through the flow field in such a way that it defines the path that a given particle of fluid has taken.

Another technique used to visualize flow patterns is to inject dye or smoke at a given point in the flow field and to observe the dye or smoke trace as it travels downstream. Such a trace is called a *streakline* (see Fig. 4.9; note the separation zone downstream of the airfoil where turbulence causes the smoke to be diffused. Separation is discussed in more detail in Sec. 4.9). In addition to dye streams or suspended particles, there are other methods to determine flow patterns. To visualize the flow next to the surface of a body being tested in a wind tunnel, one can apply a mixture of oil and pigments to the surface of the body. Droplets of the oil will be driven along the surface of the body by the wind, thus indicating the direction of the air flow next to the surface. A similar visualization method involves attaching a number of tufts of thread or yarn to the surface of the test body. Each tuft is attached only at one end. The remainder of the tuft is free to be blown by the wind flowing past it. The flow past these tufts causes the tufts to be oriented parallel to the velocity of flow past them, thus indicating the flow pattern.

The use of lasers and optical methods have led to more sophisticated techniques for flow visualization. The imaging of particles in thin sheets of laser light provide an

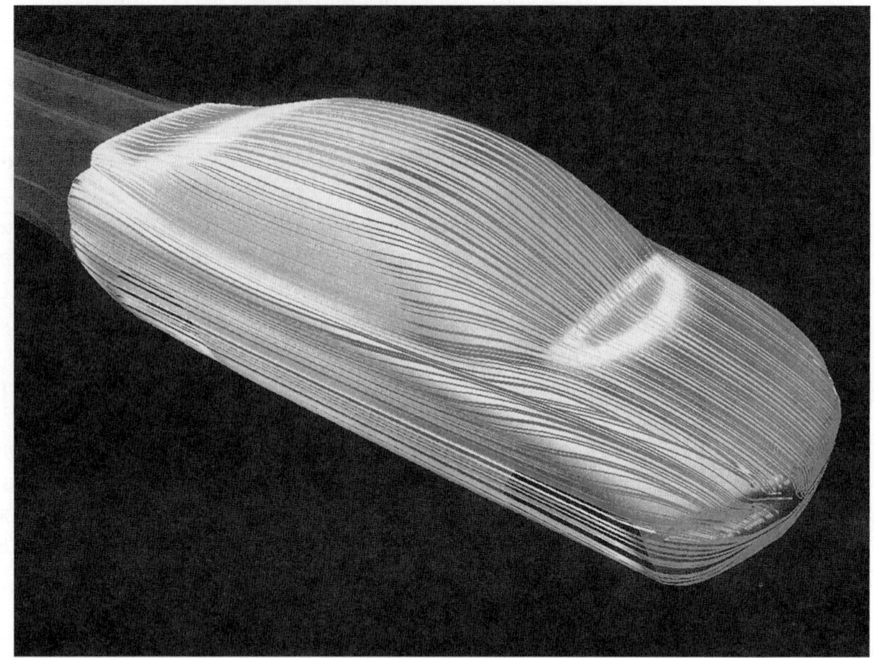

FIGURE 4.7

Predicted streamline pattern over the Volvo ECC prototype. (Courtesy of J. Michael Summa, Analytical Methods Inc.)

FIGURE 4.8

Pathlines of floating particles.

instantaneous measure of the flow velocity in a field. The measurement of all three components of the velocity vector is possible. Other techniques include the use of phosphorescent particles or gases, which emit light when subjected to a laser beam. A review of flow visualization techniques can be found in *Flow Visualization* by Merzkirch (2). The *Journal of Flow Visualization and Image Processing*, published quarterly by Begell House,

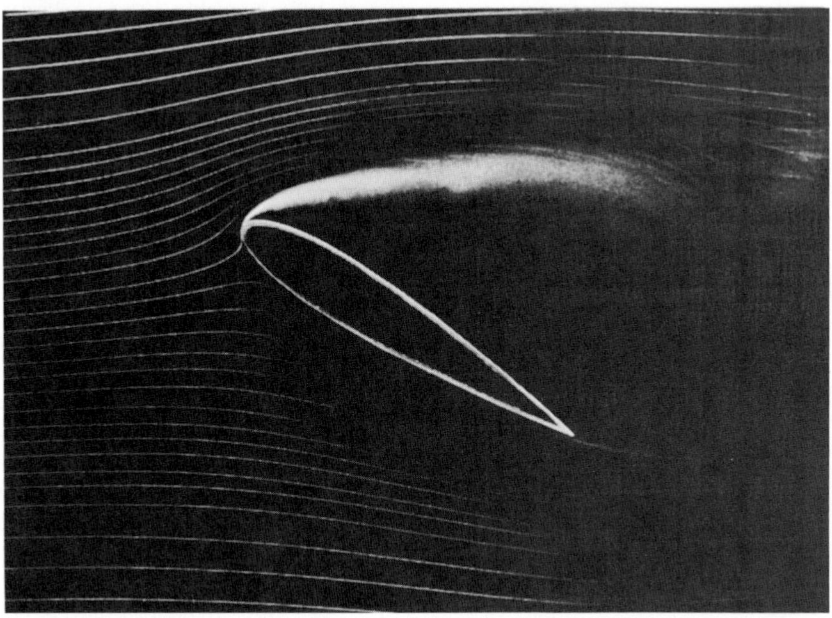

FIGURE 4.9

Smoke traces about an airfoil with a large angle of attack. (Courtesy of Education Development Center, Inc., Newton, MA.)

reports current activities in this rapidly developing area of technology. The book *An Album of Fluid Motion* by Van Dyke (3) provides a treasury of photographs of moving fluids.

Pathline, Streakline, and Streamline

Understanding the distinction between the pathline, streakline, and streamline is important in flow visualization. In steady flow all three lines are coincident if they start from the same point. Therefore, the pathline or streakline represents streamlines in steady flow.

In unsteady flows, the pathline, streakline, and streamline can be three distinct lines. Consider a two-dimensional flow that initially has horizontal streamlines as shown in Fig. 4.10. At a given time, t_0, the flow instantly changes direction and the flow moves upward to the right at 45° with no further change. The flow is unsteady because the velocity at a point changes with time. A fluid particle is tracked from the starting point, and up to time t_0, the pathline is the horizontal line segment shown on Fig. 4.10*a*. After time t_0, the particle continues to follow the streamline and moves up the right as shown. Both line segments constitute the pathline.

Next a tracer fluid, such as a dye, is introduced at the injection point and continues to be injected. At time t_0, the dye will form a line segment as shown in Fig. 4.10*c*. At this time, there is no difference between the pathline and the streakline. Now the flow changes directions and the initial horizontal dye line is transported, in whole, in the upward 45° direction. After t_0 the dye continues to be injected and forms a new line segment along the new streamline, resulting in the streakline shown in Fig. 4.10*d*. Obviously, the pathline and streakline are very different. In general, neither pathlines nor streaklines represent streamlines in an unsteady flow. Both the pathline and streakline provide a history of the flow field, and the streamlines indicate the current flow pattern.

FIGURE 4.10

Streamlines, pathlines, and streakline for an unsteady-flow field.

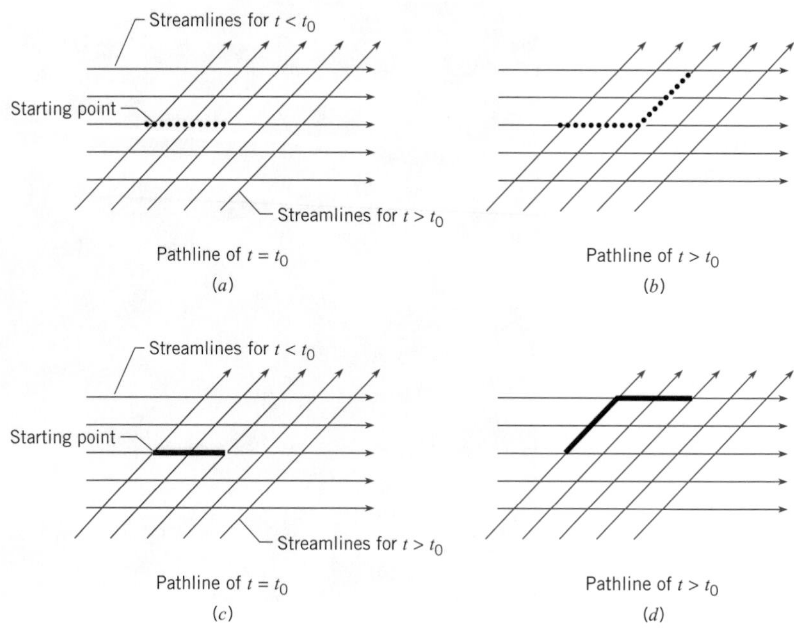

An animation of the preceding example is provided online at http://www.justask4u.com/csp/crowe. Students are encouraged to visit this site to gain a better understanding.

Acceleration

Acceleration of a fluid particle as it moves along a pathline, as shown in Fig. 4.11, is the rate of change of the particle's velocity with time. The components of the acceleration vector are shown in Fig. 4.11*b*. The normal component of acceleration a_n will be present anytime a fluid particle is moving on a curved path (i.e., centrifugal acceleration). The tangential component of acceleration a_t will be present if the particle is changing speed. In the Lagrangian formulation, the velocity is a function of time only, so the differentiation is straightforward. In the Eulerian formulation, the velocity at a point in the flow field is a function of both spatial coordinates and time, so the differentiation is somewhat more complicated.

The acceleration of a fluid particle using normal and tangential components is presented first because this approach relates to the traditional formulation introduced in dynamics. The acceleration of a fluid particle in a general Cartesian coordinate system then follows.

Normal and Tangential Components

Using normal and tangential components, the velocity of a fluid particle on a pathline (Fig. 4.11*a*) may be written as

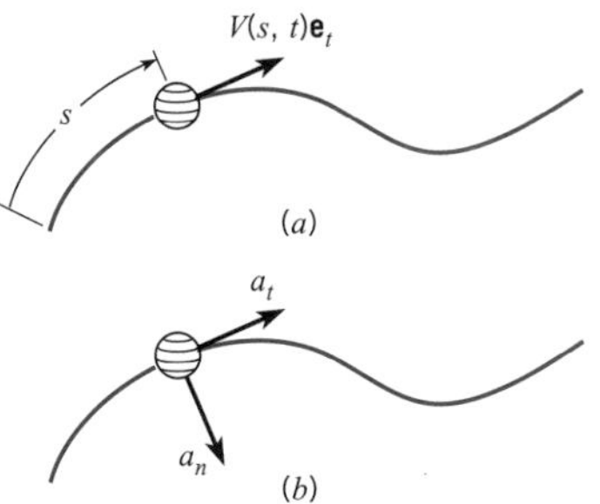

FIGURE 4.11

Fluid particle moving on a pathline. (a) Velocity. (b) Acceleration.

$$\mathbf{V} = V(s, t)\mathbf{e}_t$$

where $V(s, t)$ is the speed of the particle, which can vary with distance along the pathline, s, and time, t. The direction of the velocity vector is given by a unit vector $\mathbf{e}_t$.

Using the definition of acceleration:

$$\mathbf{a} = \frac{d\mathbf{V}}{dt} = \left(\frac{dV}{dt}\right)\mathbf{e}_t + V\left(\frac{d\mathbf{e}_t}{dt}\right) \tag{4.5}$$

To evaluate the derivative of speed in Eq. (4.5), we can use the chain rule for a function of two variables.

$$\frac{dV(s, t)}{dt} = \left(\frac{\partial V}{\partial s}\right)\left(\frac{ds}{dt}\right) + \frac{\partial V}{\partial t} \tag{4.6}$$

In a time dt, the fluid particle moves a distance ds, so the derivative ds/dt corresponds to the speed V of the particle, and Eq. (4.6) becomes

$$\frac{dV}{dt} = V\left(\frac{\partial V}{\partial s}\right) + \frac{\partial V}{\partial t} \tag{4.7}$$

In Eq. (4.5), the derivative of the unit vector $d\mathbf{e}_t/dt$ is nonzero because the direction of the unit vector changes with time as the particle moves along the pathline. The derivative is (4)

$$\frac{d\mathbf{e}_t}{dt} = \frac{V}{r}\mathbf{e}_n \tag{4.8}$$

where r is the local radius of curvature of the pathline, and $\mathbf{e}_n$ is a unit vector that is perpendicular to the pathline and pointing inward toward the center of curvature.

Substituting Eqs. (4.7) and (4.8) into Eq. (4.5) gives the acceleration of the fluid particle:

$$\mathbf{a} = \left(V\frac{\partial V}{\partial s} + \frac{\partial V}{\partial t}\right)\mathbf{e}_t + \left(\frac{V^2}{r}\right)\mathbf{e}_n \tag{4.9}$$

The first term shows that if the speed of a fluid particle is changing, there is a component of acceleration tangent to the pathline. The second term shows that a curved pathline gives rise to a component of acceleration normal to the pathline, namely, the centripetal acceleration.

Cartesian Components

It is often convenient to express acceleration using components in a Cartesian coordinate system. By the Eulerian approach, the velocity components are functions of both space and time, as given by Eq. (4.3). Thus

$$\mathbf{V} = u\mathbf{i} + v\mathbf{j} + w\mathbf{k}$$

where $u = f_1(x, y, z, t)$, $v = f_2(x, y, z, t)$, and $w = f_3(x, y, z, t)$.

The acceleration of a fluid particle in the x direction is given by

$$a_x = \frac{du}{dt}$$

where u is the x component of velocity as we follow the particle. By using the chain rule for differentiation of a multivariable function, we can express this as

$$a_x = \frac{\partial u}{\partial x}\frac{dx}{dt} + \frac{\partial u}{\partial y}\frac{dy}{dt} + \frac{\partial u}{\partial z}\frac{dz}{dt} + \frac{\partial u}{\partial t} \tag{4.10}$$

In a time dt, the fluid particle moves in the x direction a distance $dx = u\,dt$, so

$$u = \frac{dx}{dt}$$

and similarly $\quad v = \dfrac{dy}{dt} \quad$ and $\quad w = \dfrac{dz}{dt}$

Thus the acceleration component a_x of the particle is given by

$$a_x = u\frac{\partial u}{\partial x} + v\frac{\partial u}{\partial y} + w\frac{\partial u}{\partial z} + \frac{\partial u}{\partial t} \tag{4.11}*$$

Similarly, for the y and z components, we obtain

$$a_y = u\frac{\partial v}{\partial x} + v\frac{\partial v}{\partial y} + w\frac{\partial v}{\partial z} + \frac{\partial v}{\partial t} \tag{4.12}$$

$$a_z = u\frac{\partial w}{\partial x} + v\frac{\partial w}{\partial y} + w\frac{\partial w}{\partial z} + \frac{\partial w}{\partial t} \tag{4.13}$$

In summary, the acceleration components of a fluid particle in Cartesian coordinates are given by Eqs. (4.11) to (4.13). The Eulerian formulation includes more terms because the velocity is a function of both space and time whereas in the Lagrangian formulation the velocity is a function of time only.

* The acceleration of a fluid particle in the x direction is often written as $a_x = Du/Dt$, where the operator D/Dt is called the "substantive" or "material" derivative, that is, the derivative based on the moving fluid particle and defined as

$$\frac{D}{Dt} = \frac{\partial}{\partial t} + u\frac{\partial}{\partial x} + v\frac{\partial}{\partial y} + w\frac{\partial}{\partial z}$$

example 4.1

The velocity field for a fluid is given by

$$\mathbf{V} = 2x^2t\mathbf{i} + 3xy^2\mathbf{j} + 2xz\mathbf{k}$$

Find the acceleration in the x direction at the point (1,2,2) when $t = 1$. The coefficients in the equation have dimensions such that when the position is expressed in meters and time in seconds, the velocity is in m/s.

Solution The acceleration in the x direction is given by

$$a_x = \frac{\partial u}{\partial t} + u\frac{\partial u}{\partial x} + v\frac{\partial u}{\partial y} + w\frac{\partial u}{\partial z}$$

The velocity components are

$$u = 2x^2t$$

$$v = 3xy^2$$

$$w = 2xz$$

Carrying out the derivatives of the velocities and substituting into Eq. (4.11) gives

$$a_x = 2x^2 + 2x^2t \times 4xt + 3x^2y \times 0 + 2xz \times 0$$

Substituting in the values and carrying out the calculations gives

$$a_x = 2 + 8 = 10 \text{ m/s}^2$$

◁

Equations (4.12) and (4.13) can be used in the same fashion to find the components of acceleration in the y and z directions.

Convective and Local Acceleration

Inspection of Eqs. (4.11) to (4.13) reveals that some terms involve derivatives with respect to time, e.g., $\partial u/\partial t$. These terms are called *local accelerations*. Local acceleration terms occur only when a flow field is unsteady. In a steady flow, the local acceleration is zero. The remaining terms (e.g., $u(\partial u/\partial x)$, $v(\partial v/\partial y)$, etc.) are called *convective accelerations*. Convective accelerations occur when velocity is a function of position in a flow field. In uniform flows, the convective acceleration is zero.

The local and convective accelerations are also apparent in Eq. (4.10). The centrifugal acceleration is also a convective acceleration because it occurs as a result of velocity change (in direction) along the pathline.

The difference in the Eulerian and Lagrangian approaches to measuring acceleration is illustrated in the accompanying cartoon. In the Lagrangian approach, the acceleration is measured by moving with the cart (fluid particle), whereas in the Eulerian approach, the acceleration is obtained by noting the change in velocity as the cart passes by and then calculating the acceleration.

example 4.2

A nozzle is designed such that the velocity in the nozzle varies as

$$u = \frac{u_0}{1.0 - 0.5x/L}$$

where the velocity u_0 is the entrance velocity and L is the nozzle length. The entrance velocity is 10 m/s and the length is 0.5 m. The velocity is uniform across each section. Find the acceleration at the station halfway through the nozzle $(x/L = 0.5)$.

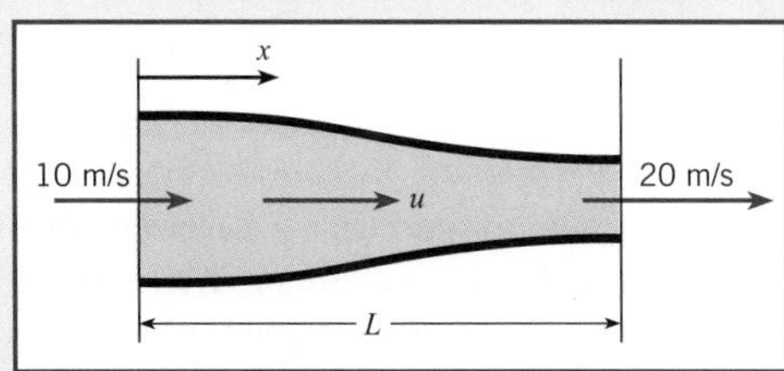

Solution It is obvious that there is an acceleration of the flow because the fluid enters at 10 m/s and exits at 20 m/s. In this case, there is only one component of acceleration. There is no local acceleration because the flow is steady, so the acceleration is due only to the convective contribution:

$$a_x = u\frac{du}{dx}$$

where the total derivative has been used because u is a function of x only.
Taking the derivative

$$\frac{du}{dx} = -\frac{u_0}{(1-0.5x/L)^2}\times\left(-\frac{0.5}{L}\right)$$
$$= \frac{1}{L}\frac{0.5u_0}{(1-0.5x/L)^2}$$

and multiplying by u gives

$$u\frac{du}{dx} = 0.5\frac{u_0^2}{L}\frac{1}{(1-0.5x/L)^3}$$

Evaluating the acceleration at $x/L = 0.5$ gives

$$a_x = 1.185\frac{u_0^2}{L}$$
$$= 1.185\times\frac{10^2}{0.5}$$
$$= 237\ \text{m/s}^2$$

◁

Since a_x is positive, the direction of the acceleration is positive, that is, in the positive x direction. This is reasonable because the velocity increases in the x direction.

Euler's Equation

In Chapter 3, we learned that the forces acting on a static fluid particle are the pressure and gravitational force (weight). With no acceleration, the sum of these forces is zero. In this section, we extend the analysis to an accelerating fluid particle.

From dynamics we know that the motion of a body is governed by Newton's second law, $F = ma$. The forces acting on a fluid mass are due to pressure and gravity (weight). For the present we are neglecting the forces due to viscous effects. Consider the cylindrical fluid element* situated between two streamlines shown in Fig. 4.12. We can regard this element as a "free body" in which the presence of the surrounding fluid is replaced by pressure forces acting on the element. Here the element is being accelerated in the ℓ direction, and the direction of ℓ is arbitrary. Note that the coordinate axis z is vertically upward and that the pressure varies along the length of the element. Applying Newton's second law in the ℓ direction, we have

$$\sum F_\ell = ma_\ell$$

$$F_{\text{pressure}} + F_{\text{gravity}} = ma_\ell \tag{4.14}$$

The mass of the fluid element is

$$m = \rho \Delta A \Delta \ell$$

Substituting the forces due to pressure and gravity (weight) into Eq. (4.14), we have

$$p\Delta A - (p + \Delta p)\Delta A - \Delta W \sin\alpha = \rho \Delta A \Delta \ell a_\ell \tag{4.15}$$

Notice that the pressure force acting on the sides of the cylindrical element do not contribute to the force in the ℓ direction. However, $\Delta W = \gamma \Delta\ell \Delta A$, so Eq. (4.15) reduces to

$$-\frac{\Delta p}{\Delta \ell} - \gamma \sin\alpha = \rho a_\ell \tag{4.16}$$

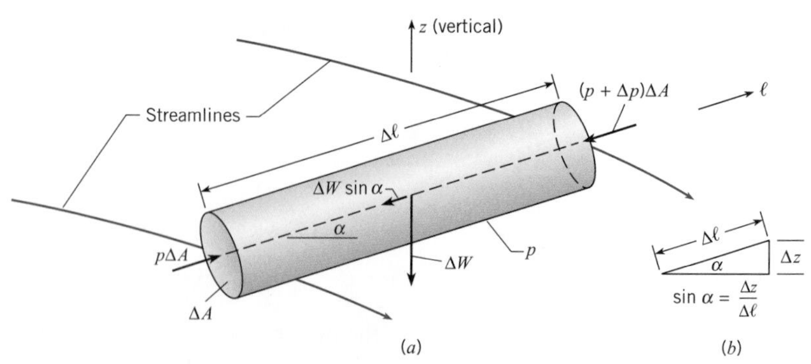

FIGURE 4.12

"Free body" diagram for fluid element accelerating in the ℓ direction. (a) Fluid element. (b) Trigometric relation.

* In this case, the words "fluid element" and "fluid particle" are synonymous. The word "element" is used here because it is more descriptive of the shape.

Pressure is a function of both position and time. Taking the limit of $\Delta p/\Delta,$ at a given time as $\Delta\ell$ approaches zero yields the partial derivative

$$\lim_{\Delta, \to 0} \left.\frac{\Delta p}{\Delta,}\right|_t = \frac{\partial p}{\partial,}$$

Figure 4.12*b* also shows that sin α is equal to $\Delta z/\Delta,$. Taking the limit as $\Delta\ell$ approaches zero at a given time yields

$$\sin\alpha = \lim_{\Delta, \to 0} \frac{\Delta z}{\Delta,} = \frac{\partial z}{\partial,}$$

Thus the limiting form Eq. (4.16) when $\Delta\ell$ approaches zero is

$$-\frac{\partial p}{\partial,} - \gamma\frac{\partial z}{\partial,} = \rho a_,$$

or, taking γ as a constant,

$$-\frac{\partial}{\partial,}(p + \gamma z) = \rho a_, \qquad (4.17)$$

Equation (4.17) is *Euler's equation* for motion of a fluid. It is interesting to note that when the acceleration is zero, Eq. (4.17) reduces to $\partial/\partial,(p + \gamma z) = 0$, which corresponds to the familiar hydrostatic equation $p + \gamma z = \text{const.}$ In other words, along a direction in which there is no acceleration the pressure distribution is hydrostatic. For example, in a flow with straight, parallel streamlines, the pressure in the direction normal to the streamlines is hydrostatic because there is no acceleration in this direction. Again, this assumes that the gravity and pressure forces are the only forces acting. When the flow is static, there is no motion (or acceleration), so the viscous stresses are zero and Euler's equation reduces to the hydrostatic equation.

An example application of Euler's equation is to the uniform acceleration of liquid in a tank.

Assume that the open tank of liquid shown in Fig. 4.13 is accelerated to the right, the positive *x* direction, at a rate of a_x. For this to occur, a net force must act on the liquid in the *x* direction; this is accomplished when the liquid redistributes itself in the tank as shown by *A′ B′ CD*. Under this condition the hydrostatic force at the left end is greater than the hydrostatic force at the right, which is consistent with the requirement of $F = Ma$.

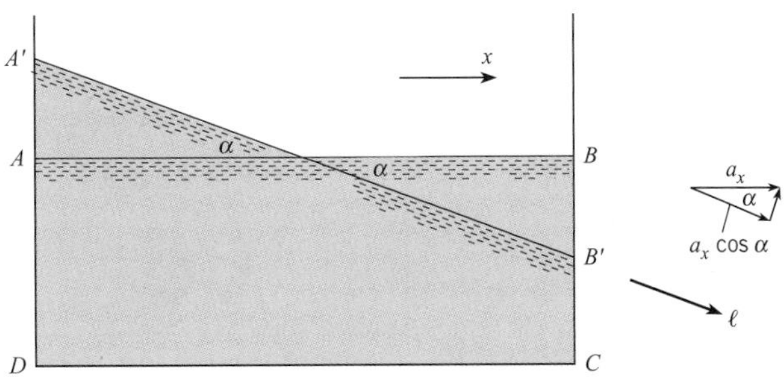

FIGURE 4.13

Uniform acceleration of a tank of liquid

Further quantitative analysis of the acceleration of the tank of liquid is made with Eq. (4.17). First consider application of the equation along the liquid surface $A'B'$. Here the pressure is constant, $p = p_{atm}$. Consequently, $\partial p/\partial, = 0$. The acceleration along $A'B'$ is given by $a_, = a_x \cos\alpha$. Hence, Euler's equation reduces to

$$\frac{d}{d,}(\gamma z) = -\rho a_x \cos\alpha \tag{4.18}$$

where the total derivative is used because the variables do not change with time. The specific weight in Eq. 4.18 is constant. Therefore, Eq. (4.18) becomes

$$\frac{dz}{d,} = -\frac{a_x \cos\alpha}{g}$$

But $dz/d, = -\sin\alpha$. Thus we obtain

$$\sin\alpha = \frac{a_x \cos\alpha}{g}$$

or

$$\tan\alpha = \frac{a_x}{g} \tag{4.19}$$

Still further analysis can be made if Euler's equation is applied along a horizontal plane in the liquid, such as at the bottom of the tank. Now z is constant and Euler's equation reduces to $\partial p/\partial, = -\rho a_x$, which shows that the pressure must decrease in the direction of acceleration. The change in pressure is consistent with the change in depth of the liquid because hydrostatic pressure variation prevails in the vertical direction, since there is no component of acceleration in that direction. Thus as the depth decreases in the direction of acceleration, the pressure along the bottom of the tank must also decrease. Another case of uniform acceleration is given in the following example.

example 4.3

The tank on a tank truck is filled completely with gasoline, which has a specific weight of 42 lbf/ft^3 (6.60 kN/m^3).

a. If the tank on the trailer is 20 ft (6.1 m) long and if the pressure at the top rear end of the tank is atmospheric, what is the pressure at the top front when the truck decelerates at a rate of 10 ft/s^2 (3.05 m/s^2)?

b. If the tank is 6 ft (1.83 m) high, what is the maximum pressure in the tank?

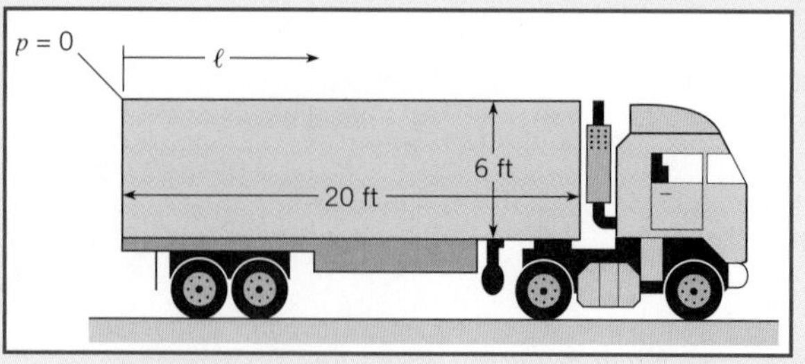

Solution Apply Euler's equation along the top of the tank. Here z is constant and the pressure does not vary with time during this phase of deceleration. Therefore, one may write

$$\frac{dp}{d,} = -\rho a_,$$

Integrating, one obtains

$$p = -\rho,a_, + C$$

When , = 0, $p = 0$; hence, $C = 0$ and $p = -\rho,a_,$.

Now substituting $-10\ \text{ft/s}^2$ ($-3.05\ \text{m/s}^2$) for a_ℓ, 20 ft (6.1 m) for ℓ, and 1.30 slugs/ft^3 (672 kg/m^3) for ρ, which is equal to γ/g, one obtains

$$p = -1.30\ \text{slugs/ft}^3 \times (-10\ \text{ft/s}^2) \times 20\ \text{ft} = 260\ \text{psfg} \qquad \triangleleft$$

SI units
$$p = -673\ \text{kg/m}^3 \times (-3.05\ \text{m/s}^2) \times 6.1\ \text{m}$$
$$= 12{,}500\ \text{N/m}^2 = 12{,}500\ \text{Pa gage} \qquad \triangleleft$$

The maximum pressure in the tank will occur at the front end of the tank bottom. Since the pressure variation is hydrostatic in the vertical direction, one obtains $p + \gamma z =$ constant, or

$$p_{\text{bottom}} + \gamma z_{\text{bottom}} = p_{\text{top}} + \gamma z_{\text{top}}$$

Solving yields

$$p_{\text{bottom}} = 260 + (42)(6)$$

$$p_{\text{max}} = p_{\text{bottom}} = 512\ \text{psfg} \qquad \triangleleft$$

SI units
$$p_{\text{max}} = p_{\text{bottom}} = 12{,}500\ \text{N/m}^2 + 6.6\ \text{kN/m}^3 \times 1.83\ \text{m}$$
$$= 24.6\ \text{kPa gage} \qquad \triangleleft$$

4·5

The Bernoulli Equation

The Bernoulli Equation along a Streamline

From the dynamics of particles in solid-body mechanics, we know that integrating Newton's second law for particle motion along a pathline provides a relationship between the change in kinetic energy and the work done on the particle. Integrating Euler's equation along a pathline in the steady flow of an incompressible fluid yields an equivalent relationship called the Bernoulli equation.

We now develop the Bernoulli equation by applying Euler's equation along a pathline with the direction ℓ replaced by s, the distance along the pathline, and the acceleration a_ℓ replaced by a_t, the direction tangent to the pathline. Euler's equation becomes

$$-\frac{\partial}{\partial s}(p+\gamma z) = \rho a_t \tag{4.20}$$

The tangential component of acceleration is given by Eq. (4.9), namely,

$$a_t = V\frac{\partial V}{\partial s} + \frac{\partial V}{\partial t} \tag{4.21}$$

For a steady flow, the local acceleration is zero and the pathline becomes a streamline. Also, the properties along a streamline depend only on the distance s, so the partial derivatives become ordinary derivatives. Euler's equation now becomes

$$-\frac{d}{ds}(p+\gamma z) = \rho V\frac{dV}{ds} = \rho\frac{d}{ds}\left(\frac{V^2}{2}\right) \tag{4.22}$$

Moving all the terms to one side, we have

$$\frac{d}{ds}\left(p+\gamma z+\rho\frac{V^2}{2}\right) = 0 \tag{4.23}$$

or

$$p+\gamma z+\rho\frac{V^2}{2} = C \tag{4.24a}$$

where C is a constant. This is known as the *Bernoulli equation,* which states that the sum of the piezometric pressure $(p+\gamma z)$ and kinetic pressure $(\rho V^2/2)$* is constant along a streamline for the *steady* flow of an *incompressible, inviscid* fluid. Dividing Eq. (4.24a) by the specific weight yields the equivalent form of the Bernoulli equation along a streamline:

$$\frac{p}{\gamma}+z+\frac{V^2}{2g} = h+\frac{V^2}{2g} = C \tag{4.24b}$$

in terms of the piezometric head (h) and velocity head $(V^2/2g)$.

Application of the Bernoulli Equation

Stagnation Tube

Consider a curved tube such as that shown in Fig. 4.14. When the Bernoulli equation is written between points 1 and 2, we note that $z_1 = z_2$. Therefore, the Bernoulli equation reduces to

* Another term, dynamic pressure, which is closely associated with kinetic pressure, is equal to the difference between the total pressure (pressure at a point of stagnation) and the static pressure. Under conditions where the Bernoulli equation applies, kinetic and dynamic pressure are equal. However, in high-speed gas flow, where compressibility effects are important, the two may have significantly different values.

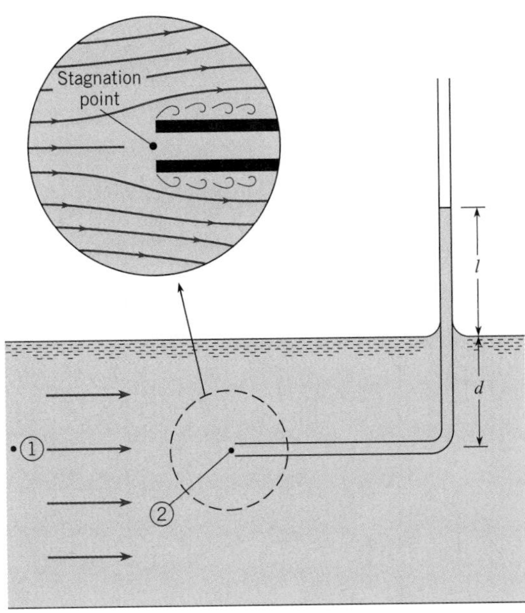

FIGURE 4.14

Stagnation tube.

$$p_1 + \frac{\rho V_1^2}{2} = p_2 + \frac{\rho V_2^2}{2} \tag{4.25}$$

Also note that the velocity at point 2 is zero (a *stagnation point*). Hence, Eq. (4.25) reduces to

$$V_1^2 = \frac{2}{\rho}(p_2 - p_1) \tag{4.26}$$

By the equations of hydrostatics (there is no acceleration normal to the streamlines where the streamlines are straight and parallel), $p_1 = \gamma d$ and $p_2 = \gamma(\ell + d)$. Therefore, Eq. (4.26) can be written as

$$V_1^2 = \frac{2}{\rho}(\gamma(\ell + d) - \gamma d)$$

which reduces to

$$V_1 = \sqrt{2g\ell} \tag{4.27}$$

Thus it is seen that a very simple device such as this curved tube can be used to measure the velocity of flow.

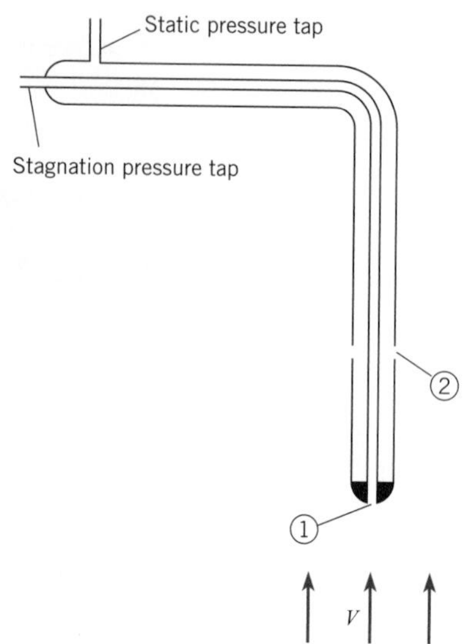

FIGURE 4.15
Pitot tube.

Pitot Tube

The Pitot tube, named after the eighteenth-century French hydraulic engineer who invented it, is based on the same principle as the stagnation tube, but it is much more versatile than the stagnation tube. The Pitot tube has a pressure tap at the upstream end of the tube for sensing the *stagnation pressure.* There are also ports located several tube diameters downstream of the front end of the tube for sensing the *static pressure* in the fluid where the velocity is essentially the same as the approach velocity. When the Bernoulli equation is applied between points 1 and 2 in Fig. 4.15, we have

$$\frac{p_1}{\gamma} + \frac{V_1^2}{2g} + z_1 = \frac{p_2}{\gamma} + \frac{V_2^2}{2g} + z_2$$

But $V_1 = 0$, so solving that equation for V_2 gives the *pitot tube equation*

$$V_2 = \left[\frac{2}{\rho}(p_{z,1} - p_{z,2})\right]^{1/2} \tag{4.28}$$

Here $V_2 = V$, where V is the velocity of the stream and $p_{z,1}$ and $p_{z,2}$ are the piezometric pressures at points 1 and 2, respectively.

By connecting a pressure gage or manometer between taps that lead to points 1 and 2, we can easily measure the flow velocity with the Pitot tube. A major advantage of the Pitot tube is that it can be used to measure velocity in a pressurized pipe; a simple stagnation tube is not convenient to use in such a situation.

If a differential pressure gage is connected across the taps, Eq. (4.28) simplifies to $V = \sqrt{2\Delta p/\rho}$, where Δp is the pressure difference measured by the gage.

example 4.4

A mercury–kerosene manometer is connected to the Pitot tube as shown. If the deflection on the manometer is 7 in., what is the kerosene velocity in the pipe? Assume that the specific gravity of the kerosene is 0.81.

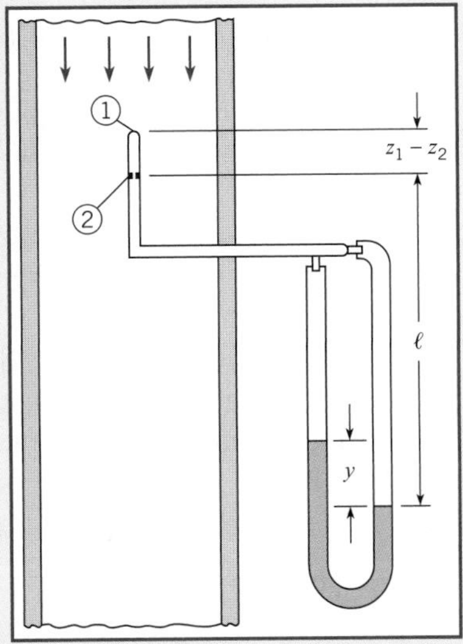

Solution We need to know the difference in piezometric pressure between points 1 and 2. We evaluate this difference by applying the manometer equation.

$$p_1 + (z_1 - z_2)\gamma_{\text{kero}} + \ell\gamma_{\text{kero}} - y\gamma_{\text{Hg}} - (\ell - y)\gamma_{\text{kero}} = p_2$$

or

$$p_1 + \gamma_{\text{kero}} z_1 - (p_2 + \gamma_{\text{kero}} z_2) = y(\gamma_{\text{Hg}} - \gamma_{\text{kero}})$$

$$p_{z,1} - p_{z,2} = y(\gamma_{\text{Hg}} - \gamma_{\text{kero}})$$

Then using the Pitot tube equation,

$$V = \left[\frac{2}{\rho_{\text{kero}}} y(\gamma_{\text{Hg}} - \gamma_{\text{kero}})\right]^{1/2}$$

$$= \left[2gy\left(\frac{\gamma_{\text{Hg}}}{\gamma_{\text{kero}}} - 1\right)\right]^{1/2}$$

From Table A.4 in the Appendix, the specific gravity of mercury is 13.55. Substituting in the values gives

$$V = \left[2 \times 32.2 \times \frac{7}{12}\left(\frac{13.55}{0.81} - 1\right)\right]^{1/2}$$

$$= \left[2 \times 32.2 \times \frac{7}{12}(16.7 - 1)\right]^{1/2}$$

$$= 24.3 \text{ ft/s}$$

◁

Note: The −1 in the quantity $(16.7 - 1)$ reflects the effect of the column of kerosene in the right leg of the manometer, which tends to counterbalance the mercury in the left leg. Thus if we have a gas–liquid manometer, the counterbalancing effect is negligible.

example 4.5

A differential pressure gage is connected across the taps of a Pitot tube. When this Pitot tube is used in a wind tunnel test, the gage indicates a Δp of 730 Pa. What is the air velocity in the tunnel? The pressure and temperature in the tunnel are 98 kPa absolute and 20°C.

Solution

$$V = \sqrt{2\Delta p/\rho}$$

where

$$\rho = \frac{p}{RT} = \frac{98 \times 10^3 \text{ N/m}^2}{(287 \text{ J/kg K}) \times (20 + 273 \text{ K})} = 1.17 \text{ kg/m}^3$$

$$\Delta p = 730 \text{ Pa}$$

Therefore

$$V = \sqrt{(2 \times 730 \text{ N/m}^2)/(1.17 \text{ kg/m}^3)} = 35.3 \text{ m/s}$$

◁

Pressure Coefficient

It is often convenient to express the pressure in terms of a nondimensional quantity called the *pressure coefficient*, which is defined as

$$C_p = \frac{p - p_0}{\frac{1}{2}\rho V_0^2} \tag{4.29a}$$

where p_0 is a reference pressure and V_0 is a reference velocity. In many applications, the reference pressure and velocity are the free-stream values. A negative pressure coefficient means that the local pressure is less than the reference pressure. This form of the pressure coefficient is useful in gas flows, where changes in hydrostatic pressures are usually negligible. For liquid flows the following form for the pressure coefficient is recommended:

$$C_p = \frac{h - h_0}{V_0^2/(2g)} \tag{4.29b}$$

where h is the piezometric head.

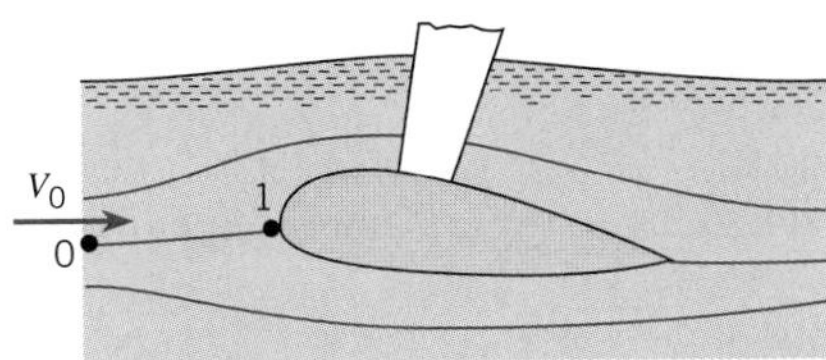

FIGURE 4.16

Flow over a hydrafoil.

The pressure coefficient is often used to reduce data from experimental pressure measurements. For example, the pressures measured on an airfoil surface are generally expressed in terms of pressure coefficients.

By applying the Bernoulli equation, we can show that the pressure coefficient at the stagnation point is unity. Consider the flow over the hydrofoil section shown in Fig. 4.16. Point 0 on the dividing streamline is far upstream from the hydrofoil section, where "free-stream" conditions prevail. Point 1 is the stagnation point. The Bernoulli equation states

$$h_0 + \frac{V_0^2}{2g} = h_1 + \frac{V_1^2}{2g}$$

But at the stagnation point, $V_1 = 0$, so

$$C_p = \frac{h_1 - h_0}{V_0^2/2g} = 1$$

example 4.6

An airfoil section is mounted in a steady air flow as shown. The upstream static pressure and free-stream velocity at position 1 on the separating streamline is 14 psia and the velocity is 300 ft/s. At point 2 on the airfoil (upper surface) the velocity is 330 ft/s, and at point 3 the velocity is 270 ft/s. the flow is incompressible and inviscid. The air density is 0.07 lbm/ft^3. Find the pressure difference between points 2 and 3 and evaluate the pressure coefficient at the same two locations.

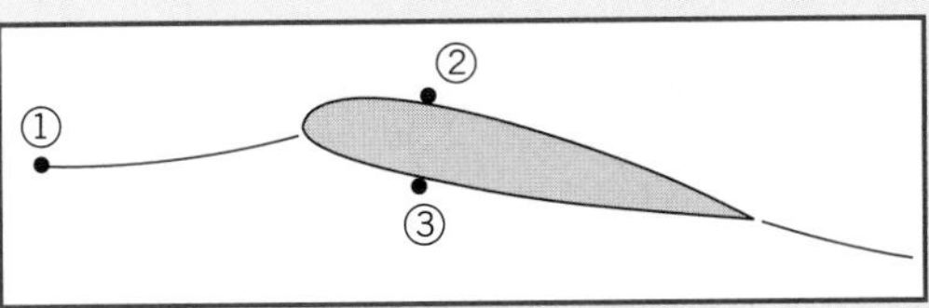

Solution Since the flow is incompressible, steady, and inviscid, the Bernoulli equation is valid along a streamline. Also, neglect the pressure change due to elevation difference (γz) since the fluid is air. Applying the Bernoulli equation from point 1 to 2, we have

$$p_1 + \rho\frac{V_1^2}{2} = p_2 + \rho\frac{V_2^2}{2}$$

Then using the Bernoulli equation between points 1 and 3,

$$p_1 + \rho\frac{V_1^2}{2} = p_3 + \rho\frac{V_3^2}{2}$$

so

$$p_2 + \rho\frac{V_2^2}{2} = p_3 + \rho\frac{V_3^2}{2}$$

Solving for the pressure difference,

$$p_3 - p_2 = \frac{\rho}{2}\left(V_2^2 - V_3^2\right)$$

The specific weight is 0.07 lbf/ft^3, so the density is

$$\rho = \frac{0.07\ \text{lbm}/\text{ft}^3}{32.2\ \text{lbm}/\text{slug}} = 0.0022\ \text{slug}/\text{ft}^3$$

Substituting in the values, we have

$$\begin{aligned} p_3 - p_2 &= \frac{0.0022}{2}(330^2 - 270^2) \\ &= 39.6\ \text{psf} \end{aligned}$$

The pressure coefficient at point 2 is

$$C_{p_2} = \frac{p_2 - p_1}{(1/2)\rho V_1^2}$$

From the Bernoulli equation between points 1 and 2,

$$p_2 - p_1 = \frac{\rho}{2}(V_1^2 - V_2^2)$$

Then

$$\begin{aligned} C_{p_2} &= 1 - \left(\frac{V_2}{V_1}\right)^2 \\ &= 1 - \left(\frac{330}{300}\right)^2 \\ &= -0.21 \end{aligned}$$

The pressure coefficient at point 3 is

$$C_{p_3} = 1 - \left(\frac{V_3}{V_1}\right)^2$$

$$= 1 - \left(\frac{270}{300}\right)^2$$

$$= 0.19 \qquad \triangleleft$$

The negative C_{p_2} indicates that the local static pressure at point 2 is less than the free-stream value and the positive C_{p_3} shows a larger pressure at point 3 than the free-stream value.

4.6 Rotation and Vorticity

Concept of Rotation

Consider a tank of liquid that is being rotated about a vertical axis. A plan view of such a tank is given in Fig. 4.17. If we focus on a given element, it can be seen that this element will rotate but not deform as time passes. In this process, all lines drawn through the element, such as *a-a* and *b-b* in Fig. 4.17, will rotate at the same rate. This is unquestionably a case of fluid rotation. Now consider fluid flow between two horizontal plates, Fig. 4.18, where the bottom plate is stationary and the top is moving to the right with a velocity V. The velocity distribution is linear; therefore, an element of fluid will deform as shown. Here we see that the element face that initially vertical rotates clockwise, whereas the horizontal face does not. It is not clear whether this is a case of rotational motion or not.

Rotation is defined as the average rotation of two initially mutually perpendicular faces of a fluid element. The test is to look at the rotation of the line that bisects both faces (*a-a* and *b-b* in Fig. 4.18*a*). The angle between this line and the horizontal axis is θ. If this line rotates, the flow is rotational. Obviously, in this case, there is rotation because the bisector does rotate. If the bisector does not rotate, the flow is irrotational. The rotation can be monitored by inserting a cruciform (cross) shape in the flow, as shown in Fig. 418*b*, and checking if it rotates. The cross will rotate with the bisector. If there is no rotation, the flow is irrotational.

We will now derive an expression that will give the rate of rotation of the bisector in terms of the velocity gradients in the flow. Consider the element shown in Fig. 4.19. The sides of the element are initially perpendicular. Then the element moves with time and deforms as shown. After time Δt the horizontal side has rotated counterclockwise by $\Delta\theta_A$ and the vertical side clockwise by $\Delta\theta_B$. By definition, counterclockwise rotation is positive. The rotational rate of the bisector is half the sum of the rotational rate of each side, so

$$\dot{\theta} = \frac{1}{2}(\dot{\theta}_A - \dot{\theta}_B)$$

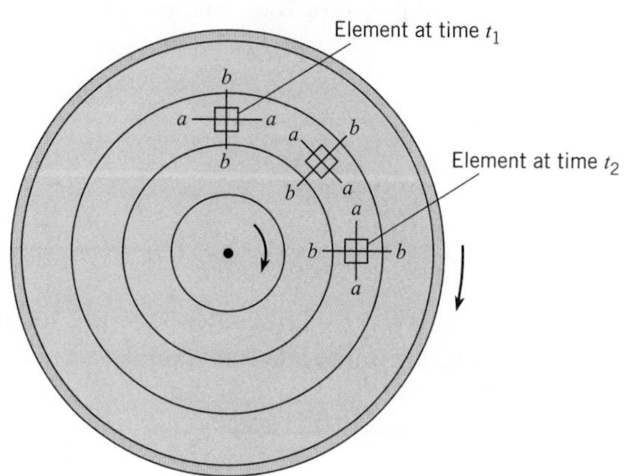

FIGURE 4.17

Rotation of a fluid element in a rotating tank of fluid.

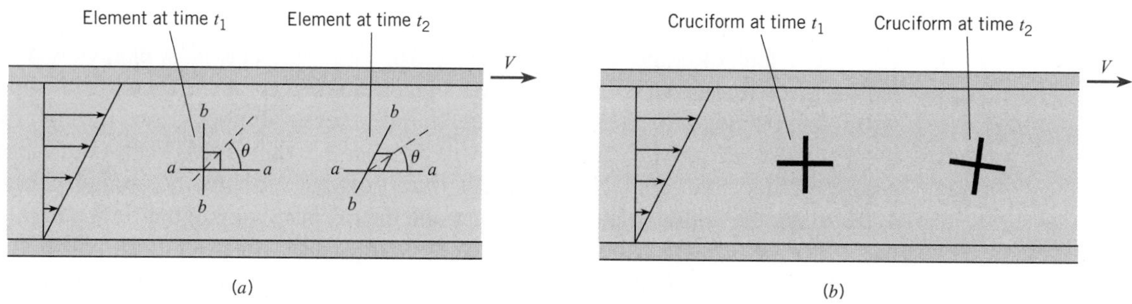

FIGURE 4.18

Rotation of fluid element and cruciform in flow between moving and stationary parallel plates.

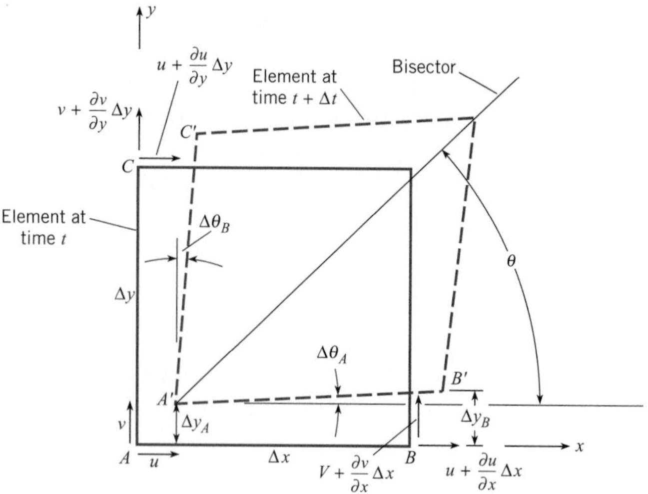

FIGURE 4.19

Translation and deformation of a fluid element.

The rotational rate of the element sides is related to the velocity gradients.

Referring to Fig. 4.19, the angle $\Delta\theta_A$ is given by

$$\Delta\theta_A \cdot \frac{\Delta y_B - \Delta y_A}{\Delta x}$$

$$\cdot \frac{\left(v + \dfrac{\partial v}{\partial x}\right)\Delta t - v\Delta t}{\Delta x}$$

$$\cdot \frac{\partial v}{\partial x}\Delta t$$

or, in the limit as $\Delta t \longrightarrow 0$,

$$\dot{\theta}_A = \frac{\partial v}{\partial x}$$

Similarly, we can show

$$\dot{\theta}_B = \frac{\partial u}{\partial y}$$

so the rotation rate of the element about the z-axis (normal to the page) is

$$\dot{\theta} = \frac{1}{2}\left(\frac{\partial v}{\partial x} - \frac{\partial u}{\partial y}\right)$$

This component of rotational velocity is defined as Ω_z, so

$$\Omega_z = \frac{1}{2}\left(\frac{\partial u}{\partial y} - \frac{\partial v}{\partial x}\right) \tag{4.30a}$$

Likewise, the rotation rates about the other axes are

$$\Omega_x = \frac{1}{2}\left(\frac{\partial w}{\partial y} - \frac{\partial v}{\partial z}\right) \tag{4.30b}$$

$$\Omega_y = \frac{1}{2}\left(\frac{\partial u}{\partial z} - \frac{\partial w}{\partial x}\right) \tag{4.30c}$$

The rate-of-rotation vector is

$$\mathbf{V} = \Omega_x\mathbf{i} + \Omega_y\mathbf{j} + \Omega_z\mathbf{k} \tag{4.31}$$

An irrotational flow $(\mathbf{V} = 0)$ requires that

$$\frac{\partial v}{\partial x} = \frac{\partial u}{\partial y} \tag{4.32a}$$

$$\frac{\partial w}{\partial y} = \frac{\partial v}{\partial z} \tag{4.32b}$$

$$\frac{\partial u}{\partial z} = \frac{\partial w}{\partial x} \tag{4.32c}$$

The most extensive application of these equations is in ideal flow theory. An ideal flow is the flow of an irrotational, incompressible fluid. Flow fields in which viscous effects are small can often be regarded as irrotational. In fact, if a flow of an incompressible, inviscid fluid is initially irrotational, it will remain irrotational.

Vorticity

Another property used frequently in fluid mechanics is vorticity. The vorticity is twice the rate-of-rotation vector, so the *vorticity equation* is

$$\begin{aligned}\omega &= 2\mathbf{V}\\ &= \left(\frac{\partial w}{\partial y}-\frac{\partial v}{\partial z}\right)\mathbf{i}+\left(\frac{\partial u}{\partial z}-\frac{\partial w}{\partial x}\right)\mathbf{j}+\left(\frac{\partial u}{\partial y}-\frac{\partial v}{\partial x}\right)\mathbf{k} \qquad (4.33)\\ &= \nabla\times\vec{V}\end{aligned}$$

An irrotational flow signifies that the vorticity vector is everywhere zero.

example 4.7

The vector $V = 10x\mathbf{i} - 10y\mathbf{j}$ represents a two-dimensional velocity field. Is the flow irrotational?

Solution In a two-dimensional flow in the *xy*-plane, the flow is irrotational if

$$\frac{\partial v}{\partial x}=\frac{\partial u}{\partial y}$$

The velocity components and derivatives are

$$u = 10x \qquad \frac{\partial u}{\partial y} = 0$$

$$v = -10y \qquad \frac{\partial v}{\partial x} = 0$$

So the irrotationality condition is satisfied and the flow is irrotational.

example 4.8

A fluid exists between stationary and moving parallel flat plates, and the velocity is linear as shown. The distance between the plates is 1 cm and the upper plate moves at 2 cm/s. Find the amount of rotation that fluid elements located at 0.25 cm, 0.5 cm, and 0.75 cm will undergo after they have traveled a distance of 1 cm.

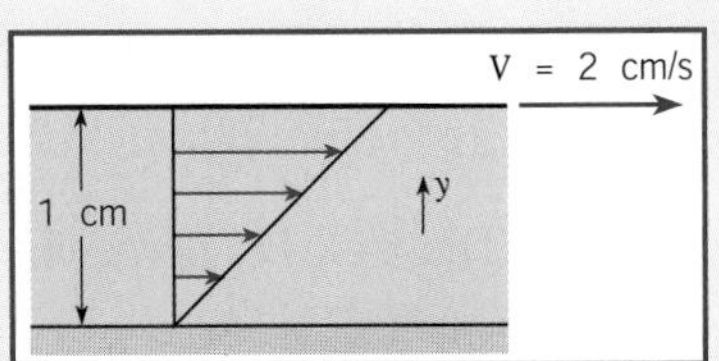

Solution Since the plates are parallel, there is no cross-stream component of velocity; i.e., $v = 0$. The rotation rate is

$$\Omega_z = \frac{1}{2}\left(\frac{\partial v}{\partial x} - \frac{\partial u}{\partial y}\right)$$

$$= -\frac{1}{2}\frac{\partial u}{\partial y}$$

The linear velocity profile is given by

$$u = 2y$$

so

$$\Omega_z = -1(\text{rad/s})$$

The rotation rate is uniform in the channel. The time to travel 1 cm is obtained from

$$\Delta t = \frac{1}{u} = \frac{1}{2y}$$

The total rotation is

$$\Delta\theta = \Omega_z \Delta t$$

$$= -\frac{1}{2y}$$

The total rotations for the three starting positions are

y	$\Delta\theta$
0.25	–2.0
0.5	–1.0
0.75	–0.67

An animation of the rotation in this example is shown in http://www.justask4u.com/csp/crowe.

Rotation in Flows with Concentric Streamlines

It is interesting to realize that a flow field rotating with circular streamlines can be irrotational; that is, the fluid elements do not rotate. Consider the two-dimensional flow field shown in Fig. 4.20. The circumferential velocity on the circular streamline is V and its radius is r. The x-axis is perpendicular to the page. As before, the rotation of the element is quantified by the rotation of the bisector, which is

$$\dot{\theta} = \frac{1}{2}(\dot{\theta}_A + \dot{\theta}_B)$$

From geometry, the angle $\Delta\theta_B$ is equal to the angle $\Delta\phi$. The rotational rate of angle ϕ is V/r, so

$$\dot{\theta}_B = \frac{V}{r}$$

The rate of change of the angle θ_A is

$$\dot{\theta}_A = \frac{\partial V}{\partial r}$$

Since V is a function of r only, the partial derivative can be replaced by the total derivative. Therefore the rotational rate about the z-axis is

$$\Omega_z = \frac{1}{2}\left(\frac{dV}{dr} + \frac{V}{r}\right) \tag{4.34}$$

For a flow rotating as a solid body, the velocity distribution is $V = \omega r$, so the rate of rotation is

$$\Omega_z = \frac{1}{2}\left[\frac{d}{dr}(\omega r) + \omega\right]$$
$$= \omega$$

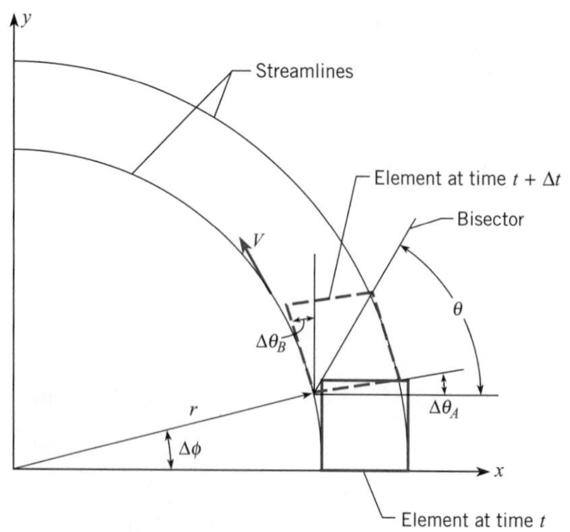

FIGURE 4.20

Deformation of element in flow with concentric, circular streamlines.

as expected. This type of circular motion is called a "forced" vortex.

If the flow is irrotational, then

$$\frac{dV}{dr} = -\frac{V}{r}$$

or

$$\frac{dV}{V} = -\frac{dr}{r}$$

Integrating this equation leads to

$$V = \frac{C}{r} \tag{4.35}$$

where C is a constant. In this case, the circumferential velocity varies inversely with r, so the velocity decreases with increasing radius. This flow field is known as a "free" vortex. The fluid elements are going around in circles, but they are not rotating.

As mentioned earlier, the rotation of a fluid can be visually monitored by inserting a cruciform shape into the flow field. If the cruciform figure does not rotate, the flow is irrotational. The student is encouraged to observe the animation in http://www.justask4u.com/csp/crowe, which demonstrates the fluid rotation in forced and free vortices.

Vortices

A vortex is defined as the motion of a multitude of fluid particles around a common center. As we have seen, fluid dynamists have identified two types of vortices with circular streamlines: the forced vortex, where the fluid rotates as a rigid body, and the free vortex, in which the velocity varies inversely with the radius. Actual vortices (for example, tornadoes and whirlpools in rivers) are approximated by a combination of forced and free vortices. as shown in Fig. 4.21, with the forced vortex at the center and the free vortex on the outside. This is called a combined vortex. The velocity distribution is shown in the figure. In reality, there would not be such a sharp demarcation between the forced and free vortices.

4·7 Pressure Distribution in Rotating Flows

A common type of rotating flow is the flow in which the fluid rotates as a rigid body. This type of flow constitutes the vortex core of the vortex system shown in Fig. 4.21. It is also encountered in centrifuges where the centrifugal force causes separation due to density difference.

To learn how pressure varies in a rotating flow, let us apply Euler's equation in the direction normal to the streamlines and outward from the center of rotation. In this case the direction of ℓ in Euler's equation (Eq. 4.17) is replaced by r, and we have

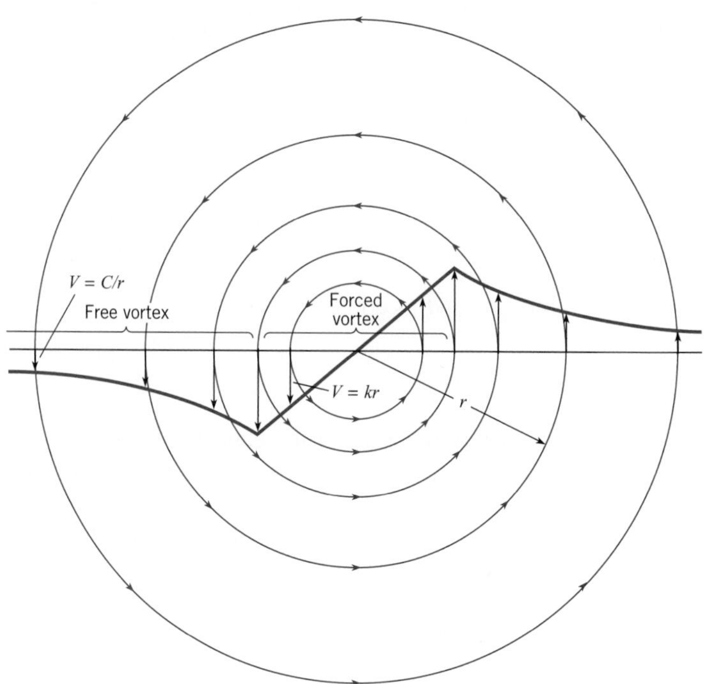

FIGURE 4.21

Velocity distribution in a vortex.

$$-\frac{d}{dr}(p + \gamma z) = \rho a_r \tag{4.36}$$

where the partial derivative has been dropped since the flow is steady and a function only of the radius r. From Eq. (4.9), the acceleration in the radial direction (away from the center of curvature) is

$$a_r = -\frac{V^2}{r}$$

and Euler's equation becomes

$$-\frac{d}{dr}(p + \gamma z) = -\rho\frac{V^2}{r} \tag{4.37}$$

For a liquid rotating as a rigid body,

$$V = \omega r$$

Substituting the velocity distribution into Euler's equation,

$$\frac{d}{dr}(p + \gamma z) = \rho r \omega^2 \tag{4.38}$$

Integrating Eq. (4.38) with respect to r then gives us

$$p + \gamma z = \frac{\rho r^2 \omega^2}{2} + \text{const.} \tag{4.39}$$

or

$$\frac{p}{\gamma}+z-\frac{\omega^2 r^2}{2g}=C \tag{4.40}$$

This equation describes *pressure variation in rotating flow.* It is important to note that this is *not* the Bernoulli equation.

example 4.9

A cylindrical tank of liquid shown in the figure is rotating as a solid body at a rate of 4 rad/s. The tank diameter is 0.5 m. The line *AA* depicts the liquid surface before rotation and the line *A′A′* shows the surface profile after rotation has been established. Find the elevation difference between the liquid at the center and the wall during rotation.

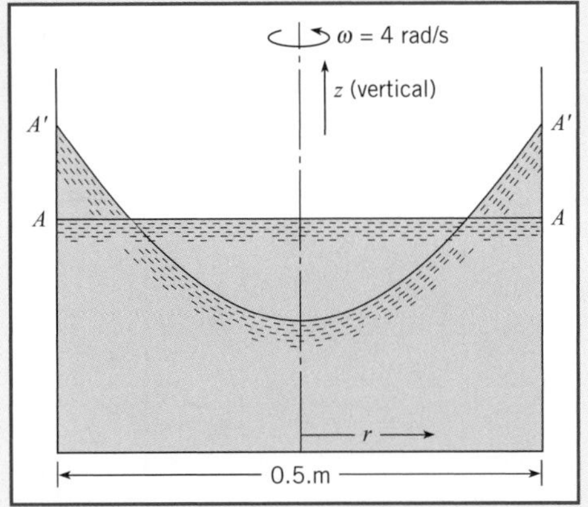

Solution At the liquid surface, the pressure is constant (atmospheric), so the equation for pressure variation in rotating flow reduces to

$$z-\frac{\omega^2 r^2}{2g}=C$$

At the tank center $r = 0$, so the surface profile varies as

$$z=z_0+\frac{\omega^2 r^2}{2g}$$

where z_0 is the elevation at the tank center. The liquid elevation at the edge of the tank is

$$z_{D/2}=z_0+\frac{\omega^2}{2g}\left(\frac{D}{2}\right)^2$$

$$=z_0+\frac{\omega^2 D^2}{8g}$$

The difference in elevation is

$$z_{D/2} - z_0 = \frac{\omega^2 D^2}{8g}$$

Substituting in the values,

$$z_{D/2} - z_0 = \frac{4^2 \times 0.5^2}{8 \times 9.81}$$

$$= 0.051 \text{ or } 5.1 \text{ cm}$$

◁

example 4.10

When the U-tube is not rotated, the water stands in the tube as shown. If the tube is rotated about the eccentric axis at a rate of 8 rad/s, what are the new levels of water in the tube?

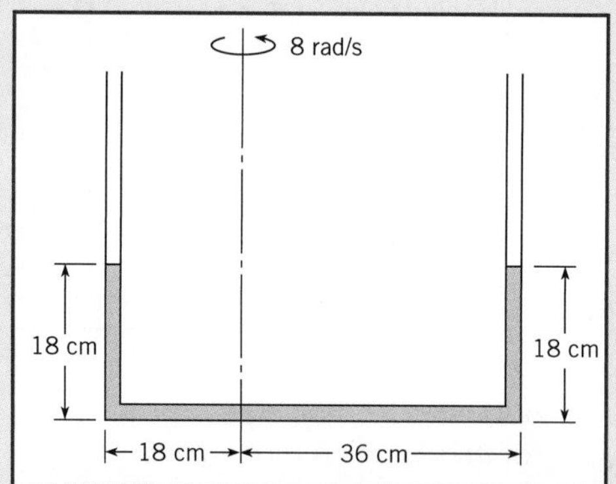

Solution The solution of this problem is based on the equation for pressure variation in rotating flow and also on the fact that the water occupies a given volume of the tube, which may be expressed in terms of a given length of tube. Let the elevation reference be at the level of the horizontal part of the tube. Then, by considering a point at the water surface in the left tube where $p = 0$ and also a point at the surface in the right tube, one can write Eq. (4.40) as follows:

$$\gamma z_l - \frac{\rho r_l^2 \omega^2}{2} = \gamma z_r - \frac{\rho r_r^2 \omega^2}{2}$$

Another independent equation involving the volume of tube occupied by the liquid may be written as

$$z_l + z_r = 0.36 \text{ m}$$

Substituting $r_l = 0.18$ m, $r_r = 0.36$ m, and $\omega = 8$ rad/s into the first equation above and solving the equations for z_l and z_r yields

$$z_l = 2.1 \text{ cm} \quad \text{and} \quad z_r = 33.9 \text{ cm}$$

◁

4.8

The Bernoulli Equation in Irrotational Flow

For a free (irrotational) vortex, the velocity distribution derives from

$$\frac{dV}{dr} = -\frac{V}{r}$$

Applying Euler's equation in the r direction gives

$$\begin{aligned} -\frac{d}{dr}(p + \gamma z) &= -\rho\frac{V^2}{r} \\ &= \rho V\frac{dV}{dr} \\ &= \frac{d}{dr}\left(\rho\frac{V^2}{2}\right) \end{aligned}$$

since the density is constant. This equation can be rewritten as

$$\frac{d}{dr}\left(p + \gamma z + \rho\frac{V^2}{2}\right) = 0$$

or

$$p + \gamma z + \rho\frac{V^2}{2} = C \tag{4.41}$$

which is the Bernoulli equation, and C is constant in the r direction (across streamlines).

The general irrotationality condition can be expressed as

$$\frac{dV}{dn} = -\frac{V}{r}$$

where n is the coordinate direction normal outward (from the center of curvature) and r is local radius of curvature. The Euler equation in the n direction becomes

$$\frac{d}{dn}\left(p + \gamma z + \rho\frac{V^2}{2}\right) = 0$$

or

$$p + \gamma z + \rho\frac{V^2}{2} = C$$

Thus for an irrotational flow, the constant C in the Bernoulli equation is the same across streamlines as well as along streamlines, so it is the same everywhere in the flow field. Equivalently, the sum of the piezometric head and velocity head

$$\frac{p}{\gamma} + z + \frac{V^2}{2g} = C$$

is constant everywhere in the flow field if the flow is steady, incompressible, inviscid, and irrotational. Thus for any two points, 1 and 2, in the flow field,

$$\frac{p_1}{\gamma} + z_1 + \frac{V_1^2}{2g} = \frac{p_2}{\gamma} + z_2 + \frac{V_2^2}{2g}$$

Note that the V in the Bernoulli equation is the speed of the fluid and not a velocity component.

The limitation for inviscid fluids follows from the inherent limitation of Euler's equation, from which the Bernoulli equation is derived. Although no fluid is inviscid, the Bernoulli equation can be applied where viscous effects are negligible. In many applications, the viscous effects are confined to narrow regions adjacent to a wall. The Bernoulli equation can be used to predict pressure distributions outside these regions if the velocity distribution is known.

example 4.11

A free vortex in air rotates in a horizontal plane and has a velocity of 40 m/s at a radius of 4 km from the vortex center. Find the velocity at 10 km from the center and the pressure difference between the two locations. The air density is 1.2 kg/m^3.

Solution The variation of velocity in a free (irrotational) vortex is

$$V = \frac{C}{r}$$

The ratio of velocities as a function of radius is

$$\frac{V_{10\text{km}}}{V_{4\text{km}}} = \frac{r_{4\text{km}}}{r_{10\text{km}}} = 0.4$$

so

$$\begin{aligned} V_{10\text{km}} &= 0.4 \times 40 \\ &= 16 \text{ m/s} \end{aligned}$$

Because the flow is irrotational, the Bernoulli equation applies. There is no change in elevation, so

$$p_{4km} + \rho\frac{V_{4km}^2}{2} = p_{10km} + \rho\frac{V_{10km}^2}{2}$$

$$p_{10km} - p_{4km} = \frac{\rho}{2}(V_{4km}^2 - V_{10km}^2)$$

$$= \frac{1.2}{2}(40^2 - 16^2)$$

$$= 806 \text{ Pa}$$

◁

Pressure Variation in a Tornado

A model for the flow field in a tornado is a forced vortex at the center surrounded by a free vortex, as shown in Fig. 4.21. This model is used in several applications of vortex flows. In practice, however, there will be no discontinuity in the slope of the velocity distribution as shown in Fig. 4.21, but rather a smooth transition between the inner forced vortex and the outer free vortex. Still, the model can be used to make reasonable predictions of the pressure field. The pressure distribution in the forced vortex is given by Eq. (4.40) while the Bernoulli equation can be used for the pressure distribution in the free vortex.

Take point 1 as the center of the forced vortex and point 2 at the junction of the forced and free vortices, where the velocity is maximum. Let point 3 be at the extremity of the free vortex, where the velocity is essentially zero and the pressure is atmospheric pressure p_0. Applying the Bernoulli equation to the free vortex, between any arbitrary point and point 3, we have

$$p + \gamma z + \rho\frac{V^2}{2} = p_3 + \gamma z_3 + \rho\frac{V_3^2}{2} \tag{4.42}$$

Neglecting any elevation change, setting $p_0 = p_3$ and taking V_3 as zero gives

$$p - p_0 = -\rho\frac{V^2}{2} \tag{4.43}$$

which shows that the pressure decreases toward the center of the vortex. This decreasing pressure provides the centripetal force to keep the flow moving along circular streamlines. The pressure at point 2 is

$$p_2 - p_0 = -\rho\frac{V_{max}^2}{2} \tag{4.44}$$

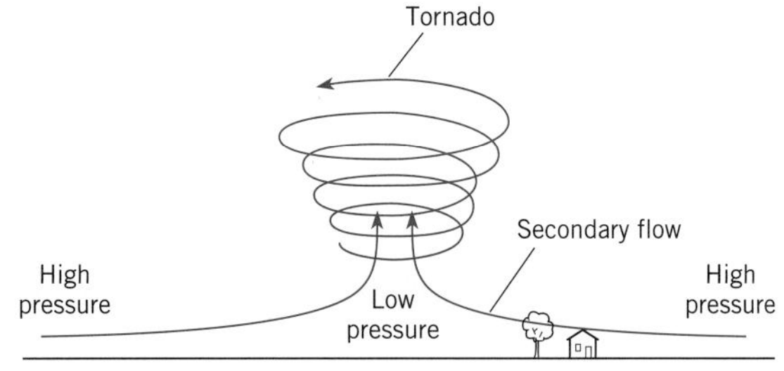

FIGURE 4.22

The "secondary" flow produced by the pressure difference in a tornado.

Applying the equation for pressure variation in rotating flow, Eq. (4.40), we have

$$p + \gamma z - \rho\frac{\omega^2 r^2}{2} = p_2 + \gamma z_2 - \rho\frac{\omega^2 r_2^2}{2} \tag{4.45}$$

Once again there is no elevation change, so $z = z_2$. At point 2, ωr_2 is the maximum speed V_{max}, and ωr is the speed of the fluid in the forced vortex. Solving for the pressure, we find

$$p = p_2 - \rho\frac{V_{max}^2}{2} + \rho\frac{V^2}{2} \tag{4.46}$$

Substituting in the expression for p_2 from Eq. (4.44) gives

$$p = p_0 - \rho V_{max}^2 + \rho\frac{V^2}{2} \tag{4.47}$$

The pressure difference between the center of the tornado where the speed is zero and the outer edge of the tornado is

$$p_1 - p_0 = -\rho V_{max}^2 \tag{4.48}$$

The minimum pressure at the vortex center can give rise to a "secondary" flow as shown in Fig. 4.22. In this case the secondary flow is produced by the pressure gradient in the primary (vortex) flow. In the region near the ground, the wind velocity is decreased due to the friction provided by the ground. However, the pressure difference in the radial direction causes a radially inward flow adjacent to the ground and an upward flow at the vortex center.

example 4.12

Assume that a tornado is modeled as the combination of a forced and a free vortex. The maximum wind speed in the tornado is 150 mph. What is the pressure difference, in inches of mercury, between the center and the outer edge of the tornado? The density of the air is 0.075 lbm/ft^3.

Solution A speed of 150 mph corresponds to 220 ft/s and a density of 0.075 lbm/ft^3 to 0.00223 slugs/ft^3. Substituting these values into Eq. (4.48) gives

$$p_1 - p_0 = -0.00223\ \frac{\text{slugs}}{\text{ft}^3} \times 220^2\ \frac{\text{ft}^2}{\text{s}^2} = -112.8\ \text{psf}$$

or

$$p_1 - p_0 = -1.60\ \text{in. Hg}$$

◁

This pressure difference causes the radially inward flow near the ground level in a tornado.

Pressure Distribution Around a Circular Cylinder—Ideal Fluid

If a fluid is nonviscous and incompressible (an *ideal* fluid) and if the flow is initially irrotational, then the flow will be irrotational throughout the entire flow field.* Then, if the flow is also steady, the Bernoulli equation will apply everywhere because all the restrictions for the Bernoulli equation will have been satisfied. The flow pattern about a circular cylinder with such restrictions is shown in Fig. 4.23.

Because the flow pattern is symmetrical with either the vertical or the horizontal axis through the center of the cylinder, the pressure distribution on the surface of the cylinder, obtained by application of the Bernoulli equation, is also symmetrical. In Fig. 4.24 the relative pressure C_p is plotted outward (negative) or inward (positive) from the surface of the cylinder, depending on the sign of the relative pressure and on a line normal to the surface of the cylinder. It should also be noted that p_0 and V_0 are the pressure and velocity

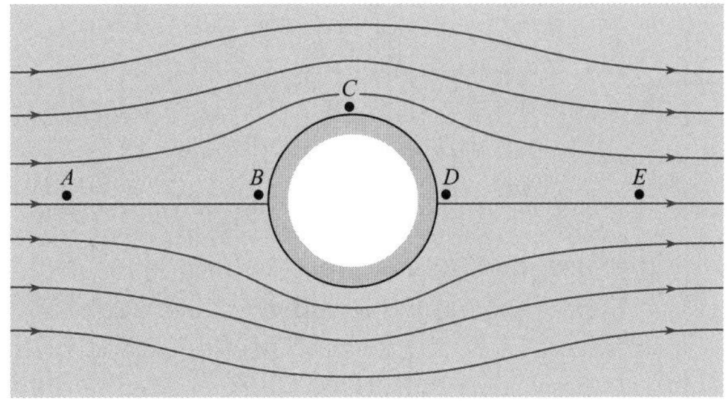

FIGURE 4.23
Irrotational flow past a cylinder.

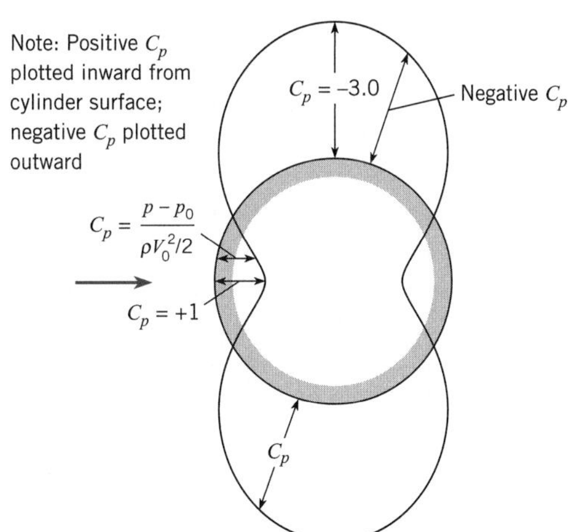

FIGURE 4.24
Pressure distribution on a cylinder—irrotational flow.

* This can be seen in a qualitative sense if one visualizes a small spherical element of fluid within a nonviscous flow field. Since pressure forces act normal to the surface of the spherical element, the element deforms if the pressure is not of equal intensity over the entire surface. However, the element cannot rotate (irrotational situation) because there is no shear stress (viscosity is zero) on the surface of the sphere to possibly cause rotation.

of the free stream far upstream or downstream of the body. Thus we see that the points at the front and rear of the cylinder are points of stagnation ($C_p = +1.0$) and that the minimum pressure ($C_p = -3.0$) occurs at the midsection, where the velocity is highest. If we visualize a fluid particle as it travels around the cylinder from *A* to *B* to *C* and finally to *D* in Fig. 4.23, we see that it first decelerates, which is consistent with the increase in pressure from *A* to *B*. Then as it passes from *B* to *C*, it is accelerated to its highest speed by the action of the pressure gradient; that is, the pressure decreases over the entire path from *B* to *C*. Next, as the particle travels from *C* to *D*, its momentum at *C* is sufficient to allow it to travel to *D* against the adverse pressure gradient (pressure increases in the direction of flow here). Finally, the particle accelerates to the free-stream velocity in its passage from *D* to *E*. Understanding this qualitative description of how the fluid travels from one point to another will be helpful when the phenomenon of separation is explained in the next section.

4.9 Separation

Consider the flow of a real (viscous) fluid past a cylinder as shown in Fig. 4.25. The flow pattern upstream of the midsection is very similar to the pattern for an ideal fluid. However, in a viscous fluid the velocity at the surface is zero (no-slip condition), whereas with the flow of an inviscid fluid the surface velocity need not be zero. Because of viscous effects, a thin layer next to the surface, called a *boundary layer*, occurs. The velocity changes from zero to the free-stream velocity across the boundary layer. Over the forward section of the cylinder, where the pressure decreases from the stagnation point to the midsection, the boundary layer is quite thin. Outside the boundary layer, viscous effects are unimportant and the flow can be treated as inviscid and irrotational.

Downstream of the midsection, the pressure increases with distance along the surface. With the low velocity in the boundary layer the fluid particles can go only so far against the adverse pressure gradient until they are forced to detour away from the surface. This is called the *separation point*. A recirculatory flow called a *wake* develops behind the cylinder. The flow in the wake region is called *separated flow*. The pressure distribution on the cylinder surface in the wake region is nearly constant, as shown in Fig. 4.26. The point of separation depends on the free-stream velocity, cylinder diameter, and fluid kinematic viscosity through a nondimensional number called the Reynolds number, defined as VD/ν. This number will de defined in more detail in Chapter 8. Separation is not observed for Reynolds numbers less than 50. The separation point is also affected by the roughness of the cylinder surface and turbulence in the free-stream flow. The wake region is also very unsteady, with the shedding of eddies causing oscillating surface pressures.

A video of separation behind a circular cylinder is available on the Web at www.wiley.com/college/crowe. This video shows pathlines in the wake region of the cylinder, and one can see the unsteadiness associated with separated flow.

Separation and the development of a wake region also occurs on blunt objects and cross sections with sharp edges as shown in Fig. 4.27. In these situations, the flow cannot negotiate the turn at the sharp edges and separates from the body, generating eddies, a separated region,

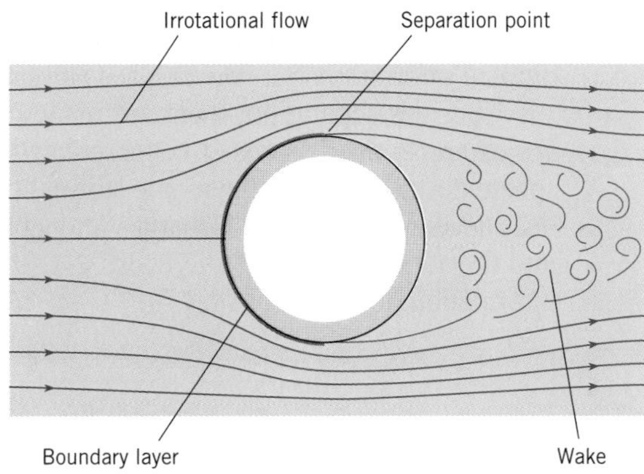

FIGURE 4.25

Flow of a real fluid past a circular cylinder.

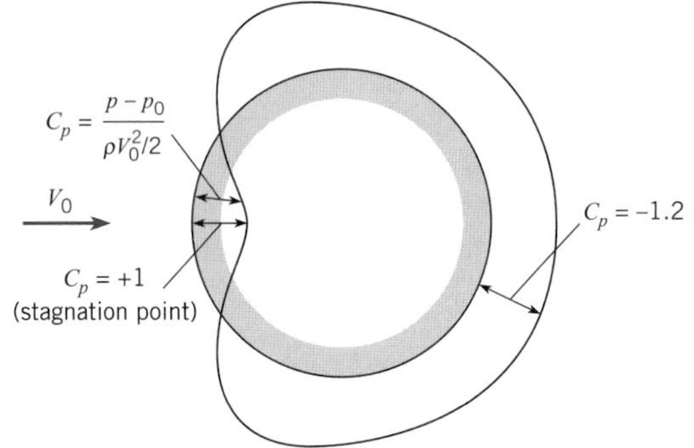

FIGURE 4.26

Pressure distribution on a circular cylinder, $Re = 10^5$; after Fage and Warsap (5).

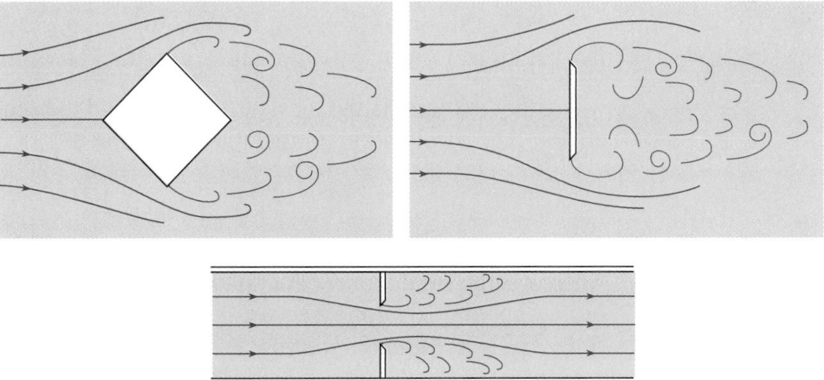

FIGURE 4.27

Flow past a square rod and a disk and through a sharp-edged orifice.

and wake flow. As we will learn in Chapter 11, the low pressure in the wake zone leads to drag (flow resistance).

The eddies developed by separation are initially large but break down to smaller eddies and are ultimately dissipated by viscosity. The process of vortex generation and decay is typical of all turbulent flows and is one of the most significant aspects of fluid mechanics. The prediction of velocities in turbulent flows generated by separation is a continuing challenge for engineers involved with computational fluid mechanics. For additional information on vortices, see Hussaini and Salas (6) and Lugt (7).

Besides separation, there are many other natural processes that generate vortices. For example, the Coriolis effect associated with low-pressure storm centers in the atmosphere develop vortices (cyclonic storms) that extend hundred of miles. Large-scale vortices develop from river discharges into a bay or ocean or when a smoke stack discharges into the atmosphere. See references (8), (9), and (10) for discussion of eddies and basic information on turbulent flow.

Summary

There are two approaches to describe the velocity of a flowing fluid. In the Lagrangian approach, the position of a specific fluid particle traveling along a pathline is recorded with time. In the Eulerian approach, the properties of fluid particles passing a given point in space are recorded with time. The Eulerian approach is generally used to analyze fluid motion.

The streamline is a curve everywhere tangent to the local velocity vector. The configuration of streamlines in a flow field is called the flow pattern. The pathline is the line traced out by a particle. A streakline is the line produced by a dye introduced at a point in the field. Pathlines, streaklines, and streamlines are coincident in steady flow but differ in unsteady flows.

In a uniform flow, the velocity does not change along a streamline. In a steady flow, the velocity does not change with time at any location.

The tangential acceleration of a fluid element along a pathline is

$$a_t = \frac{\partial V}{\partial t} + V\frac{\partial V}{\partial s}$$

where the first term is the local acceleration and the second term is the convective acceleration. The acceleration normal to the pathline is

$$a_n = \frac{V^2}{r}$$

where r is the local radius of curvature of the pathline.

Applying Newton's second law to a fluid element in an incompressible, inviscid flow results in *Euler's equation*,

$$-\frac{\partial}{\partial_{,}}(p + \gamma z) = \rho a_{,}$$

where ℓ is an arbitrary direction. Integrating Euler's equation along a streamline in steady flow results in the *Bernoulli equation*,

$$p + \gamma z + \rho\frac{V^2}{2} = C$$

where V is the speed of the fluid and C is a constant along a streamline. The value of C may vary from streamline to streamline. In an irrotational flow, the value of C is the same for every streamline. The Bernoulli equation expressed in terms of heads is

$$\frac{p}{\gamma} + z + \frac{V^2}{2g} = C$$

The *pressure coefficient* is a nondimensional pressure difference defined as

$$C_p = \frac{p - p_0}{\rho V_0^2/2}$$

or in terms of piezometric heads

$$C_p = \frac{h - h_0}{V_0^2/2g}$$

The pressure coefficient at the stagnation point is unity.

The rotation of a fluid element is defined as the average rotation of two initially perpendicular lines defining the sides of the element. If every fluid element in a flow does not rotate, the flow is irrotational.

The *vorticity vector* is defined as

$$\omega = \left(\frac{\partial w}{\partial y} - \frac{\partial v}{\partial z}\right)\mathbf{i} + \left(\frac{\partial u}{\partial z} - \frac{\partial w}{\partial x}\right)\mathbf{j} + \left(\frac{\partial u}{\partial y} - \frac{\partial v}{\partial x}\right)\mathbf{k}$$

and is equal to twice the fluid rotation vector. In an irrotational flow, the vorticity is zero.

Separation occurs when streamlines move away from the surface of the body and create a local recirculation zone or wake. Typically the pressure in the recirculation zone assumes the value at the point of separation.

References

1. Crowe, C. T., M. Sommerfeld, and Y. T. Tsuji. *Multiphase Flow with Droplets and Particles.* CRS Press, Boca Raton, Florida, 1998.

2. Merzkirch, W. *Flow Visualization*, 2nd ed. Academic Press, New York, 1987.

3. Van Dyke, M. *An Album of Fluid Motion.* Parabolic Press, Stanford, California, 1982.

4. Hibbeler, R. C. *Dynamics.* Prentice Hall, Englewood Cliffs, New Jersey, 1995.

5. Fage, A., and J. H. Warsap. "The Effects of Turbulence and Surface Roughness on the Drag of a Circular Cylinder." *Aero. Res. Comm., London, Rept. Mem.,* 1283 (1929).

6. Hussaini, M. Y., and M. D. Salas, eds. *Studies of Vortex Dominated Flows.* Springer-Verlag, New York, 1987.

7. Lugt, H. J. *Vortex Flows in Nature and Technology.* John Wiley, New York, 1983.

8. Panofsky, H. A., and J. A. Dutton. *Atmospheric Turbulence.* John Wiley, New York, 1984.

9. Vinnichenko, N. K., et al. *Turbulence in the Free Atmosphere,* 2nd ed. Consultants Bureau, a Division of Plenum Publishing Co., New York, 1980.

10. Landahl, M. T., and E. Mollo-Christensen. *Turbulence and Random Processes in Fluid Mechanics.* Cambridge University Press, New York, 1986.

11. Moran, J. M., and M. D. Morgan. *Meteorology: The Atmosphere and Science of Weather.* Prentice Hall, Englewood Cliffs, New Jersey, 1997.

Problems

4.1 In the system in the figure, the valve at C is gradually opened in such a way that a constant rate of increase in discharge is produced. How would you classify the flow at B while the valve is being opened? How would you classify the flow at A?

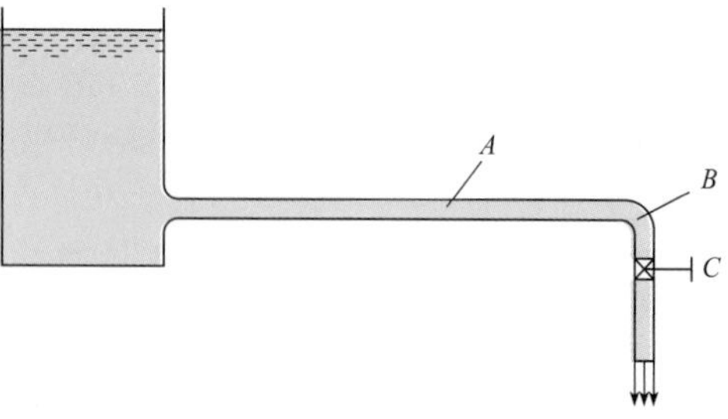

PROBLEM 4.1

4.2 a. Water flows in the passage shown. If the flow rate is decreasing with time, the flow is classified as (a) steady, (b) unsteady, (c) uniform, (d) nonuniform.
b. In the flow there is (a) local acceleration, (b) convective acceleration, (c) no acceleration.

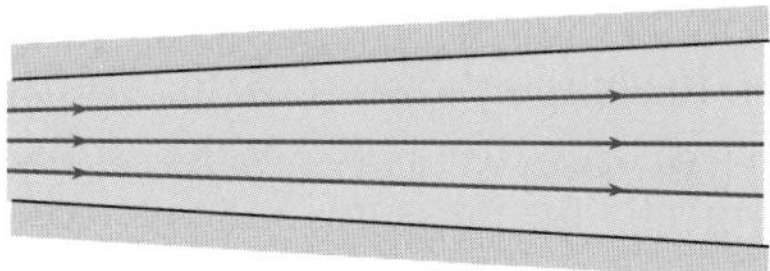

PROBLEM 4.2

4.3 If a flow pattern has converging streamlines, how would you classify the flow?

4.4 Consider flow in a straight conduit. The conduit is circular in cross section. Part of the conduit has a constant diameter, and part has a diameter that changes with distance. Then, relative to flow in that conduit, correctly match the items in column A with those in column B.

A	B
Steady flow	$V_s \partial V_s / \partial s = 0$
Unsteady flow	$V_s \partial V_s / \partial s \neq 0$
Uniform flow	$\partial V_s / \partial t = 0$
Nonuniform flow	$\partial V_s / \partial t \neq 0$

4.5 Refer to the flow as indicated by the pathlines in Fig. 4.8. Assume that the discharge is constant and the flow is nonturbulent. Which of the following statement(s) are true? (a) There is convective acceleration in the flow. (b) There is local acceleration in the flow. (c) The pathlines are also streamlines. (d) There is separation in this flow field. If you identified statement (d) as true, indicate where separation is occurring.

4.6 At time $t = 0$, dye was injected at point A in a flow field of a liquid. When the dye had been injected for 4 s, a pathline for a particle of dye that was emitted at the 4-s instant was started. The streakline at the end of 10 s is shown below. Assume that the speed (but not the velocity) of flow is the same throughout the 10-s period. Draw the pathline of the particle that was emitted at $t = 4$ s. Make your own assumptions for any missing information.

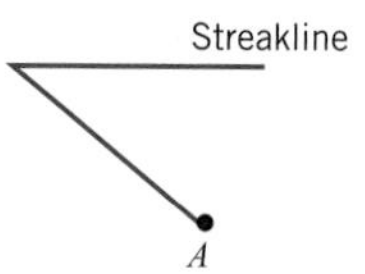

PROBLEM 4.6

4.7 For a given hypothetical flow, the velocity from time $t = 0$ to $t = 5$ s was $u = 2$ m/s, $v = 0$. Then, from time $t = 5$ s to $t = 10$ s, the velocity was $u = +3$ m/s, $v = -4$ m/s. A dye streak was started at a point in the flow field at time $t = 0$, and the path of a particle in the fluid was also traced from that same point starting at the same time. Draw to scale the streakline, pathline of the particle, and streamlines at time $t = 10$ s.

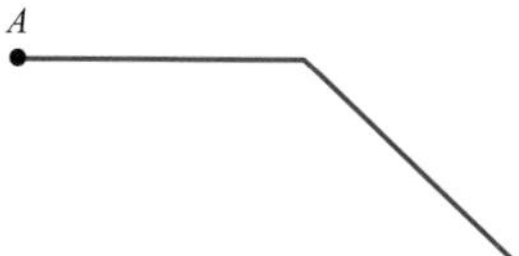

PROBLEM 4.8

4.8 At time $t = 0$, a dye streak was started at point A in a flow field of liquid. The speed of the flow is constant over a 10-s period, but the flow direction is not necessarily constant. At any particular instant the velocity in the entire field of flow is the same. The streakline produced by the dye is shown above. Draw (and label) a streamline for the flow field at $t = 8$ s.
Draw (and label) a pathline that one would see at $t = 10$ s for a particle of dye that was emitted from point A at $t = 2$ s.

4.9 A flow field is periodic in that the streamline pattern repeats at definite intervals. For the first second, the fluid is moving upward at 45° to the right and in the second second, the flow is moving downward at 45° to the right, etc. [see (*a*) below]. The speed of flow is constant at 10 m/s. After 2.5 s the pathline of a particle released at point *A* at time zero is given in (*b*). If dye is first emitted from point *A* at time zero, what will be the resulting streakline at time 2.5 s?

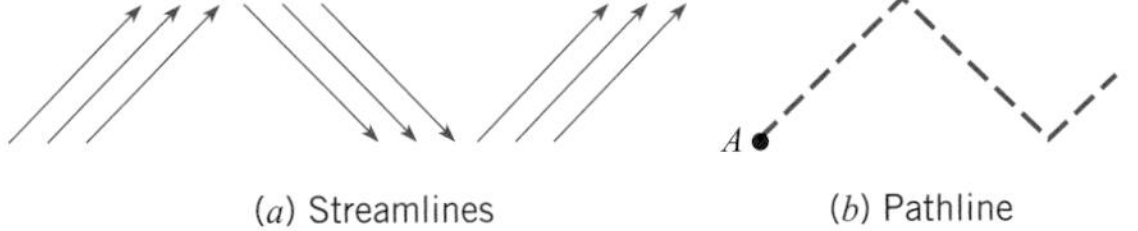

(*a*) Streamlines (*b*) Pathline

PROBLEM 4.9

4.10 The figure shows the pathline of a fluid particle that was released from point *A* at time $t = 0$. Also, the sketch shows the streakline that was developed by dye that was emitted from point *A* over the time period from $t = 0$ to $t = 5$ s. For the flow associated with this pathline and streakline, carefully sketch a streamline for $t = 0$.

For the flow associated with this pathline and streakline, one can conclude that the flow is (a) steady, (b) unsteady.

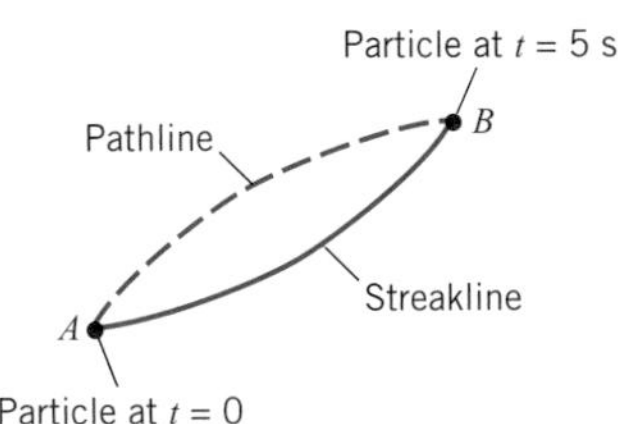

PROBLEM 4.10

4.11 For a given fluid the velocity for the entire field of flow is $u = 5$ m/s, $v = -2t$ m/s. Here t is the time in seconds. A particle is released at time $t = 0$ and a dye stream is started from the same point at the same time. Draw the streakline, the pathline of the particle, and the streamlines at time $t = 5$ s. Your solution can be checked by observing animation at http://www.justask4u.com/csp/crowe.

4.12 A flow has circular streamlines as shown. At time $t = 0$ the flow is in the clockwise direction and the angular velocity is π rad/s. This flow exists for 1 s, then reverses to the counterclockwise direction with an angular velocity of $-\pi$ rad/s and exists for 1s. For the 2 s of flow, draw the pathline for a particle of fluid that started at the point shown at time $t = 0$. Also draw the streakline for a dye stream that was emitted from the same point for the 2-s period. Carefully identify by labels the pathline and streakline. Your solution can be checked by observing animation at http://www.justask4u.com/csp/crowe.

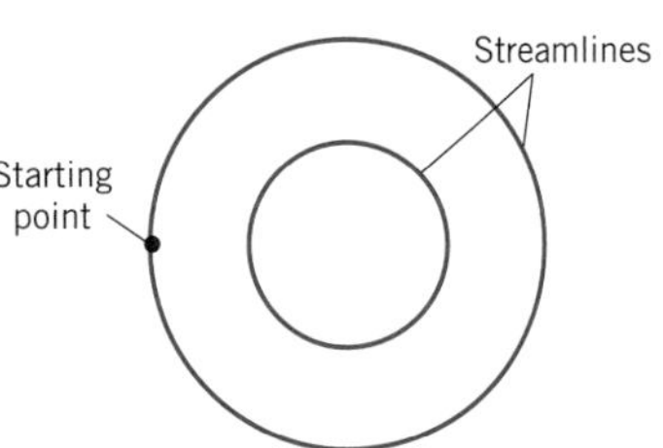

PROBLEM 4.12

4.13 In a three-dimensional flow field, the velocity for the first second is 1 m/s in the x direction. Then, for the next second, the velocity is 1 m/s in the y direction and finally, for the third second, the velocity is 1 m/s in the z direction.

a. For a particle starting at the origin, sketch the pathline on a three-dimensional coordinate system.

b. Sketch the streakline at the end of three seconds for a dye injected at the origin.

4.14 A droplet is moving across a flow field as shown. As the droplet moves it leaves a trail of vapor. The speed of the droplet is half that of the fluid. Sketch the location of the vapor trail after the droplet has moved from *A* to *B*.

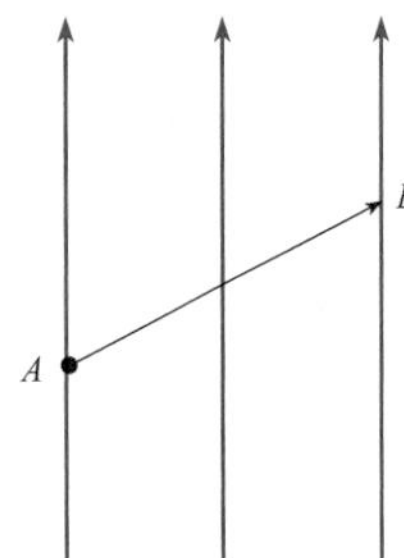

PROBLEM 4.14

4.15 Develop a program to give the coordinates of a streakline and a pathline beginning at the origin for a flow field given by

$$u = 20t^2$$

$$v = 30t^{1/2}$$

Execute the calculation for $0 \leq t \leq 1$.

4.16 Classify each of the following as a one-dimensional, two-dimensional, or three-dimensional flow.

a. Water flow over the crest of a long spillway of a dam.

b. Flow in a straight horizontal pipe.

c. Flow in a constant-diameter pipeline that follows the contour of the ground in hilly country.

d. Air flow from a slit in a plate at the end of a large rectangular duct.

e. Air flow past an automobile.

f. Air flow past a house.
g. Water flow past a pipe that is laid normal to the flow across the bottom of a wide rectangular channel.

4.17 Figure 4.23 on p. 121 shows the flow pattern for flow past a circular cylinder. Assume that the approach velocity at A is constant (does not vary with time).
a. Is the flow past the cylinder steady or unsteady?
b. Is this a case of one-dimensional, two-dimensional, or three-dimensional flow?
c. Are there any regions of the flow where local acceleration is present? If so, show where they are and show vectors representing the local acceleration in the regions where it occurs.
d. Are there any regions of flow where convective acceleration is present? If so, show vectors representing the convective acceleration in the regions where it occurs.

4.18 Given:

$$u = xt + 2y \qquad v = xt^2 - yt \qquad w = 0$$

What is the acceleration at a point $x = 1$ m, $y = 2$ m, and at time $t = 3$ s?

4.19 Tests on a sphere are conducted in a wind tunnel at an air speed of U_0. The velocity of flow toward the sphere along the longitudinal axis is found to be $u = -U_0(1 - r_0^3/x^3)$, where r_0 is the radius of the sphere and x the distance from its center. Determine the acceleration of an air particle on the x-axis upstream of the sphere in terms of x, r_0, and U_0.

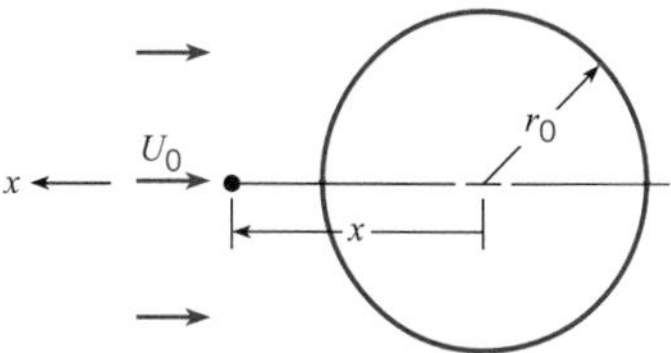

PROBLEM 4.19

4.20 Consider the flow field given in Prob. 4.12. The velocity at a radius of 10 m is given as $V_\theta = 10t$ m/s, where t is in seconds. Find the magnitude of the acceleration of a particle of fluid at $r = 10$ m when $t = 1$ s.

4.21 In this flow passage the velocity is varying with time. The velocity is given by Q/A, where Q is in m^3/s and A is in m^2. Q varies with time as

$$Q = Q_0 - Q_1\frac{t}{t_0}$$

At time $t = 0.50$ s, it is known that at section A-A the velocity gradient in the s direction is +2 m/s per meter. Given that Q_0, Q_1, and t_0 are constants with values of 0.985 m^3/s, 0.5 m^3/s, and 1 s, respectively, and assuming that one-dimensional flow applies, answer the following questions for time $t = 0.5$ s.
a. What is the velocity at A-A?
b. What is the local acceleration at A-A?
c. What is the convective acceleration at A-A?

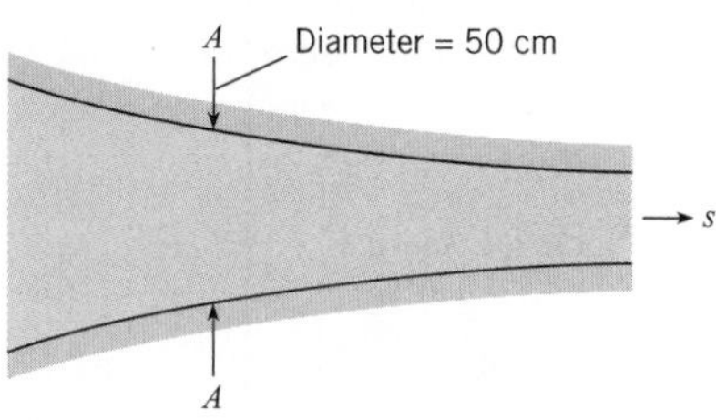

PROBLEM 4.21

4.22 The nozzle in the figure is shaped such that the velocity of flow varies linearly from the base of the nozzle to its tip. Assuming one-dimensional flow, what is the convective acceleration midway between the base and the tip if the velocity is 1 ft/s at the base and 4 ft/s at the tip? Nozzle length is 18 inches.

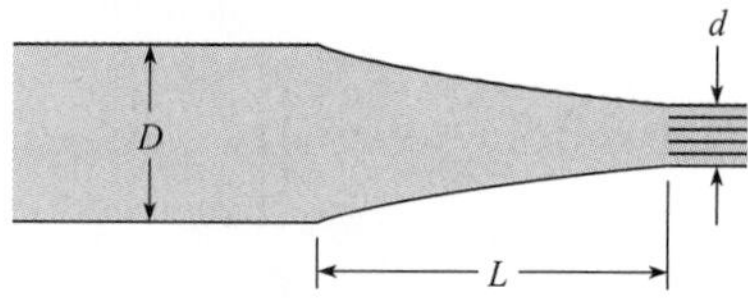

PROBLEMS 4.22, 4.23

4.23 In Prob. 4.22 the velocity varies linearly with time throughout the nozzle. The velocity at the base is $1t$ (ft/s) and at the tip is $4t$ (ft/s). What is the local acceleration midway along the nozzle when $t = 2$ s?

4.24 Liquid flows through this two-dimensional slot with a velocity of $V = 2(q_0/b)(t/t_0)$, where q_0 and t_0 are reference values. What will be the local acceleration at $x = 2B$ and $y = 0$ in terms of B, t, t_0, and q_0?

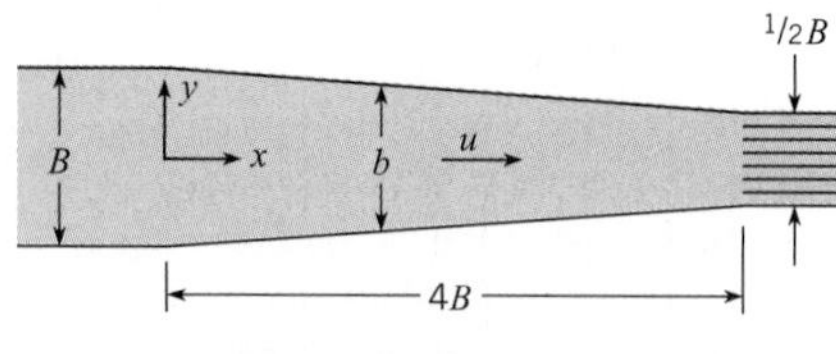

PROBLEMS 4.24, 4.25

4.25 What will be the convective acceleration for the conditions of Prob. 4.24?

4.26 The velocity of water flow in the nozzle shown is given by the following expression: $V = 2t/(1 - 0.5x/L)^2$, where V = velocity in feet per second, t = time in seconds, x = distance along the nozzle, and L = length of nozzle = 4 ft. When $x = 0.5L$ and $t = 3$ s, what is the local acceleration along the centerline? What is the convective acceleration? Assume one-dimensional flow prevails.

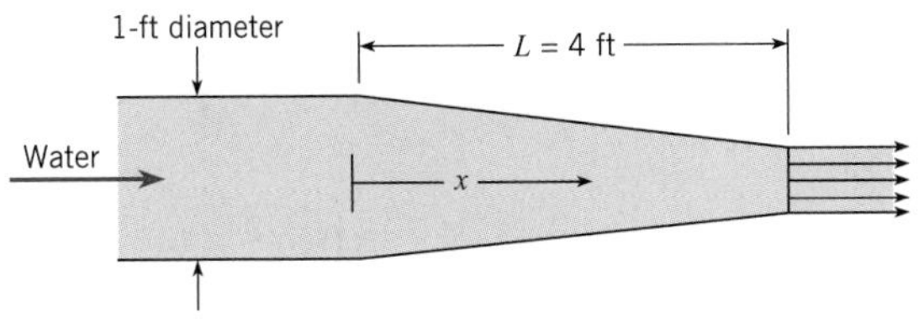

PROBLEM 4.26

4.27 A pipe slopes upward in the direction of liquid flow at an angle of 30° with the horizontal. What is the pressure gradient in the flow direction along the pipe in terms of the specific weight of the liquid if the liquid is decelerating (accelerating opposite to flow direction) at a rate of $0.3g$?

4.28 What pressure gradient is required to accelerate kerosene ($S = 0.80$) vertically upward in a vertical pipe at a rate of $0.2g$?

4.29 The hypothetical liquid in the tube shown in the figure has zero viscosity and a specific weight of 10 kN/m^3. If $p_B - p_A$ is equal to 12 kPa, one can conclude that the liquid in the tube is being accelerated (a) upward, (b) downward, (c) neither: acceleration = 0.

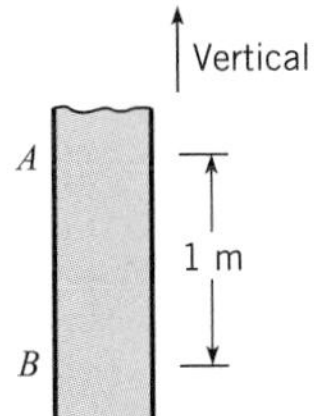

PROBLEM 4.29

4.30 If the piston and water are accelerated upward at a rate of $0.5g$, what will be the pressure at a depth of 2 ft in the water column?

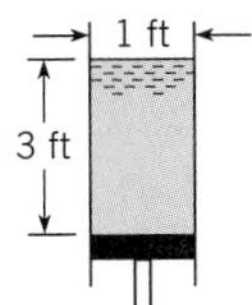

PROBLEM 4.30

4.31 Water stands at a depth of 10 ft in a vertical pipe that is open at the top and closed at the bottom by a piston. What upward acceleration of the piston is necessary to create a pressure of 9 psig immediately above the piston?

4.32 What pressure gradient is required to accelerate water in a horizontal pipe at a rate of 6 m/s^2?

4.33 Water is accelerated from rest in a horizontal pipe that is 100 m long and 30 cm in diameter. If the acceleration rate (toward the downstream end) is 6 m/s^2, what is the pressure at the upstream end if the pressure at the downstream end is 90 kPa gage?

4.34 Water stands at a depth of 10 ft in a vertical pipe that is closed at the bottom by a piston. Assuming that the vapor pressure is zero (abs), determine the maximum downward acceleration that can be given to the piston without causing the water immediately above it to vaporize.

4.35 A liquid with a specific weight of 100 lbf/ft^3 is in the conduit. This is a special kind of liquid that has zero viscosity. The pressures at points A and B are 170 psf and 100 psf, respectively. Which one (or more) of the following conclusions can one draw with certainty? (a) The velocity is in the positive ℓ direction. (b) The velocity is in the negative ℓ direction. (c) The acceleration is in the positive ℓ direction. (d) The acceleration is in the negative ℓ direction. FLUID SOLUTIONS

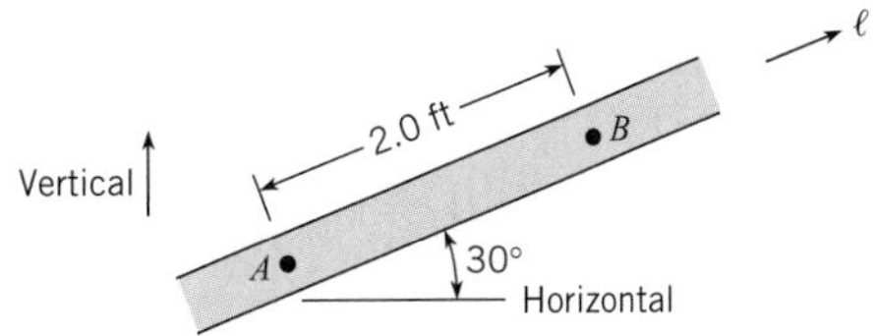

PROBLEM 4.35

4.36 If the velocity varies linearly with distance through this water nozzle, what is the pressure gradient, dp/dx, halfway through the nozzle?

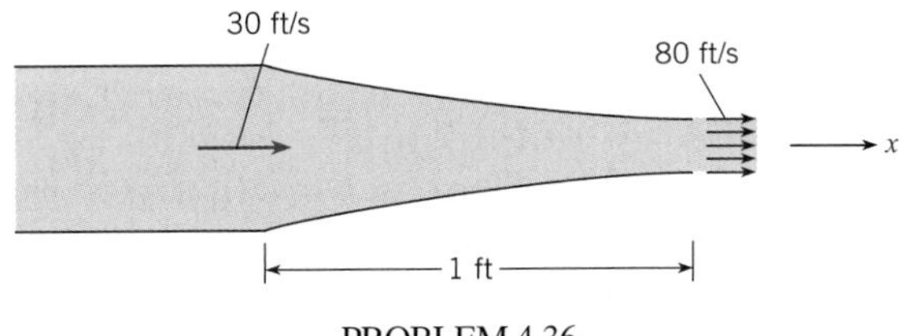

PROBLEM 4.36

4.37 This tank is accelerated in the x direction to maintain the liquid surface slope at $-3/5$. What is the acceleration of the tank?

4.38 The closed tank shown, which is full of liquid, is accelerated downward at $1.5g$ and to the right at $0.9g$. Here $L = 3$ ft, $H = 4$ ft, and the specific gravity of the liquid is 1.1. Determine $p_C - p_A$ and $p_B - p_A$.

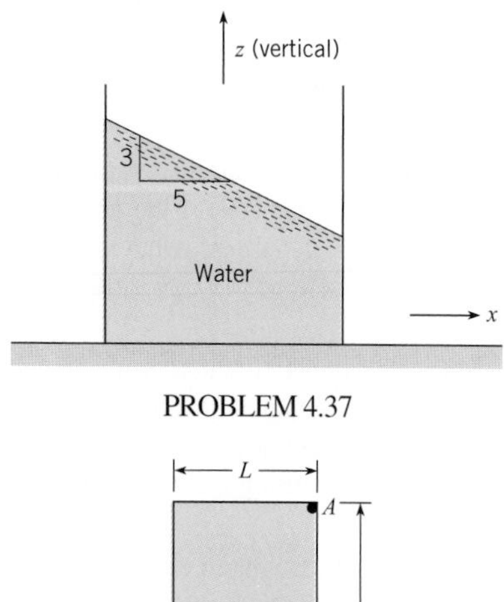

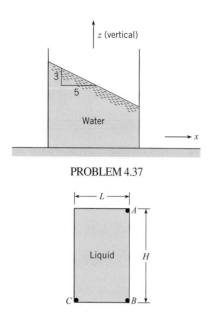

PROBLEM 4.37

PROBLEMS 4.38, 4.39

4.39 The closed tank shown, which is full of liquid, is accelerated downward at $\frac{2}{3}g$ and to the right at $1g$. Here $L = 2.5$ m, $H = 3$ m, and the liquid has a specific gravity of 1.3. Determine $p_C - p_A$ and $p_B - p_A$.

4.40 A truck carries a tank that is open at the top. The tank is 18 ft long, 6 ft wide, and 7 ft high. Assuming that the driver will not accelerate or decelerate the truck at a rate greater than 8.02 ft/s^2, to what maximum depth may the tank be filled so that water will not be spilled?

4.41 A truck carries a cylindrical tank (axis vertical) of liquid that is open at the top. Assuming that the driver will not accelerate or decelerate the truck at a rate greater than $\frac{1}{3}g$, to what maximum depth may the tank be filled so that water will not be spilled? Also, if the truck goes around an unbanked curve ($r = 50$ m), what maximum speed can it go before water will be spilled? Assume that the tank's height is the same as its diameter and that the depth for the second part of the problem is the same as that for the first.

4.42 Given: Liquid orientation in a tank as shown. The conditions could be caused by (a) constant acceleration of the tank to the right, (b) the tank being placed on a vehicle that travels at a constant speed about a circular track (center of the circle to the left of the vehicle), (c) the tank being placed on a vehicle that travels at a constant speed about a circular track (center of the circle to the right of the vehicle), (d) none of the above apply.

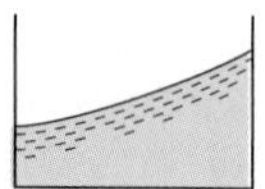

PROBLEM 4.42

4.43 The tank shown is 4 m long, 3 m high, and 3 m wide, and it is closed except for a small opening at the right end. It contains oil (S = 0.83) to a depth of 2 m in a static situation. If the tank is uniformly accelerated to the right at a rate of 19.62 m/s^2, what will be the maximum pressure intensity in the tank during acceleration?

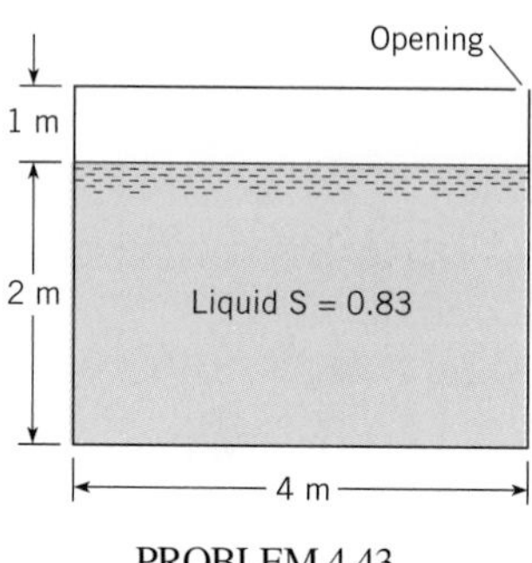

PROBLEM 4.43

4.44 A water jet issues vertically from a nozzle, as shown. The water velocity as it exits the nozzle is 20 ft/s. Calculate how high h the jet will rise. (*Hint:* Apply the Bernoulli equation along the centerline.)

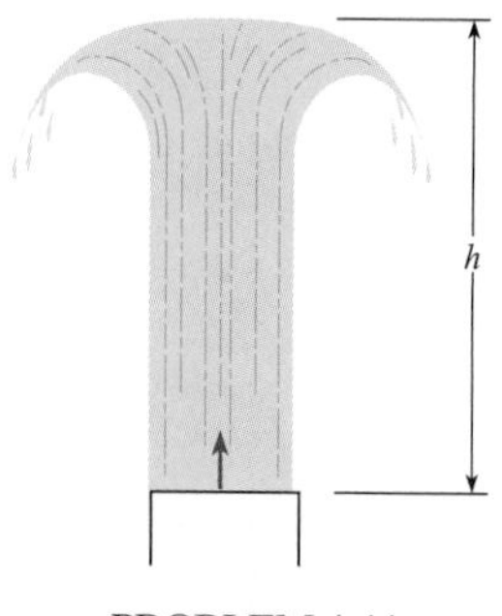

PROBLEM 4.44

4.45 A Pitot static tube is mounted on an airplane to measure airspeed. At an altitude of 10,000 ft, where the temperature is 23°F and the pressure is 10 psia, a pressure difference corresponding to 10 in. of water is measured. What is the airspeed?

4.46 An arm with a stagnation tube on the end is rotated in a horizontal plane 10 cm below a liquid surface as shown. The arm is 20 cm long, and the tube at the center of rotation extends above the liquid surface. The liquid in the tube is the same as that in the tank and has a specific weight of 10,000 N/m^3. Find the location of the liquid surface in the central tube.

4.47 A glass tube with a 90° bend is open at both ends. It is inserted into a flowing stream of water so that one opening is directed upstream and the other is vertical (see Fig. 4.14, p.101). If the water surface in the vertical tube is 10 in. higher than the stream surface, what is the velocity?

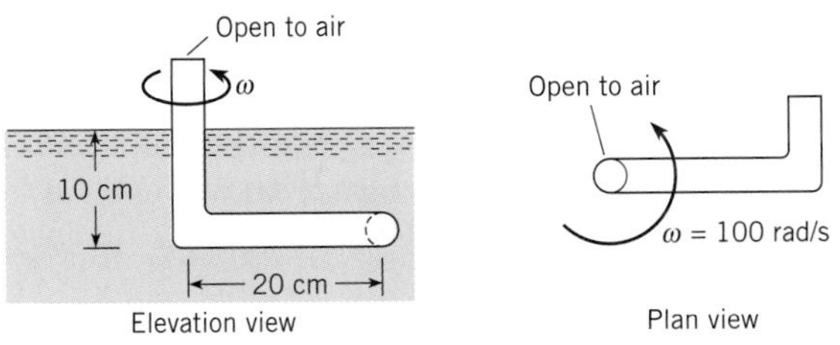

PROBLEM 4.46

4.48 A glass tube like the one described in Prob. 4.47 is inserted into a flowing stream of water with one opening directed upstream and the other end vertical (see Fig. 4.14). If the water velocity is 3 m/s, how high will the water rise in the vertical leg relative to the level of the water surface of the stream?

4.49 A Bourdon-tube gage is tapped into the center of a disk as shown. Then for a disk that is about 1 ft in diameter and for an approach velocity of air (V_0) of 40 ft/s, the gage would read a pressure intensity that is (a) less than $\rho V_0^2/2$, (b) equal to $\rho V_0^2/2$, (c) greater than $\rho V_0^2/2$.

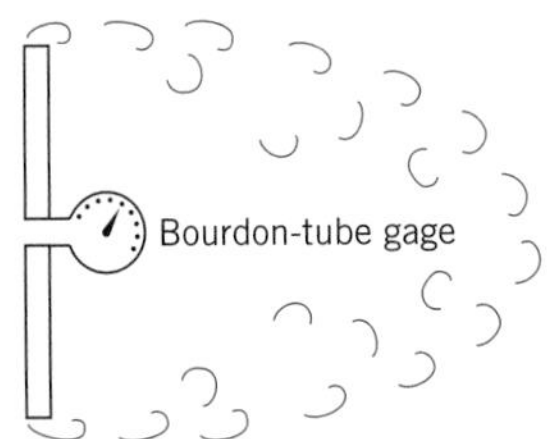

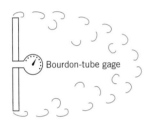

PROBLEM 4.49

4.50 An air-water manometer is connected to a Pitot tube used to measure air velocity. If the manometer deflects 2 in., what is the velocity? Assume $T = 60°\text{F}$ and $p = 15$ psia.

4.51 A flow-metering device consists of a stagnation probe at station 2 and a static pressure tap at station 1. The velocity at station 2 is twice that at station 1. Air with a density of 1.2 kg/m^3 flows through the duct. A water manometer is connected between the stagnation probe and the pressure tap, and a deflection of 10 cm is measured. What is the velocity at station 2?

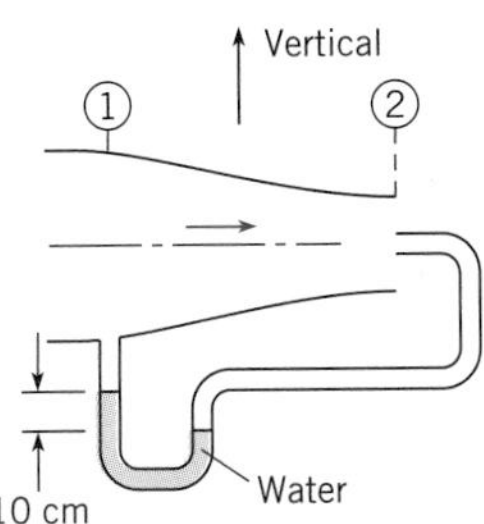

PROBLEM 4.51

4.52 A "spherical" Pitot probe is used to measure the flow velocity in water. Pressure taps are located at the forward stagnation point and at 90° from the forward stagnation point. The speed of fluid next to the surface of the sphere varies as $1.5\,V_0 \sin\theta$, where V_0 is the free-stream velocity and θ is measured from the forward stagnation point. The pressure taps are at the same level; that is, they are in the same horizontal plane. The piezometric pressure difference between the two taps is 3 kPa. What is the free-stream velocity V_0?

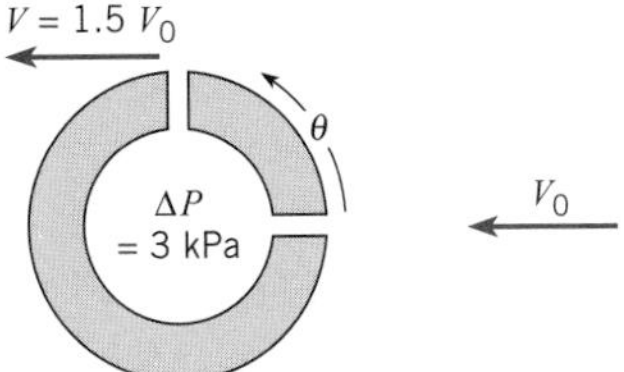

PROBLEM 4.52

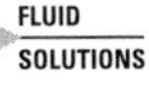

4.53 A device used to measure the velocity of fluid in a pipe consists of a cylinder, with a diameter much smaller than the pipe diameter, mounted in the pipe with pressure taps at the forward stagnation point and at the rearward side of the cylinder. Data show that the pressure coefficient at the rearward pressure tap is –0.3. Water with a density of 1000 kg/m^3 flows in the pipe. A pressure gage connected by lines to the pressure taps shows a pressure difference of 500 Pa. What is the velocity in the pipe?

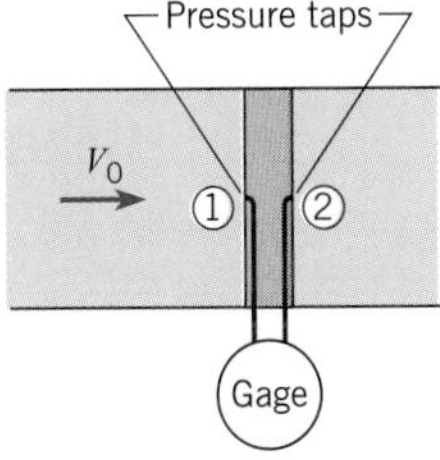

PROBLEM 4.53

4.54 A spherical Pitot probe is to be designed to measure flow velocity. One pressure tap is located at the stagnation point and another pressure tap is located at the angular position where the pressure is equal to the static pressure. The fluid velocity at the surface of a sphere is given by

$$V = 1.5 V_0 \sin\theta$$

where V_0 is the free-stream velocity and θ is the angle between the location on the sphere and the stagnation point.

a. Find the angle for the pressure tap that would yield the static pressure. The change in hydrostatic pressure is negligible.

b. Derive the equation that relates the pressure difference between the two taps to the free-stream velocity.

c. Assume the flow approaches the sphere with an angle β with respect to the axis of the sphere; that is, the stagnation point is displaced by an angle β from the pressure tap located to provide the stagnation pressure. Write a program to find the error in the velocity calculated using the equation derived in part *b* as a function of β. Plot the results for β from 0 to 10 degrees.

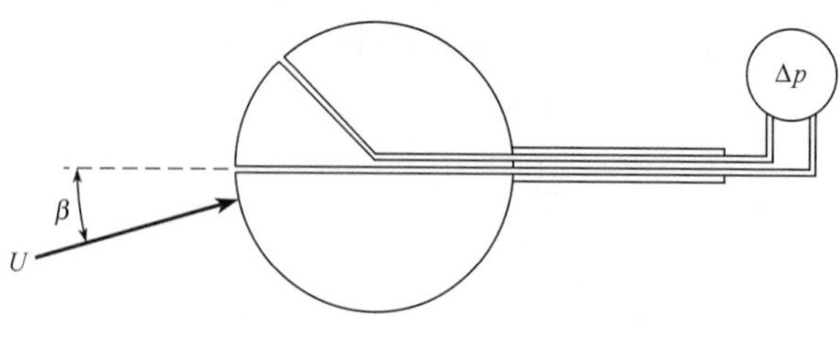

PROBLEM 4.54

4.55 Explain how you might design a spherical Pitot probe to provide the direction and velocity of a flowing stream. The Pitot probe will be mounted on a sting that can be oriented in any direction.

4.56 Two Pitot tubes are shown. The one on the top is used to measure the velocity of air, and it is connected to an air-water manometer as shown. The one on the bottom is used to measure the velocity of water, and it too is connected to an air-water manometer as shown. If the deflection h is the same for both manometers, then one can conclude that (a) $V_A = V_w$, (b) $V_A > V_w$, (c) $V_A < V_w$.

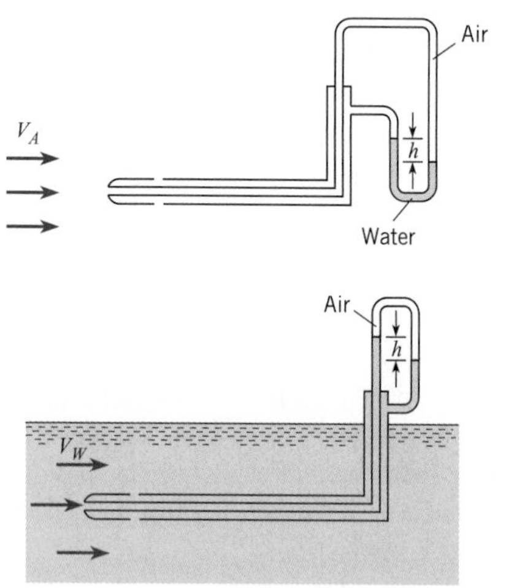

PROBLEM 4.56

4.57 A Pitot tube is used to measure the velocity at the center of a 12-in. pipe. If kerosene at 68°F is flowing and the deflection on a mercury-kerosene manometer connected to the Pitot tube is 5 in., what is the velocity?

4.58 A Pitot tube used to measure air velocity is connected to a differential pressure gage. If the air temperature is 20°C at standard atmospheric pressure at sea level, and if the differential gage reads a pressure difference of 3 kPa, what is the air velocity?

4.59 A Pitot tube used to measure air velocity is connected to a differential pressure gage. If the air temperature is 60°F at standard atmospheric pressure at sea level, and if the differential gage reads a pressure difference of 11 psf, what is the air velocity?

4.60 A Pitot tube is used to measure the gas velocity in a duct. A pressure transducer connected to the Pitot tube registers a pressure difference of 0.9 psi. The density of the gas in the duct is 0.12 lbm/ft^3. What is the gas velocity in the duct?

4.61 A sphere moves horizontally through still water at a speed of 11 ft/s. A short distance directly ahead of the sphere (call it point *A*), the velocity, with respect to the earth, induced by the sphere is 1 ft/s in the same direction as the motion of the sphere. If p_0 is the pressure in the undisturbed water at the same depth as the center of the sphere, then the value of the ratio p_A/p_0 will be (a) less than unity, (b) equal to unity, (c) greater than unity.

4.62 Body *A* travels through water at a constant speed of 13 m/s. Velocities at points *B* and *C* are induced by the moving body and are observed to have magnitudes of 5 m/s and 3 m/s, respectively. What is $p_B - p_C$?

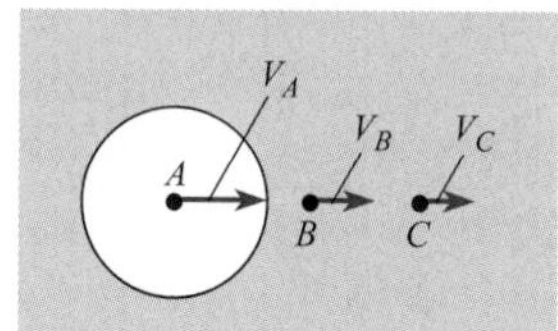

PROBLEM 4.62

4.63 Water in a flume is shown for two conditions. If the depth *d* is the same for each case, will gage *A* read greater or less than gage *B*? Explain.

4.64 The apparatus shown in the figure is used to measure the velocity of air at the center of a duct having a 10-cm diameter. A tube mounted at the center of the duct has a 2-mm diameter and is attached to one leg of a slant-tube manometer. A pressure tap in the wall of the duct is connected to the other end of the slant-tube manometer. The well of the slant-tube manometer is sufficiently large that the elevation of the fluid in it does not change significantly when fluid moves up the leg of the manometer. The air in the duct is at a temperature of 20°C, and the pressure is 150 kPa. The manometer liquid has a specific gravity of 0.7, and the slope of the leg is 30°. When there is no flow in the duct, the liquid surface in the manometer lies at 2.3 cm on the slanted scale. When there is flow in the duct, the liquid moves up to 6.7 cm on the slanted scale. Find the velocity of the air in the duct. Assuming a uniform velocity profile in the duct, calculate the rate of flow of the air.

4.65 A spherical probe is used for finding gas velocity by measuring the pressure difference between the upstream and down-

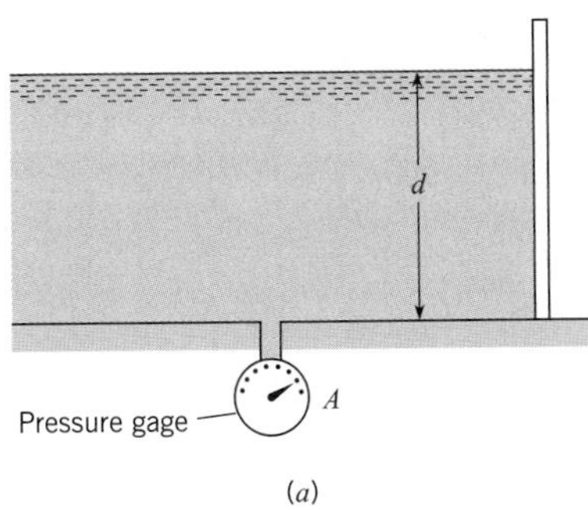

(a)

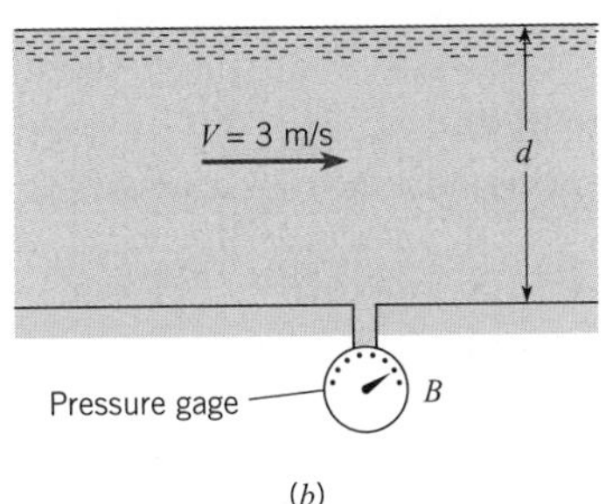

(b)

PROBLEM 4.63

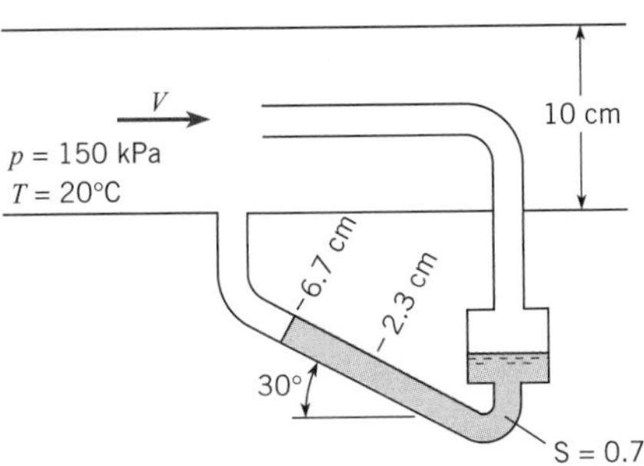

PROBLEM 4.64

stream points A and B. The pressure coefficients at points A and B are 1.0 and -0.4. The pressure difference $p_A - p_B$ is 4 kPa, and the gas density is 1.50 kg/m^3. Calculate the gas velocity.

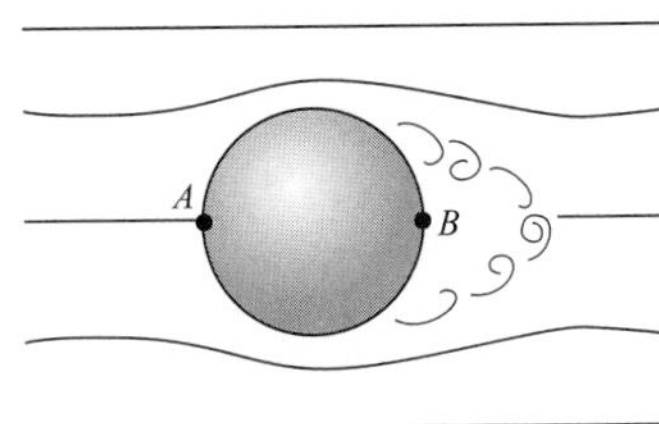

PROBLEM 4.65

4.66 A rugged instrument used frequently for monitoring gas velocity in smoke stacks consists of two open tubes oriented to the flow direction as shown and connected to a manometer. The pressure coefficient is 1.0 at A and -0.3 at B. Assume that water, at 20°C, is used in the manometer and that a 0.8-cm deflection is noted. The pressure and temperature of the stack gases are 101 kPa and 250°C. The gas constant of the stack gases is 200 J/kg K. Determine the velocity of the stack gases.

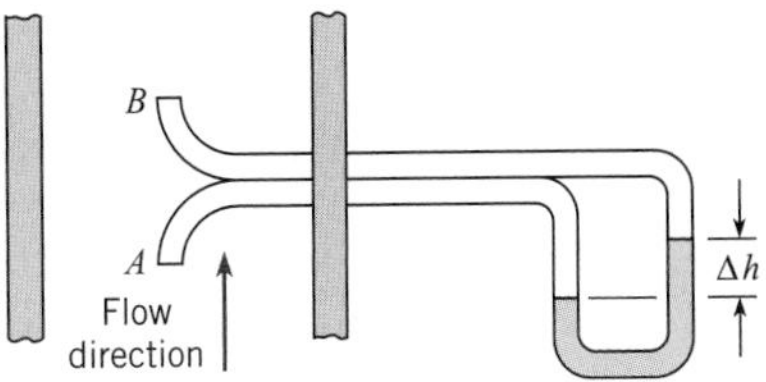

PROBLEM 4.66

4.67 A special spherical probe is used to measure the velocity of water. Pressure taps are located at the forward stagnation point and at the maximum width of the sphere. A mercury manometer is connected to the pressure taps as shown. A deflection of 5 cm is measured on the manometer. The velocity distribution around the sphere is given by

$$u = 1.5U\sin\theta$$

where U is the free-stream velocity. Find the free-stream velocity.

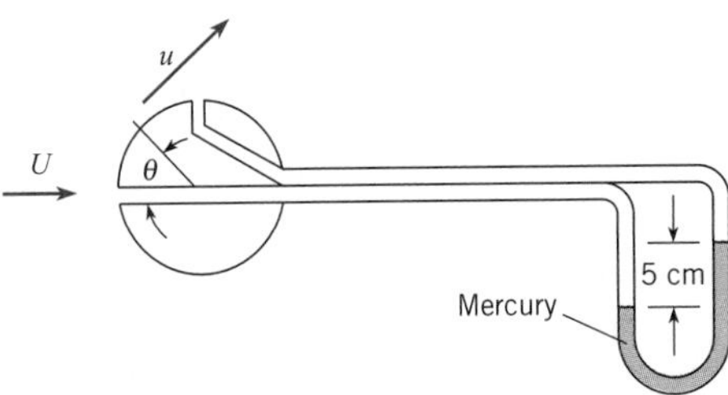

PROBLEMS 4.67, 4.68, 4.69

4.68 The pressure in the wake of a bluff body is approximately equal to the pressure at the point of separation. Assume flow past the sphere of Prob. 4.67 separates at an angle θ of 120°. If the free-stream velocity is 100 m/s and the fluid is air ($\rho = 1.2$ kg/m^3), find the pressure coefficient in the separated region next to the sphere. Also, what is the gage pressure in this region if the free-stream pressure is atmospheric?

4.69 If a pressure transducer is connected between the taps of Prob. 4.67 and it reads 120 Pa, what is the free-stream velocity? Assume the air density is 1.2 kg/m^3.

4.70 A Pitot tube is used to measure the airspeed of an airplane. The Pitot tube is connected to a pressure-sensing device calibrated to indicate the correct airspeed when the temperature is 17°C and the pressure is 101 kPa. The airplane flies at an altitude of 3000 m, where the pressure and temperature are 70 kPa and $-6.3°C$. The indicated airspeed is 70 m/s. What is the true airspeed?

4.71 The pressure coefficient distribution on a cylinder in a cross flow is given by

$$C_p = 1 - 4\sin^2\theta$$

where θ is the angular displacement from the forward stagnation point. Assume that two pressure taps are located at $\pm 30°$ as shown and connected to a water manometer. The cylinder is immersed in air with a density of 1.2 kg/m^3 and a velocity of 50 m/s in the direction shown on the figure. What will be the deflection on the manometer, in centimeters?

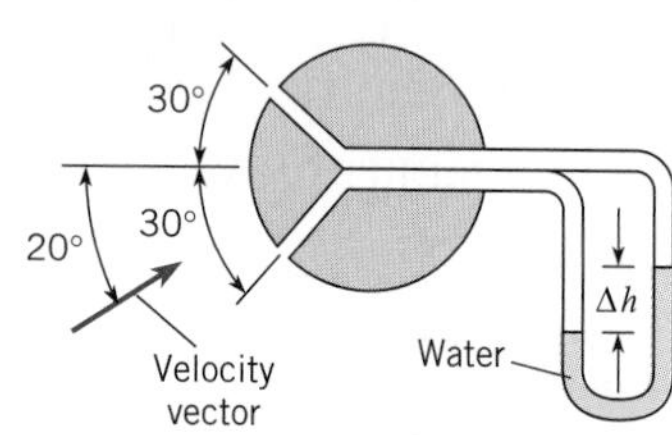

PROBLEM 4.71

FLUID SOLUTIONS **4.72** You need to measure air flow velocity. You order a commercially available Pitot tube and the accompanying instructions state that the air flow velocity is given by

$$V\ (\text{ft/min}) = 1096.7\sqrt{\frac{h_v}{d}}$$

where h_v is the "velocity pressure" in inches of water and d is the density in pounds per cubic foot. The velocity pressure is the deflection measured on a water manometer attached to the static and total pressure ports. The instructions also state the density d can be calculated using

$$d\ (\text{lbm/ft}^3) = 1.325\frac{P_a}{T}$$

where P_a is the barometric pressure in inches of mercury and T is the absolute temperature in degrees Rankine. Before you use the Pitot tube you want to confirm that the equations are correct. Determine if these equations are correct.

4.73 Consider the flow of water over the surfaces shown. For each case the depth of water at section *D-D* is the same (1 ft), and the mean velocity is the same and equal to 10 ft/s. Which of the following statements are valid?

a. $p_C > p_B > p_A$
b. $p_B > p_C > p_A$
c. $p_A = p_B = p_C$
d. $p_B < p_C < p_A$
e. $p_A < p_B < p_C$

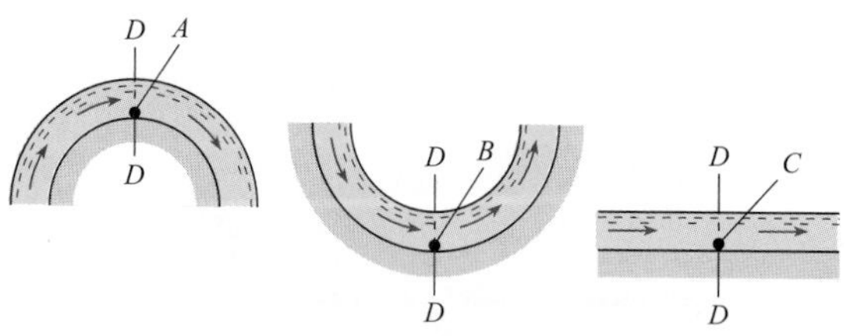

PROBLEM 4.73

4.74 The vector $\mathbf{V} = 10x\mathbf{i} - 10y\mathbf{j}$ represents a two-dimensional velocity field. Is the flow irrotational?

4.75 The u and v velocity components of a flow field are given by $u = -\omega y$ and $v = \omega x$. Determine the vorticity and the rate of rotation of flow field.

4.76 The velocity components for a two-dimensional flow are

$$u = \frac{C(y^2 - x^2)}{(y^2 + x^2)^2} \qquad v = \frac{-2Cxy}{(x^2 + y^2)^2}$$

where C is a constant. Is the flow irrotational?

4.77 A two-dimensional flow field is defined by $u = xt + 2y$ and $v = xt^2 - yt$. What is the acceleration at a point $x = 1$ m, $y = 1$ m, and at time $t = 1$ s? Is the flow rotational or irrotational?

4.78 Fluid flows between two parallel stationary plates. The distance between the plates is 1 cm. The velocity profile between the two plates is a parabola with a maximum velocity at the centerline of 2 cm/s. The velocity is given by FLUID SOLUTIONS

$$u = 2(1 - 4y^2)$$

where y is measured from the centerline. The cross-flow component of velocity, v, is zero. There is a reference line located 1 cm downstream. Find an expression, as a function of y, for the amount of rotation (in radian) a fluid element will undergo when it travels a distance of 1 cm downstream. You can check your answer by observing the animation on http://www.justask4u.com/csp/crowe.

4.79 A combination of a forced and a free vortex is represented by the velocity distribution FLUID SOLUTIONS

$$v_\theta = \frac{1}{r}[1 - \exp(-r^2)]$$

For $r \to 0$ the velocity approaches a rigid body rotation and as r becomes large, a free vortex velocity distribution is approached. Find the amount of rotation (in radians) that a fluid element will experience in completing one circuit around the center as a function of r. *Hint:* The rotation rate in a flow with concentric streamlines is given by

$$\omega_z = \frac{dv_\theta}{dr} + \frac{v_\theta}{r} = \frac{1}{r}\frac{d}{dr}(v_\theta r)$$

Evaluate the rotation for $r = 0.5, 1.0$, and 1.5. Your answer can be checked by observing the animation on http://www.justask4u.com/csp/crowe.

4.80 This closed tank, which is 4 ft in diameter, is filled with water and is spun around its vertical centroidal axis at a rate of 15 rad/s. An open piezometer is connected to the tank as shown so that it is also rotating with the tank. For these conditions, what is the pressure at the center of the bottom of the tank?

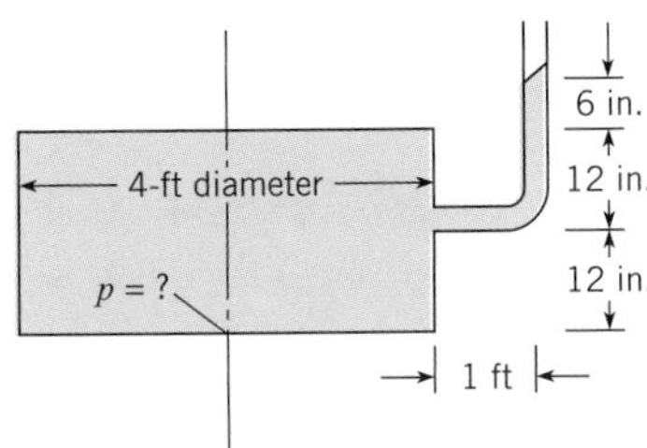

PROBLEM 4.80

4.81 A tank of liquid (S = 0.80) that is 1 ft in diameter and 1.0 ft high (h = 1.0 ft) is rigidly fixed (as shown) to a rotating arm having a 2-ft radius. The arm rotates such that the speed at point A is 20 ft/s. If the pressure at A is 25 psf, what is the pressure at B?

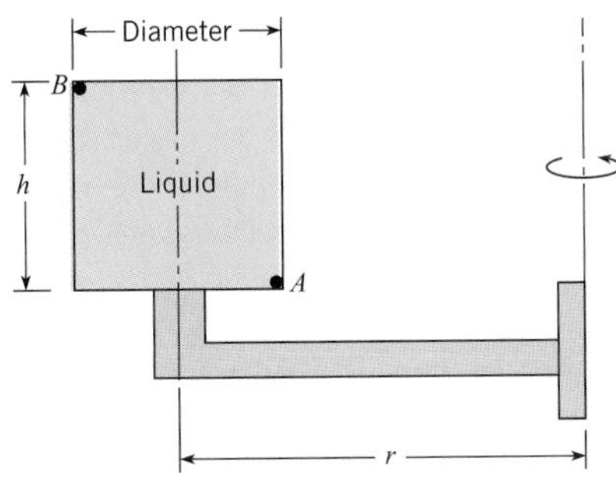

PROBLEM 4.81

4.82 A closed tank of liquid (S = 1.2) is rotated about a vertical axis (see the figure), and at the same time the entire tank is accelerated upward at 4 m/s^2. If the rate of rotation is 10 rad/s, what is the difference in pressure between points A and B $(p_B - p_A)$? Point B is at the bottom of the tank at a radius of 0.5 m from the axis of rotation, and point A is at the top on the axis of rotation.

4.83 A U-tube is rotated about one leg, as shown. Before being rotated the liquid in the tube fills 0.25 m of each leg. The length of the base of the U-tube is 0.5 m, and each leg is 0.5 m long. What would be the maximum rotation rate (in rad/s) to ensure that no liquid is expelled from the outer leg?

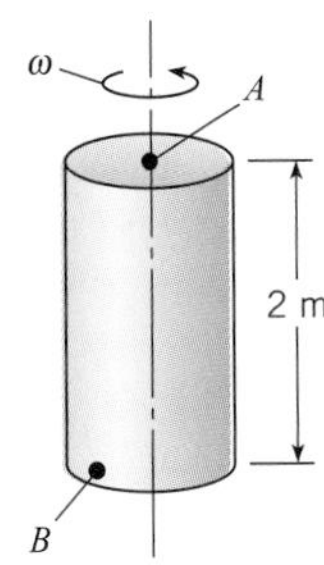

PROBLEM 4.82

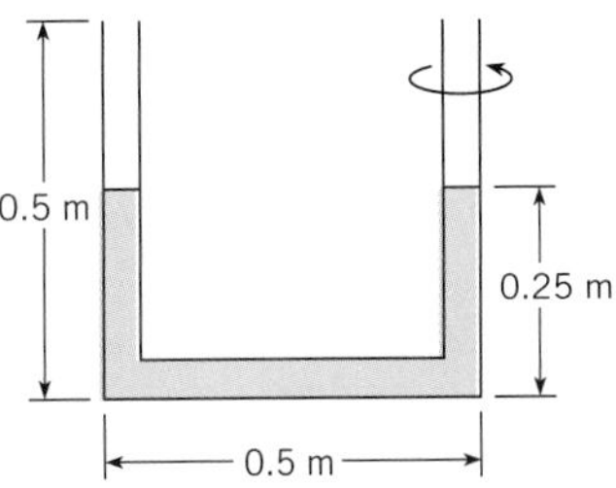

PROBLEM 4.83

4.84 A U-tube is rotated at 60 rev/min about one leg. The fluid at the bottom of the U-tube has a specific gravity of 3.0. The distance between the two legs of the U-tube is 1 ft. A 6-in. height of another fluid is in the outer leg of the U-tube. Both legs are open to the atmosphere. Calculate the specific gravity of the other fluid.

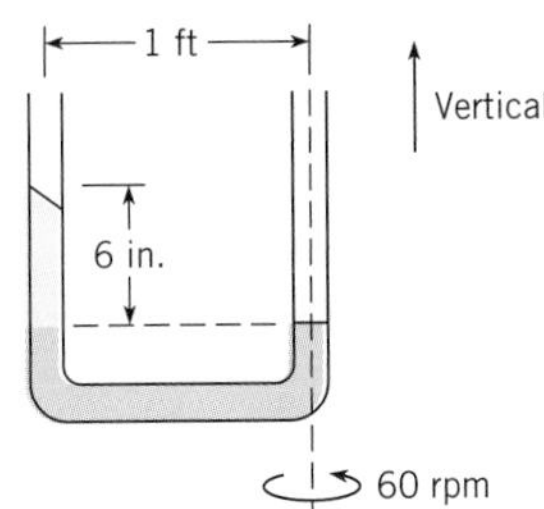

PROBLEM 4.84

4.85 If the U-tube is rotated about axis 0-0 at a rate of 32.12 rad/s, determine the new position of the water surface in the outside leg during rotation. The values of L_1, L_2, and L_3 are 1 m, 25 cm, and 30 cm, respectively.

4.86 The U-tube is attached to platform B, and the liquid levels in the U-tube are shown for conditions at rest. The platform and U-tube are then rotated about axis A-A at a rate of 4 rad/s. What will be the elevation of the liquid in the smaller leg of the U-tube after rotation?

4.87 A manometer spins about one leg as shown. The leg about which the manometer rotates contains water with a height of 10 cm. The other leg, which is 1 m from the axis of rotation, contains

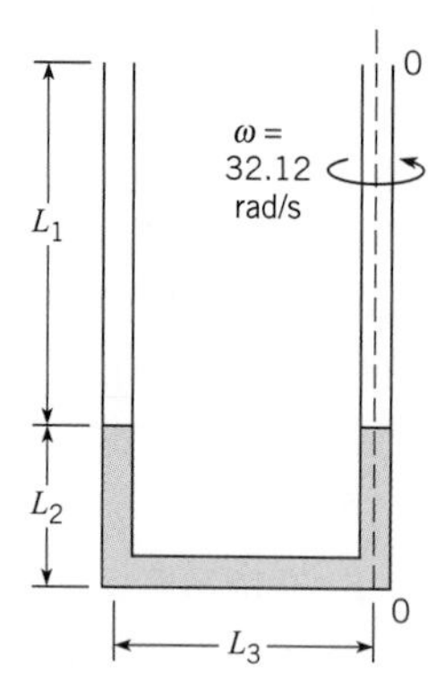

PROBLEM 4.85

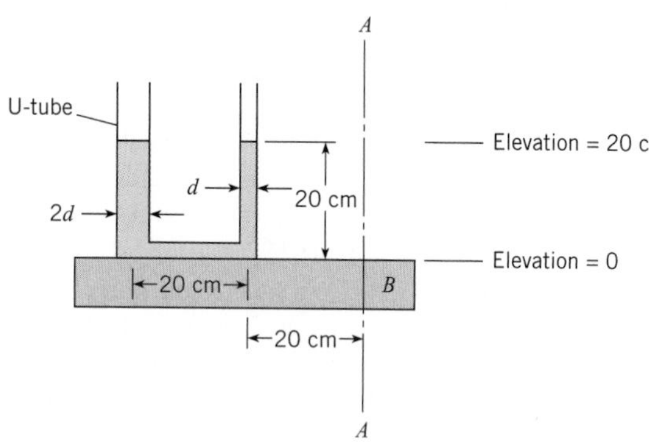

PROBLEM 4.86

mercury (S = 13.6) with a height of 1 cm. What is the rotational speed (in radians per second)?

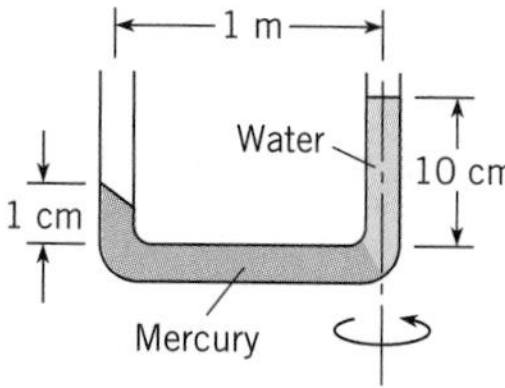

PROBLEM 4.87

4.88 A manometer is rotated around one leg, as shown. The difference in elevation between the liquid surfaces in the legs is 25 cm. The radius of the rotating arm is 10 cm. The liquid in the manometer is oil with a specific gravity of 0.8. Find the number of g's of acceleration in the leg with greatest amount of oil.

4.89 A fuel tank for a rocket in space under a zero-g environment is rotated to keep the fuel in one end of the tank. The system is rotated at 3 rev / min. The end of the tank (point A) is 1.5 m from the axis of rotation, and the fuel level is 1 m from the rotation axis. The pressure in the nonliquid end of the tank is 0.1 kPa, and the density of the fuel is 800 kg / m^3. What is the pressure at the exit (point A)?

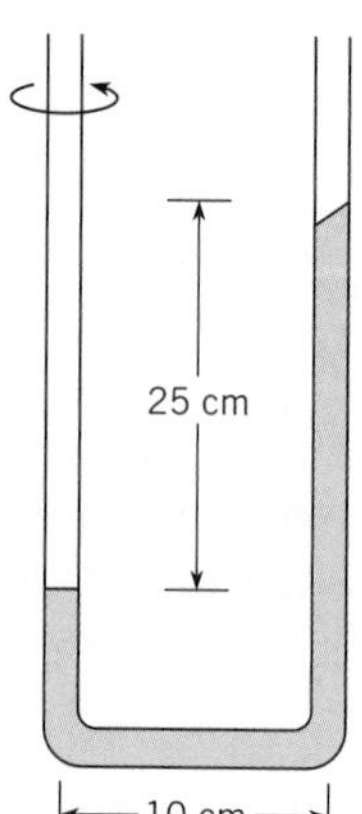

PROBLEM 4.88

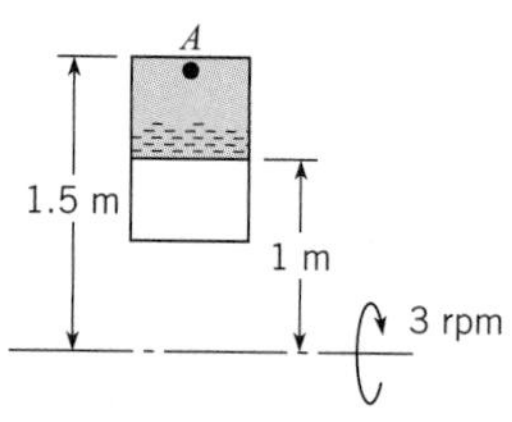

PROBLEM 4.89

4.90 Water stands in these tubes as shown when no rotation occurs. Derive a formula for the angular speed at which water will just begin to spill out of the small tube when the entire system is rotated about axis A-A.

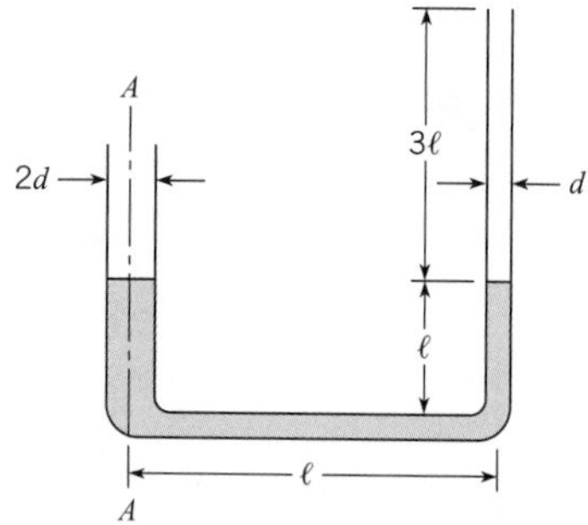

PROBLEM 4.90

4.91 Mercury is the liquid in the rotating U-tube. Determine the rate of rotation ω if $\ell = 5$ in. Then, if rotation is stopped, to what level z will the mercury level drop in the larger leg?

4.92 Water fills a slender tube 1 cm in diameter, 40 cm long, and closed at one end. When the tube is rotated in the horizontal plane about its open end at a constant speed of 60.8 rad / s, what force is exerted on the closed end?

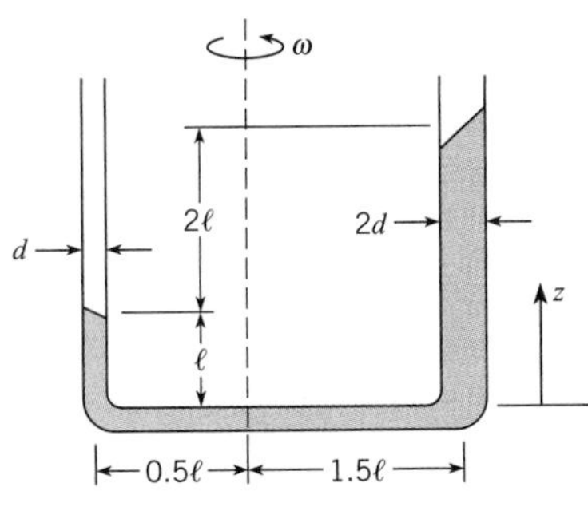

PROBLEM 4.91

4.93 Mercury maintains the position shown in the rotating tube. What is the rate of rotation in terms of g and ℓ?

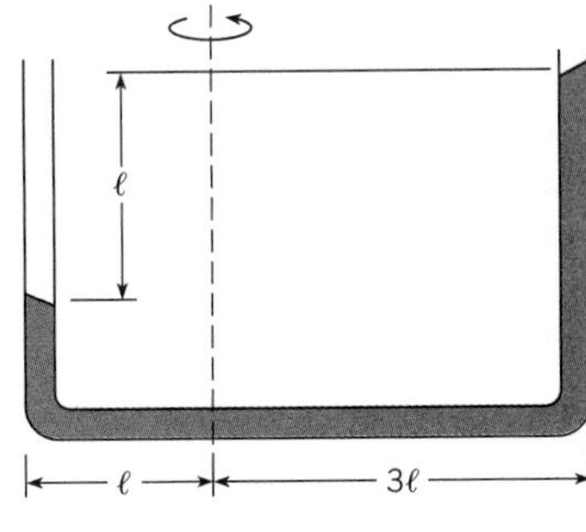

PROBLEM 4.93

4.94 The U-tube is shown with water standing in it when there is no rotation.

a. If the tube is rotated about the left leg at a rate of 5 rad/s, where will the water stand in the tube for $\ell = 25$ cm ?

b. Where will the water surface be when the rate of rotation is increased to 15 rad/s for $\ell = 30$ cm ?

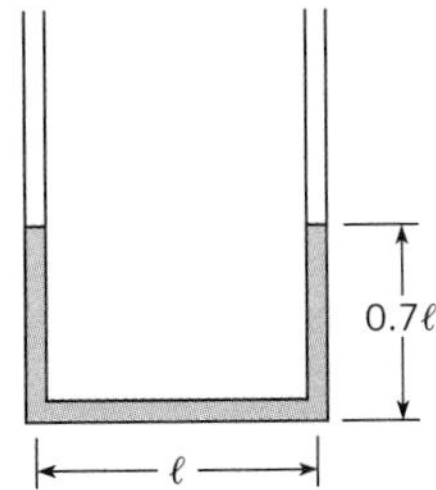

PROBLEM 4.94

4.95 This tube with liquid is rotated about a vertical axis as indicated. If the rate of rotation is 8 rad/s, what will be the pressure in the liquid at point A (at the axis of rotation) and at point B? Then, if the tube is rotated at a rate of 20 rad/s, what will the pressure be at points A and B?

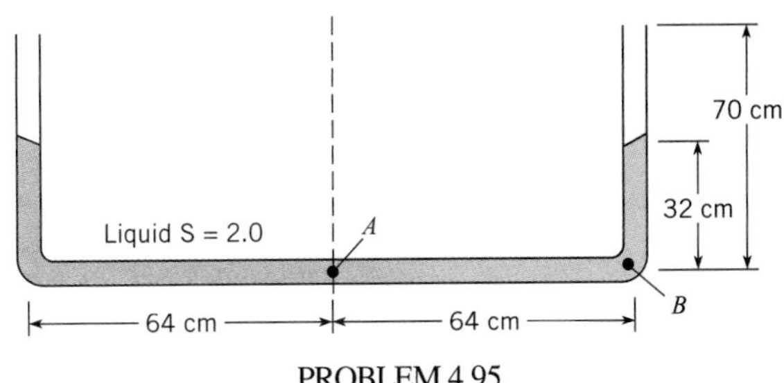

PROBLEM 4.95

4.96 Water stands in the closed-end U-tube as shown when there is no rotation. If $\ell = 10$ cm and if the entire system is rotated about axis A-A, at what angular speed will water just begin to spill out of the open tube? Assume that the temperature for the system is the same before and after rotation.

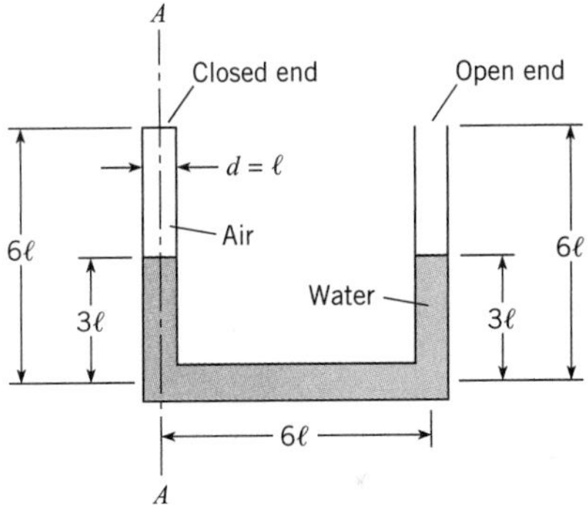

PROBLEM 4.96

4.97 A simple centrifugal pump consists of a 10-cm disk with radial ports as shown. Water is pumped from a reservoir through a central tube on the axis. The wheel spins at 2500 rev/min, and the liquid discharges to atmospheric pressure. To establish the maximum height for operation of the pump, assume that the flow rate is zero and the pressure at the pump intake is atmospheric pressure. Calculate the maximum operational height z for the pump.

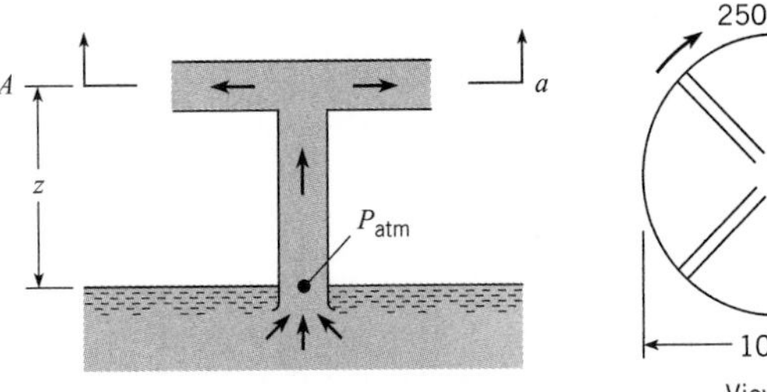

PROBLEM 4.97

4.98 A closed cylindrical tank of water is rotated about its horizontal axis as shown. The water inside the tank rotates with the tank ($V = r\omega$). Derive an equation for dp/dz along a vertical-radial line through the center of rotation. What is dp/dz along this line for $z = -1$ m, $z = 0$, and $z = +1$ m when $\omega = 5$ rad/s? Here $z = 0$ at the axis.

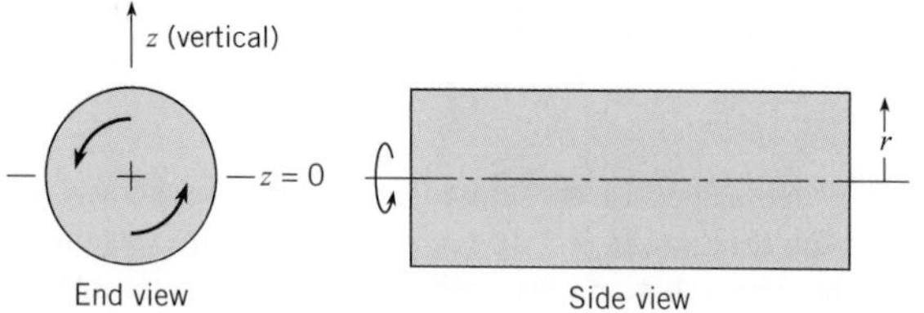

PROBLEMS 4.98, 4.99, 4.100

4.99 For the conditions of Prob. 4.98, derive an equation for the maximum pressure difference in the tank as a function of the significant variables.

4.100 The tank shown is 4 ft in diameter and 12 ft long and is closed and filled with water. It is rotated about its horizontal-centroidal axis, and the water in the tank rotates with the tank ($V = r\omega$). The maximum velocity is 20 ft/s. What is the maximum difference in pressure in the tank? Where is the point of minimum pressure?

4.101 Liquid flows with a free surface around a bend. The liquid is inviscid and incompressible, and the flow is steady and irrotational. The velocity varies with the radius across the flow as $V = 1/r$ m/s, where r is in meters. Find the difference in depth of the liquid from the inside to the outside radius. The inside radius of the bend is 1 m and the outside radius is 3 m.

4.102 The velocity in the outlet pipe from this reservoir is 16 ft/s and $h = 15$ ft. Because of the rounded entrance to the pipe, the flow is assumed to be irrotational. Under these conditions, what is the pressure at A?

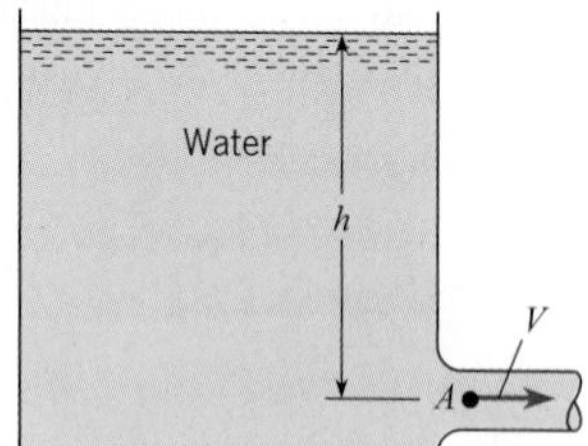

PROBLEMS 4.102, 4.103

4.103 The velocity in the outlet pipe from this reservoir is 6 m/s and $h = 15$ m. Because of the rounded entrance to the pipe, the flow is assumed to be irrotational. Under these conditions, what is the pressure at A?

4.104 The maximum velocity of the flow past a circular cylinder, as shown, is twice the approach velocity. What is Δp between the point of highest pressure and the point of lowest pressure in a 40 m/s wind? Assume irrotational flow and standard atmospheric conditions.

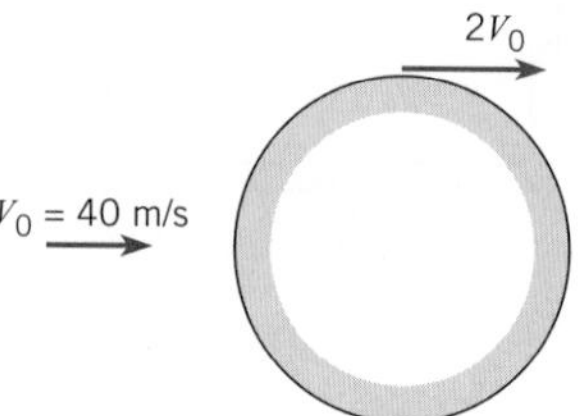

PROBLEM 4.104

4.105 The velocity and pressure are given at two points in the flow field. Assume that the two points lie in a horizontal plane and that the fluid density is uniform in the flow field and is equal to 1000 kg/m^3. Assume steady flow. Then, given these data, determine which of the following statements is true. (a) The flow in the contraction is nonuniform and irrotational. (b) The flow in the contraction is uniform and irrotational. (c) The flow in the contraction is nonuniform and rotational. (d) The flow in the contraction is uniform and rotational.

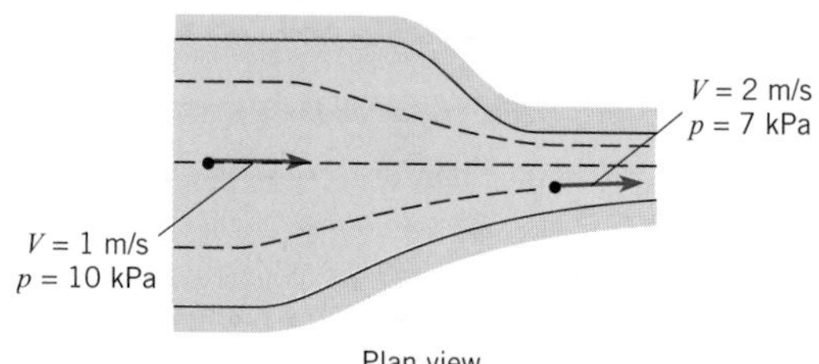

PROBLEM 4.105

4.106 Water flows from the large orifice at the bottom of the tank as shown. Assume that the flow is irrotational. Point B is at zero elevation, and point A is at 1 ft elevation. If $V_A = 8$ ft/s at an angle of 45° with the horizontal and if $V_B = 20$ ft/s vertically downward, what is the value of $p_A - p_B$?

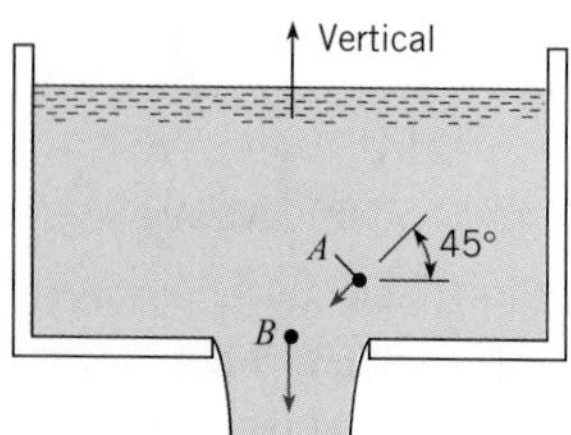

PROBLEM 4.106

4.107 Ideal-flow theory will yield a flow pattern past an airfoil similar to that shown. If the approach air velocity V_0 is 80 m/s, what is the pressure difference between the bottom and the top of this airfoil at points where the velocities are $V_1 = 85$ m/s and $V_2 = 75$ m/s ? Assume ρ_{air} is uniform at 1.2 kg/m^3.

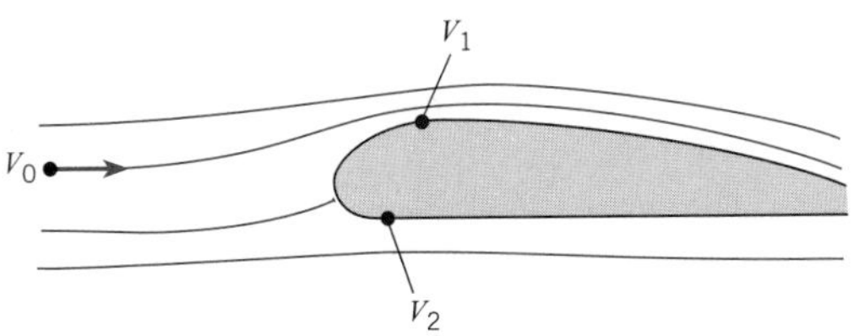

PROBLEM 4.107

4.108 Consider the flow of water between two parallel plates in which one plate is fixed. The distance between the plates is h and the speed of the moving plate is V. A person wishes to calculate the pressure difference between the plates and applies the Bernoulli equation between points 1 and 2,

$$z_1 + \frac{p_1}{\gamma} + \frac{V_1^2}{2g} = z_2 + \frac{p_2}{\gamma} + \frac{V_2^2}{2g}$$

and concludes that

$$p_1 - p_2 = \gamma(z_2 - z_1) + \rho\frac{V_2^2}{2}$$

$$= \gamma h + \rho\frac{V^2}{2}$$

Is this correct? Provide the reason for your answer.

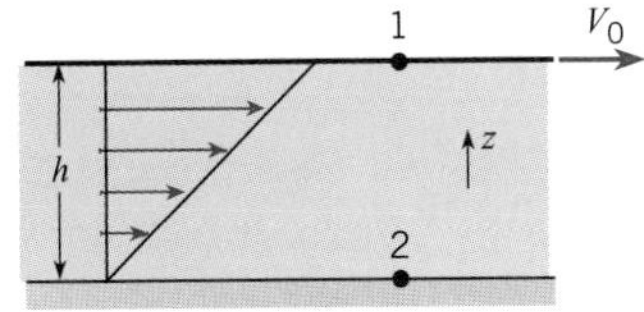

PROBLEM 4.108

4.109 Assume that the air flow in the outer portion ($r > 50$ mi) of a cyclonic storm is approximated by irrotational vortex flow. If you observed a tangential wind speed of 15 mph at a radial distance of 200 mi from the center of such a storm, what wind speeds would you predict at radial distances of 100 mi and 50 mi?

4.110 A tornado is modeled as a combined forced and free vortex. The core of the tornado (forced vortex region) has a diameter of 10 miles. The wind velocity at a distance of 50 miles from the center of the tornado is 20 mph. What is the wind velocity at the edge of the core? What is the centrifugal acceleration of the air at this location?

4.111 A whirlpool is modeled as the combination of a free and a forced vortex. The maximum velocity in the whirlpool is 10 m/s and the radius at the juncture of the free and the forced vortex is 10 m. The pressure is atmospheric (gage pressure = 0) at the free surface. Plot the shape of the water surface from the center to a radius of 50 m. The elevation is zero for the vortex center, where the velocity is zero.

4.112 The intensity of a tornado is measured by the Fujita Tornado Intensity Scale (F-Scale). An intense tornado with an F-scale reading of 4 has a maximum wind velocity of 350 km/hr (11). The tornado is modeled as a combination of a free and a forced vortex, and the radius of the forced vortex is 50 m. The atmospheric pressure is 100 kPa. Plot the variation of pressure with radius from the center. There is no elevation change.

4.113 Plot the pressure coefficient as a function of the radial position in a tornado. The pressure coefficient is defined as

$$C_p = \frac{p - p_0}{\frac{1}{2}\rho V_{max}^2}$$

where p_0 is the pressure outside of the tornado and V_{max} is the maximum velocity in the tornado. Plot the pressure coefficient as a function of r/r_c, where r_c is the radius of the forced vortex.

4.114 A weather balloon is caught in a tornado modeled as a combination free-forced vortex. Will it move toward the center or away from the center? Think carefully about pressure gradients and buoyancy. Provide a rationale for your answer.

4.115 The pressure distribution in a tornado is predicted using the Bernoulli equation, which is based on a constant density. However, the density will decrease as the pressure decreases in the tornado. Does the Bernoulli equation overpredict or underpredict the pressure drop in the tornado? Explain.

4.116 Euler's equations for a planar (two-dimensional) flow in the xy-plane are

$$u\frac{\partial u}{\partial x} + v\frac{\partial u}{\partial y} = -g\frac{\partial h}{\partial x} \qquad x \text{ direction}$$

$$u\frac{\partial u}{\partial x} + v\frac{\partial u}{\partial y} = -g\frac{\partial h}{\partial y} \qquad y \text{ direction}$$

a. The slope of a streamline is given by

$$\frac{dy}{dx} = \frac{v}{u}$$

Using this relation in Euler's equation, show that

$$d\left(\frac{u^2 + v^2}{2g} + h\right) = 0$$

or

$$d\left(\frac{V^2}{2g} + h\right) = 0$$

which means that $V^2/2g + h$ is constant along a streamline.

b. For an irrotational flow,

$$\frac{\partial u}{\partial y} = \frac{\partial v}{\partial x}$$

Substituting this equation into Euler's equation, show that

$$\frac{\partial}{\partial x}\left(\frac{V^2}{2g} + h\right) = 0$$

$$\frac{\partial}{\partial y}\left(\frac{V^2}{2g} + h\right) = 0$$

which means that $V^2/2g + h$ is constant in all directions.

4.117 Breathe in and out of your mouth. Try to sense the air flow patterns near your face while doing this. Discuss the type of flow associated with these flow processes. If you were to blow out a candle, you would do it while exhaling (at least most people do). Why is it easier to do this by exhaling than by inhaling?

4.118 Very high winds tend to lift roofs from buildings rather than forcing them downward. Explain why this should occur.

High-velocity water jets are often used for cutting solids and, in some cases, an abrasive is mixed with the water to enhance the cutting action. This abrasive water jet is being discharged from a vessel that is pressurized to 50,000 psi. The abrasive consists of particles of garnet and the jet is cutting through a 10-inch-thick block of titanium. (Courtesy Quest Integrated, Inc.)

CHAPTER 5

Control Volume Approach and Continuity Principle

In the previous chapter we learned about fluids in motion, the dynamics of a fluid element, and the Bernoulli equation, which relates pressure, elevation, and velocity along a streamline. To extend and expand these concepts, this chapter introduces an approach called the *control volume approach.* When engineers apply this approach, they focus on a volume in space and consider the flow that passes through the volume. The control volume approach derives from the Eulerian description of fluid motion.

In this chapter the control volume approach is applied to conservation of mass, resulting in the continuity principle. In Chapter 6, Newton's second law is formulated using the control volume approach, yielding the momentum principle. Then, in Chapter 7, the same approach is applied to conservation of energy, leading to the energy principle.

The control volume approach involves transforming the governing equations for a given mass, the Lagrangian form, into the corresponding equations for mass passing through a volume in space, the Eulerian form. The mathematical equation for carrying out this transformation is the *Reynolds transport theorem.* Understanding the Reynolds transport theorem is the basis for comprehending the control volume approach. Putting a sincere effort into learning about the control volume approach and its application to conservation of mass is time well spent.

5.1 Rate of Flow

Discharge

In the analysis of fluid systems, we often need to know the rate at which fluid is passing through a pipe or channel. The discharge, Q, often called the volume flow rate, is simply the volume of fluid that passes through an area per unit time. Typical units for discharge are ft^3/s (cfs), ft^3/min (cfm), gpm, m^3/s, and 1/s.

The discharge in a pipe is related to the flow velocity and cross-sectional area of the pipe. Consider the idealized flow of fluid in a pipe as shown in Fig. 5.1 in which the velocity is constant across the pipe section. Suppose a marker is injected at section A-A for a period of time Δt. The fluid that passes A-A in time Δt is represented by the marked volume. The length of the marked volume is $V\Delta t$ so the volume is $\Delta \forall = VA\Delta t$, where A is the cross-sectional area of the pipe. The rate of volume flow past A-A is $\Delta \forall / \Delta t = AV$, so

$$Q = AV \tag{5.1}$$

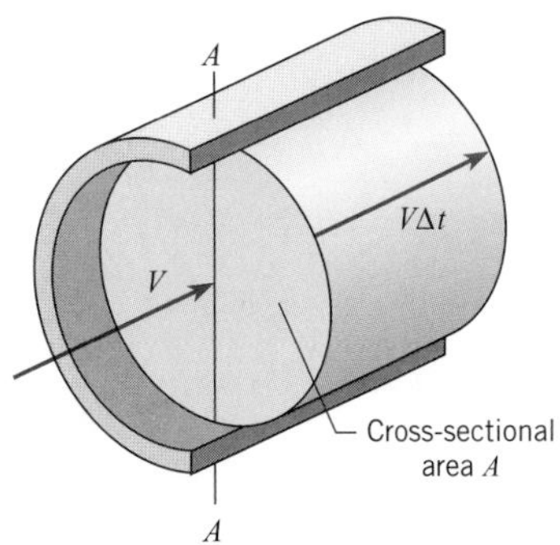

FIGURE 5.1

Volume of fluid in flow with uniform velocity distribution that passes section A-A in time Δt.

Mass Flow Rate

Another quantity frequently needed in the analysis of flow systems is the mass flow rate, $\dot{m}$, which is the mass of fluid passing a station per unit time. The usual units for mass flow rate are kg/s, lbm/s, and slugs/s. Using the same approach as for volume flow rate, the mass of the fluid in the marked volume in Fig. 5.1 is $\Delta m = rAV\Delta t$, where ρ is the density. The equation for mass flow rate is $\dot{m} = \Delta m/\Delta t$ or

$$\begin{aligned}\dot{m} &= \rho A V \\ &= \rho 0 Q\end{aligned} \tag{5.2}$$

Flow Rate with Variable Velocity Distribution

The expression $Q = AV$ is based on a uniformly distributed (constant) velocity across the pipe section. In general, the velocity varies across the pipe section as shown in Fig. 5.2. The discharge through a differential area of the section is $V\,dA$, and the total volume rate of flow Q is obtained by integration over the entire flow section:

$$Q = \int_A V\,dA \tag{5.3}$$

In a similar manner, the mass rate of flow past a section is given by

$$\dot{m} = \int_A \rho V\,dA \tag{5.4}$$

If density is constant across the flow section, the mass flow rate is given by

$$\dot{m} = \rho \int_A V\,dA = \rho Q \tag{5.5}$$

In the foregoing developments, the cross-sectional area was always oriented normal to the velocity vector. If other orientations are considered, such as in Fig. 5.3, where flow occurs past section A-A, it can be seen that only the normal component of velocity (the x component in this case) contributes to flow through the section. Therefore, to

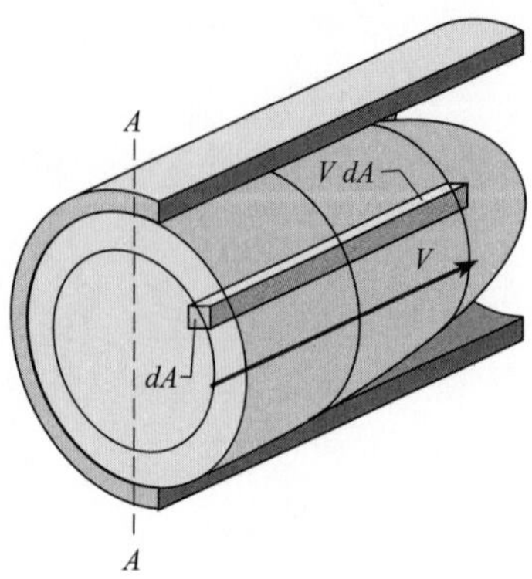

FIGURE 5.2

Volume of fluid that passes section A-A in time Δt.

evaluate flow rate, one must always consider either the area of a section normal to the total velocity or the velocity component normal to the given area. Thus the discharge for the case of Fig. 5.3 is given by

$$Q = \int_A u \, dA \quad \text{or} \quad Q = \int_A V \cos\theta \, dA$$

If we define an area vector as one that has the magnitude of the area in question and that is oriented normal to the area, then by definition, $V\cos\theta \, dA = V \cdot dA$, and the discharge can be written as

$$Q = \int_A \mathbf{V} \cdot d\mathbf{A} \tag{5.6}$$

and mass flow rate becomes

$$\dot{m} = \int_A \rho\mathbf{V} \cdot d\mathbf{A} \tag{5.7}$$

If the velocity is constant over the area, then the discharge is given as

$$Q = \mathbf{V} \cdot \mathbf{A} \tag{5.8}$$

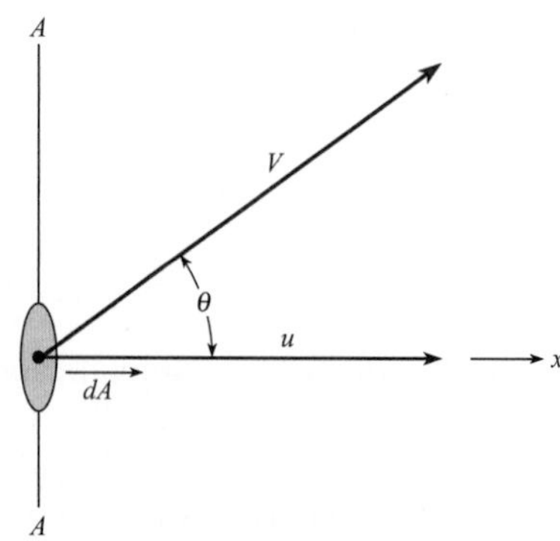

FIGURE 5.3

Velocity not normal to the section.

The scalar, or dot, product means that the velocity component normal to the area is used to obtain the discharge.

Mean Velocity

In many problems—for example, those involving flow in pipes—one may be given the discharge and need to find the mean (average) velocity without knowing the actual velocity distribution across the pipe. By definition, the mean velocity is the discharge divided by the cross-sectional area:

$$\overline{V} = \frac{Q}{A}$$

For laminar flows in circular pipes, the velocity profile is parabolic like the case shown in Fig. 5.2. In this case, the mean velocity is half the centerline velocity. However, for turbulent pipe flow as shown in Fig. 4.6*b*, the velocity profile is nearly flat, so the mean velocity is fairly close to the velocity at the pipe center. It is customary to leave the bar off the velocity symbol and simply indicate the mean velocity with *V*.

example 5.1

Air that has a mass density of 1.24 kg/m^3 (0.00241slugs/ft^3) flows in a pipe with a diameter of 30 cm (0.984 ft) at a mass rate of flow of 3 kg/s (0.206 slugs/s). What are the mean velocity of flow in this pipe and the volume rate of flow?

Solution

$$\dot{m} = \rho Q \qquad \text{or} \qquad Q = \frac{\dot{m}}{\rho} = 2.42 \text{ m}^3/\text{s (85.5 cfs)}$$

$$V = \frac{Q}{A} = \frac{2.42}{(\frac{1}{4}\pi) \times (0.30)^2} = 34.2 \text{ m/s (112 ft/s)} \qquad \triangleleft$$

example 5.2

Water flows in a channel that has a slope of 30°. If the velocity is assumed to be constant, 12 m/s, and if a depth of 60 cm measured along a vertical line, what is the discharge per meter of width of the channel?

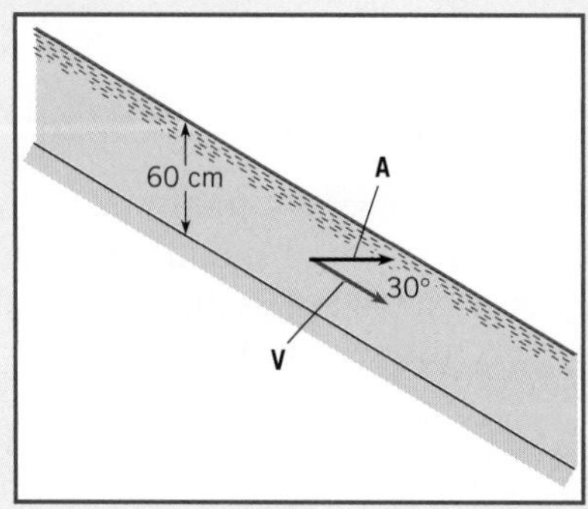

Solution The discharge in 1 meter of width is

$$\begin{aligned} Q &= V\cos 30° \times A \\ &= 12 \text{ m/s} \times \cos 30° \times (0.6 \text{ m} \times 1 \text{ m}) \\ &= 62.4 \text{ m}^3/\text{s} \end{aligned}$$

◁

The discharge per unit width is usually designated as q and is obtained by dividing through by the width, w, which in this case is 1 meter, so

$$q = \frac{Q}{w} = 62.4 \text{ m}^2/\text{s}$$

◁

example 5.3

The water velocity in the channel shown in the accompanying figure has a distribution across the vertical section equal to $u/u_{max} = (y/d)^{1/2}$. What is the discharge in the channel if the channel is 2 m deep ($d = 2$ m) and 5 m wide and the maximum velocity is 3 m/s?

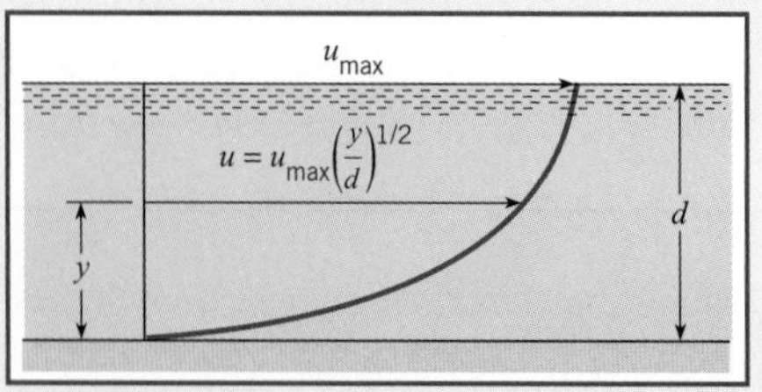

Solution The discharge is given by

$$Q = \int_0^d u \, dA$$

The channel is 5 m wide, so the differential area is 5 *dy*. Thus

$$\begin{aligned} Q &= \int_0^2 u_{\max}(y/d)^{1/2} 5 \, dy \\ &= \frac{5u_{\max}}{d^{1/2}} \int_0^2 y^{1/2} \, dy \\ &= \frac{5u_{\max}}{d^{1/2}} \frac{2}{3} y^{3/2} \Big|_0^2 \\ &= \frac{5 \times 3}{2^{1/2}} \times \frac{2}{3} \times 2^{3/2} = 20 \text{ m}^3/\text{s} \end{aligned}$$

◁

5.2 Control Volume Approach

The control volume approach is the method whereby a volume in the flow field is identified and the governing equations are solved for the flow properties associated with this volume. This is the Eulerian concept, where the control volume represents the "window" into the flow field. This is a very useful and powerful approach and is the basis for the vast majority of fluid flow analyses. In this section we will learn about the Reynolds transport theorem, which enables us to express the equations for a fluid mass in the form appropriate for a control volume or, in other words, for the control volume approach.

System, Control Volume, and Control Surface

The definitions of a system, control volume, and control surface are important to the development of the control volume approach. A fluid system is defined as a continuous mass of fluid that always contains the same fluid particles. By definition, the mass of a system is constant.

A control volume (cv) is defined as a volume in space, and the control surface (cs) is the surface enclosing the control volume. The control volume can deform with time as well as move and rotate in space. Mass can cross the control surface flowing into or out of the control volume. The mass in the control volume can be continuously changing with time.

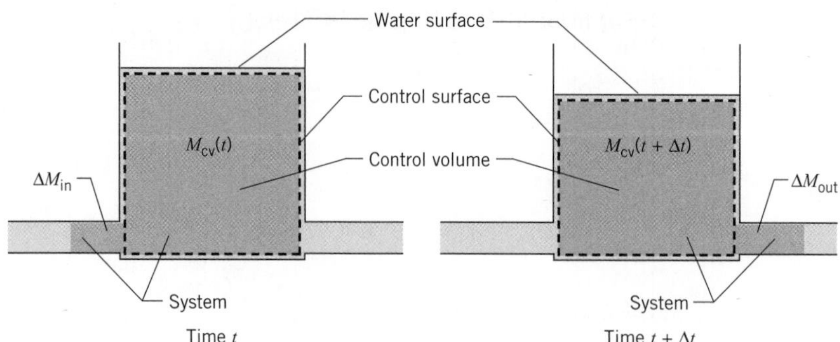

FIGURE 5.4

System, control surface, and control volume for liquid flowing through a tank.

To gain an appreciation of the relationship between system, control surface, and control volume, consider the tank of liquid shown in Fig. 5.4. There is flow into the tank from the left and flow out of the tank to the right. The level of the liquid in the tank may also be changing. The control volume is defined by the tank walls and the top of the liquid in the tank. The control surface encloses the control volume and is designated by the dashed line in the figure. There is a mass flow across the control surface as liquid enters and leaves the tank. The system selected for this illustration is the mass of fluid corresponding to the shaded area. At time t part of the system is in the tank $[M_{cv}(t)]$ and the other part $[\Delta M_{in}]$ is about to enter the tank. At this time, the mass of the system is

$$M_{sys}(t) = M_{cv}(t) + \Delta M_{in}$$

At time $t + \Delta t$, the portion ΔM_{in} has completely entered the tank and part of the system, ΔM_{out}, has left the tank. At the same time the water level in the tank has changed. Now the mass of the system is

$$M_{sys}(t + \Delta t) = M_{cv}(t + \Delta t) + \Delta M_{out}$$

By definition, the mass of the system is constant, so

$$M_{cv}(t) + \Delta M_{in} = M_{cv}(t + \Delta t) + \Delta M_{out}$$

or

$$\Delta M_{cv} = \Delta M_{in} - \Delta M_{out} \tag{5.9}$$

This equation appeals to our common sense: The change in mass in the control volume is equal to that which came in minus that which went out. Dividing Eq. (5.9) by Δt and taking the limit as $\Delta t \longrightarrow 0$ yields the rate form of the continuity principle:

$$\frac{dM_{cv}}{dt} = \dot{m}_{in} - \dot{m}_{out} \tag{5.10}$$

where $\dot{m}_{in}$ and $\dot{m}_{out}$ are the mass flow rate into and mass flow rate out of the control volume, respectively. The Reynolds transport theorem provides a more general and formal approach to deriving the continuity principle.

example 5.4

A tank with a cross-sectional area of 10 m^2 has an inflow of 7 kg/s and an outflow of 5 kg/s. Find the rate at which the water level in the tank is changing.

Solution Take the tank walls and the water level as the control surface as shown in Fig. 5.4. The volume of the control volume is

$$\forall = Ah$$

where A is the cross-sectional area of the tank and h is the elevation of the water surface. The mass in the control volume is

$$M_{\text{cv}} = \rho\forall = \rho Ah$$

The rate of change of mass in the control volume is

$$\frac{dM_{\text{cv}}}{dt} = \rho A \frac{dh}{dt}$$

By the continuity equation the rate of change of water elevation in the tank is

$$\frac{dh}{dt} = \frac{\dot{m}_{\text{in}} - \dot{m}_{\text{out}}}{\rho A}$$

$$= \frac{7 \text{ kg/s} - 5 \text{ kg/s}}{1000 \text{ kg/m}^3 \times 10 \text{ m}^2} = 0.0002 \text{ m/s or } 0.72 \text{ m/hr} \quad \triangleleft$$

Intensive and Extensive Properties

The Reynolds transport theorem involves the change of fluid properties inside the control volume and the flow of properties across the control surface. This leads to the need to distinguish between intensive and extensive fluid properties.

The extensive properties of a system are proportional to the mass of the system and include mass, M, momentum, $M\mathbf{v}$, and energy, E, where $\mathbf{v}$ is the velocity of the system with respect to inertial space. The intensive properties are independent of system mass and are obtained by dividing the extensive properties by the system mass, so the intensive property for mass is simply unity, for momentum $\mathbf{v}$, and energy e, the energy per unit mass. Because we will be developing a general equation for any property, we define a generic extensive property as B and the corresponding intensive property as b. The amount of extensive property B contained in a control volume at a given instant is

$$B_{\text{cv}} = \int_{\text{cv}} b dm = \int_{\text{cv}} b\rho \, d\forall \tag{5.11}$$

where dm and $d\forall$ are the differential mass and differential volume, respectively, and the integral is taken over the control volume.

FIGURE 5.5

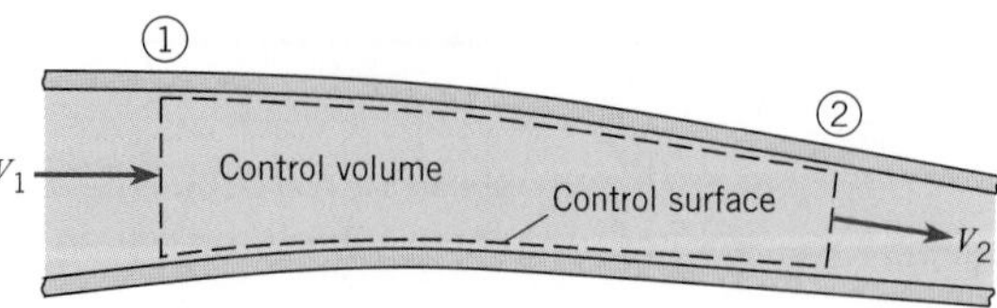

FIGURE 5.6

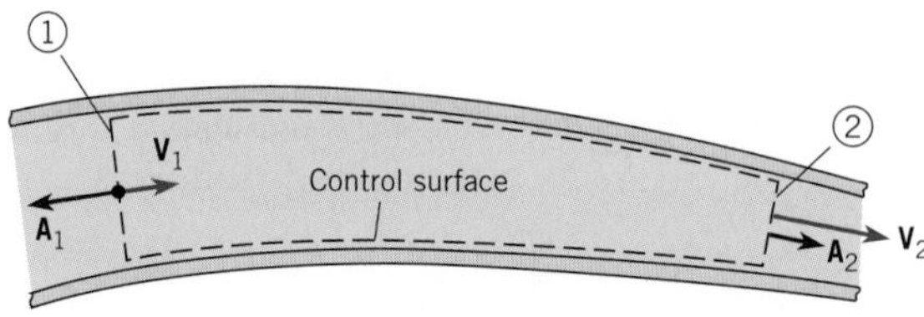

Property Transport across the Control Surface

When fluid flows across a control surface, properties such as mass, momentum, and energy are transported with the fluid either into or out of the control volume. Consider the flow through the duct in Fig. 5.5. If the velocity is uniformly distributed across a surface, the mass flow rate is given by

$$\dot{m} = \rho \mathbf{V} \cdot \mathbf{A}$$

where $\mathbf{V}$ is the velocity of the flow with respect to the surface and $\mathbf{A}$ is the area vector with magnitude equal to the area and in the direction normal to the surface. It is obvious that all surfaces have two sides, so a rule must be established as to which side the area is normal to. The sign convention we use is that the area vector always points outward from the control volume. Thus in Fig. 5.6 we show the flow in the passage with the proper orientation of the area vectors. The dot product of $\mathbf{V}_1$ and $\mathbf{A}_1$ is

$$\mathbf{V}_1 \cdot \mathbf{A}_1 = -V_1 A_1$$

because the cosine of the angle between the vectors (180°) is –1. However, the dot product of $\mathbf{V}_2$ and $\mathbf{A}_2$ is

$$\mathbf{V}_2 \cdot \mathbf{A}_2 = V_2 A_2$$

because the vectors are collinear. The net mass flow rate out of the control volume is

$$\begin{aligned} \text{net mass flow rate out} &= \rho_2 V_2 A_2 - \rho_1 V_1 A_1 \\ &= \rho_2 \mathbf{V}_2 \cdot \mathbf{A}_2 + \rho_1 \mathbf{V}_1 \cdot \mathbf{A}_1 \\ &= \sum_{\text{cs}} \rho \mathbf{V} \cdot \mathbf{A} \end{aligned} \tag{5.12}$$

Equation (5.12) states that if we sum the dot product $\rho \mathbf{V} \cdot \mathbf{A}$ for all flows into and out of the control volume, we get the net mass flow rate of the control volume, known as the net mass "efflux." If the summation is positive, the net mass flow rate is out of the control volume. If it is negative, the net mass flow rate is into the control volume. If the inflows and outflows are equal, then $\sum_{\text{cs}} \rho \mathbf{V} \cdot \mathbf{A}$ will be zero.

In a similar manner, if we want the net rate of flow of an extensive property B out of the control volume, we multiply the mass rate by the intensive property b:

$$\dot{B}_{\text{net}} = \sum_{\text{cs}} b\overbrace{\rho \mathbf{V} \cdot \mathbf{A}}^{\dot{m}} \tag{5.13}$$

To reinforce the validity of Eq. (5.13) one may consider the dimensions involved. Equation (5.13) states that the flow rate of B is given by

$$\underbrace{\left(\frac{B}{\text{mass}}\right)}_{b}\underbrace{\left(\frac{\text{mass}}{\text{time}}\right)}_{\dot{m}} = \left(\frac{B}{\text{time}}\right) = \dot{B}$$

Equation (5.13) is applicable for all one-dimensional flows. If the velocity varies across a flow section, then it becomes necessary to integrate the velocity across the section to obtain the rate of flow. A more general expression for the net rate of flow of the extensive property from the control volume is thus

$$\dot{B}_{\text{net}} = \int_{\text{cs}} b\rho \mathbf{V} \cdot d\mathbf{A} \tag{5.14}$$

Equation (5.14) will be used in the most general form of the Reynolds transport theorem.

Reynolds Transport Theorem

The Reynolds transport theorem is fundamental to the control volume approach. It is derived by considering the rate of change of an extensive property of a system of fluid as the system passes through a control volume.

A control volume with a system moving through it is shown in Fig. 5.7. The control volume is enclosed by the control surface identified by the dashed line. The system is identified by the shaded region. At time t the system consists of the material inside the control volume and the material going in, so the property B of the system at this time is

$$B_{\text{sys}}(t) = B_{\text{cv}}(t) + \Delta B_{\text{in}} \tag{5.15}$$

At time $t + \Delta t$ the system has moved and now consists of the material in the control volume and the material passing out, so B of the system is

$$B_{\text{sys}}(t + \Delta t) = B_{\text{cv}}(t + \Delta t) + \Delta B_{\text{out}} \tag{5.16}$$

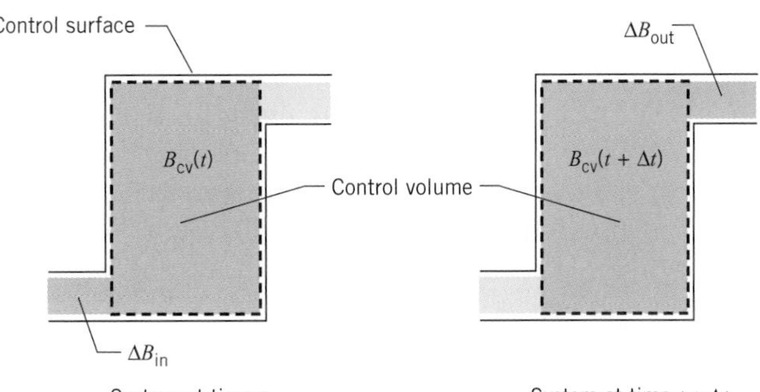

FIGURE 5.7

Progression of a system through a control volume.

The rate of change of the property B is

$$\frac{dB_{\text{sys}}}{dt} = \lim_{\Delta t \to 0}\left[\frac{B_{\text{sys}}(t+\Delta t) - B_{\text{sys}}(t)}{\Delta t}\right] \tag{5.17}$$

Substituting in Eqs. (5.15) and (5.16) results in

$$\frac{dB_{\text{sys}}}{dt} = \lim_{\Delta t \to 0}\left[\frac{B_{\text{cv}}(t+\Delta t) - B_{\text{cv}}(t) + \Delta B_{\text{out}} - \Delta B_{\text{in}}}{\Delta t}\right] \tag{5.18}$$

Rearranging terms yields

$$\frac{dB_{\text{sys}}}{dt} = \lim_{\Delta t \to 0}\left[\frac{B_{\text{cv}}(t+\Delta t) - B_{\text{cv}}(t)}{\Delta t}\right] + \lim_{\Delta t \to 0}\frac{\Delta B_{\text{out}}}{\Delta t} - \lim_{\Delta t \to 0}\frac{\Delta B_{\text{in}}}{\Delta t} \tag{5.19}$$

The first term on the right side of Eq. (5.19) is the rate of change of the property B inside the control volume, or

$$\lim_{\Delta t \to 0}\left[\frac{B_{\text{cv}}(t+\Delta t) - B_{\text{cv}}(t)}{\Delta t}\right] = \frac{dB_{\text{cv}}}{dt} \tag{5.20}$$

The remaining terms are

$$\lim_{\Delta t \to 0}\frac{\Delta B_{\text{out}}}{\Delta t} = \dot{B}_{\text{out}} \qquad \text{and} \qquad \lim_{\Delta t \to 0}\frac{\Delta B_{\text{in}}}{\Delta t} = \dot{B}_{\text{in}}$$

These two terms can be combined to give

$$\dot{B}_{\text{net}} = \dot{B}_{\text{out}} - \dot{B}_{\text{in}} \tag{5.21}$$

or the net efflux of the property B through the control surface. Equation (5.19) can now be written as

$$\frac{dB_{\text{sys}}}{dt} = \frac{d}{dt}B_{\text{cv}} + \dot{B}_{\text{net}}$$

Substituting in Eq. (5.14) for $\dot{B}_{\text{net}}$ and Eq. (5.11) for B_{cv} results in the most general form of the *Reynolds transport theorem*:

$$\underbrace{\frac{dB_{\text{sys}}}{dt}}_{\text{Lagrangian}} = \underbrace{\frac{d}{dt}\int_{\text{cv}} b\rho\, d\forall + \int_{\text{cs}} b\rho\mathbf{V}\cdot d\mathbf{A}}_{\text{Eulerian}} \tag{5.22}$$

The left side is the Lagrangian form and represents the rate of change of property B of the system. The right side is the change of property B in the control volume and the net efflux of B through the control surface and corresponds to the Eulerian form. This equation, often called the control volume equation, is the basis of the control volume approach. The velocity $\mathbf{V}$ is always the velocity of the fluid with respect to the control surface.

A simplified form of the Reynolds transport theorem can be written if the mass crossing the control surface occurs through a number of inlet and outlet ports, and the velocity, density, and intensive property b are uniformly distributed (constant) across each port. Then

$$\frac{dB_{\text{sys}}}{dt} = \frac{d}{dt}\int_{\text{cv}} \rho b\, d\forall + \sum_{\text{cs}} b\rho \mathbf{V} \cdot \mathbf{A} \tag{5.23}$$

where the summation is carried out for each port crossing the control surface.

An alternative, simplified form can be written in terms of mass flow rates:

$$\frac{dB_{\text{sys}}}{dt} = \frac{d}{dt}\int_{\text{cv}} \rho b\, d\forall + \sum_{\text{cs}} \dot{m}_o b_o - \sum_{\text{cs}} \dot{m}_i b_i \tag{5.24}$$

where the subscripts i and o refer to the inlet and outlet ports, respectively, located on the control surface. This form of the equation does not require that the velocity and density be uniformly distributed across each inlet and outlet, but the property b must be.

Selection of the Control Volume

The selection of the control volume is dependent on the flow problem to be solved. After some experience, the engineer will have a good idea how to select the control volume to address the problem at hand and simplify the solution. Obviously, the control surface has to be positioned where velocity and area information is available. Sometimes it may be possible to select a control volume to simplify the analysis. For example, if a control volume can be selected such that the dB_{cv}/dt term in the Reynolds transport theorem is zero, the solution of the equations may be easier. If the flow is steady—that is, properties at a point do not change with time—then dB_{cv}/dt is zero provided the control volume is stationary and its size does not change with time.

It is interesting to see how the motion of the control volume itself can be chosen to make the term dB_{cv}/dt zero. Consider the case at which a ship is traveling at constant speed in shallow water as shown in Fig. 5.8. If the control surface is fixed to the sea bottom, then the condition inside the control volume will continuously change with time, so dB_{cv}/dt will not be zero. However, if the control volume moves at the same speed as the ship, then none of the conditions will change with time, so dB_{cv}/dt will be zero, likely simplifying the analysis.

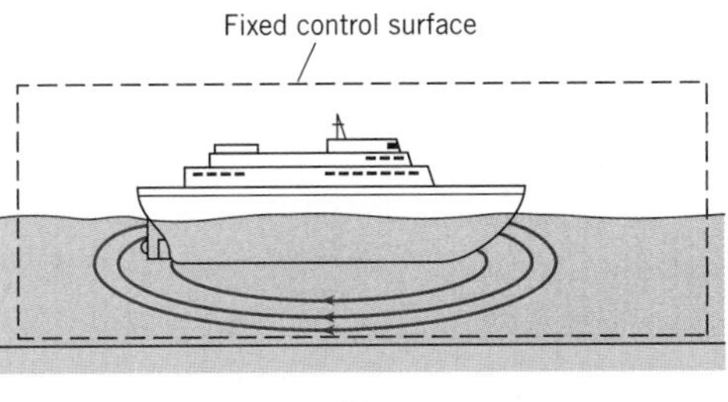

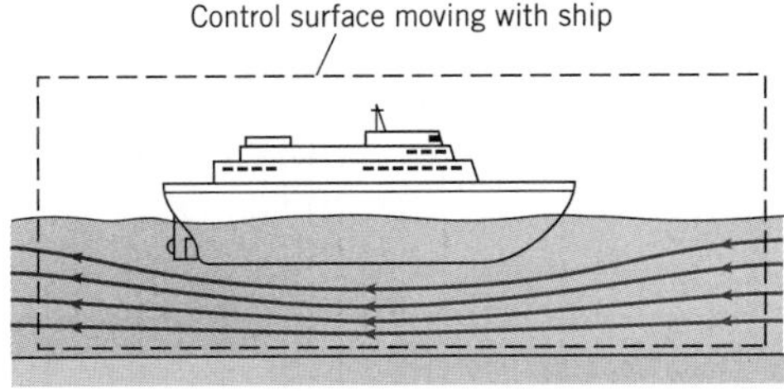

FIGURE 5.8

Change from unsteady to steady flow by change of the velocity of the control volume. (a) Unsteady flow. (b) Steady flow.

Continuity Equation

The continuity equation derives from the conservation of mass, which, in Lagrangian form, simply states that the mass of the system is constant.

$$M_{\text{sys}} = \text{const}$$

The Eulerian form is derived by applying the Reynolds transport theorem. In this case the extensive property of the system is its mass, $B_{\text{cv}} = M_{\text{sys}}$. The corresponding intensive property is the mass per unit mass, or simply unity.

$$b = \frac{M_{\text{sys}}}{M_{\text{sys}}} = 1$$

General Form of the Continuity Equation

The general form of the continuity equation is obtained by substituting the properties for mass into the Reynolds transport theorem, resulting in

$$\frac{dM_{\text{sys}}}{dt} = \frac{d}{dt}\int_{\text{cv}} \rho \, d\forall + \int_{\text{cs}} \rho \mathbf{V} \cdot d\mathbf{A}$$

However, $dM_{\text{sys}}/dt = 0$, so the general, or integral, form of the *continuity equation* is

$$\frac{d}{dt}\int_{\text{cv}} \rho \, d\forall + \int_{\text{cs}} \rho \mathbf{V} \cdot d\mathbf{A} = 0 \tag{5.25}$$

This equation states that the rate of accumulation of mass in the control volume plus the net mass efflux through the control surface is zero. The net efflux is the mass flow rate out of the control volume minus the mass flow rate in. Another way of expressing the same idea is that the increase in mass in a control volume is equal to mass entering the control volume minus the mass leaving.

If the mass crosses the control surface through a number of inlet and exit ports, the continuity equation simplifies to

$$\frac{d}{dt}M_{\text{cv}} + \sum_{\text{cs}} \dot{m}_o - \sum_{\text{cs}} \dot{m}_i = 0 \tag{5.26}$$

which is obtained by setting $b = 1$ and $B_{\text{sys}} = \text{const}$ in Eq. (5.24). M_{cv} is the mass of fluid in the control volume. Note that if there is only one inlet and outlet port, this equation reduces to

$$\frac{d}{dt}M_{\text{cv}} = \dot{m}_i - \dot{m}_o$$

which was obtained earlier, as Eq. (5.10).

example 5.5

A jet of water discharges into an open tank, and water leaves the tank through an orifice in the bottom at a rate of 0.003 m^3/s. If the cross-sectional area of the jet is 0.0025 m^2 where the velocity of water is 7 m/s, at what rate is water accumulating in (or evacuating from) the tank?

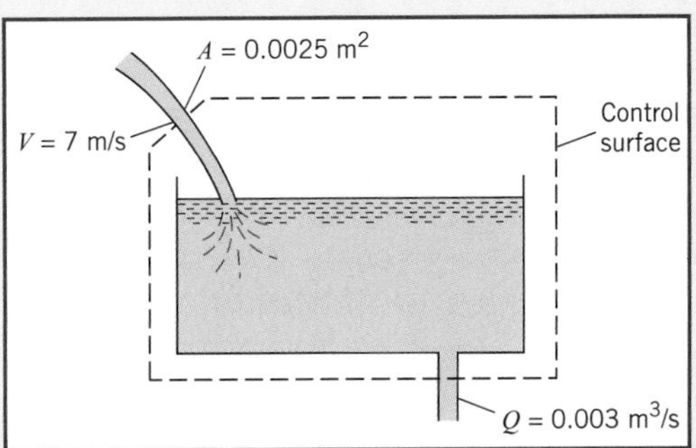

Solution By drawing a control surface to enclose the entire tank, we observe that there are two streams crossing the control surface, one entering and one leaving. Applying the continuity equation, we have

$$\frac{dM_{\text{cv}}}{dt} = \dot{m}_i - \dot{m}_o$$

The mass flow rate out is

$$\begin{aligned}\dot{m}_o = \rho Q &= 1000 \text{ kg/m}^3 \times 0.003 \text{ m}^3/\text{s} \\ &= 3 \text{ kg/s}\end{aligned}$$

The mass flow rate in is

$$\begin{aligned}\dot{m}_i &= \rho VA \\ &= 1000 \text{ kg/m}^3 \times 7 \text{ m/s} \times 0.0025 \text{ m}^2 \\ &= 17.5 \text{ kg/s}\end{aligned}$$

From the continuity equation,

$$\begin{aligned}\frac{dM_{\text{cv}}}{dt} &= 17.5 - 3 \\ &= 14.5 \text{ kg/s}\end{aligned}$$ ◁

so the mass is accumulating in the tank at the rate of 14.5 kg/s.

example 5.6

The river discharges into the reservoir shown at a rate of 400,000 ft^3/s (cfs), and the outflow rate from the reservoir through the flow passages in the dam is 250,000 cfs. If the reservoir surface area is 40 mi^2, what is the rate of rise of water in the reservoir?

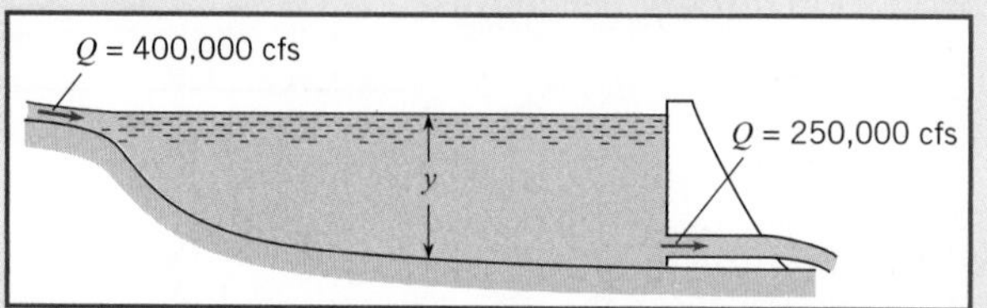

Solution We first choose a control volume as shown in the figure.

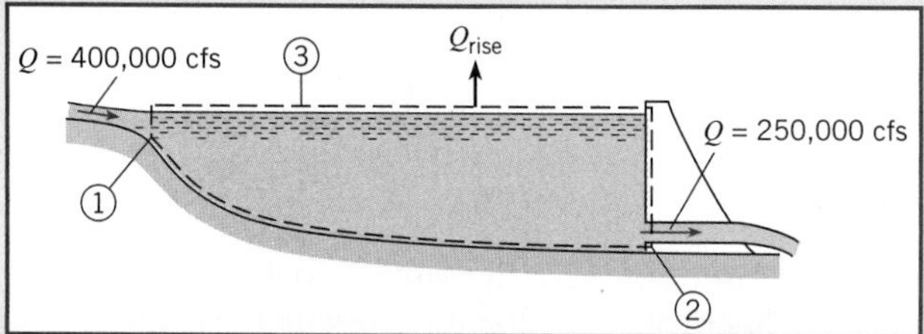

The control surface selected at section 3 is stationary. As the reservoir rises, there is a flow through the control surface with a velocity equal to the rise rate of the reservoir. The mass within the control volume does not change, so the continuity equation reduces to

$$\sum_{cs} \dot{m}_o - \sum_{cs} \dot{m}_i = 0$$

and because ρ is constant,

$$\sum_{cs} Q_o - \sum_{cs} Q_i = 0$$

Applied to this example, we have

$$Q_3 + Q_2 - Q_1 = 0$$

$$Q_{rise} + 250{,}000 \text{ ft}^3/\text{s} - 400{,}000 \text{ ft}^3/\text{s} = 0$$

$$Q_{rise} = 150{,}000 \text{ ft}^3/\text{s}$$

Q_{rise} is related to V_{rise} by $V_{rise} A_R$, where A_R is the area of the reservoir. Then we have

$$V_{rise} = \frac{150{,}000 \text{ ft}^3/\text{s}}{40 \text{ mi}^2 \times (5280)^2 \text{ft}^2/\text{mi}^2}$$

$$\text{Rate of rise} = 1.34 \times 10^{-4} \text{ft/s}$$

or

$$V_{rise} = 0.484 \text{ ft/hr}$$

◁

example 5.7

A 10-cm jet of water issues from a 1-m diameter tank, as shown. Assume that the velocity in the jet is $\sqrt{2gh}$ m/s. How long will it take for the water surface in the tank to drop from $h_0 = 2$ m to $h_f = 0.50$ m?

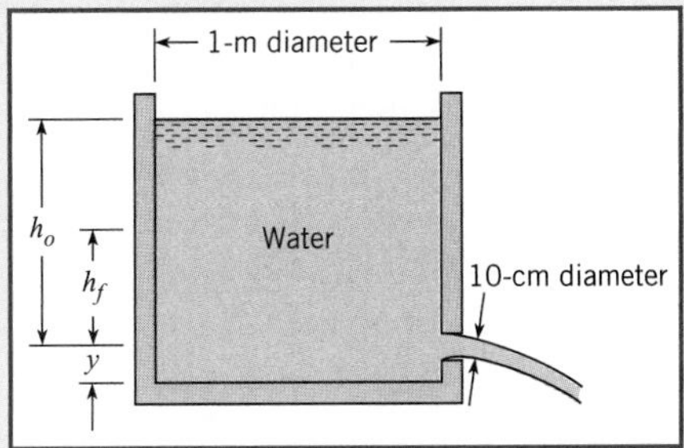

Solution In Example 5.6 we arbitrarily established a fixed control volume. In this example, however, we will let the control surface surround the volume of liquid within the tank at all times, so that the control volume will decrease in size as time passes. In other words, the control surface coincident with the liquid surface in the tank will drop right along with the liquid surface. Water is discharged from the control volume at section 1. We write the continuity equation as

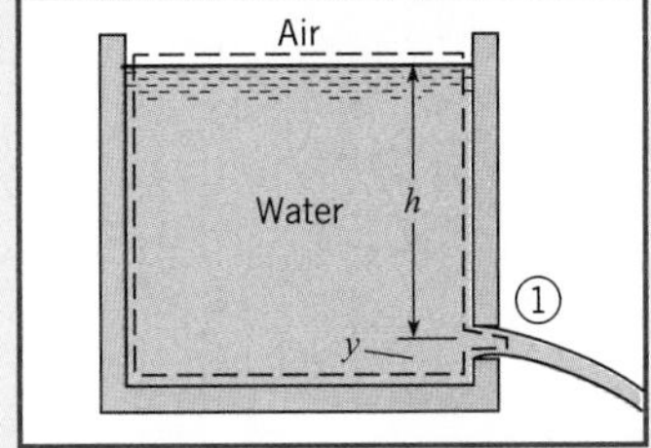

$$\sum_{\text{cs}} \dot{m}_i - \sum_{\text{cs}} \dot{m}_o = \frac{dM_{\text{cv}}}{dt}$$

$$-\rho_1 V_1 A_1 = \frac{dM_{\text{cv}}}{dt}$$

where M_{cv} represents the total mass of water in the control volume, so we write it as $\rho A_T(h + y)$. The continuity equation then becomes

$$\rho_1 V_1 A_1 = -\frac{d}{dt}[\rho A_T(h + y)]$$

where $\rho_1 = \rho = \rho_{water}$, A_T is the cross-sectional area of tank, h is the depth of water in tank above outlet, and y is the depth of water below the outlet. Since ρ in the foregoing equation is constant, we cancel it out, and when the differentiation is carried out on the right-hand side (remember that A_T and y are constant), the equation reduces to

$$V_1 A_1 = -A_T \frac{dh}{dt}$$

It was already noted that $V_1 = \sqrt{2gh}$, so substitution for V_1 yields

$$\sqrt{2gh} A_1 = -A_T \frac{dh}{dt}$$

Separating variables, we obtain

$$dt = \frac{-A_T}{\sqrt{2g}A_1} \frac{dh}{\sqrt{h}} \quad \text{or} \quad dt = \frac{-A_T}{\sqrt{2g}A_1} h^{-1/2} dh$$

Noting now that $A_T / \sqrt{2g}A_1$ is constant, we integrate the differential equation and get

$$t = \frac{-2A_T}{\sqrt{2g}A_1} h^{1/2} + C$$

The constant of integration is evaluated by arbitrarily letting $t = 0$ when $h = h_0$. Then

$$C = +\frac{2A_T}{\sqrt{2g}A_1} h_0^{1/2}$$

So we have

$$t = \frac{2A_T}{\sqrt{2g}A_1} (h_0^{1/2} - h^{1/2})$$

Thus, for this particular example, the elapsed time for the water level to drop from $h_0 = 2$ m to $h_f = 0.50$ m will be

$$t = \frac{2A_T}{\sqrt{2g}A_1} (2^{1/2} - 0.5^{1/2})$$

But
$$A_T = \frac{\pi}{4} D^2 = \frac{\pi}{4} \times 1^2 = \frac{\pi}{4} \text{ m}^2$$

$$A_1 = \frac{\pi}{4}(0.10)^2 = 0.01\left(\frac{\pi}{4}\right) \text{ m}^2$$

Hence
$$t = \frac{2\pi/4}{\sqrt{2g}(\pi/4 \times 0.01)}(1.414 - 0.707) = 31.9 \text{ s}$$ ◁

example 5.8

Methane escapes through a small (10^{-7} m^2) hole in a 10-m^3 tank. The methane escapes so slowly that the temperature in the tank remains constant at 23°C. The mass flow rate of methane through the hole is given by $\dot{m} = 0.66pA/\sqrt{RT}$, where p is the pressure in the tank, A is the area of the hole, R is the gas constant, and T is the temperature in the tank. Calculate the time required for the absolute pressure in the tank to decrease from 500 to 400 kPa.

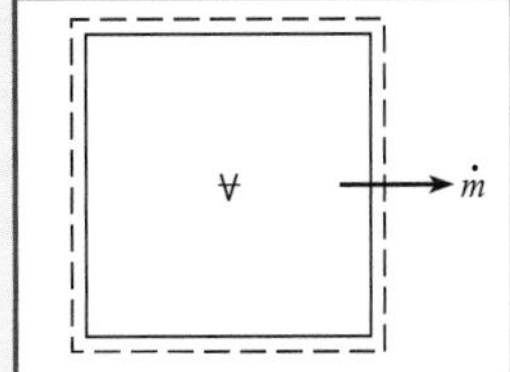

Solution Consider a control surface that just encloses the tank. Writing the continuity equation for the control volume, we have

$$\frac{d}{dt}\int_{cv} \rho \, d\mathcal{V} + \dot{m}_o - \dot{m}_i = 0$$

Since the density is uniform throughout the tank and mass flows outward from the tank at the location of the leak only,

$$\frac{d}{dt}(\rho \mathcal{V}) + \dot{m} = 0$$

The tank volume is constant, so

$$d\rho/dt = -\dot{m}/\mathcal{V}$$

The ideal-gas law provides the following relationship between pressure and density: $\rho = p/RT$. Since T is constant, $d\rho = dp/RT$. Substituting this and the mass-flow expression into the equation above gives

$$\frac{dp}{p} = -0.66\frac{A\sqrt{RT}dt}{\mathcal{V}}$$

Integrating, we obtain

$$t = \frac{1.52\mathcal{V}}{A\sqrt{RT}}\ln\frac{p_0}{p}$$

where p_0 is the initial pressure. Substituting in the appropriate values, we calculate

$$t = \frac{1.52(10 \text{ m}^3)}{(10^{-7} \text{ m}^2)\left(518\frac{\text{J}}{\text{kg}\cdot\text{K}}300 \text{ K}\right)^{1/2}}\ln\frac{500}{400} = 8.6\times 10^4 \text{ s}$$

or approximately 1 day. ◁

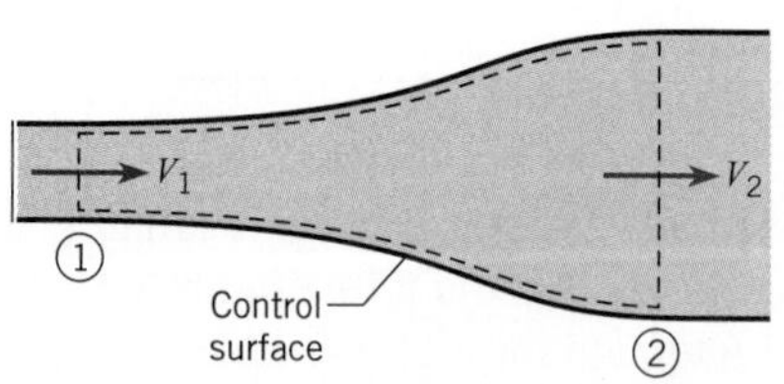

FIGURE 5.9

Flow through a pipe section.

Continuity Equation for Flow in a Pipe

The continuity equation for flow in a pipe is developed by positioning a control volume inside a pipe, as shown in Fig. 5.9. Mass enters through station 1 and exits through station 2. The control volume is fixed to the pipe walls and its volume is constant. If the flow is steady, them M_{cv} is constant so the mass flow formulation of the continuity equation becomes

$$\dot{m}_2 = \dot{m}_1$$

For flow with a uniform velocity and density distribution, the continuity equation for steady flow in a pipe is

$$\rho_2 A_2 V_2 = \rho_1 A_1 V_1 \tag{5.27}$$

If the flow is incompressible, then

$$A_2 V_2 = A_1 V_1 \tag{5.28}$$

or, equivalently,

$$Q_2 = Q_1$$

This equation is valid for both steady and unsteady incompressible flow.

Equations (5.27) and (5.28) are very common forms of the continuity equation and are used in numerous applications. If the flow is not uniformly distributed, the mass flow must be calculated using Eq. (5.4).

If there are more than two ports, then the general form of the continuity equation for steady flow is

$$\sum_{cs} \dot{m}_i = \sum_{cs} \dot{m}_o \tag{5.29}$$

If the flow is incompressible, Eq. (5.29) can be written in terms of discharge:

$$\sum_{cs} Q_i = \sum_{cs} Q_o \tag{5.30}$$

example 5.9

A 120-cm pipe is in series with a 60-cm pipe. The rate of flow of water in the system of pipes is 2 m^3/s. What is the velocity of flow in each pipe?

Solution

$$Q = V_{120}A_{120}$$

Here $\quad Q = 2\ \text{m}^3/\text{s} \quad$ and $\quad A_{120} = \frac{\pi}{4} \times (1.20)^2 = 1.13\ \text{m}^2$

Therefore $\quad V_{120} = \frac{Q}{A_{120}} = \frac{2\ \text{m}^3/\text{s}}{1.13\ \text{m}^2} = 1.77\ \text{m/s} \quad \triangleleft$

Also $\quad V_{120}A_{120} = V_{60}A_{60}$

So $\quad V_{60} = V_{120}\left(\frac{A_{120}}{A_{60}}\right) = V_{120}\left(\frac{120}{60}\right)^2 = 7.08\ \text{m/s} \quad \triangleleft$

example 5.10

As shown in the accompanying figure, water flows steadily into a tank through pipes 1 and 2 and discharges at a steady rate out of the tank through pipes 3 and 4. The mean velocity of inflow and outflow in pipes 1, 2, and 3 is 50 ft/s, and the hypothetical outflow velocity in pipe 4 varies linearly from zero at the wall to a maximum at the center of the pipe. What are the mass rate of flow and the discharge from pipe 4, and what is the maximum velocity in pipe 4?

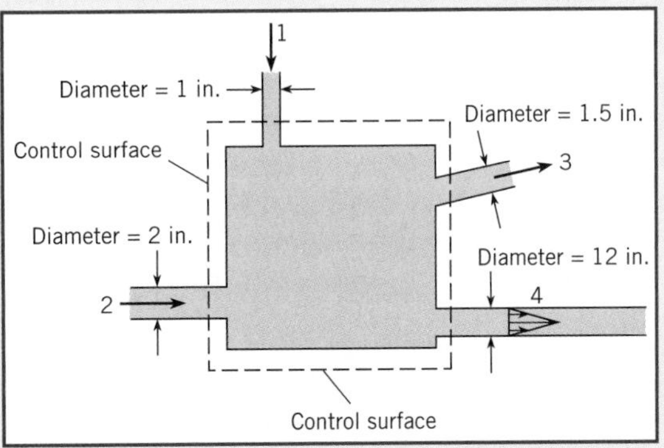

Solution The control volume chosen for this problem is shown in the figure. There are two inlet ports and two outlet ports. The mass in the control volume is constant and the flow is incompressible (water), so the continuity equation simplifies to

$$\sum_{\text{cs}} Q_o = \sum_{\text{cs}} Q_i$$

or

$$Q_3 + Q_4 = Q_1 + Q_2$$

Solving for Q_4, we have

$$\begin{aligned} Q_4 &= Q_1 + Q_2 - Q_3 \\ &= V_1A_1 + V_2A_2 - V_3A_3 \\ &= 50 \times \frac{\pi}{4}(D_1^2 - D_2^2 + D_3^2) \\ &= 50 \times \frac{\pi}{4} \times \frac{1}{144}(1 + 4 - 2.25) \\ Q_4 &= 0.750 \text{ ft}^3/\text{s} \end{aligned}$$

Now we can solve for V_{max} in pipe 4. The velocity is linearly distributed from zero at the wall to maximum at the center as shown.

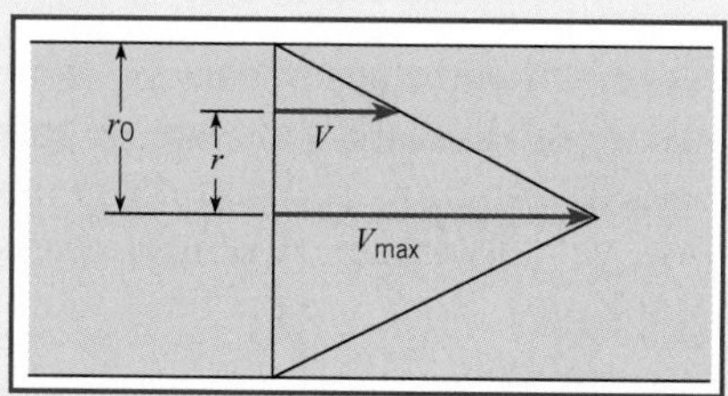

Therefore, by proportions we have

$$\frac{V}{r_0 - r} = \frac{V_{max}}{r_0} \quad \text{or} \quad V = V_{max}\left(1 - \frac{r}{r_0}\right)$$

Thus

$$Q = \int_A V\,dA = \int_0^{r_0} V_{max}\left(1 - \frac{r}{r_0}\right) 2\pi r\,dr$$

$$Q = 2\pi V_{max} \int_0^{r_0} \left(1 - \frac{r}{r_0}\right) r\,dr = 2\pi V_{max} r_0^2 \left(\tfrac{1}{2} - \tfrac{1}{3}\right) = \tfrac{1}{3}\pi r_0^2 V_{max}$$

Thus $$V_{max} = \frac{Q}{\frac{1}{3}\pi r_0^2} = \frac{0.75 \text{ ft}^3/\text{s}}{\frac{1}{3}\pi \times (\frac{1}{2})^2 \text{ ft}^2} = 2.86 \text{ ft/s}$$ ◁

example 5.11

Air flows at a steady rate in a pipe in which the cross-sectional area doubles between stations 1 and 2 and the velocity is reduced by a factor of 1.95. Find the percentage change in air density.

Solution Assume a one-dimensional flow. The continuity equation between the two stations is

$$\rho_1 V_1 A_1 = \rho_2 V_2 A_2$$

The density ratio can be expressed as

$$\frac{\rho_2}{\rho_1} = \frac{V_1}{V_2}\frac{A_1}{A_2}$$

Substituting in the values,

$$\frac{\rho_2}{\rho_1} = \frac{1.95}{1}\frac{1}{2} = 0.975$$

The density is reduced by 0.025 or 2.5%.

example 5.12

Water with a density of 1000 kg/m^3 flows through a vertical venturimeter as shown. A pressure gage is connected across two taps in the pipe (1) and the throat (2). The area ratio A_{throat}/A_{pipe} is 0.5. The velocity in the pipe is 10 m/s. Find the pressure difference recorded by the pressure gage. Assume the flow has a uniform velocity distribution and that viscous effects are not important.

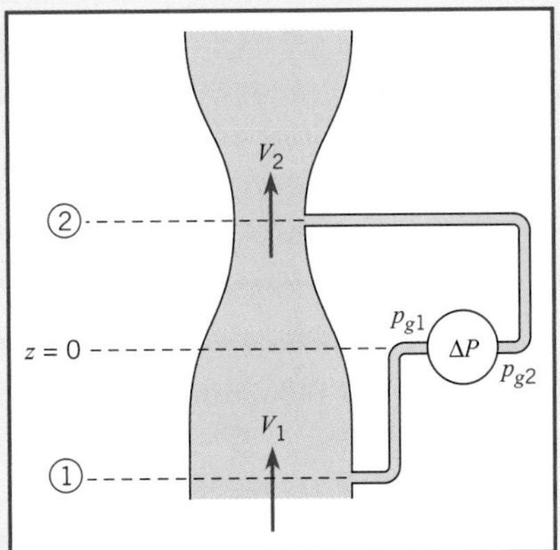

Solution The Bernoulli equation is used to relate the pressures at stations (1) and (2):

$$p_1 + \gamma z_1 + \rho\frac{V_1^2}{2} = p_2 + \gamma z_2 + \rho\frac{V_2^2}{2}$$

The steady-flow continuity equation is used to find the velocity ratio between taps 2 and 1:

$$\frac{V_2}{V_1} = \frac{A_1}{A_2}$$

Define the zero elevation at the gage location. The water in the lines from the pressure taps to the gage is static, so the pressure at the upstream connection to the gage, p_{g_1}, is

$$p_{g_1} = p_1 + \gamma z_1$$

and the pressure at the downstream connection, p_{g_2}, is

$$p_{g_2} = p_2 + \gamma z_2$$

The Bernoulli equation simplifies to

$$p_{g_1} + \rho\frac{V_1^2}{2} = p_{g_2} + \rho\frac{V_2^2}{2}$$

and the pressure across the gage is

$$p_{g_1} - p_{g_2} = \frac{\rho}{2}(V_2^2 - V_1^2)$$

$$\Delta p_g = \rho\frac{V_1^2}{2}\left(\frac{V_2^2}{V_1^2} - 1\right)$$

Using the steady-flow continuity equation,

$$\Delta p_g = \rho\frac{V_1^2}{2}\left(\frac{A_1^2}{A_2^2} - 1\right)$$

$$= \frac{1000\ \text{kg/m}^3 \times 10^2(\text{m}^2/\text{s}^2)}{2}(2^2 - 1)$$

$$= 150\ \text{kPa}$$

◁

5·4

Cavitation

Cavitation is the phenomenon that occurs when the fluid pressure is reduced to the local vapor pressure and boiling occurs. Under such conditions vapor bubbles form, grow, and then collapse, producing shock waves, noise, and dynamic effects that lead to decreased equipment performance and, frequently, equipment failure. Engineers are often concerned about the possibility of cavitation, and they must design flow systems to avoid potential problems.

FIGURE 5.10

Flow through pipe restriction: Variation of piezometric head.

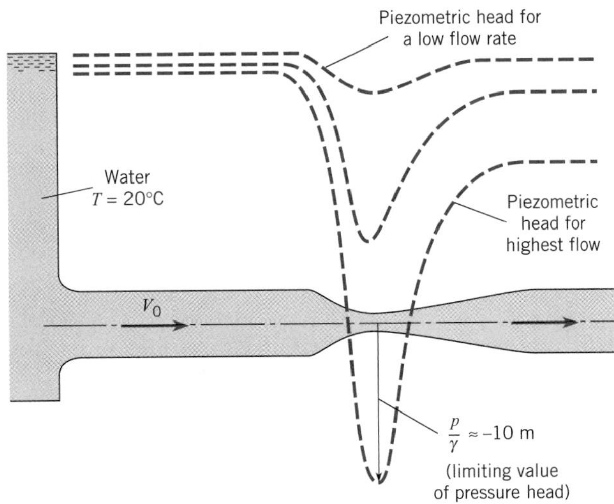

Cavitation typically occurs at locations where the velocity is high. Consider the water flow through the pipe restriction shown in Fig. 5.10. The pipe area is reduced, so the velocity is increased according to the continuity equation and, in turn, the pressure is reduced as dictated by the Bernoulli equation. The physical configuration and the plots of piezometric head along the wall of the pipe are shown in Fig. 5.10. For low flow rates, there is a relatively small drop in pressure at the restriction, so the water remains well above the vapor pressure and boiling does not occur. This is indicated in Fig. 5.10 where the piezometric head lies above the centerline, indicating a positive pressure. However, as the flow rate increases, the pressure at the restriction can become sub-atmospheric. The pressure can drop no lower than the vapor pressure of the liquid because, at this point, the liquid will boil and cavitation ensues.

The formation of vapor bubbles at the restriction of a venturimeter is shown in Fig. 5.11*a*. The vapor bubbles form and then collapse as they move into a region of higher pressure and are swept downstream with the flow. When the flow velocity is increased further, the minimum pressure is still the local vapor pressure, but the zone of bubble formation is extended as shown in Fig. 5.11*b*. In this case, the entire vapor pocket may intermittently grow and collapse, producing serious vibration problems. Severe damage that occurred on a centrifugal pump impeller is shown in Fig. 5.12, and serious erosion produced by cavitation in a spillway tunnel of Hoover Dam is shown in Fig. 5.13. Obviously, cavitation should be avoided or minimized by proper design of equipment and structures and by proper operational procedures.

A video of cavitation occurring in the region of a marine propeller is available on the net at www.wiley.com/college/crowe. In this situation, the high liquid velocities produced by the rotating propeller cause a low pressure and cavitation. The bubbles observed in the video indicate cavitation.

The cavitation number, σ, is defined as the pressure coefficient where cavitation occurs. Ideally, the cavitation number is the pressure coefficient based on the vapor pressure,

$$\sigma = \frac{(p_v/g) - h_0}{V_0^2/2g}$$

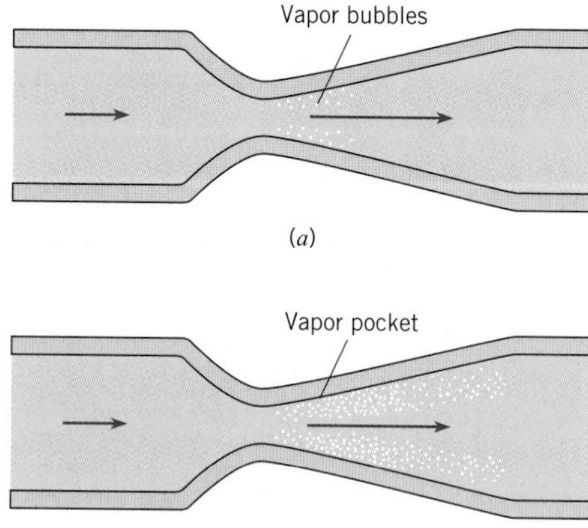

FIGURE 5.11

Formation of vapor bubbles in the process of cavitation. (a) Cavitation. (b) Cavitation—higher flow rate.

FIGURE 5.12

Cavitation damage to impeller of a centrifugal pump.

However, cavitation is also affected by contaminant gases, turbulence, and viscosity.

Experimental studies reveal that very high intermittent pressure, as high as 800 MPa (115,000 psi), develops in the vicinity of the bubbles when they collapse (1). Therefore, if bubbles collapse close to boundaries such as pipe walls, pump impellers, valve casings, and dam slipway floors, they can cause considerable damage. Usually this damage occurs in the form of fatigue failure brought about by the action of millions of bubbles impacting (in effect, imploding) against the material surface over a long period of time, thus producing a material pitting in the zone of cavitation.

Background and detailed discussions of cavitation can be found in Brennen (2) and Young (3). Typical papers on cavitation are given in references 4, 5, and 6.

The world's largest and most technically advanced water tunnel for studying cavitation is located in Memphis, Tennessee. This facility is used to test large-scale models

FIGURE 5.13

Cavitation damage to a power dam spillway tunnel.

of submarine systems and full-scale torpedoes as well as applications in the maritime shipping industry. Basic research on cavitation has focused on topics such as bubble dynamics, boundary layer effects, and surface roughness. Current information on cavitation research is found in publications of the International Symposia on Cavitation, which are held periodically at different locations around the world.

5·5

Differential Form of the Continuity Equation

In the analysis of fluid flows one of the fundamental independent equations needed is the differential form of the continuity equation. This equation is derived by applying the continuity equation, Eq. (5.25), to an infinitesimal control volume and taking the limit as the volume approaches zero, that is, as the volume is reduced to a point. Consider the cubical elemental control volume shown in Fig. 5.14. The sides of the volume are oriented normal to the x, y, and z coordinate directions.

The general form of the continuity equation is

$$\int_{\mathrm{cs}} \rho \mathbf{V} \cdot d\mathbf{A} + \frac{d}{dt}\int_{\mathrm{cv}} \rho \, d\forall = 0$$

where $\mathbf{V}$ is the velocity measured with respect to the local control surface. Applying the Leibnitz theorem for differentiation of an integral allows the unsteady term to be expressed as

$$\frac{d}{dt}\int_{\mathrm{cv}} \rho \, d\forall = \int_{\mathrm{cv}} \frac{\partial \rho}{\partial t} d\forall + \int_{\mathrm{cs}} \rho \mathbf{V}_s \cdot d\mathbf{A}$$

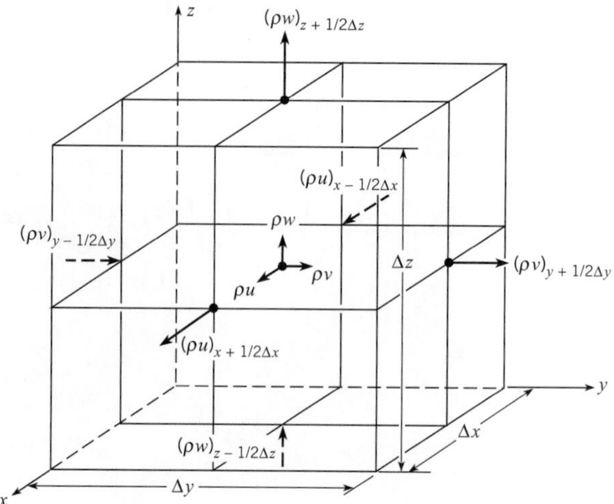

FIGURE 5.14

Elemental control volume.

where $\mathbf{V}_s$ is the local velocity of the control surface with respect to the reference frame. For a control volume with stationary sides, as shown in Fig. 5.14, $\mathbf{V}_s = 0$, so the continuity equation for the control volume can be written as

$$\int_{cs} \rho\mathbf{V} \cdot d\mathbf{A} + \int_{cv} \frac{\partial \rho}{\partial t} d\text{¥} = 0$$

The element is infinitesimal, so we can assume that the velocity and densities are uniformly distributed across each face (control surface) and the mass flux term becomes

$$\int_{cs} \rho\mathbf{V} \cdot d\mathbf{A} = \sum_{cs} \rho\mathbf{V} \cdot \mathbf{A}$$

and the continuity equation assumes the form

$$\sum_{cs} \rho\mathbf{V} \cdot \mathbf{A} + \int_{cv} \frac{\partial \rho}{\partial t} d\text{¥} = 0$$

Considering the flow rates through the six faces of the cubical element and applying those to the foregoing form of the continuity equation, we obtain

$$\begin{aligned}
&[(\rho u)_{x+(1/2)\Delta x}]\Delta y\Delta z - [(\rho u)_{x-(1/2)\Delta x}]\Delta y\Delta z \\
&\quad + [(\rho v)_{y+(1/2)\Delta y}]\Delta z\Delta x - [(\rho v)_{y-(1/2)\Delta y}]\Delta z\Delta x \\
&\quad\quad + [(\rho w)_{z+(1/2)\Delta z}]\Delta x\Delta y - [(\rho w)_{z-(1/2)\Delta z}]\Delta x\Delta y \\
&\quad\quad\quad + \frac{\partial \rho}{\partial t}\Delta x\Delta y\Delta z = 0
\end{aligned} \tag{5.31}$$

Dividing Eq. (5.31) by the volume of the element ($\Delta x\Delta y\Delta z$) yields

$$\frac{[(\rho u)_{x+(1/2)\Delta x}] - [(\rho u)_{x-(1/2)\Delta x}]}{\Delta x} + \frac{[(\rho v)_{y+(1/2)\Delta y}] - [(\rho v)_{y-(1/2)\Delta y}]}{\Delta y} + \frac{[(\rho w)_{z+(1/2)\Delta z}] - [(\rho w)_{z-(1/2)\Delta z}]}{\Delta z} + \frac{\partial \rho}{\partial t} = 0$$

Taking the limit as the volume approaches zero (that is, as Δx, Δy, and Δz uniformly approach zero) yields the differential form of the continuity equation

$$\frac{\partial}{\partial x}(\rho u) + \frac{\partial}{\partial y}(\rho v) + \frac{\partial}{\partial z}(\rho w) = -\frac{\partial \rho}{\partial t} \tag{5.32}$$

If the flow is steady, we obtain

$$\frac{\partial}{\partial x}(\rho u) + \frac{\partial}{\partial y}(\rho v) + \frac{\partial}{\partial z}(\rho w) = 0 \tag{5.33}$$

And if the fluid is incompressible, we have

$$\frac{\partial u}{\partial x} + \frac{\partial v}{\partial y} + \frac{\partial w}{\partial z} = 0 \tag{5.34a}$$

for both steady and unsteady flow.

In vector notation, Eq. (5.34*a*) is given as

$$\nabla \cdot \mathbf{V} = 0 \tag{5.34b}$$

where ∇ is the del operator, defined as

$$\nabla = \mathbf{i}\frac{\partial}{\partial x} + \mathbf{j}\frac{\partial}{\partial y} + \mathbf{k}\frac{\partial}{\partial z}$$

example 5.13

The expression $\nabla = 10x\mathbf{i} - 10y\mathbf{j}$ is said to represent the velocity for a two-dimensional incompressible flow. Check it to see whether it satisfies continuity.

Solution

$$u = 10x \quad \text{so} \quad \frac{\partial u}{\partial x} = 10$$

$$v = -10y \quad \text{so} \quad \frac{\partial v}{\partial y} = -10$$

$$\frac{\partial u}{\partial x} + \frac{\partial v}{\partial y} = 10 - 10 = 0$$

Continuity is satisfied.

Summary

Flow rate refers to either the volume per unit time or the mass per unit time passing through a surface. The volume flow rate, or discharge, is given by

$$Q = \int_A \mathbf{V} \cdot d\mathbf{A}$$

where **A** is the surface area vector and **V** is the velocity vector. If the area vector and velocity vector are aligned, then

$$Q = \int_A V\, dA \quad = \overline{V}A$$

where $\overline{V}$ is the average velocity. The corresponding mass flow rate is

$$\dot{m} = \int_A \rho V\, dA$$

If the density is uniformly distributed across the area,

$$\dot{m} = \rho Q$$

A fluid system is a given quantity of matter consisting always of the same matter. A control volume (cv) is a geometric volume defined in space and enclosed by a control surface (cs). Mass can cross the control surface.

The Reynolds transport theorem relates the time rate of change of a property of a system to the rate of change of the property in the control volume and the net efflux of the property across the control surface. It provides a relationship between the Lagrangian and Eulerian descriptions of fluid property changes.

The continuity equation derives from the conservation of mass principle and is expressed as

$$\frac{d}{dt}\int_{\text{cv}} \rho\, d\forall + \int_{\text{cs}} \rho \mathbf{V} \cdot d\mathbf{A} \quad = 0$$

where **V** is the velocity with respect to the control surface and $d\mathbf{A}$ is the differential area directed outward from the control volume. An alternative form of the continuity equation is

$$\frac{d}{dt}M_{\text{cv}} = \sum \dot{m}_i - \sum \dot{m}_o$$

where M_{cv} is the mass in the control volume and $\dot{m}_i$ and $\dot{m}_o$ are the mass flow rates of flows entering and leaving the control volume, respectively.

For steady, one-dimensional flow in a pipe, the continuity equation reduces to

$$\rho_1 A_1 V_1 = \rho_2 A_2 V_2$$

where the subscripts 1 and 2 refer to the inlet and outlet of the pipe. If, in addition, the flow is incompressible,

$$A_1V_1 = A_2V_2$$

Cavitation occurs when the pressure drops to the local vapor pressure of the liquid and bubbles appear due to liquid boiling. The presence of cavitation can cause serious equipment failures.

References

1. Knapp, R.T., J. W. Daily, and F. G. Hammitt. *Cavitation.* McGraw-Hill, New York, 1970.
2. Brennen, C. E. *Cavitation and Bubble Dynamics.* Oxford University Press, New York, 1995.
3. Young, F. R. *Cavitation.* McGraw-Hill, New York, 1989.
4. Falvey, H. T. "Predicting Cavitation in Spillway Tunnels." *Water Power and Dam Construction* (August 1982).
5. Rayan, M. A., et al. "Cavitation Erosive Wear in Centrifugal Pump Impeller." pp. 92–95, *Cavitation and Multiphase Flow Forum—1987.* ASME, 345 East 47th St., New York, 1987.
6. Thomas, H. A., and E. P. Schuleen. "Cavitation in Outlet Conduits of High Dams." *Trans. Am. Soc. Civil Eng.*, 107 (1942).

Problems

5.1 The discharge of water in a 25-cm pipe is 0.04 m^3/s. What is the mean velocity?

5.2 A pipe with a 16-in. diameter carries water having a velocity of 3 ft/s. What is the discharge in cubic feet per second and in gallons per minute (1 cfs is equivalent to 449 gpm)?

5.3 A pipe with a 2-m diameter carries water having a velocity of 4 m/s. What is the discharge in cubic meters per second and in cubic feet per second?

5.4 A pipe whose diameter is 8 cm transports air with a temperature of 20°C and pressure of 200 kPa absolute at 20 m/s. Determine the mass-flow rate.

5.5 Natural gas (methane) flows at 20 m/s through a pipe with a 1-m diameter. The temperature of the methane is 15°C, and the pressure is 150 kPa gage. Determine the mass-flow rate.

5.6 An aircraft engine test pipe is capable of providing a flow rate of 200 kg/s at altitude conditions corresponding to an absolute pressure of 50 kPa and a temperature of –18°C. The velocity of air through the duct attached to the engine is 240 m/s. Calculate the diameter of the duct.

5.7 A heating and air-conditioning engineer is designing a system to move 1100 m^3 of air per hour at 100 kPa abs, and 30°C. The duct is rectangular with cross-sectional dimensions of 1 m by 20 cm. What will be the air velocity in the duct?

5.8 The hypothetical velocity distribution in a circular duct is $v/V_0 = 1 - r/R$, where r is the radial location in the duct, R is the duct radius, and V_0 is the velocity on the axis. Find the ratio of the mean velocity to the velocity on the axis.

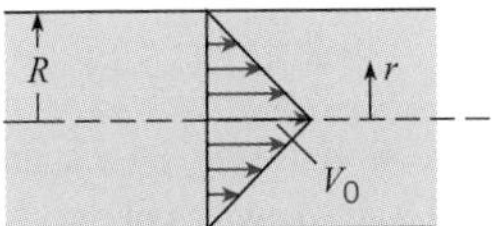

PROBLEM 5.8

5.9 Water flows in a two-dimensional channel of width W and depth D as shown in the diagram. The hypothetical velocity profile for the water is

$$V(x, y) = V_s\left(1 - \frac{4x^2}{W^2}\right)\left(1 - \frac{y^2}{D^2}\right)$$

where V_s is the velocity at the water surface midway between the channel walls. The coordinate system is as shown; x is measured from the center plane of the channel and y downward from the water surface. Find the discharge in the channel in terms of V_s, D, and W.

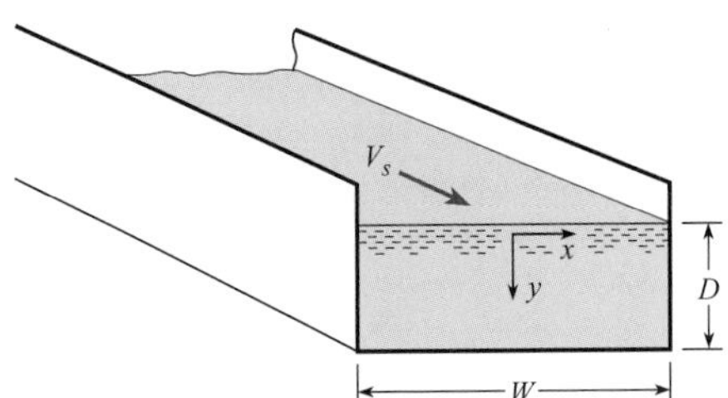

PROBLEM 5.9

5.10 Water flows in a pipe that has a 4-ft diameter and the following hypothetical velocity distribution: The velocity is maximum at the centerline and decreases linearly with r to a minimum at the pipe wall. If $V_{max} = 15$ ft/s and $V_{min} = 12$ ft/s, what is the discharge in cubic feet per second and in gallons per minute?

5.11 In Prob. 5.10, if $V_{max} = 8$ m/s, $V_{min} = 6$ m/s, and $D = 2$ m, what is the discharge in cubic meters per second and the mean velocity?

FLUID SOLUTIONS **5.12** Air enters this square duct at section 1 with the velocity distribution as shown. Note that the velocity varies in the y direction only (for a given value of y, the velocity is the same for all values of z).

a. What is the volume rate of flow?
b. What is the mean velocity in the duct?
c. What is the mass rate of flow if the mass density of the air is 1.2 kg/m^3?

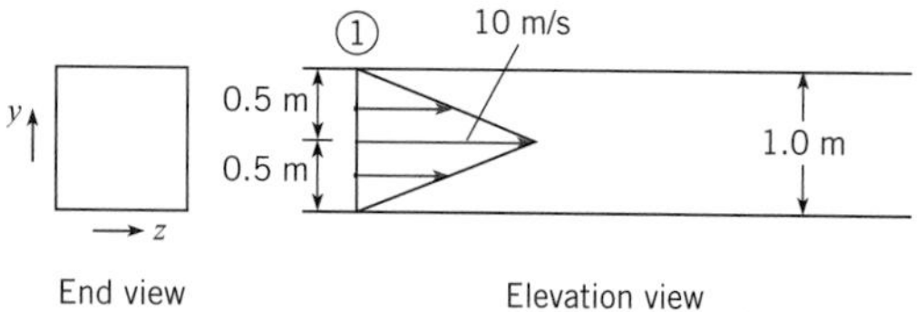

PROBLEM 5.12

5.13 The velocity at section *A-A* is 18 ft/s, and the vertical depth y at the same section is 4 ft. If the width of the channel is 25 ft, what is the discharge in cubic feet per second?

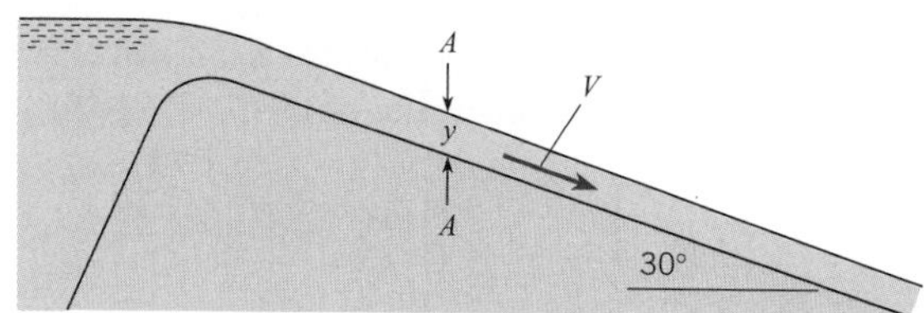

PROBLEM 5.13

5.14 The rectangular channel shown is 1.5 m wide. What is the discharge in the channel?

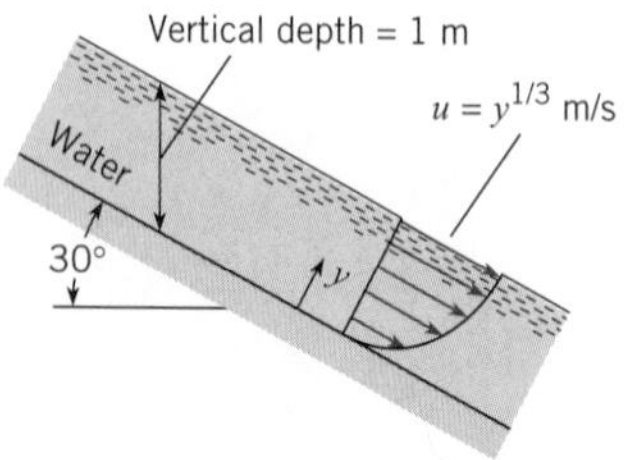

PROBLEM 5.14

5.15 If the velocity in the channel of Prob. 5.14 is given as $u = 10[\exp(y) - 1]$ m/s and the channel width is 2 m, what is the discharge in the channel and what is the mean velocity?

5.16 Water from a pipe is diverted into a weigh tank for exactly 15 min. The increased weight in the tank is 20 kN. What is the discharge in cubic meters per second? Assume $T = 20°C$.

5.17 Water enters the lock of a ship canal through 180 ports, each port having a 2 ft by 2 ft cross section. The lock is 900 ft long and 105 ft wide. The lock is designed so that the water surface in it will rise at a maximum rate of 6 ft/min. For this condition, what will be the mean velocity in each port?

5.18 An empirical equation for the velocity distribution in a horizontal, rectangular, open channel is given by $u = u_{max}(y/d)^n$, where u is the velocity at a distance y feet above the floor of the channel. If the depth d of flow is 1.2 m, $u_{max} = 3$ m/s, and $n = 1/6$, what is the discharge in cubic meters per second per meter of width of channel? What is the mean velocity?

5.19 The hypothetical water velocity in a V-shaped channel (see the accompanying figure) varies linearly with depth from zero at the bottom to maximum at the water surface. Determine the discharge if the maximum velocity is 6 ft/s.

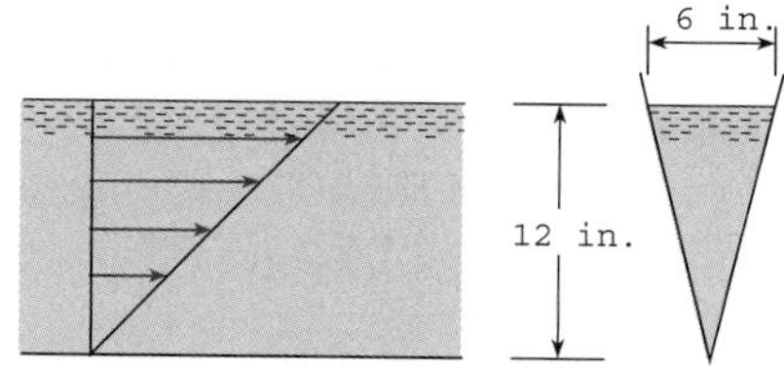

PROBLEM 5.19

5.20 The velocity of flow in a circular pipe varies according to the equation $V/V_C = (1 - r^2/r_0^2)^n$, where V_C is the centerline velocity, r_0 is the pipe radius, and r is the radial distance from the centerline. The exponent n is general and is chosen to fit a given profile ($n = 1$ for laminar flow). Determine the mean velocity as a function of V_c and n.

5.21 Plot the velocity distribution across the pipe, and determine the discharge of a fluid flowing through a pipe 1 m in diameter that has a velocity distribution given by $V = 12(1 - r^2/r_0^2)$ m/s. Here r_0 is the radius of the pipe and r is the radial distance from the centerline. What is the mean velocity?

5.22 Water flows through a 1.5-in.-diameter pipeline at 80 lbm/min. Calculate the mean velocity. Assume $T = 60°F$.

5.23 Water flows through a 20-cm pipeline at 1000 kg/min. Calculate the mean velocity in meters per second if $T = 20°C$.

5.24 Water from a pipeline is diverted into a weigh tank for exactly 10 min. The increased weight in the tank is 4765 lbf. What is the average flow rate in gallons per minute and in cubic feet per second? Assume $T = 60°F$.

5.25 The mean velocity of water in a 4-in. pipe is 8 ft/s. Determine the flow in slugs per second, gallons per minute, and cubic feet per second if $T = 60°\text{F}$.

5.26 Two parallel disks of diameter D are brought together, each with a normal speed of V. When their spacing is h, what is the radial component of convective acceleration at the section just inside the edge of the disk (section A) in terms of V, h, and D? Assume uniform velocity distribution across the section.

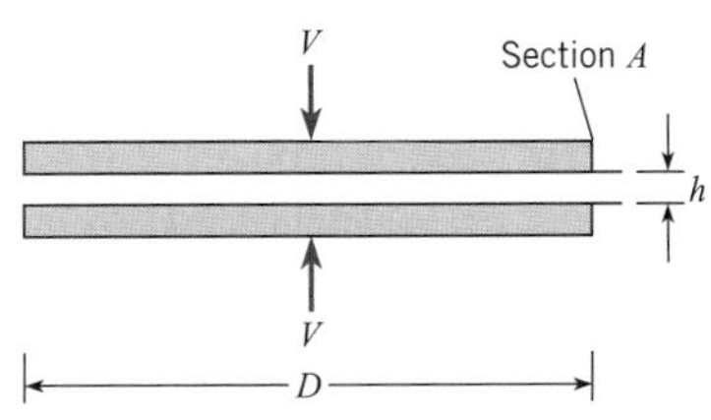

PROBLEM 5.26

5.27 For the conditions given in Prob. 5.26, find the radial component of local acceleration at A in terms of D, V, and h.

5.28 Two streams discharge into a pipe as shown. The flows are incompressible. The volume flow rate of stream A into the pipe is given by $Q_A = 0.02t\ \text{m}^3/\text{s}$ and that of stream B by $Q_B = 0.008t^2\ \text{m}^3/\text{s}$, where t is in seconds. The exit area of the pipe is $0.01\ \text{m}^2$. Find the velocity and acceleration of the flow at the exit at $t = 1$ s.

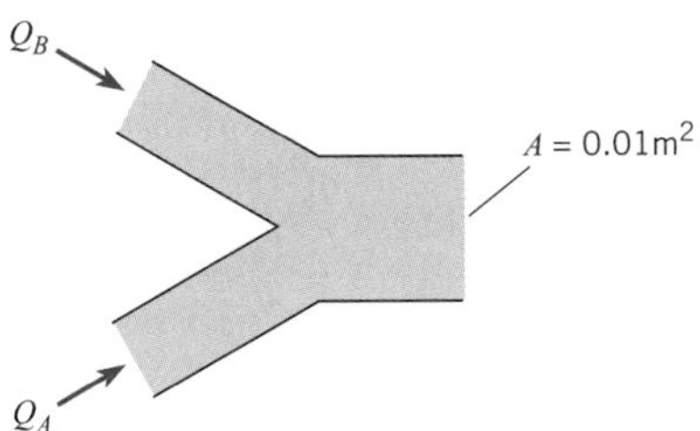

PROBLEM 5.28

5.29 Air discharges downward in the pipe and then outward between the parallel disks. Assuming negligible density change in the air, derive a formula for the acceleration of air at point A, which is a distance r from the center of the disks. Express the acceleration in terms of the constant air discharge Q, the radial distance r, and the disk spacing h. If $D = 10$ cm, $h = 0.5$ cm, and $Q = 0.380\ \text{m}^3/\text{s}$, what are the velocity in the pipe and the acceleration at point A where $r = 20$ cm?

5.30 All the conditions of Prob. 5.29 are the same except that $h = 1$ cm and the discharge is given as $Q = Q_0(t/t_0)$, where $Q_0 = 0.1\ \text{m}^3/\text{s}$ and $t_0 = 1$ s. For the additional conditions, what will be the acceleration at point A when $t = 2$ s and $t = 3$ s?

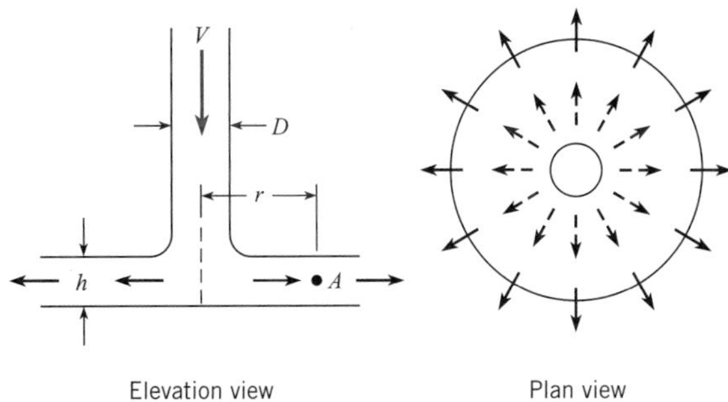

PROBLEMS 5.29, 5.30

5.31 A tank has a hole in the bottom with a cross-sectional area of $0.0025\ \text{m}^2$ and an inlet line on the side with a cross-sectional area of $0.0025\ \text{m}^2$, as shown. The cross-sectional area of the tank is $0.1\ \text{m}^2$. The velocity of the liquid flowing out the bottom hole is $V = \sqrt{2gh}$, where h is the height of the water surface in the tank above the outlet. At a certain time the surface level in the tank is 1 m and rising at the rate of 0.1 cm/s. The liquid is incompressible. Find the velocity of the liquid through the inlet.

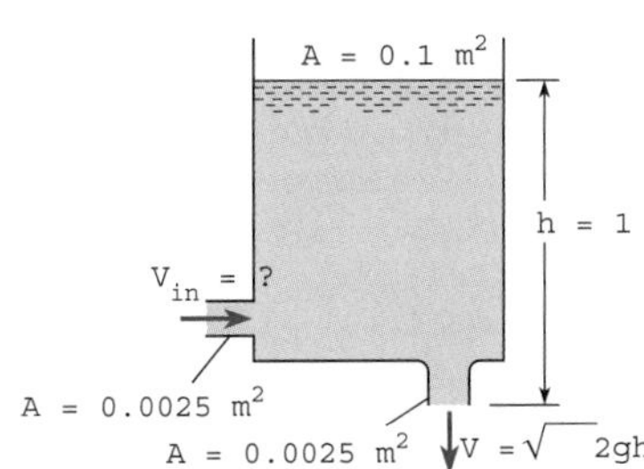

PROBLEM 5.31

5.32 A mechanical pump is used to pressurize a bicycle tire. The inflow to the pump is 1 cfm. The density of the air entering the pump is $0.075\ \text{lbm/ft}^3$. The inflated volume of a bicycle tire is $0.04\ \text{ft}^3$. The density of air in the inflated tire is $0.4\ \text{lbm/ft}^3$. How many seconds does it take to pressurize the tire if there initially was no air in the tire?

5.33 For the conditions in each of the flow cases (a and b) shown, respond to the following questions and statements concerning the application of the control-volume equation to the continuity principle

a. What is the value of b?
b. Determine the value of dB_{sys}/dt.
c. Determine the value of

$$\sum_{\text{cs}} b\rho\mathbf{V}\cdot\mathbf{A}.$$

d. Determine the value of $d/dt \int_{\text{cv}} b\rho\, d\mathcal{V}$.

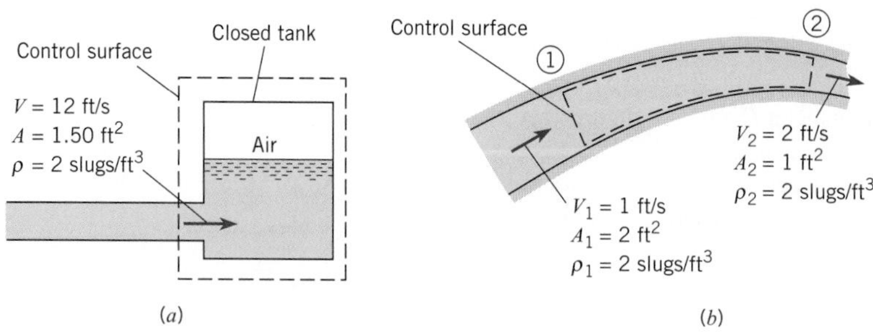

PROBLEM 5.33

5.34 Gas flows into and out of the chamber as shown. For the conditions shown, which of the following statement(s) are true of the application of the control-volume equation to the continuity principle?

a. $B_{sys} = 0$

b. $dB_{sys}/dt = 0$

c. $\sum_{cs} b\rho \mathbf{V} \cdot \mathbf{A} = 0$

d. $\frac{d}{dt}\int_{cv} \rho \, d\forall = 0$

e. $b = 0$

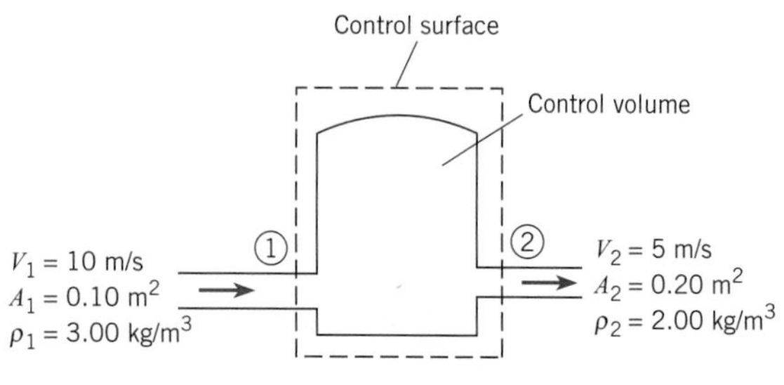

PROBLEM 5.34

FLUID SOLUTIONS **5.35** Both pistons are moving to the left, but piston A has a speed twice as great as that of piston B. Then the water level in the tank is (a) rising, (b) not moving up or down, (c) falling

5.36 The piston in the cylinder is moving up. Assume that the control volume is the volume inside the cylinder above the piston (the control volume changes in size as the piston moves). A gaseous mixture exists in the control volume. For the given conditions, indicate which of the following statements are true.

a. $\sum_{cs} \rho \mathbf{V} \cdot \mathbf{A}$ is equal to zero.

b. $\frac{d}{dt}\int_{cv} \rho \, d\forall$ is equal to zero.

c. The mass density of the gas in the control volume is increasing with time.

d. The temperature of the gas in the control volume is increasing with time.

e. The flow inside the control volume is unsteady.

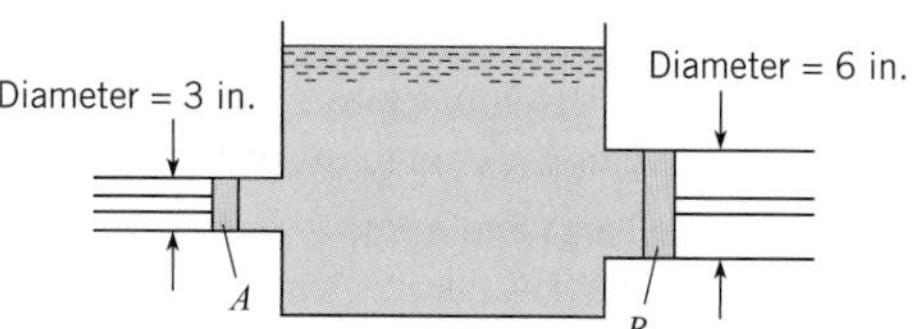

PROBLEM 5.35

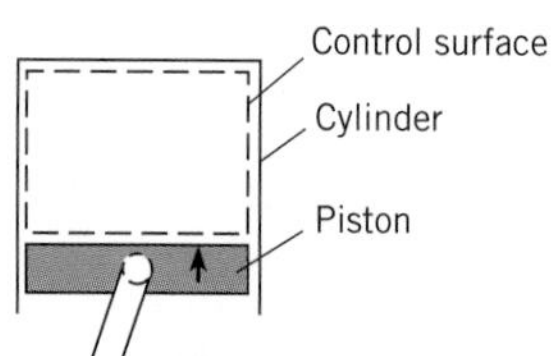

PROBLEM 5.36

5.37 For the conditions shown, respond to the following questions and statements concerning application of the control-volume equation to the continuity principle.

a. What is the value of b?

b. Determine the value of dB_{sys}/dt.

c. Determine the value of

$$\sum_{cs} b\rho \mathbf{V} \cdot \mathbf{A} .$$

d. Determine the value of $d/dt\int_{cv} b\rho \, d\forall$.

5.38 This plunger moves downward in the conical receptacle, which is filled with oil. At what level (y in terms of d) above the bottom of the receptacle will the mean upward velocity of the oil (between the plunger and the receptacle wall) be exactly the same magnitude as the downward velocity of the plunger?

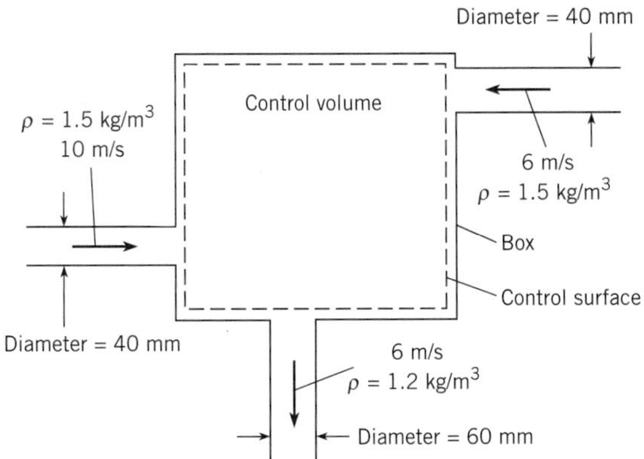

PROBLEM 5.37

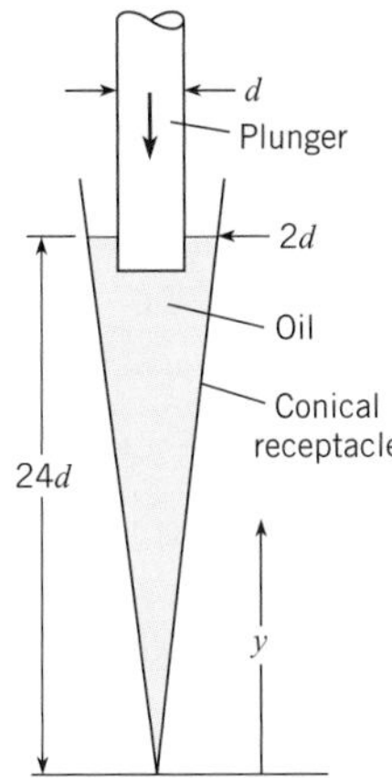

PROBLEM 5.38

5.39 A 6-in.-diameter cylinder falls at a rate of 3 ft/s in an 8-in.-diameter tube containing an incompressible liquid. What is the mean velocity of the liquid (with respect to the tube) in the space between the cylinder and the tube wall?

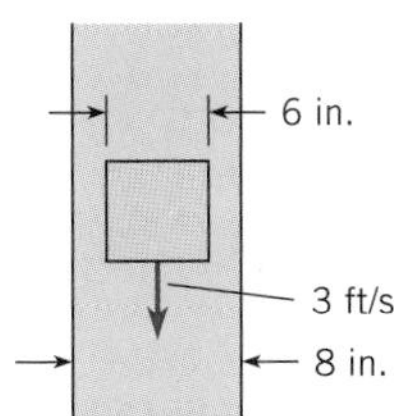

PROBLEM 5.39

5.40 This circular tank of water is being filled from a pipe as shown. The velocity of flow of water from the pipe is 10 ft/s. What will be the rate of rise of the water surface in the tank?

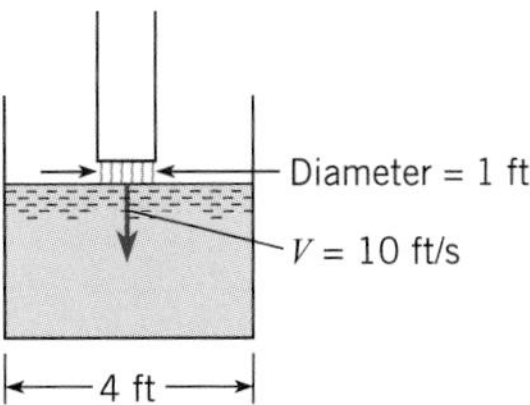

PROBLEM 5.40

5.41 A sphere 8 inches in diameter falls at 4 ft/s downward axially through water in a 1-ft-diameter container. Find the upward speed of the water with respect to the container wall at the midsection of the sphere.

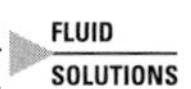

5.42 A rectangular air duct 20 cm by 50 cm carries a flow of 1.44 m^3/s. Determine the velocity in the duct. If the duct tapers to 10 cm by 40 cm, what is the velocity in the latter section? Assume constant air density.

5.43 A 30-cm pipe divides into a 20-cm branch and a 15-cm branch. If the total discharge is 0.30 m^3/s and if the same mean velocity occurs in each branch, what is the discharge in each branch?

5.44 The conditions are the same as in Prob. 5.43 except that the discharge in the 20-cm branch is twice that in the 15-cm branch. What is the mean velocity in each branch?

5.45 Water flows in an 8-in. pipe that is connected in series with a 6-in. pipe. If the rate of flow is 898 gpm (gallons per minute), what is the mean velocity in each pipe?

5.46 What is the velocity of the flow of water in leg *B* of the tee shown in the figure?

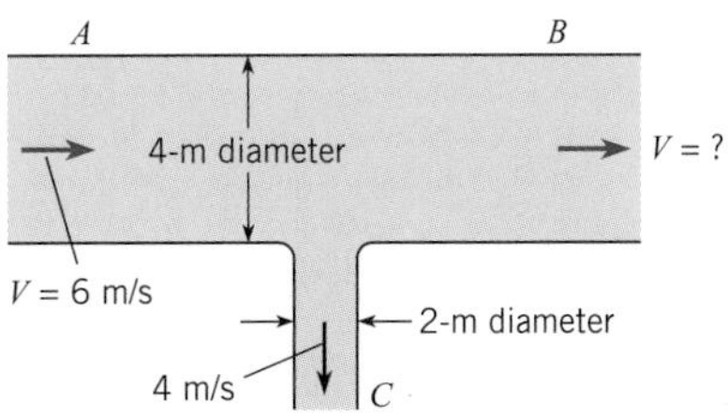

PROBLEM 5.46

5.47 For a steady flow of gas in the conduit shown, what is the mean velocity at section 2?

5.48 Two pipes are connected to an open water tank. The water is entering the bottom of the tank from pipe *A* at 8 cfm. The water level in the tank is rising at 1.0 in./min, and the surface area of the

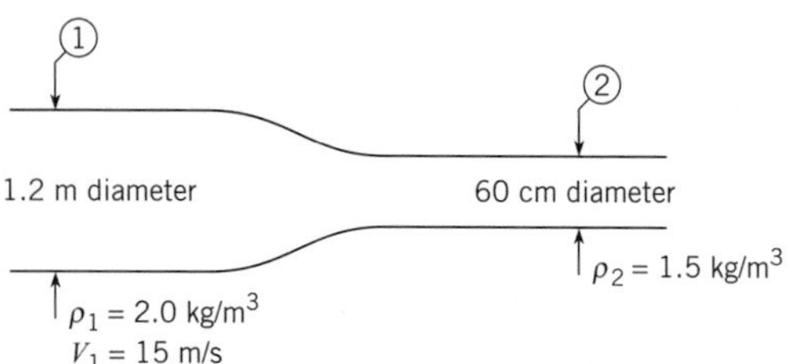

PROBLEM 5.47

tank is 80 ft^2. Calculate the discharge in a second pipe, pipe B, that is also connected to the bottom of the tank. Is the flow entering or leaving the tank from pipe B?

5.49 Is the tank in the figure filling or emptying? At what rate is the water level rising or falling in the tank?

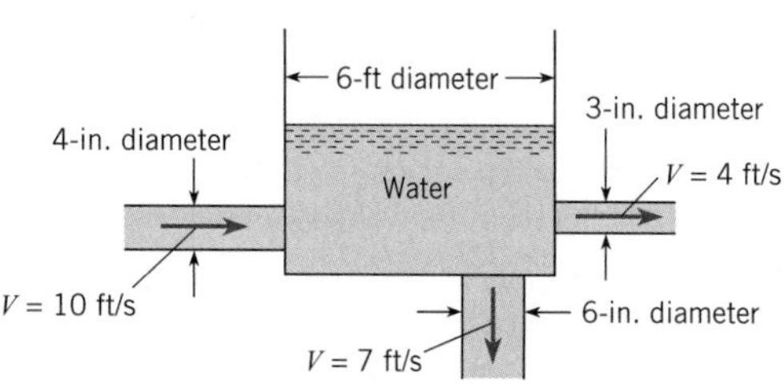

PROBLEM 5.49

5.50 Given: Flow velocities as shown in the figure and water surface elevation (as shown) at $t = 0$ s. At the end of 22 s, will the water surface in the tank be rising or falling, and at what speed?

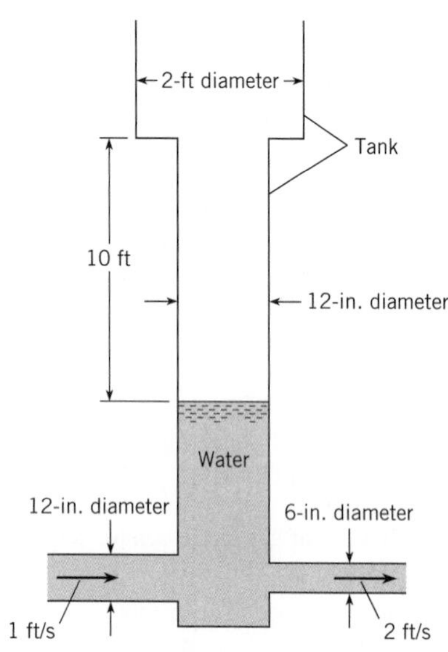

PROBLEM 5.50

5.51 A lake with no outlet is fed by a river with a constant flow of 1000 ft^3/s. Water evaporates from the surface at a constant rate of 13 ft^3/s per square mile surface area. The area varies with depth h (feet) as A (square miles) = 4.5 + 5.5h. What is the equilibrium depth of the lake? Below what river discharge will the lake dry up?

5.52 A stationary nozzle discharges water against a plate moving toward the nozzle at half the jet velocity. When the discharge from the nozzle is 5 cfs, at what rate will the plate deflect water?

5.53 The open tank shown has a constant inflow discharge of 20 ft^3/s. A 1.0-ft-diameter drain provides a variable outflow velocity V_{out} equal to $\sqrt{(2gh)}$ ft/s. What is the equilibrium height h_{eq} of the liquid in the tank?

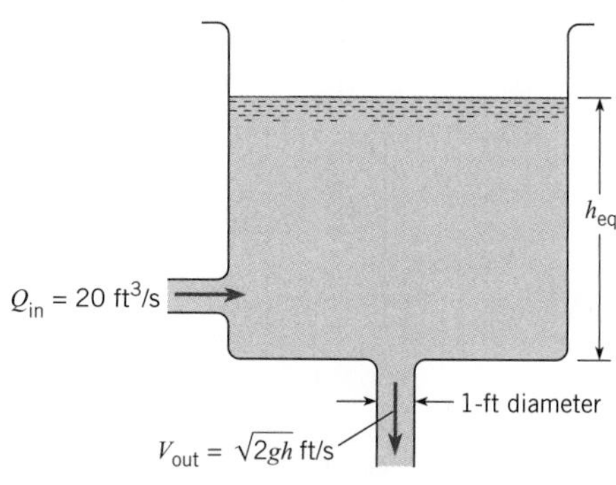

PROBLEM 5.53

5.54 Assuming that complete mixing occurs between the two inflows before the mixture discharges from the pipe at C, find the mass rate of flow, the velocity, and the specific gravity of the mixture in the pipe at C.

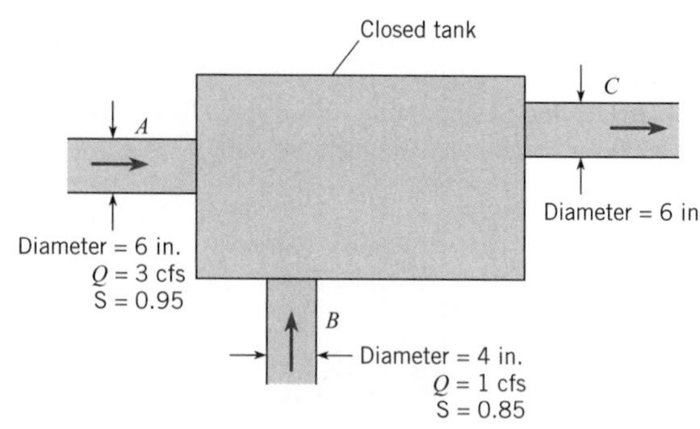

PROBLEM 5.54

5.55 Oxygen and methane are mixed at 200 kPa absolute pressure and 100°C. The velocity of the gases into the mixer is 5 m/s. The density of the gas leaving the mixer is 2.2 kg/m^3. Determine the exit velocity of the gas mixture.

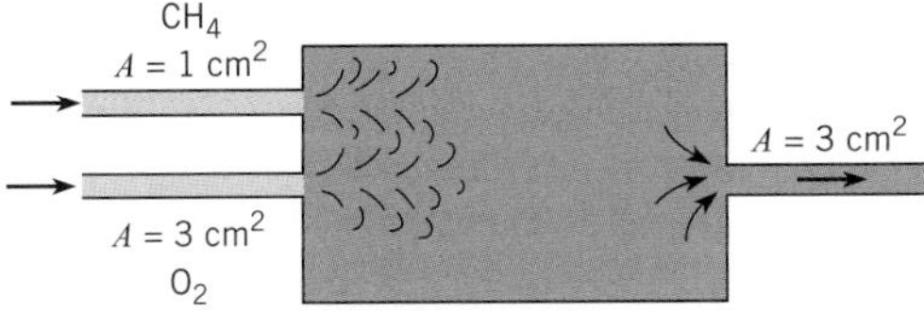

PROBLEM 5.55

5.56 A compressor supplies gas to a 10-m^3 tank. The inlet mass flow rate is given by $\dot{m} = 0.5\rho_0/\rho$ (kg/s) where ρ is the density in the tank and ρ_0 is the initial density. Find the time it would take to increase the density in the tank by a factor of 2 if the initial density is 2 kg/m^3. Assume the density is uniform throughout the tank.

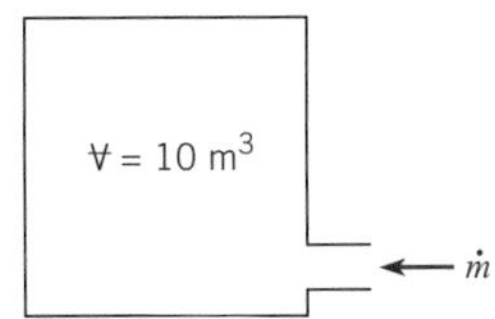

PROBLEM 5.56

5.57 A slow leak develops in a tire (assume constant volume), in which it takes 3 hr for the pressure to decrease from 30 psig to 25 psig. The air volume in the tire is 0.5 ft^3, and the temperature remains constant at 60°F. The mass-flow rate of air is given by $\dot{m} = 0.68pA/\sqrt{RT}$. Calculate the area of the hole in the tire. Atmospheric pressure is 14 psia.

5.58 Oxygen leaks slowly through a small orifice in an oxygen bottle. The volume of the bottle is 0.1 m^3, and the diameter of the orifice is 0.15 mm. The temperature in the tank remains constant at 18°C, and the mass-flow rate is given by $\dot{m} = 0.68pA/\sqrt{RT}$. How long will it take the absolute pressure to decrease from 10 to 5 MPa?

5.59 How long will it take the water surface in the tank shown to drop from $h = 3$ m to $h = 50$ cm?

5.60 A cylindrical drum of water, lying on its side, is being emptied through a 2-in.-diameter short pipe at the bottom of the drum. The velocity of the water out of the pipe is $V = \sqrt{2gh}$, where g is the acceleration due to gravity and h is the height of the water surface above the outlet of the tank. The tank is 4 ft long and 2 ft in diameter. Initially the tank is half-full. Find the time for the tank to empty.

5.61 A pipe is supplying fluid to a funnel as shown. The discharge from the pipe is 0.03 ft^3/s. The exit velocity V_e varies with the height of the liquid in the funnel as $V_e = \sqrt{2gh}$ where g is the acceleration due to gravity. The cross-sectional area of the tank varies as A_s (ft^2) $= 0.01h^2$, where h is in feet. The exit diameter of the pipe is 1 in. Find the level of the fluid in the funnel for which steady-state conditions are achieved.

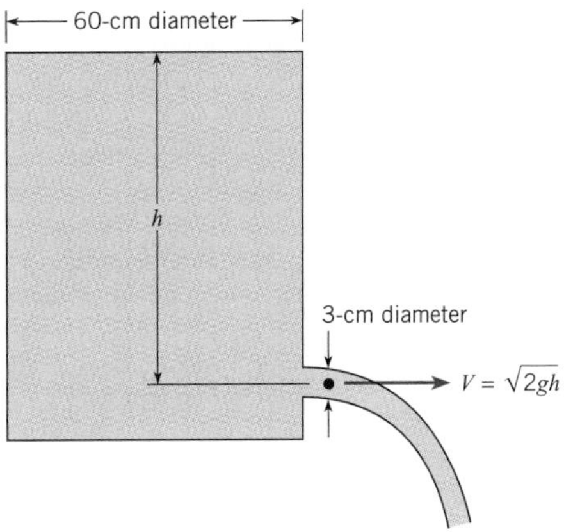

PROBLEM 5.59

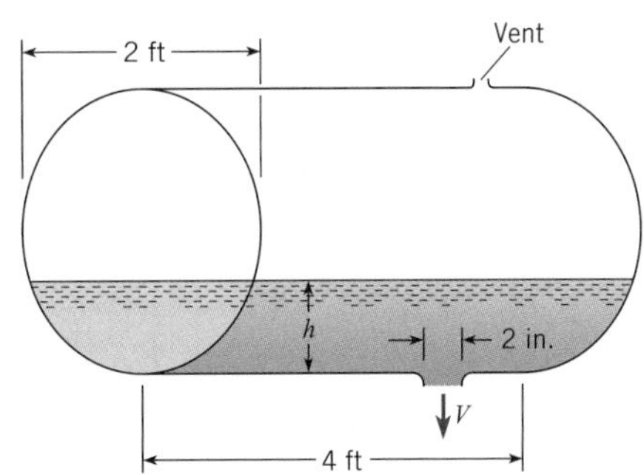

PROBLEM 5.60

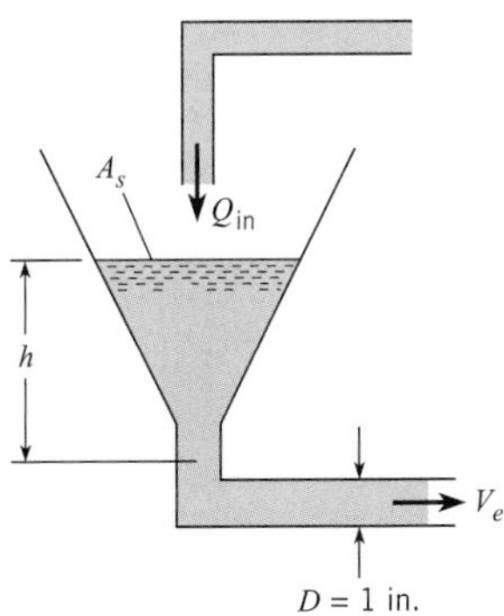

PROBLEM 5.61

5.62 Water is draining from a pressurized tank as shown in the figure. The exit velocity is given by

$$V_e = \sqrt{\frac{2p}{\rho} + 2gh}$$

where p is the pressure in the tank, ρ is the water density, and h is the elevation of the water surface above the outlet. The depth of the water in the tank is 2 m. The tank has a cross-sectional area of 1 m^2 and the exit area of the pipe is 10 cm^2. The pressure in the tank is maintained at 10 kPa. Find the time required to empty the tank. Compare this value with the time required if the tank is not pressurized.

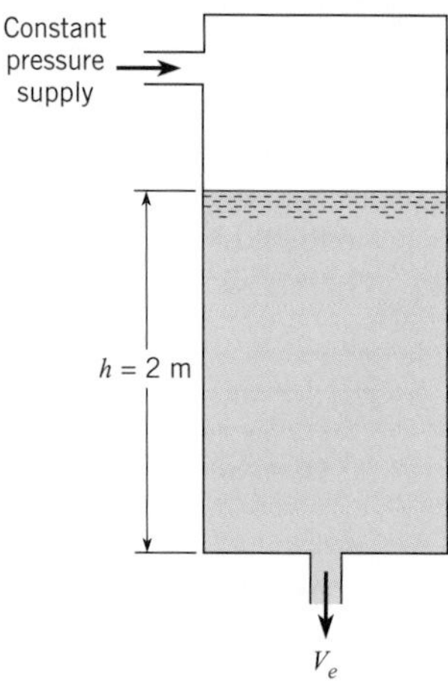

PROBLEM 5.62

FLUID SOLUTIONS **5.63** For the type of tank shown, the tank diameter is given as $D = d + C_1 h$, where d is the bottom diameter and C_1 is a constant. Derive a formula for the time of fall of liquid surface from $h = h_0$ to $h = h$ in terms of d_j, d, h_0, h, and C_1. Solve for t if $h_0 = 1$ m, $h = 20$ cm, $d = 20$ cm, $C_1 = 0.3$, and $d_j = 5$ cm. The velocity of water in the liquid jet exiting the tank is $V_e = \sqrt{2gh}$.

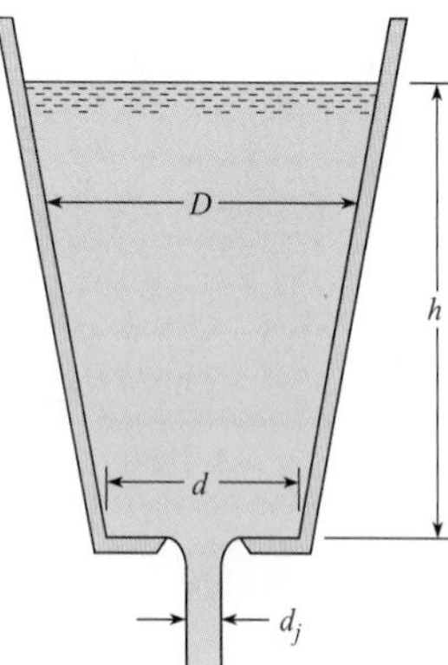

PROBLEM 5.63

5.64 Water drains out of a trough as shown. The angle with the vertical of the sloping sides is α, and the distance between the parallel sides is B. The width of the trough is $W_0 + 2h\tan\alpha$, where h is the distance from the trough bottom. The velocity of the water issuing from the opening in the bottom of the trough is equal to $V_e = \sqrt{2gh}$. The area of the water stream at the bottom is A_e. Derive an expression for the time to drain to depth h in terms of h/h_0, W_0/h_0, $\tan\alpha$, and $A_e g^{0.5}/(h_0^{1.5}B)$, where h_0 is the original depth. Find the time to drain to one-half the original depth for $W_0/h_0 = 0.2$, $\alpha = 30°$, $A_e g^{0.5}/(h_0^{1.5}B) = 0.01\ \text{s}^{-1}$.

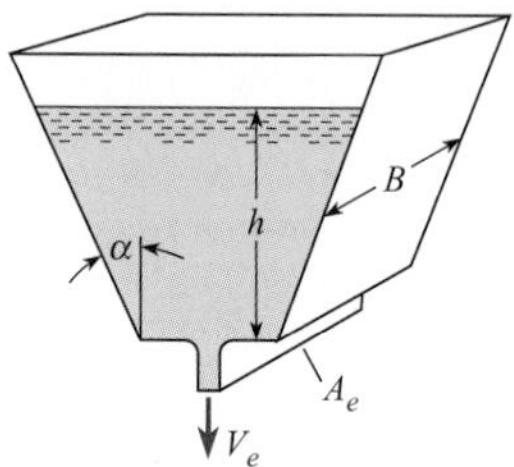

PROBLEM 5.64

5.65 A spherical tank with a diameter of 1 m is half filled with water. A port at the bottom of the tank is opened to drain the tank. The hole diameter is 1 cm and the velocity of the water draining from the hole is $V_e = \sqrt{2gh}$, where h is the elevation of the water surface above the hole. Find the time required for the tank to empty.

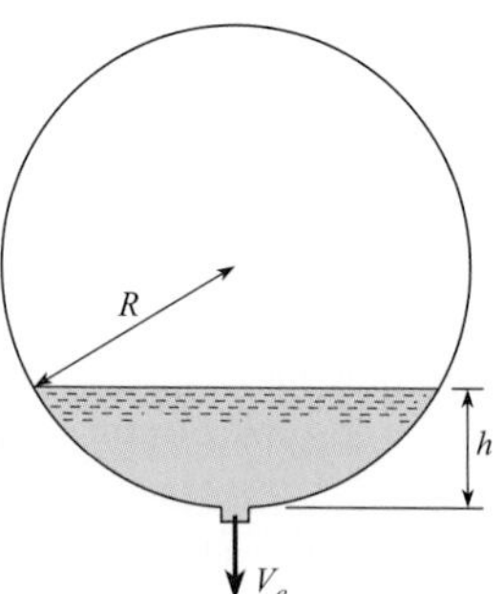

PROBLEM 5.65

5.66 A tank containing oil is to be pressurized to decrease the draining time. The tank, shown in the figure, is 2 meters in diameter and 6 meters high. The oil is originally at a level of 5 meters. The oil has a density of 880 kg/m^3. The outlet port has a diameter of 2 cm and the velocity at the outlet is given by

$$V_e = \sqrt{2gh + \frac{2p}{\rho}}$$

where p is the gage pressure in the tank, ρ is the density of the oil and h is the elevation of the surface above the hole. Assume during the emptying operation that the temperature of the air in the tank is constant. The pressure will vary as

$$p = (p_0 + p_{atm})\frac{(L-h_0)}{(L-h)} - p_{atm}$$

where L is the height of the tank, p_{atm} is the atmospheric pressure and the subscript 0 refers to the initial conditions. The initial pressure in the tank is 300 kPa gage and the atmospheric pressure is 100 kPa.

Applying the continuity equation to this problem, one finds

$$\frac{dh}{dt} = -\frac{A_e}{A_T}\sqrt{2gh + \frac{2p}{\rho}}$$

Integrate this equation numerically to predict the depth of the oil with time for a period of one hour.

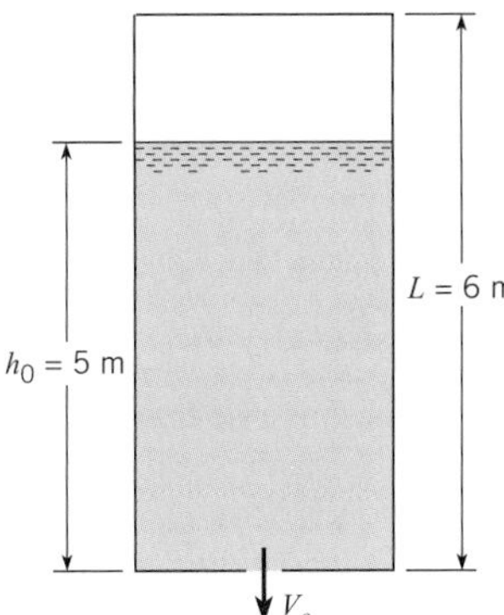

PROBLEM 5.66

5.67 An end-burning rocket motor has a chamber diameter of 10 cm and a nozzle exit diameter of 8 cm. The density of the propellant is 1750 kg/m^3 and the surface regresses at the rate of 1 cm/s. The gases crossing the nozzle exit plane have a pressure of 10 kPa abs and a temperature of 2000°C. The gas constant of the exhaust gases is 415 J/kg K. Calculate the gas velocity at the nozzle exit plane.

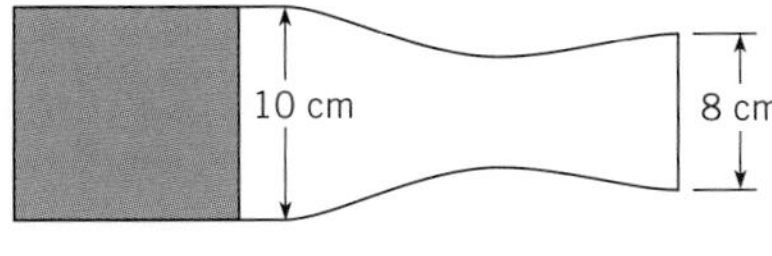

PROBLEM 5.67

5.68 A cylindrical-port rocket motor has a grain design consisting of a cylindrical shape as shown. The curved internal surface and both ends burn. The propellant surface regresses uniformly at 1.2 cm/s. The propellant density is 2000 kg/m^3. The inside diameter of the motor is 20 cm. The propellant grain is 40 cm long and has an inside diameter of 12 cm. The diameter of the nozzle exit plane is 20 cm. The gas velocity at the exit plane is 2000 m/s. Determine the gas density at the exit plane.

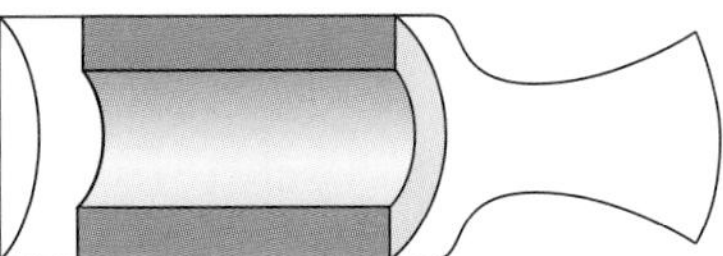

PROBLEM 5.68

5.69 The mass-flow rate through a nozzle is given by

$$\dot{m} = 0.65\frac{p_c A_t}{\sqrt{RT_c}}$$

where p_c and T_c are the pressure and temperature in the rocket chamber and R is the gas constant of the gases in the chamber. The propellant burning rate (surface regression rate) can be expressed as $\dot{r} = ap_c^n$, where a and n are two empirical constants. Show, by application of the continuity equation, that the chamber pressure can be expressed as

$$p_c = \left(\frac{a\rho_p}{0.65}\right)^{1/(1-n)}\left(\frac{A_g}{A_t}\right)^{1/(1-n)}(RT_c)^{1/[2(1-n)]}$$

where ρ_p is the propellant density and A_g is the grain surface burning area. If the operating chamber pressure of a rocket motor is 3.5 MPa and $n = 0.3$, how much will the chamber pressure increase if a crack develops in the grain, increasing the burning area by 20%?

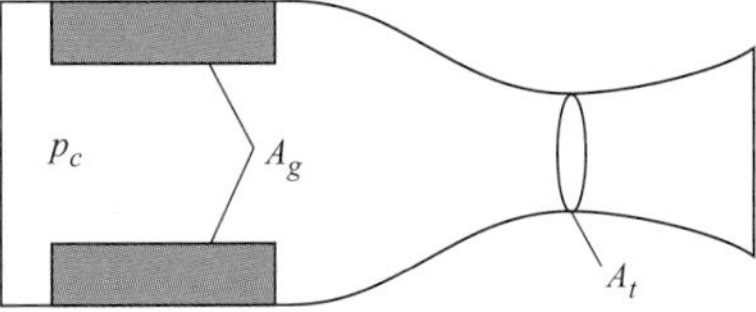

PROBLEM 5.69

5.70 A piston is moving up during the exhaust stroke of a four-cycle engine. Mass escapes through the exhaust port at a rate given by

$$\dot{m} = 0.65\frac{p_c A_v}{\sqrt{RT_c}}$$

where p_c and T_c are the cylinder pressure and temperature, A_v is the valve opening area, and R is the gas constant of the exhaust gases. The bore of the cylinder is 10 cm, and the piston is moving upward at 30 m/s. The distance between the piston and the head is 10 cm. The valve opening area is 1 cm^2, the chamber pressure is 300 kPa abs, the chamber temperature is 600°C, and the gas constant is 350 J/kg K. Applying the continuity equation, determine the rate at which the gas density is changing in the cylinder. Assume the density and pressure are uniform in the cylinder and the gas is ideal.

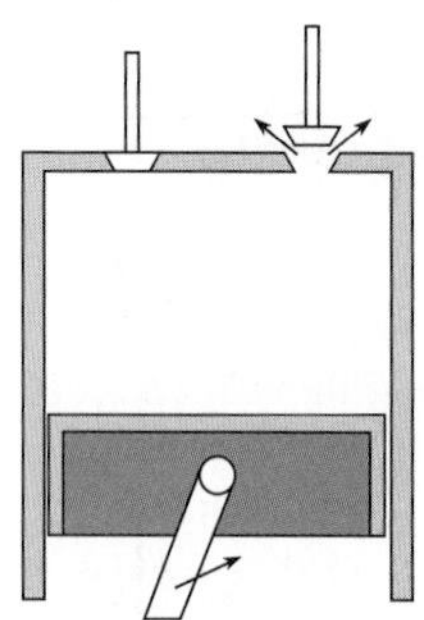

PROBLEM 5.70

FLUID SOLUTIONS **5.71** The flow pattern through the pipe contraction is as indicated, and the discharge of water is 70 cfs. For $d = 2$ ft and $D = 6$ ft, what will be the pressure at point B if the pressure at point A is 3500 psf?

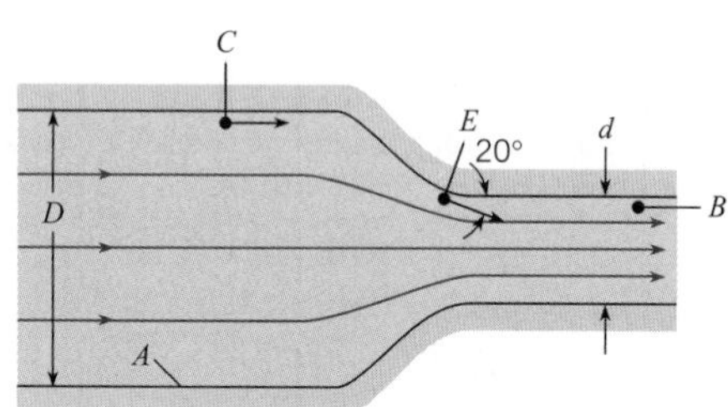

PROBLEMS 5.71, 5.72

5.72 Consider the flow of water in the pipe contraction. Assuming steady, irrotational, nonviscous flow, find the pressure at point E (same elevation as point B) if the velocity at point E is 50 ft/s and the pressure and velocity at point C are 15 psi and 10 ft/s, respectively.

5.73 The annular venturimeter is useful for metering flows in pipe systems for which upstream calming distances are limited. The annular venturimeter consists of a cylindrical section mounted inside a pipe as shown in the figure. The pressure difference is measured between the upstream pipe and at the region adjacent to the cylindrical section. Air at standard conditions flows in the system. The pipe diameter is 4 inches. The ratio of the cylindrical section diameter to the inside pipe diameter is 0.8. A pressure difference of 2 inches of water is measured. Find the volume flow rate in the system. Assume the flow is incompressible, inviscid, and steady and that the velocity is uniformly distributed across the pipe sections.

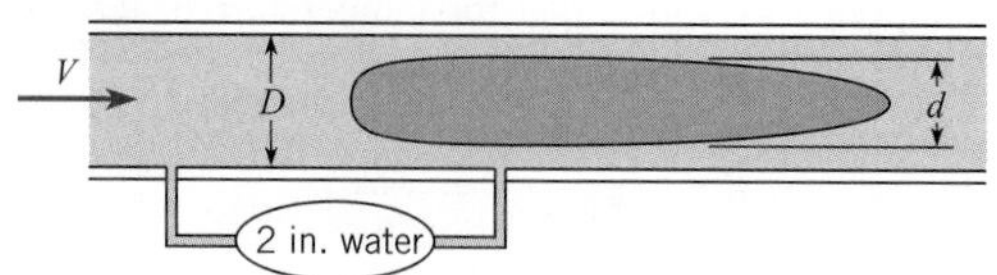

PROBLEM 5.73

5.74 Venturi-type applicators are frequently used to spray liquid fertilizers. Water flowing through the venturi creates a sub-atmospheric pressure at the throat, which in turn causes the liquid fertilizer to flow up the feed tube and mix with the water in the throat region. The venturi applicator shown in the figure uses water at 20°C to spray a liquid fertilizer with the same density. The venturi exhausts to the atmosphere, and the exit diameter is 1 cm. The ratio of exit area to throat area (A_2/A_1) is 2. The flow rate of water through the venturi is 10 lpm (liters per minute). The bottom of the feed tube in the reservoir is 5 cm below the liquid fertilizer surface and 10 cm below the centerline of the venturi. The pressure at the liquid fertilizer surface is atmospheric. The flow rate through the feed tube between the reservoir and venturi throat is

$$Q_l\,(\text{lpm}) = 0.5\sqrt{\Delta h}$$

where Δh is the drop in piezometric head (in meters) between the feed tube entrance and the venturi centerline. Find the flow rate of liquid fertilizer in the feed tube, Q_l. Also find the concentration of liquid fertilizer in the mixture, $[Q_l/(Q_l + Q_w)]$, at the end of the sprayer.

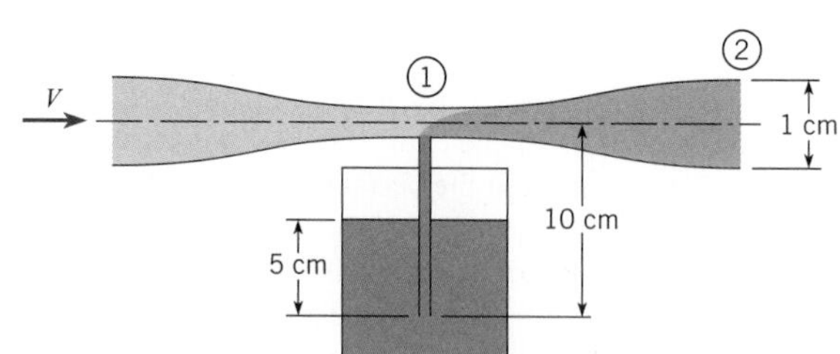

PROBLEM 5.74

5.75 When gage A indicates a pressure of 120 kPa gage, then cavitation just starts to occur in the venturi meter. If $D = 40$ cm and $d = 10$ cm, what is the water discharge in the system for this condition of incipient cavitation? The atmospheric pressure is 100 kPa gage, and the water temperature is 10°C. Neglect gravitational effects.

5.76 Air with a density of 0.0644 lbm/ft^3 is flowing upward in the vertical duct, as shown. The velocity at the inlet (station 1) is 100 ft/s, and the area ratio between station 1 and 2 is 0.5 $(A_2/A_1 = 0.5)$. Two pressure taps, 10 ft apart, are connected to a

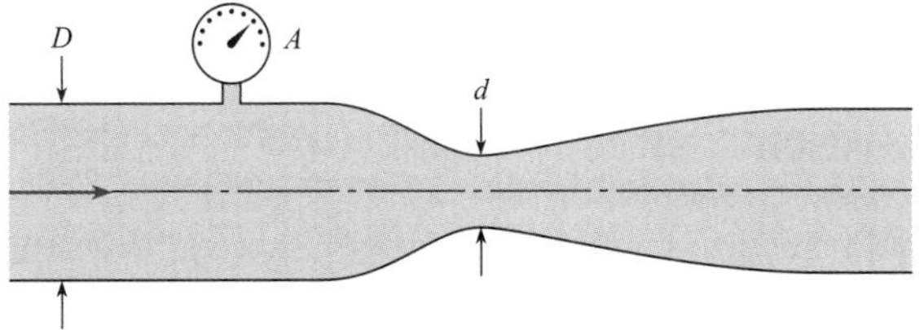

PROBLEM 5.75

manometer, as shown. The specific weight of the manometer liquid is 120 lbf/ft^3. Find the deflection, Δh, of the manometer.

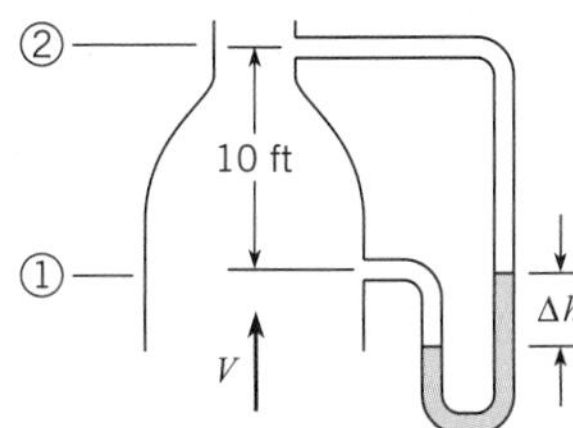

PROBLEM 5.76

5.77 An atomizer utilizes a constriction in an air duct as shown. Design an operable atomizer making your own assumptions regarding the air source.

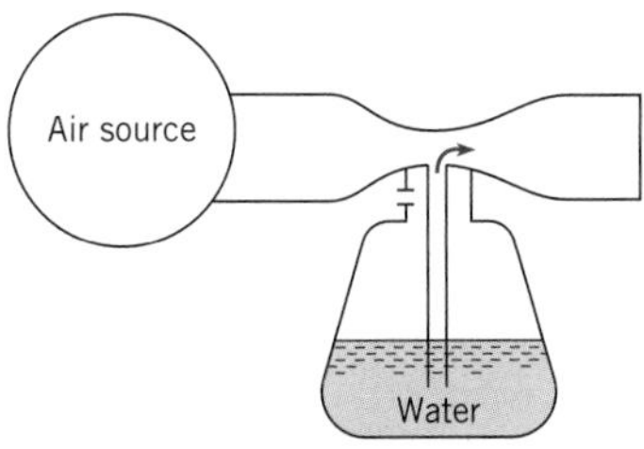

PROBLEM 5.77

5.78 A suction device is being designed based on the venturi principle to lift objects submerged in water. The operating water temperature is 15°C. The suction cup is located 1 m below the water surface, and the venturi throat is located 1 m above the water. The atmospheric pressure is 100 kPa. The ratio of the throat area to the exit area is $\frac{1}{4}$, and the exit area is 0.001 m^2. The area of the suction cup is 0.1 m^2.

a. Find the velocity of the water at the exit for maximum lift condition.

b. Find the discharge through the system for maximum lift condition.

c. Find the maximum load the suction cup can support.

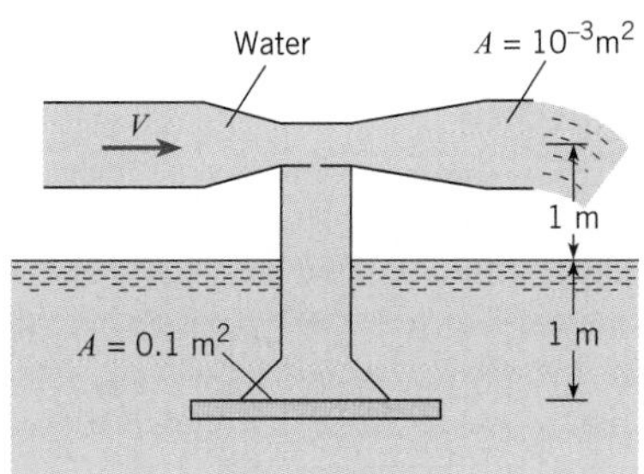

PROBLEM 5.78

5.79 A design for a hovercraft is shown in the figure. A fan brings air at 60°F into a chamber, and the air is exhausted between the skirts and the ground. The pressure inside the chamber is responsible for the lift. The hovercraft is 15 ft long and 7 ft wide. The weight of the craft including crew, fuel, and load is 2000 lbf. Assume that the pressure in the chamber is the stagnation pressure (zero velocity) and the pressure where the air exits around the skirt is atmospheric. Assume the air is incompressible, the flow is steady, and viscous effects are negligible. Find the air flow rate necessary to maintain the skirts at a height of 3 inches above the ground.

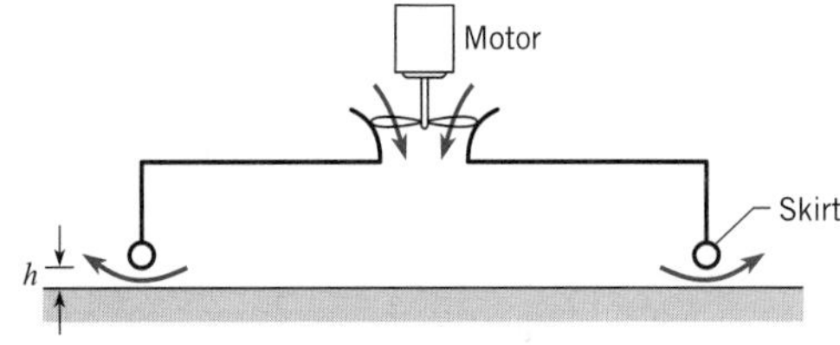

PROBLEM 5.79

5.80 Water is forced out of this cylinder by the piston. If the piston is driven at a speed of 5 ft/s, what will be the speed of efflux of the water from the nozzle if $d = 2$ in. and $D = 4$ in.? Neglecting friction and assuming irrotational flow, determine the force F that will be required to drive the piston. The exit pressure is atmospheric pressure.

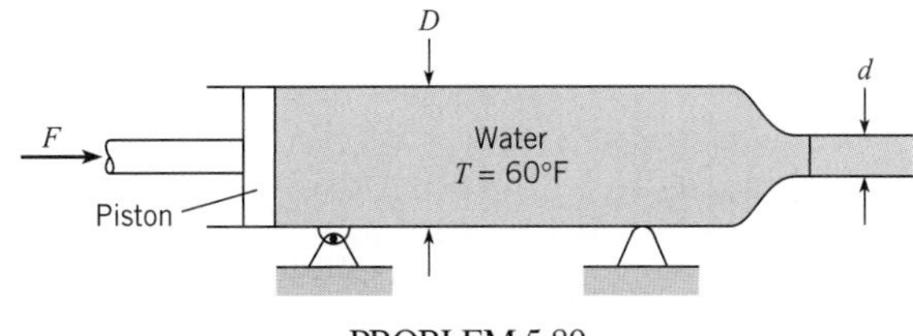

PROBLEM 5.80

5.81 Refer to the figure for Prob. 4.26 on p. 129. Assume that the jet issuing from the nozzle into the atmosphere has a diameter of 0.50 ft. Water is flowing at a constant rate of 20 ft^3/s. Assuming irrotational flow, find the gage pressure in the pipe.

5.82 Refer to Prob. 4.19 on p. 128. The air density is 1.2 kg/m^3, and $U_0 = 30$ m/s. Assume irrotational flow. Determine the pressure in the air at $x = r_0$, $x = 1.1r_0$, and $x = 2r_0$.

5.83 An engineer is designing an "elbow" meter for measuring the discharge of a liquid. The elbow meter consists of an elbow with pressure taps mounted across the meter, as shown. The elbow has a rectangular cross section. The velocity across the elbow varies as $V = K/r$, where K is a constant and r is the radius of curvature of the flow. Assume this is an irrotational flow field, so the Bernoulli equation is valid across streamlines.

a. Develop an equation for the discharge, Q, through the elbow as a function of pressure difference Δp between the taps in the form

$$Q = (2\Delta p/\rho)^{0.5} A_c f(r_2/r_1)$$

where ρ is the fluid density, A_c is the cross-sectional area of the elbow, and r_2 and r_1 are the outer and inner radii of the elbow, respectively. Find the functional relationship $f(r_2/r_1)$.

b. Evaluate the coefficient $f(r_2/r_1)$ for an elbow meter with a radius ratio of 1.5.

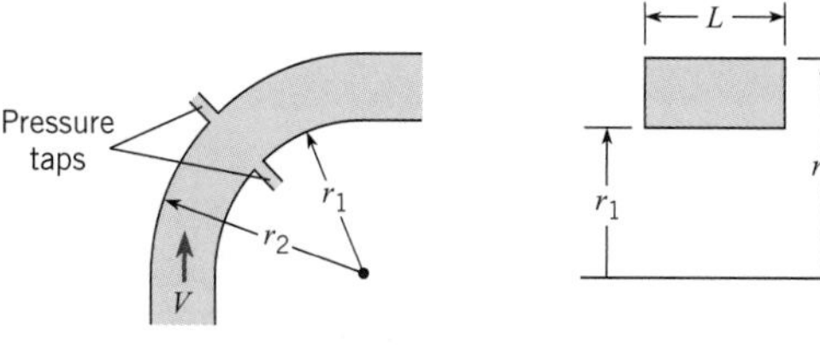

PROBLEM 5.83

5.84 A sphere 1 ft in diameter is moving at a depth of 10 ft below a water surface where the water temperature is 50°F. At what speed in still water will cavitation first occur?

5.85 When the hydrofoil shown was tested, the minimum pressure on the surface of the foil was found to be 70 kPa absolute when the foil was submerged 1.80 m and towed at a speed of 8 m/s. At the same depth, at what speed will cavitation first occur? Assume irrotational flow for both cases and $T = 10°\text{C}$.

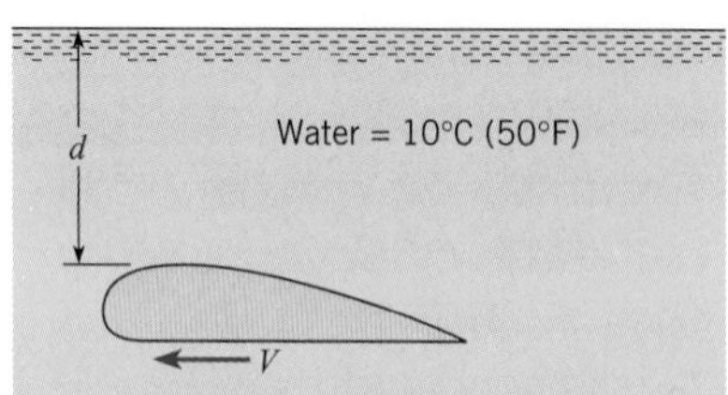

PROBLEMS 5.85, 5.86, 5.87, 5.88

5.86 For the hydrofoil of Prob. 5.85, at what speed will cavitation begin if the depth is increased to 3 m?

5.87 When the hydrofoil shown was tested, the minimum pressure on the surface of the foil was found to be 2.5 psi vacuum when the foil was submerged 4 ft and towed at a speed of 20 ft/s. At the same depth, at what speed will cavitation first occur? Assume irrotational flow for both cases and $T = 50°\text{F}$.

5.88 For the conditions of Prob. 5.87, at what speed will cavitation begin if the depth is increased to 10 ft?

5.89 A sphere is moving in water at a depth where the absolute pressure is 18 psia. The maximum velocity on a sphere occurs 90° from the forward stagnation point and is 1.5 times the free-stream velocity. The density of water is 62.4 lbm/ft^3. Calculate the speed of the sphere at which cavitation will occur. $T = 50°\text{F}$.

5.90 The minimum pressure on a cylinder moving horizontally in water ($T = 10°\text{C}$) at 5 m/s at a depth of 1 m is 80 kPa absolute. At what velocity will cavitation begin? Atmospheric pressure is 100 kPa absolute.

5.91 It is predicted that a flow field will have the following velocity components:

$$u = V(x^3 + xy^2) \qquad v = V(y^3 + yx^2) \qquad w = 0$$

V is a constant. Is such a flow field possible? (Does it satisfy continuity?)

5.92 The velocity components of a flow field are given by

$$u = \frac{y}{(x^2 + y^2)^{3/2}} \qquad v = \frac{-x}{(x^2 + y^2)^{3/2}}$$

Is continuity satisfied? Is the flow irrotational?

5.93 A u component of velocity is given by $u = Axy$, where A is a constant. What is a possible v component? What must the v component be if the flow is irrotational?

Holding and aiming a fire hose requires considerable effort. As the water exits at high velocity from the nozzle a force is needed to hold the hose in position. This force can be determined by applying the momentum principle for flowing fluids. (Corbis Digital Stock)

CHAPTER 6

Momentum Principle

In solid mechanics, Newton's second law of motion, $\mathbf{F} = m\mathbf{a}$, is applied to many problems involving force and acceleration. The law is equally useful in fluid mechanics. The analysis of forces on pipe bends, thrust produced by a turbojet, acceleration of a rocket, and torque produced by a hydraulic turbine are all examples of the application of Newton's second law in fluid mechanics. In this chapter the control volume approach is applied to $\mathbf{F} = m\mathbf{a}$ and the momentum principle, relating the change in momentum in the control volume to forces acting on the control surface, is developed. The same approach is extended to the moment-of-momentum principle. Finally, the control volume approach is used to derive the differential form of the equation of motion for a fluid, namely the Navier-Stokes equation.

Momentum Equation: Derivation

When forces act on a particle, the particle accelerates according to Newton's second law of motion:

$$\sum \mathbf{F} = m\mathbf{a} \tag{6.1}$$

By definition, the mass (m) is constant, so the equation may be written using momentum:

$$\sum \mathbf{F} = \frac{d(m\mathbf{v})}{dt} \tag{6.2}$$

Although Eqs. (6.1) and (6.2) apply to a single particle, the law can also be formulated for a system comprised of a group of particles, for example, a fluid system. In this case, the law may be written as

$$\sum \mathbf{F} = \frac{d(\mathbf{Mom}_{\text{sys}})}{dt} \tag{6.3}$$

The term $\mathbf{Mom}_{\text{sys}}$ denotes the total momentum of all mass forming the system.

Equation (6.3) is a Lagrangian equation. For analyses of fluids, it is advantageous to convert this equation to an Eulerian form, giving a result that applies to control volumes. To derive an Eulerian equation, we use the Reynolds transport theorem from Chapter 5:

$$\frac{dB_{sys}}{dt} = \frac{d}{dt}\int_{cv} b\rho \, d\forall + \int_{cs} b\rho \mathbf{V} \cdot d\mathbf{A} \tag{5.22}$$

where $\mathbf{V}$ is fluid velocity relative to the control surface at the location where the flow is crossing the surface. The extensive property B_{sys} becomes the momentum of the system: $B = \mathbf{Mom}_{sys}$. The intensive property b becomes the momentum per unit mass within the system. The momentum of an element is $m\mathbf{v}$, and so $b = \mathbf{v}$. The velocity $\mathbf{v}$ must be relative to an inertial reference frame, that is, a frame that does not rotate and can either be fixed or moving at a constant velocity. Substituting for B and b into Eq. (5.22) gives

$$\frac{d(\mathbf{Mom}_{sys})}{dt} = \frac{d}{dt}\int_{cv} \mathbf{v}\rho \, d\forall + \int_{cs} \mathbf{v}\rho \mathbf{V} \cdot d\mathbf{A} \tag{6.4}$$

Combining Eqs. (6.3) and (6.4) gives the integral form of the *momentum principle*:

$$\sum \mathbf{F} = \frac{d}{dt}\int_{cv} \mathbf{v}\rho \, d\forall + \int_{cs} \mathbf{v}\rho \mathbf{V} \cdot d\mathbf{A} \tag{6.5}$$

This equation states that the sum of external forces acting on the material in the control volume equals the rate of momentum change inside the control volume plus the net rate at which momentum flows out of the control volume.

Interpretation of the Momentum Equation

Application of the momentum equation to fluid flow problems is analogous to the use of the "free-body" approach in solid mechanics. In solid mechanics, a system of interest is isolated from its surroundings, thereby creating a free body, and forces are applied to replace the influence of the surroundings. These forces are then summed and equated with the product of mass and acceleration. In fluid mechanics, the system of interest is the material contained within the control volume, and forces are applied to this system to represent the effect of the surroundings. Forces are then summed and equated to momentum changes of the flow.

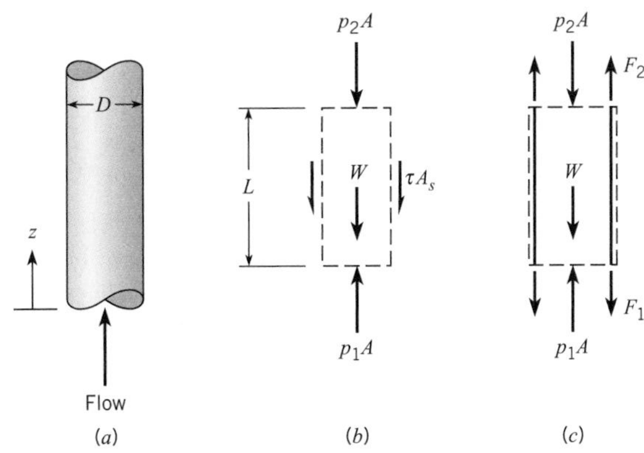

FIGURE 6.1

Forces associated with flow in a pipe: (a) pipe schematic, (b) control volume situated inside the pipe, and (c) control volume surrounding the pipe.

Force terms

In Eq. (6.5), the force term represents all external forces that act on the material within the control volume. For example, consider flow inside a vertical pipe, as shown in Fig. 6.1*a*. One possible control volume is a cylinder with diameter D and length L located just inside the pipe wall. As shown in Fig. 6.1*b*, the fluid within the control volume has been isolated from its surroundings and the effect of the surroundings are shown as forces. The effect of the wall is replaced by a force equal to the shear stress (τ) times the pipe surface area ($A_s = \pi DL$). The force due to pressure is given by pressure (p) times the section area ($A = \pi D^2/4$). The weight of the fluid is given by $W = \gamma(\pi D^2/4)L$. Thus, the net force acting in the z direction is given by

$$\sum F_z = (p_1 - p_2)\frac{\pi}{4}D^2 - \tau\pi DL - \gamma(\pi D^2/4)L \tag{6.6a}$$

Another possible control volume has a length L and a diameter that is larger than the pipe's outside diameter. As shown in Fig. 6.1*c*, this control volume cuts through the pipe wall. Comparing Figs. 6.1*b* and *c* shows that the pressure forces are the same. However, in Fig. 6.1*c*, there is no force associated with shear stress, but there are two new forces, F_1 and F_2, which represent the forces due to the pipe wall. Also, the weight of material within the control volume now includes the weight of the fluid and the pipe wall (W_p). The net z-direction force is

$$\sum F_z = (p_1 - p_2)\frac{\pi}{4}D^2 - F_1 + F_2 - (W_p + \gamma(\pi D^2/4)L) \tag{6.6b}$$

The choice of control volume depends on what information we are trying to find. To relate the pressure change between sections 1 and 2 to wall shear stress, Eq. (6.6*a*) would be best. To find the tensile force carried by the pipe wall, we would use Eq. (6.6*b*).

We identify the sketches in Fig. 6.1*b* and *c* as *force diagrams* (FD). A force diagram shows the forces acting on the material contained within a control volume. A force diagram is equivalent to a free-body diagram at the instant in time when the momentum equation is applied.

In Fig. 6.1*b*, the force of gravity (weight) acts on each mass element in the control volume (with the resultant force acting at the mass center). A force that acts on mass elements within the body is defined as a *body force*. A body force can act at a distance without any physical contact. Examples of body forces include gravitational, electrostatic, and magnetic forces.

Excepting the body force (weight), all forces shown in Figs. 6.1*b* and *c* are surface forces. A surface force is defined as a force that requires physical contact, meaning that surface forces act at the control surface. For example, p_1A_1 acts at the control surface and requires contact between the fluid outside the control volume and the fluid inside the control volume. With respect to pressure, the net force is obtained by integrating the pressure over the area of the surface. For example, if the pressure varies hydrostatically, the magnitude of the force and the line of action would be determined using the methods presented in Chapter 3. When evaluating pressure forces, engineers commonly use gage pressure. In this case, the force associated with atmospheric pressure acting over a surface is zero. In addition to pressure, surface forces can be caused by shear stress, for example the force τA_s shown in Fig. 6.1*b*.

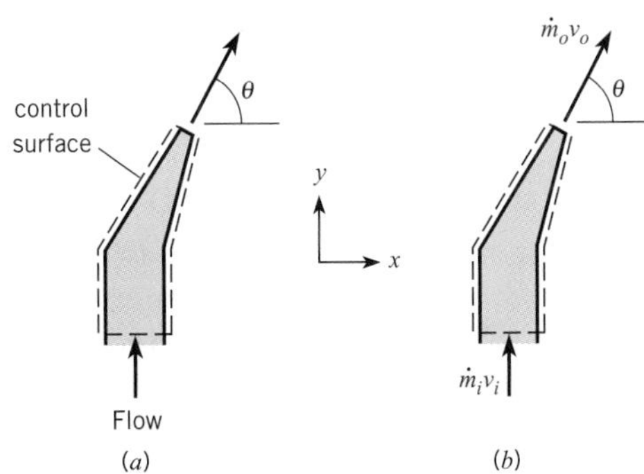

FIGURE 6.2

(a) Nozzle, and (b) momentum diagram for nozzle.

Momentum Accumulation

The first term on the right side of Eq. (6.5) represents the rate at which the momentum of the material inside the control volume is changing with time. In particular, the mass of a volume element in the control volume is $\rho\, d\mathcal{V}$, so the product $\mathbf{v}\, \rho \mathcal{V}$ is the momentum of a volume element. Integrating over the control volume gives total momentum of the material in the control volume. Taking the time derivative gives the rate at which the momentum is changing. This term may be described as the net rate of momentum accumulation, and we will refer to it as *momentum accumulation*. The units are momentum per unit time, which are equivalent to the units for force.

In many problems, the momentum accumulation is zero. For example, consider steady flow through the control volume surrounding the nozzle shown in Fig. 6.2*a*. The fluid inside the control volume has momentum because it is moving. However, the velocity and density at each point are constant with time, so the total momentum in the control volume is constant, and the momentum accumulation is zero. The evaluation of momentum accumulation is completed by considering the structural elements (i.e., the nozzle walls). Since the structural elements are stationary, there is no momentum change or momentum accumulation.

In summary, the momentum of the material inside a control volume is evaluated by integrating the momentum of each volume element over the control volume. If the momentum in each differential volume is constant with time (e.g., steady flow, a stationary structural part), the momentum accumulation is zero.

If the flow crossing the control surface occurs through a series of inlet and outlet ports and if the velocity $\mathbf{v}$ is uniformly distributed across each port, then the mass flow form of the Reynolds transport theorem, Eq. (5.24), can be used, and the *momentum principle* becomes

$$\sum \mathbf{F} = \frac{d}{dt}\int_{\text{cv}} \rho \mathbf{v}\, d\mathcal{V} + \sum_{\text{cs}} \dot{m}_o \mathbf{v}_o - \sum_{\text{cs}} \dot{m}_i \mathbf{v}_i \tag{6.7}$$

where the subscripts o and i refer to the outlet and inlet ports, respectively. Notice that the product of $\dot{m}\mathbf{v}$ corresponds to the mass per unit time times velocity, or momentum per unit time, which has the same units as force.

As long as $\mathbf{v}$ is uniform at the control surface, Eq. (6.7) applies to any control volume, including one that is moving, deforming, or both. In all cases, $\dot{m}$ is the rate at which mass is passing across the control surface, and $\mathbf{v}$ is velocity evaluated at the control surface with respect to the inertial reference frame that is selected.

Momentum Diagram

The momentum terms on the right side of Eq. (6.5) may be visualized with a *momentum diagram* (MD). The momentum diagram is created by sketching a control volume and then drawing a vector to represent the momentum accumulation term and a vector to represent momentum flow at each section where mass crosses the control surface.

Although the momentum diagram applies to the integral form of the momentum principle [Eq. (6.5)], the diagram takes on a simple form when the velocity $\mathbf{v}$ is uniformly distributed across each inlet and outlet port and Eq. (6.7) applies. For example, consider steady flow through the nozzle shown in Fig. 6.2. For the control volume indicated, the momentum accumulation term is zero, and this vector is omitted from the diagram. If the velocity is assumed to be uniform across the inlet and exit sections, the outlet momentum flow is given by $\dot{m}_o\mathbf{v}_o$, and the inlet momentum flow is given by $\dot{m}_i\mathbf{v}_i$, as shown in Fig. 6.2*b*. To evaluate the momentum flow, we can use the diagram to see that

$$\sum_{\text{cs}} \dot{m}_o\mathbf{v}_o = [\dot{m}_o v_o \cos(\theta)]\mathbf{i} + [\dot{m}_o v_o \sin(\theta)]\mathbf{j}$$

and

$$\sum_{\text{cs}} \dot{m}_i\mathbf{v}_i = [\dot{m}_i v_i]\mathbf{j}$$

Recognizing that $\dot{m}_i = \dot{m}_o = \dot{m}$, we can combine the above equations to show that the net outward flow of momentum is

$$\sum_{\text{cs}} \dot{m}_o\mathbf{v}_o - \sum_{\text{cs}} \dot{m}_i\mathbf{v}_i = [\dot{m}v_o\cos(\theta)]\mathbf{i} + [\dot{m}v_o\sin(\theta) - \dot{m}v_i]\mathbf{j}$$

Using the momentum diagram is a straightforward way to evaluate the momentum terms.

The Momentum Equation for Cartesian Coordinates

When velocity $\mathbf{v}$ varies across the control surface, the general form of momentum equation (6.5) needs to be used.

However, if the velocity $\mathbf{v}$ is uniformly distributed across each inlet and outlet port, Eq. (6.7) applies and three scalar equations can be written for the three coordinate directions:

$$x \text{ direction: } \sum F_x = \frac{d}{dt}\int_{cv} v_x \rho \, d\forall + \sum_{cs} \dot{m}_o v_{ox} - \sum_{cs} \dot{m}_i v_{ix} \tag{6.8a}$$

$$y \text{ direction: } \sum F_y = \frac{d}{dt}\int_{cv} v_y \rho \, d\forall + \sum_{cs} \dot{m}_o v_{oy} - \sum_{cs} \dot{m}_i v_{iy} \tag{6.8b}$$

$$z \text{ direction: } \sum F_z = \frac{d}{dt}\int_{cv} v_z \rho \, d\forall + \sum_{cs} \dot{m}_o v_{oz} - \sum_{cs} \dot{m}_i v_{iz} \tag{6.8c}$$

where the subscripts x, y, and z refer to the force and velocity components in the coordinate directions. The mass flows in the above equations are always scalar (not vector) quantities.

Systematic Approach

A systematic approach is recommended for using the momentum equation. One such approach is summarized below.

Problem setup

- Select an appropriate control volume. Sketch the control volume and coordinate axes. Select an inertial reference frame.
- Identify governing equations. This will include the applicable vector or scalar form of the momentum equation, and may include other equations such as the Bernoulli equation and the continuity equation.

Force analysis and diagram

- Sketch body force(s) (usually only gravitational force) on the force diagram.
- Sketch surface forces on the force diagram; these are forces caused by pressure distribution, shear stress distribution, and supports and structures.

Momentum analysis and diagram

- Evaluate the momentum accumulation term. If the flow is steady and other materials in the control volume are stationary, the momentum accumulation is zero. Otherwise, the momentum accumulation term is evaluated by integration, and an appropriate vector is added to the momentum diagram.
- Sketch momentum flow vectors on the momentum diagram. For uniform velocity, each vector is $\dot{m}\mathbf{v}$.

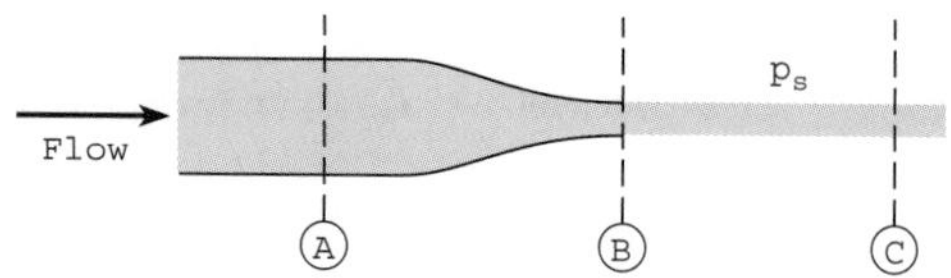

FIGURE 6.3

A fluid jet exiting a nozzle. Pressure of the ambient fluid is p_s. Letters indicate cross sections.

6.3 Typical Applications

This section discusses four common applications of the momentum equation: fluid jets, nozzles, vanes, and pipes.

Fluid Jets

A fluid jet leaving a nozzle is shown in Fig. 6.3. It is conventional to assume that pressure is constant across any cross-section of the jet, e.g., sections B, C, etc. Since pressure is uniform across the jet, the pressure will equal the pressure of the surrounding fluid, p_s.* Thus $p_B = p_C = p_s$. Because of this result, a fluid jet does not exert a surface force when it passes across a control surface. Finally, we typically assume that fluid velocity is uniform across the cross section of a jet.

example 6.1

The sketch below shows a 40-g rocket, of the type used for model rocketry, being fired on a test stand in order to evaluate thrust. The exhaust jet from the rocket motor has a diameter of $d = 1$ cm, a speed of $v = 450$ m/s, and a density of $\rho = 0.5$ kg/m^3. Assume the pressure in the exhaust jet equals ambient pressure, and neglect any momentum changes inside the rocket motor. Find the force F_b acting on the beam that supports the rocket.

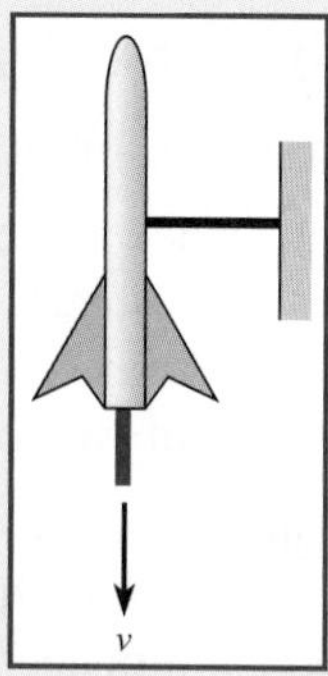

* This assumption is valid if a jet is subsonic, meaning the speed of the jet is less than the local speed of sound in the fluid. Otherwise, the exit pressure can be higher than atmospheric pressure. Supersonic jets are discussed in Chapter 12.

Solution The sketch below shows the control volume, and defines the y coordinate axis. A stationary reference frame was selected. Because this problem involves only one direction, the momentum equation in the z direction (6.8c) will be used.

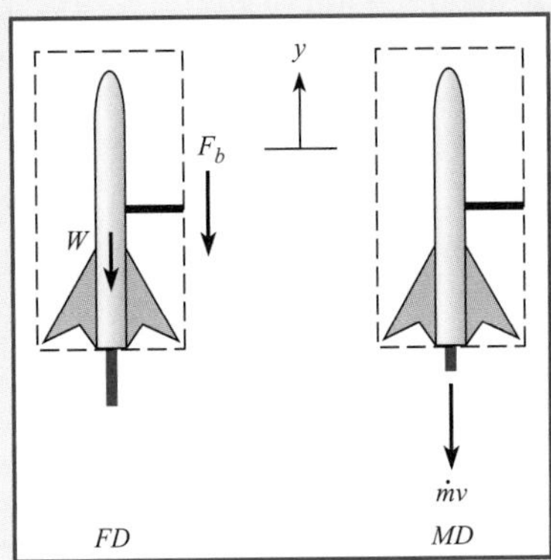

The force diagram (FD) shows a body force that is the weight W, and a surface force that is the force of the beam F_b acting at the control surface. From the force diagram,

$$\sum F_z = (-F_b - W)$$

The momentum accumulation term is zero because the structure is stationary and any momentum changes of the gases within the motor are neglected. As shown on the momentum diagram (MD), there is no inflow ($\dot{m}_i = 0$) and only one outflow of momentum. The momentum flow terms in Eq. (6.8c) may be evaluated using the momentum diagram:

$$\sum_{cs} \dot{m}_o v_{o_z} - \sum_{cs} \dot{m}_i v_{i_z} = \dot{m}(-v) = -\dot{m}v$$

where the minus sign on the right side of the equation denotes that the outward momentum flow is negative with respect to the z direction, as shown on the momentum diagram.

Substituting force and momentum terms into the momentum equation gives

$$-F_b - W = -\dot{m}v$$

The weight is

$$W = mg = (0.04 \text{ kg})(9.81 \text{ m/s}^2) = 0.392 \text{ N}$$

The magnitude of the momentum flow is

$$\dot{m}v = (\rho A v)v = (0.5 \text{ kg/m}^3)(\pi \times 0.01^2/4 \text{ m}^2)(450^2 \text{ m}^2/\text{s}^2) = 7.95 \text{ N}$$

The force is

$$F_b = \dot{m}v - W = 7.95 \text{ N} - 0.392 \text{ N} = 7.56 \text{ N} \quad \triangleleft$$

and the direction of F_b (on the beam) is upward.

The thrust force of the rocket motor is $\dot{m}v = 7.95$ N (1.79 lbf); this value is typical of a small motor used for model rocketry.

The force F_b acts downward at the control surface, yet acts upward on the support beam, as shown in the sketch below. This is an example of an action and reaction force, as described by Newton's third law of motion.

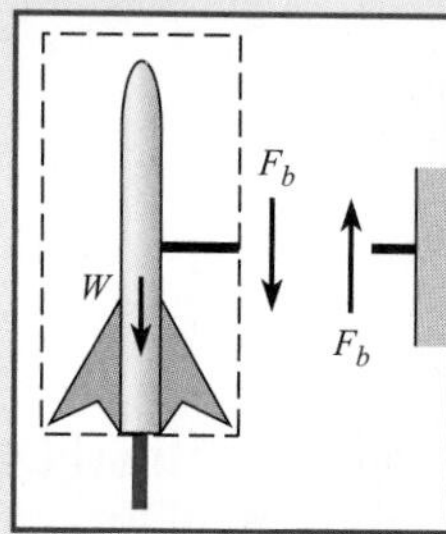

For solving this problem, we used two separate diagrams: the force and the momentum diagram. The rationale for this approach is to regard forces and momentum flows as separate phenomena. This facilitates writing the equations and provides a systematic approach to more complex problems.

example 6.2

As shown in the sketch, concrete is flowing into a cart sitting on a scale. The stream of concrete has a density of $\rho = 150 \text{ lbm/ft}^3$, an area of $A = 1 \text{ ft}^2$, and a speed of $v = 10$ ft/s. At the instant shown, the weight of the cart plus the concrete is 800 lbf. Determine the tension in the cable and the weight recorded by the scale. Assume steady flow.

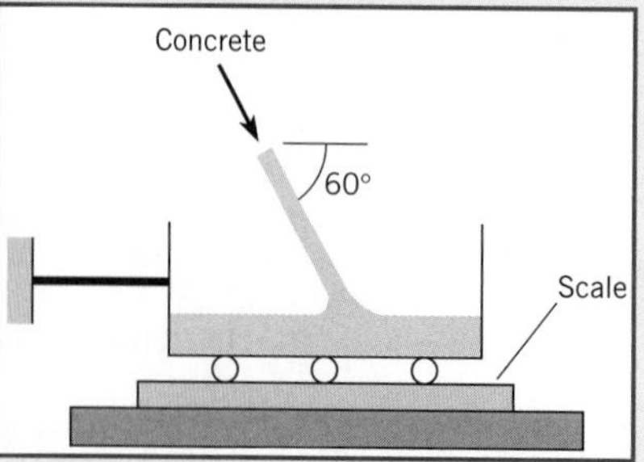

Solution The control volume is shown in the sketch below. A stationary reference frame was selected. Since this problem involves two directions, the scalar equations for momentum in the x and y directions, Eqs. (6.8*a*) and (6.8*c*), will be used.

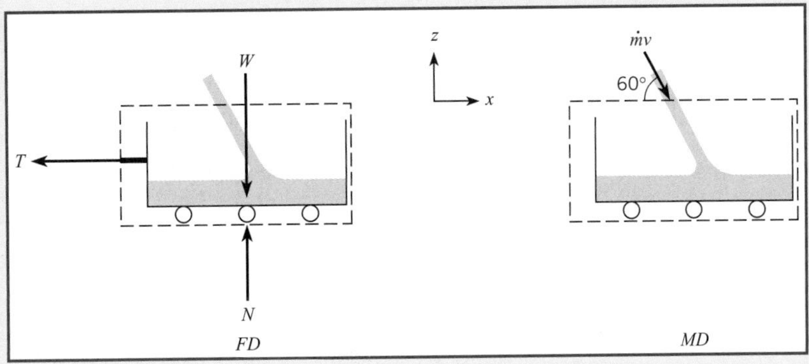

The force diagram shows the weight, the force of the cable, and the force of the scale on the cart. From the force diagram

$$\sum F_x = -T$$

$$\sum F_z = N - W$$

Since the flow is steady and the cart is stationary, the momentum accumulation term is zero. The momentum diagram shows an inward momentum flow corresponding to the flow of concrete into the cart. The momentum flow terms from Eqs. (6.8*a*) and (6.8*c*) may be evaluated using the momentum diagram:

$$\sum_{\text{cs}} \dot{m}_o v_{o_x} - \sum_{\text{cs}} \dot{m}_i v_{i_x} = -(\dot{m}v \cos 60°)$$

$$\sum_{\text{cs}} \dot{m}_o v_{o_z} - \sum_{\text{cs}} \dot{m}_i v_{i_z} = -(-\dot{m}v \sin 60°)$$

and so the momentum equation in the x direction is

$$-T = -\dot{m}v \cos 60°$$

and the momentum equation in the z direction is

$$N - W = \dot{m}v \sin 60°$$

The momentum flow rate is

$$\dot{m}v = (\rho A v)v$$
$$= (150\ \text{lbm}/\text{ft}^3)(1.0\ \text{slug}/32.2\ \text{lbm})(1\ \text{ft}^2)(10^2\ \text{ft}^2/\text{s}^2) = 466\ \text{lbf}$$

The weight recorded by the scale is

$$N = \dot{m}v \sin 60° + W$$
$$= 466\ \text{lbf} \times \sin 60° + 800\ \text{lbf}$$
$$= 1200\ \text{lbf} \quad \triangleleft$$

The tension in the cable is

$$T = \dot{m}v \cos 60° = 466\ \text{lbf} \times \cos 60° = 233\ \text{lbf} \quad \triangleleft$$

The weight recorded by the scale is larger than the weight of the cart because of the momentum carried by the fluid jet. Notice that unit conversions are usually needed when using English units.

Nozzles

When a fluid flows through a nozzle as shown in Fig. 6.3, we often assume the velocity is uniform across sections A and B. Hence, momentum flows will have magnitude $\dot{m}v$. If the nozzle exhausts into the atmosphere, the pressure at section B is atmospheric. Applying the Bernoulli equation between sections A and B will provide an equation for the pressure at section A. This pressure will exert a force of magnitude pA, where p is the pressure at the centroid of section A.

example 6.3

The sketch shows air flowing through a nozzle. The inlet pressure is $p_1 = 105$ kPa, abs., and the air exhausts into the atmosphere, where the pressure is 101.3 kPa, abs. The nozzle has an inlet diameter of 60 mm and an exit diameter of 10 mm, and the nozzle is connected to the supply pipe by flanges. Find the air speed at the exit of the nozzle and the force required to hold the nozzle stationary. Assume the air has a constant density of 1.22 kg/m^3. Neglect the weight of the nozzle.

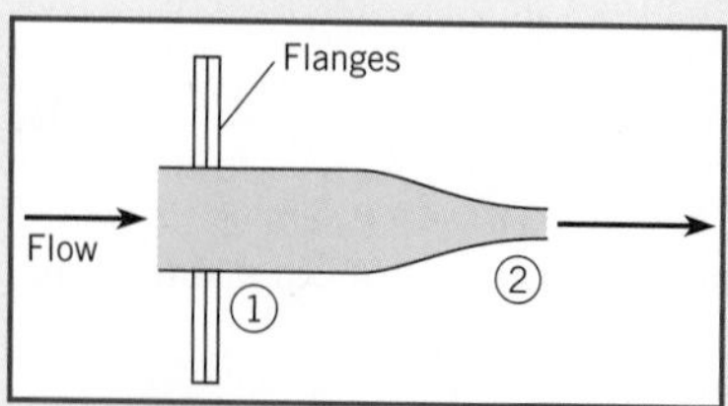

Solution The control volume is shown in the sketch below. A stationary reference frame was selected. This problem involves only one direction, the x direction, and so Eq. (6.8a) will be used. In addition, the Bernoulli equation can be applied between sections 1 and 2. Since the fluid is a gas, elevation changes are negligible. Also, $p_2 = 0$ kPa gage. Hence, the Bernoulli equation simplifies to

$$p_1 + \rho v_1^2/2 = \rho v_2^2/2$$

The continuity equation gives

$$v_1 A_1 = v_2 A_2$$

Substituting for area and combining the previous two equations gives

$$v_2 = \sqrt{\frac{2p_1}{\rho(1-(d_2/d_1)^4)}}$$

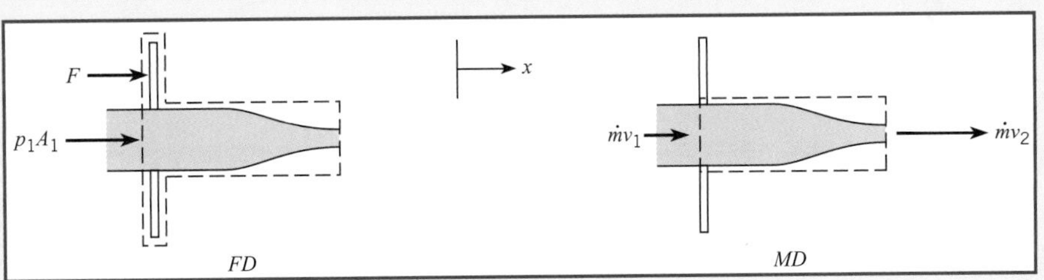

The force diagram shows a force due to pressure p_1A_1 and a force F, which is the net force acting on the flange of the nozzle. Since the direction of F is unknown, it was assumed to act in the positive direction. From the force diagram,

$$\sum F_x = F + p_1A_1$$

There is no momentum accumulation because the flow is steady and the nozzle is stationary. The mass flow rate is constant, so $\dot{m}_1 = \dot{m}_2 = \dot{m}$. The momentum diagram shows an inward and outward momentum flow. From the momentum diagram,

$$\sum_{\text{cs}} \dot{m}_o v_{ox} - \sum_{\text{cs}} \dot{m}_i v_{ix} = \dot{m}v_2 - \dot{m}v_1$$

Substituting into the momentum equation and solving for force results in

$$F = \dot{m}(v_2 - v_1) - p_1A_1$$

To begin the calculations, v_2 is found from the previously derived equation:

$$v_2 = \sqrt{\frac{2\times(105-101.3)\times 1000\ \text{Pa}}{(1.22\ \text{kg/m}^3)(1-(10/60)^4)}} = 77.9\ \text{m/s} \qquad \triangleleft$$

Then v_1 is

$$v_1 = v_2 A_2 / A_1 = (77.9\ \text{m/s})(10/60)^2 = 2.16\ \text{m/s}$$

Mass flow rate is

$$\dot{m} = \rho A_2 v_2 = (1.22\ \text{kg/m}^3)(\pi \times 0.01^2/4\ \text{m}^2)(77.9\ \text{m/s})$$
$$= 7.46 \times 10^{-3}\ \text{kg/s}$$

The net rate of momentum flow is

$$\dot{m}(v_2 - v_1) = (7.46 \times 10^{-3}\ \text{kg/s})(77.9 - 2.16)\ \text{m/s}$$
$$= 0.57\ \text{N}$$

The force due to pressure is calculated using gage pressure:

$$p_1 A_1 = [(105 - 101.3) \times 1000\ \text{Pa}][\pi \times 0.06^2/4\ \text{m}^2] = 10.46\ \text{N}$$

and the force is

$$F = -p_1 A_1 + \dot{m}(v_2 - v_1)$$
$$= -10.46\ \text{N} + 0.57\ \text{N} = -9.89\ \text{N}$$ ◁

Because F is negative, the direction is opposite to the direction assumed on the force diagram. Hence, the force acts in the negative x direction.

Vanes

A vane is a structural component, typically thin, that is used to turn a fluid jet or be turned by a fluid jet. Examples include a blade in a turbine and a sail on a ship. Figure 6.4 shows a flat vane impacted by a jet of fluid. A typical control volume is also shown. In analyzing flow over a vane, it is common to apply the Bernoulli equation and to neglect changes in elevation. Since the pressure is constant (atmospheric pressure), $v_1 = v_2 = v_3$. Another common assumption is that viscous forces are negligible compared to pressure forces. Thus when a vane is flat, as in Fig. 6.4, the force needed to hold the vane stationary is normal to the vane.

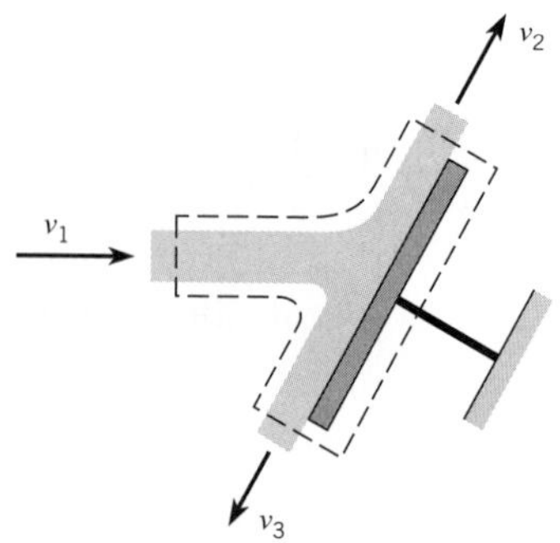

FIGURE 6.4

Fluid jet striking a flat vane.

example 6.4

A water jet is deflected 60° by a stationary vane as shown in the figure. The incoming jet has a speed of 100 ft/s and a diameter of 1 in. Find the force exerted by the jet on the vane. Neglect the influence of gravity.

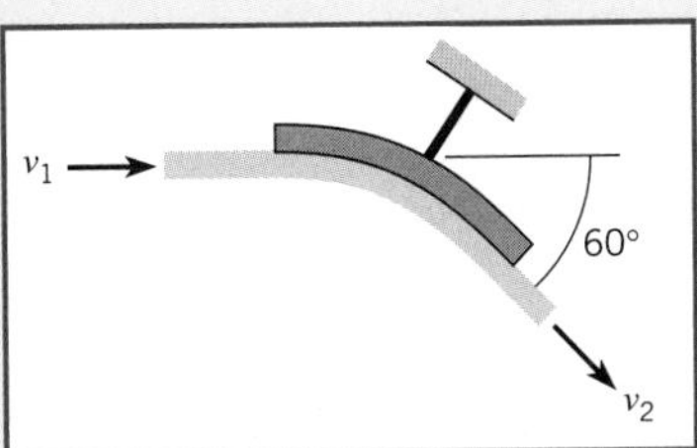

Solution The control volume is shown in the sketch below. A stationary reference frame was selected. Application of the Bernoulli equation shows that the inlet and exit speeds are equal: $v_1 = v_2$. Also, the inlet and exit mass flow rates are equal, $\dot{m}_1 = \dot{m}_2 = \dot{m}$. This problem involves two directions, the x and y directions. We can select the momentum equation as either the two component equations (6.8*a* and *b*) or the vector equation (6.7). We will use the vector form.

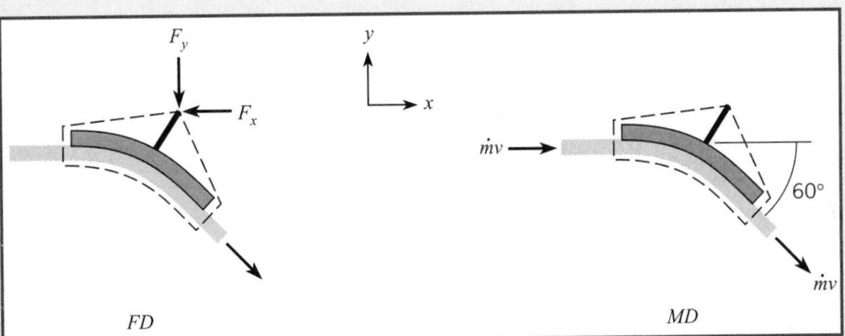

In the force diagram, the term **F** is the force needed to hold the vane stationary. From the force diagram,

$$\sum \mathbf{F} = -F_x\mathbf{i} - F_y\mathbf{j}$$

Since the flow is steady and the vane is stationary, momentum accumulation is zero. The momentum diagram shows one inward momentum flow and one outward momentum flow. From the momentum diagram,

$$\dot{m}\mathbf{v}_i = \dot{m}v\mathbf{i}$$

$$\dot{m}\mathbf{v}_o = [(\dot{m}v\cos 60°)\mathbf{i} - (\dot{m}v\sin 60°)\mathbf{j}]$$

so the net momentum flow is

$$\sum_{cs} \dot{m}_o \mathbf{v}_o - \sum_{cs} \dot{m}_i \mathbf{v}_i = (\dot{m}v\cos 60° - \dot{m}v)\mathbf{i} - (\dot{m}v\sin 60°)\mathbf{j}$$

Equating force and momentum terms gives

$$-F_x\mathbf{i} - F_y\mathbf{j} = (\dot{m}v\cos 60° - \dot{m}v)\mathbf{i} - (\dot{m}v\sin 60°)\mathbf{j}$$

so
$$-F_x = \dot{m}v\cos 60° - \dot{m}v$$

and
$$-F_y = -\dot{m}v\sin 60°$$

The variable $\dot{m}$ gives the rate at which mass is crossing the control surface:

$$\dot{m} = \rho A v$$

Evaluation of the x direction momentum equation gives

$$\begin{aligned} F_x &= \dot{m}v(1 - \cos 60°) = \rho A v^2(1 - \cos 60°) \\ &= (1.94 \text{ slug/ft}^3)(\pi \times 0.0417^2 \text{ ft}^2)(100^2 \text{ ft}^2/\text{s}^2)(1 - \cos 60°) \\ &= 53.0 \text{ lbf} \end{aligned}$$

In the y direction,

$$\begin{aligned} F_y &= \dot{m}v\sin 60° = \rho A v^2 \sin 60° \\ &= (1.94 \text{ slug/ft}^3)(\pi \times 0.0417^2 \text{ ft}^2)(100^2 \text{ ft}^2/\text{s}^2)\sin 60° \\ &= 91.8 \text{ lbf} \end{aligned}$$

The force of the jet on the vane ($\mathbf{F}_{\text{jet}}$) is opposite in direction to the force required to hold the vane stationary ($\mathbf{F}$). Therefore, $\mathbf{F}_{\text{jet}} = (53.0 \text{ lbf})\mathbf{i} + (91.8 \text{ lbf})\mathbf{j}$ ◁

example 6.5

As shown in the figure, an incident jet of water with density ρ, speed v, and area A is deflected through an angle β by a stationary, axisymmetric vane. Find the force required to hold the vane stationary. Express your answer using ρ, v, A, and β. Neglect the influence of gravity.

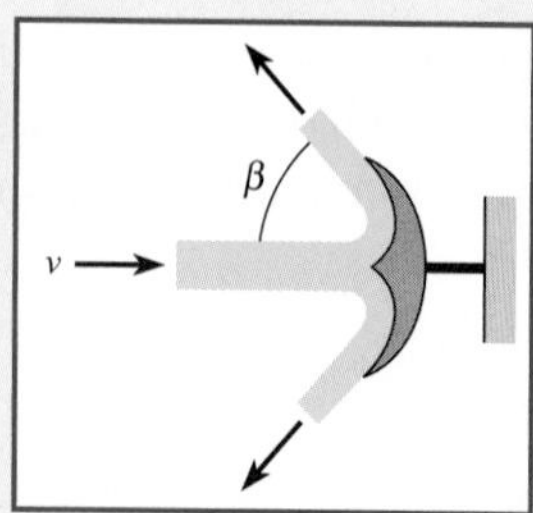

Solution The Bernoulli equation shows that the incoming jet and the deflected jet have the same speed. The control volume is shown in the sketch below. A stationary reference frame was chosen. Only the x direction is involved in this problem, so Eq. (6.8a) will be used.

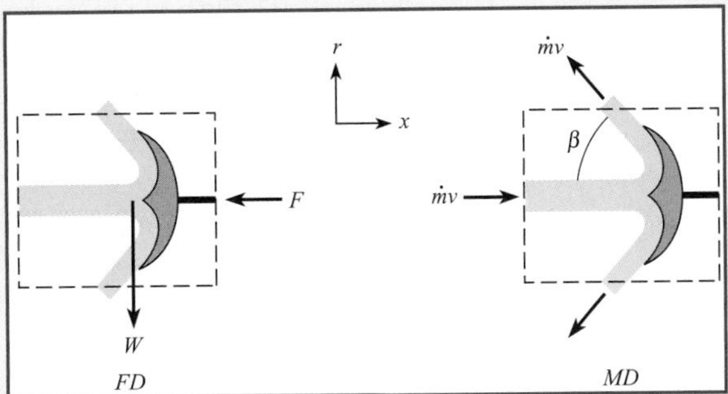

The force diagram shows a single term F, which is the force needed to hold the vane stationary. Hence

$$\sum F_x = -F$$

Flow is steady and the vane is stationary, so momentum accumulation is zero. The r direction momentum flow is zero because the vane is axisymmetric. Hence, the net flow of momentum, as shown on the momentum diagram, is

$$\sum_{\text{cs}} \dot{m}_o v_{ox} - \sum_{\text{cs}} \dot{m}_i v_{ix} = (-\dot{m}v\cos b) - (\dot{m}v) = -\dot{m}v(1 + \cos b)$$

The x direction momentum equation becomes

$$-F = -\dot{m}v(1 + \cos b)$$

Letting $\dot{m} = \rho A v$ gives

$$F = \rho A v^2 (1 + \cos b) \qquad \triangleleft$$

and the direction of this force is to the left, as shown in the force diagram.

Pipes

Because flow in a pipe is usually turbulent, it is common to assume that velocity is nearly constant across each cross section of the pipe. Also, the force acting on a pipe cross section is given by pA, where p is the pressure at the centroid of area and A is area.

example 6.6

As shown in the figure, a 1-m-diameter pipe bend is carrying crude oil (S = 0.94) with a steady flow rate of 2 m^3/s. The bend has an angle of 30° and lies in a horizontal plane. The volume of oil in the bend is 1.2 m^3, and the empty weight of the bend is 4 kN. Assume the pressure along the centerline of the bend is constant with a value of 75 kPa gage. Find the net force required to hold the bend in place.

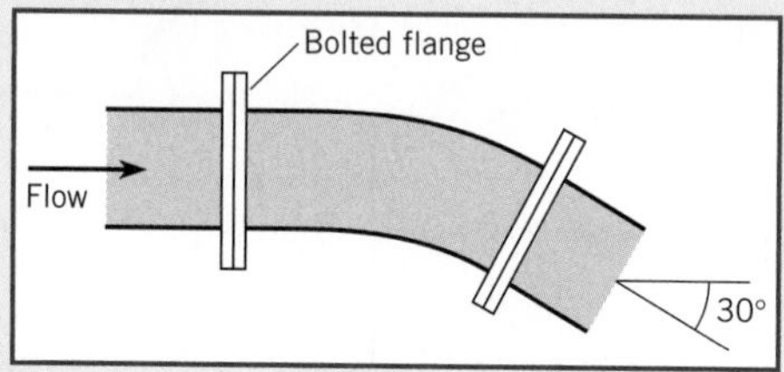

Solution The control volume is shown in the sketch below. A stationary reference frame was selected. Flow is steady, and the continuity equation gives $\dot{m}_i = \dot{m}_o = \dot{m}$. Also, the bend has a constant diameter and so $v_i = v_o = v$. This problem involves three coordinate directions, with the z direction outward from the page (right-hand coordinate system). The vector form of the momentum equation (6.7) will be used.

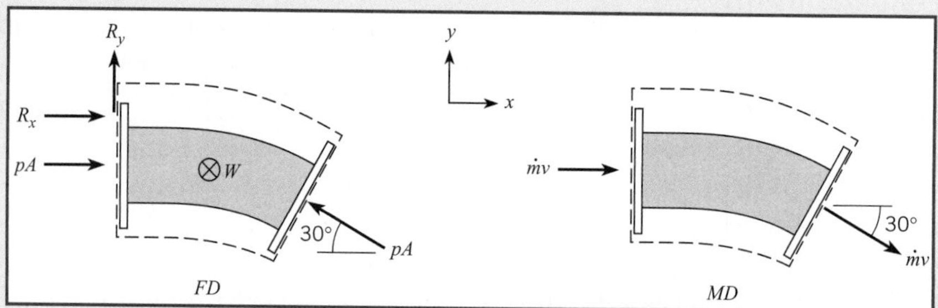

On the force diagram, pressure forces act where the flow crosses the control surfaces. The weight of the pipe and fluid therein is W and acts in the negative z direction. The resultant force (**R**) acting on the bend is due to the bolts and pipe flanges. This force has components in the x, y, and z directions, with R_z acting in the positive z direction. From the force diagram,

$$\sum \mathbf{F} = (R_x + pA - pA\cos 30°)\mathbf{i} + (R_y + pA\sin 30°)\mathbf{j} + (R_z - W)\mathbf{k}$$

Momentum accumulation is zero because the flow is steady and the bend is stationary. From the momentum diagram,

$$\sum_{\text{cs}} \dot{m}_o \mathbf{v}_o - \sum_{\text{cs}} \dot{m}_i \mathbf{v}_i = [(\dot{m}v\cos 30°)\mathbf{i} - (\dot{m}v\sin 30°)\mathbf{j}] - [(\dot{m}v)\mathbf{i}]$$

Equating force and momentum terms in each of the three coordinate directions results in

$$R_x + pA - pA\cos 30° = \dot{m}v\cos 30° - \dot{m}v$$
$$R_y + pA\sin 30° = -\dot{m}v\sin 30°$$
$$R_z - W = 0$$

The pressure force is

$$pA = (75\ \text{kN/m}^2)(\pi \times 0.5^2\ \text{m}^2) = 58.9\ \text{kN}$$

The fluid speed is

$$v = Q/A = \frac{(2\ \text{m}^3/\text{s})}{(\pi \times 0.5^2\ \text{m}^2)} = 2.55\ \text{m/s}$$

The momentum flow rate is

$$\dot{m}v = \rho Q v = (0.94 \times 1000\ \text{kg/m}^3)(2\ \text{m}^3/\text{s})(2.55\ \text{m/s}) = 4.80\ \text{kN}$$

The value of R_x is

$$R_x = -(pA + \dot{m}v)(1 - \cos 30°)$$
$$= -(58.9 + 4.80)(\text{kN})(1 - \cos 30°) = -8.53\ \text{kN}$$

The value of R_y is

$$R_y = -(pA + \dot{m}v)\sin 30°$$
$$= -(58.9 + 4.80)(\text{kN})(\sin 30°) = -31.8\ \text{kN}$$

The bend weight includes the oil plus the empty pipe:

$$W = \gamma\forall + 4\ \text{kN}$$
$$= (0.94 \times 9.81\ \text{kN/m}^3)(1.2\ \text{m}^3) + 4\ \text{kN} = 15.1\ \text{kN}$$

So $R_z = 15.1$ kN. The net force acting on the bend to hold it stationary is

$$\mathbf{R} = (-8.53\ \text{kN})\mathbf{i} + (-31.8\ \text{kN})\mathbf{j} + (15.1\ \text{kN})\mathbf{k}$$ ◁

example 6.7

Water flows through a 180° reducing bend, as shown. The discharge is 0.25 m^3/s, and the pressure at the center of the inlet section is 150 kPa gage. If the bend volume is 0.10 m^3, and it is assumed that the Bernoulli equation is valid, what force is required to hold the bend in place? The metal in the bend weighs 500 N.

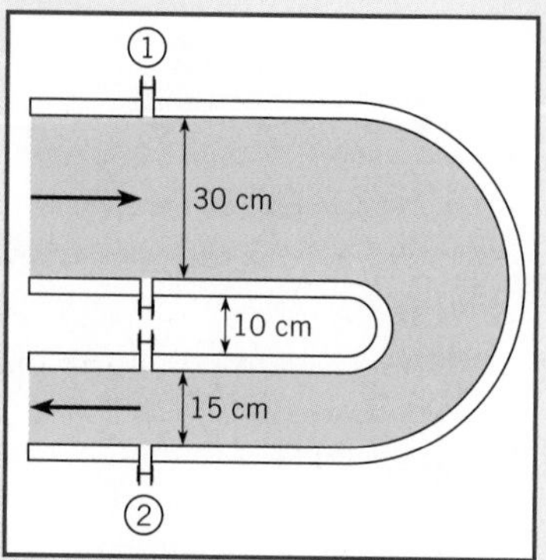

Solution The control volume is shown in the sketch below. A stationary reference frame was selected. The flow is steady, and continuity gives $\dot{m}_1 = \dot{m}_2 = \dot{m}$.

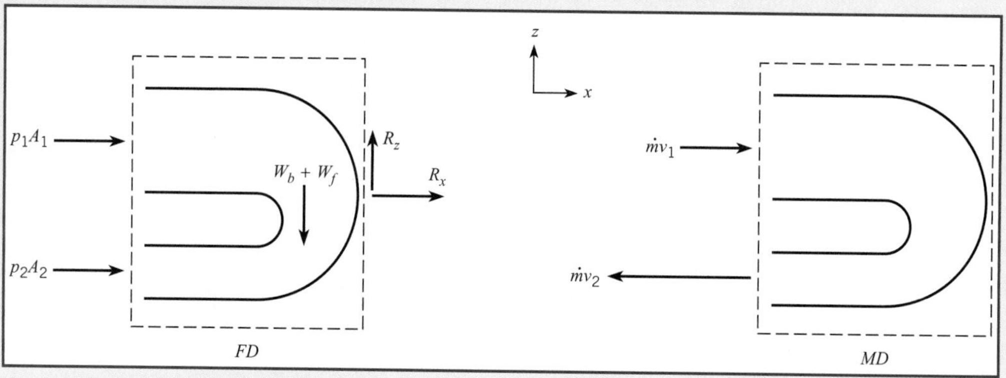

Velocities may be found using continuity:

$$Q = A_1 v_1 = A_2 v_2$$

The Bernoulli equation can be used to relate pressure at sections 1 and 2:

$$\frac{p_1}{\gamma} + \frac{v_1^2}{2g} + z_1 = \frac{p_2}{\gamma} + \frac{v_2^2}{2g} + z_2$$

Because this problem involves the x and z directions, Eqs. (6.8*a*) and (6.8*c*) will be used.

The force diagram shows weight, pressure forces, and **R**, which is the net force due to the connections at sections 1 and 2. From the force diagram,

$$\sum F_x = p_1A_1 + p_2A_2 + R_x$$

$$\sum F_z = R_z - W_b - W_f$$

where W_b is the weight of the bend and W_f is the weight of the fluid within the bend.

Momentum accumulation is zero because the flow is steady and the bend is stationary. Each momentum flow can be represented as shown on the momentum diagram. From the momentum diagram,

$$\sum_{cs} \dot{m}_o v_{ox} - \sum_{cs} \dot{m}_i v_{ix} = \dot{m}(-v_2) - \dot{m}v_1$$

$$= -\dot{m}(v_2 + v_1)$$

and the y direction momentum terms are zero. Substituting force and momentum terms into Eq. (6.8a) results in

$$p_1A_1 + p_2A_2 + R_x = -\dot{m}(v_2 + v_1)$$

Similarly, evaluation of Eq. (6.8c) yields

$$R_z - W_b - W_f = 0$$

Speeds are given by

$$v_1 = \frac{Q}{A_1} = \frac{0.25 \text{ m}^3/\text{s}}{\pi/4 \times 0.3^2 \text{ m}^2} = 3.54 \text{ m/s}$$

$$v_2 = \frac{Q}{A_2} = \frac{0.25 \text{ m}^3/\text{s}}{\pi/4 \times 0.15^2 \text{ m}^2} = 14.15 \text{ m/s}$$

Mass flow rate is given by

$$\dot{m} = \rho Q = (1000 \text{ kg/m}^3)(0.25 \text{ m}^3)$$

$$= 250 \text{ kg/s}$$

The net outward momentum flow rate is

$$\dot{m}(v_2 + v_1) = (250 \text{ kg/s})(14.15 + 3.54)(\text{m/s}) = 4420 \text{ N}$$

Pressure at section 2 is given by the Bernoulli equation:

$$p_2 = p_1 + \frac{\rho(v_1^2 - v_2^2)}{2} + \gamma(z_1 - z_2)$$

$$= 150\text{ kPa} + \frac{(1000)(3.54^2 - 14.15^2)\text{Pa}}{2} + (9810)(0.325)\text{Pa}$$

$$= 59.3\text{ kPa}$$

R_x is given by $\qquad R_x = -(p_1A_1 + p_2A_2) - \dot{m}(v_2 + v_1)$

The net pressure force is

$$p_1A_1 + p_2A_2 = (150\text{ kPa})(\pi \times 0.3^2/4\text{ m}^2) + (59.3\text{ kPa})(\pi \times 0.15^2/4\text{ m}^2)$$
$$= 11.6\text{ kN}$$

The x component of the support force is

$$R_x = -(p_1A_1 + p_2A_2) - \dot{m}(v_2 + v_1)$$
$$= -(11.6\text{ kN}) - (4.42\text{ kN})$$
$$= -16.0\text{ kN} \quad \triangleleft$$

and the z component is

$$R_z = W_b + W_f$$
$$= 500\text{ N} + (9810\text{ N/m}^3)(0.1\text{ m}^3)$$
$$= 1.48\text{ kN} \quad \triangleleft$$

6.4

Additional Applications

Nonuniform Velocity Distribution

The previous examples in this chapter were cases wherein it was assumed that the velocity across each flow section was constant. Thus, we used the simplified forms of the momentum equation, Eqs. (6.7) or (6.8). The next example involves a nonuniform velocity distribution across a flow section. Therefore, the general form of the momentum equation (6.5) must be used.

example 6.8

The drag force of a bullet-shaped device may be measured using a wind tunnel. The tunnel is round with a diameter of 1 m, the pressure at section 1 is 1.5 kPa gage, the pressure at section 2 is 1.0 kPa gage, and air density is 1.0 kg/m^3. At the inlet, the velocity is uniform with a magnitude of 30 m/s. At the exit, the velocity varies linearly as shown in the sketch. Determine the drag force on the device and support vanes. Neglect viscous resistance at the wall, and assume pressure is uniform across sections 1 and 2.

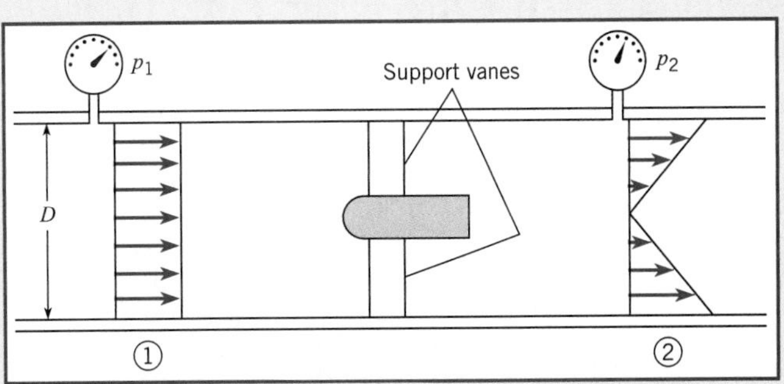

Solution A free-body diagram of the device is shown below. The net drag force F_D is balanced by forces on the support vanes, so $F_D = F_{s1} + F_{s2}$.

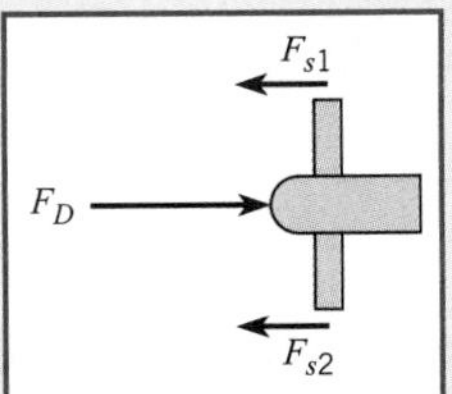

The sketch below shows the control volume. A stationary reference frame was selected. The velocity distribution across the exit section is nonuniform, so the general form of the momentum equation (6.5) will be used.

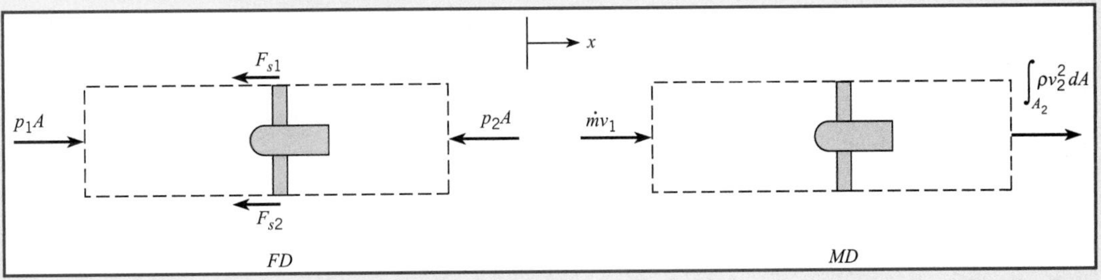

On the force diagram, shear stress and weight are omitted, and the pressure forces are as shown. At the two locations where the control surface passes through the support vanes, the surface forces are F_{s1} and F_{s2}. From the force diagram,

$$\sum F_x = p_1A - p_2A - (F_{s1} + F_{s2})$$

Letting $F_D = F_{s1} + F_{s2}$ results in

$$\sum F_x = p_1A - p_2A - F_D$$

The momentum diagram shows an inward momentum flow and an outward momentum flow. Each momentum flow term was analyzed using $\int_A v_x \rho \mathbf{V} \cdot d\mathbf{A}$. Across the inlet (section 1), variables in the integral are $v_x = v_1$, $\mathbf{V}_1 = v_1\mathbf{i}$, $\mathbf{A}_1 = (-A\mathbf{i})$. Because these variables are constants, the integral simplifies to

$$\int_{A_1} v_x \rho \mathbf{V} \cdot d\mathbf{A} = v_x \rho \mathbf{V}_1 \cdot \mathbf{A}_1 = v_x \rho(-v_x A)$$
$$= -\dot{m} v_1$$

Across the exit (section 2), variables in the integral are $v_x = v_2$, $\mathbf{V}_2 = v_2\mathbf{i}$, $d\mathbf{A} = dA\mathbf{i}$ and the outward momentum flow is

$$\int_{A_2} v_x \rho \mathbf{V} \cdot d\mathbf{A} = \int_{A_2} v_2 \rho(\mathbf{v}_2 \cdot d\mathbf{A}) = \int_{A_2} \rho v_2^2\, dA$$

where v_2 cannot be factored out of the integral because it varies across the area: $v_2 = v_2(r)$. From the momentum diagram, the net momentum flow is given by

$$\int_{cs} v_x \rho \mathbf{V} \cdot d\mathbf{A} = \int_{A_2} \rho v_2^2\, dA - \dot{m} v_1$$

Equation (6.5), when applied to this problem, simplifies to

$$\sum F_x = \int_{cs} v_x \rho \mathbf{V} \cdot d\mathbf{A}$$

Substituting forces and momentum terms into the momentum equation results in

$$p_1 A_1 - p_2 A_2 - F_D = \int_{A_2} \rho v_2^2\, dA - \dot{m} v_1$$

$$F_D = A_1(p_1 - p_2) + \dot{m} v_1 - \int_{A_2} \rho v_2^2\, dA$$

The net pressure force is

$$A_1(p_1 - p_2) = (\pi \times 0.5^2\ \mathrm{m}^2)(1.5 - 1.0)(10^3)\ \mathrm{N/m}^2$$
$$= 392.7\ \mathrm{N}$$

The inward momentum flow is

$$\dot{m} v_1 = \rho A_1 v_1^2 = (1.0\ \mathrm{kg/m}^3)(\pi \times 0.5^2\ \mathrm{m}^2)(30^2\ \mathrm{m}^2/\mathrm{s}^2) = 706.8\ \mathrm{N}$$

To evaluate the outward momentum flow, we need an equation for $v_2(r)$. Since the outlet velocity distribution is linear, $v_2(r) = v_{\max}(r/r_o)$, where $v_{\max}$ is the velocity near the wall, and r_o is the outer radius of the tunnel. To evaluate $v_{\max}$, we use continuity:

$$Q_1 = Q_2$$

$$A_1 v_1 = \int_{A_2} v_2(r)\, dA = \int_0^{r_o} v_{\max}(r/r_o) 2\pi r\, dr$$

Thus

$$(\pi \times 0.5^2\ \text{m}^2)(30\ \text{m/s}) = \int_0^{0.5} v_{max}(r/0.5)2\pi r\,dr$$

which can be solved to show that $v_{max} = 45\ \text{m/s}$. Evaluating the exit momentum flow gives

$$\begin{aligned}\int_{A_2} \rho v_2^2\,dA &= \rho\int_0^{0.5} v_{max}^2(r/0.5)^2 2\pi r\,dr \\ &= (1.0\ \text{kg/m}^3)(45^2\ \text{m}^2/\text{s}^2)(0.5^2\ \text{m}^2)(\pi/2) \\ &= 795.2\ \text{N}\end{aligned}$$

The drag force is

$$\begin{aligned}F_D &= A_1(p_1 - p_2) + \dot{m}v_1 - \int_{A_2} \rho v_2^2\,dA \\ &= 392.7\ \text{N} + 706.8\ \text{N} - 795.2\ \text{N} \\ &= 304\ \text{N}\end{aligned}$$

◁

Moving Control Volumes

When an object is moving, it is often convenient to define a control volume that moves with the object. As discussed in Section 6.1, the velocity **v** in the momentum equation must be relative to an inertial reference frame. The reference frame selected may move with the control volume, or it can be independent of the control volume. For example, a stationary reference frame may be selected for a problem involving a moving control volume.

When Eq. (6.7) or (6.8) is used, each mass flow rate is calculated using the velocity with respect to the control surface. In other words, the term $\dot{m}$ gives the rate at which mass passes in or out of the control volume. Mass flow rate is independent of the reference frame that is selected to evaluate **v.**

example 6.9

A stationary nozzle produces a jet with a speed v_j and an area A_j. The jet strikes a moving block and is deflected 90° relative to the block. The block is sliding with a constant speed v_b on a rough surface. Find the frictional force F acting on the block.

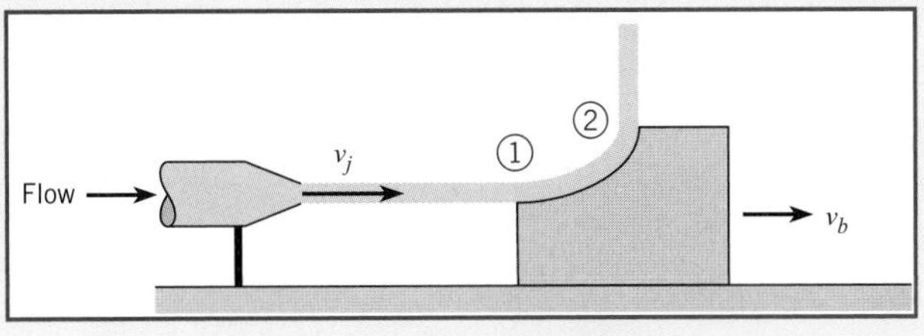

Solution We select a control volume that encloses the block and moves with it. We choose a moving reference frame that is fixed to the block. This reference frame is inertial because the block has a constant velocity. We select the vector form of the momentum equation (6.7).

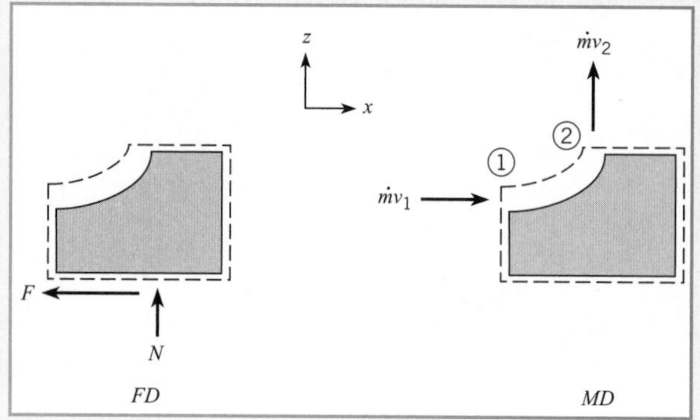

The force diagram shows the weight of the block, a normal force, and a friction force. From the force diagram,

$$\sum \mathbf{F} = -F\mathbf{i} + (N - W)\mathbf{k}$$

Momentum terms depend on the reference frame. From the selected reference frame, the momentum accumulation is zero because the flow is steady and the block is stationary. The momentum diagram shows an inward and an outward momentum flow. Notice that the outward momentum flow, relative to the reference frame, would be upward. From the momentum diagram,

$$\sum_{\text{cs}} \dot{m}_o \mathbf{v}_o - \sum_{\text{cs}} \dot{m}_i \mathbf{v}_i = \dot{m} v_2 \mathbf{k} - \dot{m} v_1 \mathbf{i}$$

Substituting force and momentum terms into Eq. (6.7) results in

$$-F\mathbf{i} + (N - W)\mathbf{k} = \dot{m} v_2 \mathbf{k} - \dot{m} v_1 \mathbf{i}$$

Equating terms in the x direction gives

$$-F = -\dot{m} v_1$$

The mass flow rate is calculated using the velocity relative to the control surface $(v_j - v_b)$, and so

$$\dot{m} = \rho A_j (v_j - v_b)$$

Notice that $\dot{m} \longrightarrow 0$ as $v_b \longrightarrow v_j$, which should be the case because $\dot{m}$ is the rate at which mass is crossing the control surface.

The speed v_1 is relative to the moving reference frame, and so

$$v_1 = v_j - v_b$$

Combining terms results in

$$F = \dot{m}v_1 = \rho A_j (v_j - v_b)^2 \quad \triangleleft$$

The problem can also be analyzed using a stationary reference frame. We select the same moving control volume, but now select a reference frame that is fixed to the support of the nozzle. The mass flow rate $\dot{m}$ is independent of the reference frame selected. Relative to the stationary reference frame, momentum accumulation is zero, and Eq. (6.7) again simplifies to

$$\sum \mathbf{F} = \dot{m}\mathbf{v}_2 - \dot{m}\mathbf{v}_1$$

Forces are independent of reference frame and so the left side of the equation is unchanged.

As shown in the following figure, the momentum diagram has changed because momentum terms depend on the reference frame. The inward momentum flow has a magnitude $\dot{m}v_1$, where v_1 is evaluated from the stationary reference frame: $v_1 = v_j$. Thus,

$$\dot{m}\mathbf{v}_1 = \dot{m}v_j\mathbf{i}$$

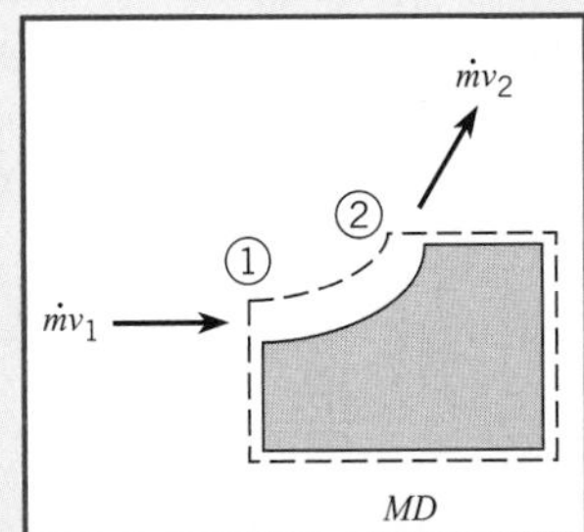

The outward momentum flow has an x and z component:

$$\dot{m}\mathbf{v}_2 = \dot{m}(v_{2x}\mathbf{i} + v_{2z}\mathbf{k})$$

The x component equals the speed of the block: $v_{2x} = v_b$. The y component can be found by applying the continuity equation to the control volume: $v_{2y} = v_j - v_b$. Combining variables gives

$$\dot{m}\mathbf{v}_2 = \dot{m}[v_b\mathbf{i} + (v_j - v_b)\mathbf{k}]$$

The net momentum flow is

$$\begin{aligned} \dot{m}\mathbf{v}_2 - \dot{m}\mathbf{v}_1 &= \dot{m}[v_b\mathbf{i} + (v_j - v_b)\mathbf{k}] - \dot{m}v_j\mathbf{i} \\ &= \dot{m}[(v_b - v_j)\mathbf{i} + (v_j - v_b)\mathbf{k}] \end{aligned}$$

Substituting force and momentum terms into the momentum equation and equating terms in the x direction results in

$$-F = \dot{m}(v_b - v_j)$$

Since $\dot{m}$ is independent of reference frame, we can use the previous result:

$$\dot{m} = \rho A_j(v_j - v_b)$$

Combining the previous two equations gives

$$F = \rho A_j(v_j - v_b)^2 \quad \triangleleft$$

which is the same solution that was obtained using a moving reference frame. Notice that analysis using the moving reference frame was easier. Usually, selection of a moving reference frame facilitates solution of a problem of this type. However, this moving reference frame needs to be inertial. Finally, it is important to stress that values of mass flow rate are independent of the choice of reference frame.

example 6.10

Consider a rocket of mass m_r traveling at a speed v_r as measured from the ground. Exhaust gases leave the engine nozzle (area A_e) at a speed V_e relative to the nozzle of the rocket, and with a pressure that is higher than local atmospheric pressure by an amount p_e. The aerodynamic drag force on the rocket is D. Derive an equation for the acceleration of the rocket.

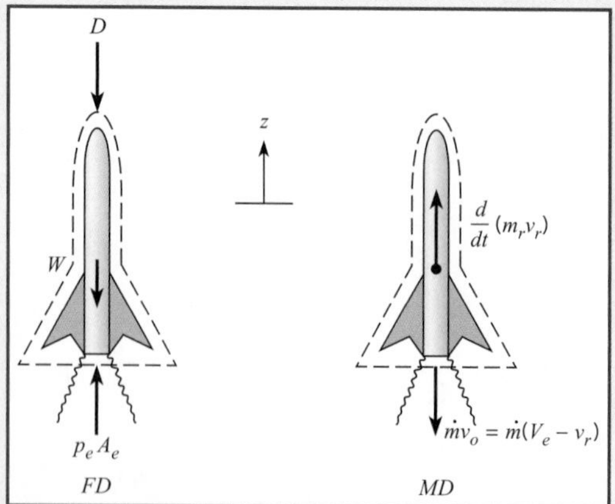

Solution We selected a control volume enclosing the rocket and moving with the rocket. Because a reference frame attached to the rocket would be noninertial, we chose a stationary reference frame fixed to the launch pad. We selected the z-direction momentum equation (6.8c).

The force diagram shows three forces: weight, aerodynamic drag force, and the pressure force acting at the nozzle exit. From the force diagram,

$$\sum F_z = p_e A_e - F - W$$

The momentum diagram shows an accumulation term and an outflow term. The momentum accumulation term is not zero because the momentum of the rocket is changing with time as the rocket accelerates. The momentum of the rocket is the mass of the rocket times the velocity of the rocket, and the accumulation term is given by $(d/dt)(m_r v_r)$. This can be formally developed by:

$$\frac{d}{dt}\int_{cv} v_z \rho \, d\forall = \frac{d}{dt}\left(v_r \int_{cv} \rho \, d\forall\right) = \frac{d}{dt}(m_r v_r)$$

In this development, we assumed that all parts of the rocket were traveling at the same speed v_r. The outward momentum flow is $\dot{m} v_o$, where v_o is relative to the stationary reference frame:

$$v_o = (V_e - v_r)$$

so

$$\dot{m} v_o = \dot{m}(V_e - v_r)$$

From the momentum diagram,

$$\frac{d}{dt}\int_{cv} v_z \rho \, d\forall + \sum_{cs} \dot{m}_o v_{oz} - \sum_{cs} \dot{m}_i v_{iz} = \frac{d(m_r v_r)}{dt} + [-\dot{m}(V_e - v_r)]$$

Substituting force and momentum terms into the momentum equation gives

$$p_e A_e - F - W = \frac{d(m_r v_r)}{dt} - \dot{m}(V_e - v_r)$$

The accumulation term can be expanded using the product rule for differentiation:

$$\frac{d}{dt}(m_r v_r) = m_r \frac{dv_r}{dt} + v_r \frac{dm_r}{dt}$$

where dm_r/dt is the rate at which the mass of the rocket is decreasing with time. This term may be evaluated by applying the continuity equation to the control volume:

$$\frac{dm_r}{dt} = -\dot{m}$$

Combining the two previous equations gives

$$\frac{d}{dt}(m_r v_r) = m_r \frac{dv_r}{dt} - \dot{m} v_r$$

Substituting this result into the momentum equation yields

$$m_r \frac{dv_r}{dt} = (\dot{m}V_e + p_e A_e) - D - W$$

The mass flow rate exiting the rocket nozzle is

$$\dot{m} = \rho_e A_e V_e$$

Combining the above two equations results in

$$m_r \frac{dv_r}{dt} = (\rho_e A_e V_e^2 + p_e A_e) - D - W \qquad \triangleleft$$

The acceleration of the rocket (dv_r/dt) depends on the instantaneous mass of the rocket and the four terms on the right side of this equation. The term $T = \rho_e A_e V_e^2 + p_e A_e$ is known as the thrust of the rocket motor. The exit pressure of the exhaust jet (p_e) is often expressed using absolute pressure, so the thrust is given by

$$T = \rho_e A_e V_e^2 + A_e(p_e - p_o)$$

where p_o is the local atmospheric pressure, which will change with altitude as a rocket ascends.

Force on a Rectangular Sluice Gate

A gate across a channel under which water flows, as shown in Fig. 6.5*a*, is known as a sluice gate. Suppose we wish to find the resultant force due to the pressure distribution acting on the left face of the sluice gate. As shown in Fig. 6.5*b*, the pressure distribution near the upper part of the sluice gate is approximately hydrostatic because fluid velocities are quite low. However, near the bottom of the sluice gate, velocity is quite large and the pressure distribution differs dramatically from a hydrostatic pressure distribution. To find the force due to pressure, we can balance the forces shown in Fig. 6.5*b*:

$$\int_{A_w} p\, dA = F_G$$

where A_w is the area on the left face of the sluice gate that is exposed to water, and F_G is the force needed to hold the sluice gate stationary. To find F_G, we can apply the momentum equation.

For the indicated control volume in Fig. 6.5*a*, the inlet speed v_1 and exit speed v_2 are assumed to be uniform. The inlet water depth is y_1 and outlet water depth is y_2. The accumulation term is zero, and the x direction momentum equation (6.8*a*) simplifies to

$$\sum F_x = \dot{m}v_2 - \dot{m}v_1$$

Substituting force and momentum terms results in

$$\bar{p}_1 A_1 - \bar{p}_2 A_2 - F_{\text{visc}} - F_G = \rho Q v_2 - \rho Q v_1$$

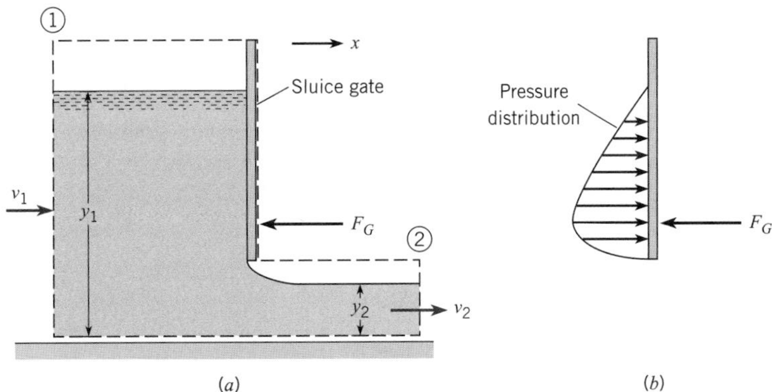

FIGURE 6.5

(a) Flow under a sluice gate. (b) Free-body diagram of the sluice gate.

Because the flow is uniform, pressure is hydrostatic. Hence, $\bar{p}_1$ is given by the pressure at depth $y_1/2$. Hence $\bar{p}_1 A_1 = (\gamma y_1/2)(y_1 b)$, where b is the width of the sluice gate and channel. Similarly, $\bar{p}_2 A_2 = (\gamma y_2/2)(y_2 b)$. The viscous force F_{visc} acts at the lower control surface. It is typically neglected because it is much smaller than the other forces. Carrying out the algebra gives

$$F_G = \left(\frac{\gamma y_1^2 b}{2} - \frac{\gamma y_2^2 b}{2}\right) + \rho Q(v_1 - v_2) \tag{6.9}$$

Water Hammer: Physical Description

Whenever a valve is closed in a pipe, a positive pressure wave is created upstream of the valve and travels up the pipe at the speed of sound. In this context a positive pressure wave is defined as one for which the pressure is greater than the steady-state pressure. This pressure wave may be great enough to cause pipe failure. Therefore, a basic understanding of this process, which is called *water hammer,* is necessary for the proper design and operation of such systems. The simplest case of water hammer will be considered here. For a more comprehensive treatment of the subject, the reader is referred to Chaudhry (1) and Streeter and Wylie (2).

Consider flow in the pipe shown in Fig. 6.6. Initially the valve at the end of the pipe is only partially open (Fig. 6.6*a*); consequently, an initial velocity V and initial pressure p_0 exist in the pipe. At time $t = 0$ it is assumed that the valve is instantaneously closed, thus creating the pressure wave that travels toward the reservoir at the speed of sound, c. All the water between the pressure wave and the upper end of the pipe will have the initial velocity V, but all the water on the other side of the pressure wave (between the wave and the valve) will be at rest. This condition is shown in Fig. 6.6*b*. Once the pressure wave reaches the upper end of the pipe (after time $t = L/c$), it can be visualized that all of the water in the pipe will be under a pressure $p_0 + \Delta p$; however, the pressure in the reservoir at the end of the pipe is only p_0. This imbalance of pressure at the reservoir end causes the water to flow from the pipe back into the reservoir with a velocity V. Thus a new pressure wave is formed that travels toward the valve end of the pipe (Fig. 6.6*c*), and the pressure on the reservoir side of the wave is reduced to p_0. When this wave finally reaches the valve, all the water in the pipe is

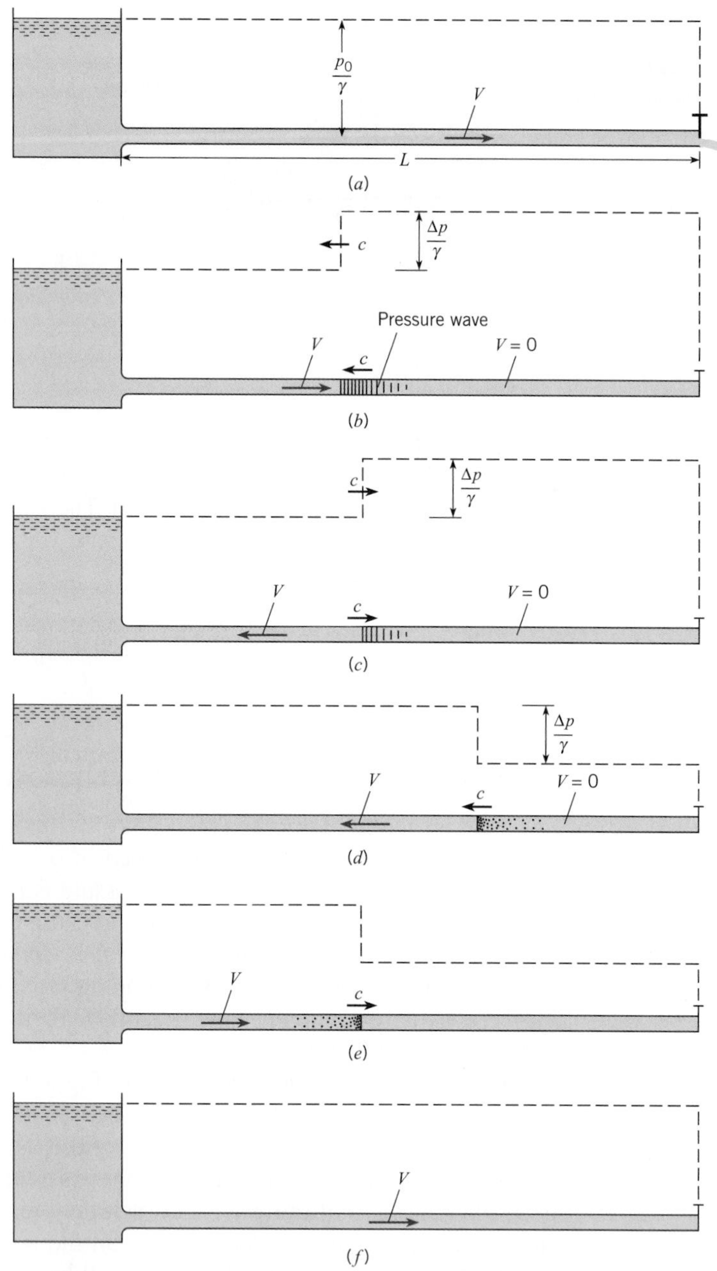

FIGURE 6.6

Water hammer process.
(a) Initial condition.
(b) Condition during time $0 < t < L/c$.
(c) Condition during time $L/c < t < 2L/c$.
(d) Condition during time $2L/c < t < 3L/c$.
(e) Condition during time $3L/c < t < 4L/c$.
(f) Condition at time $t = 4L/c$.

flowing toward the reservoir with a velocity V. This condition is only momentary, however, because the closed valve prevents any sustained flow.

Next, during time $2L/c < t < 3L/c$, a rarefied wave of pressure ($p < p_0$) travels up to the reservoir, as shown in Fig. 6.6*d*. When the wave reaches the reservoir, all the water in the pipe has a pressure less than that in the reservoir. This imbalance of pressure causes flow to be established again in the entire pipe, as shown in Fig. 6.6*f*,

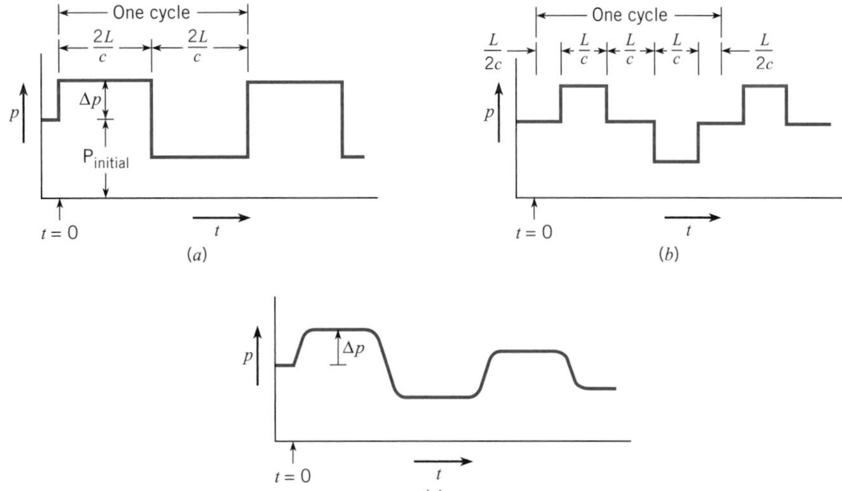

FIGURE 6.7

Variation of water-hammer pressure with time at two points in a pipe.
(a) Location: adjacent to valve.
(b) Location: at midpoint of pipe.
(c) Actual variation of pressure near valve.

and the condition is exactly the same as in the initial condition (Fig. 6.6*a*). Hence the process will repeat itself in a periodic manner.

From this description, it may be seen that the pressure in the pipe immediately upstream of the valve will be alternately high and low, as shown in Fig. 6.7*a*. A similar observation for the pressure at the midpoint of the pipe reveals a more complex variation of pressure with time, as shown in Fig. 6.7*b*. Obviously, a valve cannot be closed instantaneously, and viscous effects, which were neglected here, will have a damping effect on the process. Therefore, a more realistic pressure–time trace for the point just upstream of the valve is given in Fig. 6.7*c*. The finite time of closure erases the sharp discontinuities in the pressure trace that were present in Fig. 6.7*a*. However, it should be noted that the maximum pressure developed at the valve will be virtually the same as for instantaneous closure if the time of closure is less than $2L/c$. That is, the change in pressure will be the same for a given change in velocity unless the negative wave from the reservoir mitigates the positive pressure, and it takes a time $2L/c$ before this negative wave can reach the valve. The value $2L/c$ is called the *critical time of closure* and is given the symbol t_c.

Magnitude of Water-Hammer Pressure and Speed of Pressure Wave

We can analyze the quantitative relations for water hammer with the momentum equation by letting the control volume either move with the pressure wave, thus creating steady motion, or be fixed, thus retaining the inherently unsteady character of the process. To illustrate the use of the momentum equation with unsteady motion, let us take the latter approach. Consider a pressure wave in a rigid pipe, as shown in Fig. 6.8. The density, pressure, and velocity of the fluid on the reservoir side of the pressure wave are ρ, p, and V, respectively, and the similar quantities on the valve side of the wave are $\rho + \Delta\rho$, $p + \Delta p$, and 0. Because the wave in this case is traveling from the valve to the reservoir, its distance from the valve at any time t is given as ct. We can now apply the

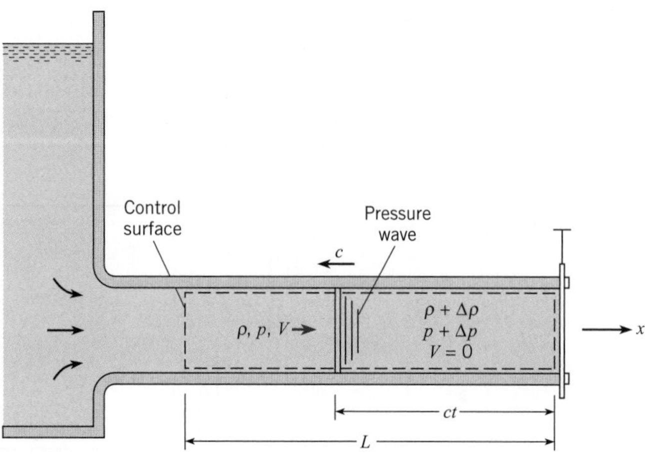

FIGURE 6.8

Pressure wave in a pipe.

momentum equation to the flow in the control volume. Let the x direction be along the pipe. The equation for x momentum (Eq. 6.8a) simplifies to

$$\sum F_x = \frac{d}{dt}\int_{\text{cv}} v_x \rho \, d\forall - \dot{m}v_i$$

The force terms are given by

$$\sum F_x = pA - (p + \Delta p)A$$

The inlet momentum flow is given by $\dot{m}v_i = \rho A V^2$. The momentum within the control volume decreases with time because fluid that is in motion stops as the pressure wave passes by. Evaluation of the momentum accumulation term gives

$$\begin{aligned}\frac{d}{dt}\int_{\text{cv}} v_x \rho \, d\forall &= \frac{d}{dt}[V\rho(L - ct)A] \\ &= -V\rho cA\end{aligned}$$

When forces and momentum terms are substituted into the momentum equation, we obtain

$$pA - (p + \Delta p)A = -\rho V^2 A - \rho V c A$$

This reduces to

$$\Delta p = \rho V^2 + \rho V c$$

In this equation the first term on the right-hand side is usually negligible with respect to the second term on the right, because for liquids c is much greater than V. Consequently, the equation reduces to

$$\Delta p = \rho V c \tag{6.10}$$

We will now determine the speed of the pressure wave by applying the continuity principle to the control volume in Fig. 6.8. The *continuity principle* is

$$0 = \sum \dot{m}_o - \sum \dot{m}_i + \frac{d}{dt}\int_{cv} \rho \, d\forall$$

and applied to Fig. 6.8, we have

$$0 = \dot{m}_i + \frac{d}{dt}[\rho(L-ct)A + (\rho+\Delta\rho)ctA]$$

because there is no mass flow out of the control volume. The mass flow rate is given by $\dot{m}_i = \rho VA$, so the continuity principle reduces to

$$\frac{\Delta\rho}{\rho} = \frac{V}{c}$$

or

$$c = \frac{V}{\Delta\rho/\rho} \tag{6.11}$$

However, by definition $E_v = (\Delta p/(\Delta\rho/\rho))$. Therefore,

$$\frac{\Delta\rho}{\rho} = \frac{\Delta p}{E_v} \tag{6.12}$$

Now when $\Delta\rho/\rho$ is eliminated between Eqs. (6.11) and (6.12), we have

$$c = \frac{VE_v}{\Delta p} \tag{6.13}$$

From Eq. (6.10), $\Delta p = \rho Vc$. Therefore, Eq. (6.13) becomes

$$c = \sqrt{\frac{E_v}{\rho}} \tag{6.14}$$

Thus, by application of the momentum and continuity equations, we have derived equations for both Δp and c.

example 6.11

A rigid pipe leading from a reservoir is 3000 ft long, and water is flowing through it with a velocity of 4 ft/s. If the initial pressure at the downstream end is 40 psig, what maximum pressure will develop at the downstream end when a rapid-acting valve at that end is closed in 1 s?

Solution

$$c = \sqrt{\frac{E_v}{\rho}} = \sqrt{\frac{320{,}000 \text{ lbf/in.}^2 \times 144 \text{ in.}^2/\text{ft}^2}{1.94 \text{ slugs/ft}^3}} = 4874 \text{ ft/s}$$

Next, determine whether the closure time is greater or less than the critical closure time, t_c.

$$t_c = 2L/c$$
$$= 2(3000\ \text{ft}/4874\ \text{ft/s}) = 1.23\ \text{s}$$

Since the closure time is less than t_c, the maximum value of Δp will be equal to ρVc.

$$\Delta p = 1.94\ \text{slugs/ft}^3 \times 4\ \text{ft/s} \times 4874\ \text{ft/s}$$
$$= 37{,}820\ \text{lbf/ft}^2 = 263\ \text{psi}$$

The maximum pressure is the initial pressure plus the pressure change, which is

$$p_{max} = 40 + 263 = 303\ \text{psig}$$ ◁

As indicated by Example 6.11, water-hammer pressures can be quite large. Therefore, engineers must design piping systems to keep the pressure within acceptable limits. This is done by installing an accumulator near the valve and/or operating the valve in such a way that rapid closure is prevented. Accumulators may be in the form of air chambers for relatively small systems, or surge tanks (a surge tank is a large open tank connected by a branch pipe to the main pipe) for large systems. Another way to eliminate excessive water-hammer pressures is to install pressure-relief valves at critical points in the pipe system. These valves are pressure-activated so that water is automatically diverted out of the system when the water-hammer pressure reaches excessive levels.

6.5 Moment-of-Momentum Equation

The moment-of-momentum equation is very useful for situations that involve moments. This includes analyses of rotating machinery such as pumps, turbines, fans, and blowers.

Torques acting on a control volume are related to changes in angular momentum through the moment-of-momentum equation. Development of this equation parallels the development of the momentum equation as presented in Section 6.1. When forces act on a system of particles, for example a fluid system, Newton's second law of motion can be used to derive an equation for rotational motion:

$$\sum \mathbf{M} = \frac{d(\mathbf{H}_{sys})}{dt} \tag{6.15}$$

where $\mathbf{M}$ is a moment and $\mathbf{H}_{sys}$ is the total angular momentum of all mass forming the system.

Equation (6.15) is a Lagrangian equation, which can be converted to an Eulerian equation using the Reynolds transport theorem [Eq. (5.22)]. The extensive property B_{sys} becomes the angular momentum of the system: $B_{sys} = \mathbf{H}_{sys}$. The intensive property b becomes the angular momentum per unit mass. The angular momentum of an element is $\mathbf{r} \times m\mathbf{v}$, and so $b = \mathbf{r} \times \mathbf{v}$. Substituting for B_{sys} and b into Eq. (5.22) gives

$$\frac{d(\mathbf{H}_{sys})}{dt} = \frac{d}{dt}\int_{cv} (\mathbf{r} \times \mathbf{v})\rho\, d\forall + \int_{cs} (\mathbf{r} \times \mathbf{v})\rho \mathbf{V} \cdot d\mathbf{A} \tag{6.16}$$

Combining Eqs. (6.15) and (6.16) gives the integral form of the *moment-of-momentum principle*:

$$\sum \mathbf{M} = \frac{d}{dt}\int_{\text{cv}} (\mathbf{r} \times \mathbf{v})\rho \, d\forall + \int_{\text{cs}} (\mathbf{r} \times \mathbf{v})\rho \mathbf{V} \cdot d\mathbf{A} \tag{6.17}$$

where **r** is a position vector that extends from the moment center, **V** is flow velocity relative to the control surface, and **v** is flow velocity relative to the inertial reference frame selected.

The moment-of-momentum equation has the following physical interpretation: The sum of moments acting on the material within the control volume equals the rate of change of angular momentum within the control volume plus the net rate at which angular momentum flows out of the control volume.

If the mass crosses the control surface through a series of inlet and outlet ports with uniformly distributed properties across each port, the moment-of-momentum principle becomes

$$\sum \mathbf{M} = \frac{d}{dt}\int_{\text{cv}} (\mathbf{r} \times \mathbf{v})\rho \, d\forall + \sum_{\text{cs}} \mathbf{r}_o \times (\dot{m}_o \mathbf{v}_o) - \sum_{\text{cs}} \mathbf{r}_i \times (\dot{m}_i \mathbf{v}_i) \tag{6.18}$$

The methods used for applying the moment-of-momentum equation parallel the methods described in Section 6.2. The origin for evaluating moments may be selected at any convenient location.

example 6.12

If the bend in Example 6.7 is supported at point A, find the moment the support system must resist. As shown, the weight (1418 N) acts 20 cm to the right of point A.

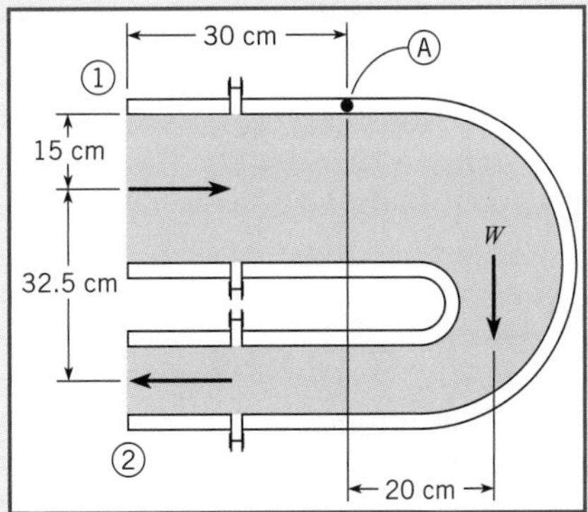

Solution The control volume is shown below. A stationary reference frame was selected. Point A was chosen as the moment center. The flow is steady, and $\dot{m}_1 = \dot{m}_2 = \dot{m}$. The vector form of the moment-of-momentum equation (6.18) will be used.

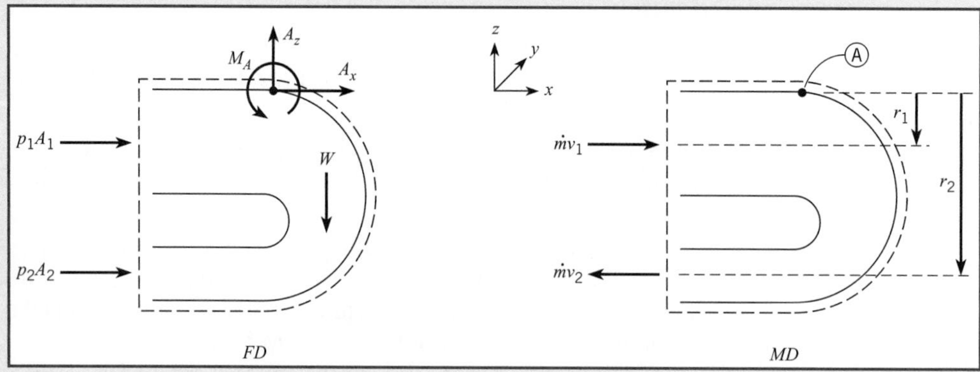

The force diagram shows pressure forces and weight. Also shown are forces applied by the support (A_x, A_z) and a moment (M_A), which is also due to the support. Using the force diagram with dimensions from the problem sketch yields

$$\sum \mathbf{M}_A = -(M_A + 0.15 p_1 A_1 + 0.475 p_2 A_2 - 0.2W)\mathbf{j}$$

The fluid within the control volume has constant angular momentum about point A, and so the accumulation term is zero. The momentum diagram shows an inward and outward momentum flow and the associated radius vectors ($\mathbf{r}_1$, $\mathbf{r}_2$). At the bend exit, the linear momentum flow rate is $\dot{m}\mathbf{v}_2$ and this term has a moment arm of $\mathbf{r}_2$. Thus, the outward flow of angular-momentum flow is

$$\mathbf{r}_2 \times (\dot{m}\mathbf{v}_2) = r_2 \dot{m} v_2 \mathbf{j}$$

Similarly, the inward angular-momentum flow is

$$\mathbf{r}_1 \times (\dot{m}\mathbf{v}_1) = r_1 \dot{m} v_1 \mathbf{j}$$

The net outward-moment-of-momentum-flow is

$$\sum_{\text{cs}} \mathbf{r}_o \times (\dot{m}_o \mathbf{v}_o) - \sum_{\text{cs}} \mathbf{r}_i \times (\dot{m}_i \mathbf{v}_i) = (r_2 \dot{m} v_2)\mathbf{j} + (r_1 \dot{m} v_1)\mathbf{j}$$

Substituting torques and momentum terms into the moment-of-momentum equation (6.18) results in

$$M_A + 0.15 p_1 A_1 + 0.475 p_2 A_2 - 0.2W = -r_2 \dot{m} v_2 - r_1 \dot{m} v_1$$

which simplifies to

$$M_A = -0.15 p_1 A_1 - 0.475 p_2 A_2 + 0.2W - \dot{m}(r_2 v_2 + r_1 v_1)$$

The torques due to pressure are

$$0.15 p_1 A_1 = (0.15 \text{ m})(150 \times 1000 \text{ N/m}^2)(\pi \times 0.3^2/4 \text{ m}^2)$$
$$= 1590 \text{ N} \cdot \text{m}$$
$$0.475 p_2 A_2 = (0.475 \text{ m})(59.3 \times 1000 \text{ N/m}^2)(\pi \times 0.15^2/4 \text{ m}^2)$$
$$= 497.8 \text{ N} \cdot \text{m}$$

The mass flow rate is

$$\dot{m} = \rho Q = (1000 \text{ kg/m}^3)(0.25 \text{ m}^3/\text{s}) = 250 \text{ kg/s}$$

The net angular-momentum outflow is

$$\dot{m}(r_2 v_2 + r_1 v_1) = (250 \text{ kg/s})(0.475 \times 14.15 + 0.15 \times 3.54)(\text{m}^2/\text{s})$$
$$= 1813 \text{ N} \cdot \text{m}$$

The moment exerted by the support is

$$M_A = -0.15 p_1 A_1 - 0.475 p_2 A_2 + 0.2W - \dot{m}(r_2 v_2 + r_1 v_1)$$
$$= -(1590 \text{ N} \cdot \text{m}) - (497.8 \text{ N} \cdot \text{m})$$
$$+ (0.2 \text{ m} \times 1481 \text{ N}) - (1813 \text{ N} \cdot \text{m})$$
$$= -3.60 \text{ kN} \cdot \text{m} \qquad \triangleleft$$

Thus, a moment of 3.60 kN · m acting in the (**j**) or clockwise direction is needed to hold the bend stationary. Stated differently, the support system must be designed to withstand a counterclockwise moment of 3.60 kN · m.

example 6.13

Determine the power produced by the sprinkler-like turbine shown in the following sketch. The turbine rotates in a horizontal plane about point O with a steady angular speed of 500 rpm. Water enters the turbine from a vertical pipe that is coaxial with the axis of rotation. Water exits the turbine through two identical nozzles, each of which has a section area of 10 cm^2. The exit speed of the water is 50 m/s relative to the nozzle. Water density is 1000 kg/m^3, and pressure at the exit of each nozzle is atmospheric.

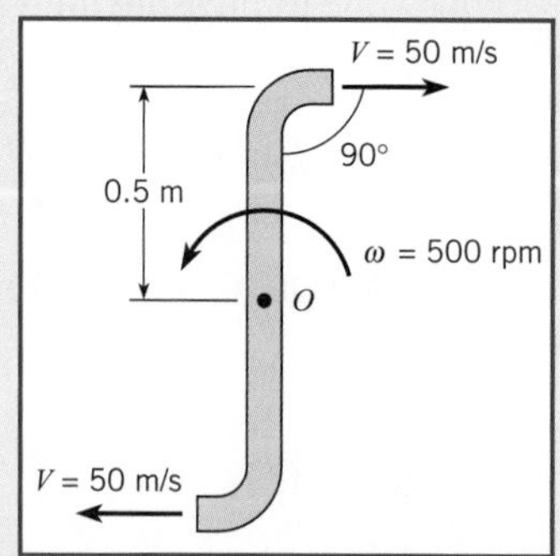

Solution We selected a moving control volume that surrounds the turbine and rotates. We chose a stationary reference frame that is fixed to point O. Note that a reference frame attached to the rotating turbine would be noninertial. We selected point O as the moment center. Equation (6.18) will be used for this problem.

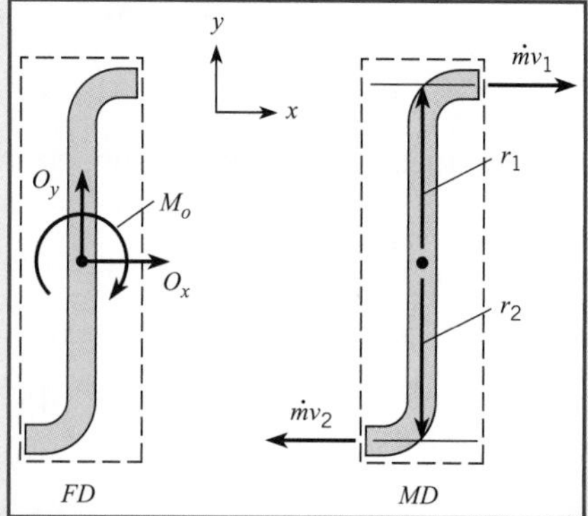

The force diagram shows a moment (M_o) due to the generator that is being used to produce power. Also shown are forces (O_x, O_y) due to the support system. From the force diagram

$$\sum \mathbf{M}_o = -M_o\mathbf{k}$$

From the selected reference frame, both the water and sprinkler arms have angular momentum about point O. However, the rotational speed is constant and so there is no change of angular momentum with time. Thus, we may set the accumulation term to zero. The momentum diagram shows two outward momentum flows and the corresponding radius vectors. There is no inflow of angular momentum because the radius vector associated with the incoming water (at point O) is zero. From the momentum diagram,

$$\mathbf{r}_1 \times (\dot{m}\mathbf{v}_1) + \mathbf{r}_2 \times (\dot{m}\mathbf{v}_2) = (-r_1\dot{m}v_1 - r_2\dot{m}v_2)\mathbf{k}$$

Substituting force and momentum terms into Eq. (6.17) gives

$$M_o = (r_1\dot{m}_1v_1 + r_2\dot{m}_2v_2) = 2(r\dot{m}v)$$

where the subscripts 1 and 2 were dropped because $r_1 = r_2$ and $v_1 = v_2$. The velocity v is relative to the selected stationary reference frame, so

$$v = V - \omega r$$

where ω is the angular speed of the turbine. The mass flow rate term uses the speed V that is relative to the control surface,

$$\dot{m} = \rho A V$$

Combining terms gives

$$M_o = 2r\rho A V(V - \omega r)$$

Mass flow rate is

$$\dot{m} = \rho A V = (1000\ \text{kg/m}^3)(0.001\ \text{m}^2)(50\ \text{m/s}) = 50\ \text{kg/s}$$

The rotational speed is $\omega = 500$ rpm $= 52.4$ rad/s, and so the moment is

$$\begin{aligned} M_o &= 2r\dot{m}[V - \omega r] \\ &= 2(0.5\ \text{m})(50\ \text{kg/s})[50\ \text{m/s} - (52.4\ \text{rad/s})(0.5\ \text{m})] \\ &= 1190\ \text{N} \cdot \text{m} \end{aligned}$$

The power produced by the turbine is the product of the moment and the angular speed:

$$\begin{aligned} P = M_o\omega &= (1190\ \text{N} \cdot \text{m})(52.4\ \text{rad/s}) \\ &= 62.4\ \text{kW} \end{aligned}$$

◁

example 6.14

A centrifugal pump is a device that takes liquid in on the pump axis and ejects the liquid at a high speed through the action of a rotating impeller. The kinetic energy is converted to an increased pressure. A centrifugal pump shown in the diagram takes water in at 20°C on the axis and ejects the water at the tip of the impeller. The diameter of the impeller is 20 cm. The impeller rotates at 1000 rpm. The flow rate through the pump is 400 lpm. Find the torque and power required to operate the pump.

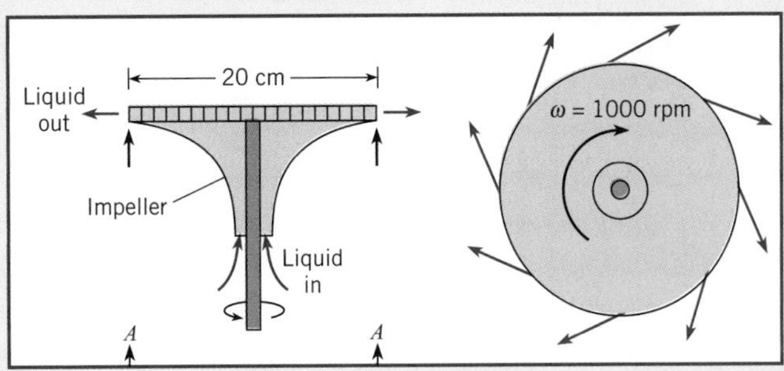

Solution The solution strategy is to first apply the moment-of-momentum equation to evaluate the torque. The power is the product of the torque and rotational speed.

Draw a stationary control volume around the impeller as shown.

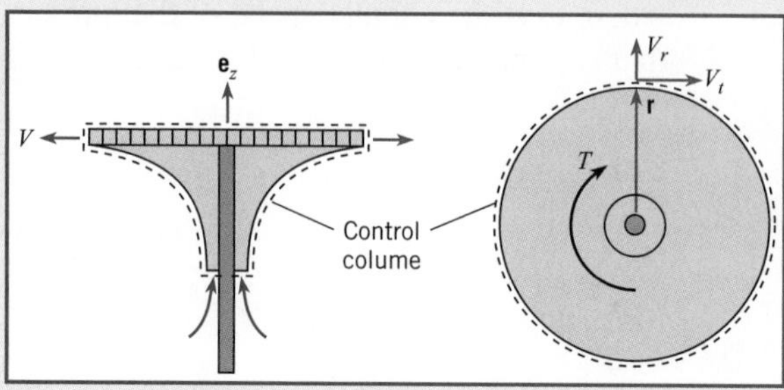

The torque, T, is applied in the positive z direction. The flow is steady, so the moment-of-momentum equation reduces to

$$T\mathbf{e}_z = \mathbf{r}_o \times (\dot{m}_o \mathbf{v}_o) - \mathbf{r}_i \times (\dot{m}_i \mathbf{v}_i)$$

At the inlet, there is no tangential component of velocity so there is no vector component in the z direction. So the equation further reduces to

$$T\mathbf{e}_z = \mathbf{r}_o \times (\dot{m}_o \mathbf{v}_o)$$

At the exit

$$\begin{aligned}\mathbf{r}_o \times \mathbf{v}_o &= \mathbf{e}_r r \times (\mathbf{e}_r V_r + \mathbf{e}_t V_t) \\ &= \mathbf{e}_z r V_t\end{aligned}$$

where $\mathbf{e}_r$ and $\mathbf{e}_t$ are the unit vectors in the radial and tangential directions, respectively. The tangential velocity is ωr, so from the moment-of momentum equation the torque is

$$\begin{aligned}T &= \dot{m}\omega r^2 \\ &= \dot{m}\omega \frac{D^2}{4}\end{aligned}$$

where D is the outlet diameter of the impeller. The mass flow rate is

$$\begin{aligned}\rho Q &= 1000 \text{ kg/m}^3 \times 400 \ l/\text{min} \times (1/60) \text{ min/s} \times 10^{-3} \text{m}^3/l \\ &= 6.67 \text{ kg/s}\end{aligned}$$

and the rotational rate is

$$\begin{aligned}\omega &= 1000 \text{ rev/min} \times 2\pi \ \text{ rad/rev} \times (1/60) \text{ min/s} \\ &= 104.7 \text{ rad/s}\end{aligned}$$

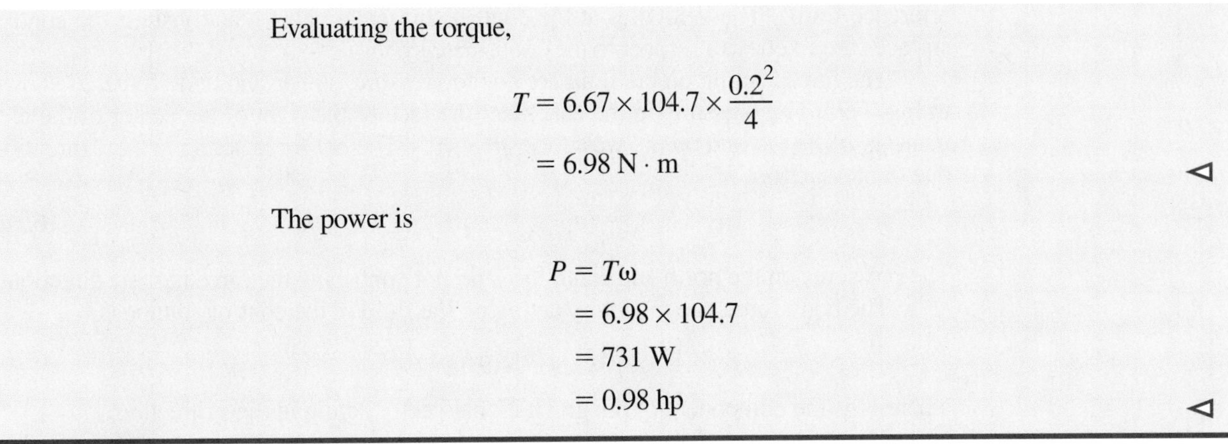

Evaluating the torque,

$$T = 6.67 \times 104.7 \times \frac{0.2^2}{4}$$
$$= 6.98\ \text{N} \cdot \text{m} \qquad \triangleleft$$

The power is

$$P = T\omega$$
$$= 6.98 \times 104.7$$
$$= 731\ \text{W}$$
$$= 0.98\ \text{hp} \qquad \triangleleft$$

6.6 Navier-Stokes Equations

In Chapter 5, the continuity equation at a point in the flow is derived using a control volume of infinitesimal size. The resulting differential equation is an independent equation in the analysis of fluid flow. The same approach can be applied to the momentum equation yielding the differential equation for momentum at a point in the flow. For simplicity, the derivation will be restricted to a two-dimensional planar flow and the extension to three dimensions will be outlined.

Consider the infinitesimal control volume shown in Fig. 6.9*a*. The dimensions of the control volume are Δx and Δy, and the dimension in the third direction (normal to page) is taken as unity. Assume that the center of the control volume is fixed with respect to the coordinate system and that the coordinate system is an inertial reference frame. Also assume that the control surfaces are fixed with respect to the coordinate system. The x-direction momentum equation is Eq. (6.8*a*), namely

$$\sum F_x = \frac{d}{dt}\int_{cv} \rho v_x \, d\forall + \sum_{cs} \dot{m}_o v_{o_x} - \sum_{cs} \dot{m}_i v_{i_x}$$

where v_x is the x component of velocity of the fluid with respect to an inertial reference frame. In this derivation, the component velocities in the x and y directions are u and v, respectively. These velocities are referenced to the coordinate system, which is an inertial

FIGURE 6.9

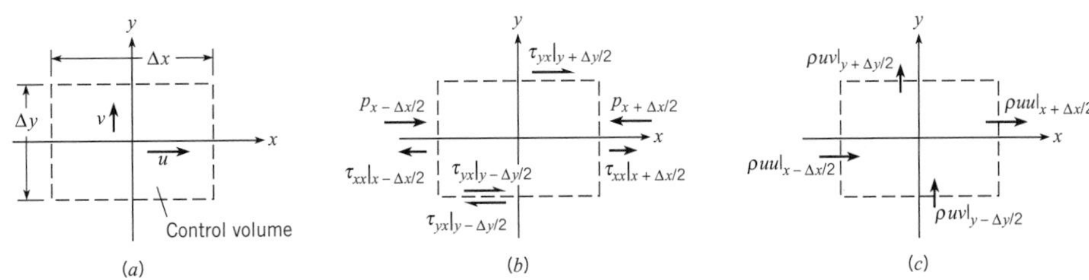

reference frame. The velocities at the control surface are also u and v since the control surfaces are fixed with respect to the coordinate system.

The forces acting on the fluid are due to pressure, stress, and body force, as shown in Fig. 6.9*b*. The pressure on the east face (the face to the right of the center of the element) is $p|_{x+\Delta x/2}$ and on the west face, $p|_{x-\Delta x/2}$. The net force acting in the x direction due to pressure is

$$F_{x,p} = (p|_{x-\Delta x/2} - p|_{x+\Delta x/2})\Delta y \tag{6.19}$$

The pressures on the north and south faces do not contribute the force in the x direction.

The body force due to gravity acting on the fluid in the control volume is

$$F_{x,g} = g_x \rho \Delta x \Delta y \tag{6.20}$$

where g_x is the component of the gravitational vector acting in the x direction.

The evaluation of the net shear stress forces is somewhat more complicated. The shear stress acting in the x direction on the north face is $\tau_{yx}|_{y+\Delta y/2}$. The subscripts on τ refer to the face on which the force acts and to the direction of the force. Thus τ_{yx} is the shear stress that acts on the y-face in the x direction. The face of a control surface is defined in the same way as the area vector in Section 4.2; that is, the y-face corresponds to the face with the area vector in the y direction. The shear stress on the south face inside the control volume is $\tau_{yx}|_{y-\Delta y/2}$. However, we are interested in the stress that acts on the south face outside the control volume, which is equal and opposite to the stress on the inside face. Thus, the stress on the fluid on the south face of the control volume is $-\tau_{yx}|_{y-\Delta y/2}$.

There is also a normal stress (other than pressure), which acts on the east and west faces. This stress is proportional to strain rate of the fluid in the control volume in the x direction. The stress on the east face is $\tau_{xx}|_{x+\Delta x/2}$. The stress acting on the outside west face of the control volume is $-\tau_{xx}|_{x-\Delta x/2}$ using the same argument as used for the shear stress on the south face. The net force in the x direction due to shear and normal stresses is

$$F_{\tau,x} = (\tau_{yx}|_{y+\Delta y/2} - \tau_{yx}|_{y-\Delta y/2})\Delta x + (\tau_{xx}|_{x+\Delta x/2} - \tau_{xx}|_{x-\Delta x/2})\Delta y \tag{6.21}$$

Applying Liebnitz theorem for the differentiation of an integral in the same fashion as done for the continuity equation, the rate of change of momentum in the x direction of the fluid in the control volume can be expressed as

$$\frac{d}{dt}\int_{cv} \rho u \, d\forall = \int_{cv} \frac{\partial}{\partial t}(\rho u)\, d\forall + \sum_{cs} \rho u(\mathbf{V}_c \cdot \mathbf{A}) \tag{6.22}$$

where $\mathbf{V}_c$ is the velocity of the control surface with respect to the coordinate system. For the control volume used in this derivation, $\mathbf{V}_c = 0$. Thus the rate of change of momentum of the fluid in the control volume becomes*

$$\frac{d}{dt}\int_{cv} \rho u \, d\forall = \int_{cv} \frac{\partial}{\partial t}(\rho u)\, dx\, dy \cong \frac{\partial}{\partial t}(\rho u)\Delta x \Delta y \tag{6.23}$$

* In the limit, as Δx and Δy approach zero,

$$\lim_{\Delta x, \Delta y \to 0} \frac{1}{\Delta x \Delta y}\int_{cv} \frac{\partial}{\partial t}(\rho u)\, dx\, dy = \frac{\partial}{\partial t}(\rho u)$$

The net efflux of momentum is obtained by summing the momentum flow from all four faces as shown in Fig. 6.9*c*. The flux of momentum outward from the east face is $\dot{m}_{x+\Delta x/2}u_{x+\Delta x/2}$ or $(\rho u_{x+\Delta x/2}\Delta y)u_{x+\Delta x/2}$. The momentum flux outward from the north face is $\dot{m}_{y+\Delta y/2}u_{x+\Delta y/2}$ or $(\rho v_{y+\Delta y/2}\Delta x)u_{y+\Delta y/2}$. The moment flux inward from the west and south faces is calculated in the same fashion. Finally, the net efflux of momentum is

$$\sum_{\text{cs}} \dot{m}_o v_{o_x} - \sum_{\text{cs}} \dot{m}_i v_{i_x} = (\rho uu|_{x+\Delta x/2} - \rho uu|_{x-\Delta x/2})\Delta y + (\rho vu|_{y+\Delta y/2} - \rho vu|_{y-\Delta y/2})\Delta x \tag{6.24}$$

Collecting all the terms that comprise Eq. (6.8*a*) and dividing through by the product $\Delta x \Delta y$ results in

$$\frac{1}{\Delta x \Delta y}\int_{\text{cv}} \frac{\partial}{\partial t}(\rho u)dxdy + \frac{\rho uu|_{x+\Delta x/2} - \rho uu|_{x-\Delta x/2}}{\Delta x} + \frac{\rho vu|_{y+\Delta y/2} - \rho vu|_{y-\Delta y/2}}{\Delta y} = \frac{p|_{x+\Delta x/2} - p|_{x+\Delta x/2}}{\Delta x} + \frac{\tau_{yx}|_{y+\Delta y/2} - \tau_{yx}|_{y-\Delta y/2}}{\Delta y} + \frac{\tau_{xx}|_{x+\Delta x/2} - \tau_{xx}|_{x-\Delta x/2}}{\Delta x} + \rho g_x \tag{6.25}$$

Taking the limit as Δx and Δy approach zero yields the differential form of the momentum equation in the x direction:

$$\frac{\partial}{\partial t}(\rho u) + \frac{\partial}{\partial x}(\rho uu) + \frac{\partial}{\partial y}(\rho uv) = -\frac{\partial p}{\partial x} + \frac{\partial \tau_{xx}}{\partial x} + \frac{\partial \tau_{yx}}{\partial y} + \rho g_x \tag{6.26}$$

One further step is to use the differential form of the continuity equation for two-dimensional flow [Eq. (5.33) with $\partial(\rho w)/\partial z = 0$] to convert the equation to a different form. Through differentiation by parts, the left side of Eq. (6.26) can be written as

$$\frac{\partial}{\partial t}(\rho u) + \frac{\partial}{\partial x}(\rho uu) + \frac{\partial}{\partial y}(\rho uv) = \rho\frac{\partial u}{\partial t} + \rho u\frac{\partial u}{\partial x} + \rho v\frac{\partial u}{\partial y} + u\left[\frac{\partial \rho}{\partial t} + \frac{\partial}{\partial x}(\rho u) + \frac{\partial}{\partial y}(\rho v)\right] \tag{6.27}$$

One notes that the last term is zero because of the continuity equation (Eq. 5.33), so the momentum equation in the x direction at a point becomes

$$\rho\frac{\partial u}{\partial t} + \rho u\frac{\partial u}{\partial x} + \rho v\frac{\partial u}{\partial y} = -\frac{\partial p}{\partial x} + \frac{\partial \tau_{xx}}{\partial x} + \frac{\partial \tau_{yx}}{\partial y} + \rho g_x \tag{6.28}$$

The left side of this equation is the product of the density and the acceleration of a fluid element in the x direction (see Section 4.3), so the equation can be written more simply as

$$\rho\frac{Du}{Dt} = -\frac{\partial p}{\partial x} + \frac{\partial \tau_{xx}}{\partial x} + \frac{\partial \tau_{yx}}{\partial y} + \rho g_x \tag{6.29}$$

By applying Eq. (6.8*b*) to the same control volume, the momentum equation in the y direction at a point can be derived. The result is

$$\rho\frac{\partial v}{\partial t}+\rho u\frac{\partial v}{\partial x}+\rho v\frac{\partial v}{\partial y}=\rho\frac{Dv}{Dt}=-\frac{\partial p}{\partial y}+\frac{\partial \tau_{yy}}{\partial y}+\frac{\partial \tau_{xy}}{\partial x}+\rho g_y \tag{6.30}$$

The same approach can be used to derive the momentum equation in three dimensions. In this case the two-dimensional infinitesimal volume is extended to a three-dimensional volume with depth Δz and with a velocity component w in the z direction.

In order to complete the development of these equations, a relationship is needed between the shear and normal stresses and the rate of strain of the fluid elements. These relationships are called the "constitutive equations." For a Newtonian fluid, by definition, the stress is proportional to the *rate* of strain. The rate of shear strain of a fluid element can be related to the gradients in velocity in the same way as the rate of rotation in Section 4.6. With reference to Fig. 4.19, the shear strain of an element is $\Delta\theta_B-\Delta\theta_C$, so the rate of shear strain is

$$\lim_{\Delta t\to 0}\left(\frac{\Delta\theta_B}{\Delta t}-\frac{\Delta\theta_C}{\Delta t}\right)=\omega_{AB}-\omega_{AC}=\frac{1}{2}\left(\frac{\partial u}{\partial y}+\frac{\partial v}{\partial x}\right) \tag{6.31}$$

The constant of proportionality between the shear stress and the rate of shear strain is 2μ, where μ is the coefficient of dynamic viscosity, so the shear stress τ_{yx} is related to the velocity gradients by

$$\tau_{yx}=\mu\left(\frac{\partial u}{\partial y}+\frac{\partial v}{\partial x}\right) \tag{6.32}$$

The normal stress in the x direction for an incompressible fluid* is given by

$$\tau_{xx}=2\mu\frac{\partial u}{\partial x} \tag{6.33}$$

Substituting the above constitutive relations into Eq. (6.30) yields

$$\rho\frac{Du}{Dt}=-\frac{\partial p}{\partial x}+\frac{\partial}{\partial x}\left(2\mu\frac{\partial u}{\partial x}\right)+\frac{\partial}{\partial y}\left[\mu\left(\frac{\partial u}{\partial y}+\frac{\partial v}{\partial x}\right)\right]+\rho g_x \tag{6.34}$$

If the dynamic viscosity is constant, the normal and shear stress terms can be written as

$$\frac{\partial}{\partial x}\left(2\mu\frac{\partial u}{\partial x}\right)+\frac{\partial}{\partial y}\left[\mu\left(\frac{\partial u}{\partial y}+\frac{\partial v}{\partial x}\right)\right]=\mu\left[\frac{\partial^2 u}{\partial x^2}+\frac{\partial^2 u}{\partial y^2}\right]+\mu\frac{\partial}{\partial x}\left(\frac{\partial u}{\partial x}+\frac{\partial v}{\partial y}\right) \tag{6.35}$$

From the continuity equation for the planar flow of an incompressible fluid, the last term is zero, so Eq. (6.34) reduces to

$$\rho\frac{Du}{Dt}=-\frac{\partial p}{\partial x}+\mu\left[\frac{\partial^2 u}{\partial x^2}+\frac{\partial^2 u}{\partial y^2}\right]+\rho g_x \tag{6.36}$$

* An additional term is needed for the normal stress equation for a compressible fluid, which is discussed in Schlichting (3).

The corresponding equation in the y direction is

$$\rho\frac{Dv}{Dt} = -\frac{\partial p}{\partial y} + \mu\left[\frac{\partial^2 v}{\partial x^2} + \frac{\partial^2 v}{\partial y^2}\right] + \rho g_y \tag{6.37}$$

The three-dimensional form for these equations can be found in Schlichting (3). These are called the Navier-Stokes equations after L. M. Navier (1785–1836) and G. G. Stokes (1819–1903), who are credited with their development.

An application of the Navier-Stokes equations will be presented in Chapter 9.

Summary

The momentum principle is used to analyze problems involving forces and flow. It is expressed as

$$\sum_{cs}\mathbf{F} = \frac{d}{dt}\int_{cv}\mathbf{v}\rho\, d\forall + \int_{cs}\mathbf{v}\rho\mathbf{V}\cdot d\mathbf{A}$$

where $\mathbf{V}$ is flow velocity relative to the control surface, and $\mathbf{v}$ is flow velocity relative to an inertial (nonaccelerating) reference frame.

The physical interpretation of the momentum principle is that the sum of forces equals the rate of momentum change inside the control volume plus the net rate at which momentum flows out of the control volume. To apply the momentum equation, one selects a control volume and then evaluates the forces, momentum accumulation, and momentum flow terms. These terms may be represented visually by using a force diagram and a momentum diagram.

The force term represents all external forces that act on the material inside the control volume. These forces can be either body forces or surface forces. For most problems, the only body force is weight. There are three common types of surface forces: those caused by structural elements, and those caused by pressure and shear stress distributions.

The momentum accumulation term $(d/dt\int_{cv}\mathbf{v}\rho\, d\forall)$ gives the rate at which the momentum inside the control volume is changing with time. If flow is steady, and other mass in the control volume is stationary, the momentum accumulation is zero. Otherwise, the momentum accumulation is evaluated by integration.

The momentum flow term $(\int_{cs}\mathbf{v}\rho\mathbf{V}\cdot d\mathbf{A})$ gives the net rate at which momentum is flowing outward across the control surface. If velocity varies across the control surface, integration is needed to find the momentum flow.

If mass enters and leaves the control volume through a number of ports and if the velocity $\mathbf{v}$ is uniformly distributed across each port, the momentum equation simplifies to

$$\sum_{cs}\mathbf{F} = \frac{d}{dt}\int_{cv}\mathbf{v}\rho\, d\forall + \sum_{cs}\dot{m}_o\mathbf{v}_o - \sum_{cs}\dot{m}_i\mathbf{v}_i$$

where subscripts o and i denote out and in, respectively. In this equation, $\dot{m}$ is the rate at which mass is crossing the control surface and $\mathbf{v}$ is flow velocity evaluated at the control surface with respect to the inertial reference frame selected.

Water hammer is due to a pressure wave in a duct that travels at the speed of sound. The pressure rise across the wave is

$$\Delta p = \rho V_c$$

where V is the duct flow velocity and c is the speed of sound, given by

$$c = \sqrt{\frac{E_v}{\rho}}$$

Problems involving moments may be analyzed by applying the moment-of-momentum principle. If mass crosses the control surface through a number of inlet and outlet ports and if the properties are uniformly distributed across each port, the moment-of-momentum principle is expressed as

$$\sum_{cs} \mathbf{M} = \frac{d}{dt}\int_{cv} \rho \mathbf{v}\, d\forall + \sum_{cs} \mathbf{r}_o \times (\dot{m}_o \mathbf{v}_o) - \sum_{cs} \mathbf{r}_i \times (\dot{m}_i \mathbf{v}_i)$$

where $\mathbf{r}$ is a position vector from the moment center. The physical interpretation is that the sum of moments equals the rate of angular-momentum change inside the control volume plus the net rate at which angular momentum flows out of the control volume. Application of the moment-of-momentum equation parallels the approaches used for the momentum equation.

The Navier-Stokes equation is a differential form of Newton's second law that applies to motion of a fluid element. The Navier-Stokes equation is a nonlinear, partial-differential equation that is widely used in advanced studies of fluid mechanics phenomena.

References

1. Chaudhry, M. H. *Applied Hydraulic Transients,* 2nd ed. Van Nostrand Reinhold, New York, 1987.

2. Streeter, V. L., and E. B. Wylie. *Fluid Transients.* FEB Press, Ann Arbor, Michigan, 1983.

3. Schlichting, H. *Boundary Layer Theory.* McGraw-Hill, New York, 1979.

Problems

6.1 A "balloon rocket" is a balloon suspended from a taut wire by a hollow tube (drinking straw) and string. The nozzle is formed of a 1-cm-diameter tube, and an air jet exits the nozzle with a speed of 40 m/s and a density of 1.2 kg/m^3. Find the force F needed to hold the balloon stationary. Neglect friction.

6.2 The balloon rocket is held in place by a force F. The pressure inside the balloon is 8 in.-H_2O, the nozzle diameter is 1.0 cm, and the air density is 1.2 kg/m^3. Find the exit velocity v and the force F. Neglect friction and assume the air flow is inviscid and irrotational.

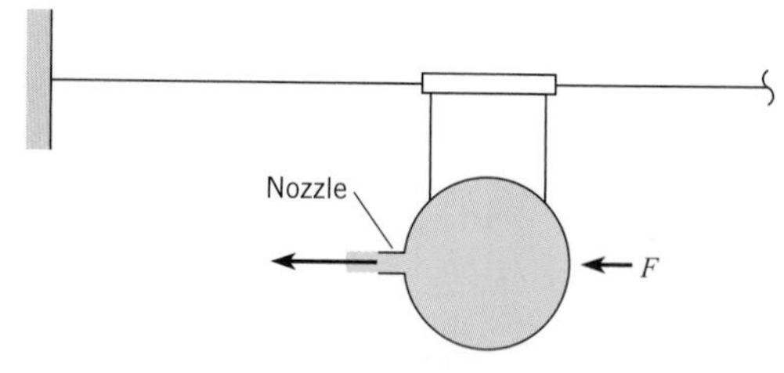

PROBLEMS 6.1, 6.2

6.3 A water jet of diameter 30 mm and speed $v = 20 \text{ m/s}$ is filling a tank. The tank has a mass of 20 kg and contains 20 liters of water at the instant shown. The water temperature is 15°C. Find the force acting on the bottom of the tank and the force acting on the stop block. Neglect friction.

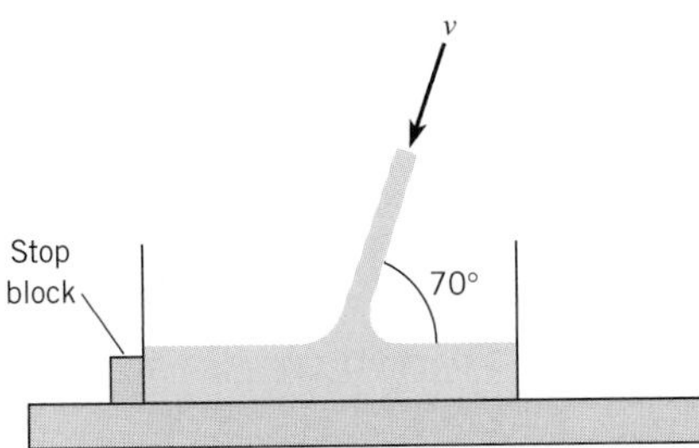

PROBLEMS 6.3, 6.4

6.4 A water jet of diameter 2 inches and speed $v = 50 \text{ ft/s}$ is filling a tank. The tank has a mass of 25 lbm and contains 5 gallons of water at the instant shown. The water temperature is 70°F. Find the minimum coefficient of friction such that the force acting on the stop block is zero.

6.5 A design contest features a submarine that will travel at a steady speed of $V_{\text{sub}} = 1.5 \text{ m/s}$ in 15°C water. The sub is powered by a water jet. This jet is created by drawing water from an inlet of diameter 25 mm, passing this water through a pump and then accelerating the water through a nozzle of diameter 5 mm to a speed of V_{jet}. The hydrodynamic drag force (F_D) can be calculated using

$$F_D = C_D\left(\frac{\rho V_{\text{sub}}^2}{2}\right)A_p$$

where the coefficient of drag is $C_D = 0.3$ and the projected area is $A_p = 0.28 \text{ m}^2$. Specify an acceptable value of V_{jet}.

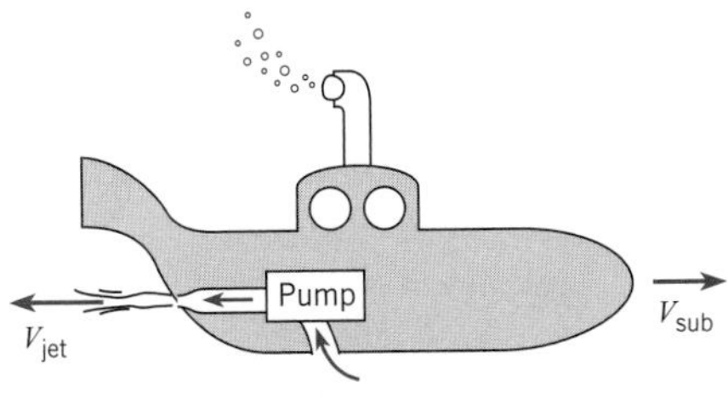

PROBLEM 6.5

6.6 A horizontal water jet at 70°F impinges on a vertical-perpendicular plate. The discharge is 2 cfs. If the external force required to hold the plate in place is 200 lbf, what is the velocity of the water?

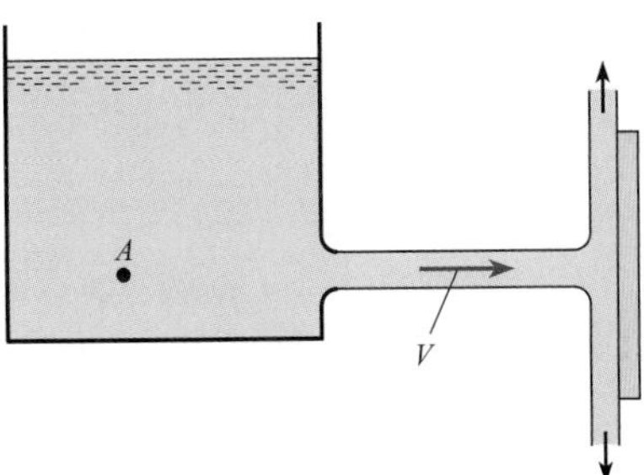

PROBLEMS 6.6, 6.7

6.7 A horizontal water jet at 70°F issues from a circular orifice in a large tank. The jet strikes a vertical plate that is normal to the axis of the jet. A force of 500 lbf is needed to hold the plate in place against the action of the jet. If the pressure in the tank is 25 psig at point A, what is the diameter of the jet just downstream of the orifice?

6.8 An engineer, who is designing a water toy, is making preliminary calculations. A user of the product will apply a force F_1 that moves a piston ($D = 80$ mm) at a speed of $V_{\text{piston}} = 300 \text{ mm/s}$. Water at 20°C jets out of a converging nozzle of diameter $d = 15$ mm. To hold the toy stationary, the user applies a force F_2 to the handle. Which force (F_1 versus F_2) is larger? Explain your answer using concepts of the momentum principle. Then calculate F_1 and F_2. Neglect friction between the piston and the walls.

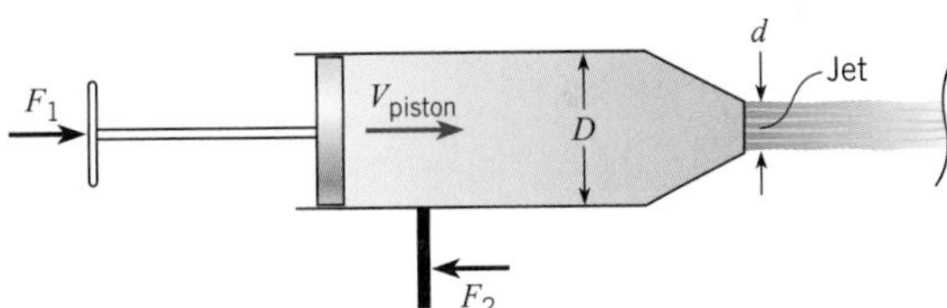

PROBLEM 6.8

6.9 A firehose on a boat is producing a 3-in.-diameter water jet with a speed of $V = 70$ mph. The boat is held stationary by a cable attached to a pier, and the water temperature is 50°F. Calculate the tension in the cable.

6.10 A boat is held stationary by a cable attached to a pier. A firehose directs a spray of 5°C water at a speed of $V = 50 \text{ m/s}$. If the allowable load on the cable is 5 kN, calculate the mass flow rate of the water jet. What is the corresponding diameter of the water jet?

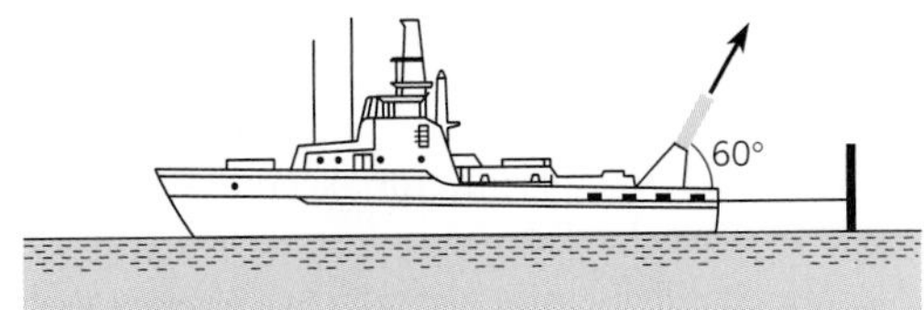

PROBLEMS 6.9, 6.10

FLUID SOLUTIONS **6.11** Water at 60°F flows through a nozzle that contracts from a diameter of 3 in. to 1 in. The pressure at section 1 is 2000 psfg, and atmospheric pressure prevails at the exit of the jet. Calculate the speed of the flow at the nozzle exit and the force required to hold the nozzle stationary. Neglect weight.

6.12 Water at 15°C flows through a nozzle that contracts from a diameter of 10 cm to 2 cm. The exit speed is $v_2 = 25$ m/s, and atmospheric pressure prevails at the exit of the jet. Calculate the pressure at section 1 and the force required to hold the nozzle stationary. Neglect weight.

6.13 Write a computer program that models the nozzle described in Prob. 6.12. Apply your computer program to solve Prob. 6.12 and 6.14. Compare your computer program prediction with the answers given in the back of the text.

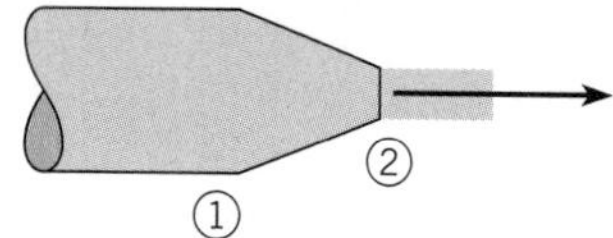

PROBLEMS 6.11, 6.12, 6.13

6.14 A tank of water (15°C) with a total weight of 200 N (water plus the container) is suspended by a vertical cable. Pressurized air drives a water jet ($d = 12$ mm) out the bottom of the tank such that the tension in the vertical cable is 10 N. If $H = 425$ mm, find the required air pressure in units of atmospheres (gage). Assume the flow of water is irrotational.

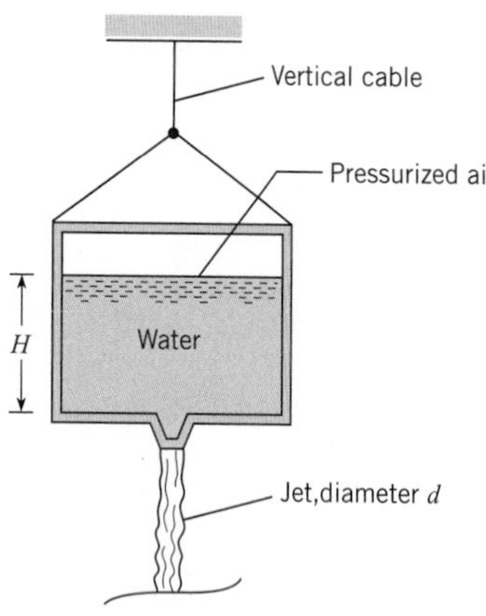

PROBLEM 6.14

6.15 A jet of water (60°F) is discharging at a constant rate of 2.0 cfs from the upper tank. If the jet diameter at section 1 is 4 in., what forces will be measured by scales A and B? Assume the empty tank weighs 300 lbf, the cross-sectional area of the tank is 4 ft^2, $h = 1$ ft, and $H = 9$ ft.

6.16 A conveyor belt discharges gravel into a barge as shown at a rate of 50 yd^3/min. If the gravel weighs 120 lbf/ft^3, what is the tension in the hawser that secures the barge to the dock?

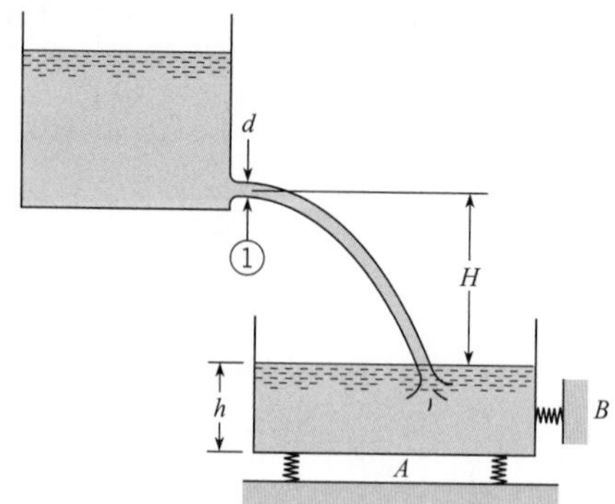

PROBLEM 6.15

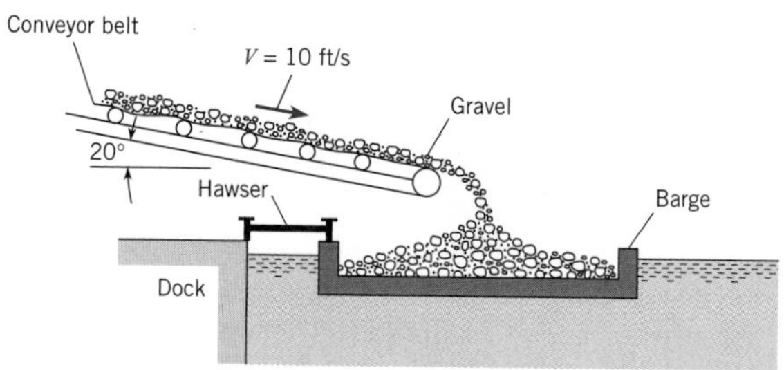

PROBLEM 6.16

6.17 Determine the external reactions in the x and y directions needed to hold this fixed vane, which turns the oil jet in a horizontal plane. Here V_1 is 18 m/s, $V_2 = 17$ m/s, and $Q = 0.15$ m^3/s.

6.18 Solve Prob. 6.17 for $V_1 = 90$ ft/s, $V_2 = 85$ ft/s, and $Q = 2$ cfs.

6.19 This planar water jet (60°F) is deflected by a fixed vane. What are the x and y components of force per unit width needed to hold the vane stationary? Neglect gravity.

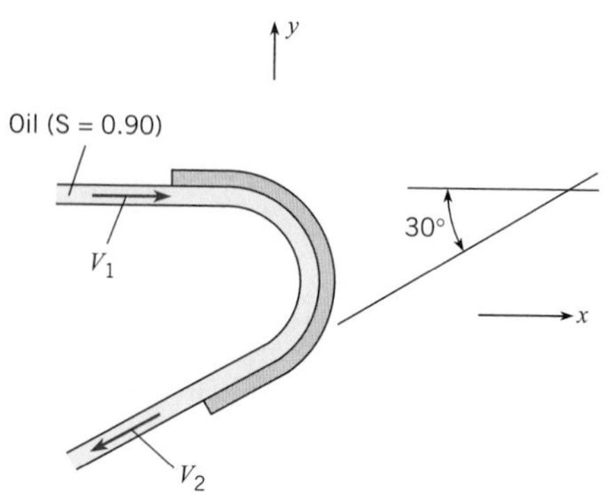

PROBLEMS 6.17, 6.18

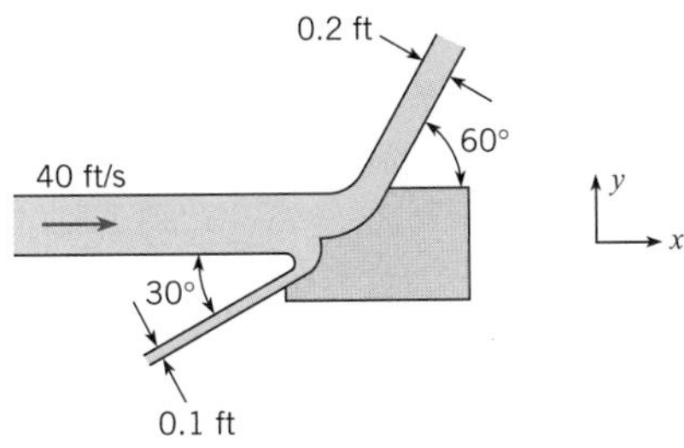

PROBLEM 6.19

6.20 A water jet with a speed of 20 ft / s and a mass flow rate of 25 lbm / s is turned 30° by a fixed vane. Find the force of the water jet on the vane. Neglect gravity.

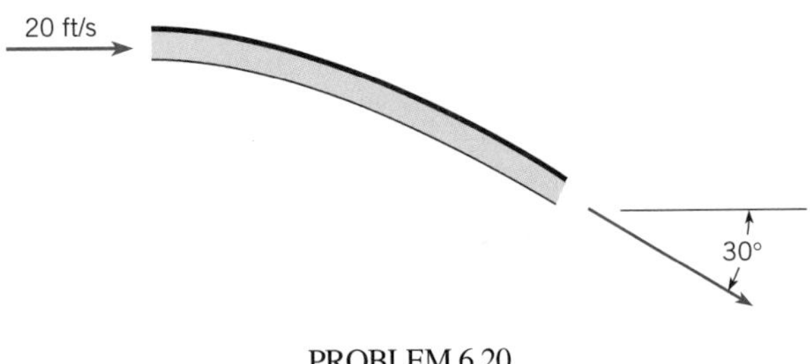

PROBLEM 6.20

6.21 Water strikes a block as shown and is deflected 30°. The flow rate of the water is 1 kg / s, and the inlet velocity is $V = 10$ m / s. The mass of the block is 1 kg. The coefficient of static friction between the block and the surface is 0.1 (friction force / normal force). If the force parallel to the surface exceeds the frictional force, the block will move. Determine the force on the block and whether the block will move. Neglect the weight of the water.

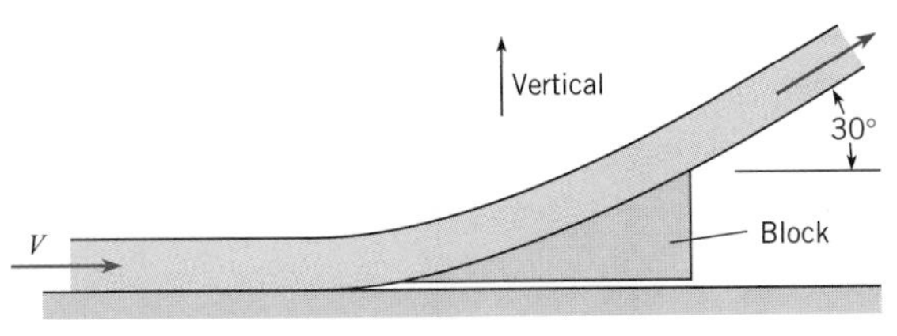

PROBLEMS 6.21, 6.22

6.22 For the situation described in Prob. 6.21, find the maximum inlet velocity (V) such that the block will not slip.

6.23 Plate A is 50 cm in diameter and has a sharp-edged orifice at its center. A water jet strikes the plate concentrically with a speed of 30 m / s. What external force is needed to hold the plate in place if the jet issuing from the orifice also has a speed of 30 m / s? The diameters of the jets are $D = 5$ cm and $d = 2$ cm.

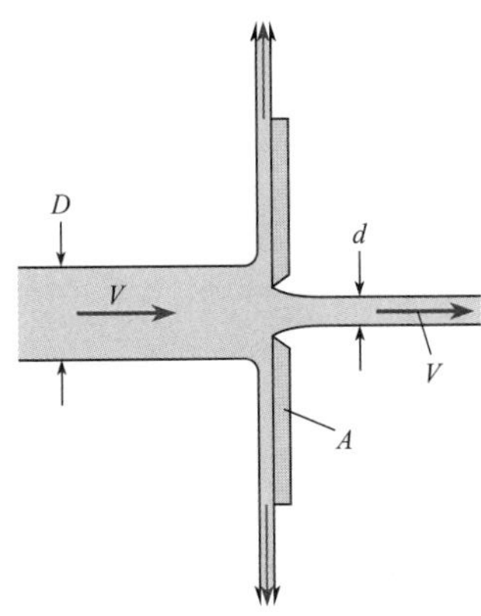

PROBLEM 6.23

6.24 This two-dimensional liquid jet impinges on the horizontal floor. Derive formulas for d_2 and d_3 as functions of b_1 and θ. Assume that the jet speed is large enough to cause the gravitational effects to be negligible. Also assume that the speed of the liquid is the same at sections 1, 2, and 3.

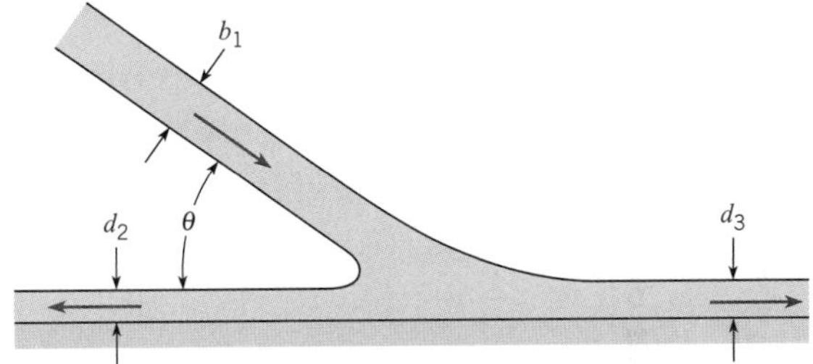

PROBLEM 6.24

6.25 A two-dimensional liquid jet impinges on a vertical wall. Assuming that the incoming jet speed is the same as the exiting jet speed ($V_1 = V_2$), derive an expression for the force per unit width of jet exerted on the wall. What form do you think the upper liquid surface will take next to the wall? Sketch the shape you think it will take, and explain your reasons for drawing it that way.

6.26 A ramjet operates by taking in air at the inlet, providing fuel for combustion, and exhausting the hot air through the exit. The mass flow at the inlet and outlet of the ramjet is 50 kg / s (the mass flow rate of fuel is negligible). The inlet velocity is 225 m / s. The density of the gases at the exit is 0.25 kg / m^3, and the exit area is 0.5 m^2. Calculate the thrust delivered by the ramjet. The ramjet is not accelerating and the flow within the ramjet is steady.

6.27 A discharge Q of a liquid of very low viscosity drops vertically into a short horizontal-rectangular channel of width B as shown. The depth at section 2 is y_2. Derive an equation that gives y_1 in terms of y_2, Q, B, and γ.

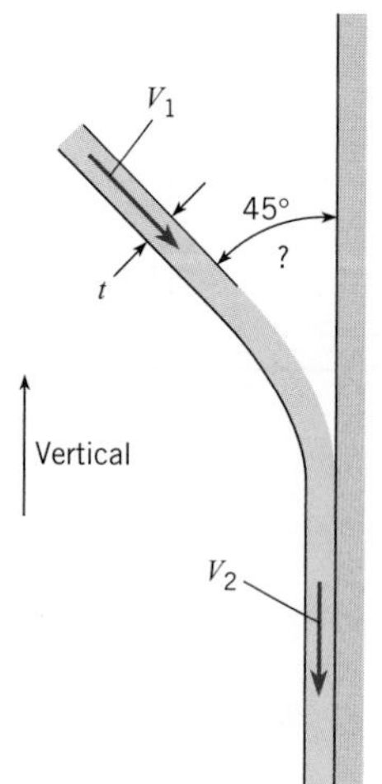

PROBLEM 6.25

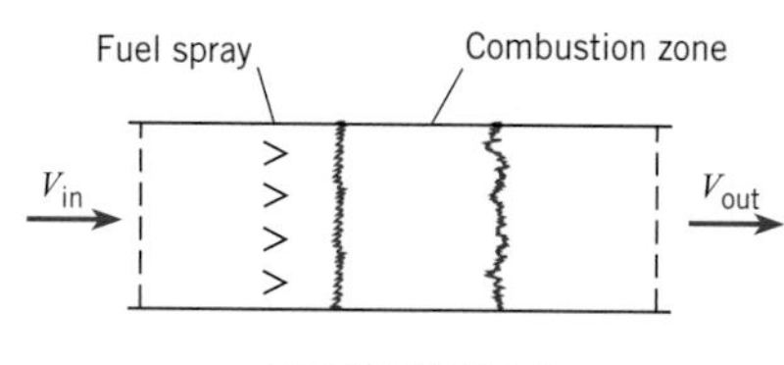

PROBLEM 6.26

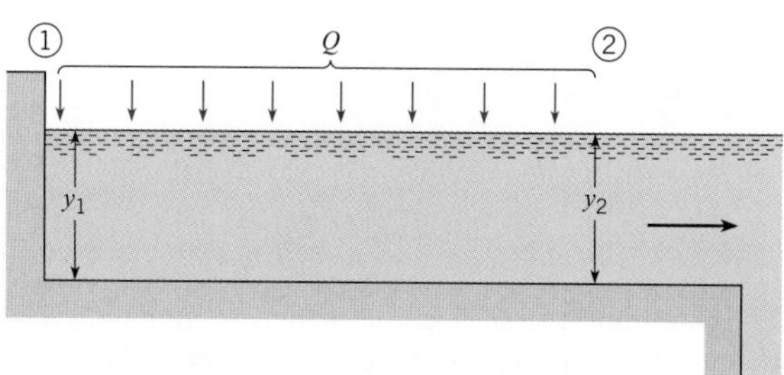

PROBLEM 6.27

6.28 The end section of a pipe has a slot cut in it so that liquid will discharge laterally as shown. The liquid is discharged from the slot at a rate given by

$$dQ = \Delta y \sqrt{2p/\rho}\, dx$$

where dQ is the discharge per differential length dx along the slot, Δy is the thickness of the jet issuing from the slot, and p is the gage pressure in the pipe. Tell qualitatively how the pressure will change in the pipe from $x = 0$ to $x = L$. Also devise a way of solving for the pressure distribution, given p_{end} (pressure at the end of the pipe) and the slot dimensions. Neglect viscous resistance in developing your solution procedure.

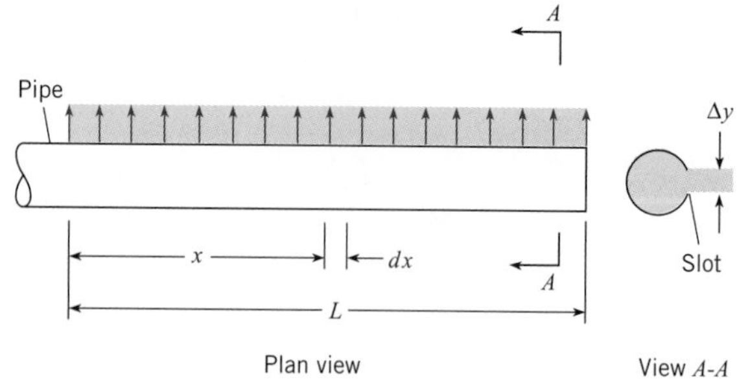

PROBLEM 6.28

6.29 A cone that is held stable by a wire is free to move in the vertical direction and has a jet of water striking it from below. The cone weighs 30 N. The initial speed of the jet as it comes from the orifice is 15 m/s, and the initial jet diameter is 2 cm. Find the height to which the cone will rise and remain stationary. *Note:* The wire is only for stability and should not enter into your calculations.

6.30 A 6-in. horizontal pipe has a 180° bend in it. If the rate of flow of water (60°F) in the bend is 6 cfs and the pressure therein is 20 psi, what external force in the original direction of flow is required to hold the bend in place?

6.31 A hot gas stream enters a uniform-diameter return bend as shown. The entrance velocity is 100 ft/s, the gas density is 0.02 lbm/ft^3, and the mass flow rate is 1 lbm/s. Water is sprayed into the duct to cool the gas down. The gas exits with a density of 0.06 lbm/ft^3. The mass flow of water into the gas is negligible. The pressures at the entrance and exit are the same and equal to the atmospheric pressure. Find the force required to hold the bend.

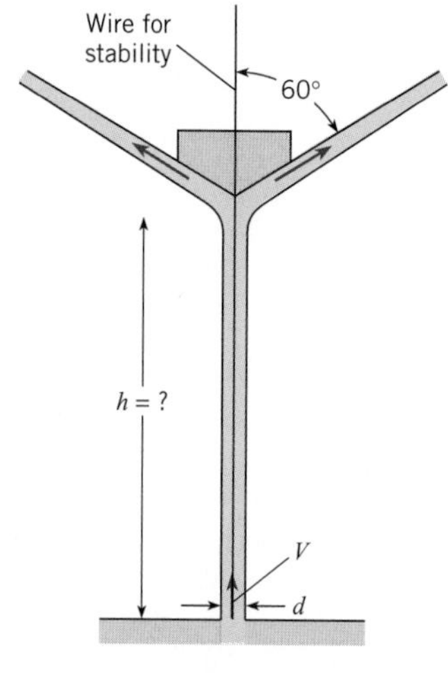

PROBLEM 6.29

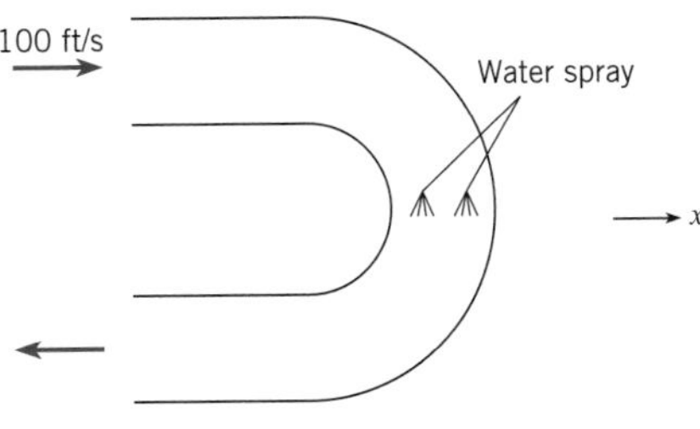

PROBLEM 6.31

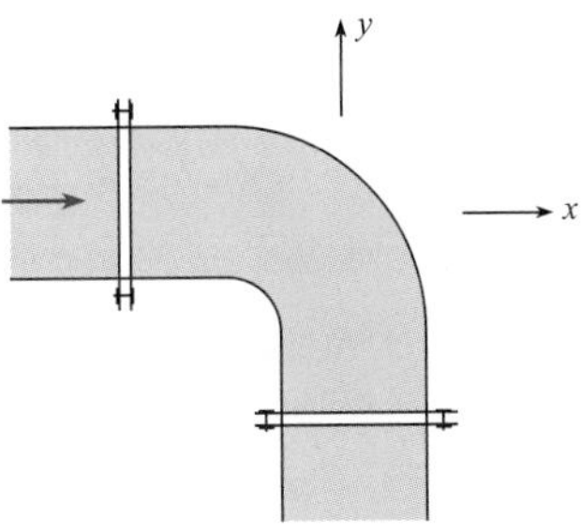

PROBLEMS 6.35, 6.36

6.32 Assume that the gage pressure p is the same at sections 1 and 2 in this horizontal bend. The fluid flowing in the bend has density ρ, discharge Q, and velocity V. The cross-sectional area of the pipe is A. Then the magnitude of the force (neglecting gravity) required at the flanges to hold the bend in place will be (a) pA, (b) $pA + \rho QV$, (c) $2pA + \rho QV$, (d) $2pA + 2\rho QV$.

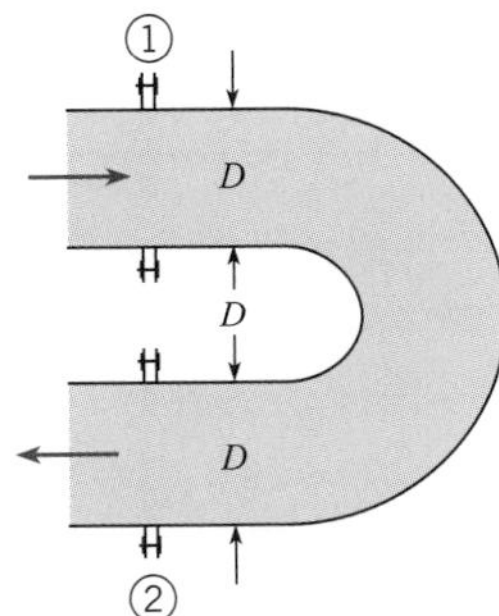

PROBLEMS 6.32, 6.33, 6.34

6.33 This pipe has a 180° vertical bend in it. The diameter D is 1 ft, and the pressure at the center of the upper pipe is 15 psig. If the flow in the bend is 20 cfs, what external force will be required to hold the bend in place against the action of the water? The bend weighs 200 lbf, and the volume of the bend is 3 ft^3. Assume the Bernoulli equation applies.

6.34 This pipe has a 180° horizontal bend in it as shown, and D is 20 cm. The discharge of water in the pipe and bend is 0.30 m^3/s, and the pressure in the pipe and bend is 100 kPa gage. If the bend volume is 0.10 m^3 and the bend itself weighs 500 N, what force must be applied at the flanges to hold the bend in place?

6.35 Water flows in the horizontal bend at a rate of 10 cfs and discharges into the atmosphere past the downstream flange. The pipe diameter is 1 ft. What force must be applied at the upstream flange to hold the bend in place? Assume that the volume of water downstream of the upstream flange is 4 ft^3 and that the bend and pipe weigh 100 lbf. Assume the pressure at the inlet section is 4 psig.

6.36 The gage pressure throughout the horizontal 90° pipe bend is 300 kPa. If the pipe diameter is 1 m and the water flow rate is 10 m^3/s, what x component of force must be applied to the bend to hold it in place against the water action?

6.37 This 30° vertical bend in a pipe with a 2-ft diameter carries water at a rate of 31.4 cfs. If the pressure p_1 is 10 psi at the lower end of the bend, where the elevation is 100 ft, and p_2 is 8.5 psi at the upper end, where the elevation is 103 ft, what will be the vertical component of force that must be exerted by the "anchor" on the bend to hold it in position? The bend itself weighs 300 lb, and the length L is 4 ft.

6.38 This bend discharges water into the atmosphere. Determine the force components at the flange required to hold the bend in place. The bend lies in a horizontal plane. Assume viscous forces are negligible. The interior volume of the bend is 0.25 m^3, D_1 = 60 cm, D_2 = 30 cm, and V_2 = 10 m/s. The mass of the bend material is 250 kg.

6.39 This nozzle bends the flow from vertically upward to 30° with the horizontal and discharges water (γ = 62.4 lbf/ft^3) at a speed of V = 130 ft/s. The volume within the nozzle itself is 1.8 ft^3, and the weight of the nozzle is 100 lbf. For these conditions, what *vertical* force must be applied to the nozzle at the flange to hold it in place.

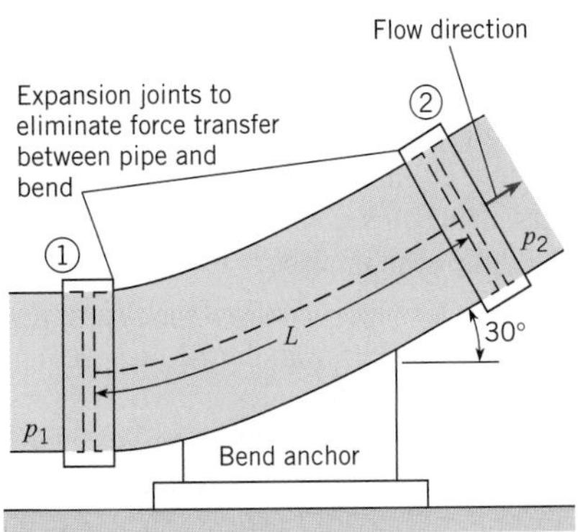

PROBLEM 6.37

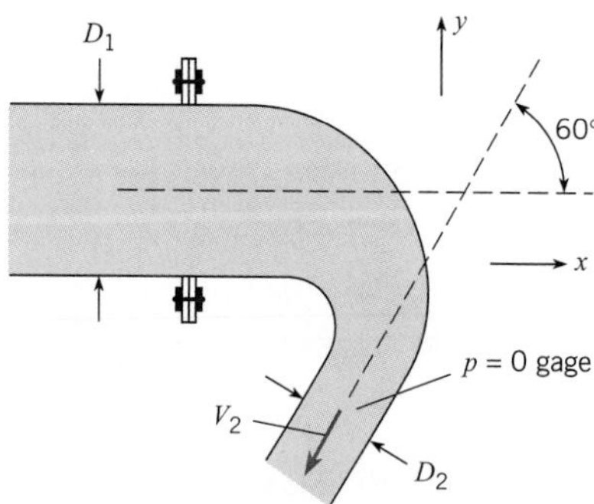

PROBLEM 6.38

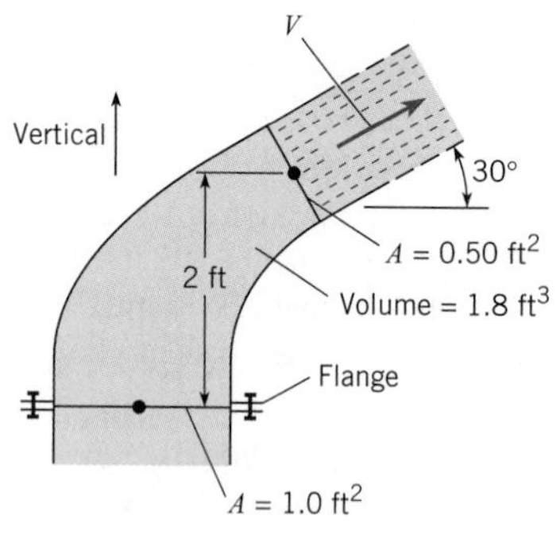

PROBLEM 6.39

6.40 A pipe 1 ft in diameter bends through an angle of 135°. The velocity of flow of gasoline (S = 0.8) is 15 ft / s, and the pressure is 10 psi in the bend. What external force is required to hold the bend against the action of the gasoline? Neglect the gravitational force.

6.41 A pipe 15 cm in diameter bends through 135°. The velocity of flow of gasoline (S = 0.8) is 8 m / s, and the pressure is 100 kPa gage throughout the bend. Neglecting gravitational force, determine the external force required to hold the bend against the action of the gasoline.

6.42 A horizontal reducing bend turns the flow of water through 60°. The inlet area is 0.001 m^2, and the outlet area is 0.0001 m^2. The water from the outlet discharges into the atmosphere with a velocity of 50 m / s. What horizontal force (parallel to the initial flow direction) acting through the metal of the bend at the inlet is required to hold the bend in place?

6.43 This 130-cm-diameter overflow pipe from a small hydroelectric plant conveys water from the 70-m elevation to the 40-m elevation. The pressures in the water at the bend entrance and exit are 20 kPa and 25 kPa, respectively. The bend interior volume is 3 m^3, and the bend itself weighs 10 kN. Determine the force that a thrust block must exert on the bend to secure it if the discharge is 16 m^3 / s.

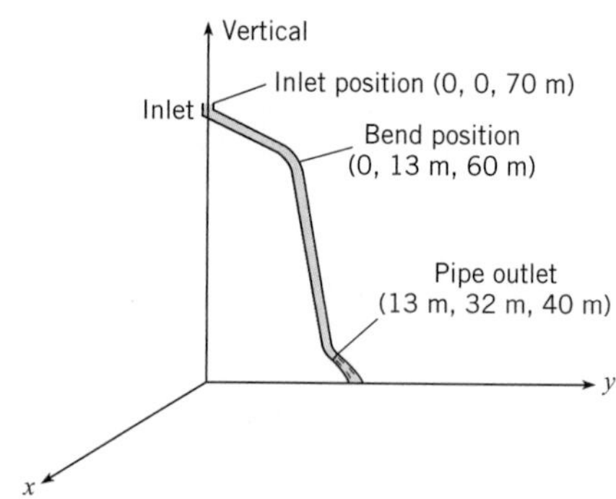

PROBLEM 6.43

6.44 Water flows in a duct as shown. The inlet water velocity is 10 m / s. The cross-sectional area of the duct is 0.1 m^2. Water is injected normal to the duct wall at the rate of 500 kg / s midway between stations 1 and 2. Neglect frictional forces on the duct wall. Calculate the pressure difference $(p_1 - p_2)$ between stations 1 and 2.

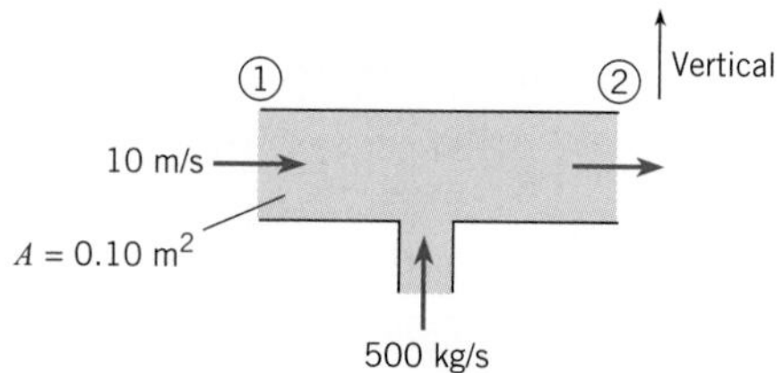

PROBLEM 6.44

6.45 For this wye fitting, which lies in a horizontal plane, the cross-sectional areas at sections 1, 2, and 3 are 1 ft^2, 1 ft^2, and 0.25 ft^2, respectively. At these same respective sections the pressures are 1000 psfg, 900 psfg, and 0 psfg, and the water discharges are 20 cfs to the right, 12 cfs to the right, and 8 cfs. What x component of force would have to be applied to the wye to hold it in place?

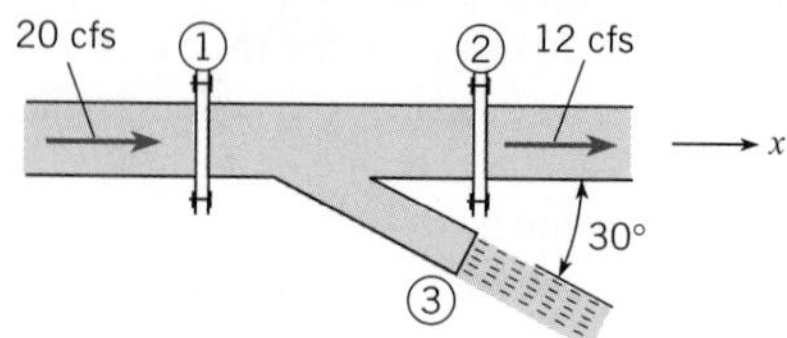

PROBLEM 6.45

6.46 Water flows through a horizontal bend and T section as shown. The mass flow rate entering at section a is 10 lbm / s, and those exiting at sections b and c are 5 lbm / s each. The pressure at section a is 5 psig. The pressure at the two outlets is atmospheric. The cross-sectional areas of the pipes are the same: 5 in^2. Find the x component of force necessary to restrain the section.

6.47 Water flows through a horizontal bend and T section as shown. At section a the flow enters with a velocity of 6 m / s, and the pressure is 4.8 kPa. At both sections b and c the flow exits the device with a velocity of 3 m / s, and the pressure at these sections is atmospheric ($p = 0$). The cross-sectional areas at a, b, and c are all the same: 0.20 m^2. Find the x and y components of force necessary to restrain the section.

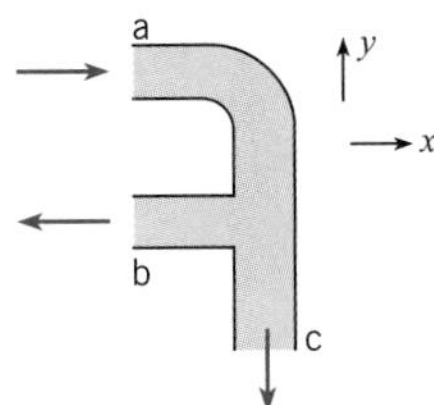

PROBLEMS 6.46, 6.47

6.48 For this horizontal T through which water ($\rho = 1000$ kg / m^3) is flowing, the following data are given: $Q_1 = 0.25$ m^3 / s, $Q_2 = 0.10$ m^3 / s, $p_1 = 100$ kPa, $p_2 = 70$ kPa, $p_3 = 80$ kPa, $D_1 = 15$ cm, $D_2 = 7$ cm, and $D_3 = 15$ cm. For these conditions, what external force in the x-y plane (through the bolts or other supporting devices) is needed to hold the T in place?

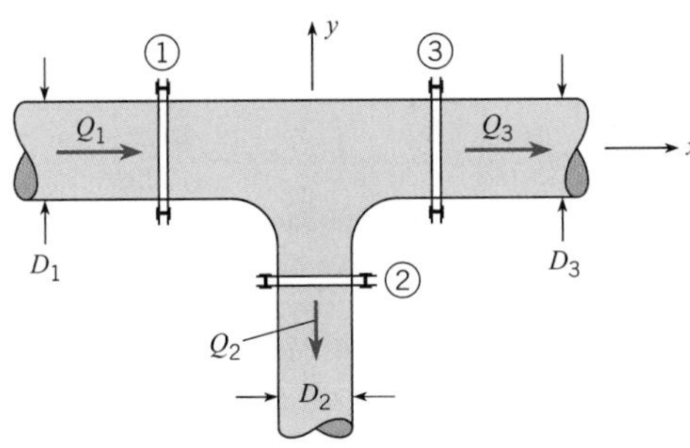

PROBLEM 6.48

6.49 For this weird nozzle, what force would have to be applied through the bolts in the flange to hold the nozzle in place? Assume irrotational flow. Water is flowing, and the nozzle itself weighs 200 N.

6.50 Water flows through this nozzle at a rate of 15 cfs and discharges into the atmosphere. $D_1 = 12$ in. and $D_2 = 9$ in. Determine the force required at the flange to hold the nozzle in place. Assume irrotational flow. Neglect gravitational forces.

6.51 Solve Prob. 6.50 using the following values: $Q = 0.30$ m^3 / s, $D_1 = 30$ cm, and $D_2 = 10$ cm.

6.52 This "double" nozzle discharges water into the atmosphere at a rate of 16 cfs. If the nozzle is lying in a horizontal plane, what

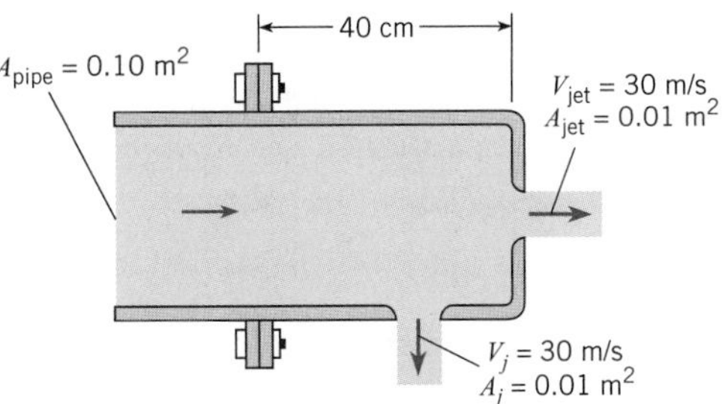

PROBLEM 6.49

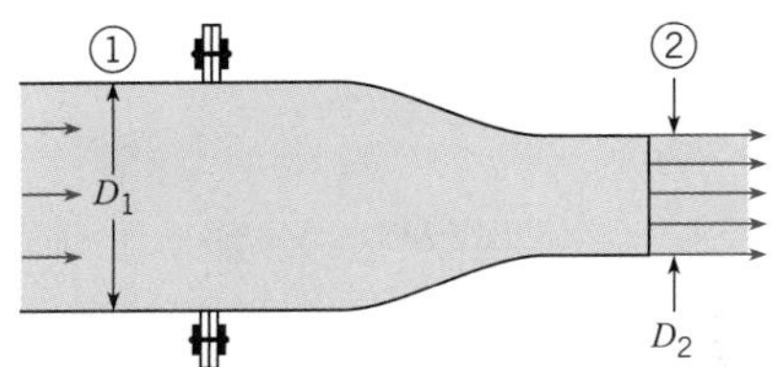

PROBLEMS 6.50, 6.51

x component of force acting through the flange bolts is required to hold the nozzle in place? *Note:* Assume irrotational flow, and assume the water speed in each jet to be the same. Jet A is 4 in. in diameter, jet B is 4.5 in. in diameter, and the pipe is 1 ft in diameter.

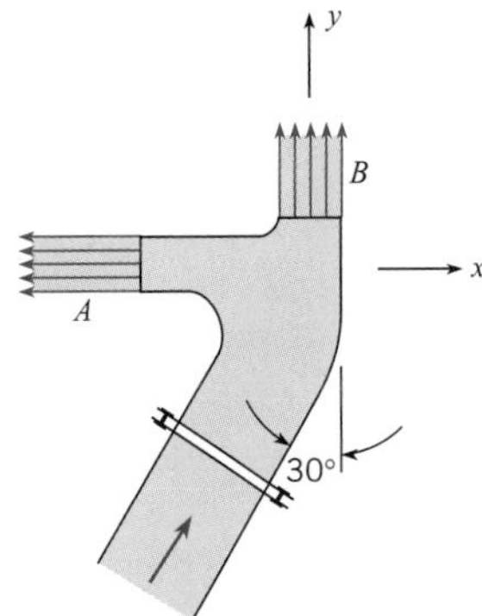

PROBLEMS 6.52, 6.53

6.53 This "double" nozzle discharges water into the atmosphere at a rate of 0.50 m^3 / s. If the nozzle is lying in a horizontal plane, what x component of force acting through the flange bolts is required to hold the nozzle in place? *Note:* Assume irrotational flow, and assume the water speed in each jet to be the same. Jet A is 10 cm in diameter, jet B is 12 cm in diameter, and the pipe is 30 cm in diameter.

6.54 A 15-cm nozzle is bolted with six bolts to the flange of a 30-cm pipe. If water discharges from the nozzle into the atmosphere, calculate the tension load in each bolt when the pressure in the pipe is 200 kPa. Assume irrotational flow.

FLUID SOLUTIONS **6.55** Water is discharged from the two-dimensional slot shown at the rate of 5 cfs per foot of slot. Determine the pressure p at the gage and the water force per foot on the vertical end plates A and C. The slot and jet dimensions B and b are 8 in. and 4 in., respectively.

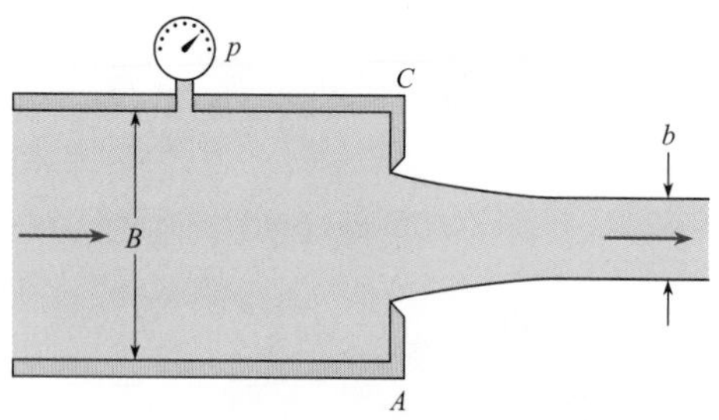

PROBLEMS 6.55, 6.56

6.56 Water is discharged from the two-dimensional slot shown at the rate of 0.40 m^3/s per meter of slot. Determine the pressure p at the gage and the water force per meter on the vertical end plates A and C. The slot and jet dimensions B and b are 20 cm and 7 cm, respectively.

6.57 This spray head discharges water at a rate of 3 ft^3/s. Assuming irrotational flow and an efflux speed of 65 ft/s in the free jet, determine what force acting through the bolts of the flange is needed to keep the spray head on the 6 in. pipe. Neglect gravitational forces.

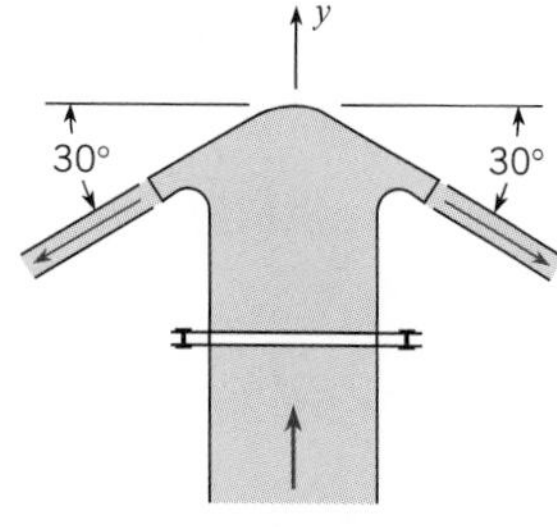

PROBLEM 6.57

6.58 Two circular water jets of 1-in. diameter (d = 1 in.) issue from this unusual nozzle. If the efflux speed is 80.2 ft/s, what force is required at the flange to hold the nozzle in place? The pressure in the 4-in. pipe (D = 4 in.) is 43 psig.

FLUID SOLUTIONS **6.59** Liquid (S = 1.2) enters the "black sphere" through a 2-in. pipe with velocity of 50 ft/s and a pressure of 60 psig. It leaves the sphere through two jets as shown. The velocity in the vertical jet is 100 ft/s, and its diameter is 1 in. The other jet's diameter is also 1 in. What force through the 2-in. pipe wall is required in the x and y directions to hold the sphere in place? Assume the sphere plus the liquid inside it weighs 200 lbf.

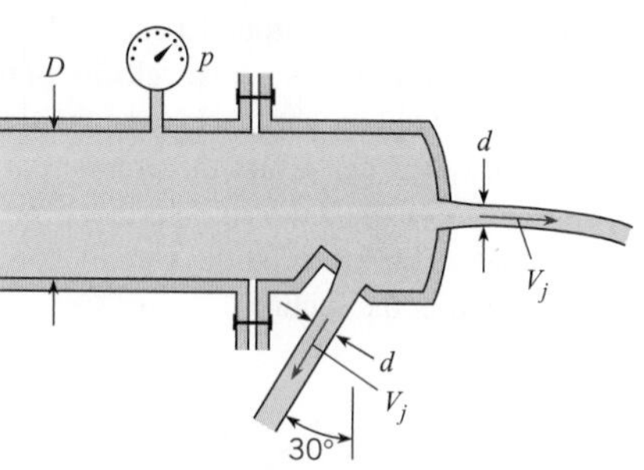

PROBLEM 6.58

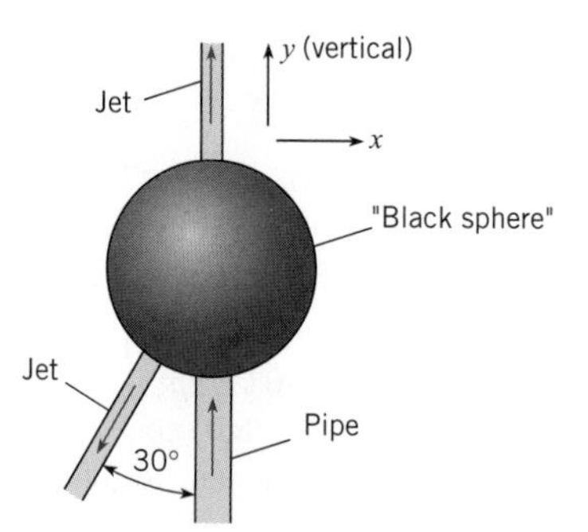

PROBLEMS 6.59, 6.60

6.60 Liquid (S = 1.5) enters the "black sphere" through a 5-cm pipe with a velocity of 10 m/s and a pressure of 400 kPa. It leaves the sphere through two jets as shown. The velocity in the vertical jet is 30 m/s, and its diameter is 25 mm. The other jet's diameter is also 25 mm. What force through the 5-cm pipe wall is required in the x and y directions to hold the sphere in place? Assume the sphere plus the liquid inside it weighs 600 N.

6.61 What force is required to hold this "black box" in place if gravitational forces are neglected and if the liquid's specific gravity is 2.0?

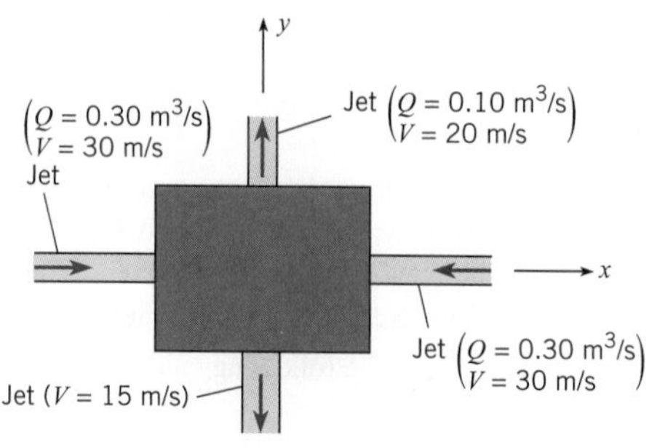

PROBLEM 6.61

6.62 Assume that you have access to a laboratory that includes a sluice gate as shown in Fig. 6.5. Design an experiment utilizing the sluice gate to verify the validity of Eq. (6.9). In your design you may attach instruments or drill holes in the sluice gate, etc., to facilitate the taking of data. In any case describe what is to be done to achieve a workable experiment.

6.63 Neglecting viscous resistance, determine the force of the water per unit of width acting on a sluice gate for which the upstream depth is 3 ft and the downstream depth is 0.6 ft.

6.64 For laminar flow in a pipe, wall shear stress (τ_0) causes the velocity distribution to change from uniform to parabolic as shown. At the fully developed section (section 2), the velocity is distributed as follows: $u = u_{max}[1 - (r/r_0)^2]$. Derive a formula for the resisting shear force F_τ as a function of U (the mean velocity in the pipe), ρ, p_1, p_2, and D (the pipe diameter).

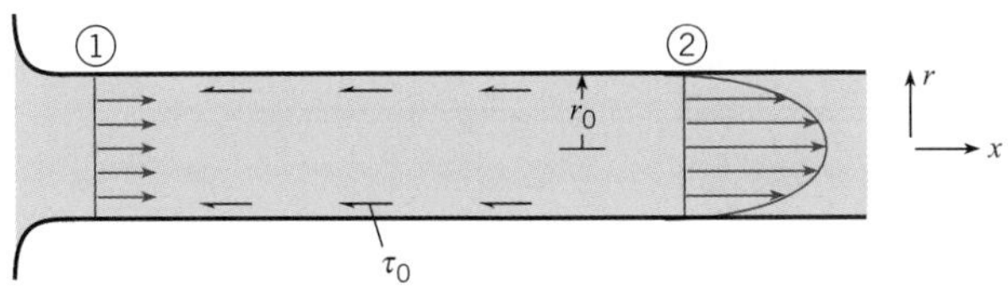

PROBLEM 6.64

6.65 The propeller on a swamp boat produces a slipstream 3 ft in diameter with a velocity relative to the boat of 90 ft / s. If the air temperature is 80°F, what is the propulsive force when the boat is not moving and also when its forward speed is 30 ft / s? *Hint:* Assume that the pressure, except in the immediate vicinity of the propeller, is atmospheric.

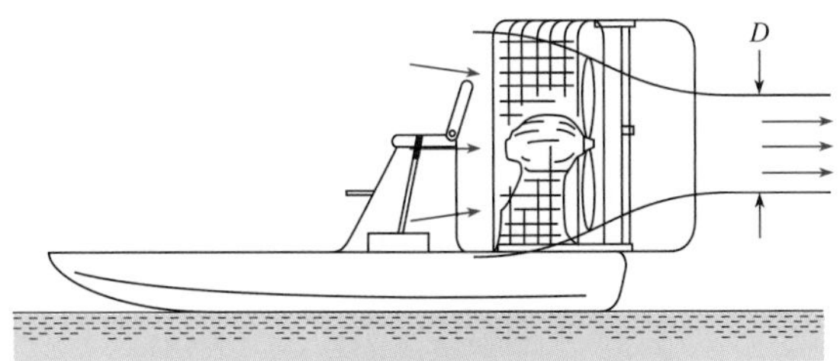

PROBLEM 6.65

6.66 A windmill is operating in a 10-m / s wind that has a density of 1.2 kg / m^3. The diameter of the windmill is 4 m. The constant-pressure (atmospheric) streamline has a diameter of 3 m upstream of the windmill and 4.5 downstream. Assume that the velocity distributions are uniform and the air is incompressible. Determine the thrust on the windmill.

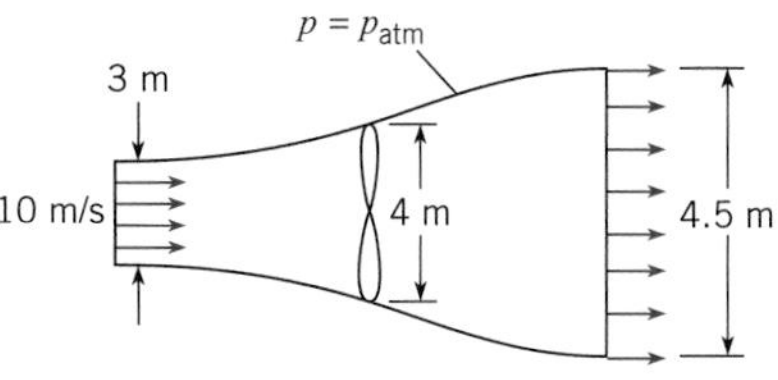

PROBLEM 6.66

6.67 The figure illustrates the principle of the jet pump. Derive a formula for $p_2 - p_1$ as a function of D_j, V_j, D_0, V_0, and ρ. Assume that the fluid from the jet and the fluid initially flowing in the pipe are the same, and assume that they are completely mixed at section 2, so that the velocity is uniform across that section. Also assume that the pressures are uniform across both sections 1 and 2. What is $p_2 - p_1$ if the fluid is water, $A_j/A_0 = 1/3$, $V_j = 15$ m / s, and $V_0 = 2$ m / s? Neglect shear stress.

6.68 Jet-type pumps are sometimes used for special purposes, such as to circulate the flow in basins in which fish are being reared. The use of a jet-type pump eliminates the need for mechanical machinery that might be injurious to the fish. The accompanying figure shows the basic concept for this type of application. For this type of basin the jets would have to increase the water surface elevation by an amount equal to $6V^2/2g$, where V is the average velocity in the basin (1 ft / s as shown in this example). Propose a basic design for a jet system that would make such a recirculating system work for a channel 8 ft wide and 4 ft deep. That is, determine the speed, size, and number of jets.

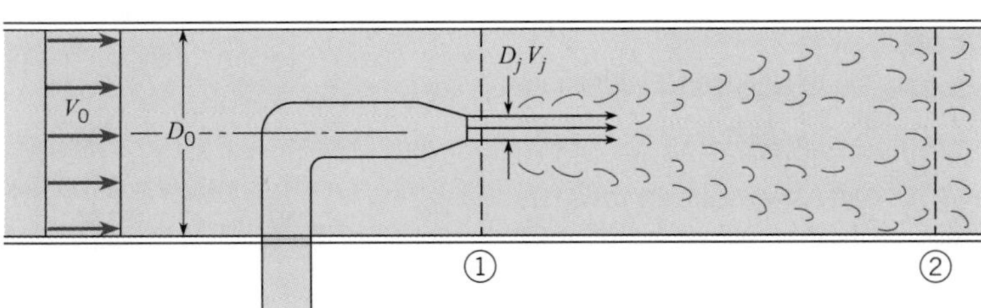

PROBLEM 6.67

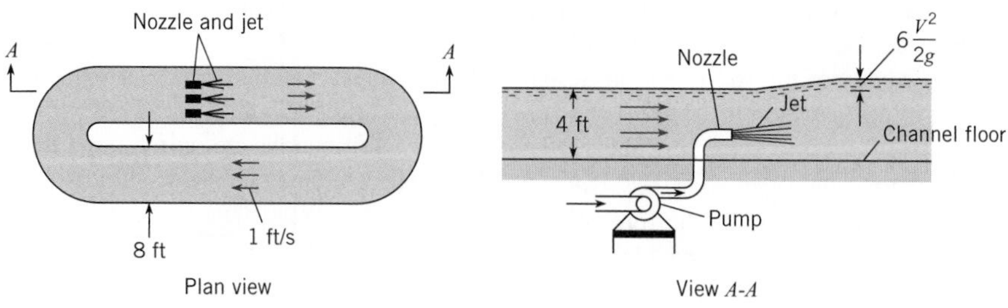

PROBLEM 6.68

FLUID SOLUTIONS **6.69** An engineer is measuring the lift and drag on an airfoil section mounted in a two-dimensional wind tunnel. The wind tunnel is 0.5 m high and 0.5 m deep (into the paper). The upstream wind velocity is uniform at 10 m / s, and the downstream velocity is 12 m / s and 8 m / s as shown. The vertical component of velocity is zero at both stations. The test section is 1 m long. The engineer measures the pressure distribution in the tunnel along the upper and lower walls and finds

$$p_u = 100 - 10x - 20x(1 - x)(\text{Pa, gage})$$
$$p_l = 100 - 10x + 20x(1 - x)(\text{Pa, gage})$$

where x is the distance in meters measured from the beginning of the test section. The gas density is homogeneous throughout and equal to 1.2 kg / m^3. The lift and drag are the vectors indicated on the figure. The forces acting on the fluid are in the opposite direction to these vectors. Find the lift and drag forces acting on the airfoil section.

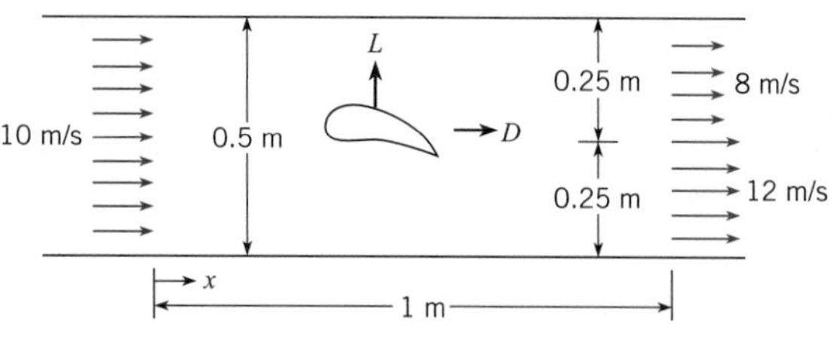

PROBLEM 6.69

6.70 A torpedo-like device is tested in a wind tunnel with an air density of 0.0026 slugs / ft^3. The tunnel is 3 ft in diameter, the upstream pressure is 0.24 psig, and the downstream pressure is 0.10 psig. If the mean air velocity V is 120 ft / s, what are the mass rate of flow and the maximum velocity at the downstream section at C? If the pressure is assumed to be uniform across the sections at A and C, what is the drag of the device and support vanes? Assume viscous resistance at the wall is negligible.

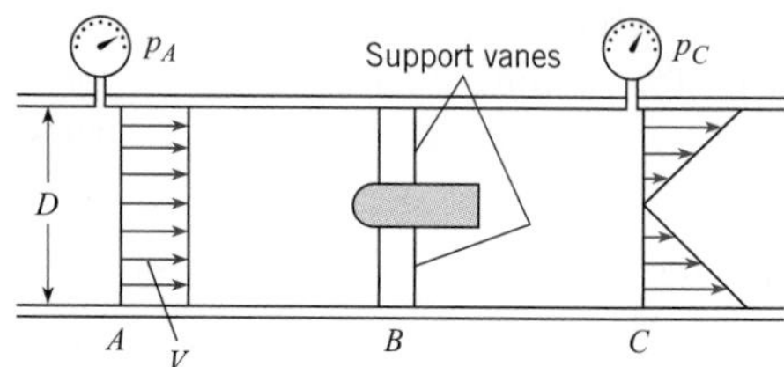

PROBLEM 6.70

6.71 Consider a tank of water in a container that rests on a sled. A high pressure is maintained by a compressor so that a jet of water leaving the tank horizontally from an orifice does so at a constant speed of 25 m / s relative to the tank. If there is 0.10 m^3 of water in the tank at time t and the diameter of the jet is 15 mm, what will be the acceleration of the sled at time t if the empty tank and compressor have a weight of 350 N and the coefficient of friction between the sled and the ice is 0.05?

6.72 A horizontal jet of water that is 6 cm in diameter and has a velocity of 20 m / s is deflected by the vane as shown. If the vane is moving at a rate of 7 m / s in the x direction, what components of force are exerted on the vane by the water in the x and y directions? Assume negligible friction between the water and the vane.

6.73 A vane on this moving cart deflects a 10-cm water jet as shown. The initial speed of the water in the jet is 20 m / s, and the cart moves at a speed of 3 m / s. If the vane splits the jet so that half goes one way and half the other, what force is exerted on the vane by the jet?

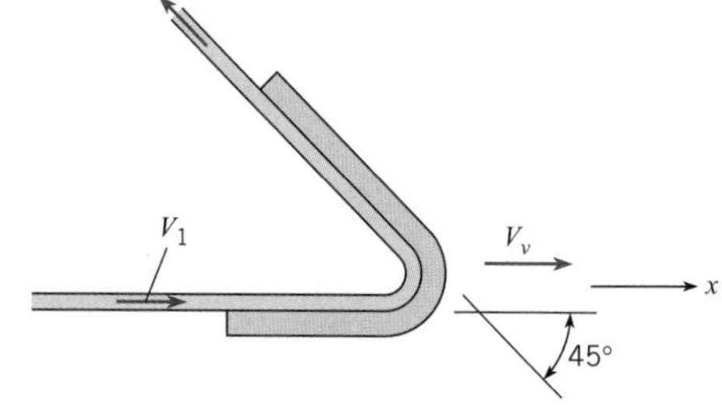

PROBLEM 6.72

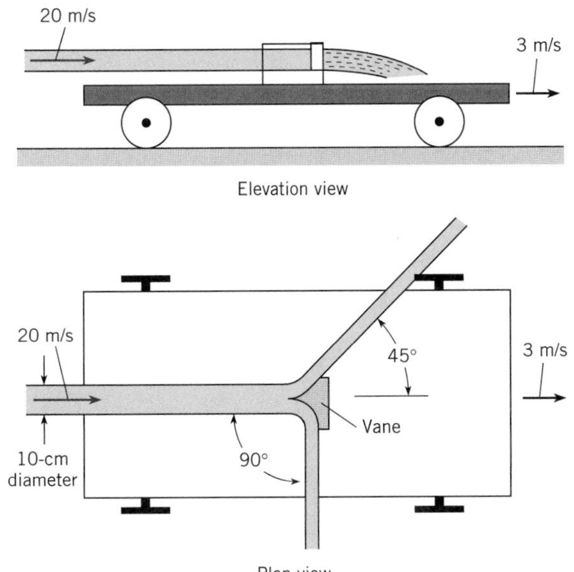

PROBLEMS 6.73, 6.74

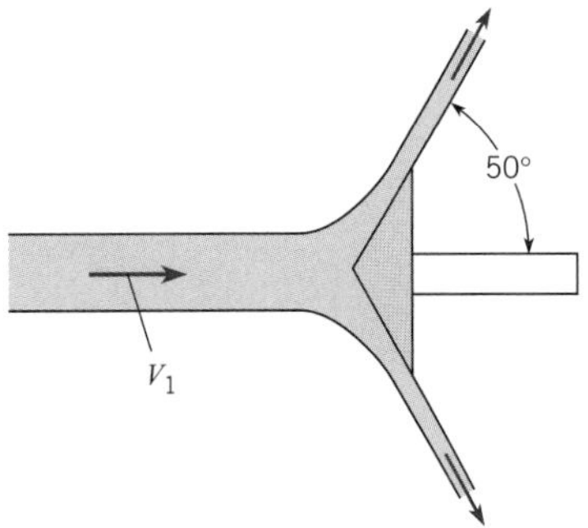

PROBLEMS 6.75, 6.76

6.74 Refer to the cart of Prob. 6.73. If the cart speed is constant at 5 ft / s and if the initial jet speed is 60 ft / s and jet diameter = 0.1 ft, what is the rolling resistance of the cart?

6.75 The water in this jet has a speed of 25 m / s to the right and is deflected by a cone that is moving to the left with a speed of 13 m / s. The diameter of the jet is 10 cm. Determine the external horizontal force needed to move the cone. Assume negligible friction between the water and the vane.

6.76 This two-dimensional water jet is deflected by the two-dimensional vane, which is moving to the right with a speed of 60 ft / s. The initial jet is 0.30 ft thick (vertical dimension), and its speed is 100 ft / s. What power per foot of the jet (normal to the page) is transmitted to the vane?

6.77 Assume that the scoop shown, which is 20 cm wide, is used as a braking device for studying deceleration effects, such as those on space vehicles. If the scoop is attached to a 1000-kg sled that is initially traveling horizontally at the rate of 100 m / s, what will be the initial deceleration of the sled? The scoop dips into the water 8 cm ($d = 8$ cm).

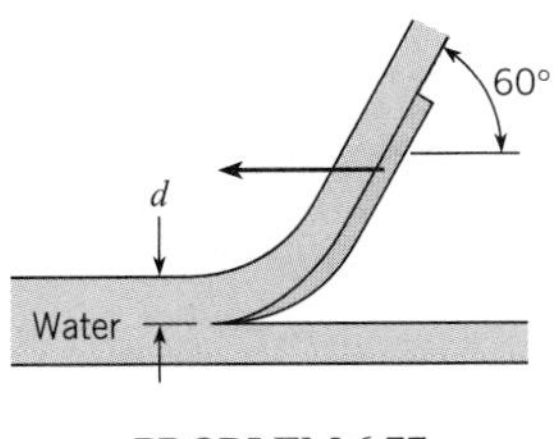

PROBLEM 6.77

6.78 This snowplow "cleans" a swath of snow that is 3 in. deep ($d = 3$ in.) and 2 ft wide ($B = 2$ ft). The snow leaves the blade in the direction indicated in the sketches. Neglecting friction between the snow and the blade, estimate the power required for just the snow removal if the speed of the snowplow is 40 ft / s.

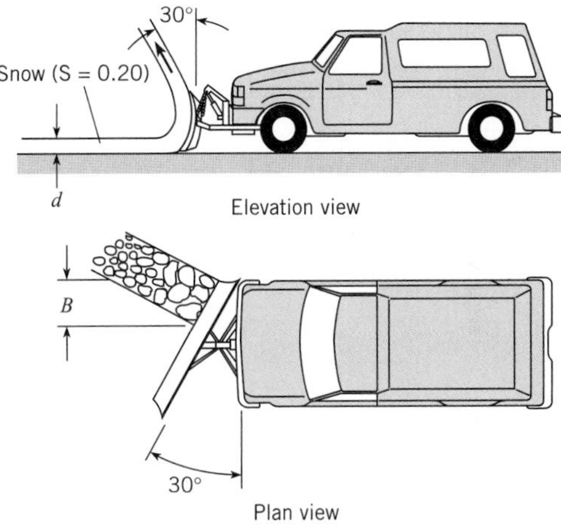

PROBLEM 6.78

6.79 A large tank of liquid is resting on a frictionless plane as shown. Explain in a qualitative way what will happen after the cap is removed from the short pipe.

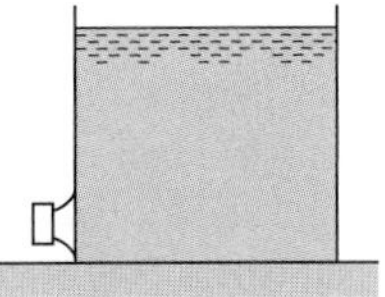

PROBLEM 6.79

6.80 A cart is moving along a track at a constant velocity of 5 m / s as shown. Water (ρ = 1000 kg/m^3) issues from a nozzle at 10 m / s and is deflected through 180° by a vane on the cart. The cross-sectional area of the nozzle is 0.0012 m^2. Calculate the resistive force on the cart.

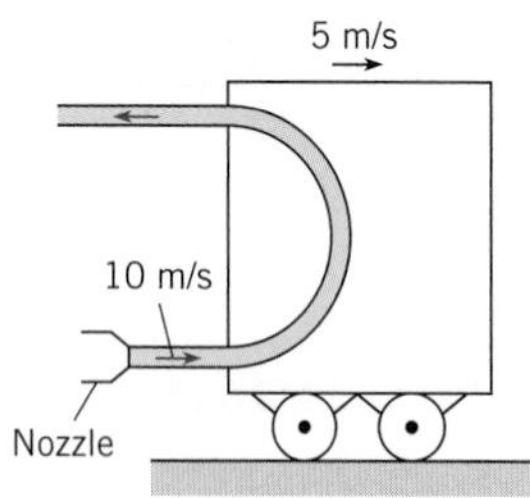

PROBLEM 6.80

6.81 A water jet is used to accelerate a cart as shown. The discharge (Q) from the jet is 0.1 m^3 / s and the velocity of the jet (V_j) is 10 m / s. When the water hits the cart, it is deflected normally as shown. The mass of the cart (M) is 10 kg. The density of water (ρ) is 1000 kg / m^3. There is no resistance on the cart, and the initial velocity of the cart is zero. The mass of the water in the jet is much less than the mass of the cart. Derive an equation for the acceleration of the cart as a function of Q, ρ, V_c, M, and V_j. Evaluate the acceleration of the cart when the velocity is 5 m / s.

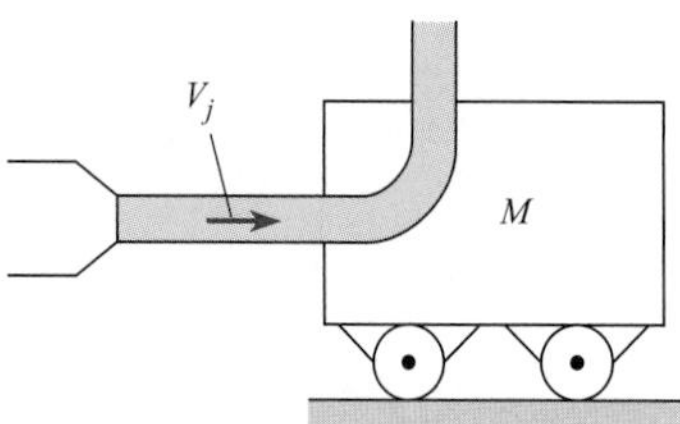

PROBLEM 6.81

6.82 The hemicircular nozzle sprays a sheet of liquid through 180° of arc as shown. The velocity is V at the efflux section where the sheet thickness is t. Derive a formula for the external force F (in the y direction) required to hold the nozzle system in place. This force should be a function of ρ, V, r, and t.

6.83 A planar nozzle is shown in the figure. The gases exit from the fuel chamber and through the nozzle with a velocity V along a radial direction as shown. The density is uniform across the exit. The pressure at the exit is equal to p_e, and the back pressure (pressure of the surroundings) is p_0. The half-angle of the nozzle is θ. Find the expression for the thrust of the nozzle in the form $T = \dot{m}\, Vf(\theta) + A_e(p_e - p_0)\,\lambda(\theta)$. Find the expressions for $f(\theta)$ and $\lambda(\theta)$.

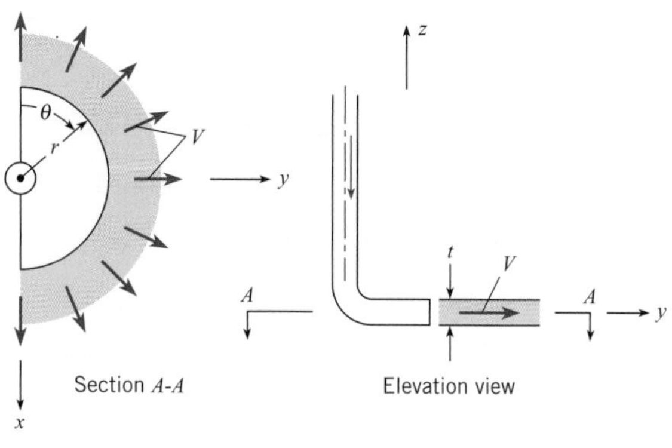

PROBLEM 6.82

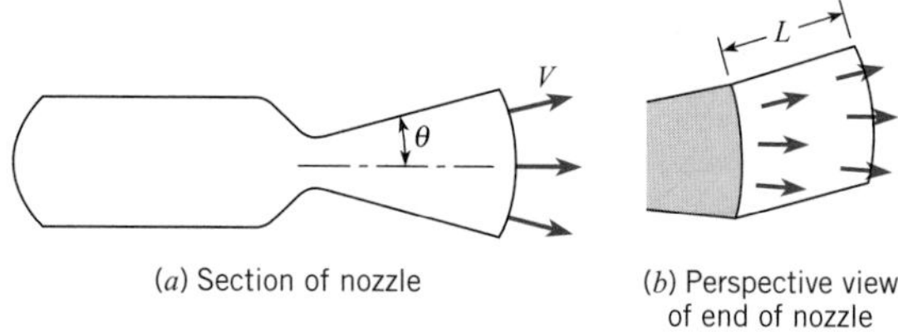

PROBLEM 6.83

6.84 A modern turbofan engine in a commercial jet takes in air, part of which passes through the compressors, combustion chambers, and turbine, and the rest of which bypasses the compressor and is accelerated by the fans. The mass flow rate of bypass air to the mass flow rate through the compressor-combustor-turbine path is called the "bypass ratio." The total flow rate of air entering a turbofan is 300 kg / s with a velocity of 300 m / s. The engine has a bypass ratio of 2.5. The bypass air exits at 600 m / s, whereas the air through the compressor-combustor-turbine path exits at 1000 m / s. What is the thrust of the turbofan engine? Clearly show your control volume and application of momentum equation.

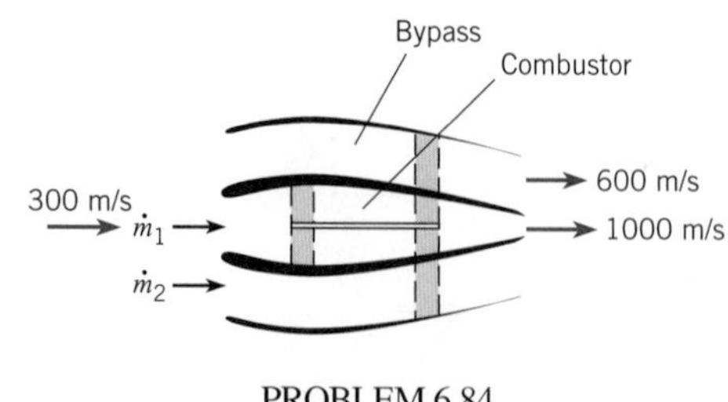

PROBLEM 6.84

6.85 It is common practice in rocket trajectory analyses to neglect the body-force term and drag, so the velocity at burnout is given by

$$v_{\text{bo}} = \frac{T}{\lambda}\ln\frac{M_0}{M_f}$$

Assuming a thrust-to-mass-flow ratio of 3000 N · s/kg and a final mass of 50 kg, calculate the initial mass needed to establish the rocket in an earth orbit at a velocity of 7200 m / s.

6.86 A very popular toy on the market several years ago was the water rocket. Water was loaded into a plastic rocket and pressurized with a hand pump. The rocket was released and would travel a considerable distance in the air. Assume that a water rocket has a mass of 50 g and is charged with 100 g of water. The pressure inside the rocket is 100 kPa gage. The exit area is one-tenth of the chamber cross-sectional area. The inside diameter of the rocket is 5 cm. Assume that Bernoulli's equation is valid for the water flow inside the rocket. Neglecting air friction, calculate the maximum velocity it will attain.

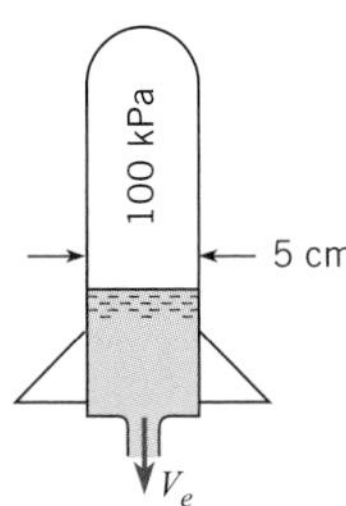

PROBLEM 6.86

6.87 A rocket is designed to have four nozzles, each canted at 30° with respect to the rocket's centerline. The gases exit at 2000 m / s through an exit area of 1 m^2. The density of the exhaust gases is 0.3 kg / m^3, and exhaust pressure is 50 kPa. The atmospheric pressure is 10 kPa. Determine the thrust on the rocket in newtons.

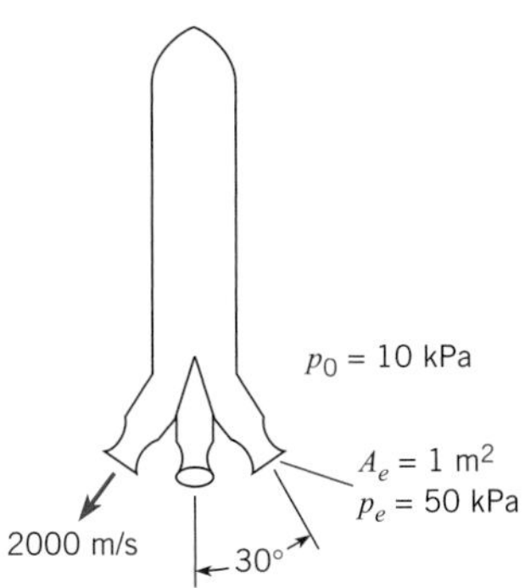

PROBLEM 6.87

6.88 A rocket nozzle designer is concerned about the force required to hold the nozzle section on the body of a rocket. The nozzle section is shaped as shown in the figure. The pressure and velocity at the entrance to the nozzle are 1.5 MPa and 100 m / s. The exit pressure and velocity are 80 kPa and 2000 m / s. The mass flow through the nozzle is 220 kg / s. The atmospheric pressure is 100 kPa. The rocket is not accelerating. Calculate the force on the nozzle-chamber connection. *Note:* The given pressures are absolute.

FLUID SOLUTIONS

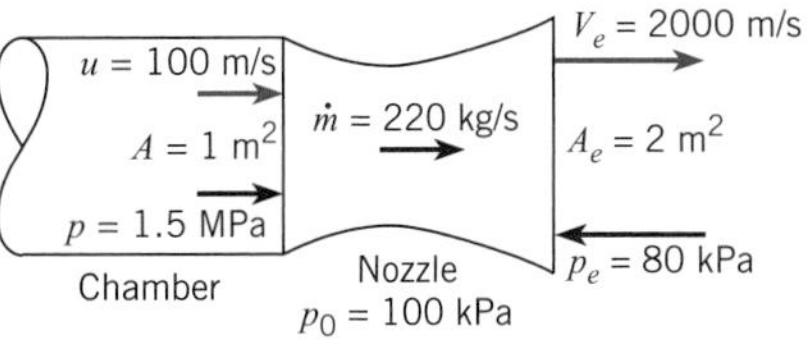

PROBLEM 6.88

6.89 The expansion section of a rocket nozzle is often conical in shape; and because the flow diverges, the thrust derived from the nozzle is less than it would be if the exit velocity were everywhere parallel to the nozzle axis. By considering the flow through the spherical section suspended by the cone and assuming that the exit pressure is equal to the atmospheric pressure, show that the thrust is given by

$$T = \dot{m}V_e\frac{(1+\cos\alpha)}{2}$$

where $\dot{m}$ is the mass flow through the nozzle, V_e is the exit velocity, and α is the nozzle half-angle.

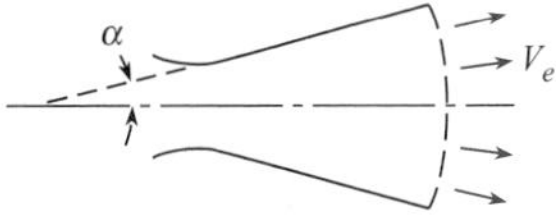

PROBLEM 6.89

6.90 A valve at the end of a gasoline pipeline is rapidly closed (assume it is closed instantaneously). If the gasoline velocity was initially 10 m / s, what will be the water-hammer pressure rise? The bulk modulus of elasticity of the gasoline is 715 MPa and the density of gasoline is 680 kg / m^3.

6.91 Estimate the maximum water-hammer pressure that is generated in a rigid pipe if the initial water velocity is 4 m / s and the pipe is 10 km long with a valve at the downstream end that is closed in 10 s.

6.92 The length of a 20-cm rigid pipe carrying 0.15 m^3 / s of water is estimated by instantaneously closing a valve at the downstream end and noting the time required for the pressure fluctuation to complete a cycle. If the time interval is 3 s, what is the pipe length?

FLUID SOLUTIONS **6.93** Estimate the maximum water-hammer pressure that is generated in a rigid pipe if the initial water velocity is 8 ft/s and the pipe is 5 mi long with a valve at the downstream end that is closed in 10 s.

6.94 A rigid pipe 4 km long and 12 cm in diameter discharges water at the rate of 0.03 m^3/s. If a valve at the end of the pipe is closed in 3 s, what is the maximum force that will be exerted on the valve as a result of the pressure rise? Assume that the water temperature is 10°C.

6.95 By letting the control volume move with the water-hammer wave, steady-flow conditions are established. Using the momentum and continuity equations and the steady-flow approach, derive Eq. (6.10).

6.96 The 60-cm pipe carries water with an initial velocity, V_0, of 0.10 m/s. If the valve at C is instantaneously closed at time $t = 0$, what will the pressure-versus-time trace look like at point B for the next 5 s? Graph your results and indicate significant quantitative relations or values from $t = 0$ to $t = 5$ s. What does the pressure versus the position along the pipe look like at $t = 1.5$ s? Plot your results and indicate the velocity or velocities in the pipe.

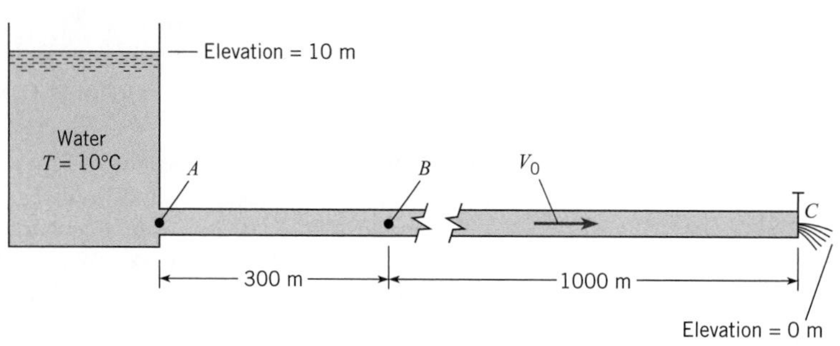

PROBLEM 6.96

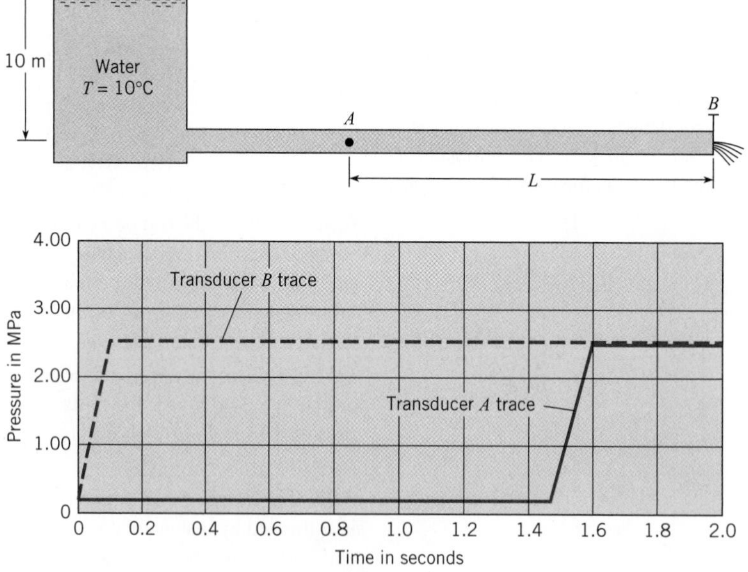

PROBLEM 6.97

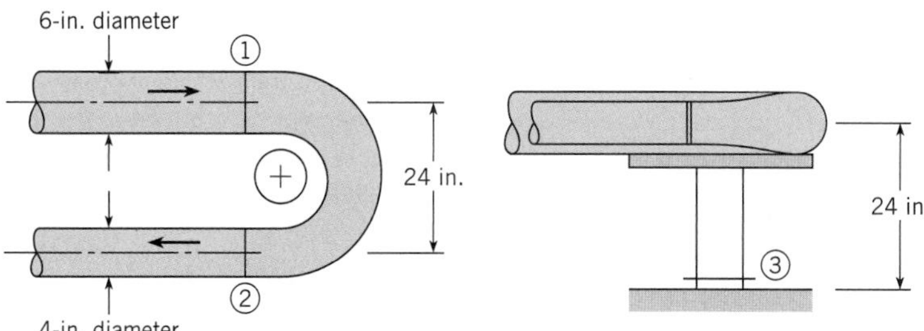

PROBLEM 6.101

6.97 Steady flow initially occurs in this 1-m steel pipe. There is a rapid-acting valve at the end of the pipe at point *B*, and there are pressure transducers at both points *A* and *B*. If the valve is closed at *B* and the *p*-versus-*t* traces are made as shown, estimate the initial discharge and the length *L* from *A* to *B*.

6.98 Water is discharged from the slot in the pipe as shown. If the resulting two-dimensional jet is 100 cm long and 15 mm thick, and if the pressure at section *A*-*A* is 30 kPa, what is the reaction at section *A*-*A*? In this calculation, do not consider the weight of the pipe.

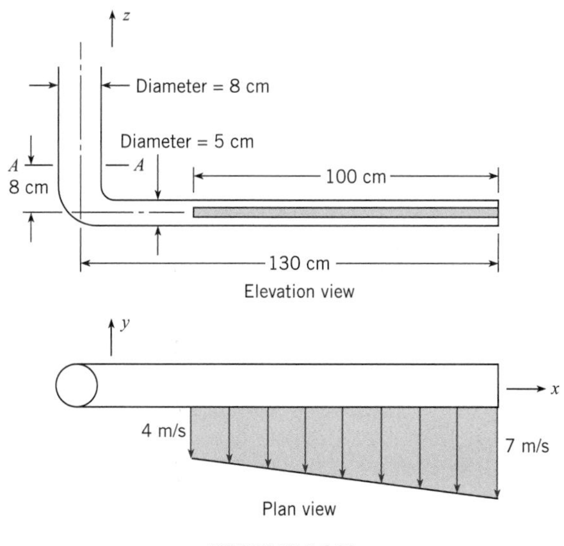

PROBLEM 6.98

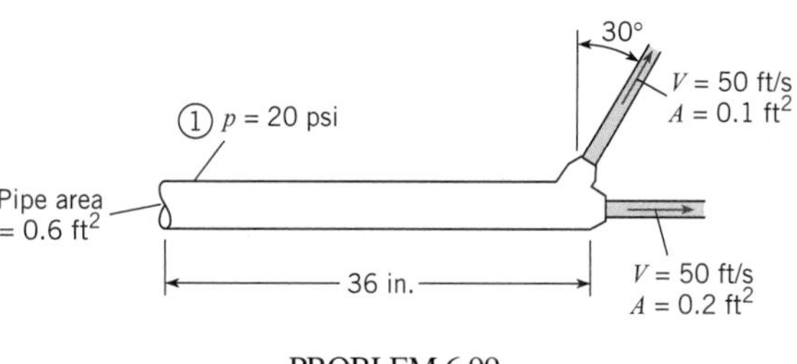

PROBLEM 6.99

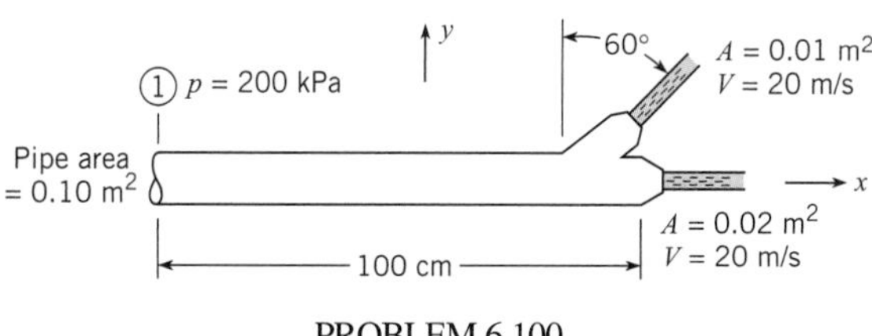

PROBLEM 6.100

6.99 What is the force and moment reaction at section 1? Water is flowing in the system. Neglect gravitational forces.

6.100 What is the reaction at section 1? Water is flowing, and the axes of the two jets lie in a vertical plane. The pipe and nozzle system weighs 90 N.

6.101 A reducing pipe bend is held in place by a pedestal as shown. There are expansion joints at sections 1 and 2, so no force is transmitted through the pipe past these sections. The pressure at section 1 is 20 psig and the rate of flow of water is 2 cfs. Find the force and moment that must be applied at section 3 to hold the bend stationary. Assume the flow is irrotational and neglect the influence of gravity. FLUID SOLUTIONS

6.102 Show how the momentum equation can be applied to derive Euler's equation for the flow of inviscid fluids. *Hint:* Select an arbitrary control volume of length Δs enclosed by a stream tube in an unsteady, nonuniform flow as shown. The volume of the control volume is $[A + (\partial A/\partial s)(\Delta s/2)]\Delta s$. First derive the continuity equation by applying the continuity principle to the flow through the control volume. Then apply the momentum equation along the stream-tube direction, and use the continuity equation to reduce it to Euler's equation.

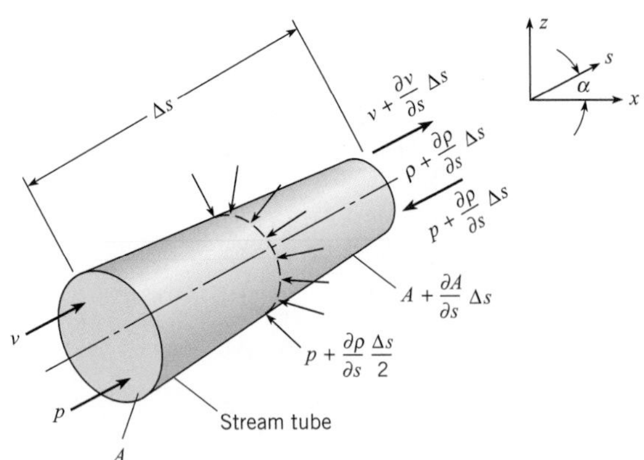

PROBLEM 6.102

6.103 Two small liquid-propellant rocket motors are mounted at the tips of a helicopter rotor to augment power under emergency conditions. The diameter of the helicopter rotor is 7 m, and it rotates at 1 rev / s. The air enters at the tip speed of the rotor, and exhaust gases exit at 500 m / s with respect to the rocket motor. The intake area of each motor is 20 cm^2, and the air density is 1.2 kg / m^3. Calculate the power provided by the rocket motors. Neglect the mass rate of flow of fuel in this calculation.

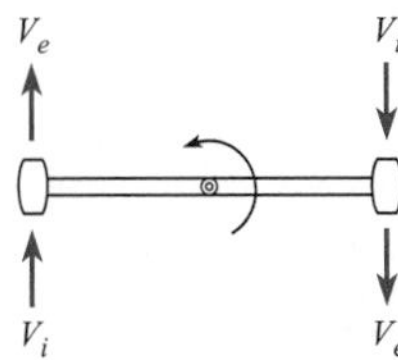

PROBLEM 6.103

6.104 Design a rotating lawn sprinkler to deliver 0.25 in. of water per hour over a circle of 50-ft radius. Make the simplifying assumptions that the pressure to the sprinkler is 50 psig and that frictional effects involving the flow of water through the sprinkler flow passages are negligible (the Bernoulli equation is applicable). However, do not neglect the friction between the rotating element and the fixed base of the sprinkler.

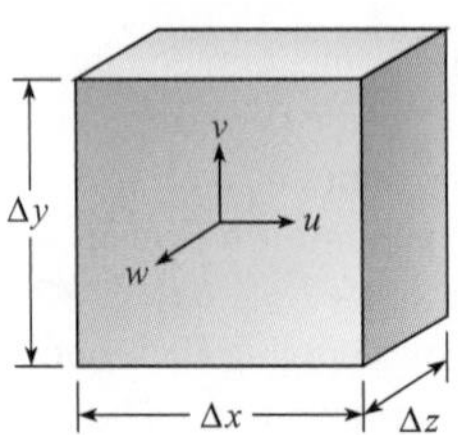

PROBLEM 6.105

6.105 Using a three-dimensional, infinitesimal parallelepiped with dimensions Δx, Δy, and Δz and velocity components u, v, and w, show that, as the volume approaches zero, (a) the rate of momentum change in the x direction per unit volume in the control volume plus the next efflux of momentum from the six surfaces per unit volume is

$$\frac{\partial}{\partial t}(\rho u) + \frac{\partial}{\partial x}(\rho uu) + \frac{\partial}{\partial y}(\rho uv) + \frac{\partial}{\partial z}(\rho uw)$$

(b) the forces due to pressure, normal, and shear stresses per unit volume are

$$\frac{\partial p}{\partial x} + \frac{\partial \tau_{xx}}{\partial x} + \frac{\partial \tau_{yx}}{\partial y} + \frac{\partial \tau_{zx}}{\partial z}$$

and (c) the body force per unit volume is ρg_x. Assemble these three components to obtain the momentum equation in the x direction at a point and use the continuity equation to arrive at the form

$$\rho\frac{\partial u}{\partial t} + u\frac{\partial u}{\partial x} + v\frac{\partial u}{\partial y} + w\frac{\partial u}{\partial z}$$
$$= -\frac{\partial p}{\partial x} + \frac{\partial \tau_{xx}}{\partial x} + \frac{\partial \tau_{yx}}{\partial y} + \frac{\partial \tau_{zx}}{\partial z} + \rho g_x$$

6.106 Using the constitutive relations (stress proportional to rate of strain) for an incompressible liquid,

$$\tau_{xx} = 2\mu\frac{\partial u}{\partial x};$$
$$\tau_{yx} = \mu\left(\frac{\partial u}{\partial y} + \frac{\partial v}{\partial x}\right);$$
$$\tau_{zx} = \mu\left(\frac{\partial u}{\partial z} + \frac{\partial w}{\partial x}\right);$$

show for a liquid with constant dynamic viscosity that

$$\frac{\partial \tau_{xx}}{\partial x} + \frac{\partial \tau_{yx}}{\partial y} + \frac{\partial \tau_{zx}}{\partial z} = \mu\left(\frac{\partial^2 u}{\partial x^2} + \frac{\partial^2 u}{\partial y^2} + \frac{\partial^2 u}{\partial z^2}\right).$$

Grand Coulee Dam in Washington State has a power-generating capacity of about 5600 MW. At this generating level the turbines discharge water at 250,000 cfs (7,080 m^3/s) and operate under a head of about 308 ft (94 m). To generate significant power during dry seasons, water is stored in Lake Roosevelt (shown in this photo) as well as in three other reservoirs in Canada. (Courtesy U.S. Bureau of Reclamation.).

C H A P T E R 7

Energy Principle

To this point, we have been concerned with the mechanical forces (pressure, gravity, shear stress) on a fluid. The energy equation allows us to incorporate the thermal energies as well. In the early nineteenth century, J. P. Joule carried out a number of tests that verified the general principle of the conservation of energy that had been previously hypothesized. From this was developed the *first law of thermodynamics,* which can be written for a given system (given quantity of matter) as follows (1):

$$\Delta E = Q - W$$

Here Q is the heat transferred to the system in a given time t, and W is the work done by the system on its surroundings in this same interval of time.* The energy E of a system can take a variety of forms, such as kinetic and potential energy of the system as a whole and energy associated with motion of the molecules. The latter includes energy involved with the structure of the atom, chemical energy, and electrical energy. It is convenient to consider kinetic energy E_k and potential energy E_p separately and to lump all other energies into a single term called *internal energy* E_u. Thus the total energy of the system is

$$E = E_u + E_k + E_p$$

Now, if we want the rate of change of E with time, the differential form of the first law of thermodynamics is given as follows (2):

$$\frac{dE}{dt} = \dot{Q} - \dot{W}$$

In the next section the first law of thermodynamics, along with the Reynolds transport theorem, will be used to develop the energy equation for fluid flow.

*Heat transferred to the system and work done by the system are defined, by convention, to be positive quantities. Heat transferred from the system and work done on the system are negative quantities.

7.1

Derivation of the Energy Equation

Reynolds Transport Theorem Applied to the First Law of Thermodynamics

The energy E introduced above refers to the total energy of the system; thus E is an extensive property of the system. Then the corresponding intensive property (energy per unit of mass) is given by e, which is made up of e_k, e_p, and u.

In applying the Reynolds transport theorem [Eq. (5.22)], we let $B_{sys} = E$ and $b = e$ to obtain

$$\frac{dE}{dt} = \frac{d}{dt}\int_{cv} e\rho \, d\Psi + \int_{cs} e\rho \mathbf{V} \cdot d\mathbf{A} \tag{7.1}$$

However, from the first law of thermodynamics, $dE/dt = \dot{Q} - \dot{W}$. Consequently, substitution is made for dE/dt in Eq. (7.1) to yield

$$\dot{Q} - \dot{W} = \frac{d}{dt}\int_{cv} e\rho \, d\Psi + \int_{cs} e\rho \mathbf{V} \cdot d\mathbf{A} \tag{7.2}$$

When we replace e by its equivalent, $e_k + e_p + u$, we obtain

$$\dot{Q} - \dot{W} = \frac{d}{dt}\int_{cv} (e_k + e_p + u)\rho \, d\Psi + \int_{cs} (e_k + e_p + u)\rho \mathbf{V} \cdot d\Psi \tag{7.3}$$

Several terms on the right of Eq. (7.3) are too general for practical application. Therefore, let us examine these carefully and express them in terms of variables associated with the flow of fluids.

The kinetic energy per unit of mass, e_k, is given by the total kinetic energy of mass having velocity V* divided by the mass itself, or

$$e_k = \frac{\Delta M V^2/2}{\Delta M} = \frac{V^2}{2} \tag{7.4}$$

The potential energy per unit of mass, e_p, is given by $E_p/\Delta M$, where E_p is the product of weight and the elevation of the centroid of the incremental mass. In this case, the potential energy is referenced to the datum from which elevation is measured. Then

$$e_p = \frac{\gamma \Delta \Psi z}{\Delta M} = \frac{\gamma \Delta \Psi z}{\rho \Delta \Psi} = gz \tag{7.5}$$

When Eqs. (7.4) and (7.5) are substituted into Eq. (7.3), we obtain

*It is assumed that the control surface is not accelerating, so V, which is referenced to the control surface, is also referenced to an inertial reference frame.

$$\dot{Q} - \dot{W} = \frac{d}{dt}\int_{cv}\left(\frac{V^2}{2} + gz + u\right)\rho\, d\forall + \int_{cs}\left(\frac{V^2}{2} + gz + u\right)\rho\mathbf{V}\cdot d\mathbf{A} \tag{7.6}$$

On the left side of this equation are the two terms $\dot{Q}$ and $\dot{W}$, which are the rate of flow of heat into the system and the rate of work done by the system on its surroundings, respectively. For convenience of analysis, work is divided into shaft work W_s and flow work W_f. These are discussed next.

Flow Work

Flow work is the work done by pressure forces as the system moves through space. Let us consider the basic figure (Fig. 7.1) depicting the system and control volume to get a better understanding of flow work. The velocity and pressure are uniformly distributed (constant) across stations 1 and 2. Consider the area A_2, which is the right end of the fluid system. The force acting on the surrounding fluid will be p_2A_2, and the distance traveled by the area in the time Δt will be $\Delta \ell_2 = V_2\Delta t$. The work done on the surrounding fluid because of this force in time Δt will be the product of the component of force in the direction of motion (p_2A_2) and the distance traveled by the area ($V_2\Delta t$). Hence the flow work done by the system on the surrounding fluid in time Δt by the downstream end of the system will be

$$\Delta W_{f2} = V_2p_2A_2\Delta t$$

The rate at which flow work is done on the area is obtained by dividing through by Δt:

$$\dot{W}_{f2} = V_2p_2A_2 \tag{7.7}$$

It can also be seen from Fig. 7.1 that $V_2A_2 = \mathbf{V}_2 \cdot \mathbf{A}_2$. Consequently, Eq. (7.7) reduces to

$$\dot{W}_{f2} = p_2\mathbf{V}_2 \cdot \mathbf{A}_2$$

In a similar manner it can be shown that the flow work done on the surrounding fluid by the upstream end of the system will be $-p_1V_1A_1$. A negative sign occurs here because the pressure force on the surrounding fluid acts in a direction opposite to the motion of the system boundary. The rate at which work is done on the surrounding fluid by the upstream end of the system can also be given in terms of the scalar product:

$$\dot{W}_{f1} = p_1\mathbf{V}_1 \cdot \mathbf{A}_1$$

A negative rate of work results from this product because the velocity vector $\mathbf{V}_1$ and the area vector $\mathbf{A}_1$ have opposite sense; thus the scalar product has a negative sign. Hence all system surfaces that are moving (represented by streams of fluid passing across the control surface) do work on the surrounding fluid according to the expression:

$$\dot{W}_f = p\mathbf{V} \cdot \mathbf{A} \tag{7.8}$$

Then the rate at which flow work is done on the system's surroundings is obtained by summing Eq. (7.8) for all streams passing through the control surface:

FIGURE 7.1

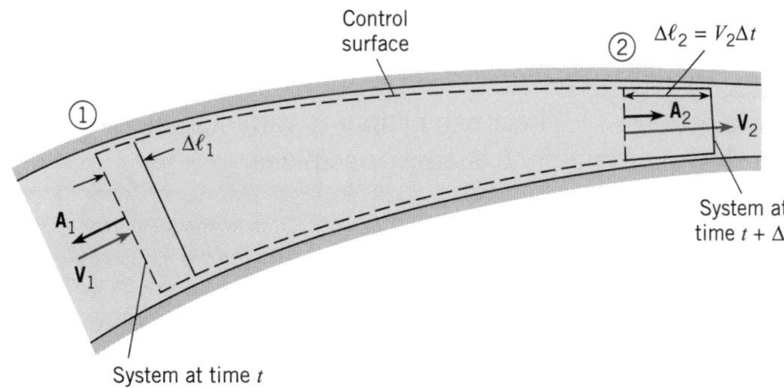

$$\dot{W}_f = \sum_{\text{cs}} p\mathbf{V} \cdot \mathbf{A} \tag{7.9}$$

If the velocity and pressure vary across the control surface, the equation can be generalized to

$$\dot{W}_f = \int_{\text{cs}} p\mathbf{V} \cdot d\mathbf{A}$$

Multiplying and dividing by the density yields the form for the flow work equation, which is incorporated into the energy equation:

$$\dot{W}_f = \int_{\text{cs}} \frac{p}{\rho} \rho\mathbf{V} \cdot d\mathbf{A} \tag{7.10}$$

Shaft Work

Shaft work is defined as work other than flow work. It is usually the work done through a shaft (from which the term originates) and is commonly associated with a pump or turbine. A pump is machine that does work on the flow, thereby increasing the energy of the flow. From the thermodynamic definition of work, this is a negative work. A turbine, on the other hand, extracts energy from the flow, doing work on the surroundings. This is a positive work.

Basic Form of the Energy Principle

If we substitute for $\dot{W}$ in Eq. (7.6) the sum of the shaft-work rate $\dot{W}_s$ and the flow-work rate, Eq. (7.10), the following equation results:

$$\begin{aligned}\dot{Q} - \dot{W}_s - &\int_{\text{cs}} \frac{p}{\rho} \rho\mathbf{V} \cdot d\mathbf{A} \\ &= \frac{d}{dt}\int_{\text{cv}} \left(\frac{V^2}{2} + gz + u\right)\rho \, d\forall + \int_{\text{cs}} \left(\frac{V^2}{2} + gz + u\right)\rho\mathbf{V} \cdot d\mathbf{A}\end{aligned} \tag{7.11}$$

The last term on each side of Eq. (7.11) has the same form. Hence these two terms may be combined as follows:

$$\dot{Q} - \dot{W}_s = \frac{d}{dt}\int_{\text{cv}}\left(\frac{V^2}{2} + gz + u\right)\rho\, d\forall + \int_{\text{cs}}\left(\frac{V^2}{2} + gz + u + \frac{p}{\rho}\right)\rho\mathbf{V}\cdot d\mathbf{A} \tag{7.12}$$

The combination of variables $p/\rho + u$ is the specific enthalpy, h, so the integral form of the energy principle is

$$\dot{Q} - \dot{W}_s = \frac{d}{dt}\int_{\text{cv}}\left(\frac{V^2}{2} + gz + u\right)\rho\, d\forall + \int_{\text{cs}}\left(\frac{V^2}{2} + gz + h\right)\rho\mathbf{V}\cdot d\mathbf{A} \tag{7.13}$$

7.2

Simplified Forms of the Energy Principle

Steady-Flow Energy Equation

For steady flow, the energy accumulation term is zero. If the flow crosses the control surface through various inlet and outlet ports and if the properties are uniformly distributed across each port, Eq. (7.13) simplifies to

$$\dot{Q} - \dot{W}_s = \sum_{\text{cs}} \dot{m}_o\left(\frac{V^2}{2} + gz + h\right)_o - \sum_{\text{cs}} \dot{m}_i\left(\frac{V^2}{2} + gz + h\right)_i \tag{7.14}$$

example 7.1

A steam turbine receives superheated steam at 1.4 MPa absolute and 400°C, which corresponds to a specific enthalpy of 3121 kJ/kg. The steam leaves the turbine at 101 kPa absolute and 100°C, for which the specific enthalpy is 2676 kJ/kg. The steam enters the turbine at 15 m/s and exits at 60 m/s. The elevation difference between entry and exit ports is negligible. The heat lost through the turbine wall is 7600 kJ/h. Calculate the power output if the mass flow through the turbine is 0.5 kg/s.

Solution First sketch the general layout of the turbine, indicating the inlet and outlet velocities, shaft work, and heat transfer, as shown.

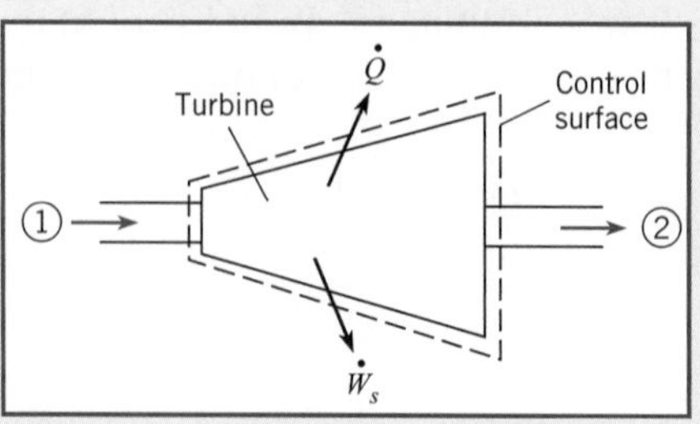

Mass crosses the control surface at two stations, entering at station 1 and exiting at station 2. The flow is steady. Writing down the energy principle for steady flow, neglecting elevation terms, gives

$$\dot{Q} - \dot{W}_s = \dot{m}_2\left(h_2 + \frac{V_2^2}{2}\right) - \dot{m}_1\left(h_1 + \frac{V_1^2}{2}\right)$$

The flow is steady, so

$$\dot{m}_1 = \dot{m}_2 = \dot{m}$$

from the conservation of mass. Thus we have

$$\dot{W}_s = \dot{Q} + \dot{m}\left(\frac{V_1^2}{2} - \frac{V_2^2}{2} + h_1 - h_2\right)$$

We must be careful to check units before substituting numbers into the equation and evaluating $\dot{W}_s$. The units for enthalpy are joules per kilogram; for velocity squared, they are meters squared per second squared. One finds that meters squared per second squared is equivalent to joules per kilogram:

$$\frac{\text{m}^2}{\text{s}^2} = \frac{\text{kg}\cdot\text{m}^2}{\text{kg}\cdot\text{s}^2} = \frac{\text{kg}\cdot\text{m}}{\text{s}^2}\frac{\text{m}}{\text{kg}} = \frac{\text{N}\cdot\text{m}}{\text{kg}} = \frac{\text{J}}{\text{kg}}$$

Substituting numbers into the energy equation, while realizing that $\dot{Q}$ (the heat transfer) has a negative value and using the correct units, gives

$$\dot{W}_s = \frac{-7600}{3600}\frac{\text{kJ}}{\text{h}}\frac{\text{h}}{\text{s}} + 0.5\frac{\text{kg}}{\text{s}}\left[\frac{15^2 - 60^2}{2\times 10^3}\frac{\text{kJ}}{\text{kg}} + (3121 - 2676)\frac{\text{kJ}}{\text{kg}}\right]$$

$$= -2.11 + 0.5(-1.69 + 445) = 220\frac{\text{kJ}}{\text{s}} = 220\text{ kW} \quad \triangleleft$$

The kinetic-energy term in this type of problem is usually negligible compared with the enthalpy difference, which is the case in this example.

Energy Equation for Steady Flow of an Incompressible Fluid in a Pipe

Consider flow through the pipe system shown in Fig. 7.2. Here the magnitude of the velocity is variable across the flow sections; thus we use the more general form of the energy principle, Eq. (7.12). When the steady-flow form $[d(\ )/dt = 0]$ of Eq. (7.12) is written between sections 1 and 2 and when the flow quantities relating to section 1 are transferred to the left side of the equation, we obtain

FIGURE 7.2

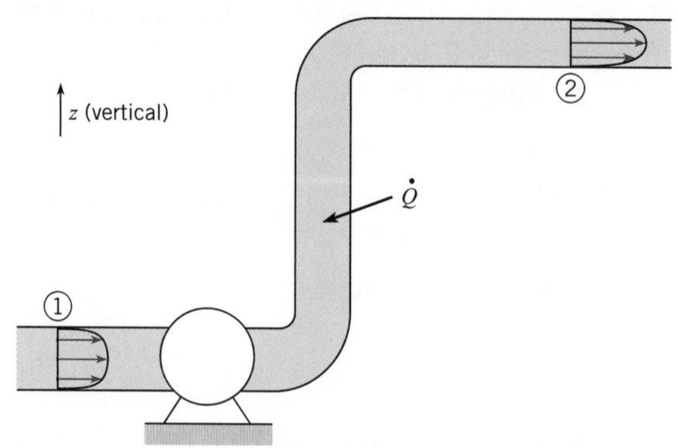

$$\dot{Q} - \dot{W}_s + \int_{A_1}\left(\frac{p_1}{\rho} + gz_1 + u_1\right)\rho V_1\, dA_1 + \int_{A_1}\frac{\rho V_1^3}{2}\, dA_1$$
$$= \int_{A_2}\left(\frac{p_2}{\rho} + gz_2 + u_2\right)\rho V_2\, dA_2 + \int_{A_2}\frac{\rho V_2^3}{2}\, dA_2 \quad (7.15)$$

At sections 1 and 2 the streamlines are parallel to the wall so there is no acceleration normal to the streamlines. Therefore, $p/\rho + gz$ is constant across the section. At any section, the internal energy is usually constant across the section. Therefore, $p/\rho + gz + u$ can be taken outside the integral to yield

$$\dot{Q} - \dot{W}_s + \left(\frac{p_1}{\rho} + gz_1 + u_1\right)\int_{A_1}\rho V_1\, dA_1 + \int_{A_1}\rho\frac{V_1^3}{2}\, dA_1$$
$$= \left(\frac{p_2}{\rho} + gz_2 + u_2\right)\int_{A_2}\rho V_2\, dA_2 + \int_{A_2}\rho\frac{V_2^3}{2}\, dA_2 \quad (7.16)$$

It can be seen that $\int \rho V\, dA = \rho \overline{V} A = \dot{m}$, the mass rate of flow. Consequently, some simplification will result if we divide through by $\dot{m}$. However, $\dot{m}$ does not appear as a factor of $\int(\rho V^3/2)\, dA$; so it is common to express $\int(\rho V^3/2)\, dA$ as $\alpha(\rho\overline{V}^3/2)A$. Then when we factor out $\dot{m}$ from each of these terms, Eq. (7.16) becomes

$$\dot{Q} - \dot{W}_s + \left(\frac{p_1}{\rho} + gz_1 + u_1 + \alpha_1\frac{\overline{V}_1^2}{2}\right)\dot{m} = \left(\frac{p_2}{\rho} + gz_2 + u_2 + \alpha_2\frac{\overline{V}_2^2}{2}\right)\dot{m} \quad (7.17)$$

When we divide through by $\dot{m}$, we get

$$\frac{1}{\dot{m}}(\dot{Q} - \dot{W}_s) + \frac{p_1}{\rho} + gz_1 + u_1 + \alpha_1\frac{\overline{V}_1^2}{2} = \frac{p_2}{\rho} + gz_2 + u_2 + \alpha_2\frac{\overline{V}_2^2}{2} \quad (7.18)$$

The coefficients α_1 and α_2 are *kinetic-energy correction factors* and are evaluated by the original expressions in which they were introduced:

$$\alpha \frac{\rho \overline{V}^3 A}{2} = \int_A \frac{\rho V^3 dA}{2} \tag{7.19}$$

or

$$\alpha = \frac{1}{A} \int_A \left(\frac{V}{\overline{V}} \right)^3 dA \tag{7.20}$$

Thus $\alpha = 1$ when the velocity is uniform across the section, and $\alpha > 1$ for nonuniform velocity distributions. Computations show that $\alpha = 2$ for laminar flow in a pipe where the velocity has a parabolic distribution across the section. For most cases of turbulent flow, $\alpha \approx 1.05$. Because this is quite close to unity, it is common practice in engineering applications to let $\alpha_1 = \alpha_2 = 1$. A similar correction factor could be used with the momentum flux terms in the momentum equation for one-dimensional flow. However, it deviates even less from unity than does α for a given velocity distribution.

example 7.2

The velocity distribution for laminar flow in a pipe is given by the equation

$$V = V_{max} \left[1 - \left(\frac{r}{r_0} \right)^2 \right]$$

Here r_0 is the radius of the pipe and r is the radial distance from the center. Determine $\overline{V}$ in terms of V_{max} and evaluate the kinetic-energy correction factor α.

Solution A sketch for the velocity distribution is shown in the figure.

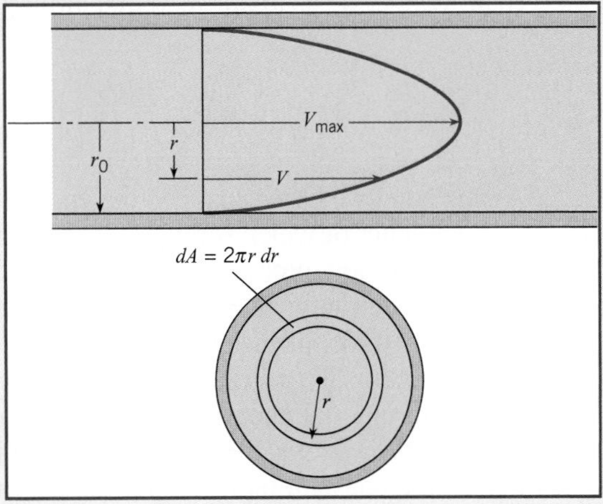

The discharge is given by $Q = \int V\, dA$, or

$$Q = \overline{V} A = \int_0^{r_0} V_{max} \left(1 - \frac{r^2}{r_0^2} \right) 2\pi r\, dr$$

or
$$\bar{V} = \frac{-\pi}{A} r_0^2 V_{max} \frac{(1 - r^2/r_0^2)^2}{2}\Bigg|_0^{r_0} = \frac{1}{2} V_{max}$$

The mean velocity is one-half the maximum velocity. To evaluate α we apply Eq. (7.20):

$$\alpha = \frac{1}{\pi r_0^2} = \int_0^{r_0} \frac{V_{max}^3 (1 - r^2/r_0^2)^3}{(\frac{1}{2})^3 V_{max}^3} 2\pi r\, dr = 2$$ ◁

The shaft-work term in Eq. (7.18) is usually the result of a turbine or pump in the flow system. When fluid passes through a turbine, the fluid system is doing shaft work on the surroundings; on the other hand, a pump does work on the fluid. It is convenient, then, to represent the shaft-work term as

$$\dot{W}_s = \dot{W}_t - \dot{W}_p \tag{7.21}$$

where $\dot{W}_t$ and $\dot{W}_p$ are magnitudes of power (work per unit time) delivered by a turbine or supplied to a pump. Substituting this expression for shaft work into Eq. (7.18) and dividing by g results in

$$\frac{\dot{W}_p}{\dot{m}g} + \frac{p_1}{\gamma} + z_1 + \alpha_1 \frac{\bar{V}_1^2}{2g} = \frac{\dot{W}_t}{\dot{m}g} + \frac{p_2}{\gamma} + z_2 + \alpha_2 \frac{\bar{V}_2^2}{2g} + \frac{u_2 - u_1}{g} - \frac{\dot{Q}}{\dot{m}g} \tag{7.22}$$

All of the terms of Eq. (7.22) have one dimension: length. Hence we designate the first term involving shaft work as h_p, head supplied by a pump, and the second term involving shaft work as h_t, head given up to a turbine. Equation (7.22) is then written as

$$\frac{p_1}{\gamma} + \alpha_1 \frac{\bar{V}_1^2}{2g} + z_1 + h_p = \frac{p_2}{\gamma} + \alpha_2 \frac{\bar{V}_2^2}{2g} + z_2 + h_t + \left[\frac{1}{g}(u_2 - u_1) - \frac{\dot{Q}}{\dot{m}g}\right] \tag{7.23}$$

At this point we have separated the energy equation into a mechanical part and a thermal part, the latter of which is represented by the last term enclosed in the brackets.

In the flow process, some of the system's mechanical energy is converted to thermal energy through viscous action between fluid particles. This energy cannot be recovered in the form of mechanical energy. It is identified as an energy loss. The effect of viscous action is to heat the fluid. If there were no heat transfer ($\dot{Q} = 0$), then the change in internal energy (which is proportional to temperature for an incompressible liquid) would be positive and the term would be positive. If there were heat transfer to maintain a constant temperature, the heat transfer would have to be negative (from the pipe), so the term would still be positive. It can be shown by application of the second law of thermodynamics that the term in brackets is always positive. It is simply referred to as the *head loss* and is represented by h_L. Thus Eq. (7.23) becomes the *energy equation*

$$\frac{p_1}{\gamma} + \alpha_1 \frac{\bar{V}_1^2}{2g} + z_1 + h_p = \frac{p_2}{\gamma} + \alpha_2 \frac{\bar{V}_2^2}{2g} + z_2 + h_t + h_L \tag{7.24}$$

In Chapter 10 we shall consider specific practical ways to estimate h_L. However, for the present we shall use it only in a general way.

In most applications of the energy equation it is understood that the kinetic-energy term $\alpha\overline{V}^2/2g$ involves the mean velocity $\overline{V}$. Hence the bar over the V is usually omitted. The coefficient α is often also omitted; it usually has a value very near unity because most flows are turbulent. In addition, it is often necessary to convert the total power of a pump or turbine to h_p or h_t, respectively, or vice versa. This is done using the original definition of h_p and h_t.

The *power equation* for power supplied to the flow by a pump is

$$\dot{W}_p = \gamma Q h_p = \dot{m} g h_p$$

and for the power delivered by a turbine

$$\dot{W}_t = \gamma Q h_t = \dot{m} g h_t$$

Both pumps and turbines lose energy due to factors such as mechanical friction, viscous dissipation, and leakage. These losses are accounted for by the mechanical efficiency. If the mechanical efficiency of the pump is η_p, the power delivered by the pump to the flow is

$$\dot{W}_p = \eta_p \dot{W}_{p,\text{act}}$$

where $\dot{W}_{p,\text{act}}$ is the actual power supplied to the pump. If the mechanical efficiency of the turbine is η_t, the actual power delivered by the turbine is

$$\dot{W}_{t,\text{act}} = \eta_t \dot{W}_t$$

It must be cautioned that Eq. (7.24) is not valid when compressibility effects are significant. The energy equation for compressible flow will be covered in Chapter 12.

example 7.3

A horizontal pipe carries cooling water for a thermal power plant from a reservoir as shown. The head loss in the pipe is given as

$$\frac{0.02(L/D)V^2}{2g}$$

where L is the length of the pipe from the reservoir to the point in question, V is the mean velocity in the pipe, and D is the diameter of the pipe. If the pipe diameter is 20 cm and the rate of flow is 0.06 m^3/s, what is the pressure in the pipe at $L = 2000$ m? Assume $\alpha = 1$.

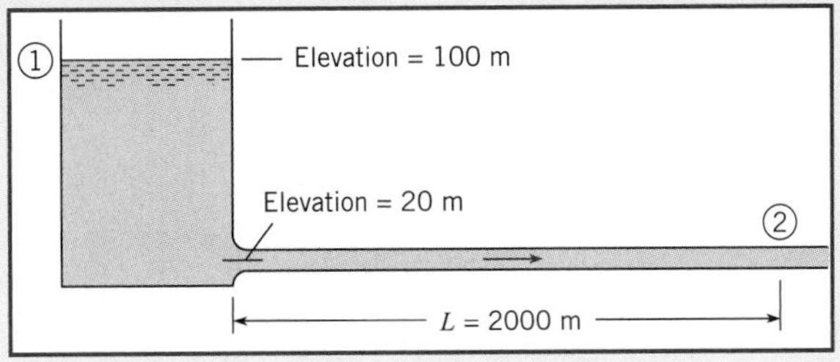

Solution Write the energy equation between the water surface in the reservoir and section 2 in the pipe:

$$\frac{p_1}{\gamma} + \frac{V_1^2}{2g} + z_1 = \frac{p_2}{\gamma} + \frac{V_2^2}{2g} + z_2 + h_L$$

where $p_1 = 0$, $V_1 = 0$, $z_1 = 100$ m, $z_2 = 20$ m, $V_2 = Q/A = 0.06/[(\pi/4) \times 0.2^2] = 1.91$ m/s, and $h_L = 0.02(L/0.2) \times 1.91^2/(2 \times 9.81) = 0.0186L$ m. Then

$$\frac{p_2}{\gamma} = 100 - 20 - 0.186 - 0.0186L = 79.8 - 0.0186L$$

At $L = 2000$ m

$$\frac{p_2}{\gamma} = 79.8 - 37.2 = 42.6 \text{ m}$$

$$p_2 = 417.9 \text{ kPa}$$ ◁

example 7.4

The pipe in Fig. 7.2 is 50 cm in diameter and carries water at a rate of 0.5 m^3/s. Also, $z_2 = 40$ m, $z_1 = 30$ m, and $p_1 = 70$ kPa gage. What power in kilowatts and in horsepower must be supplied to the flow by the pump if the gage pressure at section 2 is to be 350 kPa? Assume $h_L = 3$ m of water and $\alpha_1 = \alpha_2 = 1$.

Solution Write the energy equation between sections 1 and 2:

$$\frac{p_1}{\gamma} + \frac{V_1^2}{2g} + z_1 + h_p = \frac{p_2}{\gamma} + \frac{V_2^2}{2g} + z_2 + h_L$$

where $p_1 = 70$ kPa, $\gamma = 9.81$ kN/m^3, $V_1 = Q/A_1 = (0.50 \text{ m}^3/\text{s})/[\pi(0.25^2)\text{m}^2] = 2.55$ m/s, $V_2 = 2.55$ m/s, $z_1 = 30$ m, $z_2 = 40$ m, $p_2 = 350$ kPa, $h_L = 3$ m, and $g = 9.81$ m/s^2. Then

$$\begin{aligned} h_p &= \frac{p_2 - p_1}{\gamma} + \frac{V_2^2 - V_1^2}{2g} + z_2 - z_1 + h_L \\ &= \frac{(350 - 70) \text{ kN/m}^2}{9.81 \text{ kN/m}^3} + \frac{(2.55^2 - 2.55^2) \text{ m}^2/\text{s}^2}{9.81 \text{ m/s}^2} + 40 \text{ m} - 30 \text{ m} + 3 \text{ m} \\ &= 28.5 + 0 + 10 + 3 = 41.5 \text{ m} \end{aligned}$$

The head supplied by the pump is 41.5 m. Therefore, we obtain the total power supplied by the pump by taking the product of h_p and the weight rate of flow.

$$P = \dot{W}_P = Q\gamma h_p = 0.5 \text{ m}^3/\text{s} \times 9.81 \text{ kN/m}^3 \times 41.5 \text{ m}$$
$$= 204 \text{ m} \cdot \text{kN/s}$$
$$= 204 \text{ kJ/s} = 204 \text{ kW}$$
$$= 273 \text{ hp}$$

◁

example 7.5

At the maximum rate of power generation this hydroelectric power plant takes a discharge of 141 m^3/s. If the head loss through the intakes, penstock, and outlet works is 1.52 m, what is the rate of power generation?

Solution Assume $\alpha_1 = \alpha_2 = 1$. Then, for this application, the energy equation reduces to

$$\frac{p_1}{\gamma} + \frac{V_1^2}{2g} + z_1 = \frac{p_2}{\gamma} + \frac{V_2^2}{2g} + h_L + h_t$$

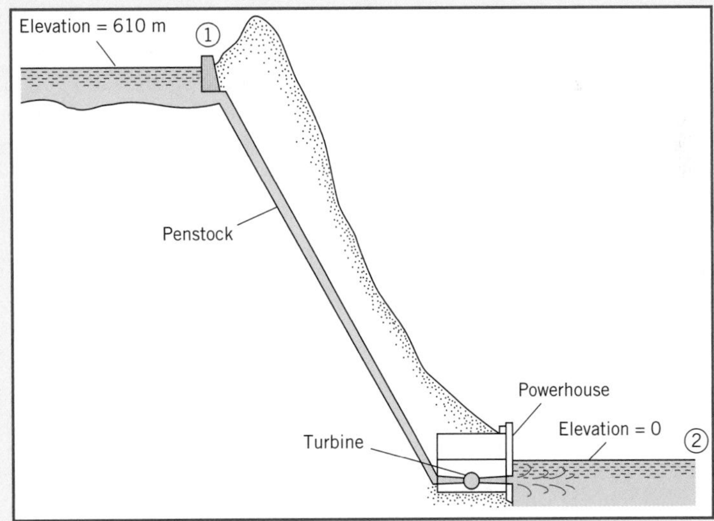

If section 1 is at the upstream reservoir and section 2 is at sea level, then $p_1 = 0$, $V_1 = 0$, $z_1 = 610$ m, $p_2 = 0$, $V_2 = 0$, and $h_L = 1.52$ m. Then

$$h_t = 610 - 1.52 = 608.5 \text{ m}$$

$$P = \dot{W}_t = Q\gamma h_t = 141 \times 9810 \times 608.5 = 842 \text{ MW}$$

◁

The Bernoulli Equation and the Energy Equation

There is often confusion over the difference between the Bernoulli equation and the energy equation. The Bernoulli equation was developed for the relationship between velocity and piezometric pressure along a streamline in a steady, incompressible, and inviscid flow, namely

$$\frac{V_1^2}{2} + \frac{p_1}{\gamma} + z_1 = \frac{V_2^2}{2} + \frac{p_2}{\gamma} + z_2$$

The energy equation was developed for viscous, incompressible flow in a pipe with additional energy being added through a pump or extracted through a turbine:

$$\alpha\frac{V_1^2}{2} + \frac{p_1}{\gamma} + z_1 + h_p = \alpha\frac{V_2^2}{2} + \frac{p_2}{\gamma} + z_2 + h_t + h_L$$

The applications of the respective equations must not be confused.

Under special circumstances the energy equation can be reduced to the Bernoulli equation. If the flow is inviscid, there is no head loss; $h_L = 0$. If the "pipe" is regarded as a small stream tube enclosing a streamline, then $\alpha = 1$. There is no pump or turbine along a streamline, so $h_p = h_t = 0$. In this case the energy equation is identical to the Bernoulli equation.

Often one finds the statement that the Bernoulli equation is used to find the pressure change in a pipe with varying cross-sectional area. However, it is really the energy equation that is being applied with the assumptions that there are no head losses ($h_L = 0$) due to viscosity and the velocity profiles at each cross-section are uniformly distributed ($\alpha = 1$). With these restrictions for this application, the Bernoulli equation and energy equation yield the same relationship between velocity and pressure. However, the energy equation is *not* the Bernoulli equation.

7·3 Application of the Energy, Momentum, and Continuity Principles in Combination

The energy, momentum, and continuity equations are independent equations. They can therefore be used together to solve a variety of problems. To illustrate the use of these equations in combination, we shall consider flow through an abrupt expansion and forces on transitions.

Abrupt Expansion

Consider the flow from a small pipe into a larger pipe, as shown in Fig. 7.3. We want to solve for the head loss due to the expansion as a function of the flow velocities in the two pipes. Normally such a problem is not amenable to analytic solution; however, one can solve this problem with a reasonable assumption about the pressure distribution at the change in section. Experience tells us that when flow occurs past an abrupt enlargement such as this, the flow separates from the boundary. Hence, in effect, a jet of fluid from the

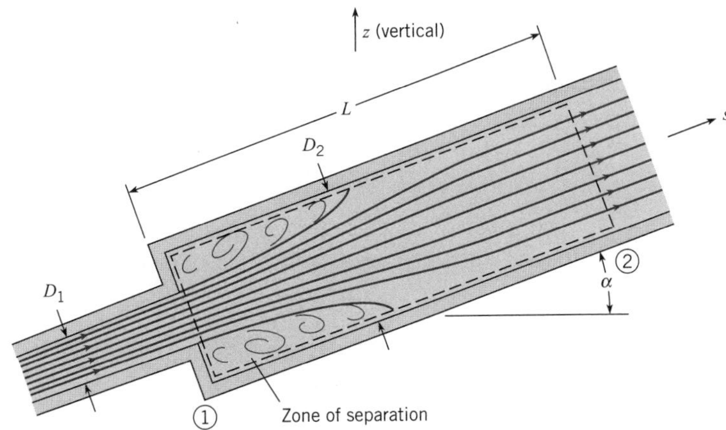

FIGURE 7.3

Flow through an abrupt expansion.

smaller pipe discharges into the larger pipe. Because the streamlines in the jet are initially straight and parallel, the pressure distribution across the jet will be simply hydrostatic. Because the fluid in the zone of separation at section 1 has such low velocity, it can be likewise assumed that the pressure in this zone is the same as that in the jet except for the hydrostatic variation. With this basic assumption for the pressure at section 1, we can apply the momentum and energy equations to solve for the head loss due to the expansion. The energy equation written between sections 1 and 2 is

$$\frac{p_1}{\gamma} + \alpha_1 \frac{V_1^2}{2g} + z_1 = \frac{p_2}{\gamma} + \alpha_2 \frac{V_2^2}{2g} + z_2 + h_L \tag{7.25}$$

We are assuming turbulent flow conditions here, so we may assume $\alpha_1 = \alpha_2 = 1$. The momentum equation for the fluid in the large pipe between section 1 and section 2, written for the s direction, is

$$\sum F_s = \dot{m}V_2 - \dot{m}V_1$$

Neglecting the force due to shear stress, we have

$$p_1 A_2 - p_2 A_2 - \gamma A_2 L \sin \alpha = \rho V_2^2 A_2 - \rho V_1^2 A_1$$

or

$$\frac{p_1}{\gamma} - \frac{p_2}{\gamma} - (z_2 - z_1) = \frac{V_2^2}{g} - \frac{V_1^2}{g}\frac{A_1}{A_2} \tag{7.26}$$

The continuity equation, $V_1 A_1 = V_2 A_2$, is also a valid independent equation for this problem. Therefore, when Eqs. (7.25) and (7.26), along with the continuity equation, are used to solve for h_L, the following equation for *sudden-expansion head loss* is obtained:

$$h_L = \frac{(V_1 - V_2)^2}{2g} \tag{7.27}$$

If a pipe discharges liquid into a reservoir, then $V_2 = 0$ and the sudden-expansion head loss simplifies to

$$h_L = \frac{V^2}{2g}$$

which is the velocity head of the liquid in the pipe. This energy is dissipated by the viscous action of the liquid in the reservoir.

Forces on Transitions

The method for determining the forces required to hold a transition or any other flow passage in place is presented in the form of an example.

example 7.6

Water flows through the contraction at a rate of 0.707 m^3/s. The head loss due to this particular transition is given by the empirical equation

$$h_L = 0.1\frac{V_2^2}{2g}$$

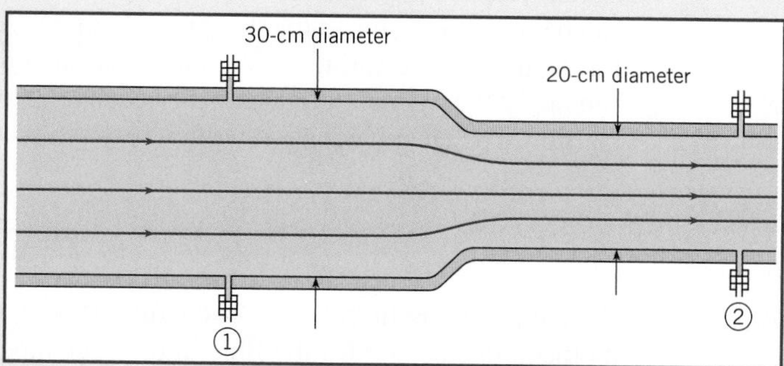

Here V_2 is the velocity in the 20-cm pipe. What horizontal force is required to hold the transition in place if $p_1 = 250$ kPa? Assume $\alpha_1 = \alpha_2 = 1$.

Solution Write the momentum equation for the transition between sections 1 and 2. The control surface is drawn so that it encloses the fluid and the transition itself; consequently, the control volume is as shown. The pressures are gage pressures, so the pressure on the exterior of the transition is zero. Hence the force on the exterior is zero. Then, for this control volume, the momentum equation is

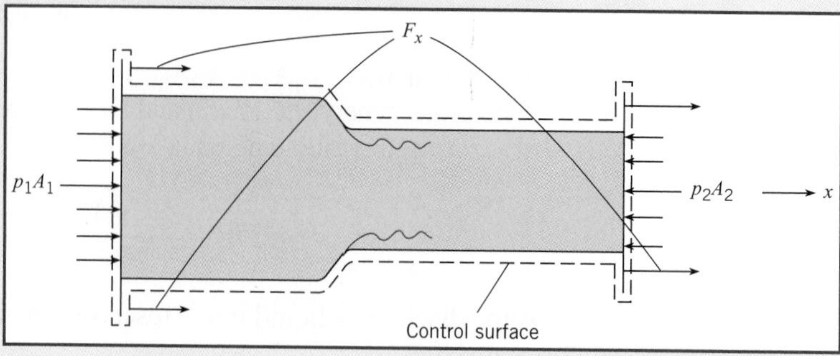

$$p_1A_1 - p_2A_2 + F_x = \dot{m}_2V_2 - \dot{m}_1V_1$$
$$p_1A_1 - p_2A_2 + F_x = \rho V_2^2A_2 - \rho V_1^2A_1$$
$$F_x = \rho Q(V_2 - V_1) + p_2A_2 - p_1A_1$$

Here Q and the velocities are known, and F_x is the unknown. Also, p_2 is still unknown. We obtain p_2 by applying the energy equation between sections 1 and 2:

$$\frac{p_1}{\gamma} + \frac{V_1^2}{2g} + z_1 = \frac{p_2}{\gamma} + \frac{V_2^2}{2g} + z_2 + h_L$$

Here we have assumed $\alpha_1 = \alpha_2 = 1$. For our problem, $z_1 = z_2$, so the foregoing equation reduces to

$$\frac{p_1}{\gamma} + \frac{V_1^2}{2g} = \frac{p_2}{\gamma} + \frac{V_2^2}{2g} + h_L$$
$$p_2 = p_1 - \gamma\left(\frac{V_2^2}{2g} - \frac{V_1^2}{2g} + h_L\right)$$

Substituting this expression for p_2 into the equation for the force on the transition yields

$$F_x = \rho Q(V_2 - V_1) + A_2\left[p_1 - \gamma\left(\frac{V_2^2}{2g} - \frac{V_1^2}{2g} + h_L\right)\right] - p_1A_1$$

$$V_1 = \frac{Q}{A_1} = \frac{0.707}{(\pi/4)\times 0.3^2} = 10 \text{ m/s}$$
$$V_2 = \frac{Q}{A_2} = \frac{0.707}{(\pi/4)\times 0.2^2} = 22.5 \text{ m/s}$$
$$h_L = \frac{0.1V_2^2}{2g} = \frac{0.1\times 22.5^2}{2\times 9.81} = 2.58 \text{ m}$$

Then

$$F_x = 1000\times(0.707)(22.5 - 10) + \frac{\pi}{4}\times 0.2^2$$
$$\times\left[250{,}000 - 9810\left(\frac{22.5^2}{2\times 9.81} - \frac{10^2}{2\times 9.81} + 2.58\right)\right]$$
$$-250{,}000\times(\pi/4)\times 0.3^2$$
$$= -8.15 \text{ kN}$$ ◁

Thus a force of 8.15 kN must be applied in the negative x direction to hold the transition in place for the given conditions.

Concept of the Hydraulic and Energy Grade Lines

The units of Eq. (7.24) are meters or feet, and we can attach a useful physical relationship to them. Consider the flow in the pipe shown in Fig. 7.4. The sum of the terms on the left side of Eq. (7.24) represents the total mechanical energy (stated in energy per unit weight of flowing liquid, $N \cdot m/N$) plus the flow-work term in the fluid at the upstream section plus the energy supplied by a pump. The sum on the right-hand side is the total energy per unit weight at the downstream section plus the head loss and energy per unit weight given up to a turbine between the two sections. For the case shown in Fig. 7.4, the velocity of flow in the reservoir is negligible. Hence the total energy at the surface is potential energy and, in terms of energy per unit weight, is simply z. At the downstream section the liquid is at a different elevation than in the reservoir and it has significant velocity. Therefore, the "energy per unit weight" here is given in terms of (p_2/γ), z_2, and $\alpha_2 V_2^2/2g$. It is common practice to lump all these terms together as *total head* in feet or meters. Note that the sum of these terms plus the head loss between sections 1 and 2 is equal to the total head at section 1. By analyzing a sketch such as Fig. 7.4 it is possible at a glance to ascertain the pressure at any point in the pipeline: one simply picks off values of p/γ and multiplies by γ to obtain p.

If one were to tap a piezometer into the pipe in Fig. 7.4, the liquid would rise to a height p/γ above the pipe. Hence the height of the p/γ line is called the *hydraulic grade line* (HGL). The total head $(p/\gamma + \alpha V^2/2g + z)$ in the system is greater than $p/\gamma + z$ by an amount $\alpha V^2/2g$. Consequently, the *energy grade line* (EGL) is above the HGL by a distance $\alpha V^2/2g$. The engineer who develops a visual concept of the energy equation as explained above will find it much easier to sense trouble spots in the system (usually points of low pressure) and to devise ways of solving the problem.

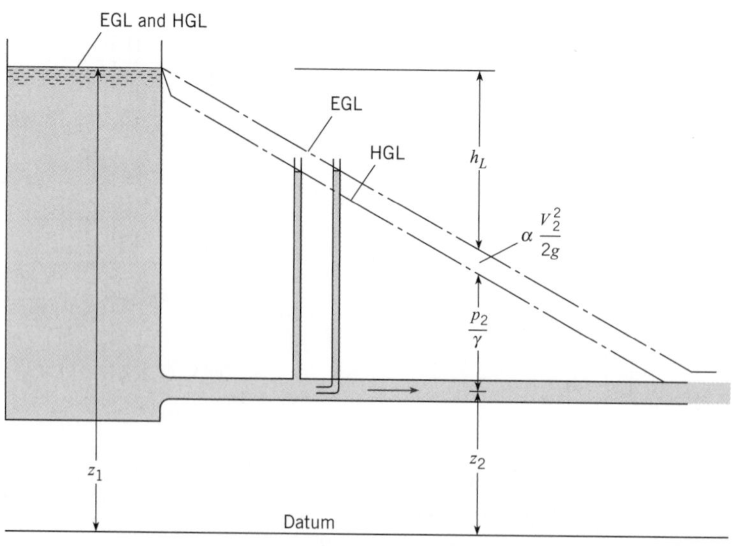

FIGURE 7.4

EGL and HGL in a straight pipe.

Here are some other helpful hints for drawing hydraulic grade lines and energy grade lines.

1. By definition, the EGL is positioned above the HGL by an amount $\alpha V^2/2g$. Thus if the velocity is zero, as in a lake or reservoir, the HGL and EGL will coincide (see Fig. 7.4) with the liquid surface.

2. Head loss for flow in a pipe or channel always means that the EGL will slope downward in the direction of flow (see Fig. 7.4). The only exception to this rule occurs when a pump supplies energy (and pressure) to the flow. Then an abrupt rise in the EGL (and the HGL) occurs from the upstream side to the downstream side of the pump (see Fig. 7.5).

3. In point 2 above, it was noted that a pump can cause an abrupt rise in the EGL (and the HGL) because energy is introduced into the flow by the pump. Similarly, if energy is abruptly taken out of the flow—by a turbine, for example—then the EGL and the HGL will drop abruptly, as in Fig. 7.6. In Fig. 7.6 and other figures (Fig. 7.7 through Fig. 7.9), it is assumed that $\alpha = 1.0$. Figure 7.6 also shows that much of the kinetic energy can be converted to pressure if there is a gradual expansion such as at the outlet. Thus the head loss at the outlet is reduced, making the turbine installation more efficient. If the outlet to a reservoir is an abrupt expansion as in Fig. 7.8, all the kinetic energy is lost. Thus the EGL drops an amount $\alpha V^2/2g$ at the outlet.

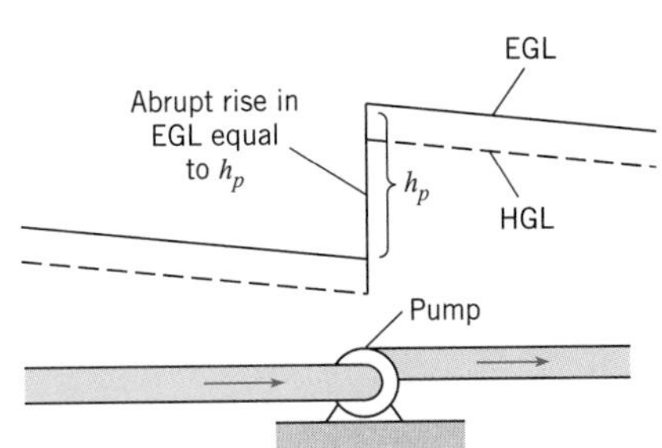

FIGURE 7.5

Rise in EGL and HGL due to pump.

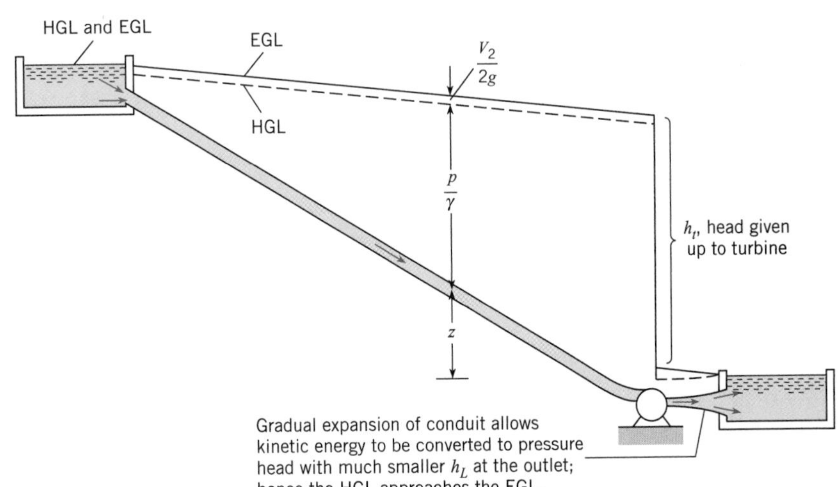

FIGURE 7.6

Drop in EGL and HGL due to turbine.

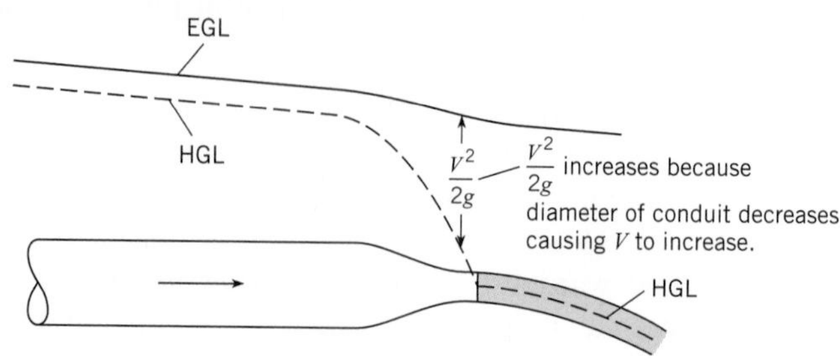

FIGURE 7.7

Change in HGL and EGL due to flow through a nozzle.

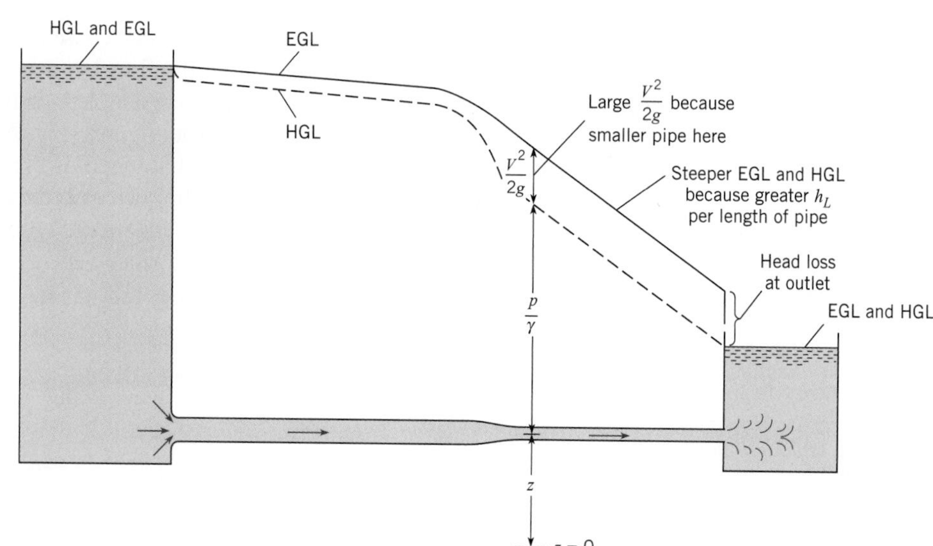

FIGURE 7.8

Change in EGL and HGL due to change in diameter of pipe.

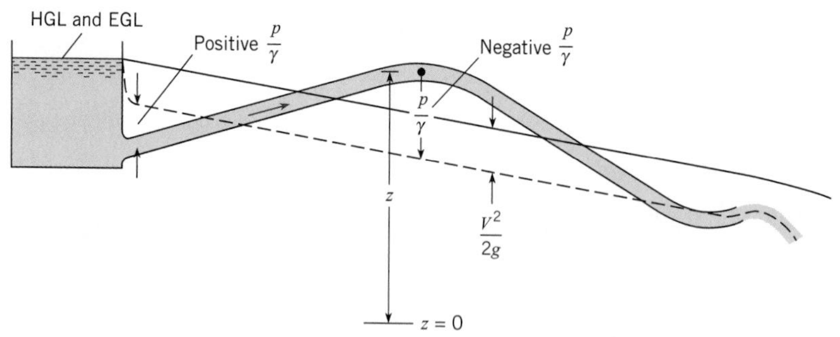

FIGURE 7.9

Subatmospheric pressure when pipe is above HGL.

4. In a pipe or channel where the pressure is zero, the HGL is coincident with the system because $p/\gamma = 0$ at these points. This fact can be used to locate the HGL at certain points in a physical system, such as at the outlet end of a pipe where the liquid discharges into the atmosphere or at the upstream end where the gage pressure is zero at the reservoir surface (see Fig. 7.4).

5. For steady flow in a pipe that has uniform physical characteristics (diameter, roughness, shape, and so on) along its length, the head loss per unit length will be constant. Thus the slope $(\Delta h_L/\Delta L)$ of the EGL and the HGL will be constant along the length of pipe (see Fig. 7.4).

6. If a flow passage changes diameter, such as in a nozzle or by means of a change in pipe size, the velocity therein will also change. Hence, the distance between the EGL and the HGL will change (see Fig. 7.7). In addition, the slope on the EGL will change because the head loss per length will be larger in the conduit with the larger velocity (see Fig. 7.8). You will learn more about this latter point in Chapter 10 when head loss in pipes is considered in more detail.

7. If the HGL falls below the pipe, then p/γ is negative, indicating subatmospheric pressure (see Fig. 7.9).

example 7.7

A pump draws water from a reservoir, where the water-surface elevation is 520 ft, and forces the water through a pipe 5000 ft long and 1 ft in diameter. This pipe then discharges the water into a reservoir with water-surface elevation of 620 ft. The flow rate is 7.85 cfs, and the head loss in the pipe is given by $0.01(L/D)(V^2/2g)$. Determine the head supplied by the pump, h_p, and the power supplied to the flow, and draw the HGL and EGL for the system. Assume that the pipe is horizontal and is 510 ft in elevation.

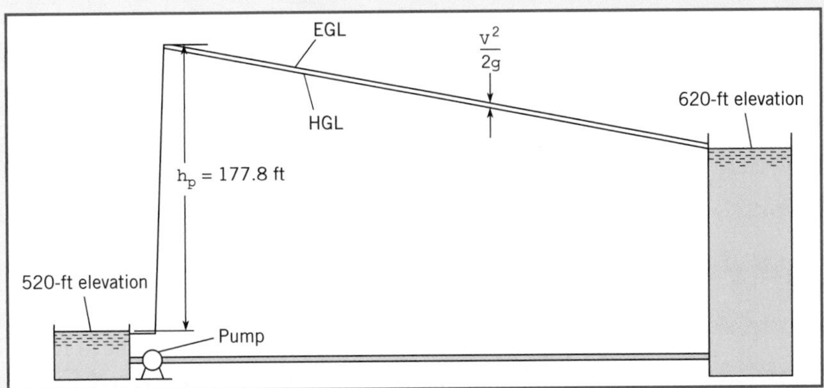

Solution First solve for h_p by using the energy equation (one-dimensional flow assumed) written from the water surface in the lower reservoir to the water surface in the upper reservoir.

$$\frac{p_1}{\gamma} + \frac{V_1^2}{2g} + z_1 + h_p = \frac{p_2}{\gamma} + \frac{V_2^2}{2g} + z_2 + h_L$$

In this example, p_1/γ, p_2/γ, V_1, and V_2 are all zero, but $z_1 = 520$ ft and $z_2 = 620$ ft.

$$h_L = 0.01 \times \frac{5000}{1}\frac{V^2}{2g} \quad \text{and} \quad V = \frac{Q}{A_p} = 10 \text{ ft/s}$$

Then

$$h_p = 620 - 520 + 0.01 \times \frac{5000}{1}\frac{100}{64.4} \text{ ft-lbf/lbf} = 178 \text{ ft}$$

However, the product of the flow rate Q and specific weight will give us the weight rate of flow. Then the power supplied by the pump will be $h_p \times$ weight rate of flow, or

$$P = h_p Q\gamma \text{ ft-lbf/s} = \frac{h_p Q\gamma}{550} = 158 \text{ hp} \quad \triangleleft$$

Summary

The energy equation relates the rate of change of energy of a system to the rate of heat transfer to the system and the rate at which the system does work on the surroundings. Applying the energy equation to a control volume with steady flow and with uniform flow where mass crosses the control surface results in

$$\dot{Q} - \dot{W}_s = \sum_{\text{cs}} \dot{m}_o \left(\frac{V^2}{2} + h + gz\right)_o - \sum_{\text{cs}} \dot{m}_i \left(\frac{V^2}{2} + h + gz\right)_i$$

where $\dot{Q}$ is the rate of heat transfer to the control volume, $\dot{W}_s$ is the rate at which shaft work is done on the surroundings, and h is the enthalpy of the fluid.

Further simplification of the energy equation for the flow of an incompressible fluid in a pipe yields

$$\frac{p_1}{\gamma} + \alpha_1 \frac{V_1^2}{2g} + z_1 + h_p = \frac{p_2}{\gamma} + \alpha_2 \frac{V_2^2}{2g} + z_2 + h_t + h_L$$

where α is the kinetic-energy correction factor, h_p is the head provided by a pump, h_t is the head removed by a turbine, and h_L is the head loss. Station 1 is always upstream and station 2 is downstream. For a laminar flow, $\alpha = 2$ and for a turbulent flow, $\alpha \simeq 1$. The increase in head across a pump is related to the pump power by

$$\dot{W}_P = \gamma Q h_p$$

and the power delivered by a turbine is given by

$$\dot{W}_t = \gamma Q h_t$$

The actual power required by the pump is $\dot{W}_{p,\text{act}} = \dot{W}_P / \eta_p$ and the actual power delivered by a turbine is $\dot{W}_{t,\text{act}} = \eta_t \dot{W}_t$, where η_p and η_t are the pump and turbine efficiencies.

The head loss is always positive and represents the irreversible conversion of mechanical energy to thermal energy through the viscous action of the fluid. The head loss due to a sudden expansion is

$$h_L = \frac{(V_1 - V_2)^2}{2g}$$

where V_1 and V_2 are the upstream and downstream velocities.

The hydraulic grade line (HGL) is the profile of the piezometric head, $p/\gamma + z$, along a pipe. The energy grade line (EGL) is a plot of the total head, $V^2/2g + p/\gamma + z$, along a pipe. If the hydraulic grade line falls below the elevation of a pipe, subatmospheric pressure exists in the pipe at that location, giving rise to the possibility of cavitation or leakage into the pipe.

References

1. Cengel, Y. A., and M. A. Bolos. *Thermodynamics: An Engineering Approach.* McGraw-Hill, New York, 1998.

2. Moran, M. J., and H. N. Shapiro. *Fundamentals of Engineering Thermodynamics.* John Wiley, New York, 1992.

Problems

7.1 The sketch shows a common consumer product called the Water Pik. This device uses a motor to drive a piston pump that produces a jet of water (d = 1/8 in., T = 10°C) with a speed of 40 m/s. Estimate the minimum electrical power in watts that is required by the device.

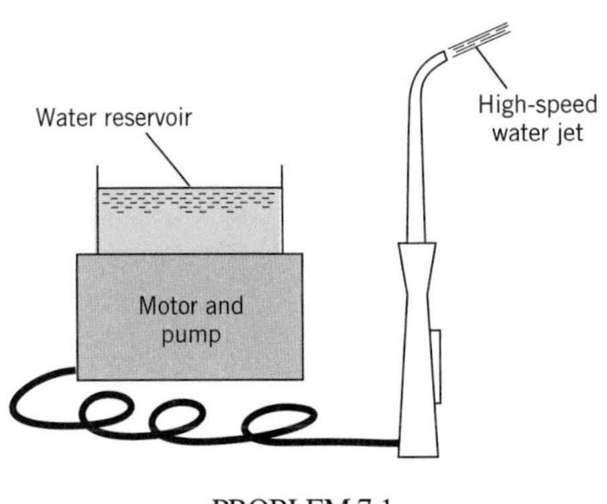

PROBLEM 7.1

7.2 A turbine receives steam at 2.0 MPa, 500°C (enthalpy = 3062 kJ/kg) at a velocity of 10 m/s. The steam leaves the turbine as a gas–liquid mixture at 101 kPa, 373 K, with an enthalpy of 2621 kJ/kg. The exit velocity is 50 m/s. Thermal energy is lost through the turbine walls at a rate of 10 kJ/h. The potential energy due to the elevation difference between inlet and exit ports can be neglected. Calculate the power if 4000 kg/h of steam pass through the turbine.

7.3 An engineer is considering the development of a small wind turbine (D = 1.0 m) for home applications. The design wind speed is 15 mph at T = 50°F and p = 0.9 bar. The efficiency of the turbine is η = 20%, meaning that 20% of the kinetic energy in the wind can be extracted. Estimate the power in watts that can be produced by the turbine.

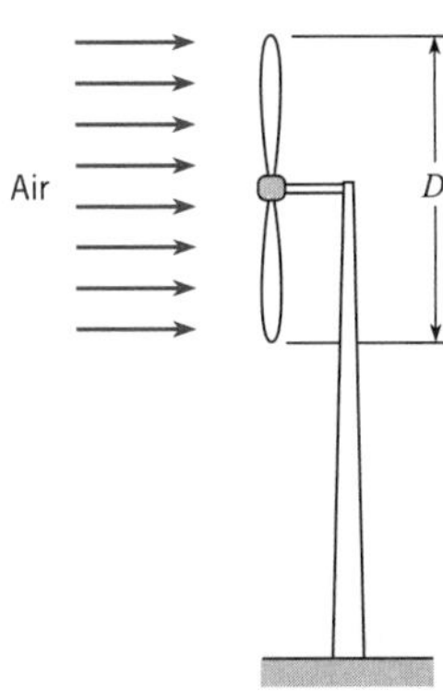

PROBLEM 7.3

7.4 A compressor is used to supply high-pressure air for a supersonic wind tunnel. The enthalpy of the air is 300 kJ/kg at the entrance and 500 kJ/kg at the compressor outlet. The outlet velocity is 200 m/s, and the inlet velocity is negligible. The mass flow through the compressor is 1.5 kg/s. The system is adiabatic. Find the power (in kilowatts) required to operate the compressor.

7.5 Air flows in a pipe 1200 m long and 8 cm in diameter at a rate of 0.5 kg/s. The pressure and temperature at the upstream end of the pipe are 50 kPa gage and 20°C. The pipe is well insulated, so that a negligible amount of heat is transferred to or from the air in the pipe. What are the velocity and temperature in the stream of air at the outlet as it discharges into the atmosphere? Assume that atmospheric pressure is 100 kPa. Assume that air is an ideal gas and that the specific enthalpy is given by h (kJ/kg) = $1.004T$ (K).

7.6 A hypothetical velocity distribution in a pipe has a maximum velocity of V_{max} at the center and a minimum velocity of $0.5V_{max}$ at the wall. If the velocity is linearly distributed from the center to the wall, what is α? What is the mean velocity V in terms of V_{max}?

FLUID SOLUTIONS **7.7** For this hypothetical velocity distribution in a wide rectangular channel, evaluate the kinetic-energy correction factor α.

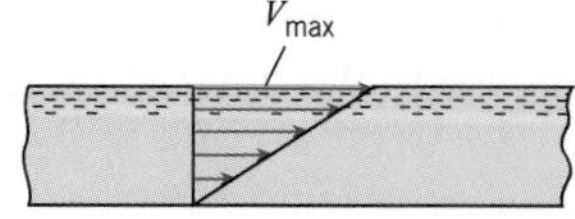

PROBLEM 7.7

7.8 For these velocity distributions in a pipe, indicate whether the kinetic-energy correction factor α is greater than, equal to, or less than unity.

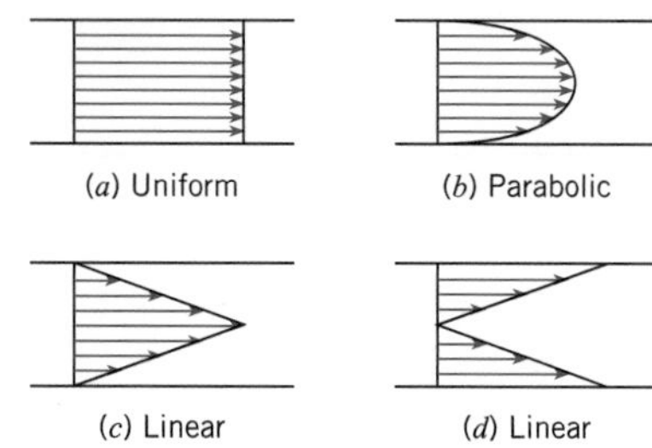

PROBLEMS 7.8, 7.9, 7.10

7.9 Calculate α for case (*c*) in Prob. 7.8.

7.10 Calculate α for case (*d*) in Prob. 7.8.

7.11 If the value of α (kinetic-energy correction factor) for flow in a pipe is 1.08, then the flow must be (a) laminar, (b) turbulent.

7.12 An approximate equation for the velocity distribution in a pipe with turbulent flow is

$$\frac{V}{V_{max}} = \left(\frac{y}{r_0}\right)^n$$

where V_{max} is the centerline velocity, y is the distance from the wall of the pipe, r_0 is the radius of the pipe, and n is an exponent that depends on the Reynolds number and varies between 1/6 and 1/8 for most applications. Derive a formula for α as a function of n. What is α if $n = 1/6$?

FLUID SOLUTIONS **7.13** An approximate equation for the velocity distribution in a rectangular channel with turbulent flow is

$$\frac{V}{V_{max}} = \left(\frac{y}{d}\right)^n$$

where V_{max} is the velocity at the surface, y is the distance from the floor of the channel, d is the depth of flow, and n is an exponent that varies from about 1/6 to 1/8 depending on the Reynolds number. Derive a formula for α as a function of n. What is the value of α for $n = 1/7$?

7.14 The following data were taken for turbulent flow in a circular pipe with a radius of 3.5 cm. Evaluate the kinetic energy correction factor. The velocity at the pipe wall is zero.

r (cm)	*V* (m/s)	*r* (cm)	*V* (m/s)
0.0	32.5	2.8	22.03
0.5	32.44	2.9	21.24
1.0	32.27	3.0	20.49
1.5	31.22	3.1	19.6
2.0	28.21	3.2	18.69
2.25	26.51	3.25	18.16
2.5	24.38	3.3	17.54
2.6	23.7	3.35	17.02
2.7	22.88	3.4	16.14

7.15 Water flows from a pressurized tank as shown. The pressure in the tank above the water surface is 100 kPa gage, and the water surface level is 12 m above the outlet. The water exit velocity is 10 m/s. The head loss in the system varies as $h_L = K_L V^2/2g$, where K_L is the head-loss coefficient. Find the value for K_L. Assume α = 1.0 at all locations.

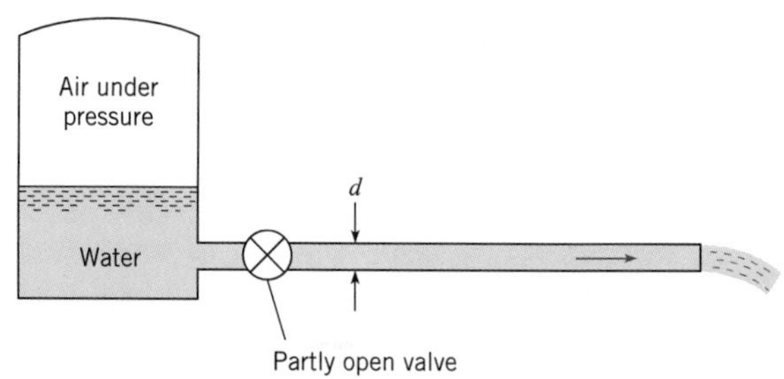

PROBLEMS 7.15, 7.16

7.16 A reservoir with water is pressurized as shown. The pipe diameter is 1 in. The head loss in the system is given by $h_L = 5V^2/2g$. The height between the water surface and the pipe outlet is 10 ft. A discharge of 0.10 ft^3/s is needed. What must the pressure in the tank be to achieve such a flow rate? Assume α = 1.0 at all locations.

7.17 A pipe drains a tank as shown. If $x = 10$ ft, $y = 4$ ft, and head losses are neglected, what is the pressure at point A and what is the velocity at the exit? Assume α = 1.0 at all locations.

7.18 A pipe drains a tank as shown. If $x = 8$ m, $y = 2$ m, and head losses are neglected, what is the pressure at point A and what is the velocity at the exit? Assume α = 1.0 at all locations.

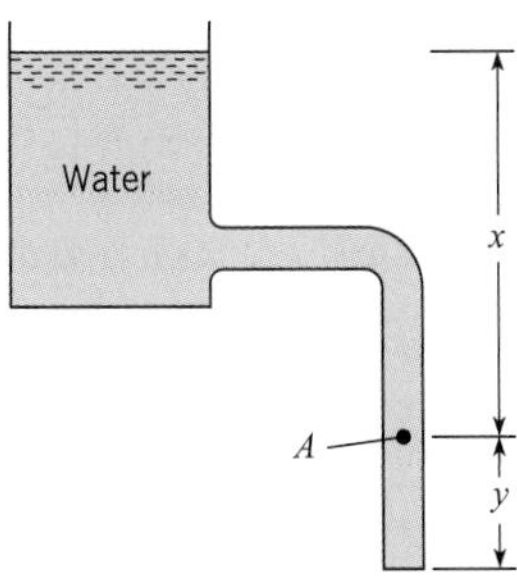

PROBLEMS 7.17, 7.18

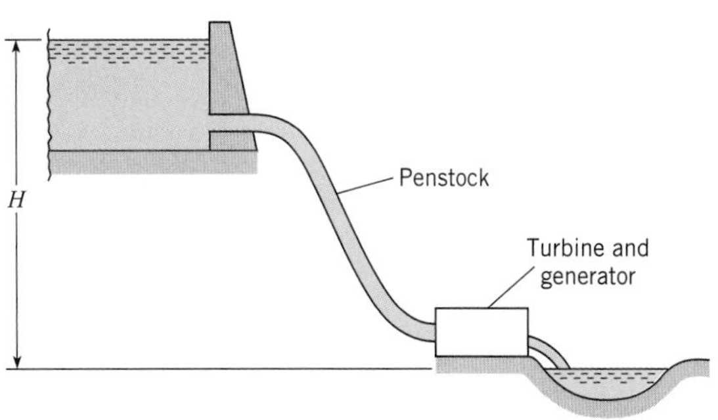

PROBLEM 7.21

7.19 If $D_A = 20$ cm, $D_B = 12$ cm, and $L = 1$ m, and if crude oil ($S = 0.90$) is flowing at a rate of $0.06\ \text{m}^3/\text{s}$, determine the difference in pressure between sections A and B. Neglect head losses.

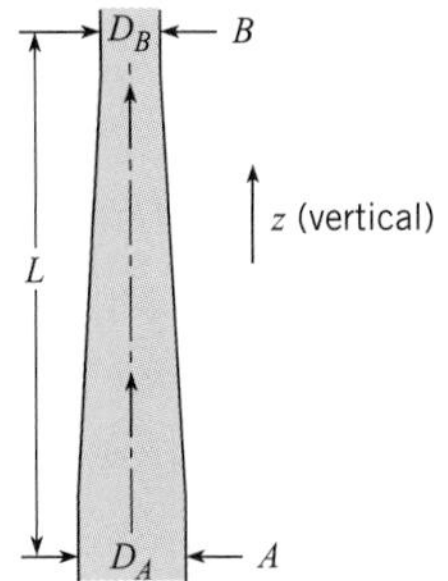

PROBLEM 7.19

7.20 In the figure for Probs. 7.15 and 7.16, suppose that the reservoir is open to the atmosphere at the top. The valve is used to control the flow rate from the reservoir. The head loss across the valve is given as $h_L = 5V^2/2g$, where V is the velocity in the pipe. The cross-sectional area of the pipe is 9 cm^2. The head loss due to friction in the pipe is negligible. The elevation of the water level in the reservoir above the pipe outlet is 11 m. Find the discharge in the pipe. Assume $\alpha = 1.0$ at all locations.

7.21 An engineer is making an estimate for a home owner. This owner has a small stream ($Q = 1.4$ cfs, $T = 40°\text{F}$) that is located at an elevation $H = 34$ ft above the owner's residence. The owner is proposing to dam the stream, diverting the flow through a pipe (penstock). This flow will spin a hydraulic turbine, which in turn will drive a generator to produce electrical power. Estimate the maximum power in kilowatts that can be generated if there is no head loss and both the turbine and generator are 100% efficient. Also, estimate the power if the head loss is 5.5 ft, the turbine is 70% efficient, and the generator is 90% efficient.

7.22 Gasoline having a specific gravity of 0.8 is flowing in the pipe shown at a rate of 6 cfs. What is the pressure at section 2 when the pressure at section 1 is 15 psig and the head loss is 6 ft between the two sections? Assume $\alpha = 1.0$ at all locations.

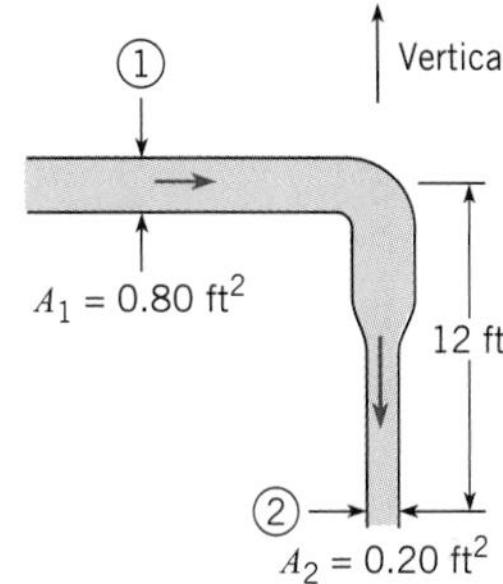

PROBLEM 7.22

7.23 Determine the discharge in the pipe and the pressure at point B. Neglect head losses. Assume $\alpha = 1.0$ at all locations.

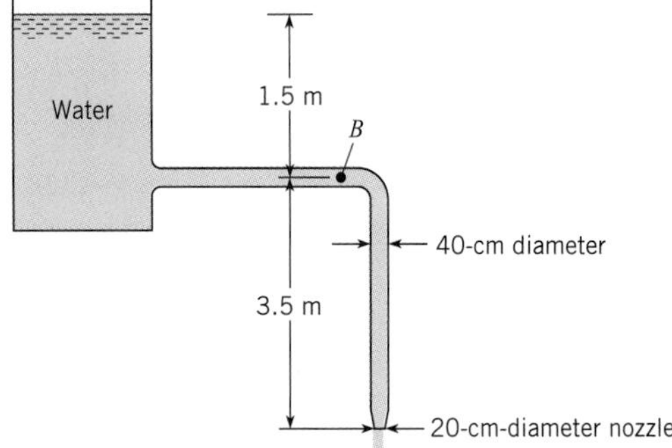

PROBLEM 7.23

7.24 A microchannel is being designed to transfer fluid in a MEMS (micro electrical mechanical system) application. The channel is 200 micrometers in diameter and is 5 cm long. Ethyl alcohol is driven through the system at the rate of 0.1 microliters/s (μl/s) with a syringe pump, which is essentially a moving piston. The pressure at the exit of the channel is atmospheric. The flow is laminar, so $\alpha = 2$. The head loss in the channel is given by

$$h_L = \frac{32\mu L V}{\gamma D^2}$$

where L is the channel length, D the diameter, V the mean velocity, μ the viscosity of the fluid, and γ the specific weight of the fluid. Find the pressure in the syringe pump. The velocity head associated with the motion of the piston in the syringe pump is negligible.

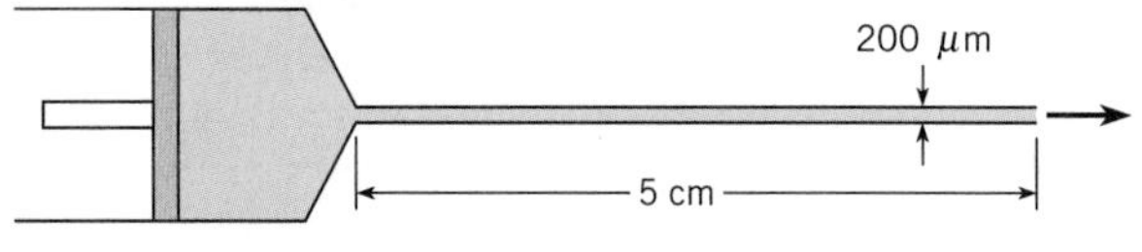

PROBLEM 7.24

7.25 Fire-fighting equipment requires that the exit velocity of the fire hose be 40 m/s at an elevation of 50 m above the hydrant. The nozzle at the end of the hose has a contraction ratio of 4:1 ($A_e/A_{hose} = 1/4$). The head loss in the hose is $10V^2/2g$, where V is the velocity in the hose. What must the pressure be at the hydrant to meet this requirement? The pipe supplying the hydrant is much larger than the fire hose.

7.26 The discharge in the siphon is 2.80 cfs, $D = 8$ in., $L_1 = 3$ ft, and $L_2 = 3$ ft. Determine the head loss between the reservoir surface and point C. Determine the pressure at point B if three-quarters of the head loss (as found above) occurs between the reservoir surface and point B. Assume $\alpha = 1.0$ at all locations.

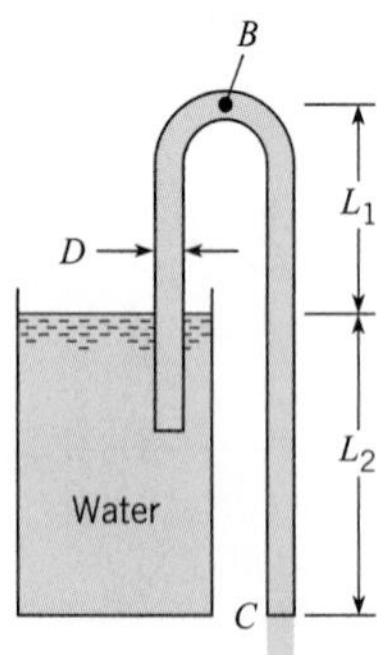

PROBLEM 7.26

7.27 Water ($\gamma = 62.4$ lbf/ft^3) flows through a horizontal constant diameter pipe with a cross-sectional area of 9 in^2. The velocity in the pipe is 15 ft/s, and the water discharges to the atmosphere. The head loss between the pipe joint and the end of the pipe is 3 ft. Find the force on the joint to hold the pipe. The pipe is mounted on frictionless rollers. Assume $\alpha = 1.0$ at all locations.

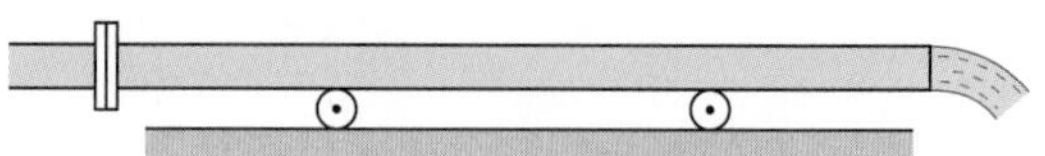

PROBLEM 7.27

7.28 For this siphon the elevations at A, B, C, and D are 30 m, 32 m, 27 m, and 26 m, respectively. The head loss between the inlet and point B is three-quarters of the velocity head, and the head loss in the pipe itself between point B and the end of the pipe is one-quarter of the velocity head. For these conditions, what is the discharge and what is the pressure at point B? The pipe diameter = 25 cm. Assume $\alpha = 1.0$ at all locations.

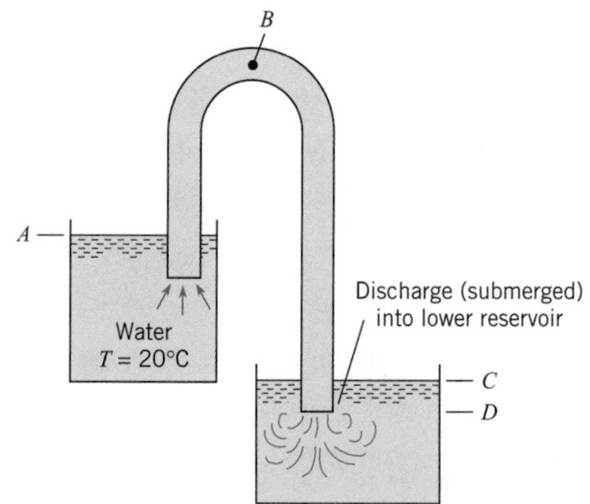

PROBLEMS 7.28, 7.29

7.29 For this system, point B is 10 m above the bottom of the upper reservoir. The head loss from A to B is $1.8V^2/2g$, and the pipe area is 10^{-4} m^2. Assume a constant discharge of 8×10^{-4} m^3/s. For these conditions, what will be the depth of water in the upper reservoir for which cavitation will begin at point B? Vapor pressure = 1.23 kPa and atmospheric pressure = 100 kPa. Assume $\alpha = 1.0$ at all locations.

7.30 Water flows at a steady rate in this vertical pipe. The pressure at A is 10 kPa, and at B it is 98.1 kPa. Then the flow in the pipe is (a) upward, (b) downward, (c) no flow.

7.31 In this system, $d = 6$ in., $D = 12$ in., $\Delta z_1 = 6$ ft, and $\Delta z_2 = 12$ ft. The discharge of water in the system is 10 cfs. Is the machine a pump or a turbine? What are the pressures at points A and B? Neglect head losses. Assume $\alpha = 1.0$ at all locations.

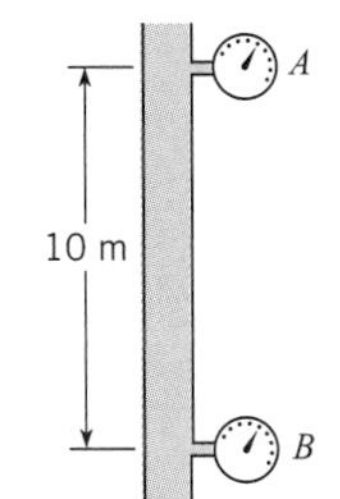

PROBLEM 7.30

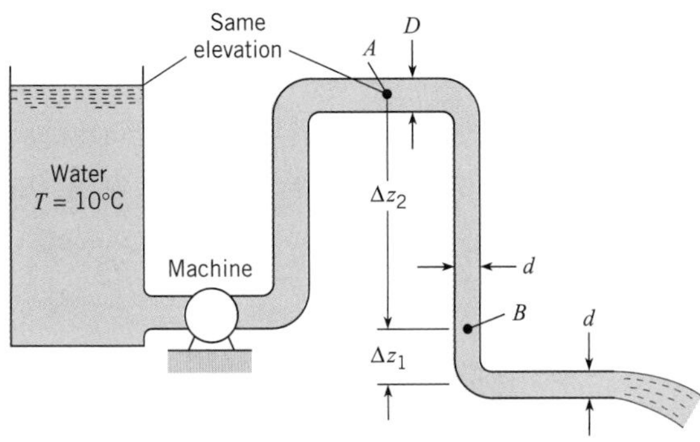

PROBLEM 7.31

7.32 For this system, the discharge of water is 0.1 m^3/s, $x = 1.0$ m, $y = 2.0$ m, $z = 7.0$ m, and the pipe diameter is 30 cm. Neglecting head losses, what is the pressure head at point 2 if the jet from the nozzle is 10 cm in diameter? Assume $\alpha = 1.0$ at all locations.

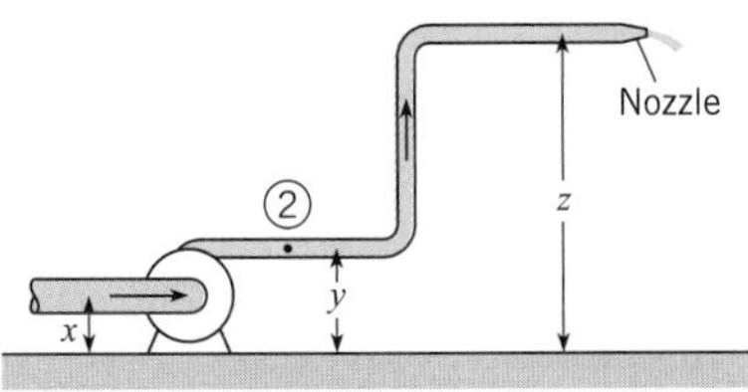

PROBLEM 7.32

7.33 A pump draws water through an 8-in. suction pipe and discharges it through a 4-in. pipe in which the velocity is 12 ft/s. The 4-in. pipe discharges horizontally into air at *C*. To what height *h* above the water surface at *A* can the water be raised if 25 hp is delivered to the pump? Assume that the pump operates at 60% efficiency and that the head loss in the pipe between *A* and *C* is equal to $2V_C^2/2g$. Assume $\alpha = 1.0$ at all locations.

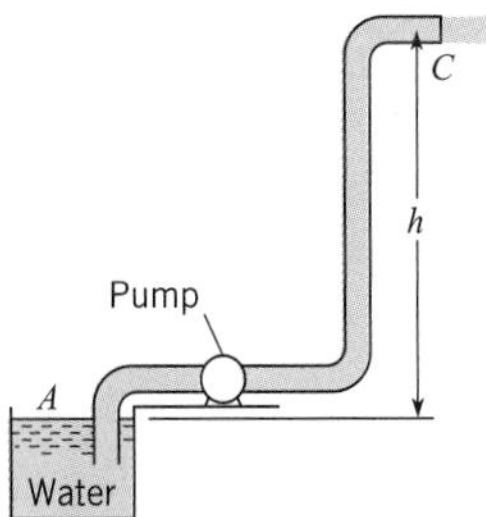

PROBLEMS 7.33, 7.34

7.34 A pump draws water through a 20-cm suction pipe and discharges it through a 10-cm pipe in which the velocity is 3 m/s. The 15-cm pipe discharges horizontally into air at point *C*. To what height *h* above the water surface at *A* can the water be raised if 25 kW is delivered to the pump? Assume that the pump operates at 60% efficiency and that the head loss in the pipe between *A* and *C* is equal to $2V_C^2/2g$. Assume $\alpha = 1.0$ at all locations.

7.35 As shown in the figure, the pump supplies energy to the flow such that the upstream pressure (12-in. pipe) is 5 psi and the downstream pressure (6-in. pipe) is 60 psi when the flow of water is 3.0 cfs. What horsepower is delivered by the pump to the flow? Assume $\alpha = 1.0$ at all locations.

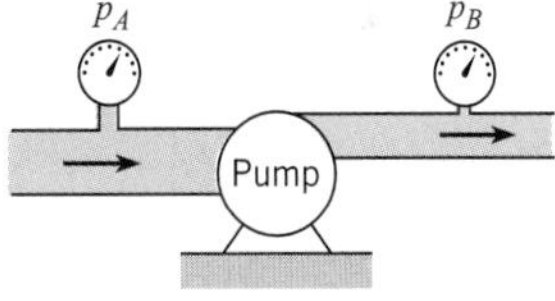

PROBLEM 7.35

7.36 A water discharge of 8 m^3/s is to flow through this horizontal pipe, which is 1 m in diameter. If the head loss is given as $7V^2/2g$ (*V* is velocity in the pipe), how much power will have to be supplied to the flow by the pump to produce this discharge? Assume $\alpha = 1.0$ at all locations.

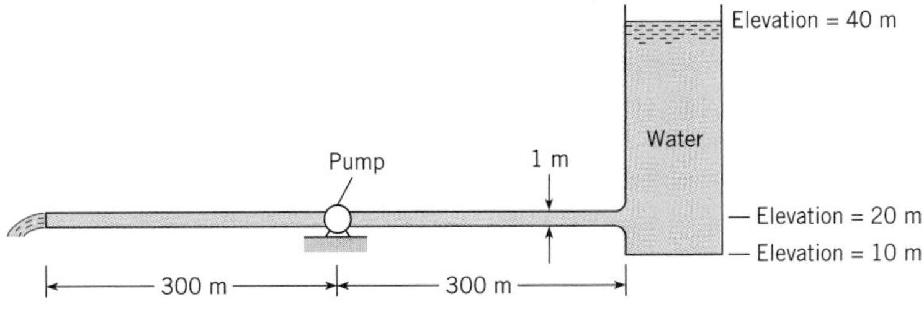

PROBLEM 7.36

7.37 Water is flowing at a rate of $0.25\ \text{m}^3/\text{s}$, and it is assumed that $h_L = 2V^2/2g$ from the reservoir to the gage, where V is the velocity in the 30-cm pipe. What power must the pump supply? Assume $\alpha = 1.0$ at all locations.

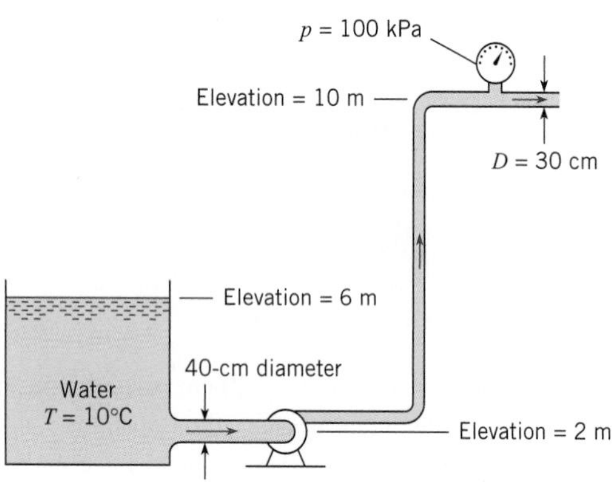

PROBLEM 7.37

7.38 In the pump test shown, the rate of flow is 6 cfs of oil (S = 0.88). Calculate the horsepower that the pump supplies to the oil if there is a differential reading of 46 in. of mercury in the U-tube manometer. Assume $\alpha = 1.0$ at all locations.

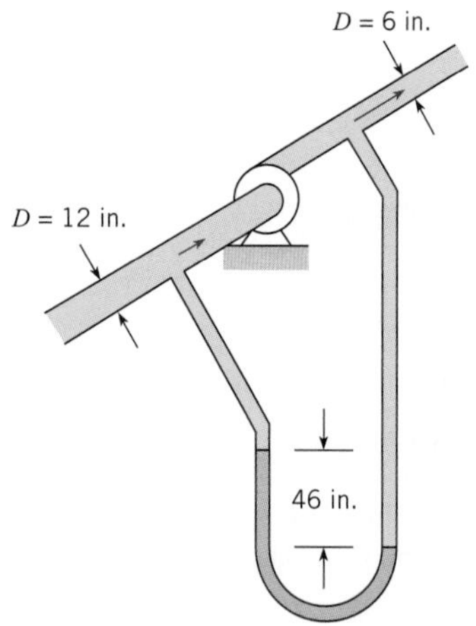

PROBLEM 7.38

7.39 If the discharge is 400 cfs, what power output may be expected from the turbine? Assume that the turbine efficiency is 90% and that the overall head loss is $1.5V^2/2g$, where V is the velocity in the 7-ft penstock Assume $\alpha = 1.0$ at all locations.

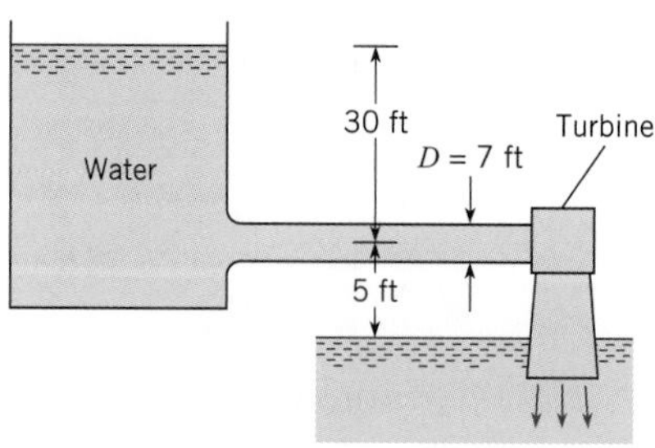

PROBLEM 7.39

7.40 A small-scale hydraulic power system is shown. The elevation difference between the reservoir water surface and the pond water surface downstream of the reservoir, H, is 15 m. The velocity of the water exhausting into the pond is 5 m/s, and the discharge through the system is $1\ \text{m}^3/\text{s}$. The head loss due to friction in the penstock is negligible. Find the power produced by the turbine in kilowatts.

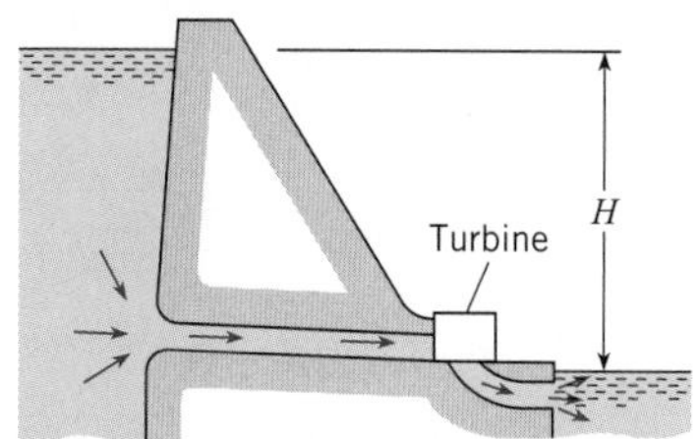

PROBLEMS 7.40, 7.41

7.41 The discharge of water through this turbine is 1000 cfs. What power is generated if the turbine efficiency is 85% and the total head loss is 4 ft? $H = 100$ ft. Also, carefully sketch the energy grade line (EGL) and the hydraulic grade line (HGL).

7.42 Neglecting head losses, determine what power the pump must deliver to produce the flow as shown. Here the elevations at points A, B, C, and D are 40 m, 65 m, 35 m, and 30 m, respectively. The nozzle area is $25\ \text{cm}^2$.

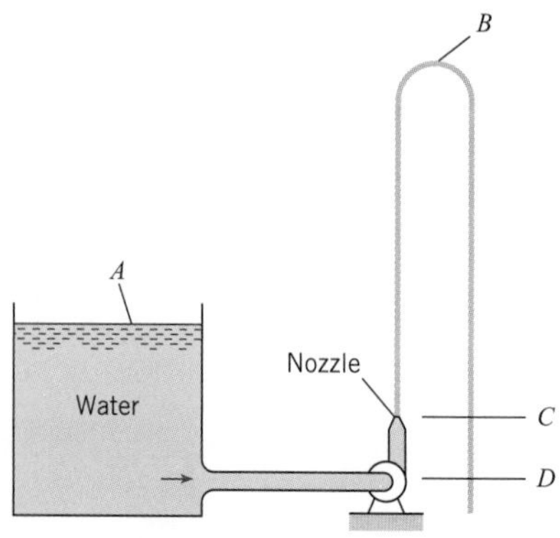

PROBLEMS 7.42, 7.43

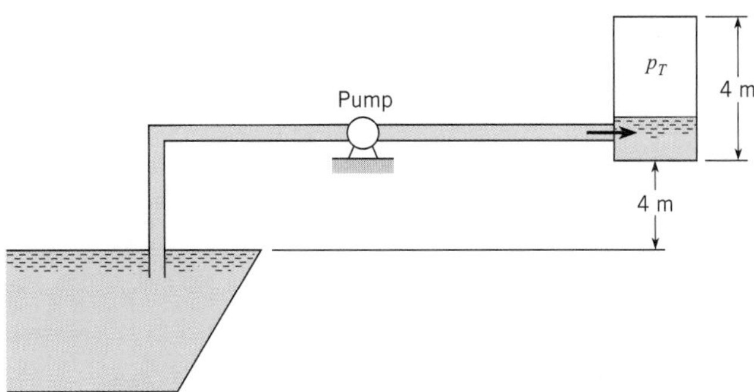

PROBLEM 7.46

7.43 Neglecting head losses, determine what horsepower the pump must deliver to produce the flow as shown. Here the elevations at points A, B, C, and D are 110 ft, 200 ft, 110 ft, and 90 ft, respectively. The nozzle area is 0.10 ft^2.

7.44 A pumping system is to be designed to pump crude oil a distance of one mile in a 1-foot-diameter pipe at a rate of 3500 gpm. The pressures at the entrance and exit of the pipe are atmospheric, and the exit of the pipe is 200 feet higher than the entrance. The pressure loss in the system due to pipe friction is 60 psi. The specific weight of the oil is 53 lbf/ft^3. Find the power, in horsepower, required for the pump.

7.45 A pump is used to transfer SAE 30 oil from tank A to tank B as shown. The tanks have a diameter of 12 meters. The initial depth of the oil in tank A is 20 m and in tank B the depth is one meter. The pump delivers a constant head of 60 m. The connecting pipe has a diameter of 20 cm and the head loss due to friction in the pipe is $20V^2/2g$. Find the time required to transfer the oil from tank A to B; that is, the time required to fill tank B to 20 m depth.

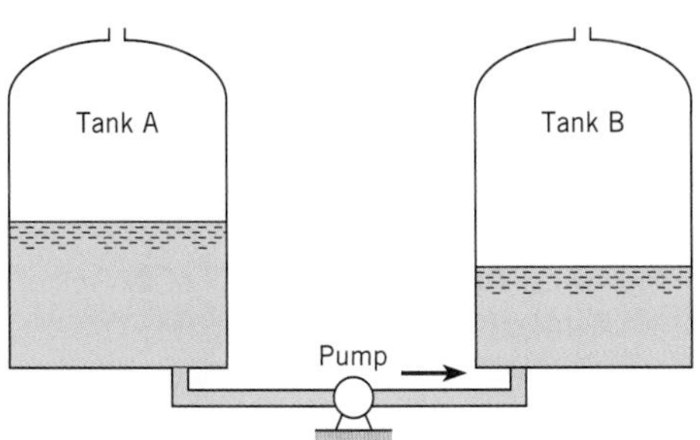

PROBLEM 7.45

7.46 A pump is used to pressurize a tank to 300 kPa, abs. The tank has a diameter of 2 m and a height of 4 m. The initial level of water in the tank is 1 m and the pressure at the water surface is 0 kPa, gage. The atmospheric pressure is 100 kPa. The pump operates with a constant head of 50 m. The water is drawn from a source that is 4 m below the tank bottom. The pipe connecting the source and the tank is 4 cm in diameter and the head loss, including the expansion loss at the tank, is $10V^2/2g$. The flow is turbulent.

Assume the compression of the air in the tank takes place isothermally, so the tank pressure is given by

$$p_T = \frac{3}{4 - z_t} p_0$$

where z_t is the depth of fluid in the tank in meters. Write a computer program that will show how the pressure varies in the tank with time, and find the time to pressurize the tank to 300 kPa, abs.

7.47 Water is draining from tank A to tank B. The elevation difference between the two tanks is 10 m. The pipe connecting the two tanks has a sudden-expansion section as shown. The cross-sectional area of the pipe from A is 10 cm^2, and the area of the pipe into B is 20 cm^2. Assume the head loss in the system consists only of that due to the sudden-expansion section and the loss due to flow into tank B. Find the discharge between the two tanks.

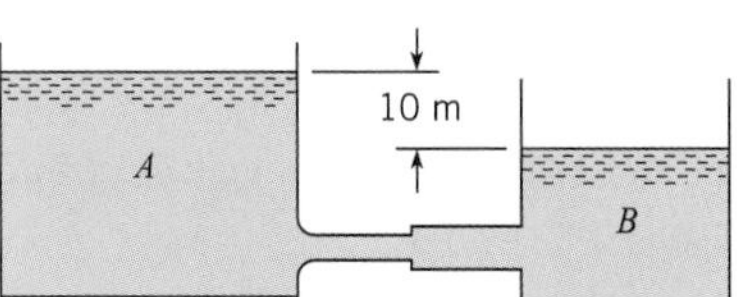

PROBLEM 7.47

7.48 A 40-cm pipe abruptly expands to a 60-cm size. These pipes are horizontal, and the discharge of water from the smaller size to the larger is 1.0 m^3/s. What horizontal force is required to hold the transition in place if the pressure in the 40-cm pipe is 70 kPa gage? Also, what is the head loss? Assume $\alpha = 1.0$ at all locations.

7.49 An 8-cm pipe carries water with a mean velocity of 4 m/s. If this pipe abruptly expands to a 15-cm pipe, what will be the head loss due to the abrupt expansion?

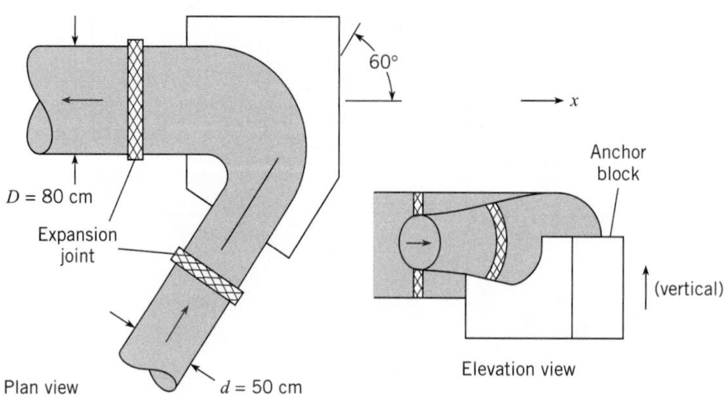

PROBLEM 7.53

7.50 A 6-in. pipe abruptly expands to a 12-in. size. If the discharge of water in the pipes is 5 cfs, what is the head loss due to abrupt expansion?

FLUID SOLUTIONS **7.51** This abrupt expansion is to be used to dissipate the high-energy flow of water in the 5-ft-diameter penstock. Assume $\alpha = 1.0$ at all locations.

a. What power (in horsepower) is lost through the expansion?

b. If the pressure at section 1 is 5 psig, what is the pressure at section 2?

c. What force is needed to hold the expansion in place?

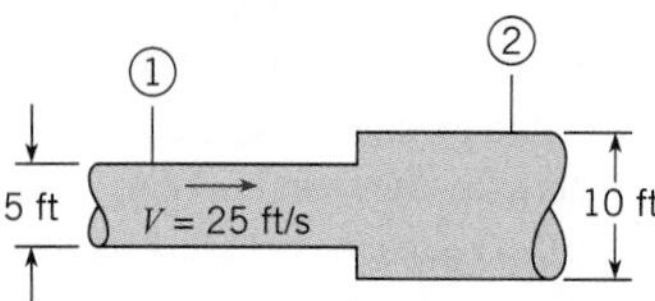

PROBLEM 7.51

7.52 This rough aluminum pipe is 6 in. in diameter. It weighs 1.5 lb per foot of length, and the length L is 50 ft. If the discharge of water is 6 cfs and the head loss due to friction from section 1 to the end of the pipe is 10 ft, what is the longitudinal force transmitted across section 1 through the pipe wall?

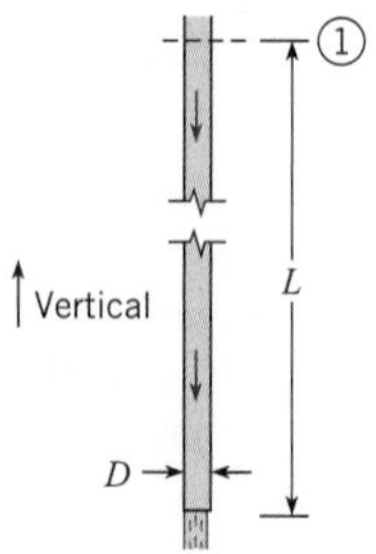

PROBLEM 7.52

7.53 Water flows in this bend at a rate of 5 m^3/s, and the pressure at the inlet is 650 kPa. If the head loss in the bend is 10 m, what will the pressure be at the outlet of the bend? Also estimate the force of the anchor block on the bend in the x direction required to hold the bend in place. Assume $\alpha = 1.0$ at all locations.

7.54 What is the head loss at the outlet of the pipe that discharges water into the reservoir at a rate of 10 cfs if the diameter of the pipe is 12 in.?

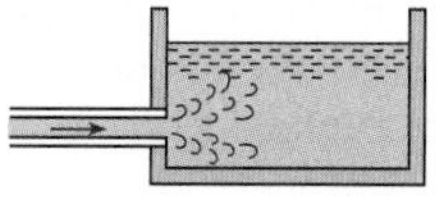

PROBLEMS 7.54, 7.55

7.55 What is the head loss at the outlet of the pipe that discharges water into the reservoir at a rate of 0.5 m^3/s if the diameter of the pipe is 50 cm?

7.56 The pipe diameter D is 30 cm, d is 15 cm, and the atmospheric pressure is 100 kPa. What is the maximum allowable discharge before cavitation occurs at the throat of the venturi meter if $H = 5$ m? Assume $\alpha = 1.0$ at all locations.

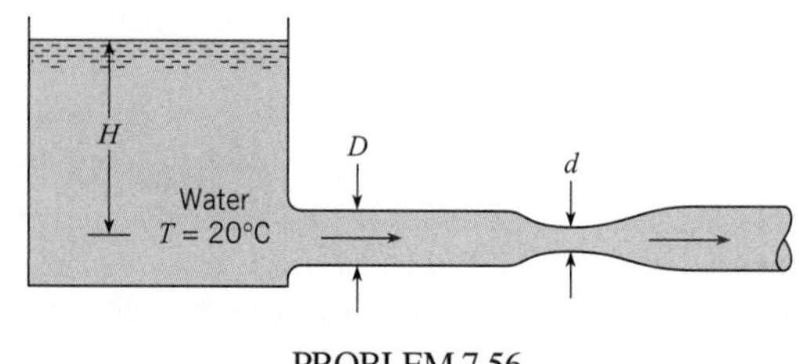

PROBLEM 7.56

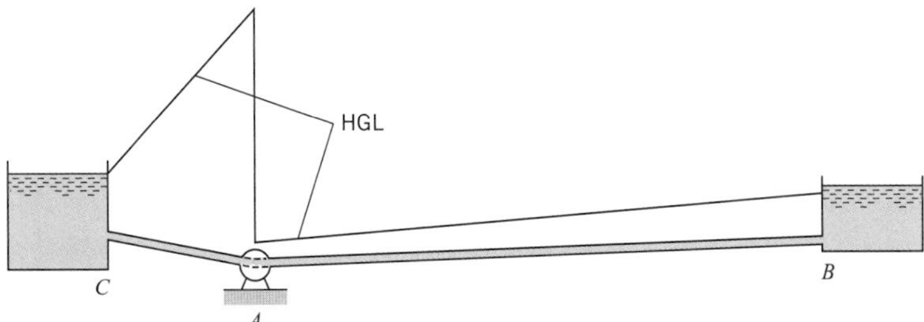

PROBLEM 7.58

7.57 In this system $d = 25$ cm, $D = 40$ cm, and the head loss from the venturi meter to the end of the pipe is given by $h_L = 0.9V^2/2g$, where V is the velocity in the pipe. Neglecting all other head losses, determine what head H will first initiate cavitation if the atmospheric pressure is 100 kPa absolute. What will be the discharge at incipient cavitation? Assume $\alpha = 1.0$ at all locations.

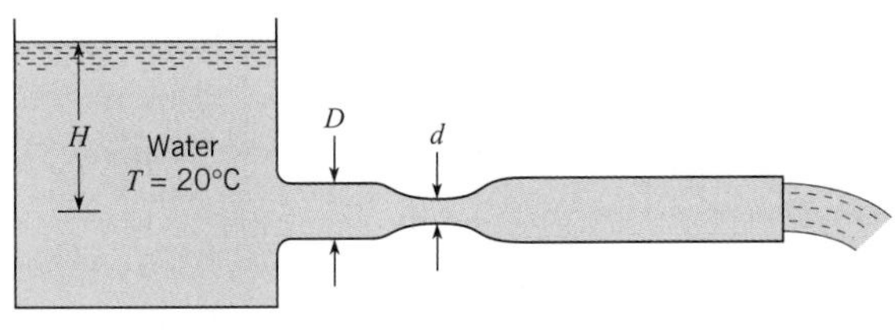

PROBLEM 7.57

7.58 For the system shown in the figure,

a. What is the flow direction?
b. What kind of machine is at A?
c. Do you think both pipes, AB and CA, are the same diameter?
d. Sketch in the EGL for the system.
e. Is there a vacuum at any point or region of the pipes? If so, identify the location.

7.59 Water flows from the reservoir through a pipe and then discharges from a nozzle as shown. The head loss in the pipe itself is given as $h_L = 0.025(L/D)(V^2/2g)$, where L and D are the length and diameter of the pipe and V is the velocity in the pipe. What is the discharge of water? Also draw the HGL and EGL for the system. Assume $\alpha = 1.0$ at all locations.

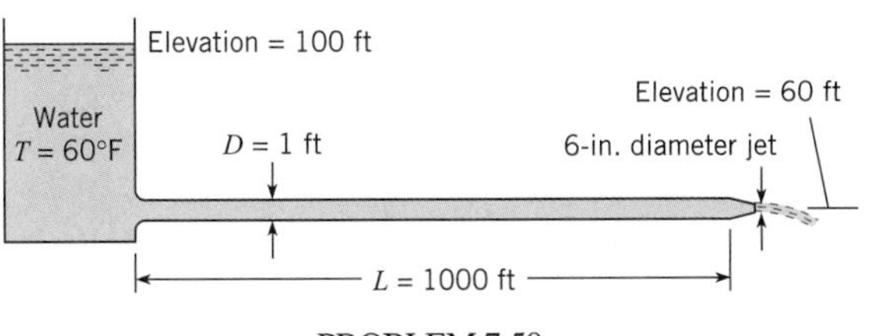

PROBLEM 7.59

7.60 Sketch the HGL and the EGL for the system of Example 7.5.

7.61 Carefully sketch the HGL and the EGL for the flow system of Prob. 7.57.

7.62 Sketch the HGL and the EGL for the reservoir and pipe of Example 7.3.

7.63 The energy grade line for steady flow in a uniform-diameter pipe is shown. Which of the following could be in the "black box"? (a) a pump, (b) a partially closed valve, (c) an abrupt expansion, (d) a turbine. Choose valid answer(s).

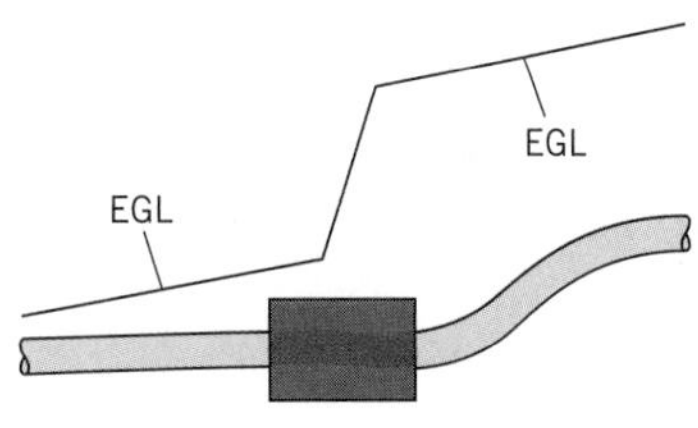

PROBLEM 7.63

7.64 If the pipe shown has constant diameter, is this type of HGL possible? If so, under what additional conditions? If not, why not?

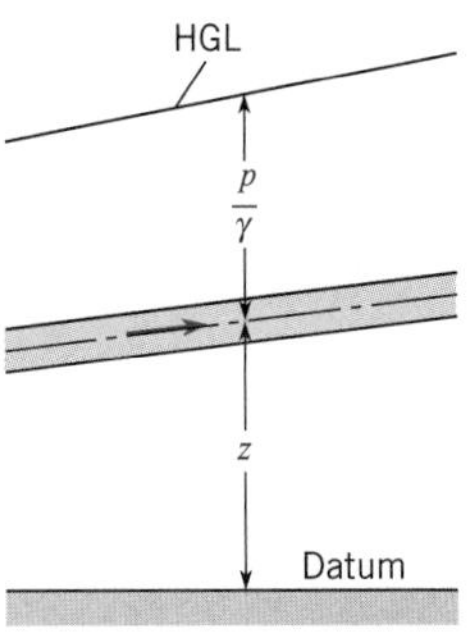

PROBLEM 7.64

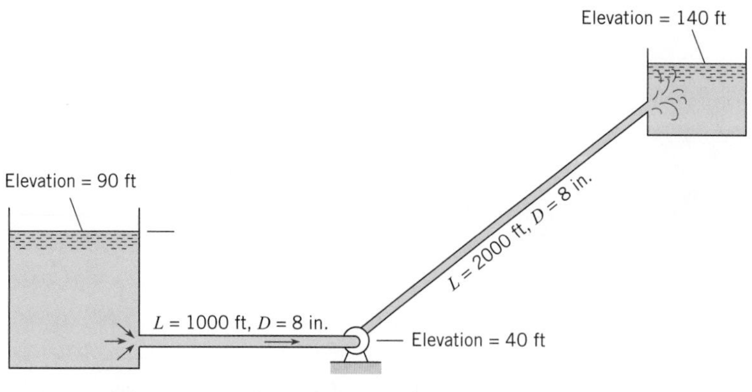

PROBLEM 7.68

7.65 Sketch the HGL and the EGL for this conduit, which tapers uniformly from the left end to the right end.

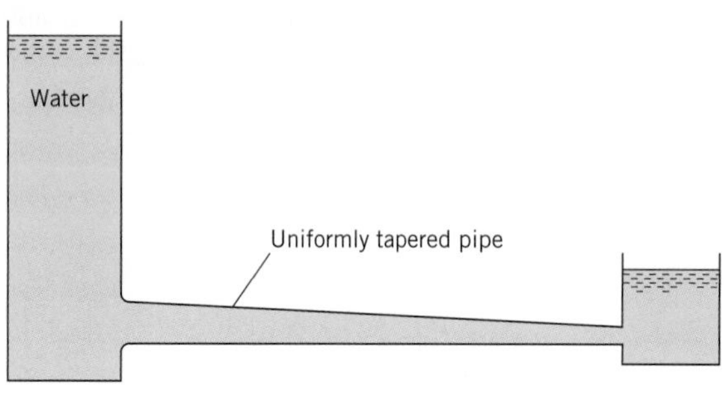

PROBLEM 7.65

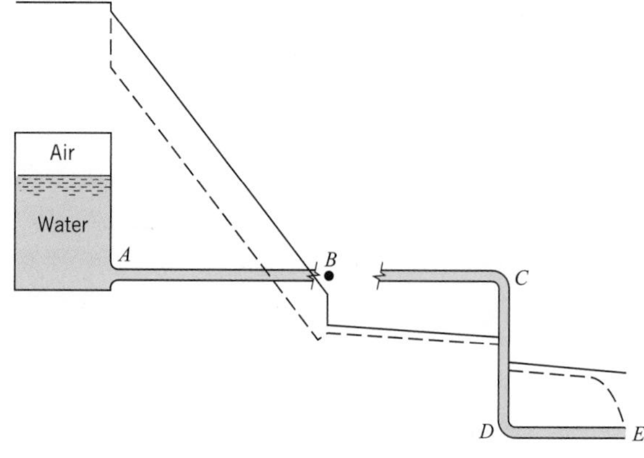

PROBLEM 7.66

7.66 The HGL and the EGL for a pipeline are shown in the figure.

a. Indicate which is the HGL and which is the EGL.
b. Are all pipes the same size? If not, which is the smallest?
c. Is there any region in the pipes where the pressure is below atmospheric pressure? If so, where?
d. Where is the point of maximum pressure in the system?
e. Where is the point of minimum pressure in the system?
f. What do you think is located at the end of the pipe at point *E*?
g. Is the pressure in the air in the tank above or below atmospheric pressure?
h. What do you think is located at point *B*?

FLUID SOLUTIONS **7.67** Refer to Fig. 7.8. Assume that the head loss in the pipes is given by $h_L = 0.02(L/D)(V^2/2g)$, where *V* is the mean velocity in the pipe, *D* is the pipe diameter, and *L* is the pipe length. The water surface elevations of the upper and lower reservoirs are 100 m and 70 m, respectively. The respective dimensions for upstream and downstream pipes are D_u = 30 cm, L_u = 200 m, and D_d = 15 cm, L_d = 100 m. Determine the discharge of water in the system.

7.68 What horsepower must be supplied to the water to pump 3.0 cfs at 68°F from the lower to the upper reservoir? Assume that the head loss in the pipes is given by $h_L = 0.018(L/D)(V^2/2g)$, where *L* is the length of the pipe in feet and *D* is the pipe diameter in feet. Sketch the HGL and the EGL.

7.69 Water flows from reservoir *A* to reservoir *B*. The water temperature in the system is 10°C, the pipe diameter *D* is 1 m, and the pipe length *L* is 300 m. If *H* = 16 m, *h* = 2 m, and the pipe head loss is given by $h_L = 0.01(L/D)(V^2/2g)$, where *V* is the velocity in the pipe, what will be the discharge in the pipe? In your solution, include the head loss at the pipe outlet, and also sketch the HGL and the EGL. What will be the pressure at point *P* halfway between the two reservoirs? Assume $\alpha = 1.0$ at all locations.

7.70 Water flows from the reservoir on the left to the reservoir on the right at a rate of 16 cfs. The formula for the head losses in the pipes is $h_L = 0.02(L/D)(V^2/2g)$. What elevation in the

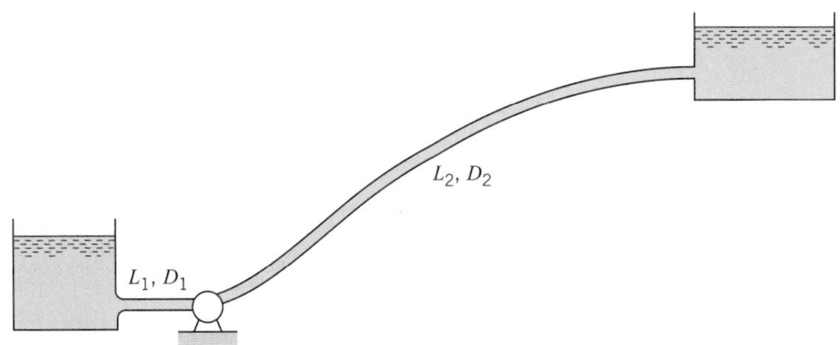

PROBLEM 7.71

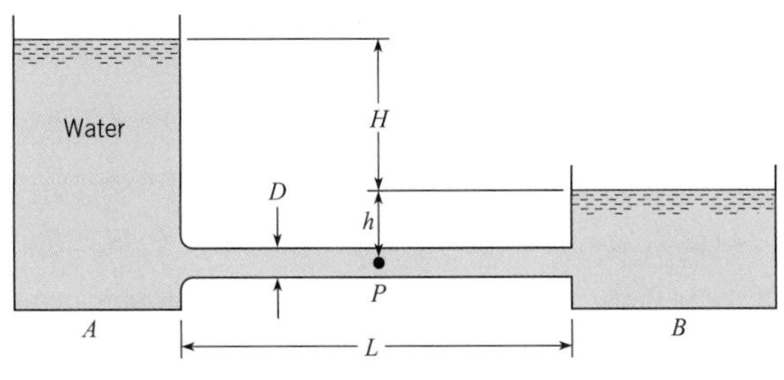

PROBLEM 7.69

left reservoir is required to produce this flow? Also carefully sketch the HGL and the EGL for the system. *Note:* Assume the head-loss formula can be used for the smaller pipe and also for the larger pipe. Assume $\alpha = 1.0$ at all locations.

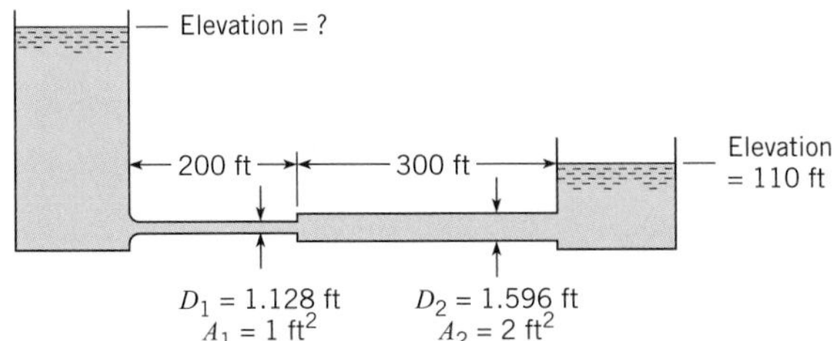

PROBLEM 7.70

7.71 What power is required to pump water at a rate of 3 m^3/s from the lower to the upper reservoir? Assume the pipe head loss is given by $h_L = 0.018(L/D)(V^2/2g)$, where L is the length of pipe, D is the pipe diameter, and V is the velocity in the pipe. The water temperature is 10°C, the water surface elevation in the lower reservoir is 150 m, and the surface elevation in the upper reservoir is 250 m. The pump elevation is 100 m, $L_1 = 100$ m, $L_2 = 1000$ m, $D_1 = 1$ m, and $D_2 = 50$ cm. Assume the pump and motor efficiency is 74%. In your solution, include the head loss at the pipe outlet and sketch the HGL and the EGL. Assume $\alpha = 1.0$ at all locations.

7.72 Refer to Fig. 7.9. Assume that the head loss in the pipe is given by $h_L = 0.02(L/D)(V^2/2g)$, where V is the mean velocity in the pipe, D is the pipe diameter, and L is the pipe length. The elevations of the reservoir water surface, the highest point in the pipe, and the pipe outlet are 200 m, 200 m, and 185 m, respectively. The pipe diameter is 30 cm, and the pipe length is 200 m. Determine the water discharge in the pipe, and, assuming that the highest point in the pipe is halfway along the pipe, determine the pressure in the pipe at that point. Assume $\alpha = 1.0$ at all locations.

7.73 A pump is used to fill a tank 5 m in diameter from a river as shown. The water surface in the river is 2 m below the bottom of the tank. The pipe diameter is 5 cm, and the head loss in the pipe is given by $h_L = 10V^2/2g$, where V is the mean velocity in the pipe. The flow in the pipe is turbulent, so $\alpha = 1$. The head provided by the pump varies with discharge through the pump as $h_p = 20 - 4 \times 10^4 Q^2$, where the discharge is given in cubic meters per second (m^3/s) and h_p is in meters. How long will it take to fill the tank to a depth of 10 m? FLUID SOLUTIONS

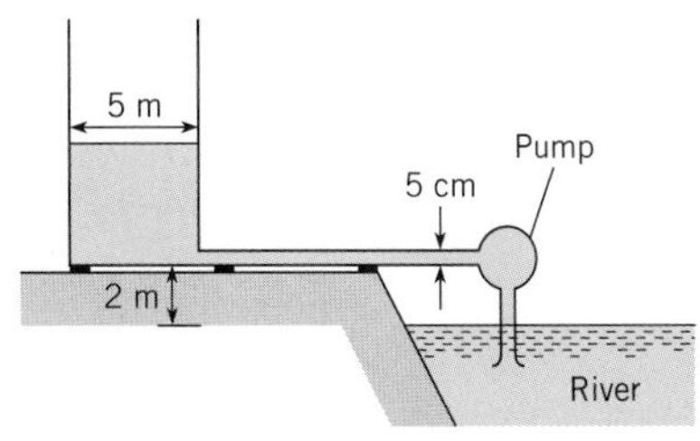

PROBLEM 7.73

7.74 The HGL and the EGL are as shown for a certain flow system.

a. Is flow from A to E or from E to A?
b. Does it appear that a reservoir exists in the system?
c. Does the pipe at E have a uniform or a variable diameter?
d. Is there a pump in the system?
e. Sketch the physical setup that could yield the conditions shown between C and D.
f. Is anything else revealed by the sketch?

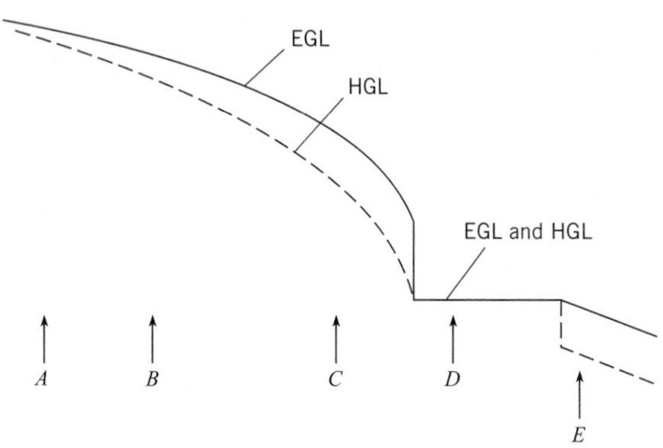

PROBLEM 7.74

FLUID SOLUTIONS **7.75** Assume that the head loss in the pipe is given by $h_L = 0.014(L/D)(V^2/2g)$, where L is the length of pipe in feet and D is the pipe diameter in feet. Assume $\alpha = 1.0$ at all locations.

a. Determine the discharge of water through this system.
b. Draw the HGL and the EGL for the system.
c. Locate the point of maximum pressure.
d. Locate the point of minimum pressure.
e. Calculate the maximum and minimum pressures in the system.

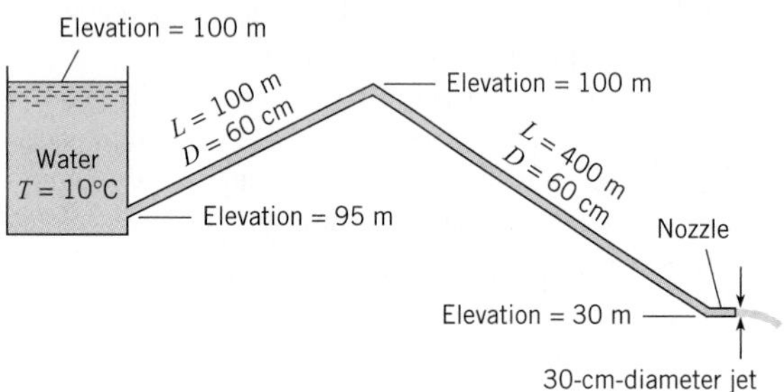

PROBLEM 7.75

7.76 In Prob. 6.66, what power is developed by the windmill? Assume $\alpha = 1.0$ at all locations.

7.77 An engineer is designing a subsonic wind tunnel. The test section is to have a cross-sectional area of 4 m^2 and an airspeed of 60 m/s. The air density is 1.2 kg/m^3. The area of the tunnel exit is 10 m^2. The head loss through the tunnel is given by $h_L = (0.025)(V_T^2/2g)$, where V_T is the airspeed in the test section. Calculate the power needed to operate the wind tunnel. *Hint:* Assume negligible energy loss for the flow approaching the tunnel in region A, and assume atmospheric pressure at the outlet section of the tunnel. Assume $\alpha = 1.0$ at all locations.

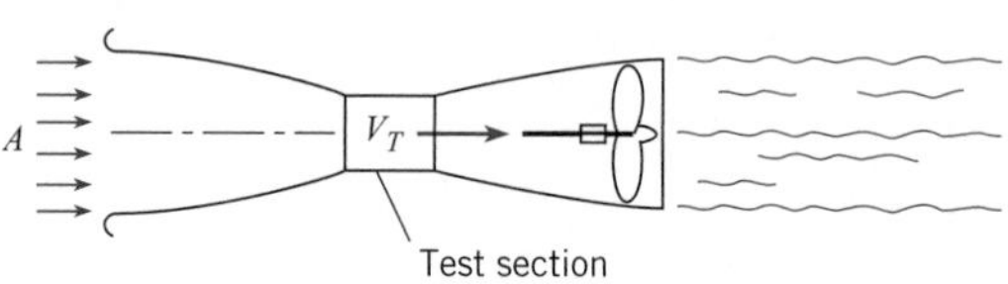

PROBLEM 7.77

7.78 Fluid flowing along a pipe of diameter D accelerates around a disk of diameter d as shown in the figure. The velocity far upstream of the disk is U, and the fluid density is ρ. Assuming incompressible flow and that the pressure downstream of the disk is the same as that at the plane of separation, develop an expression for the force required to hold the disk in place in terms of U, D, d, and ρ. Using the expression you developed, determine the force when $U = 10$ m/s, $D = 5$ cm, $d = 4$ cm, and $\rho = 1.2$ kg/m^3. Assume $\alpha = 1.0$ at all locations.

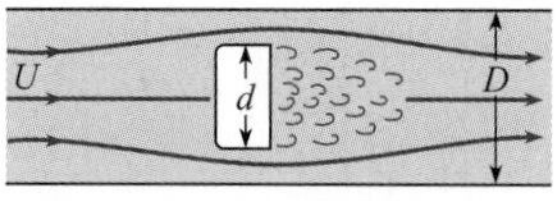

PROBLEM 7.78

Smoke traces emitted from small ports of this model office building allow engineers to evaluate the air circulation in the vicinity of the structure. The building and adjacent structures are mounted on a turntable on the floor of the wind tunnel so that the effects of different wind directions can be studied. (Courtesy CALSPAN.)

CHAPTER 8

Dimensional Analysis and Similitude

The solutions of most engineering problems involving fluid mechanics rely on data acquired by experimental means. In many cases the empirical data are general enough that engineers have need for them in their normal design practice. Consequently, they are made available by publication in journals and textbooks. Examples of such data are the resistance coefficients for pipes and the drag coefficients for blunt bodies.* For many problems, however, either the geometry of the structure that guides the flow or the flow conditions themselves are so unique that special tests on a replica of the structure at a different scale are required to predict the flow patterns and pressure variation. When such tests are performed, the replica of the structure on which the tests are made is called the *model,* and the full-scale structure employed in the actual engineering design is called the *prototype.* The model is usually made much smaller than the prototype for economic reasons.

8.1 The Need for Dimensional Analysis

Fluid mechanics is more heavily involved with empirical work than is structural engineering, machine design, or electrical engineering because the analytical tools presently available are not capable of yielding exact solutions to many of the problems in fluid mechanics. It is true that exact solutions are obtainable for all hydrostatic problems and for many laminar-flow problems. However, the most general equations solved on the largest computers yield only fair approximations for turbulent-flow problems—hence the need for experimental evaluation and verification.

For analyzing model studies and for correlating the results of experimental research, it is essential that researchers employ *dimensionless parameters.* To appreciate the advantages of using dimensionless parameters, let us consider the flow of water through the unusual orifice illustrated in Fig. 8.1. Actually, this is much like a flow nozzle, which will be presented in Chapter 13, except that the flow is in the opposite direction to that in a nozzle operating in normal fashion. It is true that an orifice operating in such a manner will have a much different performance than a flow nozzle. However, it is not unlikely that a firm or city water department might have such a situation where the flow may occur the "right way" most of the time and the "wrong way" part of the time—

*These coefficients will be presented in Chapters 10 and 11.

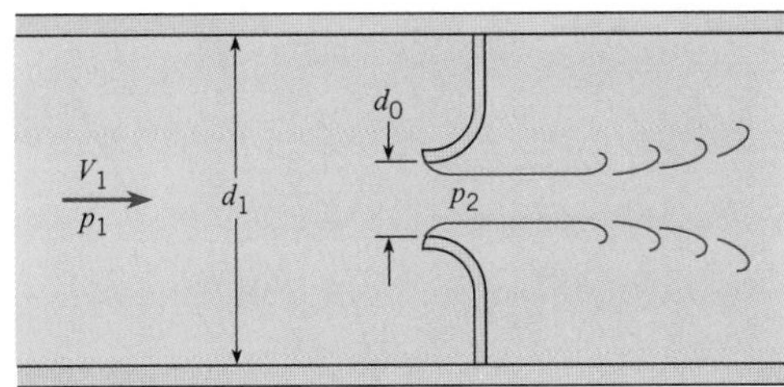

FIGURE 8.1

Flow through inverted flow nozzle.

hence the need for such knowledge. The test procedure involves testing several orifices, each with a different throat diameter d_0. We want to measure the pressure difference $p_1 - p_2$ as a function of the velocity V_1, density ρ, and diameter d_0. We think that by carrying out numerous measurements at different values of V_1, d_0, and ρ, we can plot our data as shown in Fig. 8.2*a*. Soon, however, we realize that our first plan will involve a terrific amount of work, so we look for a better scheme. We then think that the Bernoulli equation must have relevance here in a manner similar to that shown in Chapter 4. It was noted in Chapter 4 that the conditions for application of the Bernoulli equation are closely approximated if the fluid has fairly low viscosity, as does water, if the streamlines converge in the direction of flow, and if the flow is steady. These conditions prevail between sections 1 and 2 in our problem. In Section 4.5 we knew by the character of the flow passages how V_1 and V_2 were related. Thus we could write the Bernoulli equation between two points in the flow field and solve for Δp directly. However, in the orifice in Fig. 8.1, it should be expected that separation will occur downstream of the throat. Thus the smallest flow section is smaller than the throat section. Therefore, it is necessary to determine experimentally the relationship between Δp and the other variables.

We use a general form of the Bernoulli equation, much as we did in Chapter 4, but we insert coefficients for unknown relations. In other words, we write the Bernoulli equation as

$$p_1 + \rho\frac{V_1^2}{2} = p_2 + \rho\frac{V_2^2}{2}$$

or, in dimensionless form, as

$$\frac{p_1 - p_2}{\rho V_1^2/2} = \frac{V_2^2}{V_1^2} - 1 \tag{8.1}$$

This dimensionless form for pressure was identified in Chapter 4 as the pressure coefficient. In Eq. (8.1) we know that $V_2/V_1 = A_1/A_2$. However, we do not know the value of A_2, and we are most interested in relating the velocity ratio to the ratio of A_1 to the orifice area A_0. Therefore, let us simply express A_1/A_0 in functional form: $V_2/V_1 = f(A_1/A_0) = f_1(d_1/d_0)^2$. Then Eq. (8.1) can be rewritten as

$$C_p = \frac{p_1 - p_2}{\rho V_1^2/2} = [f_1(d_1/d_0)^2]^2 - 1 \tag{8.2}$$

Since the right-hand side is solely a function of d_1/d_0, we can write

$$C_p = f_2\left(\frac{d_1}{d_0}\right) \tag{8.3}$$

FIGURE 8.2

Relations of pressure, velocity, and diameter.

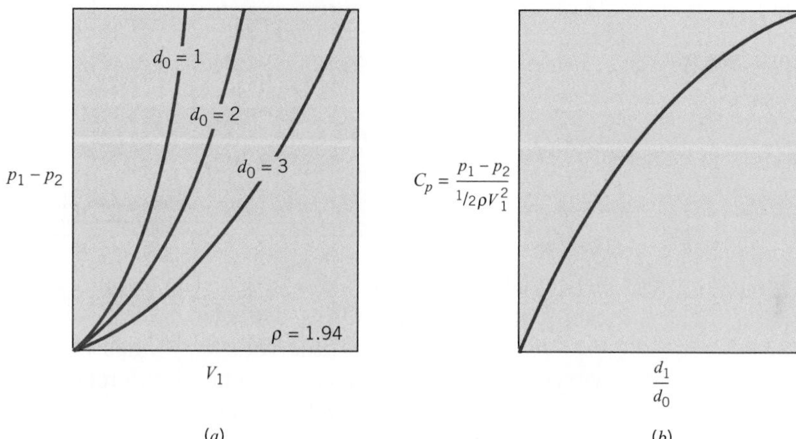

If we plot the pressure coefficient C_p (instead of the pressure difference) versus d_1/d_0, we will get a single curve, as shown in Fig. 8.2*b*. Inspection will convince the reader that everything given in Fig. 8.2*a* is included in Fig. 8.2*b*.

Thus by considering a nondimensional form of Bernoulli's equation, we have made a tremendous reduction in the experimental work that would have been required had we not considered the nondimensional form. The process of nondimensionalizing the equation reduced the correlating parameters from five (Δp, ρ, V, d_1, d_0) to two $[\Delta p/(\rho V^2/2), (d_1/d_0)]$. To collect the data for Fig. 8.2*a* would require, say, 20 data points for each d_0 value. If five d_0's were tested, then 100 data points would be needed. On the other hand, only about 20 would be required to yield the curve in Fig. 8.2*b*, so a considerable amount of time is saved.

In the foregoing development we had a clue about the governing equation. By considering a dimensionless form of that equation, we were able to obtain a set of dimensionless parameters with which to correlate our data. In many cases, however, the governing equation is not available, and we must seek the dimensionless parameters by using a formal procedure called *dimensional analysis.* Sections 8.2 and 8.3 present some basic material leading up to dimensional analysis, and Section 8.4 contains an explanation of the actual procedure of dimensional analysis.

Dimensions and Equations

All variables used in science or engineering are expressed in terms of a limited number of basic dimensions. For most engineering problems the basic dimensions are mass, length, and time.

The units of all other dimensional variables can be expressed in terms of these units. For example, from Newton's second law, force is equal to mass times acceleration, so the units for force can be expressed in terms of M, L, and T as

$$[F] = M\frac{L}{T^2} \tag{8.4}$$

Here the brackets mean "dimension of." Therefore, Eq. (8.4) reads as follows: "The dimensions of F are mass times length per time squared." Another example is viscosity, which has the units of $\mathrm{N \cdot s/m^2}$, so the dimensions of viscosity are

$$[\mu] = \frac{FT}{L^2} = \frac{M}{LT} \tag{8.5}$$

Pressure, by definition, is force per unit area, so the dimensions of pressure can be expressed as

$$[p] = \left[\frac{F}{A}\right] = \frac{M}{LT^2} \tag{8.6}$$

It goes without saying that all equations must balance in magnitude. However, all rational equations (those developed from basic laws of physics) must also be dimensionally homogeneous. That is, the left-hand side of the equation must have the same dimensions as the right-hand side. Moreover, every term in the equation must have the same dimensions.

8.3 The Buckingham Π Theorem

In 1915 Buckingham (1) showed that the number of independent dimensionless groups of variables (dimensionless parameters) needed to correlate the variables in a given process is equal to $n - m$, where n is the number of variables involved and m is the number of basic dimensions included in the variables.

Buckingham referred to the dimensionless parameters as Π, which is the reason the theorem is called the "Π theorem." We will refer to the dimensionless parameters as "π-groups." If the equation describing a physical system has n dimensional variables and is expressed as

$$y_1 = f(y_2, y_3, \ldots y_n)$$

then it can be rearranged and expressed in terms of $n - m$ dimensionless parameters (π-groups) as

$$\pi_1 = \varphi(\pi_2, \pi_3, \ldots \pi_{n-m})$$

Thus if the drag force F of a fluid flowing past a sphere is known to be a function of the velocity V, mass density ρ, viscosity μ, and diameter D, then five variables (F, V, ρ, μ, and D) and three basic dimensions (L, M, and T) are involved. We will have $5 - 3 = 2$ basic groupings of variables that can be used to correlate experimental results in the form

$$\pi_1 = \varphi(\pi_2)$$

Note that the same reduction in correlating parameters occurred in our discussion concerning the reverse-flow nozzle in Section 8.1. In the next section we will show in a step-by-step process how variables can be combined to form the dimensionless parameters.

Dimensional Analysis

The Step-by-Step Method

Several methods may be used to carry out the process of finding the π-groups, but the step-by-step approach, very clearly presented by Ipsen (2), is one of the easiest and reveals much about the process. The basic objective in dimensional analysis, as already noted, is to reduce the number of separate variables involved in a problem to a smaller number of independent dimensionless groups of variables (π-groups). The reason for this is that all rational equations can be nondimensionalized, as illustrated in Section 8.1, and that all rational equations have a certain number of independent terms in them. Thus, by the procedure of dimensional analysis, we will be simply arranging variables into a dimensionless equation. In the process of combining the variables to form dimensionless groups or π-terms, the number of independent groups thus obtained is less than the number of original variables.

We start the process by identifying only those variables that are significant to the problem. Then we include these variables in a functional equation and note their dimensions. As an example, let us consider a simple problem we are already familiar with so that we can get a feel for the process.

example 8.1

Suppose that we want to consider the velocity of a falling body in a vacuum, and we know that this velocity V is a function of the acceleration due to gravity, g, and the distance through which the body falls, h. However, let us say that we do not know how to derive the equation for the fall velocity but want to obtain the proper relationship by dimensional analysis and experimentation. Thus in this example we want to determine the dimensionless parameter(s) that apply.

Solution We include the significant variables in a functional equation:

$$V = f(g, h)$$

where

$$[V] = L/T$$

$$[g] = L/T^2$$

$$[h] = L$$

The goal now is to combine the dimensional variables in such a way as to eliminate the basic dimensions and come up with π-groups. One approach is to form a table as shown below.

Variable	[ ]	Variable	[ ]	Variable	[ ]
V	$\frac{L}{T}$	$\frac{V}{h}$	$\frac{1}{T}$	$\frac{V}{\sqrt{gh}}$	0
g	$\frac{L}{T^2}$	$\frac{g}{h}$	$\frac{1}{T^2}$		
h	L				

The variables are listed in the first column with their corresponding dimensions. We note that h has the dimensions of length, so we can use h to eliminate the length dimension in the other two variables. Dividing both V and g by h gives the variables in the second column with their corresponding dimensions. Now we note that the square root of g/h has the dimensions of $1/T$ so we can divide V/h by $\sqrt{g/h}$ to obtain the dimensionless group in the third column. The zero dimension indicates a dimensionless parameter.

The functional equation can now be written in dimensionless form as

$$\frac{V}{\sqrt{gh}} = C \qquad \triangleleft$$

Of course, our experience in basic physics tells us that $C = \sqrt{2}$.

In the foregoing illustration of the step-by-step method of dimensional analysis, it should be noted that the remaining single dimensionless parameter was consistent with the number of π-groups, $n - m$, that the Buckingham Π theorem predicts. That is, we had three variables $(n = 3)$ and two dimensions [L and T $(m = 2)$]. Hence we should have one $(3 - 2)$ π-group, and we do. In fact, the step-by-step method of dimensional analysis will always yield the correct number of parameters.*

We will take another example that is somewhat more involved than the first. However, the procedure is the same as before.

example 8.2

If the drag F_D of a sphere in a fluid flowing past the sphere is a function of the viscosity μ, the mass density ρ, the velocity of flow V, and the diameter of the sphere D, what dimensionless parameters are applicable to the flow process?

*Note that, in rare instances, the number of parameters may be one more than predicted by the Buckingham Π theorem. This anomaly can occur because it is possible that two-dimensional categories can be eliminated when dividing (or multiplying) by a given variable. See Ipsen (2, page 172) for an example of this.

Solution

$$F_D = f(V, \rho, \mu, D) \tag{8.7}$$

In this problem we have five dimensional parameters, namely F_D, V, ρ, μ, and D. From the Buckingham Π theorem there should be two π-groups. We approach this problem in the same way as used in Example 8.1, by using a table and eliminating the basic dimensions step-by-step.

Variable	[ ]	Variable	[ ]	Variable	[ ]	Variable	[]
F_D	$\frac{ML}{T^2}$	$\frac{F_D}{D}$	$\frac{M}{T^2}$	$\frac{F_D}{\rho D^4}$	$\frac{1}{T^2}$	$\frac{F_D}{\rho V^2 D^2}$	0
V	$\frac{L}{T}$	$\frac{V}{D}$	$\frac{1}{T}$	$\frac{V}{D}$	$\frac{1}{T}$		
ρ	$\frac{M}{L^3}$	ρD^3	M				
μ	$\frac{M}{LT}$	μD	$\frac{M}{T}$	$\frac{\mu}{\rho D^2}$	$\frac{1}{T}$	$\frac{\mu}{\rho VD}$	0
D	L						

We can start by using the sphere diameter D to eliminate the length dimension in the other variables. For example, dividing the force by the diameter removes the length dimension. Also, multiplying the density by the diameter cubed removes the length dimension. The resulting variables are shown in the second column with their corresponding dimensions. We can use ρD^3 to eliminate the mass dimension in the other variables, and now only three variables remain with dimensions of time only. The result is shown in the third column. The V/D grouping in the second row did not change because there was no mass dimension in this term. Finally, the V/D grouping is used to eliminate the time dimension, resulting in the fourth column. There are two π-groups remaining, which leads to an equation in the form

$$\frac{F_D}{\rho V^2 D^2} = f\left(\frac{\mu}{\rho VD}\right) \tag{8.8}$$

Although Eq. (8.8) still has the same variables as Eq. (8.7), the number of terms has been reduced from five to two. This makes the correlation of data much simpler.

The actual functional relationship between the parameters is determined by running a series of tests to observe values of $F_D/\rho V^2 D^2$ for different values of the ratio $\mu/\rho VD$. The results of these tests are plotted to yield a usable functional relationship between the two parameters. Since the relationship is experimentally determined, it is just as valid to express Eq. (8.8) with the right-hand side inverted:

$$\frac{F_D}{\rho V^2 D^2} = g\left(\frac{VD\rho}{\mu}\right) \tag{8.9}$$

Now the combination on the right is the Reynolds number, which was introduced in Chapter 4. Hence the parameter $F_D/\rho V^2D^2$ is a function of the Reynolds number, $VD\rho/\mu$.

If, in Example 8.2, we had used the grouping $\mu/\rho D^2$ to eliminate the time dimension, the two π-groups would have been $F_D\rho/\mu^2$ and $\rho VD/\mu$, so the functional relationship would be

$$\frac{F_D\rho}{\mu^2} = \phi\left(\frac{\rho VD}{\mu}\right)$$

where ϕ represents the function. This is also a valid result. However, we can always form another π-group by combining two π-groups. Thus if we formed a new dimensionless parameter by multiplying through by $(\mu/\rho VD)^2$, we would have

$$\frac{F_D\rho}{\mu^2}\left(\frac{\mu}{\rho VD}\right)^2 = \phi\left(\frac{\rho VD}{\mu}\right)\left(\frac{\mu}{\rho VD}\right)^2$$

The left side of the equation becomes $F_D/\rho V^2D^2$ and the right side can be written as

$$\phi\left(\frac{\rho VD}{\mu}\right)\left(\frac{\mu}{\rho VD}\right)^2 = f\left(\frac{\mu}{\rho VD}\right)$$

where f is the same function as in Eq. (8.8). The form of the dimensionless equation to be used depends on the application.

The Exponent Method

An alternative method for finding the dimensionless parameters, once the significant variables have been identified, is to solve a set of algebraic simultaneous equations that derive from the requirement of dimensional homogeneity for equations describing physical systems. This will be referred to as the exponent method. The method is best explained by example.

The problem described in Example 8.2 is solved here using the exponent method. The dimensions of the variables are

$$[F] = \frac{ML}{T^2}$$

$$[V] = \frac{L}{T}$$

$$[\rho] = \frac{M}{L^3}$$

$$[\mu] = \frac{M}{LT}$$

$$[D] = L$$

Dimensional homogeneity requires that each term making up the function f in Eq. (8.7) have the same dimensions as force. Thus the variables in each term must combine in such a way that the combination has the dimensions of force (ML/T^2). We assume that the variables in f_1 combine as the product

$$V^a \rho^b \mu^c D^d$$

where a, b, c, and d are exponents to be selected that yield the dimensions of force. Dimensional homogeneity requires

$$[F] = [V^a \rho^b \mu^c D^d]$$

or

$$\frac{ML}{T^2} = \left(\frac{L}{T}\right)^a \left(\frac{M}{L^3}\right)^b \left(\frac{M}{LT}\right)^c L^d = \frac{L^{a-3b-c+d} M^{b+c}}{T^{a+c}}$$

Equating the powers of M, L, and T on each side of the equation results in three algebraic equations,

$$L:\ \ a - 3b - c + d = 1$$

$$M:\ \ b + c = 1$$

$$T:\ \ a + c = 2$$

We have three equations and four unknowns. However, we can solve for three of the unknowns in terms of the fourth unknown. The solution procedure is simplified if we select as the fourth unknown the exponent that appears most frequently in the equations. Let us take a, b, and d as the exponents to be solved for in terms of c. The equations can be written in matrix form as

$$\begin{pmatrix} 1 & -1 & 1 \\ 0 & 1 & 0 \\ 1 & 0 & 0 \end{pmatrix} \begin{pmatrix} a \\ b \\ d \end{pmatrix} = \begin{pmatrix} 1 + c \\ 1 - c \\ 2 - c \end{pmatrix}$$

If the determinant of the matrix is not zero, a unique solution for a, b, and d is obtainable.* The result is

$$a = 2 - c$$

$$b = 1 - c$$

$$d = 2 - c$$

These exponents are now substituted back into the combination of the physical variables, and the result is

*The choice of the three unknowns is not necessarily arbitrary. If the determinant is zero, then a different set of the three unknowns must be chosen. If all the possible combinations of the three unknowns yield a zero determinant, then two unknowns must be chosen and solved for in terms of the remaining unknowns. In this situation, the number of dimensionless parameters will be two (not three) less than the number of variables.

$$V^a \rho^b \mu^c D^d = \rho V^2 D^2 \left(\frac{\mu}{\rho V D}\right)^c$$

The factor $\rho V^2 D^2$ has the dimensions of force and will be common to every term in the function. The combination $(\mu/\rho VD)$ is dimensionless. Thus Eq. (8.7) can be rewritten as

$$F = \rho V^2 D^2 f\left(\frac{\mu}{\rho V D}\right)$$

Dividing through by $\rho V^2 D^2$ yields the same dimensionless equation as Eq. (8.8).

$$\frac{F}{\rho V^2 D^2} = f\left(\frac{\mu}{\rho V D}\right)$$

Note that in the process of dimensional analysis, we did not need to evaluate the exponent c. We sought only the form of the dimensionless parameter. The functional form of f must be obtained from experiment.

Recapitulation of the Process of Dimensional Analysis

In the foregoing paragraphs, the concept and procedure of dimensional analysis have been presented. For review purposes and handy reference, the procedure of dimensional analysis can be summarized as follows:

1. Identify all significant variables associated with the problem and include these in a functional equation, such as

$$Z = f(U, V, W, X, Y)$$

2. First combine one variable that includes a given dimension with other variables having that same dimension in such a way as to eliminate that dimension. Then choose a second variable to combine with other variables and groups of variables to eliminate a second dimension. Repeat the process until the entire equation consists of dimensionless groups or parameters. If the exponent method is used, a set of algebraic equations for exponents of the physical variables that satisfy dimensionless homogeneity is developed. Three exponents (for systems with three dimensions) are selected and solved for in terms of the remaining exponents. Substituting the three exponents back into the combination of physical variables yields the dimensionless parameters.

3. The final form will appear something like the following:

$$\pi_1 = f(\pi_2, \pi_3, \ldots, \pi_{n-m})$$

where $\pi_1, \pi_2, \ldots, \pi_{n-m}$ represent the dimensionless parameters obtained, n is the number of variables, and m is the number of dimensions included in the list of variables.

Limitations of Dimensional Analysis

All the foregoing procedures deal with straightforward situations. However, some problems do occur. In order to apply dimensional analysis we must first decide which variables are significant. If we do not understand the problem well enough to make a good initial choice of variables, dimensional analysis seldom provides clarification.

One error might be inclusion of variables whose influence is already accounted for. For example, one might tend to include two or three length variables in a scale-model test where only one may be sufficient.

Another serious error might be the omission of a significant variable. If this is done, one of the significant dimensionless parameters will likewise be missing. In this regard, we may sometimes identify a list of variables that we think are significant to a problem and find that one dimensional category (such as M or L or T) is included in only one variable. When this occurs, we know that an error in choice of variables has been made because it will not be possible to combine two variables to eliminate the lone dimension. Either the variable with the lone dimension should not have been included in the first place (it is not significant), or another one should have been included (a significant variable was omitted).

How do we know whether a variable is significant for a given problem? Probably the truest answer is by experience. After working in the field of fluid mechanics for several years, one develops a feel for the significance of variables to certain kinds of applications. However, even the inexperienced engineer will appreciate the fact that free-surface effects have no significance in closed-conduit flow; consequently, surface tension, σ, would not be included as a variable. In closed-conduit flow, if the velocity is less than approximately one-third the speed of sound, compressibility effects are usually negligible. Such guidelines, which have been observed by previous experimenters, help the beginning engineer develop confidence in her or his application of dimensional analysis and similitude.

8.5 Common Dimensionless Numbers

In Example 8.2, one of the π-groups obtained was a form of the Reynolds number. This parameter and others recur repeatedly in fluid-flow studies. By considering all variables that might be significant in a general flow situation, we can derive these π-groups or *dimensionless numbers* by means of dimensional analysis.

Visualize a hypothetical flow condition in which the pressure difference between two points in the flow field is expected to be a function of V, L, ρ, μ, E_v, σ, and $\Delta\gamma$. Respectively, these variables are velocity, characteristic length, mass density of the fluid, viscosity of the fluid, bulk modulus of elasticity of the fluid, surface tension of the fluid, and the difference in specific weight between the flowing fluid and the fluid above the free surface. The latter variable, $\Delta\gamma$, is significant when free-surface phenomena such as waves exist. The functional relationship for these variables is given as

$$\Delta p = f(V, L, \rho, \mu, E_v, \sigma, \Delta\gamma)$$

where

$$[\Delta p] = M/LT^2$$
$$[V] = L/T$$
$$[L] = L$$
$$[\rho] = M/L^3$$
$$[\mu] = M/LT$$
$$[E_v] = M/LT^2$$
$$[\sigma] = M/T^2$$
$$[\Delta\gamma] = M/L^2T^2$$

When we make a dimensional analysis on these variables using ρ, V, and L as combining variables, we obtain the following functional relationship between the dimensionless parameters:

$$\frac{\Delta p}{\rho V^2} = f\left(\frac{VL\rho}{\mu}, \frac{V}{\sqrt{E_v/\rho}}, \frac{\rho L V^2}{\sigma}, \frac{V^2}{L\Delta\gamma/\rho}\right)$$

Without affecting the results with respect to dimensional considerations, we can rewrite the left-hand side of this equation as $\Delta p/(1/2)\rho V^2$. We do this to put the term in the same form as the pressure coefficient that was introduced in Chapter 4. The speed of sound c is given by $\sqrt{E_v/\rho}$. We then have

$$C_p = f\left(\frac{VL\rho}{\mu}, \frac{V}{c}, \frac{\rho L V^2}{\sigma}, \frac{V^2}{L\Delta\gamma/\rho}\right) \tag{8.10}$$

Thus the pressure coefficient is dependent on the dimensionless parameters (π-groups) on the right-hand side of the equation. The first group is the *Reynolds number*,

$$\text{Re} = \frac{\rho VL}{\mu}$$

which has been referred to earlier. It is named after Osborne Reynolds, a pioneer in fluid mechanics, who discovered in 1883 that this number was related to the occurrence of turbulence in pipe flow. The second group is the Mach number,

$$\text{M} = \frac{V}{c}$$

which is important in compressible flows and was first identified by Ernst Mach, an Austrian physicist, in his study of ballistics. The third group is the Weber number,

$$\mathrm{We} = \frac{\rho L V^2}{\sigma}$$

which is important in liquid atomization and attributed to Moritz Weber (1871–1951), a German professor. The last group is the Froude (rhymes with "food") number squared. The Froude number is associated with William Froude (1810–1879), who was an English naval architect. It is customary to replace ρ by γ/g, so the Froude number is

$$\mathrm{Fr} = \frac{V}{\sqrt{(\Delta\gamma/\gamma)gL}}$$

In this particular form it is called the densimetric Froude number, and it is applied in studying the motion of fluids in which there is density stratification, such as between salt water and fresh water in an estuary. It also has application in the study of thermal plumes from stacks or from heated-water effluents associated with thermal power plants. However, in most applications of the Froude number, the flowing fluid is usually a liquid such as water and the adjacent fluid at the interface is a gas such as air. Thus for all practical purposes, $\Delta\gamma = \gamma_{\text{liquid}}$. Then $\Delta\gamma/\gamma = 1$. Hence the Froude number is most often given as

$$\mathrm{Fr} = \frac{V}{\sqrt{gL}} \tag{8.11}$$

We can then write Eq. (8.10) as

$$C_p = f(\mathrm{Re, M, We, Fr}) \tag{8.12}$$

In Chapter 4 it was noted that C_p is a dimensionless parameter derived from the Bernoulli equation. One can view it as a ratio of a pressure force relative to an inertial reaction (mass times acceleration) of the fluid. The pressure force, F_p, is simply ΔpA, or, from a dimensional point of view, $F_p \propto \Delta pL^2$. The inertial, or Ma, force is expressed as

$$F_i = Ma$$

$$F_i \propto \rho L^3 V\frac{dV}{ds} \propto \tfrac{1}{2}\rho V^2 L^2$$

where M represents the mass of the fluid, and VdV/ds is acceleration in the form of convective acceleration. The ratio of the pressure force to inertial reaction is then given as

$$\frac{F_p}{F_i} \propto \frac{\Delta pL^2}{\frac{1}{2}\rho V^2 L^2} = \frac{\Delta p}{\frac{1}{2}\rho V^2} = C_p$$

Similarly, the other dimensionless numbers can be viewed as relative quantities—that is, inertial-force reaction relative to forces relating to fluid characteristics. The next paragraphs demonstrate these relations.

The Reynolds number can be viewed as a ratio of inertial to viscous forces. The viscous force acting on a surface is $F_v = \tau A = (\mu\, du/dy)A$. However, we are primarily interested in dimensional considerations; hence F_v can be written as

$$F_v \propto \mu \frac{V}{L} A \propto \mu \frac{V}{L} L^2$$

$$F_v \propto \mu V L$$

Thus the ratio of inertial to viscous force in terms of the basic variables is

$$\frac{F_i}{F_v} \propto \frac{\rho L^2 V^2}{\mu V L} = \frac{V L \rho}{\mu} = \text{Re}$$

Because the magnitude of the Reynolds number is inversely proportional to the shear force, low Reynolds numbers imply relatively large shear forces and high Reynolds numbers imply relatively low shear forces. This concept of the Reynolds number and its relationship to viscous stresses is often very helpful in understanding certain flow phenomena and in developing approximate solutions to them. For example, in Section 8.9 it will be shown that because the viscous effects are relatively insignificant at very high Reynolds numbers, one can assume that C_p is virtually constant for problems involving convective acceleration.

Expressed in terms of the basic variables involved, the other forces (surface tension, gravity, and elastic) that influence the flow are as follows:

$$F_\sigma \propto \sigma L \qquad \text{(surface-tension force)}$$

$$F_g \propto \Delta\gamma L^3 \qquad \text{(gravity force)}$$

$$F_c \propto \rho V c L^2 \qquad \text{(elastic or compressibility force)}$$

The elastic-force relation follows from $F_c = \Delta p_c L^2$, where $\Delta p_c = \rho V c$ (the pressure change that occurs in water hammer or shock waves in supersonic flow). Thus the Mach number is a ratio of inertial to elastic force,

$$\text{M} = \frac{\rho L^2 V^2}{\rho V c L^2} = \frac{V}{c}$$

the Weber number is a ratio of inertial to surface-tension force,

$$\text{We} = \frac{\rho L^2 V^2}{\sigma L} = \frac{\rho V^2 L}{\sigma}$$

and the Froude number squared is the ratio of inertial to gravity force,

$$(\text{Fr})^2 = \frac{\rho L^2 V^2}{\Delta\gamma L^3} = \frac{V^2}{L \Delta\gamma/\rho}$$

The Froude number is unimportant when gravity causes only a hydrostatic pressure distribution, such as in a closed conduit. However, if the gravitational force influences the pattern of flow, such as in flow over a spillway or in the formation of waves created by a ship as it cruises over the sea, the Froude number is a most significant parameter.

The Mach number is an indicator of how important compressibility effects are in a fluid flow. If the Mach number is small, then the inertial force associated with the fluid motion does not cause a significant density change, and the flow can be treated as incompressible (constant density). On the other hand, if the Mach number is large, there are often appreciable density changes that must be considered in model studies. The significance of the Mach number in compressible flow studies will be discussed in more detail in Chapter 12.

The Weber number is an important parameter in liquid atomization. The surface tension of the liquid at the surface of a droplet is responsible for maintaining the droplet's shape. If a droplet is subjected to an air jet and there is a relative velocity between the droplet and the gas, inertial forces due to this relative velocity cause the droplet to deform. If the Weber number is too large, the inertial force overcomes the surface-tension force to the point that the droplet shatters into even smaller droplets. Thus a Weber-number criterion can be useful in predicting the droplet size to be expected in liquid atomization. The size of the droplets resulting from liquid atomization is a very significant parameter in gas-turbine and rocket combustion.

8.6 Similitude

Scope of Similitude

Whenever it is necessary to perform tests on a model to obtain information that cannot be obtained by analytical means alone, the rules of similitude must be applied. *Similitude* is the theory and art of predicting prototype performance from model observations. We shall see that the theory of similitude involves the application of dimensionless numbers, such as the Reynolds number or the Froude number, to predict prototype performance from model tests. The art of similitude enters the problem when the engineer must make decisions about model design, model construction, performance of tests, or analysis of results that are not included in the basic theory.

Present engineering practice makes use of model tests more frequently than most people realize. For example, whenever a new airplane is being designed, tests are made not only on the general scale model of the prototype airplane but also on various components of the plane. Numerous tests are made on individual wing sections as well as on the engine pods and tail sections.

Models of automobiles and high-speed trains are also tested in wind tunnels to predict the drag and flow patterns for the prototype. Information derived from these model studies often indicates potential problems that can be corrected before the prototype is built, thereby saving considerable time and expense in development of the prototype.

In civil engineering, model tests are always used to predict flow conditions for the spillways of large dams. In addition, river models assist the engineer in the design of flood-control structures as well as in the analysis of sediment movement in the river. Marine engineers make extensive tests on model ship hulls to predict the drag of the ships. Much of this type of testing is done at the David Taylor Model Basin, Naval Surface Warfare Center, Carderock Division, near Washington, D.C. (see Fig. 8.3). Tests are also regularly performed on models of tall buildings to help predict the wind loads on the buildings, the stability characteristics of the buildings, and the air-flow patterns in their vicinity. The latter information is used by the architects to design walkways and passageways that are safer and more comfortable for pedestrians to use.

FIGURE 8.3

Ship-model test at the David Taylor Model Basin, Naval Surface Warfare Center, Carderock Division.

Geometric Similitude

The basic and perhaps the most obvious requirement of similitude is that the model be an exact geometric replica of the prototype.* Consequently, if a 1:10 scale model is specified, all linear dimensions of the model must be $1/10$ of those of the prototype. In Fig. 8.4 if the model and prototype are geometrically similar, the following equalities hold:

$$\frac{\ell_m}{\ell_p} = \frac{w_m}{w_p} = \frac{c_m}{c_p} = L_r \tag{8.13}$$

Here ℓ, w, and c are specific linear dimensions associated with the model and prototype, and L_r is the scale ratio between model and prototype. It follows that the ratio of corresponding areas between model and prototype will be the square of the length ratio: $A_r = L_r^2$. The ratio of corresponding volumes will be given by $\mathcal{V}_m/\mathcal{V}_p = L_r^3$.

Dynamic Similitude

The basic requirement for dynamic similitude is that the forces that act on corresponding masses in the model and prototype must be in the same ratio $(F_m/F_p = \text{constant})$

*For most model studies this is a basic requirement. However, for certain types of problems, such as river models, distortion of the vertical scale is often necessary to obtain meaningful results.

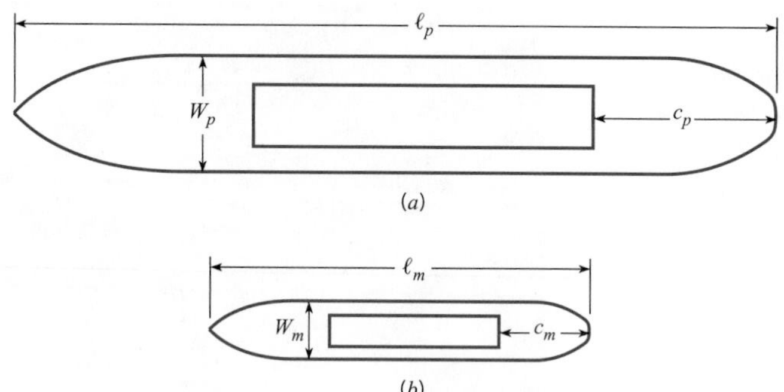

FIGURE 8.4

(a) Prototype. (b) Model.

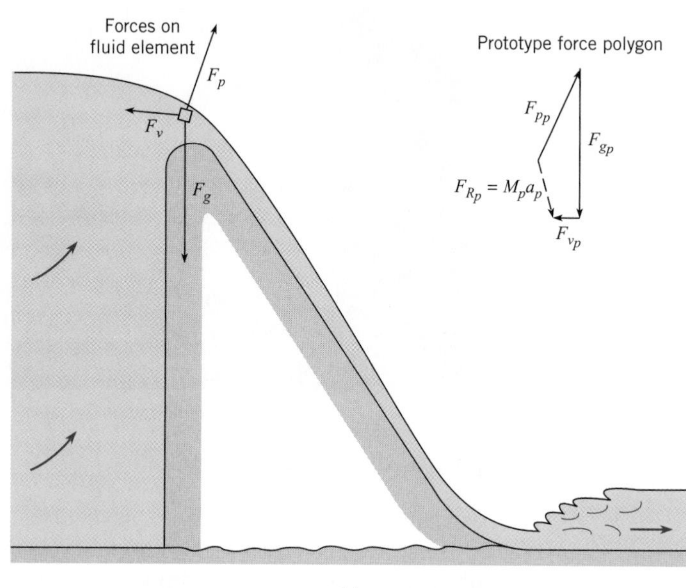

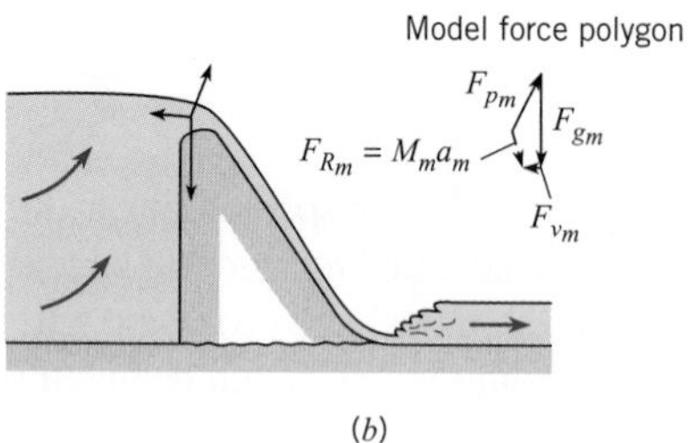

FIGURE 8.5

Model–prototype relations: view (a) and view (b).

throughout the entire flow field. Since the forces acting on the fluid elements will thus control the motion of those elements, it follows that dynamic similarity will yield similarity of flow patterns. Consequently, the flow patterns will be the same in the model as in the prototype if geometric similarity is satisfied and if the relative forces acting on the fluid are the same in the model as in the prototype. This latter condition requires that we have equality of the appropriate dimensionless numbers introduced in Section 8.5, because these dimensionless numbers are indicators of relative forces within the fluid.

Now we shall give a more physical interpretation to the foregoing developments. Consider flow over the spillway shown in Fig. 8.5*a*. Here corresponding masses of fluid in the model and prototype are acted on by corresponding forces. These forces are the force of gravity F_g, the pressure force F_p, and the viscous resistance force F_v. These forces add vectorially as shown in Fig. 8.5 to yield a resultant force F_R, which will in turn produce an acceleration of the volume of fluid in accordance with Newton's second law. Hence, because the force polygons in the prototype and model are similar, the magnitudes of the forces in the prototype and model will be in the same ratio as the magnitude of the *Ma* vectors:

$$\frac{M_m a_m}{M_p a_p} = \frac{F_{g_m}}{F_{g_p}}$$

or

$$\frac{\rho_m L_m^3 (V_m/t_m)}{\rho_p L_p^3 (V_p/t_p)} = \frac{\gamma_m L_m^3}{\gamma_p L_p^3}$$

which reduces to

$$\frac{V_m}{g_m t_m} = \frac{V_p}{g_p t_p}$$

But

$$\frac{t_m}{t_p} \propto \frac{L_m/V_m}{L_p/V_p}$$

so

$$\frac{V_m^2}{g_m L_m} = \frac{V_p^2}{g_p L_p} \tag{8.14}$$

Taking the square root of each side of Eq. (8. 14), we obtain

$$\frac{V_m}{\sqrt{g_m L_m}} = \frac{V_p}{\sqrt{g_p L_p}} \tag{8.15}$$

or

$$\mathrm{Fr}_m = \mathrm{Fr}_p$$

Thus it has been shown that the Froude number in the model must be equal to the Froude number in the prototype. However, we have looked only at the *Ma*, or inertial, forces and the gravity forces to derive Eq. (8.15). If we equate the ratio of inertial forces to the ratio of viscous forces, we obtain

$$\frac{M_m a_m}{M_p a_p} = \frac{F_{v_m}}{F_{v_p}} \tag{8.16}$$

Here $F_v \propto \mu VL$, so with a little algebra, Eq. (8.16) reduces to

$$\mathrm{Re}_m = \mathrm{Re}_p$$

And finally
$$\frac{M_m a_m}{M_p a_p} = \frac{F_{p_m}}{F_{p_p}} \tag{8.17}$$

where
$$F_p \propto \Delta p L^2$$

which yields
$$C_{p_m} = C_{p_p}$$

The foregoing development then means that dynamic similitude (similarity of pressure coefficients) will be completely achieved for flow over a spillway if the Froude number and the Reynolds number are the same in the model as in the prototype. As will be shown in Section 8.7, the latter part of this requirement can be relaxed if the model is reasonably large. In the force polygon shown in Fig. 8.5, it can be seen that the polygon can be completed with only three of the forces; hence one of them is dependent on the others. Thus, if we think of the pressure force as a dependent one, it follows that the pressure coefficient is dependent on the other parameters. In other words, if we have equality of Fr and Re, then we automatically have equality of C_p, between model and prototype. This is exactly the conclusion arrived at in Section 8.5, reached here in a different manner. If other forces, such as surface-tension or elastic forces, were significant in establishing the flow pattern, then equality of additional dimensionless parameters, such as the Weber number and the Mach number, respectively, would be required.

We conclude from the foregoing developments that *the requirement for similarity of flow between model and prototype is that the significant π-groups must be equal for model and prototype.*

For example, if we were model-testing a valve (enclosed flow), the only forces that might affect the flow pattern would be the shear forces produced by viscous effects. Consequently, dynamic similarity would prevail, with equality of Reynolds numbers. That is, the model Reynolds number would have to be the same as the prototype Reynolds number. For flow over a spillway, the gravitational force is predominant in establishing the flow pattern. Therefore, the Froude number in the model must be the same as the Froude number in the prototype for dynamic similarity in this type of application. The significance of these requirements and their actual use in predicting prototype performance will be considered in the next two sections.

8.7 Model Studies for Flows without Free-Surface Effects

Free-surface effects are absent in the flow of liquids or gases in closed conduits, including control devices such as valves, or in the flow about bodies (for example, aircraft) that travel through air or are deeply submerged in a liquid such as water (submarines). Free-surface effects are also absent where a structure such as a building is stationary and wind flows past it. In all these cases, given relatively low Mach numbers, it is the Reynolds-number criterion that must be used for dynamic similarity. That is, the Reynolds number in the model must equal the Reynolds number in the prototype. The following examples illustrate the application.

example 8.3

The drag characteristics of a blimp 5 m in diameter and 60 m long are to be studied in a wind tunnel. If the speed of the blimp through still air is 10 m/s, and if a 1/10 scale model is to be tested, what airspeed in the wind tunnel is needed for dynamically similar conditions? Assume the same air pressure and temperature for both model and prototype.

Solution For dynamic similarity, the Reynolds number of the model must equal the Reynolds number of the prototype, or

$$\text{Re}_m = \text{Re}_p$$

Hence

$$\frac{V_m L_m \rho_m}{\mu_m} = \frac{V_p L_p \rho_p}{\mu_p}$$

From this we can solve for V_m:

$$V_m = V_p \frac{L_p}{L_m} \frac{\rho_p}{\rho_m} \frac{\mu_m}{\mu_p} = 10 \times 10 \times 1 \times 1 = 100 \text{ m/s}$$

Here, we get the result that the airspeed in the wind tunnel must be 100 m/s for true Reynolds-number similitude. This speed is quite large, and in fact Mach-number effects may start to become important at such a speed. However, we will see in Section 8.9 that it is not always necessary to operate models at true Reynolds-number criteria to obtain useful results.

However, if the engineer thinks that it is essential to test models at the same Reynolds number as the prototype, then only a few alternatives are available. One way to produce high Reynolds numbers at nominal airspeeds is to increase the density of the air. A NASA wind tunnel at the Ames Research Center at Moffett Field in California is one such facility. It has a 12-ft-diameter test section; it can be pressurized up to 90 psia (620 kPa); it can be operated to yield a Reynolds number per foot up to 1.2×10^7, and the maximum Mach number at which a model can be tested in this wind tunnel is 0.6. The air flow in this wind tunnel is produced by a single-stage, 20-blade axial-flow fan, which is powered by a 15,000-horsepower, variable-speed, synchronous electric motor (3). There are several problems that are peculiar to a pressurized tunnel. First, a shell (essentially a pressurized bottle) must surround the entire tunnel and its components, and this adds to the cost of the tunnel. Second, it takes a long time to pressurize the tunnel in preparation for operation, increasing the time from the start to the finish of runs. In this regard it should be noted that the original pressurized wind tunnel at the Ames Research Center was built in 1946; however, because of extensive use, the tunnel's pressure shell began to deteriorate, so a new facility (the one previously described) was built and put in operation in 1995. Improvements over the old facility include a better data collection system, very low turbulence, and capability of depressurizing only the test section instead of the entire 620,000-ft^3 wind tunnel circuit when installing and removing models. The original pressurized wind tunnel was used to test most models of U.S. commercial aircraft over the past half century, including the Boeing 737, 757, and 767; Lockheed L-1011; and McDonnell Douglas DC-9 and DC-10.

The Boeing 777 was tested in the low-speed, pressurized 5 m–by–5 m tunnel in Farnborough, England. This tunnel, operated by the Defence Evaluation and Research Agency (DERA) of Great Britain, can operate at three atmospheres with Mach numbers up to 0.2. Approximately 15,000 hours of total testing time was required for the Boeing 777 (4).

Another method of obtaining high Reynolds numbers is to build a tunnel in which the test medium (gas) is at a very low temperature, thus producing a relatively high-density–low-viscosity fluid. NASA has built such a tunnel and operates it at the Langley Research Center. This tunnel, called the National Transonic Facility, can be pressurized up to 9 atmospheres. The test medium is nitrogen, which is cooled by injecting liquid nitrogen into the system. In this wind tunnel it is possible to reach Reynolds numbers of 10^8 based on a model size of 0.25 m (5). Because of its sophisticated design, its initial cost of approximately $100,000,000 (6), and its operating expenses are high.

Information on NASA wind tunnels can be obtained from the web site http://aocentral.arc.nasa.gov.

Another modern approach in wind-tunnel technology is the development of magnetic or electrostatic suspension of models. The use of the magnetic suspension with model airplanes is being researched (6), and the electrostatic suspension for the study of single-particle aerodynamics has already been reported (7).

The use of wind tunnels for aircraft design has grown significantly as the size and sophistication of aircraft have increased. For example, in the 1930s the DC-3 and B-17 each had about 100 hours of wind-tunnel tests at a rate of $100 per hour of run time. By contrast the F-15 fighter required about 20,000 hours of tests at a cost of $20,000 per hour (6). The latter test time is even more staggering when one realizes that a much greater volume of data per hour at higher accuracy is obtained from the modern wind tunnels because of the high-speed data acquisition made possible by computers.

example 8.4

The valve shown is the type used in the control of water in large conduits. Model tests are to be done, using water as the fluid, to determine how the valve will operate under wide-open conditions. The prototype size is 6 ft in diameter at the inlet. What flow rate is required for the model if the prototype flow is 700 cfs? Assume that the temperature for model and prototype is 60°F and that the model inlet diameter is 1 ft.

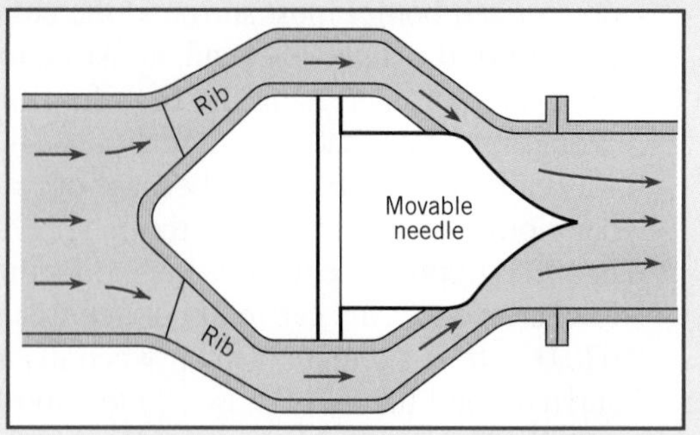

Solution Again we use the Reynolds-number criterion. Therefore

$$\text{Re}_m = \text{Re}_p$$

$$\frac{V_m L_m}{\nu_m} = \frac{V p L_p}{\nu_p}$$

$$\frac{V_m}{V_r} = \frac{L_p}{L_m}\frac{\nu_m}{\nu_p}$$

However, we are interested in the total rate of flow. Therefore, if we multiply both sides of the foregoing equation by the area ratio A_m/A_p, the resulting product on the left-hand side yields the desired discharge ratio:

$$\frac{A_m V_m}{A_p V_p} = \frac{A_m}{A_p}\frac{L_p}{L_m}\frac{\nu_m}{\nu_p}$$

or

$$\frac{Q_m}{Q_p} = \frac{A_m}{A_p}\frac{L_p}{L_m}\frac{\nu_m}{\nu_p}$$

Since $A_m/A_p = L_r^2 = (L_m/L_p)^2$, we obtain

$$\frac{Q_m}{Q_p} = \frac{L_m^2}{L_p^2}\frac{L_p}{L_m}\frac{\nu_m}{\nu_p}$$

But $\nu_m = \nu_p$, or $\nu_m/\nu_p = 1$, because water is used for both model and prototype, so

$$\frac{Q_m}{Q_p} = \frac{L_m}{L_p}$$

Finally,

$$Q_m = \tfrac{1}{6}Q_p = \tfrac{1}{6} \times 700 = 116.7 \text{ cfs}$$

◁

Note: This discharge is very large indeed, and in fact it serves to emphasize that very few model studies are made that completely satisfy the Reynolds-number criterion. This subject will be discussed further in the next sections.

8.8 Significance of the Pressure Coefficient

In the foregoing examples it was demonstrated that dynamic similarity between model and prototype exists when the significant π-groups are the same in the model and the prototype. Since none of the parameters we have considered (Re, Fr, M, or W) explicitly includes Δp, one may wonder how Δp in the model is related to Δp in the prototype. It is related by means of the pressure coefficient. If we refer to Eq. (8.12),

$$C_p = f(\text{Re, M, We, Fr})$$

we see that the pressure coefficient, $\Delta p/(\rho V^2/2)$, is a function of the basic parameters of similitude. Consequently, when dynamic similarity exists—that is, when the significant π-groups are the same in the model and the prototype—then it follows that the pressure coefficient will also be the same in the model and the prototype. Thus, when we have dynamic similarity,

$$C_{p,\,\text{model}} = C_{p,\,\text{prototype}} \tag{8.18}$$

or

$$\frac{\Delta p_m}{\frac{1}{2}\rho_m V_m^2} = \frac{\Delta p_p}{\frac{1}{2}\rho_p V_p^2} \tag{8.19}$$

The pressure coefficient can be used like any of the other basic parameters for model analysis. It is useful not only in relating pressure changes in the model to those in the prototype but also in relating total forces in model and prototype. The latter is accomplished by multiplying the pressure ratio by the area ratio.

Analogous to the pressure coefficient, the shear stress, τ, can also be expressed as a π-group by

$$c_f = \frac{\tau}{\frac{1}{2}\rho V^2}$$

where c_f is the local shear stress coefficient, a parameter that will be addressed in Chapter 9. The shear stress coefficient is also a function of the basic parameters of similitude. So when the significant dimensionless numbers are the same in the model and the prototype, the shear stress coefficient will also be the same in the model and the prototype.

example 8.5

For the given conditions of Example 8.3, if the pressure difference between two points on the surface of the model blimp is measured to be 17.8 kPa, what will be the pressure difference in the prototype for dynamically similar conditions?

Solution From Eq. (8.19), the prototype pressure difference can be given as

$$\Delta p_p = \Delta p_m \frac{\rho_p}{\rho_m}\frac{V_p^2}{V_m^2}$$

However,

$$\rho_p = \rho_m$$

and using the Reynolds similarity criterion, we get

$$\frac{V_p}{V_m} = \frac{L_m}{L_p} = \frac{1}{10}$$

Therefore,

$$\Delta p_p = \Delta p_m\left(\frac{1}{10}\right)^2 = \frac{1}{100}\Delta p_m = \frac{17{,}800}{100} = 178 \text{ Pa}$$

◁

example 8.6

For the given conditions of Example 8.3, if the drag force on the model blimp is measured to be 1530 N, what corresponding force could be expected in the prototype?

Solution Again using Eq. (8.19), we can write

$$\frac{\Delta p_p}{\Delta p_m} = \frac{\rho_p}{\rho_m}\frac{V_p^2}{V_m^2}$$

Also,

$$\text{Re}_p = \text{Re}_m$$

$$\frac{V_p L_p}{\nu_p} = \frac{V_m L_m}{\nu_m}$$

or

$$\frac{V_p}{V_m} = \frac{\nu_p L_m}{\nu_m L_p}$$

However, for this example, $\rho_p = \rho_m$ and $\nu_p = \nu_m$. Thus when we let $\rho_p/\rho_m = 1$, we obtain from the first equation in this example the following:

$$\frac{\Delta p_p}{\Delta p_m} = \frac{V_p^2}{V_m^2}$$

From the Reynolds-number relationship, when we set $\nu_p/\nu_m = 1$, we get

$$\frac{V_p}{V_m} = \frac{L_m}{L_p} \qquad \frac{V_p^2}{V_m^2} = \frac{L_m^2}{L_p^2}$$

Thus when we substitute L_m^2/L_p^2 for V_p^2/V_m^2, we get

$$\frac{\Delta p_p}{\Delta p_m} = \frac{L_m^2}{L_p^2}$$

Now, multiplying both sides of the foregoing equation by the area ratio A_p/A_m, which is the same as L_p^2/L_m^2, we obtain

$$\frac{\Delta p_p L_p^2}{\Delta p_m L_m^2} = \frac{L_m^2 L_p^2}{L_p^2 L_m^2}$$

The left-hand side of this equation is the ratio of the prototype force to the model force, and the right-hand side is unity. Hence we arrive at the interesting result that the model force is the same as the prototype force. When we use the Reynolds-number criterion and use the same fluid for model and prototype, the forces on the model will always be the same as the forces on the prototype. Consequently, for this example the force on the prototype blimp would be 1530 N.

8.9

Approximate Similitude at High Reynolds Numbers

The primary justification for model tests is that it is more economical to get answers needed for engineering design by such tests than by any other means. However, as revealed by Examples 8.3, 8.4, and 8.6, true similarity according to the Reynolds-number criterion yields quantities for the model that would require very costly model setups. Consider the size and power required for wind-tunnel tests of the blimp in Example 8.3. The wind tunnel would probably require a section at least 2 m by 2 m to accommodate the model blimp. With a 100-m/s airspeed in the tunnel, the power required for producing continuously a stream of air of this size and velocity is in the order of 4 MW. Such a test is not prohibitive, but it is very expensive. It is also conceivable that the 100-m/s airspeed would introduce Mach number effects not encountered with the prototype, thus generating concern over the validity of the model data. Furthermore, a force of 1530 N is indeed larger than that usually associated with model tests. Therefore, especially in the study of problems involving non–free-surface flows, it is desirable to perform model tests in such a way that large magnitudes of forces or pressures are not encountered.

For many cases, it is possible to obtain all the needed information from abbreviated tests. Often the Reynolds-number effect (relative viscous effect) either becomes insignificant at high Reynolds numbers or becomes independent of the Reynolds number. The point where testing can be stopped often can be detected by inspection of a graph of the pressure coefficient C_p versus the Reynolds number Re. Such a graph for a venturi meter in a pipe is shown in Fig. 8.6. In this meter, Δp is the pressure difference between the points shown, and V is the velocity in the restricted section of the venturi meter. Here it is seen that viscous forces affect the value of C_p below a Reynolds number of approximately 50,000. However, for higher Reynolds numbers, C_p is virtually constant. Physically this means that at low Reynolds numbers (relatively high viscous forces), a significant part of the change in pressure comes from viscous resistance, and the remainder comes from the inertial reaction (change in kinetic energy) of the fluid as it passes through the venturi meter. However, with high Reynolds numbers (equivalent to either small viscosity or a large product of V, D, and ρ), the viscous resistance is negligible compared with that required to overcome the inertial reaction of the fluid. Since the ratio of Δp to the inertial reaction does not change (constant C_p) for high Reynolds numbers, there is no need to carry out tests at higher Reynolds numbers. This is true in general, so long as the flow pattern does not change with the Reynolds number.

In a practical sense, whoever is in charge of the model test will try to predict from previous works approximately what maximum Reynolds number will be needed to reach the point of insignificant Reynolds-number effect and then will design the model accordingly. After a series of tests has been made on the model, C_p versus Re will be plotted to see whether the range of constant C_p has indeed been reached. If so, then no more data are needed to predict the prototype performance. However, if C_p has not reached a constant value, the test program has to be expanded or results extrapolated. Thus the results of some model tests can be used to predict prototype performance even though the Reynolds numbers are not the same for the model and the prototype. This is especially valid for angular-shaped bodies, such as model buildings, tested in wind tunnels.

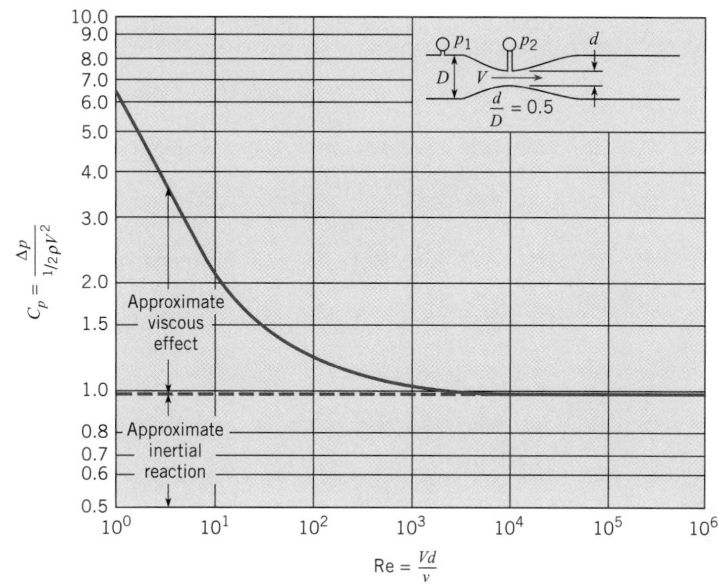

FIGURE 8.6

C_p for a venturi meter as a function of the Reynolds number.

example 8.7

Tests were made by Roberson and Crowe (8) on model-building shapes such as that shown in part (*a*) in the figure to determine the effects of free-stream turbulence and angle of incidence on the surface pressure distribution. Part (*b*) in the figure is an example of the temporal-mean pressure distribution obtained at a relative elevation on the building of $z/H = 0.7$ for an angle of incidence α of 8° and a wind speed V of 20 m/s. The air temperature in the wind tunnel was 20°C. If it is assumed that C_p does not change with higher Re, what would be the maximum and the minimum temporal-mean pressures on a similar building 100 m high from a 50-m/s wind (10°C) with the same angle of incidence? If the instantaneous magnitude of the lowest pressure is three times more negative than the lowest temporal-mean pressure (this is due to turbulence), and if the pressure inside the building is the same as the mean pressure on the leeward wall, then what will be the total wind force in the lowest pressure zone on a window that is 2 m by 2 m?

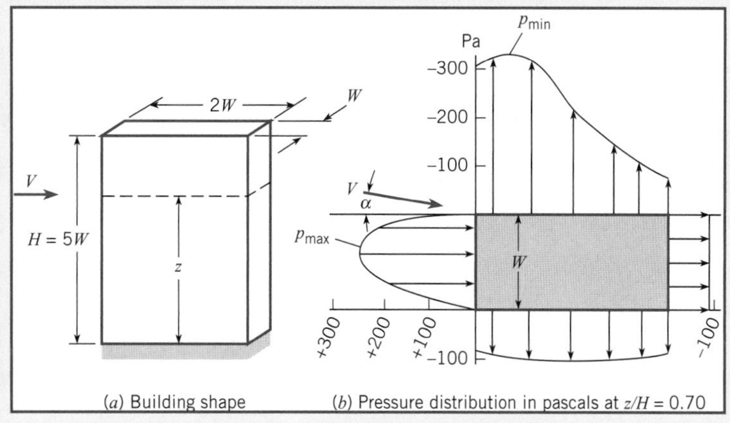

Solution We are assuming that C_p is the same for the model as for the full-scale prototype building. Therefore, we have

$$C_{p_m} = C_{p_p}$$

$$\frac{\Delta p_m}{\rho_m V_m^2/2} = \frac{\Delta p_p}{\rho_p V_p^2/2}$$

$$\frac{p_m - p_0}{\rho_m V_m^2/2} = \frac{p_p - p_0}{\rho_p V_p^2/2}$$

However, our reference pressure p_0 is zero gage, so we can solve for p_p:

$$p_p = \frac{\rho_p}{\rho_m} \frac{V_p^2}{V_m^2} p_m$$

We obtain

$$p_{p_{\max}} = \frac{1.25}{1.20}\left(\frac{50}{20}\right)^2 p_{m_{\max}}$$

$$= \frac{1.25}{1.20} \times 6.25 \times 250 = 1.63 \text{ kPa} \qquad \triangleleft$$

In a similar fashion $p_{p_{\min}} = -2.12$ kPa, and the pressure on the leeward wall (and inside the building) is

$$p_{p_{\text{inside}}} = -0.58 \text{ kPa} \qquad \triangleleft$$

The force on the window is the product of the difference of pressure across the window and the area. Here the lowest outside pressure is assumed to be $3 \times p_{\min}$, so

$$F_{\text{on window}} = (p_{\text{inside}} - p_{\text{outside}}) \times A$$

$$= \{-0.58 \text{ kPa} - [3 \times (-1.95 \text{ kPa})]\} \times 4 \text{ m}^2 = 21.1 \text{ kN} \qquad \triangleleft$$

Before leaving the section on non–free-surface flows, we should note that Mach-number effects (compressibility) usually become significant for Mach numbers exceeding 0.3. It is not always possible to tell which parameter—Mach number or Reynolds number—will be the most significant in a certain situation. Which similitude parameter is chosen depends a great deal on what information the engineer is seeking. If the engineer is interested in the viscous motion of fluid near a wall in shock-free supersonic flow, then the Reynolds number should be selected as the significant similitude parameter. However, if the shock-wave pattern over a body is of interest, then the Mach number should be selected as the similitude parameter. The Mach number and its significance are discussed in more detail in Chapter 12.

8.10 Free-Surface Model Studies

Spillway Models

The major influence, besides the spillway geometry itself, on the flow of water over a spillway is the action of gravity. Hence the Froude similarity criterion is used for such model

FIGURE 8.7

Comprehensive model for Hell's Canyon Dam. Tests were made at the Albrook Hydraulic Laboratory, Washington State University.

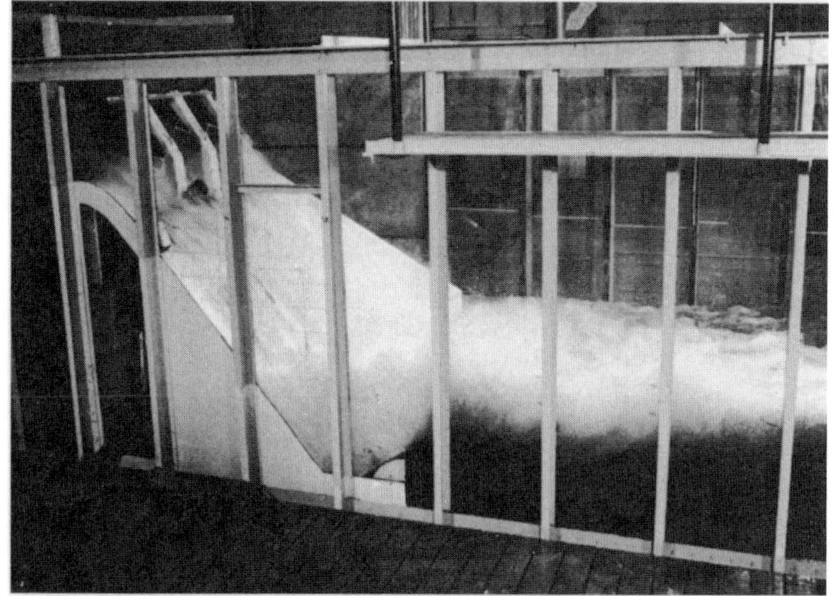

FIGURE 8.8

Spillway model for Hell's Canyon Dam. Tests were made at the Albrook Hydraulic Laboratory, Washington State University.

studies. It can be appreciated that for large spillways with depths of water on the order of 3 or 4 m and velocities on the order of 10 m/s or more, the Reynolds number is very large. At high values of the Reynolds number, the relative viscous forces are often independent of the Reynolds number, as already noted in the foregoing section (Sec. 8.9). However, if the reduced-scale model is made too small, the viscous forces as well as the surface-tension forces would have a larger relative effect on the flow in the model than in the prototype. Therefore, in practice, spillway models are made large enough that the viscous effects have about the same relative effect in the model as in the prototype (the viscous effects are nearly independent of the Reynolds number). Then the Froude number is the significant similarity parameter. Most model spillways are made at least 1 m high, and for precise studies, such as calibration of individual spillway bays, it is not uncommon to design and construct model spillway sections that are 2 or 3 m high. Figures 8.7 and 8.8 show a comprehensive model and spillway model for Hell's Canyon Dam in Idaho.

example 8.8

A 1/49 scale model of a proposed dam is used to predict prototype flow conditions. If the design flood discharge over the spillway is 15,000 m^3/s, what water flow rate should be established in the model to simulate this flow? If a velocity of 1.2 m/s is measured at a point in the model, what is the velocity at a corresponding point in the prototype?

Solution The Froude-number criterion will be used. Therefore,

$$\mathrm{Fr}_m = \mathrm{Fr}_p$$

$$\frac{V_m}{\sqrt{g_m L_m}} = \frac{V_p}{\sqrt{g_p L_p}}$$

However,

$$g_m = g_p$$

Thus we have

$$\frac{V_m}{V_p} = \sqrt{\frac{L_m}{L_p}}$$

We multiply both sides of this equation by the area ratio A_m/A_p:

$$\frac{V_m}{V_p}\frac{A_m}{A_p} = \frac{A_m}{A_p}\sqrt{\frac{L_m}{L_p}}$$

The left-hand side of the equation is the discharge ratio, and $A_m/A_p = (L_m^2/L_p^2)$. Hence we obtain

$$\frac{Q_m}{Q_p} = \left(\frac{L_m}{L_p}\right)^{5/2}$$

$$Q_m = Q_p\left(\frac{1}{49}\right)^{5/2} = 15{,}000\frac{1}{16{,}800} = 0.89\ \mathrm{m^3/s}$$

From the fourth equation in this example,

$$\frac{V_m}{V_p} = \sqrt{\frac{L_m}{L_p}}$$

Consequently,

$$V_p = V_m\sqrt{\frac{L_p}{L_m}} = 1.2 \times 7 = 8.4\ \mathrm{m/s}$$

At the given point in the prototype, we would have a velocity of 8.4 m/s.

Ship Model Tests

The largest facility for ship testing in the United States is the David Taylor Model Basin, Naval Surface Warfare Center, Carderock Division, near Washington, D.C. Two of the core facilities are the towing basins and the rotating arm facility. In the rotating arm facility, models are suspended from the end of a rotating arm in a larger circular basin. Forces and moments can be measured on ship models up to 9 m in length at steady state speeds as high as 15.4 m/s (30 knots). In the high-speed towing basin, models 1.2 m to 6.1 m can be towed at speeds up to 16.5 m/s (32 knots). Further information can be obtained from the web site http:\\www50.dt.navy.mil.

The aim of the ship model testing is to determine the resistance that the propulsion system of the ship must overcome. This resistance is the sum of the wave resistance and the surface resistance of the hull. The wave resistance is a free-surface, or Froude-number, phenomenon, and the hull resistance is a viscous, or Reynolds-number, phenomenon. Because both wave and viscous effects contribute significantly to the overall resistance, it would appear that both the Froude and Reynolds criteria should be used. However, it is impossible to satisfy both if the model liquid is water (the only practical test liquid), because the Reynolds-number criterion dictates a higher velocity for the model than for the prototype [equal to $V_p(L_p/L_m)$], whereas the Froude-number criterion dictates a lower velocity for the model [equal to $V_p(\sqrt{L_m}/\sqrt{L_p})$]. To circumvent such a dilemma, the procedure is to model for the phenomenon that is the most difficult to predict analytically and to account for the other resistance by analytical means. Since the wave resistance is the most difficult problem, the model is operated according to the Froude-number criterion and the hull resistance is accounted for analytically.

To illustrate how the test results and the analytical solutions for surface resistance are merged to yield design data, we note the necessary sequential steps:

1. First make model tests according to the Froude-number criterion, and the total model resistance is measured. This total model resistance will be equal to the wave resistance plus the surface resistance of the hull of the model.
2. Then estimate the surface resistance of the model by analytical calculations.
3. Subtract the surface resistance calculated in step 2 from the total model resistance of step 1 to yield the wave resistance of the model.
4. Using the Froude-number criterion, scale the wave resistance of the model up to yield the wave resistance of the prototype.
5. Then estimate the surface resistance of the hull of the prototype by analytical means.
6. The sum of the wave resistance of the prototype from step 4 and the surface resistance of the prototype from step 5 yields the total prototype resistance, or drag.

8.11 Summary

Dimensional analysis involves combining dimensional variables to form dimensionless groups. These groups, called "π-groups," can be regarded as the scaling parameters for fluid flow. The Buckingham Π theorem states that the number of independent π-groups is $n - m$, where n is the number of dimensional variables and m is the number of basic dimensions included in the variables. In fluid mechanics the three basic dimensions are mass (M), length (L), and time (T).

The π-groups can be found by either the step-by-step method or the exponent method. In the step-by-step method each dimension is removed by successively using a dimensional variable until the dimensionless parameters are obtained. In the exponent method, each variable is raised to a power, they are multiplied together, and three simultaneous algebraic equations formulated for dimensional homogeneity are solved to yield the dimensionless groups.

Four common dimensionless numbers (π-groups) are

$$\text{Reynolds number,} \quad \text{Re} = \frac{\rho VL}{\mu}$$

$$\text{Mach number,} \quad \text{M} = \frac{V}{c}$$

$$\text{Weber number,} \quad \text{We} = \frac{\rho V^2 L}{\sigma}$$

$$\text{Froude number,} \quad \text{Fr} = \frac{V}{\sqrt{gL}}$$

Experimental testing is often performed with a small-scale replica (model) of the full-scale structure (prototype). Similitude is the art and theory of predicting prototype performance from model observations. To achieve exact similitude, the model must be a scale model of the prototype (geometric similitude) and the values of the π-groups must be the same for the model and the prototype (dynamic similitude). In practice, it is not always possible to have complete dynamic similitude, so only the significant π-groups are matched.

References

1. Buckingham, E. "Model Experiments and the Forms of Empirical Equations." *Trans. ASME,* 37 (1915), 263.

2. Ipsen, D. C. *Units, Dimensions and Dimensionless Numbers.* McGraw-Hill, New York, 1960.

3. NASA publication available from the U.S. Government Printing Office: No. 1995-685-893.

4. Personal communication. Mark Goldhammer, Manager, Aerodynamic Design of the 777.

5. Kilgore, R. A., and D. A. Dress. "The Application of Cryogenics to High Reynolds-Number Testing in Wind Tunnels, Part 2. Development and Application of the Cryogenic Wind Tunnel Concept." *Cryogenics,* Vol. 24, no. 9, September 1984.

6. Baals, D. D., and W. R. Corliss. *Wind Tunnels of NASA.* U.S. Govt. Printing Of-fice, Washington, D.C., 1981.

7. Kale, S., et al. "An Experimental Study of Single-Particle Aerodynamics." *Proc. of First Nat. Congress on Fluid Dynamics,* Cincinnati, Ohio, July 1988.

8. Roberson, J. A., and C. T. Crowe. "Pressure Distribution on Model Buildings at Small Angles of Attack in Turbulent Flow." In *Proc. 3rd U.S. Natl. Conf. on Wind Engineering Research.* University of Florida, Gainesville, 1978.

Problems

NS **8.1** Determine which of the following equations are dimensionally homogeneous:

a.
$$Q = \tfrac{2}{3} C L \sqrt{2g} H^{3/2}$$

where Q is discharge, C is a pure number, L is length, g is acceleration due to gravity, and H is head.

b.
$$V = \frac{1.49}{n} R^{2/3} S^{1/2}$$

where V is velocity, n is length to the one-sixth power, R is length, and S is slope.

c.
$$h_f = f \frac{L}{D} \frac{V^2}{2g}$$

where h_f is head loss, f is a dimensionless resistance coefficient, L is length, D is diameter, V is velocity, and g is acceleration due to gravity.

d.
$$D = \frac{0.074}{\text{Re}^{0.2}} \frac{Bx\rho V^2}{2}$$

where D is drag force, Re is Vx/ν, B is width, x is length, ρ is mass density, and V is velocity.

8.2 Determine the dimensions of the following variables and combinations of variables in terms of the length, mass, and time system of units. (*Hint:* Convert F to M by Newton's second law.)

a. T (torque)
b. $\rho V^2/2$, where V is velocity and ρ is mass density
c. $\sqrt{\tau/\rho}$, where τ is shear stress
d. Q/ND^3, where Q is discharge, D is diameter, and N is angular speed of a pump

8.3 It takes a certain length of time for the liquid level in a tank of diameter D to drop from position h_1 to position h_2 as the tank is being drained through an orifice of diameter d at the bottom. Determine the π-groups that apply to this problem. Assume that the liquid is nonviscous. Express your answer in the functional form

$$\frac{\Delta h}{d} = f(\pi_1, \pi_2, \pi_3)$$

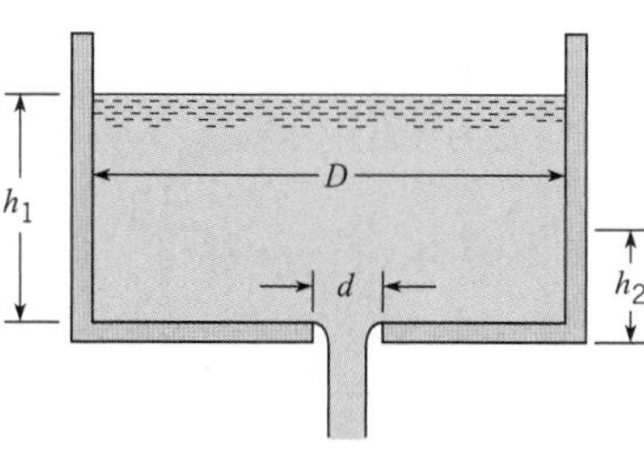

PROBLEM 8.3

8.4 The speed of small-amplitude surface waves depends on the wave amplitude h, the surface tension of the fluid σ, the specific weight of the fluid γ, and the acceleration due to gravity g. Use dimensional analysis to find a nondimensional functional form for the wave celerity (speed) V.

8.5 The maximum rise of a liquid in a small capillary tube is a function of the diameter of the tube, the surface tension, and the specific weight of the liquid. What are the significant π-groups for the problem?

8.6 For very low velocities it is known that the drag force F_D of a small sphere is a function solely of the velocity V of flow past the sphere, the diameter d of the sphere, and the viscosity μ of the fluid. Determine the π-groups involving these variables.

8.7 The drag force F_D on a very rough sphere held inside a pipe in which liquid is flowing is a function of D, ρ, μ, V, and k. D is the diameter of the sphere, ρ is mass density, μ is viscosity, V is the velocity of the liquid, and k is the height of the roughness elements on the sphere. By dimensional analysis, determine the relevant dimensionless numbers for this problem. Express your answer in the functional form

$$\frac{F_D}{\rho V^2 D^2} = f(\pi_1, \pi_2)$$

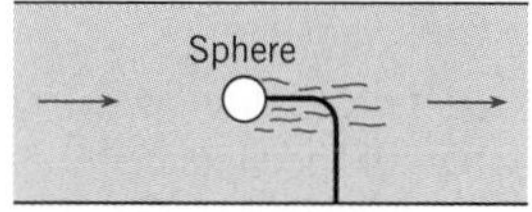

PROBLEM 8.7

8.8 Observations show that the side thrust F, for a rough spinning ball in a fluid is a function of the ball diameter, D, the free-stream velocity, V_0, the density, ρ, the viscosity, μ, the roughness height k_s, and the angular velocity of spin, ω. Determine the dimensionless parameter(s) that would be used to correlate the experimental results of a study involving the variables noted above. Express your answer in the functional form

$$\frac{F}{V_0^2 D^2} = f(\pi_1, \pi_2, \pi_3)$$

8.9 Drag tests show that the drag of a square plate placed normal to the free-stream velocity is a function of the velocity, V, the density, ρ, the plate dimensions, B, the viscosity, μ, the free-stream turbulence intensity, u', and the turbulence scale, L_x.

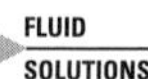

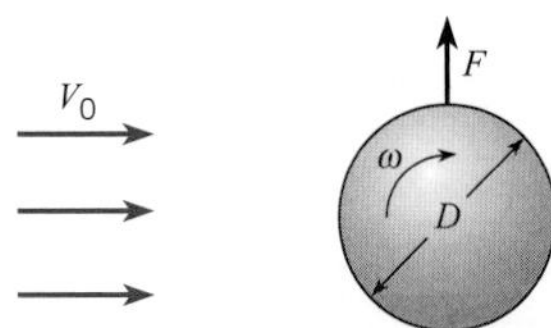

PROBLEM 8.8

Here u' and L_x are in ft/s and ft, respectively. By dimensional analysis, develop the π-groups that could be used to correlate the experimental results. Express your answer in the functional form

$$\frac{F_D}{\rho V^2 B^2} = f(\pi_1, \pi_2, \pi_3)$$

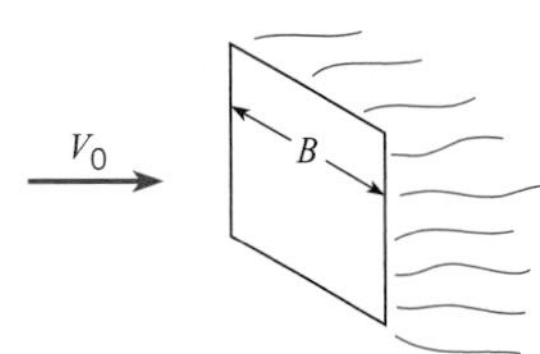

PROBLEM 8.9

8.10 Consider steady viscous flow through a small horizontal tube. For this type of flow, the pressure gradient along the tube, $\Delta p/\Delta \ell$, should be a function of the viscosity μ, the mean velocity V, and the diameter D. By dimensional analysis, derive a functional relationship relating these variables.

8.11 It is known that the pressure developed by a centrifugal pump, Δp, is a function of the diameter D of the impeller, the speed of rotation n, the discharge Q, and the fluid density ρ. By dimensional analysis, determine the π-groups relating these variables.

8.12 The frequency, f, of a bubble oscillating in a non-viscous fluid depends on the pressure in the fluid, p, the radius of the bubble, R, the density of the fluid, ρ, and the ratio of specific heats of the gas in the bubble, k. By dimensional analysis, find a nondimensional formulation for this relationship.

FLUID SOLUTIONS **8.13** The force on a satellite in the earth's upper atmosphere depends on the mean path of the molecules, λ (a length), the density, ρ, the diameter of the body, D, and the molecular speed, c: $F = f(\lambda, \rho, D, c)$. Find the nondimensional form of this equation.

8.14 The velocity V of very small ripples on a liquid surface is a function of the ripple length L, density ρ, and surface tension σ of the liquid. By dimensional analysis, find a functional expression for V.

8.15 A smooth circular plate is positioned a distance S away from a smooth boundary as shown in the figure. Oil with viscosity μ fills the space between the plate and the boundary. The torque required to rotate the plate as shown will be a function of μ, ω, S, and D, where ω is the angular velocity in radians per second. Determine the π-groups that would be involved if you were to correlate results by experimental means.

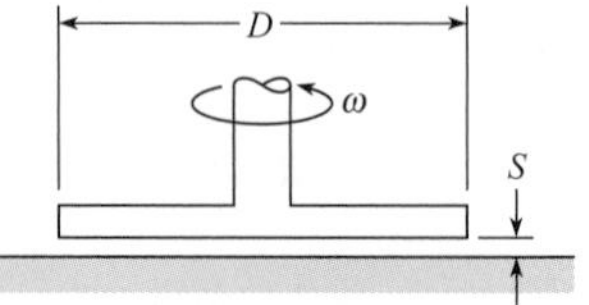

PROBLEM 8.15

8.16 A general study is to be made of the height of rise of liquid in a capillary tube as a function of time after the start of a test. Other significant variables include surface tension, mass density, specific weight, viscosity, and diameter of the tube. Determine the dimensionless parameters that apply to the problem. Express your answer in the functional form

$$\frac{h}{d} = f(\pi_1, \pi_2, \pi_3)$$

8.17 An engineer is using an experiment to characterize the power P consumed by a fan to be used in an electronics cooling application. Power depends on four variables: $P = f(\rho, D, Q, n)$, where ρ is the density of air, D is the diameter of the fan impeller, Q is the flow rate produced by the fan, and n is the rotation rate of the fan. Find the relevant π-groups and suggest a way to plot the data.

PROBLEM 8.17

8.18 The flow of a gas–particle mixture in a tube gives rise to erosion of the tube wall. That is, the collisions of particles with the wall result in removal of material from the wall. Material properties that may be of significance in this problem are modulus of elasticity, E, ultimate strength, σ, and Brinell hardness number, Br. Particle and flow properties that may be important are velocity, V, particle diameter, d, particle mass flow rate, M_p, and tube diameter, D. The Brinell hardness number is dimensionless. Perform a dimensional analysis for the erosion rate e, which has units of $\text{kg/m}^2 \cdot \text{s}$. Express your answer in the functional form

$$\frac{eV}{E} = f(\pi_1, \pi_2, \pi_3, \pi_4)$$

8.19 By dimensional analysis, determine the dimensionless relationship for the change in pressure that occurs when water or oil flows through a horizontal pipe with an abrupt contraction as shown. Express your answer in the functional form

$$\frac{\Delta p d^4}{\rho Q^2} = f(\pi_1, \pi_2)$$

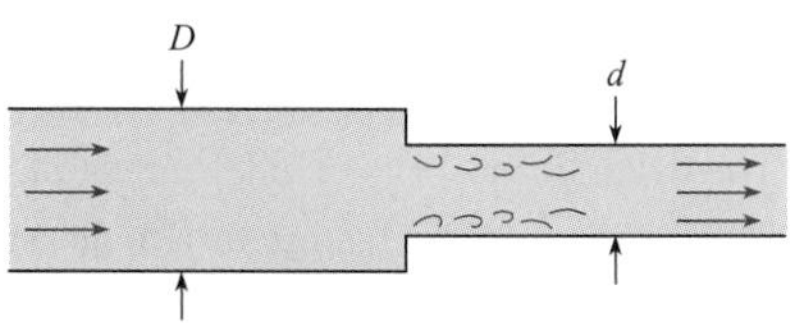

PROBLEM 8.19

8.20 For flow through a transition section from a large pipe to a small pipe, are the viscous forces relatively *large* or *small* compared to the inertial forces when the Reynolds number is very large?

8.21 A solid particle falls through a viscous fluid. The falling velocity, V, is believed to be a function of the fluid density, ρ_f, the particle density, ρ_p, the fluid viscosity, μ, the particle diameter, D, and the acceleration due to gravity, g:

$$V = f(\rho_f, \rho_p, \mu, D, g)$$

By dimensional analysis, develop the π-groups for this problem. Express your answer in the form

$$\frac{V}{\sqrt{gD}} = f(\pi_1, \pi_2)$$

8.22 A bubble is rising with a velocity V through a liquid. The rise velocity is a function of liquid density, ρ_l, bubble diameter, D, the viscosity of the liquid, μ_l, the surface tension, σ, and the acceleration due to gravity:

$$V = f(\rho_l, D, \mu_l, \sigma, g)$$

By dimensional analysis, find the dimensionless numbers for this problem. Express your answer in the form

$$\frac{V}{\sqrt{gD}} = f(\pi_1, \pi_2)$$

8.23 An experimental test program is being set up to calibrate a new flow meter. The flow meter is to measure the mass flow rate of liquid flowing through a pipe. It is assumed that the mass flow rate is a function of the following variables: FLUID SOLUTIONS

$$\dot{m} = f(\Delta p, D, \mu, \rho)$$

where Δp is the pressure difference across the meter, D is the pipe diameter, μ is the liquid viscosity, and ρ is the liquid density. Using dimensional analysis, find the π-groups. Express your answer in the form

$$\frac{\dot{m}}{\sqrt{\rho \Delta p D^4}} = f(\pi)$$

8.24 A torpedo-like device is being designed to travel just below the water surface. Which dimensionless numbers in Section 8.5 would be significant in this problem? Give a rationale for your answer. FLUID SOLUTIONS

8.25 A liquid is moving horizontally through a bed of sand. The sand grains have an average diameter D and the void fraction (volume of free space to the total volume) is α. It is assumed that the pressure drop through the bed is a function of the sand grain diameter, D, the void fraction, α, the fluid density, ρ, the fluid viscosity, μ, and the distance through the bed, Δs:

$$\Delta p = f(D, \Delta s, \alpha, \mu, \rho)$$

By dimensional analysis, find the π-groups for this problem. Express your answer in the form

$$\frac{\sqrt{\rho \Delta p} \Delta s}{\mu} = f(\pi_1, \pi_2)$$

8.26 Experiments are to be done on the drag forces on an oscillating fin in a water tunnel. It is assumed that the drag force, F_D, is a function of the liquid density, ρ, the fluid velocity, V, the surface area of the fin, S, and the frequency of oscillation, ω:

$$F_D = f(\rho, V, S, \omega)$$

By dimensional analysis, find the dimensionless parameters for this problem. Express your answer in the form

$$\frac{F_D}{\rho V^2 S} = f(\pi)$$

8.27 The discharge of a centrifugal pump is a function of the rotational speed of the pump, N, the diameter of the impeller, D, the head across the pump, h_p, the viscosity of the fluid, μ, the density of the fluid, ρ, and the acceleration due to gravity, g. The functional relationship is

$$Q = f(N, D, h_p, \mu, \rho, g)$$

By dimensional analysis, find the dimensionless parameters. Express your answer in the form

$$\frac{Q}{ND^3} = f(\pi_1, \pi_2, \pi_3)$$

8.28 The drag on a submarine moving below the free surface is to be determined by a test on a 1/15 scale model in a water tunnel. The velocity of the prototype in sea water ($\rho = 1015$ kg/m^3, $\nu = 1.4 \times 10^{-6}$ m^2/s) is 2 m/s. The test is done in pure water at 20°C. Determine the speed of the water in the water tunnel for dynamic similitude and the ratio of the drag force on the model to the drag force on the prototype.

8.29 Water with a kinematic viscosity of 10^{-6} m^2/s flows through a 5-cm pipe. What would the velocity of water have to be for the water flow to be dynamically similar to oil ($\nu = 10^{-5}$ m^2/s) flowing through the same pipe at a velocity of 0.5 m/s?

8.30 Oil with a kinematic viscosity of 4×10^{-6} m^2/s flows through a smooth pipe 15 cm in diameter at 2 m/s. What velocity should water have at 20°C in a smooth pipe 5 cm in diameter to be dynamically similar?

FLUID SOLUTIONS **8.31** A large venturi meter is calibrated by means of a 1/10 scale model using the prototype liquid. What is the discharge ratio Q_m/Q_p for dynamic similarity? If a pressure difference of 300 kPa is measured across ports in the model for a given discharge, what pressure difference will occur between similar ports in the prototype for dynamically similar conditions?

8.32 It is known that for flow past cylinders, vortices are shed alternately from one side of the cylinder and from the other. If the frequency of shedding n is a function of the approach velocity V, the diameter d of the cylinder, the mass density ρ, and the viscosity μ, what are the π-groups for this phenomenon?

8.33 A 1/5 scale model of an experimental bathosphere that will operate at great depths is to be tested to determine its drag characteristic by towing it behind a submarine. For true similitude, what should be the towing speed relative to the speed of the prototype?

8.34 A spherical balloon that is to be used in air at 60°F is tested by towing a 1/3 scale model in a lake. The model is 1 ft in diameter, and a drag of 15 lbf is measured when the model is being towed in deep water at 5 ft/s. What drag (in pounds force and newtons) can be expected for the prototype in air under dynamically similar conditions? Assume that the water temperature is 60°F.

8.35 An engineer needs a value of lift force for an airplane that has a coefficient of lift (C_L) of 0.4. The π-group is defined as

$$C_L = 2\frac{F_L}{\rho V^2 S}$$

where F_L is the lift force, ρ is the density of ambient air, V is the speed of the air relative to the airplane, and S is the area of the wings from a top view. Estimate the lift force in newtons for a speed of 80 m/s, an air density of 1.1 kg/m^3, and a wing area (planform area) of 15 m^2.

PROBLEM 8.35

8.36 An airplane travels in air (p = 100 kPa, T = 10°C) at 150 m/s. If a 1/5 scale model of the plane is tested in a wind tunnel at 25°C, what must be the density of the air in the tunnel so that both the Reynolds-number and the Mach-number criteria are satisfied? For E_v take the adiabatic relation as given in Chapter 2. *Note:* The dynamic viscosity is independent of pressure.

8.37 Flow in a given pipe is to be tested with air and then with water. Assume that the velocities (V_A and V_W) are such that the flow with air is dynamically similar to the flow with water. Then for this condition, the magnitude of the ratio of the velocities, V_A/V_W, will be (a) less than unity, (b) equal to unity, (c) greater than unity.

8.38 A smooth pipe designed to carry crude oil (diameter = 48″, ρ = 1.75 slugs/ft^3, and $\mu = 4 \times 10^{-4}$ lbf-s/ft^2) is to be modeled with a smooth pipe 4 in. in diameter carrying water (T = 60°F). If the mean velocity in the prototype is 3 ft/s, what should be the mean velocity of water in the model to ensure dynamically similar conditions?

8.39 A student is competing in a contest to design a radio-controlled blimp. The drag force acting on the blimp depends on the Reynolds number, Re = $(\rho VD)/\mu$, where V is the speed of the blimp, D is the maximum diameter, ρ is the density of air, and μ is the viscosity of air. This blimp has a coefficient of drag (C_D) of 0.3. This π-group is defined as

$$C_D = 2\frac{F_D}{\rho V^2 A_p}$$

where F_D is the drag force, ρ is the density of ambient air, V is the speed of the air relative to the blimp, and $A_p = \pi D^2/4$ is the maximum section area of the blimp from a front view. Calculate the Reynolds number, the drag force in newtons, and the power in watts required to move the blimp through the air. Blimp speed is $V = 750$ mm/s, and the maximum diameter is 475 mm. Assume that ambient air is at 20°C.

PROBLEM 8.39

8.40 Using the Reynolds-number criterion, a 1:1 scale model of a torpedo is tested in a wind tunnel. If the velocity of the torpedo in water is 10 m/s, what should be the air velocity (standard atmospheric pressure) in the wind tunnel? The temperature for both tests is 10°C.

8.41 Colonization of the moon will require an improved understanding of fluid flow under reduced gravitational forces. The gravitational force on the moon is 1/5 that on the surface of the earth. An engineer is designing a model experiment for flow in a conduit on the moon. The important scaling parameters are the Froude number and the Reynolds number. The model will be full-scale. The kinematic viscosity of the fluid to be used on the moon is 0.5×10^{-5} m^2/s. What should be the kinematic viscosity of the fluid to be used for the model on earth?

8.42 A drying tower at an industrial site is 10 m in diameter. The air inside the tower has a kinematic viscosity of 4×10^{-5} m^2/s and enters at 12 m/s. A 1/15 scale model of this tower is fabricated to operate with water that has a kinematic viscosity of 10^{-6} m^2/s. What should the entry velocity of the water be to achieve Reynolds-number scaling?

8.43 A discharge meter to be used in a 40-cm pipeline carrying oil ($\nu = 10^{-5}$ m^2/s, $\rho = 860$ kg/m^3) is to be calibrated by means of a model (1/5 scale) carrying water ($T = 20°C$ and standard atmospheric pressure). If the model is operated with a velocity of 1 m/s, find the velocity for the prototype based on Reynolds-number scaling. For the given conditions, if the pressure difference in the model was measured as 3.0 kPa, what pressure difference would you expect for the discharge meter in the oil pipeline?

8.44 Water at 10°C flowing through a rough pipe 10 cm in diameter is to be simulated by air (20°C) flowing through the same pipe. If the velocity of the water is 1.5 m/s, what will the air velocity have to be to achieve dynamic similarity? Assume the absolute air pressure in the pipe to be 150 kPa. If the pressure difference between two sections of the pipe during air flow was measured as 780 Pa, what pressure difference occurs between these two sections when water is flowing under dynamically similar conditions?

8.45 The "noisemaker" B is towed behind the mine-sweeper A to set off enemy acoustic mines such as that shown at C. The drag force of the "noisemaker" is to be studied in a water tunnel at a 1/5 scale (the model is 1/5 the size of the full scale). If the full-scale towing speed is 5 m/s, what should be the water velocity in the water tunnel for the two tests to be exactly similar? What will be the prototype drag force if the model drag force is found to be 2400 N? Assume that sea water at the same temperature is used in both the full-scale and the model tests. FLUID SOLUTIONS

8.46 An experiment is being designed to measure aerodynamic forces on a building. The model is a 1/100 scale replica of the prototype. The wind velocity on the prototype is 25 ft/s, and the density is 0.0024 slugs/ft^3. The maximum velocity in the wind tunnel is 300 ft/s. The viscosity of the air flowing for the model and the prototype is the same. Find the density needed in the wind tunnel for dynamic similarity. A force of 50 lbf is measured on the model. What will the force be on the prototype?

8.47 A 60-cm valve is designed for control of flow in a petroleum pipeline. A 1/3 scale model of the full-size valve is to be tested with water in the laboratory. If the prototype flow rate is to be 0.5 m^3/s, what flow rate should be established in the laboratory test for dynamic similitude to be established? Also, if the pressure coefficient C_p in the model is found to be 1.07, what will be the corresponding C_p in the full-scale valve? The relevant fluid properties for the petroleum are $S = 0.82$ and $\mu = 3 \times 10^{-3}$ $N \cdot s/m^2$. The viscosity of water is 10^{-3} $N \cdot s/m^2$. FLUID SOLUTIONS

8.48 The moment acting on a submarine rudder is studied by a 1/60 scale model. If the test is made in a water tunnel and if the moment measured on the model is 2 m · N when the fresh-water speed in the tunnel is 10 m/s, what are the corresponding moment and speed for the prototype? Assume the prototype operates in sea water. Assume $T = 10°C$ for both the fresh water and the sea water.

8.49 A model hydrofoil is tested in a water tunnel. For a given angle of attack, the lift of the hydrofoil is measured to be 25 kN when the water velocity is 15 m/s in the tunnel. If the prototype hydrofoil is to be twice the size of the model, what lift force would be expected for the prototype for dynamically similar conditions? Assume a water temperature of 20°C for both model and prototype.

8.50 A 1/8 scale model of an automobile is tested in a pressurized wind tunnel. The test is to simulate the automobile traveling at 80 km/h in air at atmospheric pressure and 25°C. The wind tunnel operates with air at 25°C. At what pressure in the test section must the tunnel operate to have the same Mach and Reynolds numbers? The speed of sound in air at 25°C is 345 m/s.

8.51 If the tunnel in Prob. 8.50 were to operate at atmospheric pressure and 25°C, what speed would be needed to achieve the same Reynolds number for the prototype? At this speed, would you conclude that Mach-number effects were important?

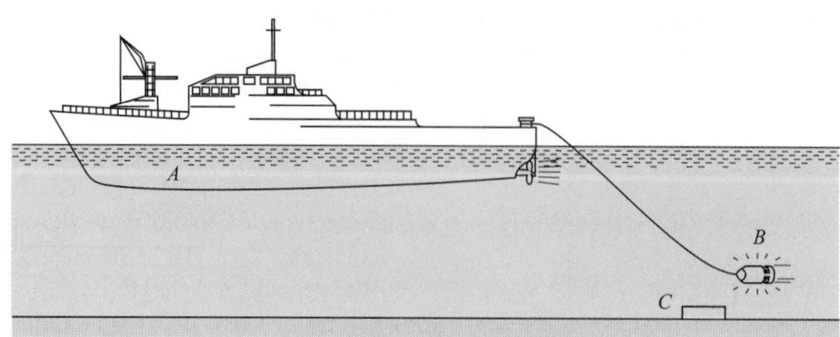

PROBLEM 8.45

8.52 An important parameter in rarefied gas dynamics is the ratio M/Re. If this ratio exceeds unity, the flow is rarefied. A spherical satellite 2 ft in diameter reenters the earth's atmosphere at 24,000 mph, where the pressure and temperature of the air are 22 psfa and –67°F. Can the flow around the satellite be classified as rarefied? The dynamic viscosity at –67°F is 3.0×10^{-7} lbf-s/ft^2, and the speed of sound is 975 ft/s.

FLUID SOLUTIONS **8.53** Experimental studies have shown that the condition for breakup of a droplet in a gas stream is

$$\mathrm{We}/\mathrm{Re}^{1/2} = 0.5$$

where Re is the Reynolds number and We is the Weber number based on the droplet diameter. What diameter water droplet would break up in a 25-m/s air stream at 20°C and standard atmospheric pressure? The surface tension of water is 7.3×10^{-2} N/m.

8.54 The critical Weber number for breakup of a liquid droplet is 6.0 based on the droplet diameter. The surface tension of heptane is 0.02 N/m. When a spray of heptane is atomized by discharging at 30 m/s into air at atmospheric pressure and 100°C, what is the expected size of the droplets?

8.55 Water is sprayed from a nozzle at 20 m/s into air at atmospheric pressure and 20°C. Estimate the size of the droplets produced if the Weber number for breakup is 6.0 based on the droplet diameter.

8.56 Determine the relationship between the kinematic viscosity ratio v_m/v_P and the scale ratio if both the Reynolds-number and the Froude-number criteria are to be satisfied in a given model test.

8.57 A hydraulic model, 1/20 scale, is built to simulate the flow conditions of a spillway of a dam. For a particular run, the waves downstream were observed to be 8 cm high. How high would be similar waves on the full-scale dam operating under the same conditions? If the wave period in the model is 2 s, what would the wave period in the prototype be?

8.58 The scale ratio between a model dam and its prototype is 1/25. In the model test, the velocity of flow near the crest of the spillway was measured to be 2.5 m/s. What is the corresponding prototype velocity? If the model discharge is 0.10 m^3/s, what is the prototype discharge?

8.59 A seaplane model is built at a 1/12 scale. To simulate take-off conditions at 125 km/h, what should be the corresponding model speed to achieve Froude-number scaling?

8.60 If the scale ratio between a model spillway and its prototype is 1/36, what velocity and discharge ratio will prevail between model and prototype? If the prototype discharge is 3000 m^3/s, what is the model discharge?

8.61 The depth and velocity at a point in a river are measured to be 20 ft and 15 ft/s, respectively. If a 1/64 scale model of this river is constructed and the model is operated under dynamically similar conditions to simulate the free-surface conditions, then what velocity and depth can be expected in the model at the corresponding point?

8.62 A 1/30 scale model of a spillway is tested in a laboratory. If the model velocity and discharge are 7.87 ft/s and 3.53 cfs, respectively, what are the corresponding values for the prototype?

8.63 Flow around a bridge pier is studied using a model at 1/10 scale. When the velocity in the model is 0.9 m/s, the standing wave at the pier nose is observed to be 2.5 cm in height. What are the corresponding values of velocity and wave height in the prototype?

8.64 A 1/25 scale model of a spillway is tested. The discharge in the model is 0.1 m^3/s. To what prototype discharge does this correspond? If it takes 1 min for a particle to float from one point to another in the model, how long would it take a similar particle to traverse the corresponding path in the prototype?

8.65 A tidal estuary is to be modeled at 1/250 scale. In the actual estuary, the maximum water velocity is expected to be 4 m/s and the tidal period is approximately 12.5 h. What corresponding velocity and period would be observed in the model?

8.66 The maximum wave force on a 1/36 model sea wall was found to be 80 N. For a corresponding wave in the full-scale wall, what full-scale force would you expect? Assume fresh water is used in the model study. Assume T = 10°C for both model and prototype water.

8.67 A model of a spillway is to be built at 1/25 scale. If the prototype has a discharge of 150 m^3/s, what must be the water discharge in the model to ensure dynamic similarity? The total force on part of the model is found to be 22 N. To what prototype force does this correspond?

8.68 A newly designed dam is to be modeled in the laboratory. The prime objective of the general model study is to determine the adequacy of the spillway design and to observe the water velocities, elevations, and pressures at critical points of the structure. The reach of the river to be modeled is 1200 m long, the width of the dam (also the maximum width of the reservoir upstream) is to be 300 m, and the maximum flood discharge to be modeled is 5000 m^3/s. The maximum laboratory discharge is limited to 0.90 m^3/s, and the floor space available for the model construction is 50 m long and 20 m wide. Determine the largest feasible scale ratio (model/prototype) for such a study.

8.69 A ship model 4 ft long is tested in a towing tank at a speed that will produce waves that are dynamically similar to those observed around the prototype. The test speed is 5 ft/s. What should the prototype speed be, given that the prototype length is 150 ft? Assume both the model and the prototype are to operate in fresh water.

8.70 The wave resistance of a model of a ship at 1/25 scale is 2 lbf at a model speed of 5 ft/s. What are the corresponding velocity and wave resistance of the prototype?

8.71 The wave resistance of a model of a ship at 1/20 scale ratio is 25 N at a speed of 4 m/s. What are the corresponding velocity and wave resistance of the prototype? Assume both the model and the prototype are to operate in fresh water.

8.72 A 1/20 scale model building that is rectangular in plan view and is three times as high as it is wide is tested in a wind tunnel. If the drag of the model in the wind tunnel is measured to be 200 N for a wind speed of 20 m/s, then the prototype building in a 40-m/s wind (same temperature) should have a drag of about (a) 40 kN, (b) 80 kN, (c) 230 kN, (d) 320 kN.

8.73 A model of a high-rise office building at 1/250 scale is tested in a wind tunnel to estimate the pressures and forces on the full-scale structure. The wind-tunnel air speed is 20 m/s at 20°C, and the full-scale structure is expected to withstand winds of 150 km/h (10°C). If the extreme values of the pressure coefficient are found to be 1.0, –2.7, and –0.8 on the windward wall, side wall, and leeward wall of the model, respectively, what corresponding pressures could be expected to act on the prototype? If the lateral wind force (wind force on building normal to wind direction) was measured as 20 N in the model, what lateral force might be expected in the prototype in the 150 km/h wind?

8.74 Experiments were carried out in a water tunnel and a wind tunnel to measure the drag force on an object. The water tunnel was operated with fresh water at 20°C, and the wind tunnel was operated at 20°C and atmospheric pressure. Three models were used with dimensions of 5 cm, 8 cm, and 15 cm. The drag force on each model was measured at different velocities. The following data were obtained.

In the water tunnel,

Model size, cm	Velocity, m/s	Force, N
5	1.0	0.064
5	4.0	0.69
5	8.0	2.20
8	1.0	0.135
8	4.0	1.52
8	8.0	4.52

and in the wind tunnel,

Model size, cm	Velocity, m/s	Force, N
8	10	0.025
8	40	0.21
8	80	0.64
15	10	0.06
15	40	0.59
15	80	1.82

The drag force is a function of the density, viscosity, velocity, and model size,

$$F_D = f(\rho, \mu, V, D)$$

Using dimensional analysis, express this equation using π-groups and then write a computer program or use a spread sheet to reduce the data. Plot the data using the dimensionless parameters.

8.75 Experiments are performed to measure the pressure drop in a pipe with water at 20°C and crude oil at the same temperature. Data are gathered with pipes of two diameters, 5 cm and 10 cm. The following data were obtained for pressure drop per unit length.

For water,

Pipe diameter, cm	Velocity, m/s	Pressure drop, N/m^3
5	1	210
5	2	730
5	5	3750
10	1	86
10	2	320
10	5	1650

and for crude oil,

Pipe diameter, cm	Velocity, m/s	Pressure drop, N/m^3
5	1	310
5	2	1040
5	5	5300
10	1	130
10	2	450
10	5	2210

The pressure drop per unit length is assumed to be a function of the pipe diameter, liquid density and viscosity, and the velocity,

$$\frac{\Delta p}{L} = f(\rho, \mu, V, D)$$

Perform a dimensional analysis to obtain the π-groups and then write a computer program or use a spreadsheet to reduce the data. Plot the results using the dimensionless parameters.

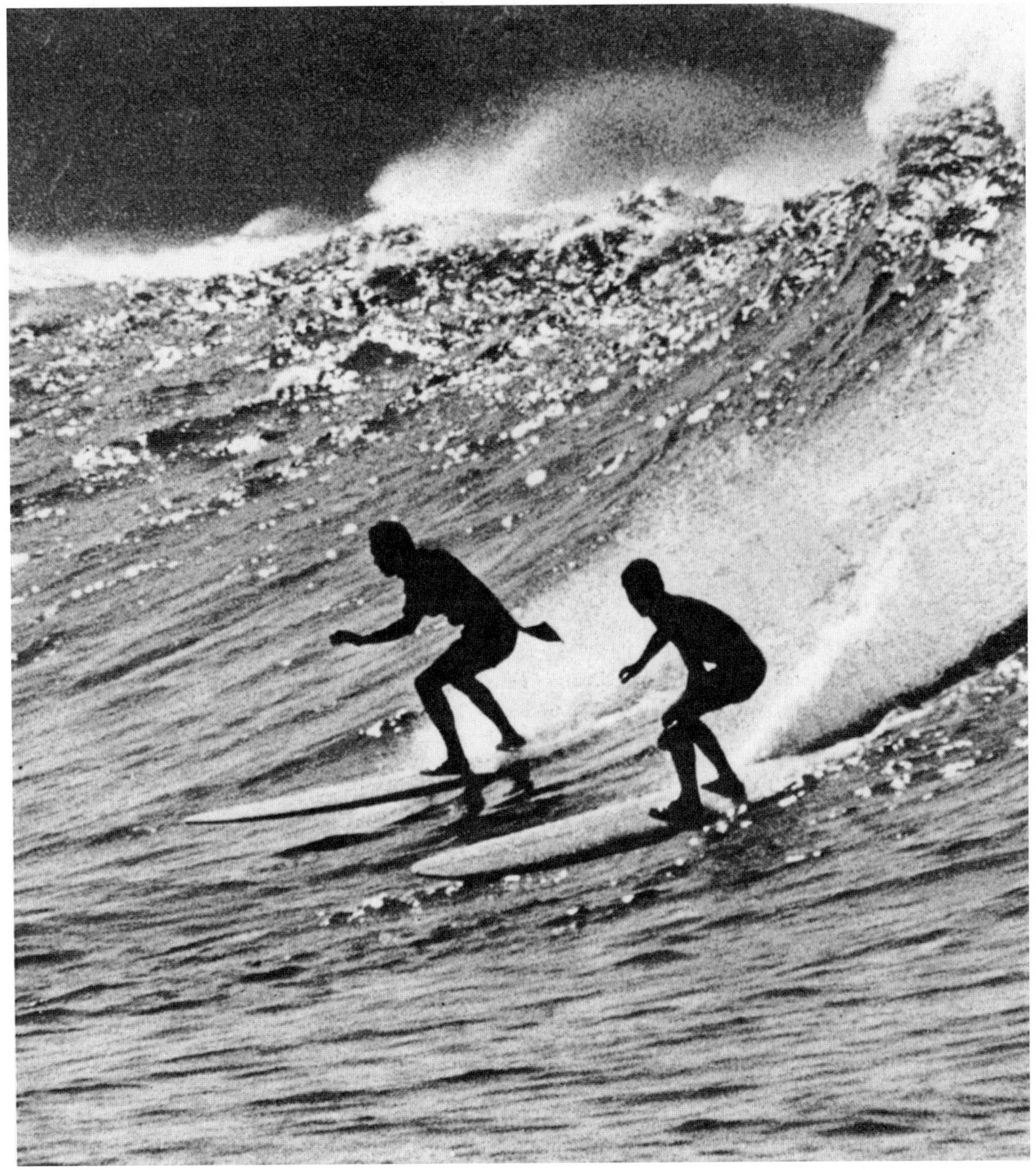

The high-speed attained by surfers is owing to the fact that "planing" of the board virtually eliminates wave drag, thus leaving only surface resistance, which is relatively small. (Courtesy Australian Tourist Commission)

CHAPTER 9

Surface Resistance

A fluid medium through which bodies such as aircraft or ships move exerts a resistance to motion on the bodies; this resistance is called *drag.* Aeronautical engineers and naval architects are vitally interested in the drag of an airplane or ship because the success or failure of the craft is directly related to its resistance to motion. If the drag is too large, the craft may be an economic failure because of the excessive costs (initial and operational) of the propulsion system.

The drag of a body is defined as the component of force in the direction of the free-stream velocity. There are two types of stresses producing drag on a body: shear stress and pressure. The difference is illustrated by comparing a flat plate oriented parallel to the flow direction and one oriented normal to the flow direction. For the plate oriented parallel to the flow direction, the only source of drag is skin friction, or shear force on the plate surface. In this orientation the pressure forces acting normal to the plate surface provide no contribution to drag. For the plate oriented normal to the flow direction, there is separation at the corner and a reduced pressure in the wake. The pressure difference across the plate produces a drag force. The shear stress on the plate surface provides no net force in the drag direction. Of course, both types of forces contribute to the drag on bodies with arbitrary shapes and orientations.

The drag, or surface resistance, from the shear stress is called *skin friction drag*, whereas the drag resulting from pressure forces is called *form drag*. In this chapter we will focus on the mechanics of skin friction drag, and in Chapter 11 we will include form drag. In this chapter we will focus on flat plates in unbounded flows, or *external flows*. We will introduce the concept of the *boundary layer*, which is a very thin layer of fluid at the plate surface over which the velocity changes from zero to the free-stream velocity. We will also address laminar and turbulent boundary layers and present some correlations that can be used to calculate boundary layer thickness, shear stress, and drag force.

9.1 Introduction

In general, the shear stress on a smooth plane surface is variable over the surface. Hence the total shear force in a given direction is obtained by integrating the component of shear stress in that direction over the total area of the surface. The shear stress on a smooth plane is a direct function of the velocity gradient next to the plane, as given by Eq. (2.6). Therefore, any problem involving shear stress also involves the flow pattern in the vicinity of the surface. The layer of fluid near the surface that has undergone a change in velocity because of the shear stress at the surface is called the *boundary layer,* and the general area of study of the flow pattern in the boundary layer, as well as of the associated shear stress at the boundary, is called *boundary layer theory.*

We will first consider the simplest cases of surface resistance resulting from uniform laminar flow in which the velocity gradient and the shear stress are constant. We will then examine a laminar boundary layer that develops from the leading edge of a smooth, flat plate. Finally, we will consider the characteristics of a turbulent boundary layer.

9.2

Surface Resistance with Uniform Laminar Flow

In this section we consider three two-dimensional laminar flows with parallel stream lines. The flows are steady and uniform; that is, there is no change in velocity along a stream line. Using the momentum equation, we will first derive a general equation for the flow velocity and then apply it to three specific problems: flow produced by a moving plate, liquid flow down an inclined plane, and flow between two stationary parallel plates.

Consider the control volume shown in Fig. 9.1, which is aligned with the flow direction s. The streamlines are inclined at an angle θ with respect to the horizontal plane. The control volume has dimensions $\Delta s \times \Delta y \times$ unity; that is, the control volume has a unit length into the page. By application of the momentum equation, the sum of the forces acting in the s direction is equal to the net outflow of momentum from the control volume. The flow is uniform, so the outflow of momentum is equal to the inflow and the momentum equation reduces to

$$\sum F_s = 0 \tag{9.1}$$

There are three forces acting on the matter in the control volume: the forces due to pressure, shear stress, and gravity. The net pressure force is

$$p\Delta y - \left(p + \frac{dp}{ds}\Delta s\right)\Delta y = -\frac{dp}{ds}\Delta s\Delta y$$

The net force due to shear stress is

$$\left(\tau + \frac{d\tau}{dy}\Delta y\right)\Delta s - \tau\Delta s = \frac{d\tau}{dy}\Delta y\Delta s$$

The component of gravitational force is $\rho g\Delta s\Delta y\sin\theta$. However, $\sin\theta$ can be related to the rate at which the elevation, z, decreases with increasing s and is given by $-dz/ds$. Thus the gravitational force becomes

$$\rho g\Delta s\Delta y \sin\theta = -\gamma\Delta y\Delta s\frac{dz}{ds}$$

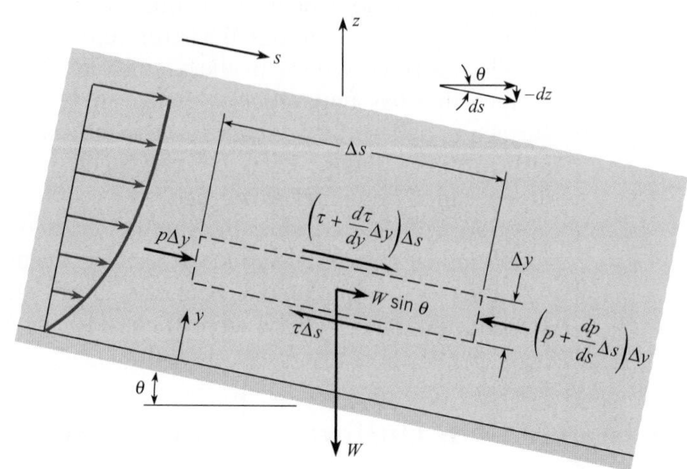

FIGURE 9.1

Control volume for analysis of uniform flow with parallel stream lines.

Summing all the forces to zero and dividing through by $\Delta s \Delta y$ results in

$$\frac{d\tau}{dy} = \frac{d}{ds}(p + \gamma z) \tag{9.2}$$

where we note that the gradient of the shear stress is equal to the change in piezometric pressure in the flow direction. The shear stress is equal to $\mu\ du/dy$, so the equation becomes

$$\frac{d^2u}{dy^2} = \frac{1}{\mu}\frac{d}{ds}(p + \gamma z) \tag{9.3}$$

We will now apply this equation to three flow configurations.

Flow Produced by a Moving Plate

Consider the flow between the two plates shown in Fig. 9.2. The lower plate is fixed and the upper plate is moving with a speed U. The plates are separated by a distance L. In this problem there is no pressure gradient in the flow direction $(dp/ds = 0)$ and the streamlines are in the horizontal direction $(dz/ds = 0)$, so Eq. (9.3) reduces to

$$\frac{d^2u}{dy^2} = 0$$

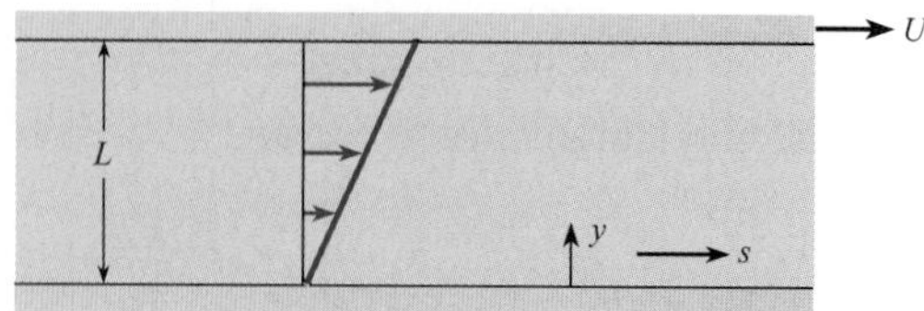

FIGURE 9.2

Flow generated by a moving plate.

The two boundary conditions are

$$u = 0 \quad \text{at} \quad y = 0$$
$$u = U \quad \text{at} \quad y = L$$

Integrating this equation twice gives

$$u = C_1 y + C_2$$

Applying the boundary conditions gives

$$u = \frac{y}{L}U \tag{9.4}$$

which shows that the velocity profile is linear between the two plates. The shear stress is constant and equal to

$$\tau = \mu\frac{du}{dy} = \mu\frac{U}{L} \tag{9.5}$$

This flow is known as a Couette flow after a French scientist, M. Couette, who did pioneering work on the flow between parallel plates and rotating cylinders.

example 9.1

If the fluid between the plates of Fig. 9.2 is SAE 30 lubricating oil at $T = 38°\text{C}$, if the plates are spaced 0.3 mm apart, and if the upper plate is moved at a velocity of 1.0 m/s, what is the surface resistance for 1.0 m^2 of the upper plate?

Solution From the Appendix (Table A.4), $\mu = 1.0 \times 10^{-1}\ \text{N} \cdot \text{s/m}^2$. Then

$$\tau = \mu\frac{du}{dy} = \mu\frac{U}{L}$$
$$= (1.0 \times 10^{-1}\ \text{N} \cdot \text{s/m}^2)(1.0\ \text{m/s})/(3 \times 10^{-4}\text{m}) = 333\ \text{N/m}^2$$
$$F_s = \tau A = 333\ \text{N/m}^2 \times 1.0\ \text{m}^2 = 333\ \text{N}$$

◁

Liquid Flow Down an Inclined Plane

Consider the flow down an inclined plane shown in Fig. 9.3. The liquid layer, which has a depth d, has a free surface where the pressure is constant, so along the surface $dp/ds = 0$. The pressure is hydrostatic across any section in the z direction but does not change in the stream line direction. The shear stress at the free surface between the air and liquid is small and therefore neglected. In this problem the gravitational force is balanced by the shear stress. The differential equation, Eq. (9.3), reduces to

$$\frac{d^2u}{dy^2} = \frac{\gamma}{\mu}\frac{dz}{ds} = -\frac{\gamma}{\mu}\sin\theta$$

The boundary conditions are

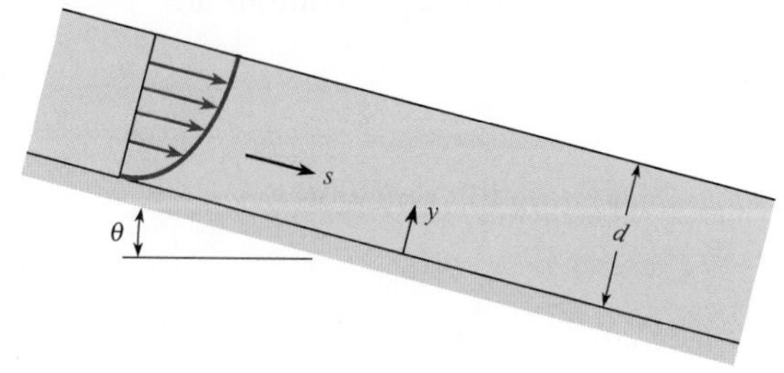

FIGURE 9.3

Uniform flow of a liquid layer on an inclined surface.

$$u = 0 \quad \text{at} \quad y = 0$$

$$\frac{du}{dy} = 0 \quad \text{at} \quad y = d$$

Integrating this equation once gives

$$\frac{du}{dy} = -\frac{y\gamma}{\mu}\sin\theta + C$$

Applying the boundary conditions at $y = d$ shows

$$C = \frac{\gamma d}{\mu}\sin\theta$$

so

$$\frac{du}{dy} = \frac{\gamma}{\mu}\sin\theta(d - y)$$

Integrating the second time results in

$$u = \frac{\gamma}{\mu}\sin\theta\left(yd - \frac{y^2}{2}\right) + C$$

and the constant of integration is set equal to zero to satisfy the boundary condition at $y = 0$. This equation can be rewritten as

$$u = \frac{\gamma\sin\theta}{2\mu}(2yd - y^2) = \frac{g\sin\theta}{2\nu}y(2d - y) \tag{9.6}$$

The resulting profile is a parabola with the maximum velocity occurring at the free surface. The maximum velocity is equal to

$$u_{\text{max}} = \frac{\gamma d^2}{2\mu}\sin\theta \tag{9.7}$$

We now obtain the discharge per unit width by integrating the velocity u over the depth of flow:

$$q = \int_0^d u\,dy = \frac{\gamma}{2\mu}\sin\theta\left[dy^2 - \frac{y^3}{3}\right]_0^d$$
$$= \frac{1}{3}\frac{\gamma}{\mu}d^3\sin\theta \tag{9.8}$$

The average velocity is now obtained by dividing Eq. (9.8) by the cross-sectional area d:

$$V = \frac{q}{d} = \frac{1}{3}\frac{\gamma}{\mu}d^2\sin\theta$$

This reduces to

$$V = \frac{gd^2}{3\nu}\sin\theta \tag{9.9a}$$

The slope, $S_0 = \tan\theta$, is approximately equal to $\sin\theta$ for small slopes. Thus Eq. (9.9*a*) can be given as

$$V = \frac{gS_0d^2}{3\nu} \tag{9.9b}$$

Experiments have shown that if the Reynolds number based on the depth of the flow, $\mathrm{Re} = Vd/\nu$, is less than 500, one can expect laminar flow in this situation. If the Reynolds number is greater than 500, the flow may become turbulent and the results of this section are no longer valid.

example 9.2

Crude oil, $\nu = 9.3 \times 10^{-5}\,\mathrm{m^2/s}$, $S = 0.92$, flows over a flat plate that has a slope of $S_0 = 0.02$. If the depth of flow is 6 mm, what is the maximum velocity and what is the discharge per meter of width of plate? Also determine the Reynolds number for this flow.

Solution First, we assume that the flow is laminar. Since $S_0 \approx \sin\theta$,

$$u = \frac{gS_0}{2\nu}y(2d - y)$$

Therefore, the maximum velocity will occur when y is maximum ($y = d$), or

$$u_{\mathrm{max}} = \frac{(9.81\ \mathrm{m/s^2})(0.02)(0.006)^2\ \mathrm{m^2}}{2 \times 9.3 \times 10^{-5}\ \mathrm{m^2/s}} = 0.038\ \mathrm{m/s} \quad \triangleleft$$

The discharge per meter of width is given by Eq. (9.3), and for a small slope we can replace $\sin\theta$ by S_0:

$$q = \frac{1}{3}\frac{\gamma}{\mu}S_0d^3 = \frac{1}{3}\frac{\gamma}{\rho\nu}S_0d^3 = \frac{1}{3}\frac{g}{\nu}S_0d^3$$

Then $$q = \frac{1}{3}\frac{(9.81\ \text{m/s}^2)}{(9.3 \times 10^{-5}\ \text{m}^2/\text{s})}(0.02)(0.006)^3\text{m}^3 = 1.52 \times 10^{-4}\ \text{m}^2/\text{s} \quad \triangleleft$$

Now we check the Reynolds number to see if our original assumption of laminar flow ($\text{Re} < 500$) was correct.

$$24\,\text{Re} = \frac{Vd}{\nu} = \frac{q}{\nu} = \frac{1.52 \times 10^{-4}\ \text{m}^2/\text{s}}{9.3 \times 10^{-5}\ \text{m}^2/\text{s}} = 1.63 \quad \triangleleft$$

The Reynolds number is less than 500. Therefore, our assumption of laminar flow was valid and the use of Eqs. (9.6) and (9.8) is justified.

Flow between Stationary Parallel Plates

Consider the two parallel plates separated by a distance B in Fig. 9.4. In this case, the flow velocity is zero at the surface of both plates, so the boundary conditions for Eq. (9.3) are

$$u = 0 \quad \text{at} \quad y = 0$$
$$u = 0 \quad \text{at} \quad y = B$$

The gradient in piezometric pressure is constant along a stream line. Integrating Eq. (9.3) twice gives

$$u = \frac{y^2}{2\mu}\frac{d}{ds}(p + \gamma z) + C_1 y + C_2$$

To satisfy the boundary condition at $y = 0$, we set $C_2 = 0$. Applying the boundary condition at $y = B$ requires that C_1 be

$$C_1 = -\frac{B}{2\mu}\frac{d}{ds}(p + \gamma z)$$

so the final equation for the velocity is

$$u = -\frac{1}{2\mu}\frac{d}{ds}(p + \gamma z)(By - y^2) = -\frac{\gamma}{2\mu}(By - y^2)\frac{dh}{ds} \tag{9.10}$$

which is a parabolic profile with the maximum velocity occurring on the centerline between the plates. The maximum velocity is

$$u_{\text{max}} = -\left(\frac{B^2}{8\mu}\right)\frac{d}{ds}(p + \gamma z) \tag{9.11a}$$

or in terms of piezometric head

$$u_{\text{max}} = -\left(\frac{B^2\gamma}{8\mu}\right)\frac{dh}{ds} \tag{9.11b}$$

The fluid always flows in the direction of decreasing piezometric pressure or piezometric head, so dh/ds is negative, giving a positive value for u_{max}.

FIGURE 9.4

Uniform flow between two stationary plates.

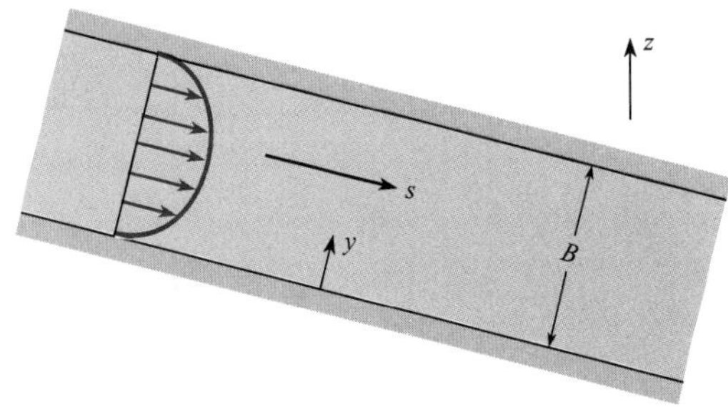

We now find the discharge per unit width by integrating the velocity over the distance between the plates:

$$q = \int_0^B u\,dy = -\left(\frac{B^3}{12\mu}\right)\frac{d}{ds}(p + \gamma z) = -\left(\frac{B^3\gamma}{12\mu}\right)\frac{dh}{ds} \tag{9.12}$$

The average velocity is

$$V = \frac{q}{B} = -\left(\frac{B^2}{12\mu}\right)\frac{d}{ds}(p + \gamma z) = \frac{2}{3}u_{max} \tag{9.13}$$

As was the case with unconfined laminar liquid flow over a plane surface, the velocity distribution is parabolic. However, in this case the maximum velocity occurs midway between the two plates. Note that flow is the result of a change of the piezometric head, not just a change of p or z alone. Experiments reveal that if the Reynolds number (VB/ν) is less than 1000, the flow is laminar. For a Reynolds number greater than 1000, the flow may be turbulent and the equations in this section invalid.

example 9.3

Oil having a specific gravity of 0.8 and a viscosity of $2 \times 10^{-2}\ \text{N}\cdot\text{s/m}^2$ flows downward between two vertical smooth plates spaced 10 mm apart. If the discharge per meter of width is $0.01\ \text{m}^2/\text{s}$, what is the pressure gradient dp/ds for this flow?

Solution First we check to see if the flow is laminar or turbulent:

$$\text{Re} = \frac{VB}{\nu} = \frac{VB\rho}{\mu} = \frac{q\rho}{\mu}$$

$$= \frac{(0.01\ \text{m}^2/\text{s}) \times 800\ \text{kg/m}^3}{0.02\ \text{N}\cdot\text{s/m}^2} = 400$$

The Reynolds number is less than 1000; therefore, the flow is laminar. The kinematic viscosity is

$$\nu = \mu/\rho = \frac{2 \times 10^{-2}\ \text{N}\cdot\text{s/m}^2}{0.8 \times 1000\ \text{kg/m}^3} = 2.5 \times 10^{-5}\ \text{m}^2/\text{s}$$

Using Eq. (9.12), the gradient in piezometric head can be written as

$$\frac{dh}{ds} = -\frac{12\mu}{B^3\gamma}q = -\frac{12\nu}{B^3 g}q$$

and gives a value of

$$\frac{dh}{ds} = -\frac{12 \times 2.5 \times 10^{-5}\text{m}^2/\text{s}}{0.01^3 \times 9.81\ \text{m}^4/\text{s}^2} \times 0.01\text{m}^2/\text{s} = -0.306$$

However,
$$\frac{dh}{ds} = \frac{d}{ds}\left(\frac{p}{\gamma} + z\right)$$

Therefore
$$\frac{d}{ds}\left(\frac{p}{\gamma} + z\right) = -0.306$$

Because the plates are vertically oriented and s is positive downward, $dz/ds = -1$. Thus

$$\frac{d(p/\gamma)}{ds} = 1 - 0.306$$

or
$$\frac{dp}{ds} = (0.8 \times 9810\ \text{N/m}^3) \times 0.694 = 5447\ \text{N/m}^2 \text{ per meter}$$ ◁

In other words, the pressure is increasing downward at a rate of 5.45 kPa per meter of length of plate.

Fully Developed Flow between Parallel Plates Using the Navier-Stokes Equations

The flow field between parallel plates will be derived here using the continuity and Navier-Stokes equations. Reference is made to Fig. 9.4, where y is the coordinate normal to the plates and x is the flow direction (same as s). The flow field is fully developed (uniform), so the derivatives $\partial u/\partial x$ and $\partial v/\partial x$ are zero. Also the flow is steady, so $\partial u/\partial t$ and $\partial v/\partial t$ are also zero. The component of the gravity vector in the x direction is $g\sin\theta$ and in the y direction, $-g\cos\theta$.

The continuity equation for planar incompressible flow is

$$\frac{\partial u}{\partial x} + \frac{\partial v}{\partial y} = 0$$

Since $\partial u/\partial x = 0$ the continuity equation reduces to

$$\frac{\partial v}{\partial y} = 0$$

or
$$v = \text{const.}$$

At the surface of the plate ($y = 0$), the velocity is zero so

$$v = 0$$

everywhere in the flow field.

The Navier-Stokes equation in the y direction is

$$\rho\frac{\partial v}{\partial t}+\rho u\frac{\partial v}{\partial x}+\rho v\frac{\partial v}{\partial y}=-\frac{\partial p}{\partial y}+\mu\left(\frac{\partial^2 v}{\partial x^2}+\frac{\partial^2 v}{\partial y^2}\right)-\rho g\cos\theta$$

Because v is zero everywhere, there is no acceleration of the fluid in the y direction, so this component of the Navier-Stokes equation reduces to

$$\frac{\partial p}{\partial y}=-\rho g\cos\theta$$

Integrating over y, one has

$$p=-y\rho g\cos\theta+p_{y=0}(x)$$

where $p_{y=0}(x)$ is the pressure distribution along the lower wall. This equation shows that the pressure decreases with elevation in the duct, as expected. In fact, $y\cos\theta$ is equal to z, so this equation can be written as the equation for hydrostatic pressure variation, namely

$$p+\gamma z=p_{y=0}(x)$$

The pressure gradient in the x direction is

$$\frac{\partial p}{\partial x}=\frac{dp_{y=0}}{dx}=\frac{dp}{dx}$$

and is the same for all values of y across the duct for any value of x.

The Navier-Stokes equation in the x direction is

$$\rho\frac{\partial u}{\partial t}+\rho u\frac{\partial u}{\partial x}+\rho v\frac{\partial u}{\partial y}=-\frac{\partial p}{\partial x}+\mu\left(\frac{\partial^2 u}{\partial x^2}+\frac{\partial^2 u}{\partial y^2}\right)+\rho g\sin\theta$$

For steady, fully developed flow the left-hand side of this equation reduces to zero (no acceleration in the x direction), and the equation becomes

$$\frac{\partial p}{\partial x}-\rho g\sin\theta=\mu\frac{\partial^2 u}{\partial y^2}$$

Because u is a function of y only $(\partial u/\partial x=0)$ and $\partial p/\partial x$ is a function only of x, this equation becomes

$$\frac{dp}{dx}-\rho g\sin\theta=\mu\frac{d^2 u}{dy^2} \tag{9.14}$$

The slope of the duct can be expressed as

$$\sin\theta=-\frac{dz}{dx}$$

which, when substituted into Eq. (9.7), yields

$$\frac{d}{dx}(p + \gamma z) = \mu \frac{d^2 u}{dy^2}$$

or

$$\gamma \frac{dh}{ds} = \mu \frac{d^2 u}{dy^2}$$

which is the same equation as used in the previous section to obtain the velocity distribution.

9·3

Qualitative Description of the Boundary Layer

Flow Pattern in a Boundary Layer

As we noted in Section 9.1, the *boundary layer* is the region next to a boundary of an object in which the fluid has had its velocity changed because of the shearing resistance created by the boundary. Outside the boundary layer the velocity is essentially the same as if an ideal (nonviscous) fluid were flowing past the object. In Section 9.2 we considered liquid flow over a plane surface and flow between parallel plates. In a narrow sense, each of these is boundary layer flow—that is, uniform flow for a fully developed laminar boundary layer. Now we will look at a boundary layer that is still developing. That is, we will look at boundary layers that grow in thickness and have significant changes of velocity with distance along the boundary. To visualize the flow pattern associated with the boundary layer, we will qualitatively analyze the interaction between a fluid and the surface of a thin, flat plate as the fluid (for example, air) passes by the plate. Figure 9.5 illustrates such a plate. Fluid passes over the top and underneath the plate, so two boundary layers are depicted in Fig. 9.5 (one above and one below the plate). In Fig. 9.5 the fluid has a uniform velocity U_0 before it reaches the vicinity of the plate. However, the fluid touching the plate has zero velocity (the velocity of the plate) because of the no-slip condition characterizing continuum flows. Therefore, a velocity gradient must exist between the fluid in the free stream and the fluid next to the plate. Consistent with this gradient, there is a shear stress at the plate surface. As the fluid particles next to the plate pass the leading edge of the plate, a retarding force (from the shear stress) begins to act on them. As these particles progress downstream, they continue to be subjected to shear stress from the plate, so they continue to decelerate. In addition, these particles (because of their lower velocity) retard other particles adjacent to them but farther out from the plate. Thus the boundary layer becomes thicker, or "grows," in the downstream direction. The broken line in Fig. 9.5 identifies the outer limit of the boundary layer. Because the boundary layer becomes thicker, the velocity gradient becomes less steep as one proceeds along the plate.

Thickening of the laminar boundary layer continues smoothly in the downstream direction until the thickness is so great that the flow becomes unstable and the boundary layer becomes turbulent. In the turbulent boundary layer, eddies mix higher-velocity fluid into the region close to the wall so that the velocity gradient du/dy at the wall becomes greater than it is at the wall in the laminar boundary layer just upstream of the transition point.

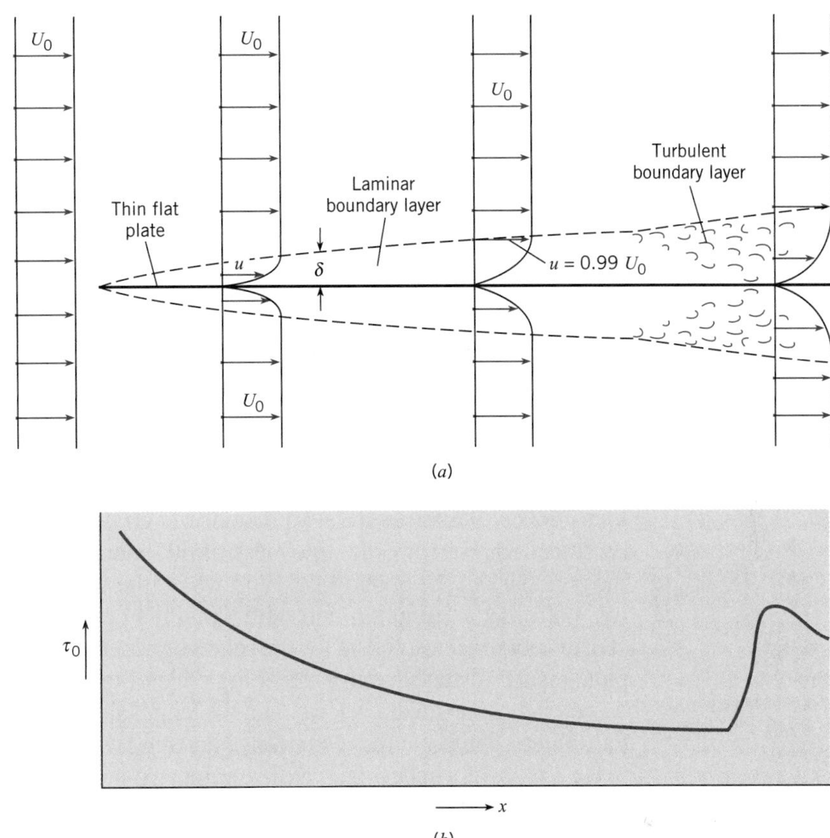

FIGURE 9.5

Development of boundary layer and distribution of shear stress along a thin, flat plate. (a) Flow pattern in boundary layers above and below the plate. (b) Shear stress distribution on either side of the plate.

If one considers an element of fluid in a boundary layer (similar to the element shown in Fig. 4.18*a*), it becomes obvious that the side normal to the boundary will rotate while the side parallel to the boundary will have virtually no rotation. Thus boundary layers, whether laminar or turbulent, are cases of rotational flow.

Shear Stress Distribution along the Boundary

In the foregoing section we noted that it is the shearing force of the plate that decelerates the fluid to produce the boundary layer, but no mention was made of the way the shear stress changes along the boundary. Because the shear stress is given by $\tau = \mu\, du/dy$, it is easy to visualize that the shear stress must be relatively large near the leading edge of the plate where the velocity gradient is steep, and that it becomes smaller as the velocity gradient becomes smaller in the downstream direction. However, where the boundary layer becomes turbulent, the shear stress at the boundary again becomes larger, consistent with the greater velocity gradient next to the wall in the turbulent boundary layer. Figure 9.5*b* depicts such a distribution of shear stress. These qualitative aspects of the boundary layer serve as a foundation for the quantitative relations presented in the next section.

9.4

Quantitative Relations for the Laminar Boundary Layer

Boundary Layer Equations

In 1904 Prandtl (1) first stated the essence of the boundary layer hypothesis, which is that viscous effects are concentrated in a thin layer of fluid (the boundary layer) next to solid boundaries. Along with his discussion of the qualitative aspects of the boundary layer, he also simplified the general equations of motion of a fluid (Navier-Stokes equations) for application to the boundary layer. Then in 1908, Blasius, one of Prandtl's students, obtained a solution for the flow in a laminar boundary layer (2). The solution was for the case of zero pressure gradient along the plate, $dp/dx = 0$, and one of Blasius's key assumptions was that the shape of the nondimensional velocity distribution did not vary from section to section along the plate. That is, he assumed that a plot of the relative velocity u/U_0 versus the relative distance from the boundary, y/δ, would be the same at each section. Here δ is the thickness of the boundary layer, defined as the distance from the boundary to the point in the fluid where the velocity is 99% of the free-stream velocity. With this assumption and with Prandtl's equations of motion for boundary layers, Blasius obtained a solution for the relative velocity distribution, shown in Fig. 9.6. In this plot, x is the distance from the leading edge of the plate, and Re_x is the Reynolds number based on the free-stream velocity and the length along the plate ($\text{Re}_x = U_0x/\nu$). In Fig. 9.6 the outer limit of the boundary layer ($u/U_0 = 0.99$) occurs at approximately $y\,\text{Re}_x^{1/2}/x = 5$. Since $y = \delta$ at this point, we have a relationship for the *boundary layer thickness* in laminar flow on a flat plate:

$$\frac{\delta}{x}\text{Re}_x^{1/2} = 5 \quad \text{or} \quad \delta = \frac{5x}{\text{Re}_x^{1/2}} \tag{9.15}$$

We can also obtain from Fig. 9.6 the inverse slope at the boundary, which is equal to 0.332, or

$$\left.\frac{d(u/U_0)}{d[(y/x)\text{Re}_x^{1/2}]}\right|_{y=0} = 0.332$$

However, at any given section, x, Re_x, and U_0 are constants. Therefore, we express the velocity gradient at the boundary as follows:

$$\left.\frac{du}{dy}\right|_{y=0} = 0.332\frac{U_0}{x}\text{Re}_x^{1/2}$$

$$\left.\frac{du}{dy}\right|_{y=0} = 0.332\frac{U_0}{x}\left(\frac{U_0x}{\nu}\right)^{1/2} \tag{9.16}$$

$$\left.\frac{du}{dy}\right|_{y=0} = 0.332\frac{U_0^{3/2}}{x^{1/2}\nu^{1/2}}$$

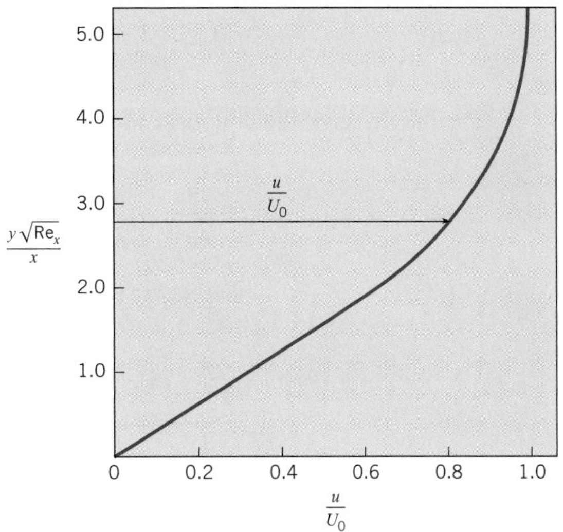

FIGURE 9.6

Velocity distribution in laminar boundary layer. [After Blasius (2)]

Equation (9.16) shows that the velocity gradient decreases as the distance x along the boundary increases.

Shear Stress

The shear stress at the boundary is obtained by multiplying the velocity gradient at the wall by the absolute viscosity:

$$\tau_0 = 0.332\mu\frac{U_0}{x}\mathrm{Re}_x^{1/2} \tag{9.17}$$

Equation (9.17) is used to obtain the local shear stress at any given section (any given value of x) for the laminar boundary layer, as shown in the following example.

example 9.4

Crude oil at 70°F ($\nu = 10^{-4}$ ft^2/s, S = 0.86) with a free-stream velocity of 10 ft/s flows past a thin, flat plate that is 4 ft wide and 6 ft long in a direction parallel to the flow. Determine and plot the boundary layer thickness and the shear stress distribution along the plate.

Solution

$$\mathrm{Re}_x = \frac{U_0 x}{\nu} = \frac{10x}{10^{-4}} = 10^5 x$$

$$\mathrm{Re}_x^{1/2} = 3.16(10^2 x^{1/2})$$

The shear stress is given by

$$\tau_0 = 0.332\mu\frac{U_0}{x}\mathrm{Re}_x^{1/2}$$

where $\mu = \rho\nu = 1.94 \times 0.86 \times 10^{-4} = 1.67 \times 10^{-4}$ lbf-s/ft^2

Then $$\tau_0 = 0.332(1.67 \times 10^{-4})\frac{10}{x}(3.16)(10^2 x^{1/2}) = \frac{0.175}{x^{1/2}} \text{ psf}$$

The thickness of the boundary layer is

$$\delta = \frac{5x}{\text{Re}_x^{1/2}} = \frac{5x}{3.16(10^2 x^{1/2})} = 1.58(10^{-2} x^{1/2}) \text{ ft}$$

$$= 1.58(12)(10^{-2} x^{1/2}) = 0.190 x^{1/2} \text{ in.}$$

The results for Example 9.4 are plotted in the accompanying figure and listed in Table 9.1.

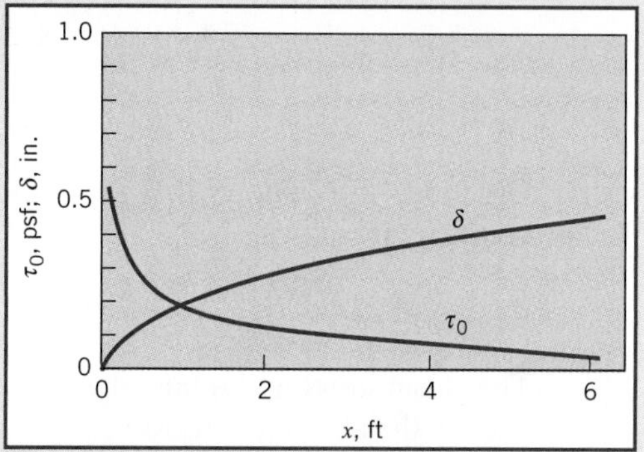

TABLE 9.1 RESULTS—δ AND τ_0 FOR DIFFERENT VALUES OF x

	x = 0.1 ft	x = 1.0 ft	x = 2 ft	x = 4 ft	x = 6 ft
$x^{1/2}$	0.316	1.00	1.414	2.00	2.45
τ_0, psf	0.552	0.174	0.123	0.087	0.071
δ, ft	0.005	0.016	0.022	0.031	0.039
δ, in.	0.060	0.189	0.270	0.380	0.466

Shearing Resistance for a Surface of Given Size

Because the shear stress at the boundary, τ_0, varies along the plate, it is necessary to integrate this stress over the entire surface to obtain the total shearing force on the surface:

$$F_s = \int_0^L \tau_0 B\, dx \tag{9.18}$$

where F_s is the surface resistance produced by viscous stresses on one side of the plate, B is the width of the plate, and L is the length. When Eq. (9.17) is substituted into Eq. (9.18), we get

$$F_s = \int_0^L 0.332 B\mu \frac{U_0 U_0^{1/2} x^{1/2}}{x\nu^{1/2}}\, dx$$
$$= 0.664 B\mu U_0 \frac{U_0^{1/2} L^{1/2}}{\nu^{1/2}} \tag{9.19}$$
$$= 0.664 B\mu U_0 \mathrm{Re}_L^{1/2}$$

In Eq. (9.19) Re_L is the Reynolds number based on the approach velocity and the length of the plate.

Shear Stress Coefficients

It is convenient to express the shear stress at the boundary, τ_0, and the total shearing force F_s in terms of π-groups involving the dynamic pressure of the free stream, $\rho U_0^2/2$. The *local sheer stress coefficient*, c_f, is defined as

$$c_f = \frac{\tau_0}{\rho U_0^2/2} \tag{9.20}$$

If one can evaluate the value of c_f, then for a given density, ρ, and free-stream velocity, U, one can calculate the shear stress at a point on the boundary where that c_f applies. In fact, by combining Eq. (9.17) with Eq. (9.20), one obtains a formula for calculating the value of the local shear stress coefficient for a given Reynolds number:

$$c_f = \frac{0.664}{\mathrm{Re}_x^{1/2}} \tag{9.21}$$

Equation (9.21) is for a laminar boundary layer. A formula for calculating the local shear stress coefficient for a *turbulent boundary layer* is developed in Section 9.5.

The total shearing force, as given by Eq. (9.18), can also be expressed as $F_s = \int_A \tau_0\, dA$. However, from Eq. (9.20), τ_0 can now be given in terms of c_f; therefore,

$$F_s = \int_A c_f(\rho U_0^2/2)\, dA$$

Here $\rho U_0^2/2$ is a constant; therefore, the above equation can be written as

$$F_s = \frac{\rho U_0^2}{2} \int_A c_f\, dA$$

Now if we divide both sides by the surface area of the plate, A, we have

$$\frac{F_s}{A} = \frac{\rho U_0^2}{2} \frac{\int_A c_f\, dA}{A}$$

The *average shear stress coefficient*, C_f, is defined as

$$C_f = \frac{F_s/A}{\rho U_0^2/2} = \frac{\int_A c_f\, dA}{A} \tag{9.22}$$

The shear resistance for a rectangular flat plate is

$$F_s = C_f \frac{\rho U_0^2}{2} BL$$

The formula for calculating the value of the average shear stress coefficient for a Reynolds number based on the plate length is obtained by combining Eq. (9.19) with Eq. (9.22):

$$C_f = \frac{1.33}{\text{Re}_L^{1/2}} \tag{9.23}$$

Equation (9.23) is for a laminar boundary layer. In Section 9.5 we develop the formula for calculating the average shear stress coefficient for a *turbulent boundary layer.*

example 9.5

For the conditions of Example 9.4, determine the resistance of one side of the plate.

Solution

$$F_s = \frac{C_f BL\rho U_0^2}{2}$$

Here

$$C_f = \frac{1.33}{\text{Re}_L^{1/2}} = \frac{1.33}{(3.16)(10^2)(6^{1/2})} = 0.0017$$

Then

$$F_s = 0.0017(4)(6)(0.86)(1.94)\left(\frac{10^2}{2}\right) = 3.40 \text{ lbf} \quad \triangleleft$$

Experiment versus Theory for the Laminar Boundary Layer

Experimental evidence indicates that the Blasius solution is valid except very near the leading edge of the plate. In the vicinity of the leading edge, an error results because of certain simplifying assumptions. However, the discrepancy is not significant for most engineering problems. For very thin, smooth plates the laminar boundary layer can be expected to change to a turbulent boundary layer at a Reynolds number of approximately 500,000. However, if the approach flow is turbulent and/or if the plate is rough, the laminar boundary layer can be expected to become turbulent at a smaller Reynolds number.

9.5 Quantitative Relations for the Turbulent Boundary Layer

Velocity Distribution in the Turbulent Boundary Layer along a Smooth Wall

The turbulent boundary layer is more complex than the laminar boundary layer. The former has three zones of flow that require different equations for the velocity distribution, as opposed to the single relationship of the laminar case. Figure 9.7 shows a portion of a turbulent boundary layer in which the three different zones of flow are identified. Each of these zones will be discussed separately.

The zone immediately adjacent to the wall is a layer of fluid that, because of the damping effect of the wall, remains relatively smooth even though most of the flow in the boundary layer is turbulent. This very thin layer is called the *viscous sublayer.* The velocity distribution in this layer is related to the shear stress and viscosity by Newton's viscosity law, which was introduced in Chapter 2. For convenience, it is repeated here as follows: $\tau = \mu \, du/dy$. In the viscous sublayer, τ is virtually constant and equal to the shear stress at the wall, τ_0. Thus $du/dy = \tau_0/\mu$, which on integration yields

$$u = \frac{\tau_0 y}{\mu} \tag{9.24}$$

If we multiply and divide the right side of Eq. (9.24) by ρ, we obtain the following:

$$\begin{aligned} u &= \frac{\tau_0/\rho}{\mu/\rho} y \\ \frac{u}{\sqrt{\tau_0/\rho}} &= \frac{\sqrt{\tau_0/\rho}}{\nu} y \end{aligned} \tag{9.25}$$

The combination of variables $\sqrt{\tau_0/\rho}$ recurs again and again in derivations involving boundary layer theory and has been given the special name *shear velocity.* The name is appropriate because the variable τ_0 relates to the shear stress and the units of the combination are those of velocity. The shear velocity (which is also sometimes called *friction velocity*) is symbolized as u_*. Thus, by definition,

$$u_* = \sqrt{\frac{\tau_0}{\rho}} \tag{9.26}$$

Now, when we substitute u_* for $\sqrt{\tau_0/\rho}$ in Eq. (9.25), we have the customary form for expressing the velocity distribution in the viscous sublayer:

$$\frac{u}{u_*} = \frac{y}{\nu/u_*} \tag{9.27}$$

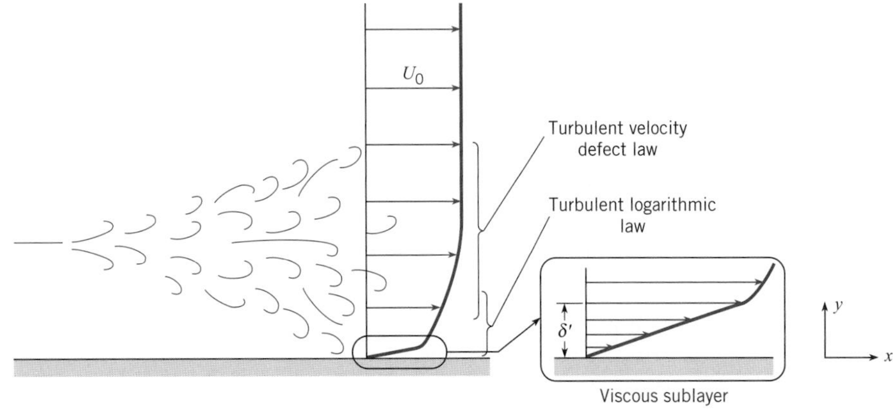

FIGURE 9.7

Sketch of zones in turbulent boundary layer.

Equation (9.27) shows that the relative velocity (velocity relative to the shear velocity) in the viscous sublayer is equal to a dimensionless distance from the wall. Experimental results show that the viscous sublayer occurs only in a film of fluid for which yu_*/ν is less than approximately 5. Consequently, the thickness of the viscous sublayer, identified by δ', is given as

$$\delta' = \frac{5\nu}{u_*} \tag{9.28}$$

It follows that the thickness of the viscous sublayer will be small for flows in which the shear stress is large, because u_* will also be large for these cases. Furthermore, the viscous sublayer will become larger along the wall in the direction of flow because the shear stress decreases along the wall in the downstream direction.

The flow zone outside the viscous sublayer is turbulent; therefore, a completely different type of flow is involved. In fact, the turbulence alters the flow regime so much that the shear stress as given by $\tau = \mu\, du/dy$ is not significant. What happens is that the mixing action of turbulence causes small fluid masses to be swept back and forth in a direction transverse to the mean flow direction. Thus, as a small mass of fluid is swept from a low-velocity zone next to the viscous sublayer into a higher-velocity zone farther out in the stream, the mass has a retarding effect on the higher-velocity stream. Through an exchange of momentum, this mass of fluid creates the effect of a retarding shear stress applied to the higher-velocity stream, much like the "conveyor belt" analogy introduced in Section 2.5. Similarly, a small mass of fluid that originates farther out in the boundary layer in a high-velocity flow zone and is swept into a region of low velocity has an effect on the low-velocity fluid much like shear stress augmenting the flow velocity. In other words, the mass of fluid with relatively higher momentum tends to accelerate the lower-velocity fluid in the region into which it moves. Although the process just described is primarily a momentum exchange phenomenon, it has the same effect as a shear stress applied to the fluid; thus in turbulent flow these "stresses" are termed *apparent shear stresses,* or *Reynolds stresses* after the British scientist-engineer who first did extensive research in turbulent flow in the late 1800s.

The mixing action of turbulence causes the velocities at a given point in a flow to fluctuate with time. If one places a velocity-sensing device, such as a hot-wire anemometer, in a turbulent flow, one can measure a fluctuating velocity, as illustrated in Fig. 9.8. It is convenient to think of the velocity as composed of two parts: a mean value, $\bar{u}$, plus a fluctuating part, u'. The fluctuating part of the velocity is responsible for the mixing action and the momentum exchange, which manifests itself as an apparent shear stress as we have noted. In fact, the apparent shear stress is related to the fluctuating part of the velocity by

$$\tau_{\text{app}} = -\rho\overline{u'v'} \tag{9.29}$$

where u' and v' refer to the x and y components of the velocity fluctuations, respectively, and the bar over these terms denotes the product of $u'v'$ averaged over a period of time.* The expression for apparent shear stress is not very useful in this form, so Prandtl

*Equation (9.29) can be derived by considering the momentum exchange that results when the transverse component of turbulent flow passes through an area parallel to the x-z plane. Or, by including the fluctuating velocity components in the Navier-Stokes equations, one can obtain the apparent shear stress terms, one of which is Eq. (9.29). Details of these derivations appear in Chapter 18 of Schlichting (3).

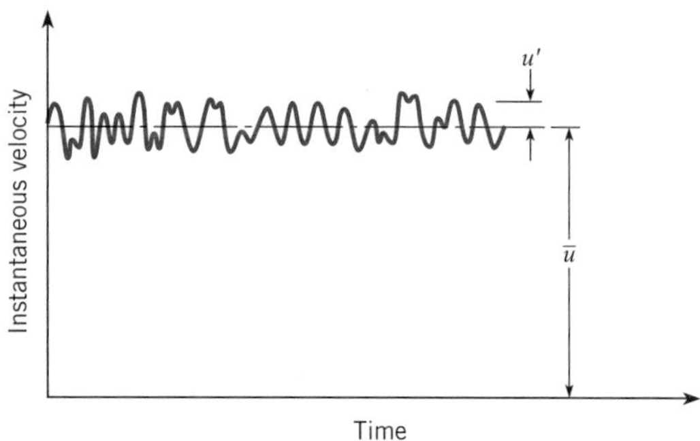

FIGURE 9.8

Velocity fluctuations in turbulent flow.

developed a theory to relate the apparent shear stress to the temporal mean velocity distribution.

Prandtl's Mixing-Length Theory

In the turbulent boundary layer, the principal flow is parallel to the boundary. However, because of turbulent eddies, there are fluctuating components transverse to the principal flow direction. By focusing on a small mass of fluid that is moving transverse to the principal flow under the action of turbulence, one can explain Prandtl's mixing-length theory. Assume that the small fluid mass is initially in a zone of relatively low mean velocity near the wall (see Fig. 9.9). Then, under the action of turbulence, it is transported to a new position farther from the wall (in the direction of increasing y). This mass of fluid in its new location will have a lower velocity (in the principal flow direction) than the surrounding fluid (it tends to retain its initial momentum), and the difference in velocity between the surrounding fluid and the small mass of fluid will be given by $\ell'\, du/dy$, where du/dy is the mean velocity gradient and ℓ' is the distance the small fluid mass traveled in the transverse direction in going from the initial position to the position farther out in the flow field. Note also that, at the point farther out in the flow field, the velocity has changed from an initial value of u to a value of $u - u'$. Therefore, Prandtl assumed that the magnitude of the fluctuating velocity component in the principal flow direction is equal to $\ell'\, du/dy$ or $|u'| = \ell'\, du/dy$. Furthermore, Prandtl assumed that the magnitude of the transverse fluctuating velocity component is proportional to the magnitude of the fluctuating component in the principal flow direction: $|v'| = K \times |u'|$, where K is a constant. This would seem to be a reasonable assumption because both components arise from the same set of eddies. Also, it should be noted that a positive v' will be associated with a negative u' (when a small mass of fluid moves away from the boundary $+v'$, it produces a negative u' in the vicinity into which it moves, as explained above). Thus $\overline{u'v'}$ will be negative, and when $\ell'\, du/dy$ is substituted for u' and $K\ell'\, du/dy$ is substituted for v' in Eq. (9.29), one obtains

$$\tau_{\text{app}} = K\rho\ell'^2\left(\frac{du}{dy}\right)^2$$

If we let $K\ell'^2$ in the foregoing equation be equal to ℓ^2, then we may write it as

$$\tau_{\text{app}} = \rho\ell^2\left(\frac{du}{dy}\right)^2 \tag{9.30}$$

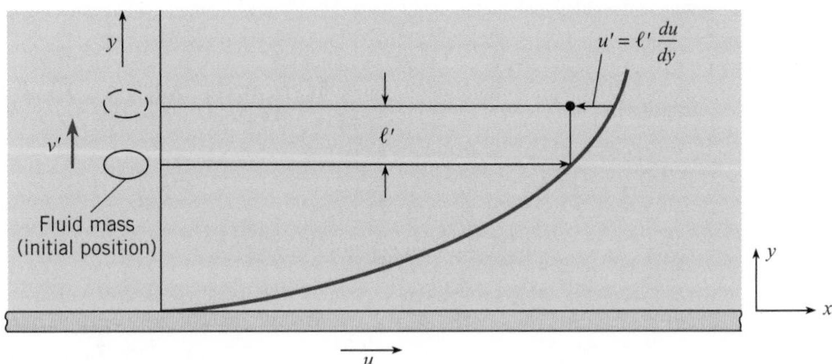

FIGURE 9.9

Concept of mixing length.

A more general form of Eq. (9.30) is

$$\tau_{\text{app}} = \rho \ell^2 \left|\frac{du}{dy}\right| \frac{du}{dy}$$

However, this form is not needed in this text.

The theory leading to Eq. (9.30) is called Prandtl's mixing-length theory and is used extensively in analyses involving turbulent flow.* The advantage of Eq. (9.30) over Eq. (9.29) is that Eq. (9.30) is expressed in terms of the temporal mean velocity distribution and can thus be integrated to obtain formulas for the velocity variation. In Eq. (9.30), ℓ is the mixing length, which was shown by Prandtl to be essentially proportional to the distance from the wall ($\ell = \kappa y$) for the region close to the wall. If we consider the velocity distribution in a boundary layer where du/dy is positive, as is shown in Fig. 9.7, and if we substitute κy for ℓ, then Eq. (9.30) reduces to

$$\tau_{\text{app}} = \rho \kappa^2 y^2 \left(\frac{du}{dy}\right)^2$$

For the zone of flow near the boundary, it is assumed that the shear stress is uniform and approximately equal to the shear stress at the wall. Thus the foregoing equation becomes

$$\tau_0 = \rho \kappa^2 y^2 \left(\frac{du}{dy}\right)^2 \tag{9.31}$$

Taking the square root of each side of Eq. (9.31) and rearranging yields

$$du = \frac{\sqrt{\tau_0/\rho}}{\kappa} \frac{dy}{y}$$

Integrating the above equation and substituting u_* for $\sqrt{\tau_0/\rho}$ gives

$$\frac{u}{u_*} = \frac{1}{\kappa} \ln y + C \tag{9.32}$$

Experiments on smooth boundaries indicate that the constant of integration C can be given in terms of u_*, ν, and a pure number as

*Prandtl published an account of his mixing-length concept in 1925. G. I. Taylor (4) published a similar concept in 1915, but the idea has been traditionally attributed to Prandtl.

$$C = 5.56 - \frac{1}{\kappa} \ln \frac{\nu}{u_*}$$

When this expression for C is substituted into Eq. (9.32), we have

$$\frac{u}{u_*} = \frac{1}{k} \ln \frac{yu_*}{\nu} + 5.56 \tag{9.33}$$

In Eq. (9.33), κ has sometimes been called the universal turbulence constant, or Karman's constant. Experiments show this constant to have a value of approximately 0.40 for the turbulent zone next to the viscous sublayer. Introducing this into Eq. (9.33) and expressing the equation in terms of the logarithm to the base 10, we get the *logarithmic velocity distribution*

$$\frac{u}{u_*} = 5.75 \log \frac{yu_*}{\nu} + 5.56 \tag{9.34}$$

This distribution is valid for values of yu_*/ν ranging from approximately 30 to 500. Thus we have a form of velocity distribution for this zone that is far different from that for the viscous sublayer. However, the relative velocity is still a function of yu_*/ν. Thus, for the range of yu_*/ν from 0 to approximately 500 (Fig. 9.10), the velocity distribution is called the *law of the wall.*

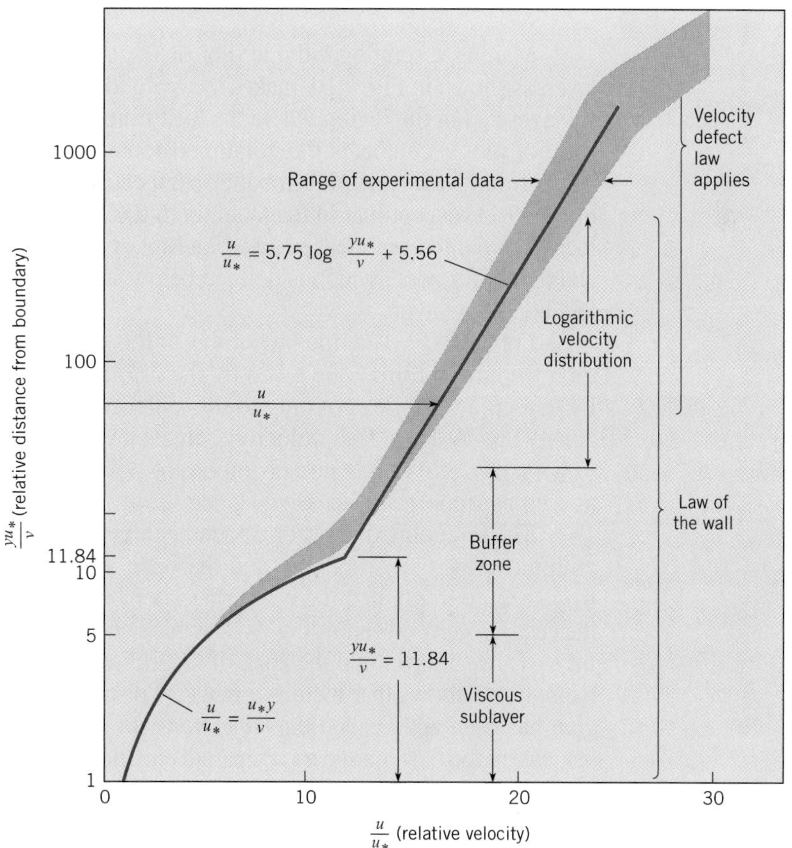

FIGURE 9.10

Velocity distribution in a turbulent boundary layer.

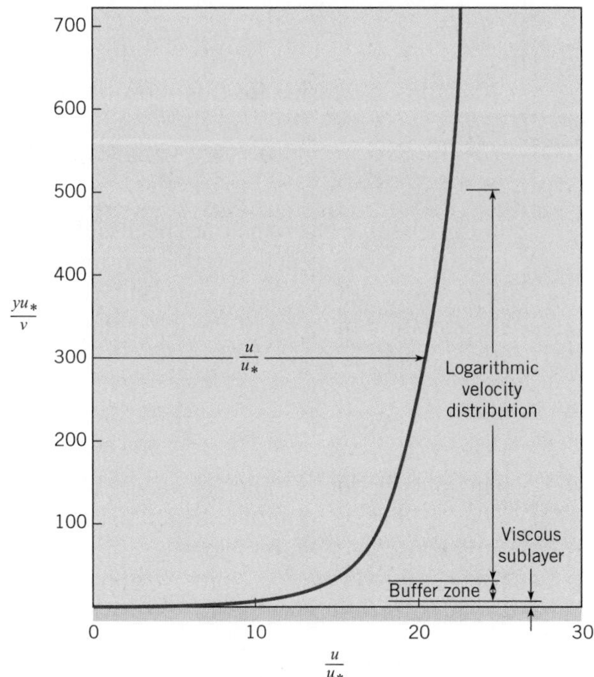

FIGURE 9.11

Velocity distribution in a turbulent boundary layer—linear scales.

Making a semilogarithmic plot of the velocity distribution in a turbulent boundary layer, as shown in Fig. 9.10, makes it easy to identify the velocity distribution in the viscous sublayer and in the region where the logarithmic equation applies. However, note that this form of plot accentuates the nondimensional distance yu_*/ν near the wall. So that the student may view this plot in better perspective, the graph shown in Fig. 9.10 is repeated in Fig. 9.11, except that in the latter, both the relative distance yu_*/ν and the relative velocity are plotted on linear scales. Figure 9.11 properly indicates that the laminar sublayer and the buffer zone, which is defined below, are a small part of the thickness of the turbulent boundary layer.

For $y/\delta > 0.15$ the law of the wall is no longer valid. Therefore, in this outer region we have a third zone given by the *velocity defect law,* Fig. 9.12. In fact, the velocity defect law applies not only to this outer region but extends well into the logarithmic zone (Fig. 9.10). The velocity defect law relates the relative defect of velocity $(U_0 - u)/u_*$ to y/δ. Another important point about this law is that it applies to rough as well as smooth surfaces.

The foregoing discussion about the three zones of flow (the viscous sublayer, the logarithmic velocity distribution, and the outer zone where the velocity defect law applies) may seem to imply that there is a sharp demarcation between zones. This is definitely not the case, as can be seen in Fig. 9.10, which shows a smooth transition of velocity between the viscous sublayer and the zone of logarithmic velocity distribution. The band of experimental data shows that there is a range of distance over which neither law applies. This region has been aptly called the *buffer zone.* In some cases it may be desirable to define the velocity in the buffer zone by a separate equation, for which the student is referred to Rouse (10). However, for some boundary layer calculations, it is convenient to ignore the precise form of velocity distribution in the buffer zone and simply let the distribution be given by

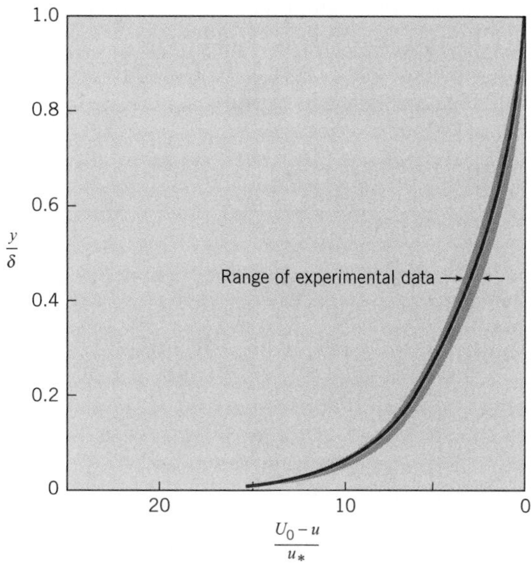

FIGURE 9.12

Velocity defect law for boundary layers. [After Rouse (5)]

the extension of the equations already cited. In this case we refer to the nominal thickness of the viscous sublayer, δ_N', which is the distance from the boundary to the point where the velocity distribution curves for the viscous sublayer and the logarithmic velocity distribution intersect. This point of intersection occurs (see Fig. 9.10) at

$$\frac{yu_*}{\nu} = 11.84 \tag{9.35}$$

In other words, for this value of yu_*/ν we can substitute δ_N' for y:

$$\delta_N' = \frac{11.84\nu}{u_*} \tag{9.36}$$

As noted in the preceding section, the actual limit of the viscous sublayer occurs at about $\delta' = 5\nu/u_*$. That is, turbulence affects the velocity distribution in the boundary layer from $\delta' = 5\nu/u_*$ outward. However, the effect is not appreciable up to the point of the intersection of the velocity distribution curve for the zone of logarithmic distribution (see Fig. 9.10). Thus, in derivations and analyses involving turbulent boundary layers, it is not uncommon to assume that the velocity distribution in the viscous sublayer extends out to $\delta_N' = 11.84\nu/u_*$.

Power-Law Formula for Velocity Distribution

Analyses have shown that for a wide range of Reynolds numbers ($10^5 < \text{Re} < 10^7$), the velocity profile in the turbulent boundary layer is reasonably approximated by the *power-law* equation

$$\frac{u}{U_0} = \left(\frac{y}{\delta}\right)^{1/7} \tag{9.37}$$

Comparisons with experimental results show that this formula conforms to those results very closely over about 90% of the boundary layer ($0.1 < y/\delta < 1$). For the inner 10% of the boundary layer, one must resort to equations for the law of the wall (see Fig. 9.10) to obtain a more precise indication of velocity. Because Eq. (9.37) is valid over the major portion of the boundary layer, it is used to advantage in deriving the overall thickness of the boundary layer as well as other relations for the turbulent boundary layer. These will be considered in the next sections.

example 9.6

Water (60°F) flows with a velocity of 20 ft/s past a flat plate. The plate is oriented parallel to the flow. At a particular section downstream of the leading edge of the plate, the shear stress next to the plate is 0.896 lbf/ft^2 and the boundary layer thickness is 0.0880 ft. Find the velocity of the water at a distance of 0.0088 ft from the plate as determined by

a. The logarithmic velocity distribution
b. The velocity defect law
c. The power-law formula

What is the nominal thickness of the viscous sublayer?

Solution Logarithmic velocity distribution:

$$u/u_* = 5.75\,\log(yu_*/\nu) + 5.56$$

where $u_* = (\tau_0/\rho)^{1/2}$, $\nu = 1.22 \times 10^{-5}\ \text{ft}^2/\text{s}$, and $y = 0.088$ ft.

$$\tau_0 = 0.896\ \text{lbf/ft}^2 \quad \text{and} \quad \rho = 1.94\ \text{slugs/ft}^3 = 1.94\ \text{lbf} \cdot \text{s}^2/\text{ft}^4$$

Thus

$$(\tau_0/\rho)^{1/2} = [(0.896\ \text{lbf/ft}^2)/(1.94\ \text{lbf} \cdot \text{s}^2/\text{ft}^4)]^{1/2} = 0.680\ \text{ft/s}$$

Also,

$$yu_*/\nu = (0.0088\ \text{ft})(0.680\ \text{ft/s})/(1.22 \times 10^{-5}\ \text{ft}^2/\text{s})$$
$$= 490$$

Then

$$u/u_* = 5.75\,\log(4.90) + 5.56$$
$$u/u_* = 21.03 \quad \text{or} \quad u = 21.03(u_*) = 21.03 \times 0.680\ \text{ft/s}$$
$$u = 14.3\ \text{ft/s}$$

◁

Velocity defect law:

$$y/\delta = 0.0088\ \text{ft}/0.088\ \text{ft} = 0.10$$

Then, from Fig. 9.12, we find that

$$(U_0 - u)/u_* \approx 8.2 \quad \text{for } y/\delta = 0.10$$

$$U_0 - u = 8.2u_*$$

or

$$u = U_0 - 8.2u_*$$
$$= 20 \text{ ft/s} - (8.2)(0.68) \text{ ft/s}$$
$$= 14.42 \text{ ft/s} \quad \triangleleft$$

Power-law formula:

$$u/U_0 = (y/\delta)^{1/7}$$
$$u = (U_0)(0.10)^{1/7}$$
$$= (20 \text{ ft/s})(0.7197)$$
$$= 14.40 \text{ ft/s}$$

Nominal viscous sublayer thickness:

$$\delta'_N = 11.84\nu/u_* = (11.84)(1.22 \times 10^{-5} \text{ ft}^2/\text{s})/(0.68 \text{ ft/s})$$
$$= 2.12 \times 10^{-4} \text{ ft} = 2.54 \times 10^{-3} \text{ in.} \quad \triangleleft$$

Momentum Equation Applied to the Boundary Layer

When the form of velocity distribution in the boundary layer is known, it is possible to derive equations for the shear stress and the thickness of the boundary layer by utilizing the momentum equation. These basic relationships involved with the momentum equation will be introduced in this section.

Consider the control volume in Fig. 9.13 for the boundary layer over a flat plate with zero pressure gradient along the plate. Flow comes into the control volume at section 1-1 and through the top of the boundary layer; it leaves the control volume at section 2-2. When we write the momentum equation for the x direction for the fluid in this control volume, we have

$$\sum F_x = \int_{cs} u\rho \mathbf{V} \cdot d\mathbf{A} \tag{9.38}$$

We assume a unit width normal to the page and designate the mass inflow through the top of the boundary layer as $\dot{m}_d$. Therefore, Eq. (9.38) becomes

$$-\tau_0 \Delta x = \int_0^{\delta_2} \rho u_2^2\, dy - \int_0^{\delta_1} \rho u_1^2\, dy - U_0 \dot{m}_\delta \tag{9.39}$$

The mass flow rate into the control volume from the top of the boundary layer can be expressed as the difference in the mass flow rates past sections 1-1 and 2-2,

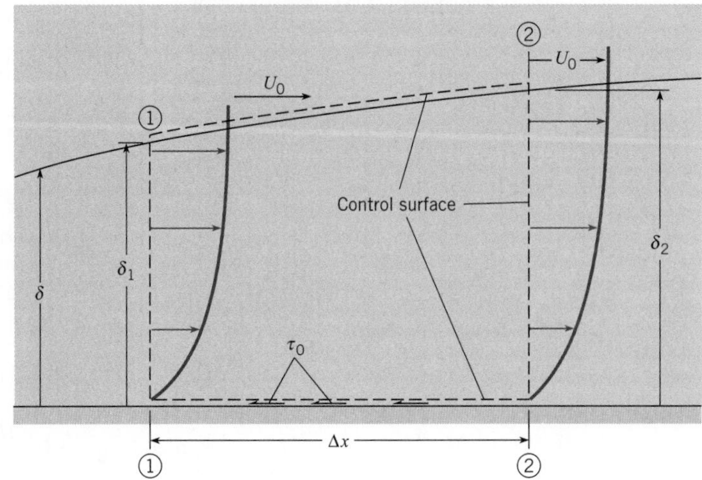

FIGURE 9.13

Control volume applied to boundary layer.

$$\dot{m}_\delta = \int_0^{\delta_2} \rho u_2\, dy - \int_0^{\delta_1} \rho u_1\, dy \tag{9.40}$$

When Eq. (9.40) is substituted into Eq. (9.39), we obtain

$$-\tau_0 \Delta x = \int_0^{\delta_2} \rho u_2^2\, dy - \int_0^{\delta_1} \rho u_1^2\, dy - U_0 \left[\int_0^{\delta_2} \rho u_2\, dy - \int_0^{\delta_1} \rho u_1 dy \right] \tag{9.41}$$

Assuming ρ is constant, we can rearrange this equation to yield

$$\begin{aligned} -\tau_0 \Delta x &= \rho \int_0^{\delta_2} (u_2^2 - U_0 u_2)\, dy - \rho \int_0^{\delta_1} (u_1^2 - U_0 u_1)\, dy \\ &= \rho U_0^2 \left\{ \int_0^{\delta_2} \left[\left(\frac{u_2}{U_0} \right)^2 - \frac{u_2}{U_0} \right] dy - \int_0^{\delta_1} \left[\left(\frac{u_1}{U_0} \right)^2 - \frac{u_1}{U_0} \right] dy \right\} \\ &= \rho U_0^2 \Delta \left\{ \int_0^{\delta} \left[\left(\frac{u}{U_0} \right)^2 - \frac{u}{U_0} \right] dy \right\} \end{aligned}$$

If we let Δx approach zero in the limit, then the foregoing equation reduces to

$$\tau_0 = \rho U_0^2 \frac{d}{dx} \int_0^{\delta} \frac{u}{U_0} \left(1 - \frac{u}{U_0} \right) dy \tag{9.42}$$

Equation (9.42) states that the shear stress at the wall is equal to ρU_0^2 times the rate of change with respect to x of an integral that is a function of the velocity distribution across the section. This equation can be used to evaluate the shear stress on a boundary indirectly by measuring the velocity distribution at various sections or, as will be shown in the next section, to derive other boundary layer equations.

Thickness of the Turbulent Boundary Layer on a Flat Plate

Using the power-law formula for velocity distribution, Eq. (9.37), in Eq. (9.42), we have

$$\frac{\tau_0}{\rho} = U_0^2 \frac{d}{dx} \int_0^{\delta} \left(\frac{y}{\delta}\right)^{1/7} \left[1 - \left(\frac{y}{\delta}\right)^{1/7}\right] dy \tag{9.43}$$

The integration in Eq. (9.43) yields

$$\frac{\tau_0}{\rho} = \frac{7}{72} U_0^2 \frac{d\delta}{dx} \tag{9.44}$$

Here we have an equation involving δ. However, it is not very useful because τ_0 and δ are both unknowns. However, another relation between the shear stress and the boundary layer thickness can be obtained by fitting more accurate analytic results for the local shear stress coefficient with a power law (3):

$$c_f = 0.020\left(\frac{\nu}{U_0 \delta}\right)^{1/6} \tag{9.45}$$

This equation provides a good fit with experimental results up to a Reynolds number of 10^{10}. Writing this equation in terms of the shear stress and density gives

$$\frac{\tau_0}{\rho} = 0.010 U_0^2 \left(\frac{\nu}{U_0 \delta}\right)^{1/6} \tag{9.46}$$

When Eq. (9.44) and Eq. (9.46) are combined, the result is

$$0.010\left(\frac{\nu}{U_0}\right)^{1/6} = \frac{7}{72} \delta^{1/6} \frac{d\delta}{dx} \tag{9.47}$$

Separating variables and integrating gives

$$\delta^{7/6} = \frac{7}{6}\left(\frac{0.010 \times 72}{7}\right)\left(\frac{\nu}{U_0}\right)^{1/6} x + C \tag{9.48}$$

We are assuming that the boundary layer starts from the leading edge, so we evaluate C by taking $x = 0$ and $\delta = 0$; therefore, $C = 0$. When Eq. (9.48) is solved for δ and simplified, we obtain the *boundary layer thickness* for turbulent flow over a flat plate:

$$\delta = \frac{0.16x}{\text{Re}_x^{1/7}} \tag{9.49}$$

Thus we have the thickness of the turbulent boundary layer as a function of both the distance along the boundary and the Reynolds number based on the distance along the boundary.

Shearing Resistance of the Turbulent Boundary Layer on a Flat Plate

When δ of Eq. (9.49) is substituted back into Eq. (9.46), we can express τ_0 in terms of the Reynolds number based on the distance along the boundary:

$$\tau_0 = \rho \frac{U_0^2}{2}\left(\frac{0.027}{\mathrm{Re}_x^{1/7}}\right) \tag{9.50}$$

Since c_f, the local shear stress coefficient, is defined as $\tau_0/(1/2)\rho U_0^2$, we can solve for c_f from Eq. (9.50):

$$c_f = \frac{\tau_0}{\rho U_0^2/2} = \frac{0.027}{\mathrm{Re}_x^{1/7}} \tag{9.51}$$

When τ_0 from Eq. (9.50) is integrated over the area of the boundary, we obtain the overall shearing resistance, which is

$$F_s = \frac{0.032BL}{\mathrm{Re}_L^{1/7}}\rho\frac{U_0^2}{2}$$

The average shear stress coefficient, C_f, is $F_s/(BL\rho U_0^2/2)$. Therefore, for a turbulent boundary layer its value may be computed from $C_f = 0.032/\mathrm{Re}_L^{1/7}$.

In the foregoing developments we have applied the integral momentum equation over the boundary layer in order to derive useful equations for local shear stress and overall plate resistance. As a point of interest, the integral momentum equation, Eq. (9.38), is also used in a variety of other applications to relate the drag of a body to the difference in pressure distribution and momentum flux between an upstream and a downstream section. Applications of this type include the determination of the drag of an airfoil section, the analysis of boundary layers over rough surfaces, the analysis of boundary layers over evaporating surfaces [see Crowe et al. (6)], and the prediction of points of separation on curved surfaces.

Even though the variation of the local skin friction coefficient with Reynolds number given by Eq. (9.51) provides a reasonably good fit with experimental data, it tends to underpredict the skin friction at higher Reynolds numbers. There are several correlations that have been proposed in the literature; see the review by Schlichting (11). A correlation proposed by White (7) that fits the data for turbulent Reynolds numbers up to 10^{10} is

$$c_f = \frac{0.455}{\ln^2(0.06\mathrm{Re}_x)} \tag{9.52a}$$

The corresponding average shear stress coefficient is

$$C_f = \frac{0.523}{\ln^2(0.06\mathrm{Re}_L)} \tag{9.52b}$$

These are the correlations for shear stress coefficients recommended here.

example 9.7

Air at a temperature of 20°C and with a free-stream velocity of 30 m/s flows past a smooth, thin plate that is 3 m wide and 6 m long in the direction of flow. Assuming that the boundary layer is forced to be turbulent from the leading edge, determine the shear stress, the thickness of the viscous sublayer, and the thickness of the boundary layer 5 m downstream of the leading edge.

Solution First compute Re_x at a distance 5 m from the leading edge:

$$\mathrm{Re}_x = U_0 \frac{x}{\nu} = \frac{(30\ \mathrm{m/s})(5\ \mathrm{m})}{1.51(10^{-5})\mathrm{m}^2/\mathrm{s}} = 10^7$$

Compute τ_0, where $\tau_0 = c_f \rho U_0^2/2$. Here the value of the local shear stress coefficient is computed from

$$c_f = \frac{0.455}{\ln^2(0.06\mathrm{Re}_x)} = \frac{0.455}{\ln^2(6\times 10^5)} = 0.0026$$

Also

$$\rho = 1.20\ \mathrm{kg/m^3} \qquad \text{(from Appendix)}$$

Then

$$\tau_0 = 0.0026 \times 1.20\ \mathrm{kg/m^3} \times \frac{30^2\ \mathrm{m^2/s^2}}{2} = 1.40\ \mathrm{N/m^2} \qquad \triangleleft$$

Now compute $u_* = \sqrt{\tau_0/\rho}$ and the thickness of the viscous sublayer:

$$u_* = \left(\frac{\tau_0}{\rho}\right)^{1/2} = \left(\frac{1.40\ \mathrm{N/m^2}}{1.20\ \mathrm{kg/m^3}}\right)^{1/2} = 1.08\ \mathrm{m/s}$$

The thickness of the viscous sublayer is given by

$$\delta' = \frac{5\nu}{u_*}$$

But $\nu = 1.49 \times 10^{-5}\ \mathrm{m^2/s}$, so

$$\delta' = \frac{5 \times 1.51 \times 10^{-5}\ \mathrm{m^2/s}}{1.08\ \mathrm{m/s}} = 6.99 \times 10^{-5}\ \mathrm{m} = 0.070\ \mathrm{mm} \qquad \triangleleft$$

Compute the boundary layer thickness:

$$\delta = \frac{0.16x}{\mathrm{Re}^{1/7}} = \frac{0.16 \times 5\ \mathrm{m}}{(10^7)^{1/7}} = 0.080\ \mathrm{m} = 80\ \mathrm{mm} \qquad \triangleleft$$

Up to this point we have developed and discussed equations for the local and average shear stress coefficients for laminar and turbulent boundary layers on a flat plate. Now we shall discuss the existence of laminar and turbulent boundary layers together on a smooth, flat plate and examine how one determines the total resistance when they both occur on the same plate. As noted in Section 9.3, the laminar boundary layer first develops on the upstream end of the plate. Then, as this layer grows in thickness, it becomes unstable and turbulence sets in. A turbulent boundary layer develops over the remainder of the plate. The onset of turbulence depends to a certain extent on the degree of smoothness of the plate and on the degree of turbulence. However, when the approach flow is nonturbulent, the transition or critical region on a smooth plate occurs at a Reynolds number, Re_x, of about 500,000. Then, when the turbulent boundary layer develops downstream of the laminar layer, we wonder whether the turbulent boundary layer will have the characteristics of one with the origin at the leading edge of the plate (same as the origin for the laminar layer) or of one with the origin at the downstream end of the laminar layer. Experiments reveal that the former is the most valid model. Thus, when calculating the overall resistance of a plate with a laminar and a turbulent boundary layer, we must evaluate the two resistances separately and sum them to obtain the total resistance. The calculation of the resistance of the laminar part is straightforward, as given in Example 9.4. However, in calculating the resistance of the turbulent part of the boundary layer, we must compute the resistance as though the entire boundary layer were turbulent and then subtract from that the resistance that would have occurred on the plate up to the transition zone.

When the foregoing model for resistance is stated mathematically, we have

$$F_s = \left(\frac{1.33}{\mathrm{Re}_{\mathrm{cr}}^{1/2}}Bx_{\mathrm{cr}} + \frac{0.523}{\ln^2(0.06\mathrm{Re}_L)}BL - \frac{0.523}{\ln^2(0.06\mathrm{Re}_{\mathrm{cr}})}Bx_{\mathrm{cr}}\right)\rho\frac{U_0^2}{2} \tag{9.53}$$

where $\mathrm{Re}_{\mathrm{cr}}$ is the Reynolds number at the transition, Re_L is the Reynolds number at the end of the plate, and x_{cr} is the distance from the leading edge of the plate to the critical or transition zone.

Because we define the average resistance coefficient as $C_f = F_s/(BL\rho U_0^2/2)$, we can see from Eq. (9.53) that

$$C_f = \frac{0.523}{\ln^2(0.06\mathrm{Re}_L)} + \frac{x_{\mathrm{cr}}}{L}\left(\frac{1.33}{\mathrm{Re}_{\mathrm{cr}}^{1/2}} - \frac{0.523}{\ln^2(0.06\mathrm{Re}_{\mathrm{cr}})}\right)$$

Here $x_{\mathrm{cr}}/L = \mathrm{Re}_{\mathrm{cr}}/\mathrm{Re}_L$. Therefore, we get

$$C_f = \frac{0.523}{\ln^2(0.06\mathrm{Re}_L)} + \frac{\mathrm{Re}_{\mathrm{cr}}}{\mathrm{Re}_L}\left(\frac{1.33}{\mathrm{Re}_{\mathrm{cr}}^{1/2}} - \frac{0.523}{\ln^2(0.06\mathrm{Re}_{\mathrm{cr}})}\right)$$

For $\mathrm{Re}_{\mathrm{cr}} = 500{,}000$, we have

$$C_f = \frac{0.523}{\ln^2(0.06\mathrm{Re}_L)} - \frac{1520}{\mathrm{Re}_L} \tag{9.54}$$

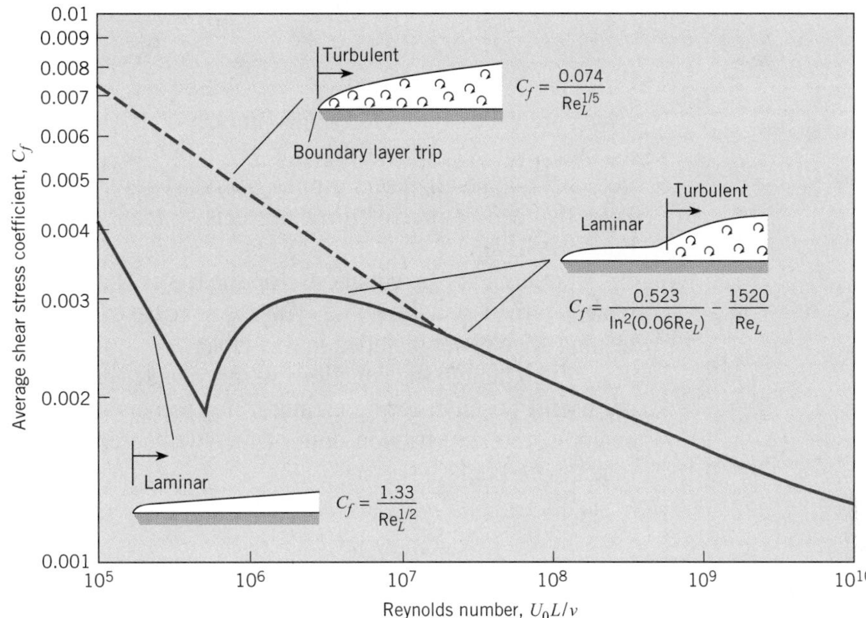

FIGURE 9.14

Average shear stress coefficients.

The variation of C_f with Reynolds number is shown by the solid line in Fig. 9.14. This curve corresponds to a boundary layer that begins as a laminar boundary layer and then changes to a turbulent boundary layer after the critical Reynolds number. This is the normal condition for a flat-plate boundary layer

TABLE 9.2 SUMMARY OF EQUATIONS FOR BOUNDARY LAYER ON A FLAT PLATE

	Laminar flow $\mathbf{Re}_x$, $\mathbf{Re}_L < 5 \times 10^5$	**Turbulent flow** $\mathbf{Re}_x$, $\mathbf{Re}_L \geq 5 \times 10^5$
Boundary layer thickness, δ	$\delta = \frac{5x}{Re_x^{1/2}}$	$\delta = \frac{0.16x}{Re_x^{1/7}}$
Local shear stress coefficient, c_f	$c_f = \frac{0.664}{Re_x^{1/2}}$	$c_f = \frac{0.455}{\ln^2(0.06Re_x)}$
Average shear stress coefficient, C_f	$C_f = \frac{1.33}{Re_L^{1/2}}$	$C_f = \frac{0.523}{\ln^2(0.06Re_L)} - \frac{1520}{Re_L}$

If the boundary layer is "tripped" by some roughness or leading-edge disturbance, the boundary layer is turbulent from the beginning. This is shown by the dashed line in Fig. 9.14. For this condition the boundary layer thickness, local shear stress coefficient, and average shear stress coefficient are fit by the empirical expressions

$$\delta = \frac{0.37x}{\text{Re}_x^{1/5}} \qquad c_f = \frac{0.058}{\text{Re}_x^{1/5}} \qquad C_f = \frac{0.074}{\text{Re}_L^{1/5}}$$

which are valid up to a Reynolds number of 10^7. For Reynolds numbers beyond 10^7, the average shear stress coefficient is given by the solid line in Fig. 9.14. It is of interest to note that marine engineers trip the boundary layer on ship models to produce a boundary layer that can be predicted more precisely than a combination of laminar and turbulent boundary layers.

A summary of equations for shear stress and boundary layer thickness on a flat plate is provided in Table 9.2. The special case of the boundary layer being tripped at the leading edge is not included in this table.

Even though the equations in this chapter have been developed for flat plates, they are useful for engineering estimates for some surfaces that are not truly flat plates. For example, the skin friction drag of the submerged part of the hull of a ship can be estimated with Eq. (9.54).

example 9.8

Assume that a boundary layer over a smooth, flat plate is laminar at first and then becomes turbulent at a critical Reynolds number of 5×10^5. If we have a plate 3 m long and 1 m wide, and if air at 20°C and normal atmospheric pressure flows past this plate with a velocity of 30 m/s, what will be the average resistance coefficient C_f for the plate? Also, what will be the total shearing resistance of one side of the plate, and what will be the resistance due to the turbulent part and the laminar part of the boundary layer?

Solution The total resistance is $F_s = C_f BL\rho(U_0^2/2)$, where from Eq. (9.54) we get C_f:

$$C_f = \frac{0.523}{\ln^2(0.06\text{Re}_L)} - \frac{1520}{\text{Re}_L}$$

In addition, $\text{Re}_L = UL/\nu$, or

$$\text{Re}_L = \frac{30 \text{ m/s} \times 3 \text{ m}}{(1.51)(10^{-5}) \text{ m}^2/\text{s}} = 5.96 \times 10^6$$

Then, solving Eq. (9.54), we have

$$C_f = 0.00320 - 0.00026 = 0.00294 \qquad \triangleleft$$

The total resistance is calculated:

$$F_s = C_f BL\rho\frac{U_0^2}{2} = 0.00294 \times 1 \times 3 \times 1.2 \times \frac{30^2}{2} = 4.76 \text{ N} \qquad \triangleleft$$

Then x_{cr} is determined:

$$\frac{Ux_{cr}}{\nu} = 500{,}000$$

or
$$x_{cr} = \frac{500{,}000 \times 1.51 \times 10^{-5}}{30} = 0.252 \text{ m}$$

Thus the laminar resistance will be

$$F_{s,\text{lam}} = \frac{1.33}{(5 \times 10^5)^{1/2}} \times 1 \times 0.252 \times 1.2 \times \frac{30^2}{2} = 0.256 \text{ N} \quad \triangleleft$$

Then
$$F_{s,\text{turb}} = 4.76 \text{ N} - 0.26 \text{ N} = 4.50 \text{ N} \quad \triangleleft$$

example 9.9

Determine the total drag of the plate given in Example 9.7.

Solution The shearing resistance of one side is given as $F_s = C_f BL\rho(U_0^2/2)$. Therefore, the total drag will be twice this for two sides of the plate:

$$F_s = C_f BL\rho U_0^2$$

We compute the Reynolds number as

$$\text{Re}_L = U_0\frac{L}{\nu} = \frac{(30 \text{ m/s})(6 \text{ m})}{(1.51)(10^{-5} \text{ m}^2/\text{s})} = 1.19(10^7)$$

From Eq. (9.54), $C_f = 0.0028$. Then

$$F_s = 0.0028(3 \text{ m})(6 \text{ m})(1.20 \text{ kg/m}^3)(30^2 \text{ m}^2/\text{s}^2) = 4.4 \text{ N} \quad \triangleleft$$

Boundary Layer Control

The preceding sections show that the boundary layer on a flat plate is initially laminar and then becomes turbulent at a distance along the plate where the Reynolds number is approximately 5×10^5. Also, it was shown (Figs. 9.5 and 9.14) that the skin friction, for a given Reynolds number, is much greater for the turbulent part of the boundary layer than for the laminar part. Thus if there were some way to prevent the boundary layer from becoming turbulent, the skin friction drag could be reduced. This is an active area of research today, and it has been for the past several decades. Procedures for producing this desirable effect are called *boundary layer control*.

One way to control the boundary layer—for example, on an airfoil—is to shape the airfoil in such a way that the pressure distribution on its surface delays the onset of turbulence. Several such airfoils have been developed, but the amount of drag reduction that it is possible to achieve in this manner is limited because alteration of the airfoil's basic shape so often decreases its lift.

Another means of boundary layer control is to make the surface of the boundary, such as an airfoil surface, somewhat porous and to apply a reduced pressure to the surface (reduced pressure or suction is maintained inside the airfoil) so that part of the boundary layer is drawn away through the porous surface (part of the air of the boundary layer is sucked into the interior of the airfoil). This keeps the boundary layer thin, so it remains stable in its laminar state and a reduced skin friction drag results. Wind tunnel tests prove that boundary layer control achieved by applying suction to a porous surface is effective in reducing drag; studies indicate that a commercial aircraft with such control would burn 20% less fuel (8)! The initial cost of the aircraft would be increased, but it is estimated that the fuel savings would pay for the additional cost within 6 months (8).

Some of the severe practical problems associated with boundary layer control by suction involve the surface finish. For example, the perforations made in the airfoil to produce the porous surface must be very small; otherwise, flow disturbances in the vicinity of each perforation will induce turbulence. One suction surface that was successfully tested at the NASA Langley Research Center in Virginia was made of titanium sheet bonded to fiberglass sandwich panel. The perforations on the sheet were 0.0025 in. in diameter with a spacing of 0.025 in. (8). These perforations were produced by electron beams.

Once such an airfoil is put in service, it must be "groomed" in such a way as to preserve its smoothness in order to prevent the development of turbulence from undesirable roughness. Insects and/or dirt cannot be allowed to build up on the surface, so an effective cleansing system is essential. For these and other details, refer to Wagner and Fisher (8).

Aircraft are not the only bodies that could benefit from boundary layer control. Bar-Hain and Weihs (9) studied submerged bodies such as submarines and torpedoes. Their study focused on the application of a distributed suction technique over a portion of an axisymmetric body.

Another boundary layer control measure involves the use of a dilute polymer solution for drag reduction in liquids. Still another drag reduction phenomenon associated with boundary layers is that of the compliant surface. Studies reveal that porpoises have a special kind of skin (its outer surface is very compliant) that delays the onset of turbulence in the boundary layer. Thus they can swim faster than any other animal of their size. Considerable research has been directed toward the development of artificial surfaces to yield reduced drag, but only limited success has been achieved. See Kramer (10) and Benjamin (11) for more information on this boundary layer phenomenon.

9·7 Summary

The shear stress in a Newtonian fluid is proportional to the local rate of strain of the fluid, and the constant of proportionality is the fluid viscosity, μ. The rate of strain is related to the velocity gradient in the fluid. The shear stress generated on a wall by fluid motion adjacent to a wall is given by

$$\tau = \mu \frac{du}{dy}$$

where u is the fluid velocity and y is the distance measured from the wall. The velocity gradient is evaluated at the wall.

The variation in velocity for a planar (two-dimensional) steady flow with parallel streamlines is governed by the equation

$$\frac{d^2u}{dy^2} = \frac{1}{\mu}\frac{d}{ds}(p + \gamma z)$$

where the distance y is normal to the streamlines and the distance s is along the streamlines. In this chapter, this equation is used to analyze three flow configurations: Couette flow (flow generated by a moving plate), flow with a free surface down an inclined plane, and flow between stationary parallel plates.

The boundary layer is the region where the viscous stresses are responsible for the velocity change between the wall and the free stream. The boundary layer thickness is the distance from the wall to the location where the velocity is 99% of the free-stream velocity. The laminar boundary layer is characterized by smooth (nonturbulent) flow where the momentum transfer between fluid layers occurs because of the fluid viscosity. As the boundary layer thickness grows, the laminar boundary layer becomes unstable and a turbulent boundary layer ensues. The transition point for a boundary layer on a flat plate occurs at a Reynolds number of 5×10^5 based on the free-stream velocity and the distance from the leading edge.

The turbulent boundary layer is characterized by an unsteady flow where the momentum exchange between fluid layers occurs because of the mixing of fluid elements normal to the direction of fluid motion. This effect, known as the Reynolds stress, significantly enhances the momentum exchange and leads to a much higher "effective" shear stress.

The local shear stress coefficient is defined as

$$c_f = \frac{\tau_0}{\frac{1}{2}\rho U_0^2}$$

where τ_0 is the wall shear stress and U_0 is the free-stream velocity. The value for the local shear stress coefficient on a flat plate depends on the Reynolds number based on the distance from the leading edge. The average shear stress coefficient is

$$C_f = \frac{F_s}{\frac{1}{2}\rho U_0^2 S}$$

where F_s is the force due to shear stress on a surface with area S. The value for the average shear stress coefficient for a flat plate depends on the nature of the boundary layer as related to the Reynolds number based on the length of the plate in the flow direction. The laminar boundary layer near the leading edge and the subsequent turbulent boundary layer contribute to the average shear stress on a flat plate. Through leading-edge roughness or other flow disturbance, the boundary layer can be "tripped" at the plate's leading edge, effecting a turbulent boundary layer over the entire plate.

References

1. Prandtl, L. "Über Flussigkeitsbewegung bei sehr kleiner Reibung." *Verhandlungen des III. Internationalen Mathematiker-Kongresses.* Leipzig, 1905.

2. Blasius, H. "Grenzschichten in Flüssigkeiten mit kleiner Reibung." *Z. Mat. Physik.* (1908), English translation in NACA TM 1256.

3. Schlichting, H. *Boundary Layer Theory.* McGraw-Hill, New York, 1979.

4. Taylor, G. I. "The Transport of Vorticity and Heat through Fluids in Turbulent Motion." *Phil. Trans A,* vol. 215 (1915), p. 1.

5. Rouse, H., et al. *Advanced Mechanics of Fluids.* John Wiley, New York, 1959.

6. Crowe, C. T., J. A. Nicholls, and R. B. Morrison. "Drag Coefficients of Inert and Burning Particles Accelerating in Gas Streams." In *Ninth International Symposium on Combustion,* pp. 395–406. Academic Press, New York, 1963.

7. White, Frank M. *Viscous Fluid Flow.* McGraw-Hill, New York, 1991.

8. Wagner, R. D., and M. C. Fisher. "Fresh Attack on Laminar Flow." *Aerospace America* (March 1984).

9. Bar-Hain, B., and D. Weihs. "Boundary-Layer Control as a Means of Reducing Drag on Fully Submerged Bodies of Revolution." *Trans. of the ASME Fluids Engineering Division,* vol. 107 (September 1985), p. 107.

10. Kramer, M. O. "Boundary Layer Stabilization by Distributed Damping." *Jour. of Aeron. Sci.,* Vol. 24, p. 459 (1957).

11. Benjamin, T. B. "Fluid Flow with Flexible Boundaries." *Proc. Intern. Congress of Applied Math,* 11th ed., p. 109. Springer-Verlag, New York, 1964.

Problems

FLUID SOLUTIONS **9.1** A cube weighing 150 N and measuring 35 cm on a side is allowed to slide down an inclined surface on which there is a film of oil having a viscosity of $10^{-2}\ \text{N} \cdot \text{s/m}^2$. What is the terminal velocity of the block if the oil has a thickness of 0.1 mm?

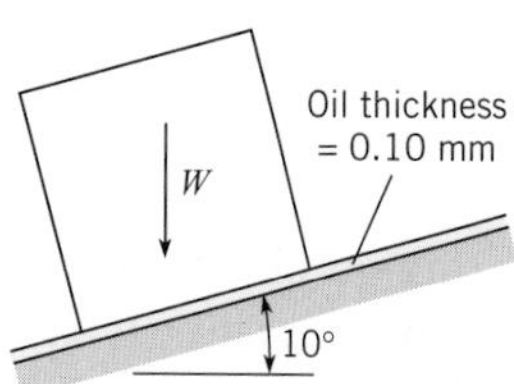

PROBLEM 9.1

9.2 A board 3 ft by 3 ft that weighs 40 lbf slides down an inclined ramp with a velocity of 0.5 fps. The board is separated from the ramp by a layer of oil 0.02 in. thick. Neglecting the edge effects of the board, calculate the approximate dynamic viscosity μ of the oil.

9.3 A board 1 m by 1 m that weighs 20 N slides down an inclined ramp with a velocity of 12 cm/s. The board is separated from the ramp by a layer of oil 0.5 mm thick. Neglecting the edge effects of the board, calculate the approximate dynamic viscosity μ of the oil.

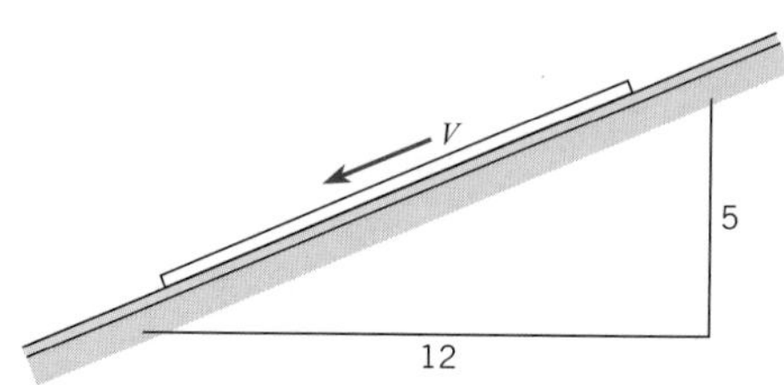

PROBLEMS 9.2, 9.3

FLUID SOLUTIONS **9.4** Uniform, steady flow is occurring between horizontal parallel plates as shown.

a. In a few words, tell what other condition must be present to cause the odd velocity distribution.

b. Where is the minimum shear stress located?

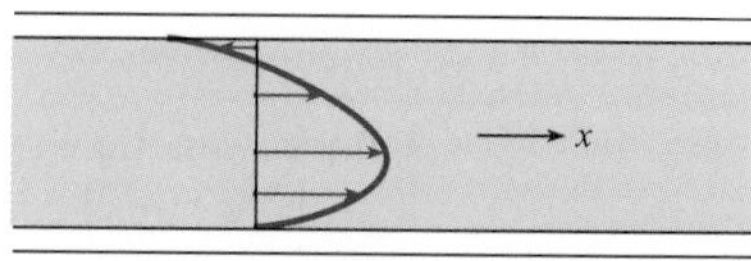

PROBLEM 9.4

9.5 Under certain conditions (pressure decreasing in the x direction and a moving plate), the laminar velocity distribution will be as shown here. For such a condition, indicate whether each of the following statements is true or false.

a. The greatest shear stress in the liquid occurs next to the fixed plate.

b. The shear stress midway between the plates is zero.
c. The minimum shear stress in the liquid occurs next to the moving plate.
d. The shear stress is greatest where the velocity is the greatest.
e. The minimum shear stress occurs where the velocity is the greatest.

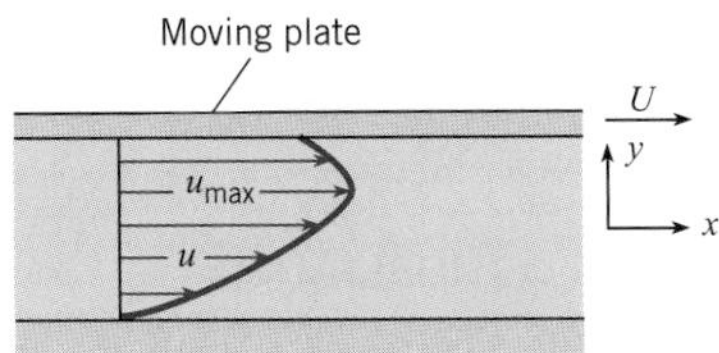

PROBLEM 9.5

9.6 A flat plate is pulled to the right at a speed of 30 cm/s. Oil with a viscosity of 4 N · s/m^2 fills the space between the plate and the solid boundary. The plate is 1 m long (L = 1 m) by 30 cm wide, and the spacing between the plate and boundary is 2.0 mm.
a. Express the velocity mathematically in terms of the coordinate system shown.
b. By mathematical means, determine whether this flow is rotational or irrotational.
c. Determine whether continuity is satisfied, using the differential form of the continuity equation.
d. Calculate the force required to produce this plate motion.

9.7 The velocity distribution that is shown represents laminar flow. Indicate whether each of the following statements is true or false.
a. The velocity gradient at the boundary is infinitely large.
b. The maximum shear stress in the liquid occurs midway between the walls.
c. The maximum shear stress in the liquid occurs next to the boundary.
d. The flow is irrotational.
e. The flow is rotational.

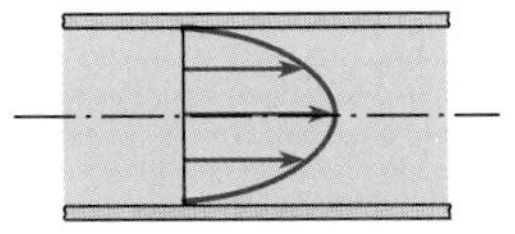

PROBLEM 9.7

9.8 A tube and a wire positioned concentrically within the tube are submerged in oil. If the wire is drawn through the tube at a constant rate, will the viscous shear stress on the wire be greater than, equal to, or less than the shear stress on the tube wall?

9.9 The upper plate shown is moving to the right with a velocity V, and the lower plate is free to move laterally under the action of the viscous forces applied to it. For steady-state conditions, derive an equation for the velocity of the lower plate. Assume that the area of oil contact is the same for the upper plate, each side of the lower plate, and the fixed boundary.

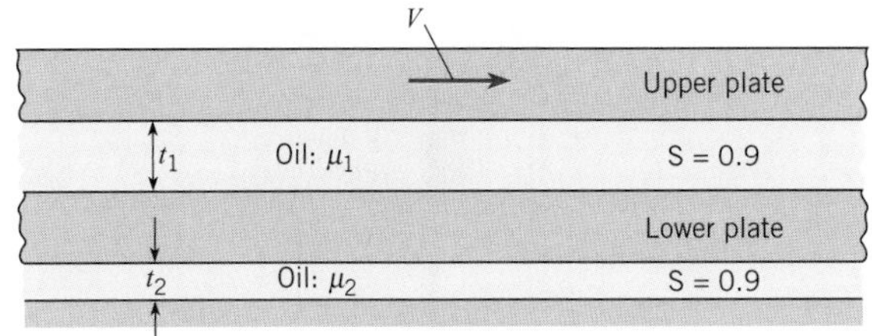

PROBLEM 9.9

9.10 A circular horizontal disk with a 12-in. diameter has a clearance of 0.001 ft from a horizontal plate. What torque is required to rotate the disk about its center at an angular velocity of 180 rpm when the clearance space contains oil (μ = 0.12 lbf-s/ft^2)?

9.11 A circular horizontal disk with a 20-cm diameter has a clearance of 2.0 mm from a horizontal plate. What torque is required to rotate the disk about its center at an angular speed of 10 rad/s when the clearance space contains oil (μ = 8 N · s/m^2)?

9.12 A movable cone fits inside a stationary conical depression as shown. When a torque is applied to the cone, the cone rotates at a speed depending on the angles θ and β, the radius r_0, and the viscosity μ of the liquid. Derive an equation for the torque in terms of the other variables, including only the viscous resistance. Assume that θ is very small.

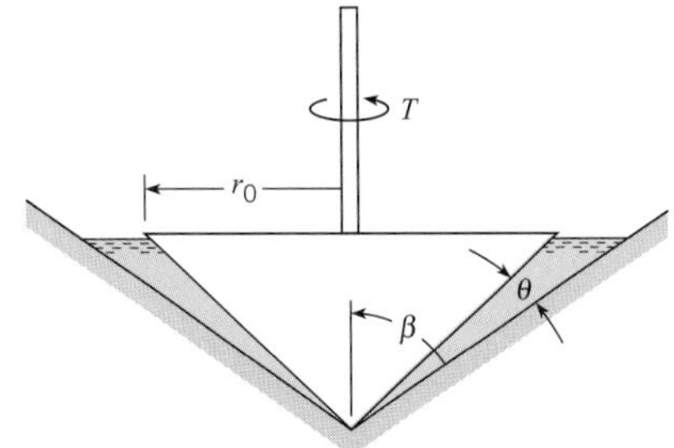

PROBLEM 9.12

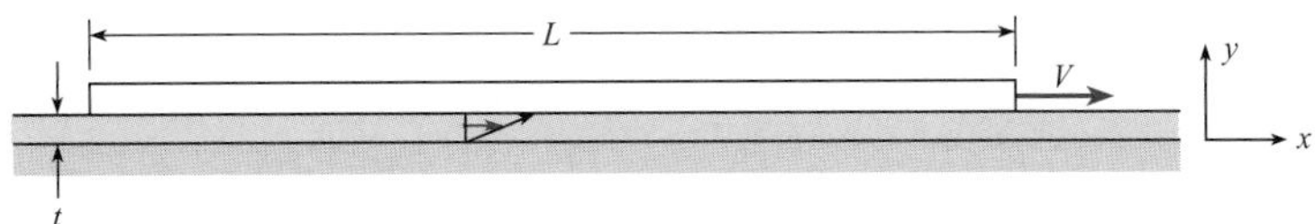

PROBLEM 9.6

9.13 A plate 2 mm thick and 1 m wide (normal to the page) is pulled between the walls shown in the figure at a speed of 0.40 m/s. Note that the space that is not occupied by the plate is filled with glycerine at a temperature of 20°C. Also, the plate is positioned midway between the walls. Sketch the velocity distribution of the glycerine at section *A-A*. Neglecting the weight of the plate, estimate the force required to pull the plate at the speed given.

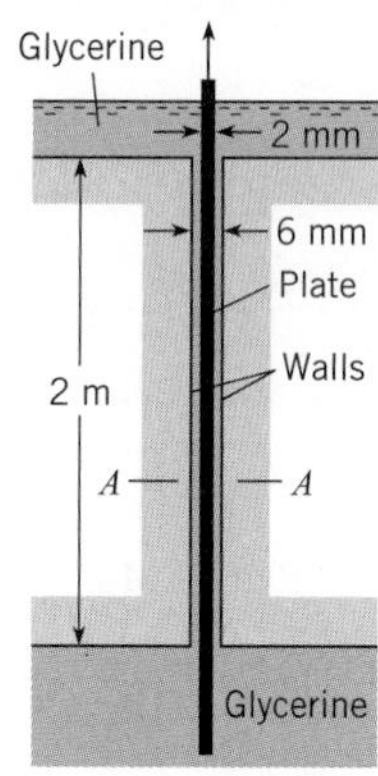

PROBLEM 9.13

FLUID SOLUTIONS **9.14** A bearing uses SAE 30 oil with a viscosity of 0.1 N · s/m². The bearing is 30 mm in diameter, and the gap between the shaft and the casing is 1 mm. The bearing has a length of 1 cm. The shaft turns at $\omega = 200$ rad/s. Assuming that the flow between the shaft and the casing is a Couette flow, find the torque required to turn the bearing.

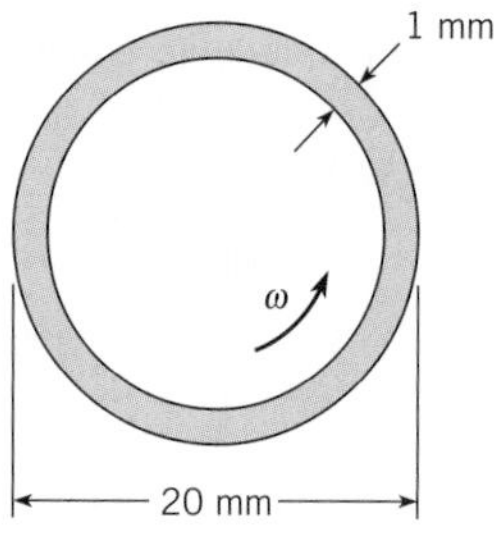

PROBLEM 9.14

9.15 An important application of surface resistance is found in lubrication theory. Consider a shaft that turns inside a stationary cylinder, with a lubricating fluid in the annular region. By considering a system consisting of an annulus of fluid of radius *r* and width Δr, and realizing that under steady-state operation the net torque on this ring is zero, show that $d(r^2\tau)/dt = 0$, where τ is the viscous shear stress. For a flow that has a tangential component of velocity only, the shear stress is related to the velocity by $\tau = \mu r d(V/r)/dr$. Show that the torque per unit length acting on the inner cylinder is given by $T = 4\pi\mu\omega r_s^2/(1 - r_s^2/r_0^2)$, where ω is the angular velocity of the shaft.

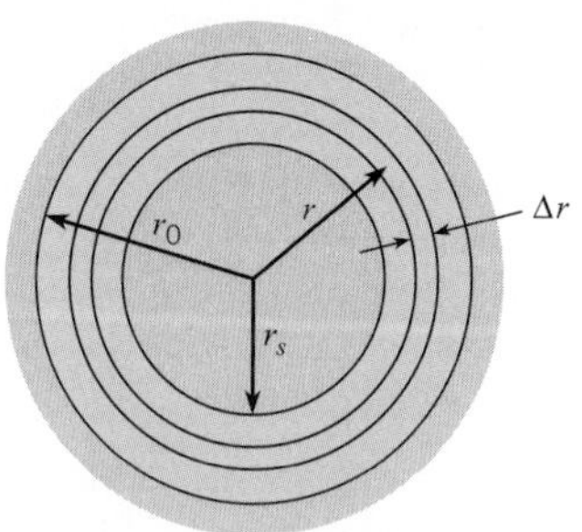

PROBLEM 9.15

9.16 Using the equation developed in Prob. 9.15, find the power necessary to rotate a 2-cm shaft at 50 rad/s if the inside diameter of the casing is 2.2 cm, the bearing is 3 cm long, and SAE 30 oil at 38°C is the lubricating fluid.

9.17 The analysis developed in Prob. 9.15 applies to a device used to measure the viscosity of a fluid. By applying a known torque to the inner cylinder and measuring the angular velocity achieved, one can calculate the viscosity of the fluid. Assume you have a 4-cm inner cylinder and a 4.5-cm outer cylinder. The cylinders are 10 cm long. When a force of 0.6 N is applied to the tangent of the inner cylinder, it rotates at 20 rpm. Calculate the viscosity of the fluid.

9.18 If a thin film of oil ($\nu = 10^{-3}$ m²/s) 2.0 mm thick flows down a surface inclined at 30° to the horizontal, what will be the maximum and mean velocity of flow?

9.19 What are the depth and discharge per unit width of oil (SAE 30 at 100°F) flowing down a 20° incline at a Reynolds number of 200?

9.20 Rain falls on a smooth roof 15 ft by 40 ft at a rate of 0.4 in./h, and the roof has a slope, in the 15-ft direction, of 10°. Estimate the depth and average velocity of flow at the lower end of the roof, assuming a temperature of 50°F.

9.21 Two horizontal parallel plates are spaced 2 mm apart. If the pressure decreases at a rate of 1.20 kPa/m in the horizontal *x* direction in the fluid between the plates, what is the maximum fluid velocity in the *x* direction? The fluid has a dynamic viscosity of 10^{-1} N · s/m² and a specific gravity of 0.90. What is the magnitude of the shearing force on the lower plate if it is 2 m long (in the direction of flow) and 1.5 m wide?

9.22 Two horizontal parallel plates are spaced 0.01 ft apart. The pressure decreases at a rate of 12 psf/ft in the horizontal *x* direction in the fluid between the plates. What is the maximum fluid velocity in the *x* direction? The fluid has a dynamic viscosity of 10^{-3} lbf-s/ft² and a specific gravity of 0.80.

9.23 A viscous fluid fills the space between these two plates, and the pressures at *A* and *B* are 150 psf and 100 psf, respectively. The fluid is not accelerating. If the specific weight of the fluid is 100 lbf/ft³, then one must conclude that (a) flow is downward, (b) flow is upward, (c) there is no flow.

9.24 Glycerine at 20°C flows downward between two vertical parallel plates separated by a distance of 0.4 cm. The ends are

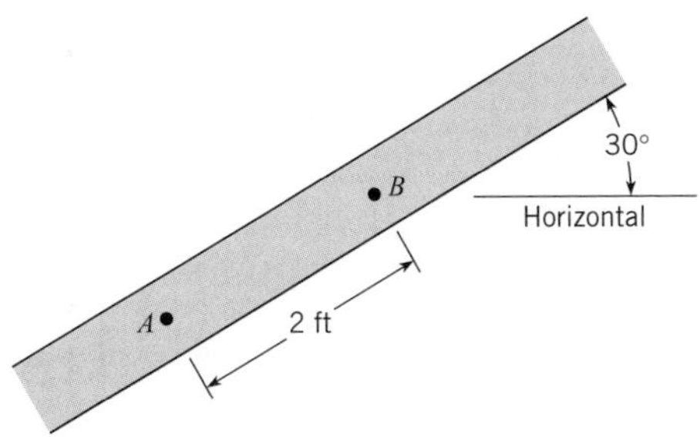

PROBLEM 9.23

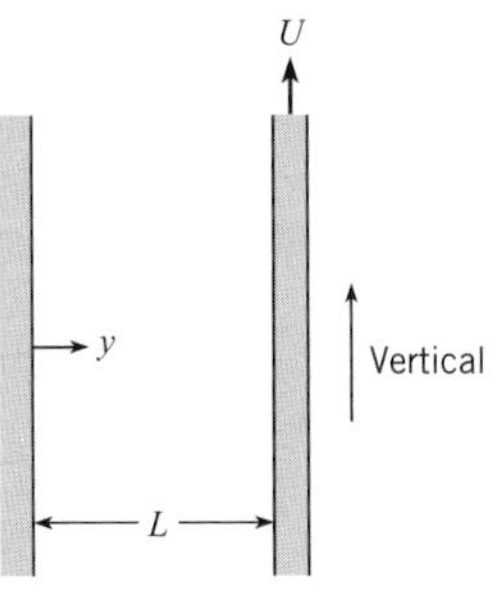

PROBLEM 9.30

open, so there is no pressure gradient. Calculate the discharge per unit width, q, in m^2/s.

9.25 Two vertical parallel plates are spaced 0.01 ft apart. If the pressure decreases at a rate of 8 psf/ft in the vertical z direction in the fluid between the plates, what is the maximum fluid velocity in the z direction? The fluid has a viscosity of 10^{-3} lbf-s/ft^2 and a specific gravity of 0.80.

9.26 Two vertical parallel plates are spaced 2 mm apart. If the pressure decreases at a rate of 10 kPa/m in the positive z direction (vertically upward) in the fluid between the plates, what is the maximum fluid velocity in the z direction? The fluid has a viscosity of 10^{-1} N · s/m^2 and a specific gravity of 0.85.

9.27 Two vertical parallel plates are spaced 0.01 ft apart. If the pressure decreases at a rate of 60 psf/ft in the vertical z direction in the fluid between the plates, what is the maximum fluid velocity in the z direction? The fluid has a viscosity of 10^{-3}lbf-s/ft^2 and a specific gravity of 0.80.

9.28 Two parallel plates are spaced 0.09 in. apart, and motor oil (SAE 30) with a temperature of 100°F flows at a rate of 0.009 cfs per foot of width between the plates. What is the pressure gradient in the direction of flow if the plates are inclined at 60° with the horizontal and if the flow is downward between the plates?

9.29 Two parallel plates are spaced 2 mm apart, and oil ($\mu = 10^{-1}$ N · s/m^2, S = 0.80) flows at a rate of 24×10^{-4} m^3/s per meter of width between the plates. What is the pressure gradient in the direction of flow if the plates are inclined at 60° with the horizontal and if the flow is downward between the plates?

9.30 A flow exists between two vertical plates as shown. The plate on the left is fixed while the plate on the right moves upward with a velocity U. The plates are separated by a distance L. There is no pressure gradient in the direction of motion. (a) Derive an expression for the velocity distribution between the plates as a function of γ, y, L, μ, and U. (b) Determine the plate velocity as a function of γ, L, and μ for which the discharge between the two plates is zero.

9.31 The flow of mud is often modeled as a Bingham plastic (see p. 20) in which the relation between stress and rate of strain is given as $\tau = \tau_0 + \eta\, du/dy$, where τ_0 is the threshold stress and η is a constant. With a Bingham plastic, there is no flow unless the threshold stress is exceeded. Consider the flow of mud down an inclined plane as shown. (a) Find the relationship between τ_0, γ, y_0, and θ for which there would be no motion. (b) Determine the velocity field $u = f(y)$ when there is flow.

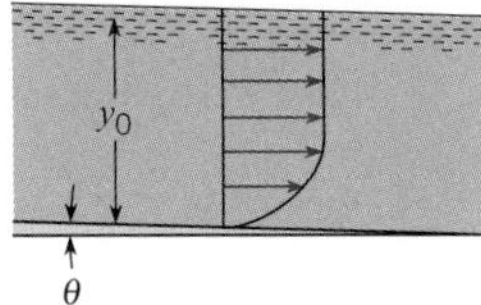

PROBLEM 9.31

9.32 Glycerine at 20°C flows downward in the annular region between two cylinders. The internal diameter of the outer cylinder is 3 cm, and the external diameter of the inner cylinder is 2.8 cm. The pressure is constant along the flow direction. The flow is laminar. Calculate the discharge. (*Hint:* The flow between the two cylinders can be treated as the flow between two flat plates.)

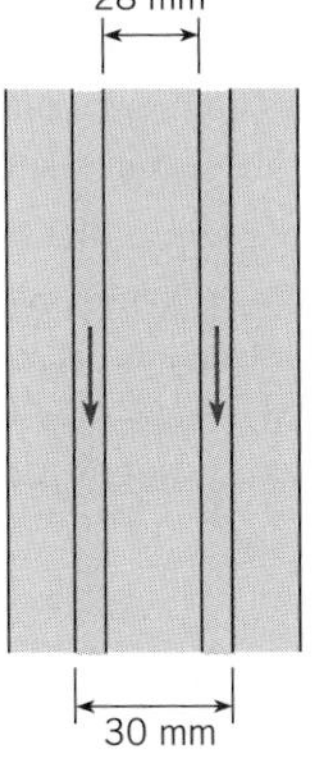

PROBLEM 9.32

FLUID SOLUTIONS **9.33** One type of bearing that can be used to support very large structures is shown in the accompanying figure. Here fluid under pressure is forced from the bearing midpoint (slot A) to the exterior zone B. Thus a pressure distribution occurs as shown. For this bearing, which is 30 cm wide, what discharge of oil from slot A per meter of length of bearing is required? Assume a 50-kN load per meter of bearing length with a clearance space t between the floor and the bearing surface of 0.60 mm. Assume an oil viscosity of 0.20 N · s/m^2. How much oil per hour would have to be pumped per meter of bearing length for the given conditions?

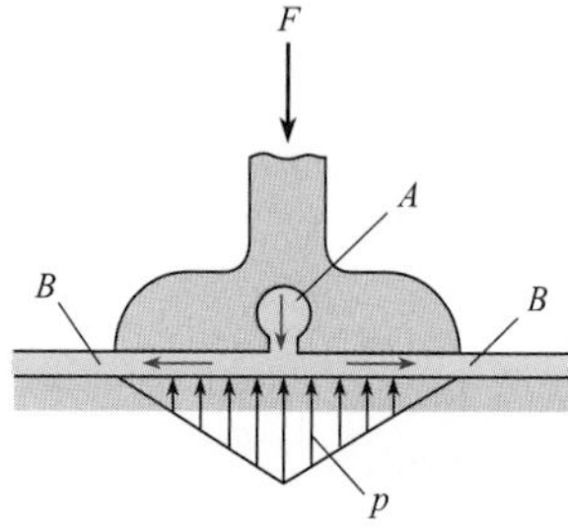

PROBLEM 9.33

9.34 Use the continuity and Navier-Stokes equations to solve for the velocity distribution in a Couette flow (see Section 9.2).

FLUID SOLUTIONS **9.35** An Eiffel-type wind tunnel operates by drawing air through a contraction, passing this air through a test section, and then exhausting the air using a large axial fan. Experimental data are recorded in the test section, which is typically a rectangular section of duct that is made of clear plastic (usually acrylic). In the test section, the velocity should have a very uniform distribution; thus, it is important that the boundary layer be very thin at the end of the test section. For the pictured wind tunnel, the test section is square with a dimension of $W = 457$ mm on each side and a length of $L = 914$ mm. Find the ratio of maximum boundary layer thickness to test section width $[\delta(x = L)/W]$ for two cases: minimum operating velocity (1 m/s) and maximum operating velocity (70 m/s). Assume air properties at 1 atm and 20°C.

9.36 A fluid flows with a speed of $U_0 = 2.4$ m/s over the top of a horizontal flat plate that has the shape of an isosceles triangle with $L = 1.5$ m. Assume that the wall shear stress decreases linearly from 10.0 to 2.0 N/m^2. For the specified shear stress distribution, find the viscous drag force in newtons on the top of the plate.

9.37 A thin plate 6 ft long and 3 ft wide is submerged and held stationary in a stream of water ($T = 60$°F) that has a velocity of 5 ft/s. What is the thickness of the boundary layer on the plate for $Re_x = 500{,}000$ (assume the boundary layer is still laminar), and at what distance downstream of the leading edge does this Reynolds number occur? What is the shear stress on the plate at this point?

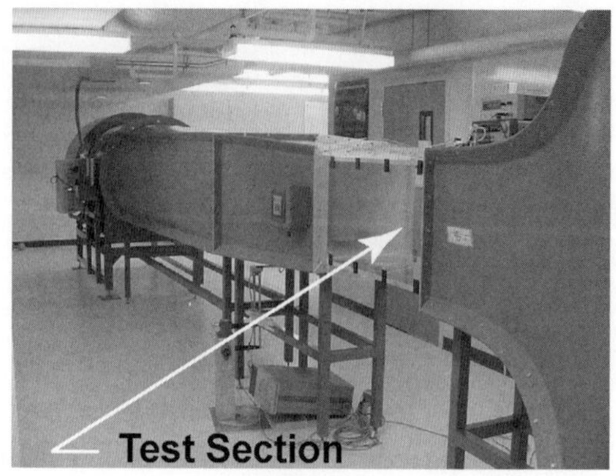

PROBLEM 9.35

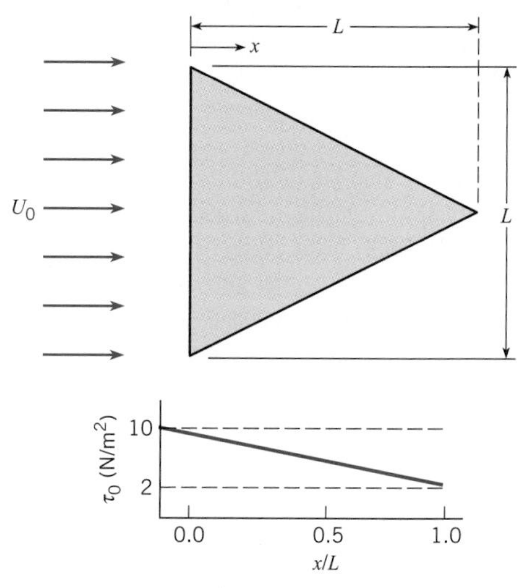

PROBLEM 9.36

9.38 What is the ratio of the boundary layer thickness on a smooth, flat plate to the distance from the leading edge just before transition to turbulent flow?

9.39 An engineer is designing a horizontal, rectangular conduit that will be part of a system that allows fish to bypass a dam. Inside the conduit, a flow of water at 40°F will be divided into two streams by a flat, rectangular metal plate. Calculate the viscous drag force on this plate, assuming boundary layer flow with free-stream velocity of 12 ft/s and plate dimensions of $L = 8$ ft and $W = 4.0$ ft.

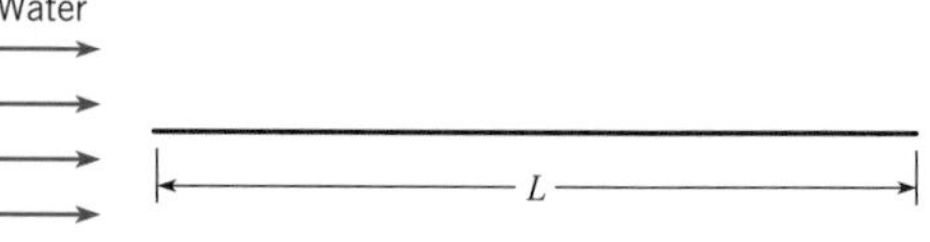

PROBLEM 9.39

9.40 Oil ($\mu = 10^{-2}$ N · s/m^2; $\rho = 900$ kg/m^3) flows past a plate in a tangential direction so that a boundary layer develops. If the velocity of approach is 4 m/s, then at a section 30 cm downstream of the leading edge the ratio of τ_δ (shear stress at the edge of the boundary layer) to τ_0 (shear stress at the plate surface) is approximately (a) 0, (b) 0.24, (c) 2.4, (d) 24.

9.41 An element for sensing local shear stress is positioned in a flat plate 1 meter from the leading edge. The element simply consists of a small plate, 1 cm × 1 cm, mounted flush with the wall, and the shear force is measured on the plate. The fluid flowing by the plate is air with a free-stream velocity of 25 m/s, a density of 1.2 kg/m^3, and a kinematic viscosity of 1.5×10^{-5} m^2/s. The boundary layer is tripped at the leading edge. What is the magnitude of the force due to shear stress acting on the element?

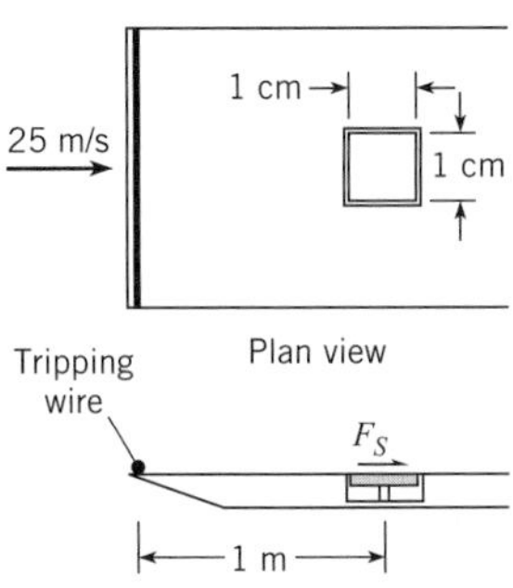

PROBLEM 9.41

9.42 You want to use the integral technique to determine the thickness of a laminar boundary layer. Assume that the velocity profile can be approximated by $u/U_0 = (y/\delta)^{1/2}$. Experimental data show that $\tau_0 = 1.66 U_0 \mu/\delta$. Use Eq. (9.42) and the foregoing relations to obtain a differential equation for δ and solve for $\delta = f(x)$. Compare your results with those of Blasius, Eq. (9.15).

9.43 A liquid ($\rho = 1000$ kg/m^3, $\mu = 2 \times 10^{-2}$ N · s/m^2, $\nu = 2 \times 10^{-5}$ m^2/s) flows tangentially past a flat plate. If the approach velocity is 1 m/s, what is the liquid velocity 1 m downstream from the leading edge of the plate and 1 mm away from the plate?

9.44 The plate of Prob. 9.43 has a total length of 3 m (parallel to the flow direction), and it is 1 m wide. What is the skin friction drag (shear force) on one side of the plate?

9.45 Oil ($\nu = 10^{-4}$ m^2/s) flows tangentially past a thin plate. If the free-stream velocity is 5 m/s, what is the velocity 1 m downstream from the leading edge and 3 mm away from the plate?

9.46 Oil ($\nu = 10^{-4}$ m^2/s, S = 0.9) flows past a plate in a tangential direction so that a boundary layer develops. If the velocity of approach is 1 m/s, what is the oil velocity 1 m downstream from the leading edge and 10 cm away from the plate?

9.47 A thin plate 0.7 m long and 1.5 m wide is submerged and held stationary in a stream of water ($T = 10$°C) that has a velocity of 1.5 m/s. What is the thickness of the boundary layer on the plate for $\text{Re}_x = 500{,}000$ (assume the boundary layer is still laminar), and at what distance downstream of the leading edge does this Reynolds number occur? What is the shear stress on the plate on this point? FLUID SOLUTIONS

9.48 For the conditions of Prob. 9.47, what is the shearing resistance on one side of the plate for the part of the plate that has a Reynolds number, Re_x, less than 500,000? What is the ratio of the laminar shearing force to the total shearing force on the plate?

9.49 An airplane wing of 2 m chord (leading edge to trailing edge distance) and 11 m span flies at 200 km/hr in air 30°C. Assume that the resistance of the wing surfaces is like that of a flat plate. FLUID SOLUTIONS

a. What is the friction drag on the wing?
b. What power is required to overcome this?
c. How much of the chord is laminar?
d. What will be the change in drag if a turbulent boundary layer is tripped at the leading edge?

9.50 A turbulent boundary layer exists in the flow of water at 20°C over a flat plate. The local shear stress measured at the surface of the plate is 0.1 N/m^2. What is the velocity at a point 1 cm from the plate surface?

9.51 A flat plate 1.5 m long and 1.0 m wide is towed in water at 20°C in the direction of its length at a speed of 15 cm/s. Determine the resistance of the plate and the boundary layer thickness at its aft end.

9.52 A liquid flows tangentially past a flat plate. The fluid properties are $\mu = 10^{-5}$ N · s/m^2 and $\rho = 1.5$ kg/m^3. Find the skin friction drag on the plate per unit width if the plate is 2 m long and the approach velocity is 20 m/s. Also, what is the velocity gradient at a point that is 1 m downstream of the leading edge and just next to the plate ($y = 0$)?

9.53 Starting with Eq. (9.44), carry out the steps leading to Eq. (9.47).

9.54 A model airplane has a wing span of 3 ft and a chord (leading edge–trailing edge distance) of 5 in. The model flies in air at 60°F and atmospheric pressure. The wing can be regarded as a flat plate so far as drag is concerned. At what speed will a turbulent boundary layer start to develop on the wing? What will be the total drag force on the wing just before turbulence appears?

9.55 For the hypothetical boundary layer on the flat plate shown, what are the skin friction drag on the top side per meter of width and the shear stress on the plate at the downstream end? Here $\rho = 1.2$ kg/m^3 and $\mu = 1.8 \times 10^{-5}$ N · s/m^2. FLUID SOLUTIONS

9.56 Starting with Eq. (9.43), perform the integration and simplify to obtain Eq. (9.44).

9.57 Assume that the velocity profile in a boundary layer is replaced by a step profile, as shown in the figure, where the velocity is zero adjacent to the surface and equal to the free-stream velocity (U) at a distance greater than δ_* from the surface. Assume also that

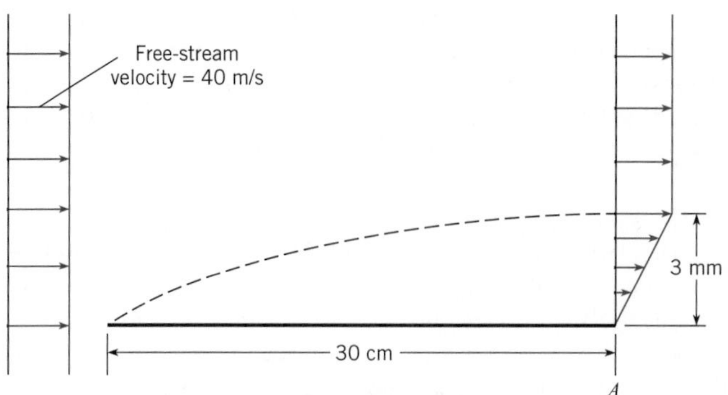

PROBLEMS 9.55, 9.58

the density is uniform and equal to the free-stream density (ρ_∞). The distance δ_* (displacement thickness) is so chosen that the mass flux corresponding to the step profile is equal to the mass flux through the actual boundary layer. Derive an integral expression for the displacement thickness as a function of ρ, ρ_∞, u, U, y, and δ.

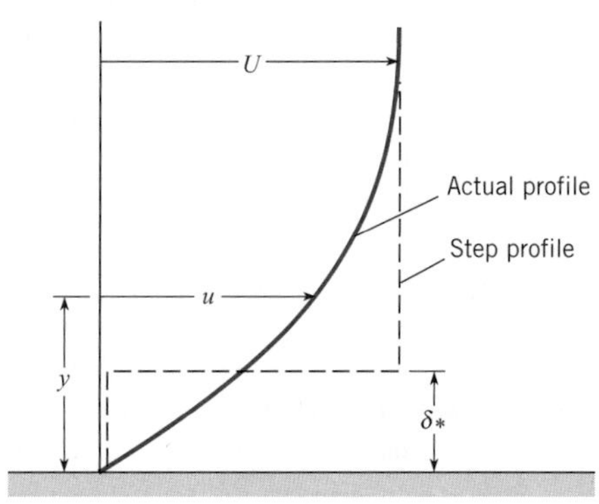

PROBLEM 9.57

9.58 Because of the reduction of velocity associated with the boundary layer, the streamlines outside the boundary layer are shifted away from the boundary. This amount of displacement of the streamlines is defined as the displacement thickness δ_*. For the boundary layer at the downstream edge of the plate in Prob. 9.55, what is the magnitude of the displacement thickness?

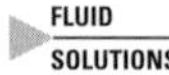

9.59 Use the integral method represented by Eq. (9.44) and the following relationship between shear stress and boundary layer thickness,

$$\frac{\tau_0}{\rho} = 0.0225 U_0^2 \left(\frac{\nu}{U_0 \delta}\right)^{1/4}$$

to find the variation of boundary layer thickness with x and Re_x, the variation of local shear stress coefficient with Re_x and the variation of average shear stress coefficient with Re_L. Assume the boundary layer profile is given by

$$\frac{u}{U_0} = \left(\frac{y}{\delta}\right)^{1/7}$$

9.60 What is the ratio of the skin friction drag of a plate 30 m long and 5 m wide to that of a plate 10 m long and 5 m wide if both plates are towed lengthwise through water ($T = 20°\mathrm{C}$) at 10 m/s?

9.61 Estimate the power required to pull the sign shown if it is towed at 35 m/s and if it is assumed that the sign has the same resistance characteristics as a flat plate. Assume standard atmospheric pressure and a temperature of 10°C.

9.62 A thin plastic panel (3 mm thick) is lowered from a ship to a construction site on the ocean floor. The plastic panel weighs 250 N in air and is lowered at a rate of 3 m/s. Assuming that the panel remains vertically oriented, calculate the tension in the cable.

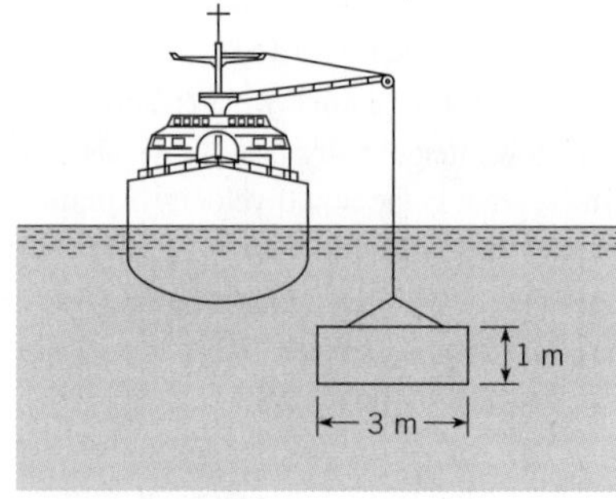

PROBLEM 9.62

9.63 The plate shown in the figure is weighted at the bottom so it will fall stably and steadily in a liquid. The weight of the plate in air is 23.5 N and the plate has a volume of 0.002 m^3. Estimate its falling speed in fresh water at 20°C. The boundary layer is normal; that is, it is not tripped at the leading edge.

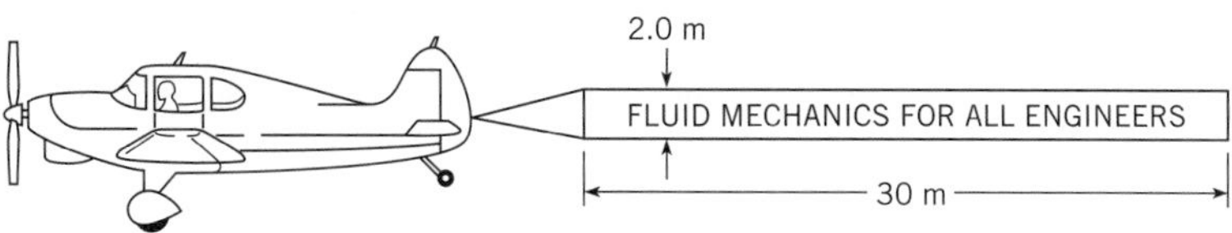

PROBLEM 9.61

In this problem, the final falling speed (terminal velocity) occurs when the weight is equal to the sum of the skin friction and buoyancy.

$$W = B + F_s = \gamma \forall + \frac{1}{2} C_f \rho U_0^2 S$$

Solve this problem using the "direct substitution method." Start by assuming a speed, calculate the Reynolds number and average shear stress coefficient, and substitute into the above equation to find the speed. Use this speed to find the new Reynolds number, skin friction, and speed. Continue until the calculated speed no longer changes.

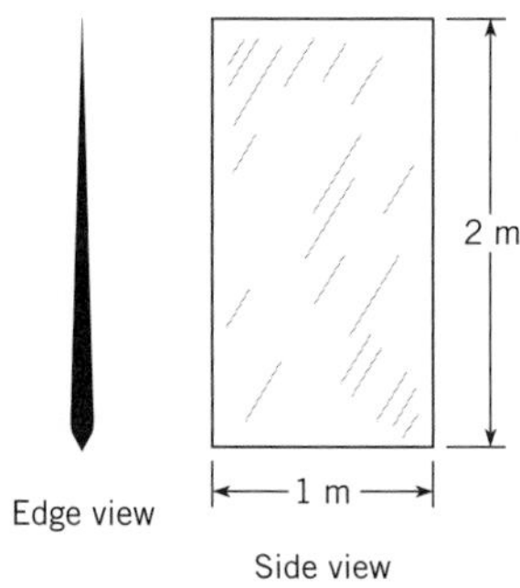

PROBLEM 9.63

9.64 A turbulent boundary layer develops from the leading edge of a flat plate with water at 20°C flowing tangentially past the plate with a free-stream velocity of 5 m/s. Determine the thickness of the viscous sublayer, δ', at a distance 1 m downstream from the leading edge. Would a roughness element 100 μm high affect the local skin friction coefficient? Why?

9.65 A model airplane descends in a vertical dive through air at standard conditions (1 atmosphere and 20°C). The majority of the drag is due to skin friction on the wing (like that on a flat plate). The wing has a span of 1 m (tip to tip) and a chord (leading edge to trailing edge distance) of 10 cm. The leading edge is rough, so the turbulent boundary layer is "tripped." The model weighs 3 N. Determine the speed (in meters per second) at which the model will fall.

9.66 A flat plate is oriented parallel to a 15 m/s air flow at 20°C and atmospheric pressure. The plate is 1 m long in the flow direction and 0.5 m wide. On one side of the plate, the boundary layer is tripped at the leading edge and on the other side there is no tripping device. Find the total drag force on the plate.

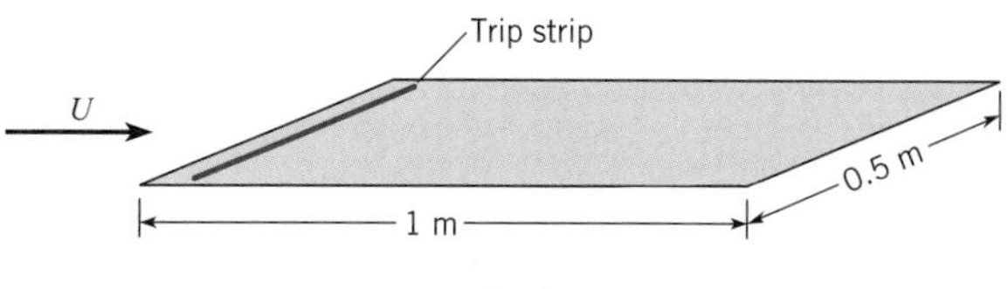

PROBLEM 9.66

9.67 A model is being developed for the entrance region between two flat plates. As shown in the figure, it is assumed that the region is approximated by a turbulent boundary layer originating at the leading edge. The system is designed such that the plates end where the boundary layers merge. The spacing between the plates is 4 mm and the entrance velocity is 10 m/s. The fluid is water at 20°C. Roughness at the leading edge trips the boundary layers. Find the length L where the boundary layers merge and find the force per unit depth (into the paper) due to shear stress on both plates.

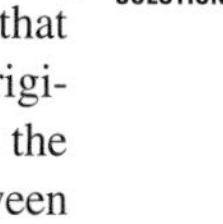

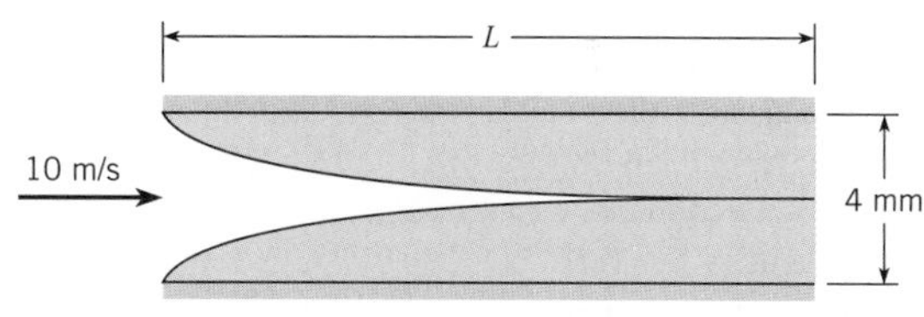

PROBLEM 9.67

9.68 Develop a computer program that requests a Reynolds number as input and information on nature of the boundary layer (tripped or natural) and provides the boundary layer thickness, the local shear stress coefficient, and the average shear stress coefficient.

9.69 An outboard racing boat "planes" at 70 mph over water at 60°F. The part of the hull in contact with the water has an average width of 3 ft and a length of 8 ft. Estimate the power required to overcome its surface resistance.

9.70 A javelin is approximately 265 cm long, has an average diameter of approximately 25 mm, and weighs 8.0 N. With a straight throw (javelin oriented parallel to the line of flight so that the relative air speed is parallel to the javelin), what will be the air drag at a speed of 30 m/s? What will be the javelin's deceleration (parallel to the line of flight) at its trajectory azimuth for the 30-m/s speed? What will be the corresponding drag and acceleration if the javelin is thrown into a 5-m/s wind and then thrown with a 5-m/s tailwind? Estimate the maximum distance of the throw if the thrower releases the javelin with a speed of 32 m/s. Assume the air temperature is 20°C.

9.71 A motor boat pulls a long, smooth, water-soaked log (0.5 m in diameter and 50 m long) at a speed of 1.7 m/s. Assuming total submergence, estimate the force required to overcome the surface resistance of the log. Assume a water temperature of 10°C.

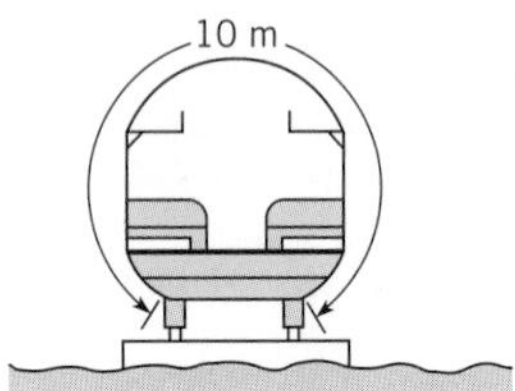

PROBLEM 9.72

9.72 Modern high-speed passenger trains are streamlined to reduce surface resistance. The cross section of a passenger car of one such train is shown. For a train 150 m long, estimate the surface resistance for a speed of 100 km/h and for one of 200 km/h. What power is required for just the surface resistance at these speeds? Assume $T = 10°C$.

FLUID SOLUTIONS **9.73** Consider the boundary layer next to the smooth hull of a ship. The ship is cruising at a speed of 45 ft/s in 60°F fresh water. Assuming that the boundary layer on the ship hull develops the same as on a flat plate, determine:

a. The thickness of the boundary layer at a distance $x = 100$ ft downstream from the bow.

b. The velocity of the water at a point in the boundary layer at $x = 100$ ft and $y/\delta = 0.50$.

c. The shear stress, τ_0, adjacent to the hull at $x = 100$ ft.

9.74 A ship 600 ft long steams at a rate of 30 ft/s through still fresh water ($T = 50°C$). If the submerged area of the ship is 50,000 ft^2, what is the skin friction drag of this ship?

9.75 A river barge has the dimensions shown. It draws 2 ft of water when empty. Estimate the skin friction drag of the barge when it is being towed at a speed of 10 ft/s through still fresh water.

9.76 A supertanker has length, breadth, and draught (fully loaded) dimensions of 325 m, 48 m, and 19 m, respectively. In open seas the tanker normally operates at a speed of 18 kt (1 kt = 0.515 m/s). For these conditions, and assuming that flat-plate boundary-layer conditions are approximated, estimate the skin friction drag of such a ship steaming in 10°C water. What power is required to overcome the skin friction drag? What is the boundary layer thickness at 300 m from the bow?

9.77 A model test is to be done to predict the wave drag on a ship. The ship is 500 ft long and operates at 30 ft/s in sea water at 10°C. The wetted area of the prototype is 25,000 ft^2. The model:prototype scale ratio is $\frac{1}{100}$. Modeling is done in fresh water at 60°F to match the Froude number. The viscous drag can be calculated by assuming a flat plate with the wetted area of the model and a length corresponding to the length of the model. A total drag of 0.1 lbf is measured in the model tests. Calculate the wave drag on the actual ship.

9.78 A ship is designed so that it is 250 m long, its beam measures 30 m, and its draft is 12 m. The surface area of the ship below the water line is 8800 m^2. A 1/30 scale model of the ship is tested and is found to have a total drag of 38.0 N when towed at a speed of 1.45 m/s. Using the methods outlined in Section 8.10, answer the following questions, assuming that model tests are made in fresh water (20°C) and that prototype conditions are sea water (10°C).

a. To what speed in the prototype does the 1.45 m/s correspond?

b. What are the model skin friction drag and wave drag?

c. What would the ship drag be in salt water corresponding to the model-test conditions in fresh water?

9.79 A hydroplane 3 m long skims across a very calm lake ($T = 20°C$) at a speed of 15 m/s. For this condition, what will be the minimum shear stress along the smooth bottom?

9.80 Estimate the power required to overcome the surface resistance of a water skier if he or she is towed at 30 mph and each ski is 4 ft by 6 in. Assume the water temperature is 60°F.

9.81 If the wetted area of an 80-m ship is 1500 m^2, approximately how great is the surface drag when the ship is traveling at a speed of 10 m/s? What is the thickness of the boundary layer at the stern? Assume seawater at $T = 10°C$.

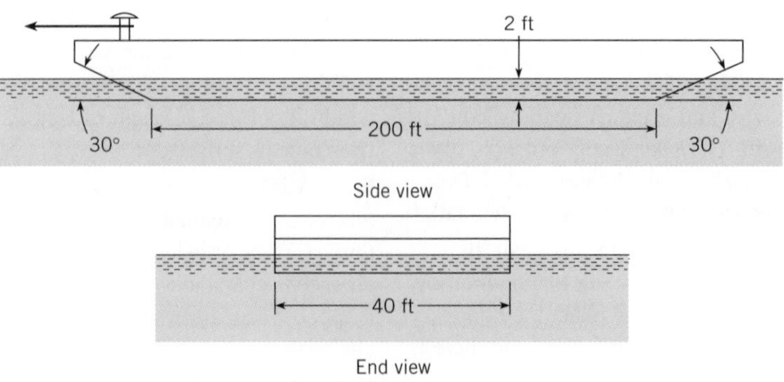

PROBLEM 9.75

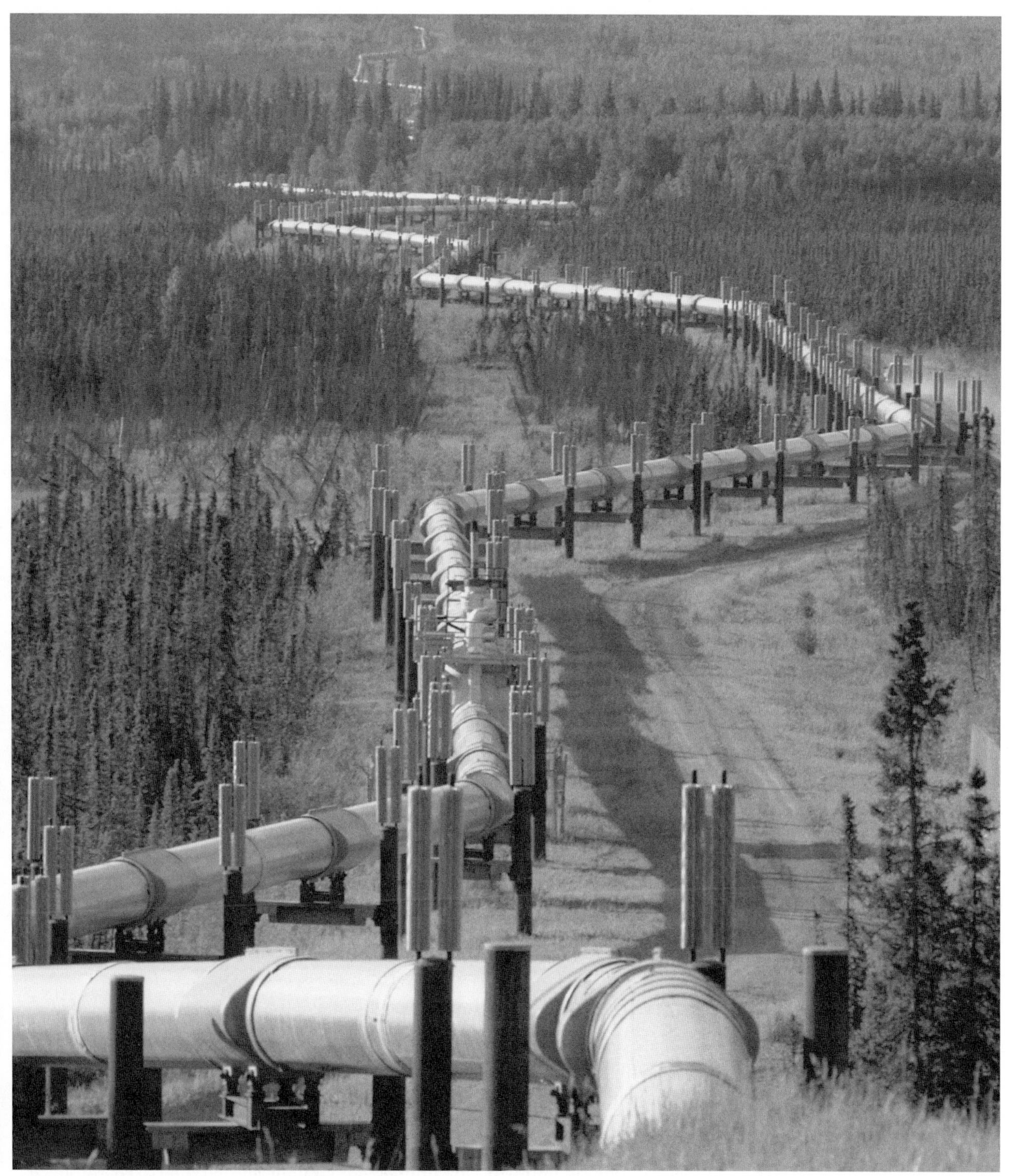

The Trans-Alaska pipeline transports crude oil over 800 miles from the North Slope of Alaska to the ice-free seaport of Valdez. The pipeline, 4 ft in diameter, crosses three mountain ranges and over 800 rivers and streams and took two years to construct. (Alaska Stock Images)

CHAPTER 10

Flow in Conduits

When one considers the conveniences and necessities of everyday life, it is truly amazing to note the role played by conduits. For example, all of the water that we use in our homes is pumped through pipes so that it will be available when and where we want it. In addition, virtually all of this water leaves our homes as dilute wastes through sewers, another type of conduit. In addition to domestic use, the consumption of water by industry is enormous, from the processing of agricultural products to the manufacturing of steel and paper. All of the water used in these manufacturing processes is transported by means of piping systems. In the United States the petroleum industry alone transports approximately 20 million barrels of liquid petroleum per day in addition to the billions of cubic feet of gas transported by pipeline.

In the foregoing examples it is the transportation of the fluid that is the primary objective. However, there are numerous applications in which flow is a necessary but secondary part of the process. For example, heating and ventilating systems, as well as electric generating stations, utilize conduit flow to circulate fluids in order to transport energy from one location to another. Piping systems are also used extensively for controlling the operation of machinery.

Thus the application of flow in conduits cuts across all fields of engineering. Consequently, every engineer should understand the basic fluid mechanics involved with such flow. In this chapter we shall introduce the fundamental theory of flow in conduits as well as basic design procedures.

Shear Stress Distribution across a Pipe Section

The velocity distribution in a pipe is directly linked to the shear stress distribution; hence it is important to understand the latter. To determine the shear stress distribution, we start with the equation of equilibrium applied to a cylindrical control volume that is oriented coaxially with the pipe, as shown in Fig. 10.1. For the conditions shown in Fig. 10.1, it is assumed that the flow is uniform (streamlines are straight and parallel). Therefore, the net momentum flow through the control volume is zero. Also, the pressure across any section of the pipe will be hydrostatically distributed. Thus the pressure force acting on an end face of the fluid element will be the product of the pressure at the center of the element (also at the center of the pipe) and the area of the face of the element. With steady

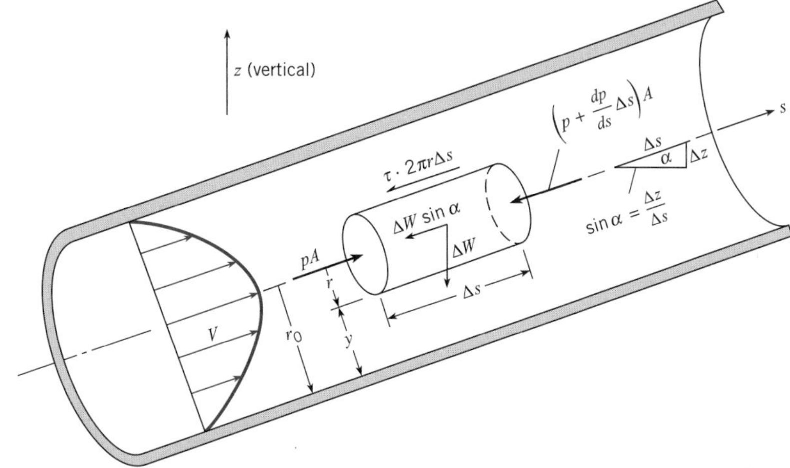

FIGURE 10.1

Flow in a pipe.

uniform flow, equilibrium between the pressure, gravity, and shearing forces acting on the fluid will prevail. Consequently, the momentum equation yields the following:

$$\sum F_s = 0$$

$$pA - \left(p + \frac{dp}{ds}\Delta s\right)A - \Delta W \sin a - t(2pr)\Delta s = 0 \tag{10.1}$$

In Eq. (10.1) $\Delta W = \gamma A \Delta s$ and $\sin a = dz/ds$. Therefore, Eq. (10.1) reduces to

$$-\frac{dp}{ds}\Delta s A - gA\Delta s\frac{dz}{ds} - t(2pr)\Delta s = 0 \tag{10.2}$$

Then, when we divide Eq. (10.2) through by ΔsA and simplify, we obtain

$$t = \frac{r}{2}\left[-\frac{d}{ds}(p + gz)\right] \tag{10.3}$$

Since the gradient itself, $d/ds\,(p + \gamma z)$, is negative (see Section 7.4) and constant across the section for uniform flow,* it follows that $-d/ds\,(p + \gamma z)$ will be positive and constant across the pipe section. Thus τ in Eq. (10.3) will be zero at the center of the pipe and will increase linearly to a maximum at the pipe wall. We will use Eq. (10.3) in the following section to derive the velocity distribution for laminar flow.

* The combination $p + \gamma z$ is constant across the section because the streamlines are straight and parallel in uniform flow, and for this condition there will be no acceleration of the fluid normal to the streamline. Thus hydrostatic conditions prevail across the flow section. For a hydrostatic condition, $p/g + z =$ constant or $p + gz =$ constant as shown in Chapter 3.

Laminar Flow in Pipes

We determine how the velocity varies across the pipe by substituting for τ in Eq. (10.3) its equivalent $\mu dV/dy$ and integrating. First, making the substitution, we have

$$m\frac{dV}{dy} = \frac{r}{2}\left[-\frac{d}{ds}(p+gz)\right] \tag{10.4}$$

Because $dV/dy = -dV/dr$, Eq. (10.4) becomes

$$\frac{dV}{dr} = -\frac{r}{2m}\left[-\frac{d}{ds}(p+gz)\right] \tag{10.5}$$

When we separate variables and integrate across the section, we obtain

$$V = -\frac{r^2}{4m}\left[-\frac{d}{ds}(p+gz)\right] + C \tag{10.6}$$

We can evaluate the constant of integration in Eq. (10.6) by noting that when $r = r_0$, the velocity $V = 0$. Therefore, the constant of integration is given by $C = (r_0^2/4\mu)[-d/ds(p + \gamma z)]$, and Eq. (10.6) then becomes

$$V = \frac{r_0^2 - r^2}{4m}\left[-\frac{d}{ds}(p+gz)\right] \tag{10.7}$$

Equation (10.7) indicates that the velocity distribution for laminar flow in a pipe is parabolic across the section with the maximum velocity at the center of the pipe. Figure 10.2 shows the variation of the shear stress and velocity in the pipe.

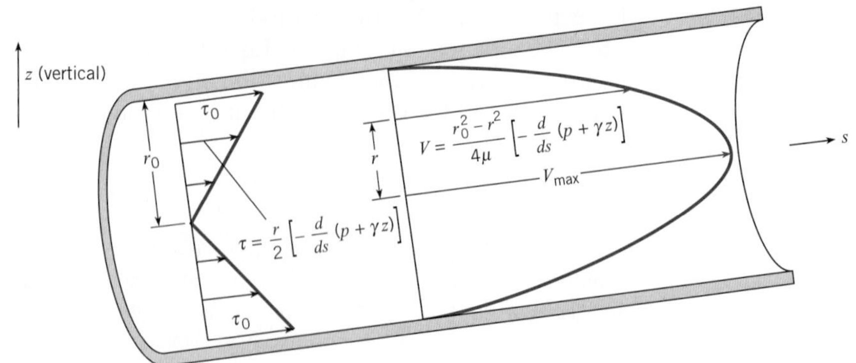

FIGURE 10.2

Distribution of shear stress and velocity for laminar flow in a pipe.

Laminar flow in a round pipe is known as *Hagen-Poiseuille flow,* named after a German, Hagen, and a Frenchman, Poiseuille, who studied low-speed flows in tubes in the 1840s.

example 10.1

Oil (S = 0.90; $m = 5 \times 10^{-1}\ \text{N} \cdot \text{s/m}^2$) flows steadily in a 3-cm pipe. The pipe is vertical, and the pressure at an elevation of 100 m is 200 kPa. If the pressure at an elevation of 85 m is 250 kPa, is the flow direction up or down? What is the velocity at the center of the pipe and at 6 mm from the center, assuming that the flow is laminar?

Solution First determine the rate of change of $p + \gamma z$. Taking s in the z direction,

$$\frac{d}{ds}(p + gz) = \frac{(p_{100} + gz_{100}) - (p_{85} + gz_{85})}{15}$$

$$= \frac{[200 \times 10^3 + 8830(100)] - [250 \times 10^3 + 8830(85)]}{15}$$

$$= \frac{(1.083 \times 10^6 - 1.00 \times 10^6)\ \text{N/m}^2}{15\ \text{m}} = 5.53\ \text{kN/m}^3$$

The quantity $p + \gamma z$ is not constant with elevation—it increases upward (decreases downward). Therefore, the direction of flow is downward. This can be seen by substituting $d(p + gz)/ds = 5.53\ \text{kN/m}^3$ into Eq. (10.7). When this is done, V is negative for all values of r in the flow. When $r = 0$ (center of the pipe), the velocity will be maximum. Thus

$$V_{\text{center}} = V_{\text{max}} = \frac{r_0^2}{4m}(-5.53\ \text{kN/m}^3)$$

$$= \frac{0.015^2\ \text{m}^2}{4(5 \times 10^{-1}\ \text{N} \cdot \text{s/m}^2)}(-5.53 \times 10^3\ \text{N/m}^3) = -0.622\ \text{m/s} \quad \triangleleft$$

At first it may seem strange that the velocity is in a direction opposite to the direction of decreasing pressure. However, it may not seem so peculiar if one realizes that in this example the pipe is vertical, so the gravitational force as well as pressure helps to establish the flow. What counts when flow is other than in the horizontal direction is how the combination $p + \gamma z$ changes with s. If $p + \gamma z$ is constant, then we have the equation of hydrostatics and no flow occurs. However, if $p + \gamma z$ is not constant, flow will occur in the direction of decreasing $p + \gamma z$.

Next determine the velocity at $r = 6\ \text{mm} = 0.006\ \text{m}$. Using Eq. (10.7), we find that

$$V = \frac{0.015^2\ \text{m}^2 - 0.006^2\ \text{m}^2}{4(5 \times 10^{-1}\ \text{N} \cdot \text{s/m}^2)}(-5.53 \times 10^3\ \text{N/m}^3) = -0.522\ \text{m/s} \quad \triangleleft$$

For many problems we wish to relate the pressure change to the rate of flow or mean velocity $\overline{V}$ in the conduit. Therefore, it is necessary to integrate $dQ = VdA$ over the cross-sectional area of flow. That is,

$$\begin{aligned} Q &= \int V\,dA \\ &= \int_0^{r_0} \frac{(r_0^2 - r^2)}{4m}\left[-\frac{d}{ds}(p + gz)\right](2pr\,dr) \end{aligned} \tag{10.8}$$

The factor $\pi[d(p + gz)/ds]/4m$ is constant across the pipe section. Therefore, upon integration, we obtain

$$Q = \frac{p}{4m}\left[\frac{d}{ds}(p + gz)\right]\frac{(r^2 - r_0^2)^2}{2}\Bigg|_0^{r_0} \tag{10.9}$$

which reduces to

$$Q = \frac{pr_0^4}{8m}\left[-\frac{d}{ds}(p + gz)\right] \tag{10.10}$$

If we divide through by the cross-sectional area of the pipe, we have an expression for the mean velocity:

$$\overline{V} = \frac{r_0^2}{8m}\left[-\frac{d}{ds}(p + gz)\right] \tag{10.11}$$

Comparing Eqs. (10.11) and (10.7) reveals that $\overline{V} = V_{max}/2$. Also, by substituting $D/2$ for r_0, we have

$$\overline{V} = \frac{D^2}{32m}\left[-\frac{d}{ds}(p + gz)\right] \tag{10.12}$$

or

$$\frac{d}{ds}(p + gz) = -\frac{32m\overline{V}}{D^2} \tag{10.13}$$

Integrating Eq. (10.13) along the pipe between sections 1 and 2, we obtain

$$p_2 - p_1 + g(z_2 - z_1) = -\frac{32m\overline{V}}{D^2}(s_2 - s_1) \tag{10.14}$$

Here $s_2 - s_1$ is the length L of pipe between the two sections. Therefore, Eq. (10.14) can be rewritten as

$$\frac{p_1}{g} + z_1 = \frac{p_2}{g} + z_2 + \frac{32mL\overline{V}}{gD^2} \tag{10.15}$$

It can be seen that when the general energy equation for incompressible flow in conduits, Eq. (7.24), is reduced to one for uniform flow in a constant-diameter pipe where $V_1 = V_2$, the result is

$$\frac{p_1}{g} + z_1 = \frac{p_2}{g} + z_2 + h_f \tag{10.16}$$

Here h_f is used instead of h_L to signify head loss due to frictional resistance of the pipe. Comparison of Eqs. (10.15) and (10.16) then shows that the head loss for laminar flow is given by

$$h_f = \frac{32mLV}{gD^2} \tag{10.17}$$

Here the bar over the V has been omitted to conform to the standard practice of denoting the mean velocity in one-dimensional flow analyses by V without the bar.

10.3 Criterion for Laminar or Turbulent Flow in a Pipe

To predict whether flow will be laminar or turbulent, it is necessary to explore the characteristics of flow in both laminar and turbulent states. Although other scientists before him had sensed the marked physical difference between laminar and turbulent flow, it was Osborne Reynolds (1) who first developed the basic laws of turbulent flow. With his analytical and experimental work he showed that the Reynolds number was a basic parameter relating to laminar as well as turbulent flow. For example, using an experimental apparatus such as that shown in Fig. 10.3, he found that the onset of turbulence in a smooth pipe was related to the Reynolds number (VDr/m) in a very interesting way. If the fluid in the upstream reservoir was not completely still or if the pipe had some vibration in it, the flow in the pipe as it was gradually increased from a low rate to higher rates was initially laminar but then changed from laminar to turbulent flow at a Reynolds number in the neighborhood of 2100. However, Reynolds found that if the fluid was initially completely motionless and if there was no vibration in the equipment while the flow was increased, it was possible to reach a much higher Reynolds number before the flow became turbulent. He also found that, when going from high-velocity turbulent flow to low-velocity flow, the change from turbulent flow always occurred at a Reynolds number of about 2000.

These experiments of Reynolds indicate that under carefully controlled conditions it is possible to have laminar flow in pipes at Reynolds numbers much higher than 2000. However, the slightest disturbances will trigger the onset of turbulence at high values of Re. Because most engineering applications involve some vibration or flow disturbance, it is reasonable to expect that pipe flow will be laminar for Reynolds numbers less than 2000 and turbulent for Reynolds numbers greater than 3000. When Re is between 2000 and 3000, the type of flow is very unpredictable and often changes back and forth between laminar and turbulent states. Fortunately, however, most engineering applications either are not in this range or are not significantly affected by the unstable flow.

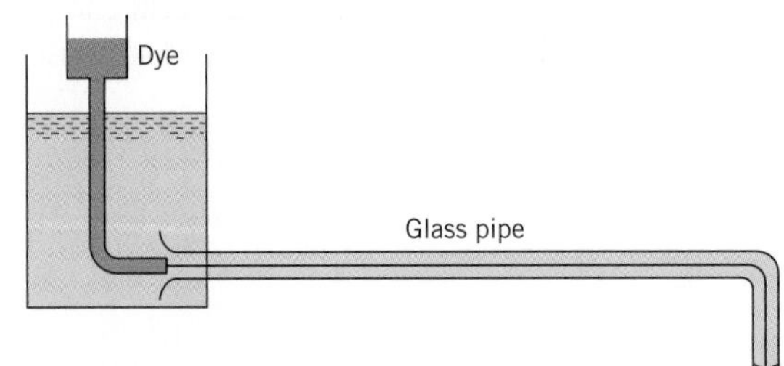

FIGURE 10.3

Schematic diagram of apparatus used by Reynolds to study laminar and turbulent flow.

example 10.2

Oil (S = 0.85) with a kinematic viscosity of $6 \times 10^{-4}\ \text{m}^2/\text{s}$ flows in a 15-cm pipe at a rate of $0.020\ \text{m}^3/\text{s}$. What is the head loss per 100-m length of pipe?

Solution First we determine whether the flow is laminar or turbulent by checking to see if the Reynolds number is below 2000 or above 3000.

$$V = \frac{Q}{A} = \frac{0.020\ \text{m}^3/\text{s}}{(p/4)D^2\ \text{m}^2} = \frac{0.020\ \text{m}^3/\text{s}}{0.785(0.15^2\ \text{m}^2)} = 1.13\ \text{m/s}$$

Then

$$\text{Re} = \frac{VD}{n} = \frac{(1.13\ \text{m/s})(0.15\ \text{m})}{6 \times 10^{-4}\ \text{m}^2/\text{s}} = 283$$

Since the Reynolds number is less than 2000, the flow is laminar. The head loss per 100 m is obtained from Eq. (10.17):

$$h_f = \frac{32mLV}{gD^2}$$

Here $m/g = n/\text{g}$; hence

$$h_f = \frac{32\nu LV}{gD^2}$$

Then

$$h_f = \frac{32(6)(10^{-4}\ \text{m}^2/\text{s})(100\ \text{m})(1.13\ \text{m/s})}{(9.81\ \text{m/s}^2)(0.15^2\ \text{m}^2)} = 9.83\ \text{m}$$

The head loss is 9.83 m/100 m of length. ◁

example 10.3

Kerosene (0°C) flows under the action of gravity in the pipe shown, which is 6 mm in diameter and 100 m long. Determine the rate of flow in the pipe.

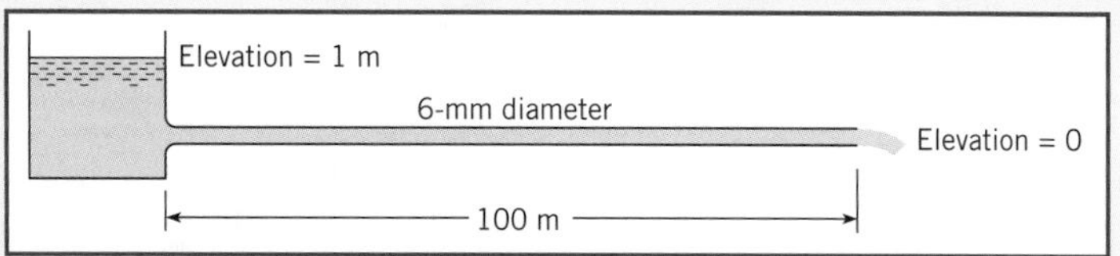

Solution Because the pipe diameter is small and because the head producing flow is also quite small, it is expected that the velocity in the pipe will be small. Hence it will be initially assumed that the flow is laminar and $V^2/2g$ is negligible. Then, to solve for the velocity, we apply the energy equation to the problem. We write this equation between a section at the upstream liquid surface and the outlet of the pipe. Thus we have

$$\frac{p_1}{g} + \frac{a_1 V_1^2}{2g} + z_1 = \frac{p_2}{g} + \frac{a_2 V_2^2}{2g} + z_2 + \frac{32 m L V}{g D^2}$$

With the assumption we have noted, this equation reduces to

$$0 + 0 + 1 = 0 + 0 + 0 + \frac{32 m L V}{g D^2}$$

or

$$\frac{32 m L V}{g D^2} = 1$$

For 0°C the viscosity (from Figs. A.2 and A.3 in the Appendix) is

$$m = 3.2 \times 10^{-3}\ \text{N} \cdot \text{s/m}^2 \qquad v = 3.9 \times 10^{-6}\ \text{m}^2/\text{s}$$

Then $$V = \frac{1 \times g D^2}{32 m L}$$

$$= \frac{1(8010\ \text{N/m}^3)(0.006^2\ \text{m}^2)}{32(3.2 \times 10^{-3}\ \text{N} \cdot \text{s/m}^2)(100\ \text{m})} = 0.0282\ \text{m/s} = 28.2\ \text{mm/s}$$

Now check Re to see if the flow is laminar, and check $V^2/2g$ to see if it is indeed negligible:

$$\text{Re} = \frac{VD}{n} = \frac{(0.0282\ \text{m/s})(0.006\ \text{m})}{3.9 \times 10^{-6}\ \text{m}^2/\text{s}} = 43.4$$

$$\frac{V^2}{2g} = \frac{(0.0282\ \text{m/s})^2}{(2)(9.81\ \text{m/s})} = 4.05 \times 10^{-5}\ \text{m} \quad \text{(negligible)}$$

Therefore, the flow is laminar and the velocity is valid. The discharge is then calculated as follows:

$$Q = VA = (0.282\ \text{m/s})\left(\frac{p}{4}\right)(0.006\ \text{m})^2 = 7.97 \times 10^{-7}\ \text{m}^3/\text{s} \qquad \triangleleft$$

FIGURE 10.4

Apparent shear stress in a pipe. [After Laufer (2)]

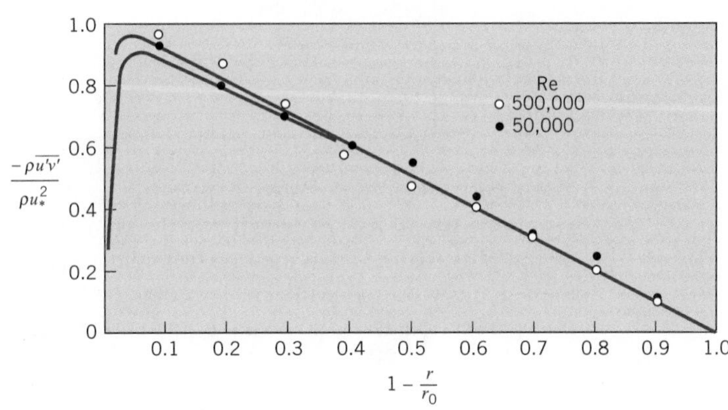

10.4

Turbulent Flow in Pipes

Turbulence and Its Influence in Pipe Flow

In the preceding section it was pointed out that pipe flow is turbulent when the Reynolds number is larger than approximately 3000. However, to say that the flow is turbulent is only a gross description of it. We can obtain a better "feel" for the flow by exploring the similarities between turbulent flow in a pipe and flow in a turbulent boundary layer and by relating the shear stress in the pipe to the level of turbulence. Once we understand these basic physical relationships, we will be better equipped to proceed to the development of equations for the velocity distribution and the resistance to turbulent flow in pipes.

The similarities between turbulent boundary-layer flow and turbulent flow in pipes are many. In fact, it is valid to think of turbulent flow in a pipe as a turbulent boundary layer that has become as thick as the radius of the pipe. With this perspective we realize that flow in a smooth pipe has a viscous sublayer just as a flat-plate boundary layer does. In addition, the velocity gradient in the viscous sublayer will be consistent with the shear stress, as given by $t = m\, du/dy$. However, outside the viscous sublayer the viscous shear stress is negligible compared with the resistance resulting from turbulence. We have already referred in Chapter 9 to the *apparent shear stress,* $t_{\text{app}} = -r\overline{u'v'}$, which involves an exchange of momentum, but its effect is like that of a true shear stress. It is zero at the pipe center and increases to a maximum near the wall, as shown in Fig. 10.4. Here it is seen that the apparent shear stress increases linearly almost to the edge of the pipe. This linear change in τ_{app} is in accordance with Eq. (10.3), which was developed in Section 10.1. Near the wall, in the viscous sublayer, τ_{app} reduces to zero because all of the shear stress there is in the form of viscous shear stress.

We have shown that there are indeed many analogies between turbulent boundary-layer flow and turbulent flow in pipes. The primary difference is that pipe flow is uniform and boundary-layer flow is not. Of course, this difference does not apply near the inlet of the pipe, where the flow is nonuniform.

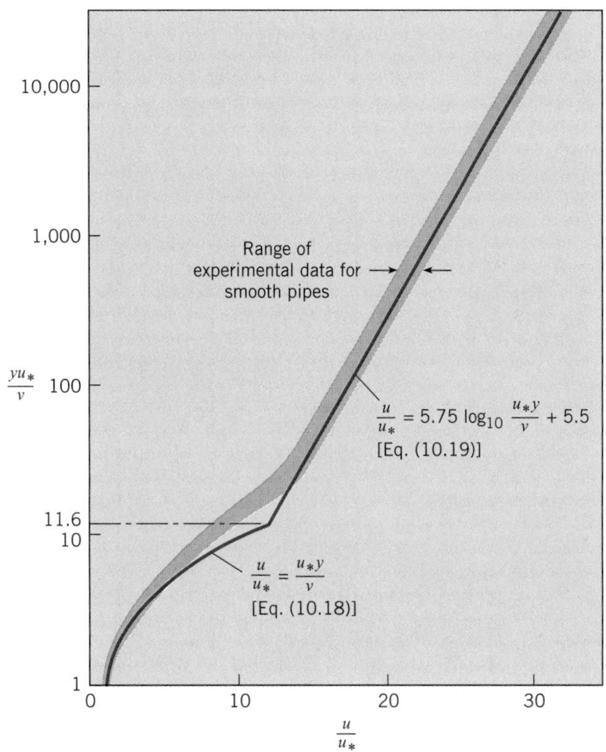

FIGURE 10.5

Velocity distribution for smooth pipes. [After Schlichting (3)]

Velocity Distribution and Resistance in Smooth Pipes

Experiments have shown that, in the viscous sublayer and in the turbulent zone near the wall, the velocity distribution equations are of the same form as those for the turbulent boundary layer. That is, for a smooth pipe,

$$\frac{u}{u_*} = \frac{u_* y}{n} \qquad \text{for } 0 < \frac{yu_*}{n} < 5 \tag{10.18}$$

$$\frac{u}{u_*} = 5.75 \log \frac{u_* y}{n} + 5.5 \qquad \text{for } 20 < \frac{yu_*}{n} \leq 10^5 \tag{10.19}$$

Figure 10.5 is a plot of Eqs. (10.18) and (10.19) as well as an indication of the range of experimental data from various sources. For flow near the center of the pipe, as for flow near the outer limit of the boundary layer, the velocity defect law is applicable, as shown in Fig. 10.6. Figure 10.6 also includes the range of experimental velocity data obtained from flow in rough conduits. Again, a power-law formula like that for the turbulent boundary layer is applicable everywhere except close to the wall. This formula is

$$\frac{u}{u_{max}} = \left(\frac{y}{r_0}\right)^m \tag{10.20}$$

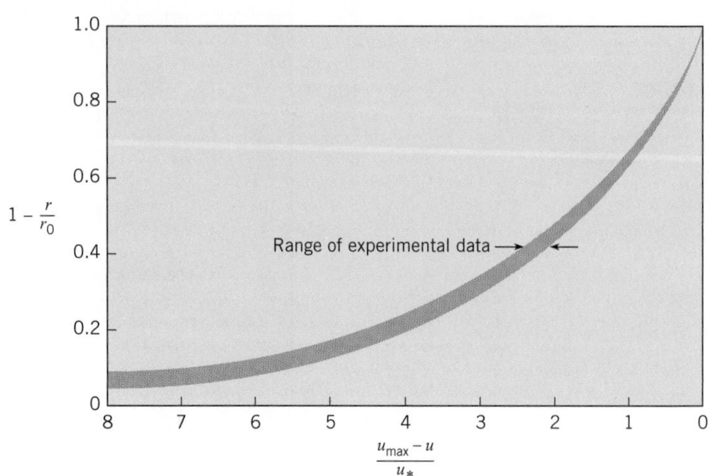

FIGURE 10.6

Velocity defect law for turbulent flow in smooth and rough pipes. [After Schlichting (3)]

Here y is the distance from the wall and m is an empirically determined quantity. Some references indicate that m has a value of $1/7$ for turbulent flow. However, Schlichting (3) shows that m varies from $1/6$ to $1/10$ depending on the Reynolds number. His values for m are given in Table 10.1.

In Chapter 9 the local shear stress on a flat plate was expressed as

$$t_0 = c_f r \frac{V_0^2}{2}$$

where c_f is a function of the character of flow (laminar or turbulent) and the Reynolds number. For pipe flow it is customary to express τ_0 in a similar manner; however, we use the mean velocity as the reference velocity, and the coefficient of proportionality is given as $f/4$ instead of c_f. Here f is called the *resistance coefficient* or *friction factor* of the pipe. Thus we have

TABLE 10.1 EXPONENTS FOR POWER-LAW EQUATION AND RATIO OF MEAN TO MAXIMUM VELOCITY

Re→	**4×10^3**	**2.3×10^4**	**1.1×10^5**	**1.1×10^6**	**3.2×10^6**
m→	$\frac{1}{6.0}$	$\frac{1}{6.6}$	$\frac{1}{7.0}$	$\frac{1}{8.8}$	$\frac{1}{10.0}$
$\bar{V}/V_{max}$→	0.791	0.807	0.817	0.850	0.865

SOURCE: Schlichting (3). Used with permission of the McGraw-Hill Companies.

$$t_0 = \frac{f}{4}r\frac{V^2}{2} \tag{10.21}$$

Or, because $\sqrt{t_0/r} = u_*$, we have

$$\frac{u_*}{V} = \sqrt{\frac{f}{8}}$$

Noting that $\tau = \tau_0$ when $r = r_0$ in Eq. (10.3), we can eliminate τ_0 between Eqs. (10.3) and (10.21). Then, by integrating between two sections along the pipe, we obtain

$$h_1 - h_2 = f\frac{L}{D}\frac{V^2}{2g}$$

$$h_f = f\frac{L}{D}\frac{V^2}{2g} \tag{10.22}$$

where h_f is the head loss created by viscous effects and is equal to the change in piezometric head along the pipe. Equation (10.22) is called the *Darcy-Weisbach equation.* It is named after Henry Darcy, a French engineer of the nineteenth century, and Julius Weisbach, a German engineer and scientist of the same era. Weisbach first proposed the use of the nondimensional resistance coefficient, and Darcy carried out numerous tests on water pipes. Brief accounts of their work are given by Rouse and Ince (4). It can be easily shown by a simultaneous solution of Eqs. (10.17) and (10.22) that the resistance coefficient for *laminar flow* is given by

$$f = \frac{64}{\text{Re}} \tag{10.23}$$

For turbulent flow, analytical and empirical results on smooth pipes yield the following approximate relation for f:

$$\frac{1}{\sqrt{f}} = 2\,\log(\text{Re}\sqrt{f}) - 0.8 \qquad \text{for Re} > 3000 \tag{10.24}$$

Equation (10.24) was first developed by Prandtl.

Velocity Distribution and Resistance—Rough Pipes

Numerous tests on flow in rough pipes all show that a semilogarithmic velocity distribution is valid over most of the pipe section (5, 4). This relationship is given in the following form:

$$\frac{u}{u_*} = 5.75\,\log\frac{y}{k} + B \tag{10.25}$$

Here y is the distance from the rough wall, k is a measure of the height of the roughness elements, and B is a function of the character of roughness. That is, B is a function of the type, concentration, and size variation of the roughness. Research by Roberson and Chen

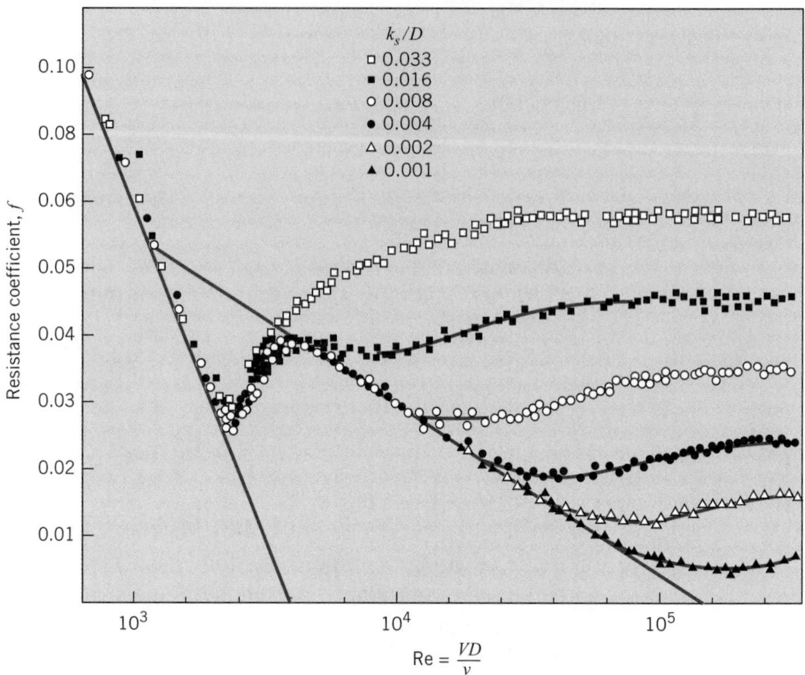

FIGURE 10.7

Resistance coefficient f versus Re for sand-roughened pipe. [After Nikuradse (6)]

(7) shows that B can be analytically determined for artificially roughened boundaries. Work by Wright (8), Calhoun (9), Kumar (10), and Eldridge (11) shows that using a numerical approach with the same theory yields solutions for B and f for natural roughness, such as found in rock-bedded streams and commercial pipes when the roughness characteristics are known. Kumar (12) has developed a simplified analytical method of solution to predict the resistance coefficient of rough boundaries. His method can be applied to both artificial and natural types of roughness.

In 1933 Nikuradse (6) carried out a number of tests on flow in pipes that were roughened with uniform-sized sand grains. From these tests he found that the value of B with this kind of roughness was 8.5. Thus, for his tests, Eq. (10.25) becomes $u/u_* = 5.75 \log(y/k_s) + 8.5$. In Nikuradse's tests the distance y was measured from the geometric mean of the wall surface, and k_s was the size of the sand grains.

Nikuradse's tests revealed two very important characteristics of rough-pipe flow. First, with low Reynolds numbers and with small-sized sand grains, the flow resistance is virtually the same as that for a smooth pipe. Second, for high values of the Reynolds number, the resistance coefficient is solely a function of the relative roughness k_s/D. These characteristics are shown in Fig. 10.7, where f is plotted as a function of Re for various values of the relative roughness k_s/D. The reason the resistance is like that of a smooth pipe for low values of k_s/D and Re is that for these conditions the roughness elements are completely within the viscous sublayer and hence have negligible influence on the main flow in the pipe. However, at high values of the Reynolds number, the viscous sublayer is so thin that the roughness elements project into the main stream of flow and the flow resistance is determined by the drag of the individual roughness elements. Hence, for relatively large values of k_s/D and for large Reynolds numbers, the resistance to flow is proportional to V^2; thus f becomes constant for these conditions. The effect of roughness can be summarized by (13)

$$\left(\frac{k_s}{D}\right)\text{Re} < 10 \qquad \text{roughness unimportant, pipe considered smooth}$$

$$\left(\frac{k_s}{D}\right)\text{Re} > 1000 \qquad \text{fully rough, } f \text{ independent of Reynolds number}$$

The region between these limits is the transitional roughness regime.

The uniform character of the sand grains used in Nikuradse's tests produced a dip in the f-versus-Re curve (Fig. 10.7) before the curve reached a constant value of f. However, tests on commercial pipes where the roughness is somewhat random reveal that no such dip occurs. Using data from commercial pipes, Colebrook (14) in 1939 developed an empirical equation, called the Colebrook-White formula, for the friction factor. Moody (4) used the Colebrook-White formula to generate a design chart similar to that shown in Fig. 10.8. This chart is now known as the *Moody diagram* for commercial pipes.

TABLE 10.2 EQUIVALENT SAND GRAIN ROUGHNESS, k_s, FOR VARIOUS PIPE MATERIALS

Boundary Material	**k_s, millimeters**	**k_s, inches**
Glass, plastic	Smooth	Smooth
Copper or brass tubing	0.0015	6×10^{-5}
Wrought iron, steel	0.046	0.002
Asphalted cast iron	0.12	0.005
Galvanized iron	0.15	0.006
Cast iron	0.26	0.010
Concrete	0.3 to 3.0	0.012–0.12
Riveted steel	0.9–9	0.035–0.35
Rubber pipe (straight)	0.025	0.001

In Fig. 10.8 the variable k_s is the symbol used to denote the *equivalent sand roughness*. That is, a pipe that has the same resistance characteristics at high Re values as a sand-roughened pipe of the same size is said to have a size of roughness equivalent to that of the sand-roughened pipe. Table 10.2 gives the equivalent sand roughness for various kinds of pipes. This table can be used to calculate the relative roughness for a given pipe diameter, which, in turn, is used in Fig. 10.8 to find the friction factor.

In Fig. 10.8 the abscissa (labeled at the bottom) is the Reynolds number Re, and the ordinate (labeled at the left) is the resistance coefficient f. Each blue curve is for a constant relative roughness k_s/D, and the values of k_s/D are given on the right at the end of each curve. To find f, given Re and k_s/D, one goes to the right to find the correct relative roughness curve. Then one looks at the bottom of the chart to find the given value of Re and, with this value of Re, moves vertically upward until the given k_s/D curve is reached. Finally, from this point one moves horizontally to the left scale to read the value of f. If the curve for the given value of k_s/D is not plotted in Fig. 10.8, then one simply

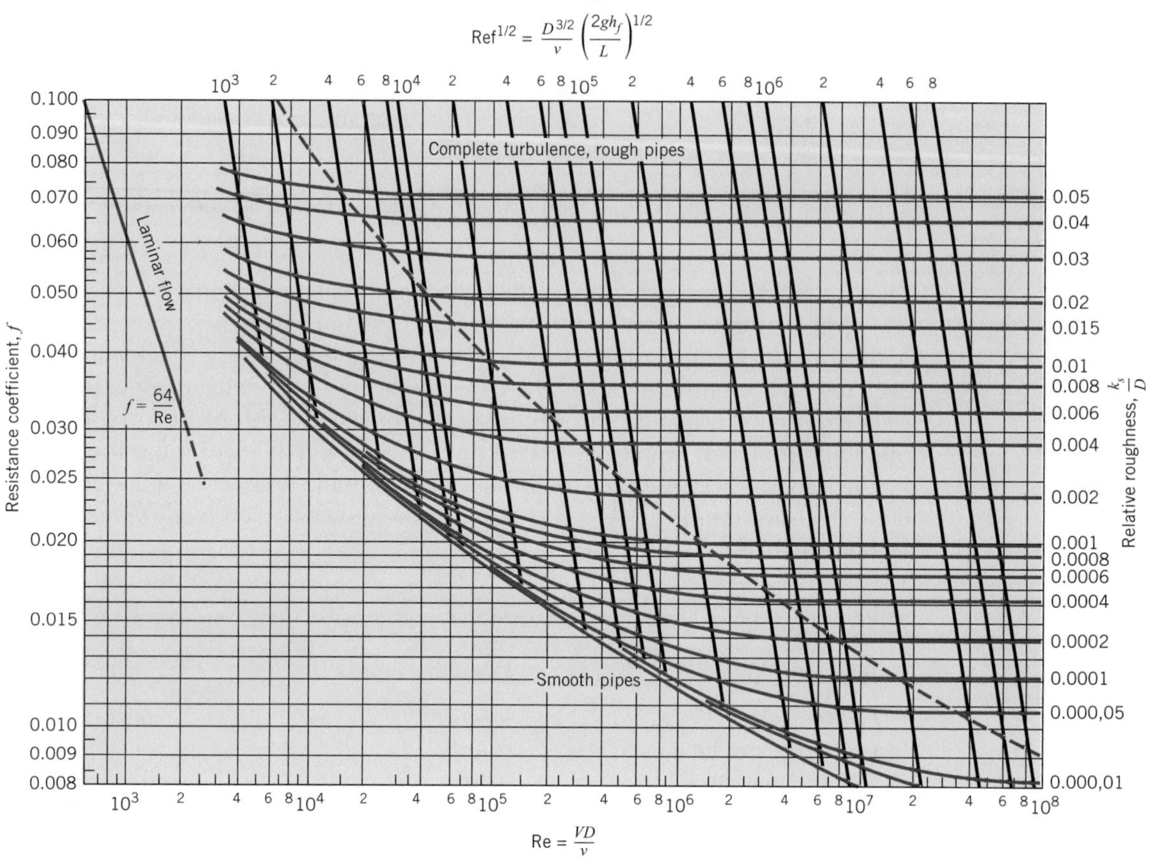

FIGURE 10.8

Resistance coefficient f versus Re. Reprinted with minor variations. [After Moody (5). Reprinted with permission from the A.S.M.E.]

finds the proper position on the graph by interpolation between the k_s/D curves that bracket the given k_s/D.

For some problems it is convenient to enter Fig. 10.8 using a value of the parameter $\text{Re} f^{1/2}$. This parameter is useful when h_f and k_s/D are known but the velocity V is not. Without V the Reynolds number cannot be computed, so f cannot be read by entering the chart with Re and k_s/D. But from $h_f = f(L/D)V^2/2g$ [or $V = (2gh_f/L)^{1/2}(D/f)^{1/2}$] and $\text{Re} = VD/n$, one can see that Re can be given as

$$\text{Re} = \frac{D^{3/2}}{f^{1/2}n}\left(\frac{2gh_f}{L}\right)^{1/2}$$

Upon multiplying both sides of the above equation by $f^{1/2}$, we get

$$\text{Re} f^{1/2} = \frac{D^{3/2}}{n}(2gh_f/L)^{1/2}$$

Thus a value of $\text{Re} f^{1/2}$ can be calculated for this type of flow problem, which, in turn, enables us to determine f directly, using Fig. 10.8, where curves of constant $\text{Re} f^{1/2}$ are plotted slanting from the upper left to lower right and the values of $\text{Re} f^{1/2}$ for each line are given at the top of the chart.

When using computers to carry out pipe flow calculations, it is much more convenient to have an equation for the friction factor as a function of Reynolds number and relative roughness. By using the Colebrook-White formula, Swamee and Jain (15) developed an explicit equation for friction factor, namely

$$f = \frac{0.25}{\left[\log_{10}\left(\frac{k_s}{3.7D} + \frac{5.74}{\text{Re}^{0.9}}\right)\right]^2} \tag{10.26}$$

It is reported that this equation predicts friction factors that differ by less than 3% from those on the Moody diagram for $4 \times 10^3 < \text{Re} < 10^8$ and $10^{-5} < k_s/D < 2 \times 10^{-2}$.

There are basically three types of problems involved with uniform flow in a single pipe. These are

1. Determine the head loss, given the kind and size of pipe and the flow rate.
2. Determine the flow rate, given the head loss, kind, and size of pipe.
3. Determine the size of pipe needed to carry the flow, given the kind of pipe, head, and flow rate.

In the first type of problem, the Reynolds number and k_s/D are first computed and then f is read from Fig. 10.8, after which the head loss is obtained by the use of Eq. (10.22).

example 10.4

Water ($T = 20°\text{C}$) flows at a rate of $0.05 \text{ m}^3/\text{s}$ in a 20-cm asphalted cast-iron pipe. What is the head loss per kilometer of pipe?

Solution First compute the Reynolds number, VD/n. Here $V = Q/A$. Thus

$$V = \frac{0.05 \text{ m}^3/\text{s}}{(p/4)(0.20^2 \text{ m}^2)} = 1.59 \text{ m/s}$$

$$n = 1.0 \times 10^{-6} \text{ m}^2/\text{s} \quad \text{(from Table A.5)}$$

Then

$$\text{Re} = \frac{VD}{n} = \frac{(1.59 \text{ m/s})(0.20 \text{ m})}{10^{-6} \text{ m}^2/\text{s}} = 3.18 \times 10^5$$

From Table 10.2, the roughness for asphalted cast-iron pipe is 0.12 mm, so the relative roughness (k_s/D) is 0.0006. Then from Fig. 10.8, using the values obtained for k_s/D and Re, we find $f = 0.019$. Finally, the head loss is computed from the Darcy-Weisbach equation:

$$h_f = f\frac{L}{D}\frac{V^2}{2g} = 0.019\left(\frac{1000\text{ m}}{0.20\text{ m}}\right)\left(\frac{1.59^2\text{ m}^2/\text{s}^2}{2(9.81\text{ m/s}^2)}\right) = 12.2\text{ m} \quad \triangleleft$$

The head loss per kilometer is 12.2 m.

In the second type of problem, k_s/D and the value of $(D^{3/2}/n)\sqrt{2gh_f/L}$ are computed so that the top scale can be used to enter the chart of Fig. 10.8. Then, once f is read from the chart, the velocity from Eq. (10.22) is solved for and the discharge is computed from $Q = VA$.

example 10.5

The head loss per kilometer of 20-cm asphalted cast-iron pipe is 12.2 m. What is the discharge of water?

Solution First compute the parameter $D^{3/2}\sqrt{2gh_f/L}/n$. Assume $T = 20°\text{C}$, so that

$$D^{3/2}\frac{\sqrt{2gh_f/L}}{n} = (0.20\text{ m})^{3/2}\frac{[2(9.81\text{ m/s}^2)(12.2\text{ m}/1000\text{ m})]^{1/2}}{1.0\times 10^{-6}\text{ m}^2/\text{s}}$$

$$= 4.38\times 10^4$$

From Table 10.2, the roughness for asphalted cast-iron pipe is 0.12 mm, so the relative roughness (k_s/D) is 0.0006. Using Fig. 10.8, we read $f = 0.019$. We use this f in the Darcy-Weisbach equation to solve for V:

$$h_f = f\frac{L}{D}\frac{V^2}{2g}$$

$$12.2\text{ m} = \frac{0.019(1000\text{ m})}{0.20\text{ m}}\frac{V^2}{2(9.81\text{ m/s}^2)}$$

$$V^2 = 2.52\text{ m}^2/\text{s}^2$$

$$V = 1.59\text{ m/s}$$

Finally we compute the discharge:

$$Q = VA = V\frac{\pi}{4}D^2$$

$$= (1.59\text{ m/s})(0.785)(0.20^2\text{ m}^2) = 0.050\text{ m}^3/\text{s} \quad \triangleleft$$

Examples 10.4 and 10.5 are good checks on the validity of the methods of solution because their basic data are exactly the same—in one case, the head loss is unknown; in the other, the discharge is unknown.

In Example 10.5 the head loss in the pipe was given. Therefore, it was possible to obtain a direct solution by entering Fig. 10.8 with a value of $\text{Re} f^{1/2}$. However, many problems for which the discharge Q is desired cannot be solved directly. For example, a problem in which water flows from a reservoir through a pipe and into the atmosphere cannot be solved directly. Here part of the available head is lost to friction in the pipe, and part of the head remains as kinetic energy in the jet as it leaves the pipe. Therefore, at the outset one does not know how much head loss occurs in the pipe itself. To effect a solution, one must iterate on f. The energy equation is written and an initial value for f is guessed. Because f tends to a constant value at high values of Re, an "educated" first guess is to use this limiting value of f. Next one solves for the velocity V. With this value of V, one then computes a Reynolds number that makes it possible to determine a better value of f using Fig. 10.8, and so on. This type of solution usually converges quite rapidly because f changes more slowly than Re. Once f and V have been determined, one calculates the discharge by using the continuity equation.

example 10.6

Determine the discharge of water through the 50-cm steel pipe shown in the figure.

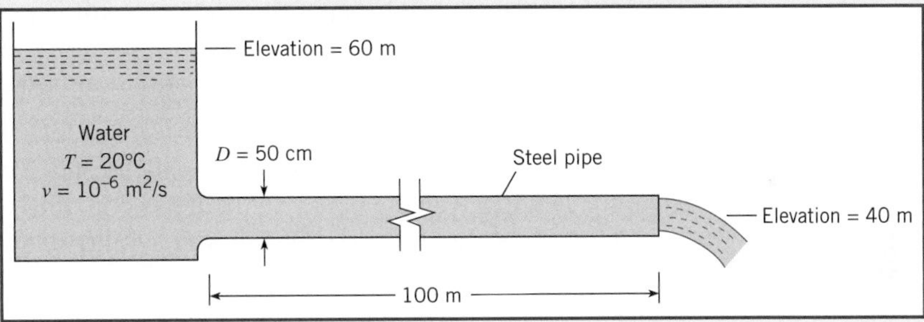

Solution From Table 10.2, the roughness for steel pipe is 0.046 mm, so the relative roughness (k_s/D) is 9.2×10^{-5}. Now write the energy equation from the reservoir water surface to the free jet at the end of the pipe:

$$\frac{p_1}{g} + \frac{V_1^2}{2g} + z_1 = \frac{p_2}{g} + \frac{V_2^2}{2g} + z_2 + h_L$$

$$0 + 0 + 60 = 0 + \frac{V_2^2}{2g} + 40 + f\frac{L}{D}\frac{V_2^2}{2g}$$

or

$$V = \left(\frac{2g \times 20}{1 + 200f}\right)^{1/2}$$

First trial: Assume $f = 0.020$; then $V = 8.86$ m/s and Re $= 4.43 \times 10^6$. With Re $= 4.43 \times 10^6$ and $k_s/D = 9.2 \times 10^{-5}$, then $f = 0.012$ (from Fig. 10.8). This f then yields $V = 10.7$ m/s.

Second trial: For $V = 10.7$ m/s, Re $= 5.35 \times 10^6$ and $f = 0.012$,

$$Q = VA = 10.7 \text{ m/s} \times (p/4) \times (0.50)^2 \text{ m}^2 = 2.10 \text{ m}^3\text{/s} \quad \triangleleft$$

This example can also be solved using a programmable calculator and the explicit equation for the friction factor, Eq. (10.26). The procedure is to write a program that, given a velocity, calculates the Reynolds number, then calculates the friction factor from Eq. (10.26) and, finally, calculates the velocity from the above equation. This velocity is then entered again and a new velocity is calculated until the velocity no longer changes. The following table gives the velocities calculated starting with an initial guess of 20 m/s and finishing when the velocity difference between iterations is less than 0.01.

1st iteration	20.0
2nd iteration	10.73
3rd iteration	10.67

The convergence is fast because the friction factor is a very weak function of Reynolds number.

In the third type of problem, it is usually best to first assume a value of f and then to solve for D, after which a better value of f is computed based on the first estimate of D. This iterative procedure is continued until a valid solution is obtained. A trial-and-error procedure is necessary because without D one cannot compute k_s/D or Re to enter Fig. 10.8.

example 10.7

What size of asphalted cast-iron pipe is required to carry water at a discharge of 3 cfs and with a head loss of 4 ft per 1000 ft of pipe?

Solution First assume $f = 0.015$. Then

$$h_f = \frac{fLV^2}{D2g} = \frac{fLQ^2/A^2}{D\ \ 2g} = \frac{fLQ^2}{2g(p/4)^2 D^5}$$

or

$$D^5 = \frac{fLQ^2}{0.785^2(2gh_f)}$$

For this example,

$$D^5 = \frac{0.015(1000\text{ ft})(3\text{ ft}^3/\text{s})^2}{0.615(64.4\text{ ft/s}^2)(4\text{ ft})} = 0.852\text{ ft}^5$$

$$D = 0.97\text{ ft}$$

Now compute a more accurate value of f:

$$\frac{k_s}{D} = 0.0004 \qquad V = \frac{Q}{A} = \frac{3\text{ ft}^3/\text{s}}{0.785(0.94\text{ ft}^2)} = 4.07\text{ ft/s}$$

Then $$\text{Re} = \frac{VD}{n} = \frac{(4.07\text{ ft/s})(0.97\text{ ft})}{1.21(10^{-5}\text{ ft}^2/\text{s})} = 3.26 \times 10^5$$

From Fig. 10.8, $f = 0.0175$. Now recompute D by applying the ratio of f's to previous calculations for D^5:

$$D^5 = \frac{0.0175}{0.015}(0.852\text{ ft}^5) = 0.994\text{ ft}^5$$

$$D = 0.999\text{ ft}$$

Use a pipe with a 12-in. diameter. ◁

Note: If a size that is not available commercially is calculated during design, it is customary practice to choose the next larger available size. The cost will be less than that for odd-sized pipe, and the pipe will be more than large enough to carry the flow.

Explicit Equations for Q and D

In the foregoing discussion, methods were presented by which Q and D can be calculated. All of these methods involve the use of the Moody diagram (Fig. 10.8).

To provide an alternative to the Moody diagram, Swamee and Jain (15) developed an explicit equation for discharge:

$$Q = -2.22D^{5/2}\sqrt{gh_f/L}\log\left(\frac{k_s}{3.7D} + \frac{1.78n}{D^{3/2}\sqrt{gh_f/L}}\right) \tag{10.27}$$

They also developed a formula for the explicit determination of D. A modified version of that formula, given by Streeter and Wylie (16), is

$$D = 0.66\left[k_s^{1.25}\left(\frac{LQ^2}{gh_f}\right)^{4.75} + nQ^{9.4}\left(\frac{L}{gh_f}\right)^{5.2}\right]^{0.04} \tag{10.28}$$

If you want to solve for head loss given Q, L, D, k_s, and ν, simply solve for f by Eq. (10.26) and compute h_f with the Darcy-Weisbach equation, Eq. (10.22). Straightforward calculations for Q and D can also be made if h_f is known. However, for problems involving head losses in addition to h_f, an iterative solution is required. For computing Q, you can assume an f and solve for Q from the energy equation after substituting Q/A in that equation. Then compute Re and use the result in Eq. (10.26) to get a better estimate of f, and so on, until Q converges analogous to the procedure for determining Q using the Moody diagram. In this case, however, Eq. (10.26) is substituted for the Moody diagram. Similarly, you can determine D if you are given Q, v, the change in pressure or head, and the geometric configuration.

example 10.8

Solve Example 10.5 using Eq. (10.27).

Solution From Table 10.2, k_s for asphalted cast-iron pipe is given as 1.2×10^{-4} m. From the given conditions, $h_f/L = 0.0122$. Assume $T = 20°\text{C}$, so $n = (10^{-6}\ \text{m}^2/\text{s})$. Then, using Eq. (10.27), we have

$$Q = -2.22(0.20\ \text{m})^{5/2}\sqrt{9.81\ \text{m/s}^2 \times 0.0122}$$

$$\times \log\left(\frac{1.2 \times 10^{-4}\ \text{m}}{3.7 \times 0.20\ \text{m}} + \frac{1.78 \times 10^{-6}\ \text{m}^2/\text{s}}{(0.20\ \text{m})^{3/2}\sqrt{9.81\ \text{m/s}^2 \times 0.0122}}\right)$$

$$= 0.050\ \text{m}^3/\text{s}$$

◁

Flow at Pipe Inlets and Losses from Fittings

In the preceding section, formulas were presented that are used to determine the head loss for uniform flow in a pipe. However, pipe systems also include inlets, outlets, bends, and other appurtenances that create additional head losses. The resulting flow separation and the generation of additional turbulence force usually cause these head losses. In this section we will consider the flow patterns and resulting head losses for some of these flow transitions.

Flow in a Pipe Inlet

If the inlet to a pipe is well rounded, as shown in Fig. 10.9, the boundary layer will develop from the inlet and grow in thickness until it extends to the center of the pipe. After that point, the flow in the pipe will be uniform. The length L_e of the developing region at the entrance is equal to approximately $0.05D\text{Re}$ for laminar flow and approximately $50D$ for turbulent flow. Velocity and pressure distribution for the inlet region of a pipe with turbulent flow are shown in Fig. 10.10. The head loss that is produced by inlets, outlets, or fittings is expressed by the equation

$$h_L = K\frac{V^2}{2g} \tag{10.29}$$

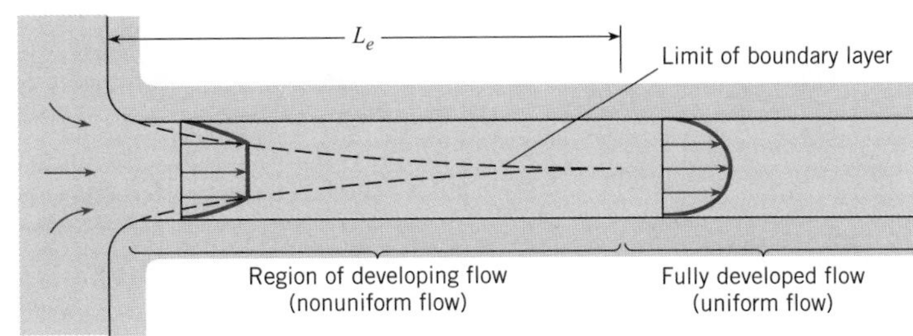

FIGURE 10.9

Flow characteristics at a pipe inlet (not to scale).

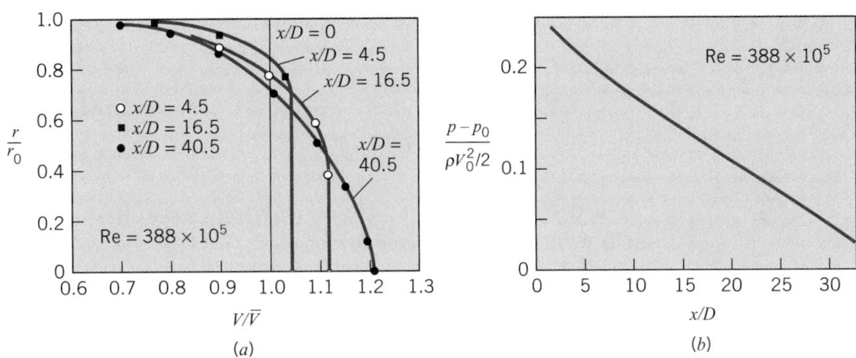

FIGURE 10.10

Distribution of velocity and pressure in the inlet region of a pipe [Barbin and Jones (17)]. (a) Velocity distribution. (b) Pressure distribution.

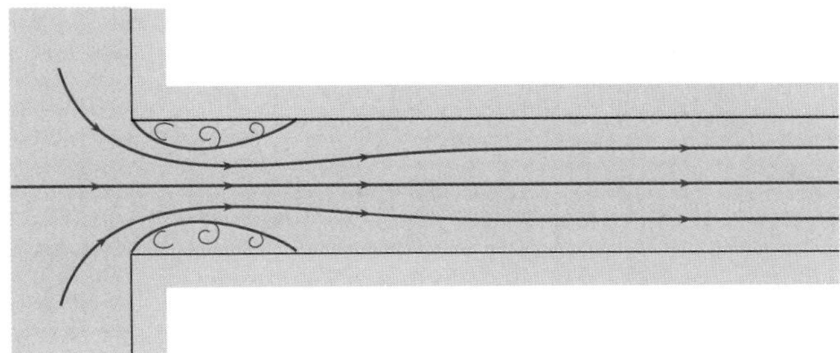

FIGURE 10.11

Flow at a sharp-edged inlet.

In Eq. (10.29) V is the mean velocity in the pipe and K is the loss coefficient for the particular fitting that is involved. For example, K for a well-rounded inlet with flow at high values of Re is approximately 0.10. Hence the head loss for such a transition is quite small compared with that for an abrupt pipe outlet for which K is 1.0.

If the pipe inlet is abrupt, or sharp-edged, as in Fig. 10.11, separation occurs just downstream of the entrance. Hence the streamlines converge and then diverge with consequent turbulence and relatively high head loss. The loss coefficient for the abrupt inlet is approximately 0.5.

Flow through an Elbow

Although the cross-sectional area of an elbow may not change from section to section, considerable head loss is produced because separation occurs near the inside of the bend and downstream of the midsection. Thus when the flow leaves the elbow, the eddies pro-

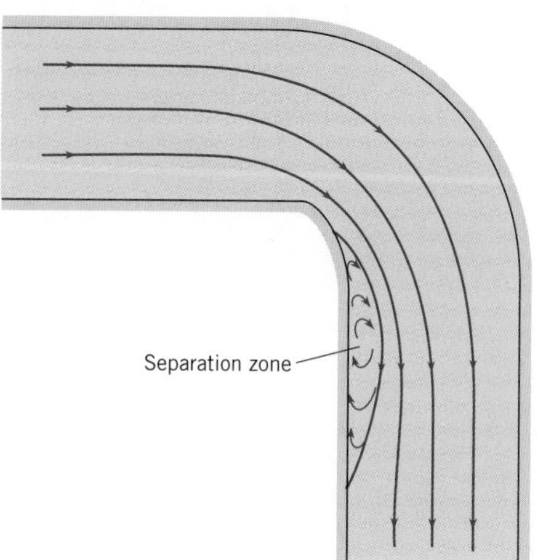

FIGURE 10.12

Flow pattern in an elbow.

duced by separation create considerable head loss. The approximate flow pattern for an elbow is shown in Fig. 10.12.

The loss coefficient for an elbow at high Reynolds numbers depends primarily on the shape of the elbow. If it is a very short-radius elbow, the loss coefficient is quite high. For larger-radius elbows, the coefficient decreases until a minimum value is found at an r/d value of about 4 (see Table 10.3). However, for still larger values of r/d, an increase in loss coefficient occurs because, for larger r/d values, the elbow itself is significantly longer than elbows with small r/d values. The greater length creates an additional head loss. The loss coefficients for various types of elbows, and for a number of other fittings and flow transitions, are given in Table 10.3.*

TABLE 10.3 LOSS COEFFICIENTS FOR VARIOUS TRANSITIONS AND FITTINGS

Description	Sketch		Additional Data	K	Source
Pipe entrance			r/d	K_e	(18)†
			0.0	0.50	
			0.1	0.12	
$h_L = K_e V^2/2g$			>0.2	0.03	
Contraction			K_C	K_C	(18)
		D_2/D_1	$u = 60°$	$u = 180°$	
		0.0	0.08	0.50	
		0.20	0.08	0.49	
		0.40	0.07	0.42	
		0.60	0.06	0.27	
		0.80	0.06	0.20	
$h_L = K_C V_2^2/2g$		0.90	0.06	0.10	

* Engineering handbooks usually include extensive tables of loss coefficients. References (18), (19), (20), (21), (22), and (23) are particularly useful in this respect.

TABLE 10.3 LOSS COEFFICIENTS FOR VARIOUS TRANSITIONS AND FITTINGS (CONTINUED)

Description	Sketch	Additional Data		K	Source
Expansion			K_E $u = 20°$	K_E $u = 180°$	(17)
		D_1/D_2			
		0.0		1.00	
		0.20	0.30	0.87	
		0.40	0.25	0.70	
		0.60	0.15	0.41	
$h_L = K_E V_1^2/2g$		0.80	0.10	0.15	
90° miter bend	Vanes	Without vanes		$K_b = 1.1$	(23)
		With vanes		$K_b = 0.2$	(23)
90° smooth bend		r/d			(24) and (17)
		1	$K_b = 0.35$		
		2	0.19		
		4	0.16		
		6	0.21		
		8	0.28		
		10	0.32		
Threaded pipe fittings	Globe valve—wide open	$K_v = 10.0$			(23)
	Angle valve—wide open	$K_v = 5.0$			
	Gate valve—wide open	$K_v = 0.2$			
	Gate valve—half open	$K_v = 5.6$			
	Return bend	$K_b = 2.2$			
	Tee				
	straight-through flow	$K_t = 0.4$			
	side-outlet flow	$K_t = 1.8$			
	90° elbow	$K_b = 0.9$			
	45° elbow	$K_b = 0.4$			

†Reprinted by permission of the American Society of Heating, Refrigerating and Air Conditioning Engineers, Atlanta, Georgia, from the 1981 *ASHRAE Handbook—Fundamentals.*

example 10.9

If oil ($n = 4 \times 10^{-5}$ m^2/s, S = 0.9) flows from the upper to the lower reservoir at a rate of 0.028 m^3/s in the 15-cm smooth pipe, what is the elevation of the oil surface in the upper reservoir?

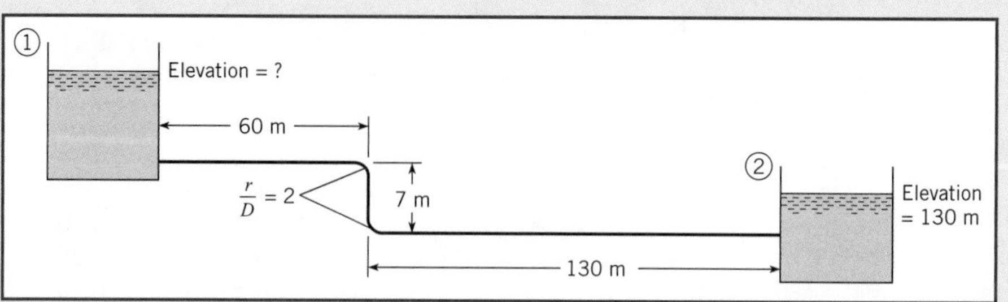

Solution Apply the energy equation between the surfaces of the upper and lower reservoirs:

$$\frac{p_1}{g} + \frac{V_1^2}{2g} + z_1 = \frac{p_2}{g} + \frac{V_2^2}{2g} + z_2 + \sum h_L$$

$$0 + 0 + z_1 = 0 + 0 + 130\text{ m} + \frac{fLV^2}{D2g} + 2K_b\frac{V^2}{2g} + K_e\frac{V^2}{2g} + K_E\frac{V^2}{2g}$$

Here K_b, K_e, and K_E are loss coefficients for bend, entrance, and outlet, respectively. These have values of 0.19, 0.5, and 1.0 (Table 10.3). To determine f, we get Re in order to enter Fig. 10.8:

$$\text{Re} = \frac{VD}{n}$$

But

$$V = \frac{Q}{A} = \frac{(0.028\text{ m}^3/\text{s})}{0.785(0.15\text{ m})^2} = 1.58\text{ m/s}$$

Then

$$\text{Re} = \frac{1.58\text{ m/s}(0.15\text{ m})}{4\times 10^{-5}\text{ m}^2/\text{s}} = 5.93\times 10^3$$

Now we read f from Fig. 10.8 (smooth pipe curve): $f = 0.035$. Then

$$z_1 = 130\text{ m} + \frac{V^2}{2g}\left[\frac{0.035(197\text{ m})}{0.15\text{ m}} + 2(0.19) + 0.5 + 1\right]$$

$$= 130\text{ m} + \left[\frac{(1.58\text{ m/s})^2}{2(9.81\text{ m/s}^2)}\right](46 + 0.38 + 0.5 + 1)$$

$$= 130\text{ m} + 6.1\text{ m} = 136.1\text{ m}$$ ◁

Transition Losses and Grade Lines

In Chapter 7 the effect on the energy and hydraulic grade lines (EGL and HGL) of the pipe head loss and head loss at an abrupt expansion were discussed in some detail. However, no mention was made of head loss due to entrances, bends, and other flow transitions. The primary effect, just as for abrupt expansions, is to cause the EGL to drop by an amount equal to the head loss produced by that transition. Generally, this drop occurs over a distance of several diameters downstream of the transition, as seen in Fig. 10.13. The HGL also drops sharply immediately downstream of the entrance because of the high-velocity flow in the contracted portion of the stream. Then, as the turbulent mixing occurs even farther downstream, energy is lost owing to viscous action occurring in the mixing process. Thus the EGL at the entrance is steeper than it is farther downstream, where the flow has become uniform. As the additional energy loss from the transition subsides, the EGL takes on the slope created by the head loss of the pipe itself.

Even though many transitions produce grade lines that have interesting local details, such as those noted for a sharp-edged entrance, it is common as a gross indication to simply show abrupt changes in the EGL and to neglect local departures between the EGL and the HGL owing to local changes in $V^2/2g$. Thus, Fig. 10.14 is a simplified plot of the EGL and the HGL for a pipe with several transitions in it.

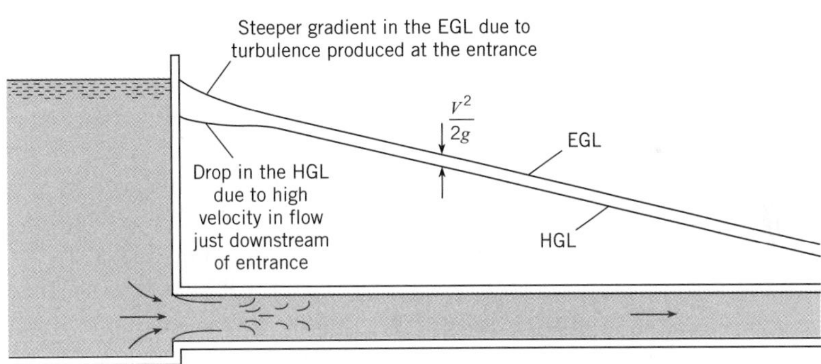

FIGURE 10.13

EGL and HGL at a sharp-edged pipe entrance.

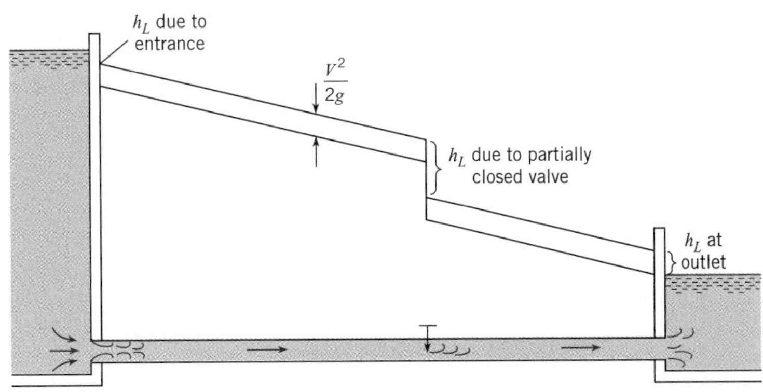

FIGURE 10.14

Head losses in a pipe.

example 10.10

Refer to Fig. 10.14. The difference in water-surface elevation between the reservoirs is 5.0 m, and the horizontal distance between them is 300 m. Using the explicit formula for f, Eq. (10.26), determine the size of steel pipe needed for a discharge of 2 m^3/s when the gate valve is wide open.

Solution First write the energy equation:

$$\frac{p_1}{g} + \frac{V_1^2}{2g} + z_1 = \frac{p_2}{g} + \frac{V_2^2}{2g} + z_2 + \sum h_L$$

$$5 \text{ m} = \sum h_L$$

$$= \frac{V^2}{2g}\left(\frac{fL}{D} + K_e + K_v + K_E\right)$$

$$= \frac{Q^2}{2gA^2}(f \times 300/D + 0.5 + 0.2 + 1.0)$$

$$= \frac{Q^2}{2g(p/4)^2 D^4}\left[300\left(\frac{f}{D}\right) + 1.7\right]$$

$$= \frac{(2 \text{ m}^3/\text{s})^2}{2 \times 9.81 \text{ m/s}^2 \times (p/4)^2 \times D^4}\left[300\left(\frac{f}{D}\right) + 1.7\right]$$

$$= \frac{0.33 \text{ m}^5}{D^4}\left(300\frac{f}{D} + 1.7\right)$$

The other equations for solving this problem are Eq. (10.26), $V = Q/A$, and $\text{Re} = VD/n$. As we did when using the Moody diagram, we make an initial assumption for D. Next we compute V and Re, after which we compute f from Eq. (10.26). Then we compute D from the energy equation (above). With this calculated value of D we go through the process again to get a better estimate of D, and so on, until the change in D is negligibly small. In this example, $n = 10^{-6}$ m^2/s and $k_s = 4.6 \times 10^{-5}$ m (from Table 10.2). We assume an initial value for D of 1 m. Then

$$V = Q/A = \frac{2 \text{ m}^3/\text{s}}{(p/4) \times (1 \text{ m})^2} = 2.55 \text{ m/s}$$

$$\text{Re} = \frac{VD}{n} = 2.55 \times \frac{1}{10^{-6}} = 2.55 \times 10^6$$

$$f = 0.25\left[\log\left(\frac{k_s}{3.70} + \frac{5.74}{\text{Re}^{0.9}}\right)\right]^{-2}$$

$$= 0.25\left[\log\left(\frac{4.6 \times 10^{-5} \text{ m}}{3.7 \times 1 \text{ m}} + \frac{5.74}{(2.55 \times 10^6)^{0.9}}\right)\right]^{-2} = 0.0116$$

With this value of f, we solve the energy equation for D to obtain

$$D = 0.79 \text{ m}$$

With $D = 0.79$ m we repeat the computational procedure again and again, until convergence. The next iteration yields $D = 0.79$ m. Since there is no significant change, we have a solution: $D = 0.79$ m. ◁

Pipe Systems

Simple Pump in a Pipeline

We have considered a number of pipe flow problems in which the head for producing the flow was explicitly given. Now we shall consider flow in which the head is developed by a pump. However, the head produced by a centrifugal pump is a function of the discharge. Hence a direct solution is usually not immediately available. The solution (that is, the flow rate for a given system) is obtained when the system equation (or curve) of head versus discharge is solved simultaneously with the pump equation (or curve) of head versus discharge. The solution of these two equations (or the point where the two curves intersect) yields the operating condition for the system. Consider flow of water in the system of Fig. 10.15. When the energy equation is written from the reservoir water surface to the outlet stream, we obtain the following equation:

$$\frac{p_1}{g} + \frac{V_1^2}{2g} + z_1 + h_p = \frac{p_2}{g} + \frac{V_2^2}{2g} + z_2 + \sum K_L \frac{V^2}{2g} + \sum \frac{fL}{D}\frac{V^2}{2g}$$

For a system with one size of pipe, this simplifies to

$$h_p = (z_2 - z_1) + \frac{V^2}{2g}\left(1 + \sum K_L + \frac{fL}{D}\right) \tag{10.30}$$

Hence, for any given discharge, a certain head h_p must be supplied to maintain that flow. Thus we can construct a head-versus-discharge curve, as shown in Fig. 10.16. Such a curve is called the *system curve.* Any given centrifugal pump has a head-versus-discharge curve that is characteristic of that pump at a given pump speed. Such curves are supplied by the pump manufacturer; a typical one for a centrifugal pump is shown in Fig. 10.16.

Figure 10.16 reveals that, as the discharge increases in a pipe, the head required for flow also increases. However, the head that is produced by the pump decreases as the discharge increases. Consequently, the two curves intersect, and the operating point is at the point of intersection—that point where the head produced by the pump is just the amount needed to overcome the head loss in the pipe.

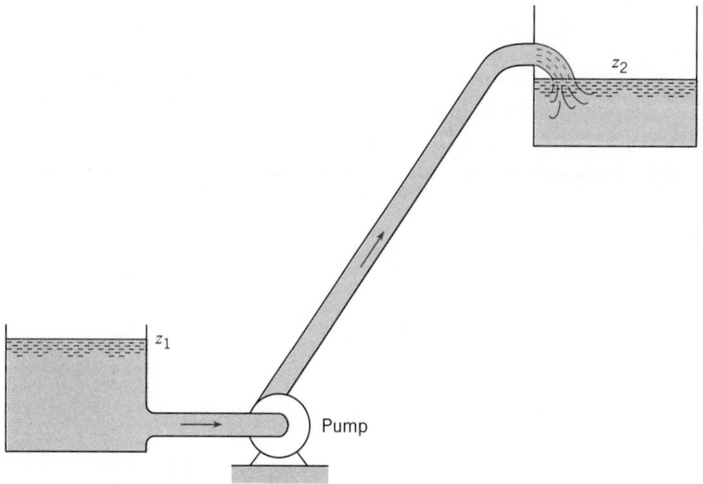

FIGURE 10.15

Pump and pipe combination.

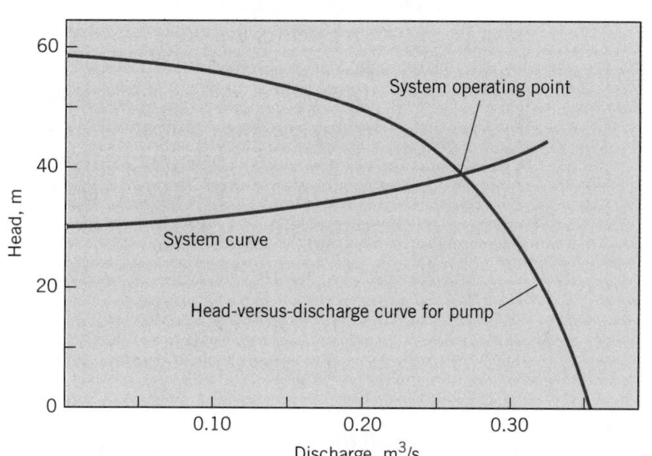

FIGURE 10.16

Pump and system curves.

example 10.11

What will be the discharge in this water system if the pump has the characteristics shown in Fig. 10.16? Assume $f = 0.015$.

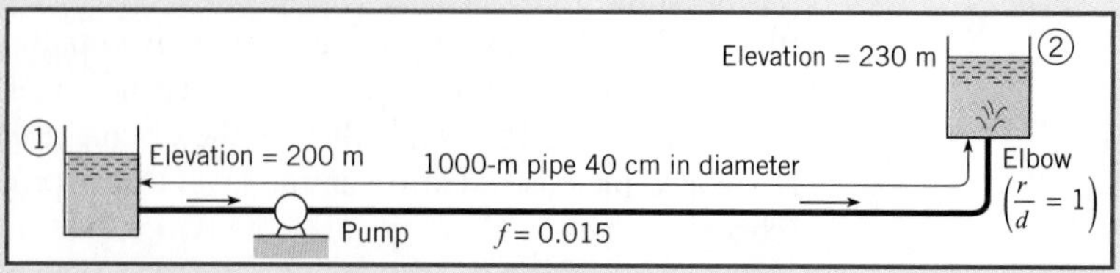

Solution First we write the energy equation from water surface to water surface:

$$\frac{p_1}{g} + \frac{V_1^2}{2g} + z_1 + h_p = \frac{p_2}{g} + \frac{V_2^2}{2g} + z_2 + \sum h_L$$

$$0 + 0 + 200 + h_p = 0 + 0 + 230 + \left(\frac{fL}{D} + K_e + K_b + K_E\right)\frac{V^2}{2g}$$

Here $K_e = 0.5$, $K_b = 0.35$, and $K_E = 1.0$. Hence

$$h_p = 30 + \frac{Q^2}{2gA^2}\left[\frac{0.015(1000)}{0.40} + 0.5 + 0.35 + 1\right]$$

$$= 30 + \frac{Q^2}{2 \times 9.81 \times [(p/4) \times 0.4^2]^2}(39.3) = 30\text{ m} + 127Q^2\text{ m}$$

Now we make a table of Q versus h_p (see below) to give values to produce a system curve that will be plotted with the pump curve. When the system curve is plotted on the same graph as the pump curve, it is seen (Fig. 10.16) that the operating condition occurs at $Q = 0.27\text{ m}^3/\text{s}$. ◁

Q, m^3/s	Q^2, m^6/s^2	$127Q^2$	$h_p = 30\text{ m} + 127Q^2\text{ m}$
0	0	0	30
0.1	1×10^{-2}	1.3	31.3
0.2	4×10^{-2}	5.1	35.1
0.3	9×10^{-2}	11.4	41.4

Pipes in Parallel

Consider a pipe that branches into two parallel pipes and then rejoins, as shown in Fig. 10.17. A problem involving this configuration might be to determine the division of flow in each pipe, given the total flow rate.

No matter which pipe is involved, the pressure difference between the two junction points is the same. Also, the elevation difference between the two junction points is the same. Because $h_L = (p_1/g + z_1) - (p_2/g + z_2)$, it follows that h_L between the two junction points is the same in both of the pipes of the parallel pipe system. Thus we can write

$$h_{L_1} = h_{L_2}$$

$$f_1\frac{L_1}{D_1}\frac{V_1^2}{2g} = f_2\frac{L_2}{D_2}\frac{V_2^2}{2g}$$

Then

$$\left(\frac{V_1}{V_2}\right)^2 = \frac{f_2L_2D_1}{f_1L_1D_2} \quad \text{or} \quad \frac{V_1}{V_2} = \left(\frac{f_2L_2D_1}{f_1L_1D_2}\right)^{1/2}$$

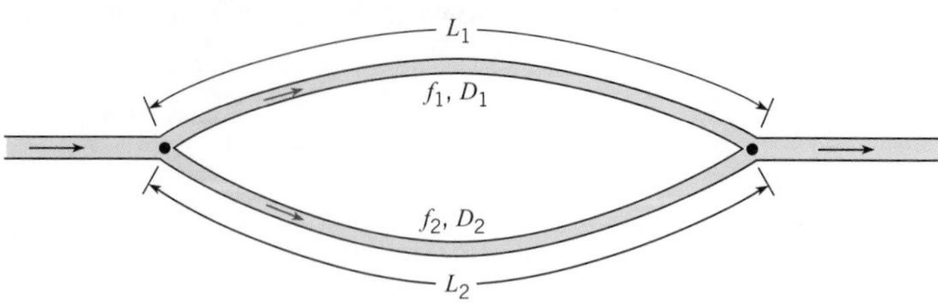

FIGURE 10.17

Flow in parallel pipes.

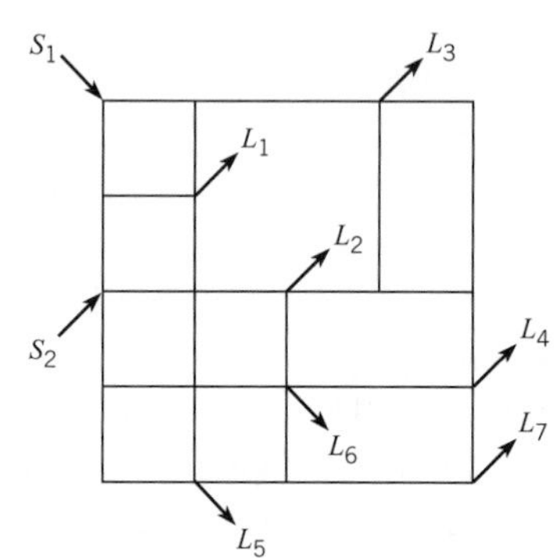

FIGURE 10.18

Pipe network.

If f_1 and f_2 are known, the division of flow can be easily determined. However, some trial-and-error analysis may be required if f_1 and f_2 are in the range where they are functions of the Reynolds number.

Pipe Networks

The most common pipe networks are the water distribution systems for municipalities. These systems have one or more sources (discharges of water into the system) and numerous loads: one for each household and commercial establishment. For purposes of simplification, the loads are usually lumped throughout the system. Figure 10.18 shows a simplified distribution system with two sources and seven loads.

The engineer is often engaged to design the original system or to recommend an economical expansion to the network. An expansion may involve additional housing or commercial developments, or it may be to handle increased loads within the existing area.

In the design of such a system, the engineer will have to estimate the future loads for the system and will need to have sources (wells or direct pumping from streams or lakes) to satisfy the loads. Also, the layout of the pipe network must be made (usually parallel to streets), and pipe sizes will have to be determined. The object of the design is to arrive at a network of pipes that will deliver the design flow at the design pressure for minimum cost. The cost will include first costs (materials and construction) as well as maintenance and operating costs. The design process usually involves a number of iterations on pipe sizes and layouts before the optimum design (minimum cost) is achieved.

So far as the fluid mechanics of the problem are concerned, the engineer must determine pressures throughout the network for various conditions—that is, for various combinations of pipe sizes, sources, and loads. The solution of a problem for a given layout and a given set of sources and loads requires that two conditions be satisfied:

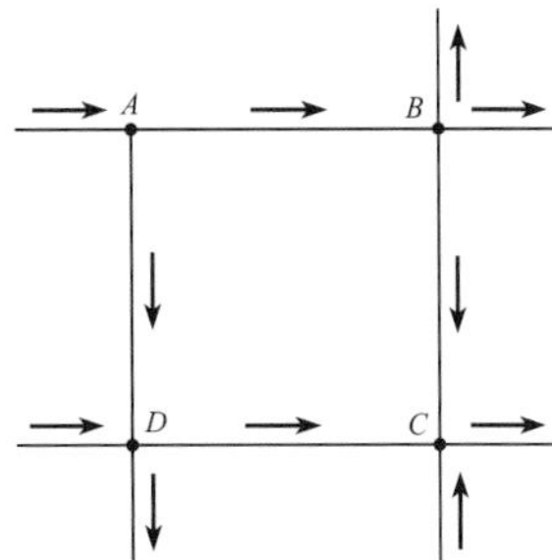

FIGURE 10.19

A typical loop of a pipe network.

1. Continuity must be satisfied. That is, the flow into a junction of the network must equal the flow out of the junction. This must be satisfied for all junctions.
2. The head loss between any two junctions must be the same regardless of the path in the series of pipes taken to get from one junction point to the other. This requirement results because pressure must be continuous throughout the network (pressure cannot have two values at a given point). This condition leads to the conclusion that the algebraic sum of head losses around a given loop must be equal to zero. Here the sign (positive or negative) for the head loss in a given pipe is given by the sense of the flow with respect to the loop, that is, whether the flow has a clockwise or counterclockwise sense.

Only a few years ago, these solutions were made by trial-and-error hand computation, but modern applications using digital computers have made the older methods obsolete. Even with these advances, however, the engineer charged with the design or analysis of such a system must understand the basic fluid mechanics of the system to be able to interpret the results properly and to make good engineering decisions based on the results. Therefore, an understanding of the original method of solution by Hardy Cross (25) may help you to gain this basic insight. The Hardy Cross method is as follows.

The engineer first distributes the flow throughout the network so that loads at various nodes are satisfied. In the process of distributing the flow through the pipes of the network, the engineer must be certain that continuity is satisfied at all junctions (flow into a junction = flow out of the junction), thus satisfying requirement 1. The first guess at the flow distribution obviously will not satisfy requirement 2 regarding head loss; therefore, corrections are applied. For each loop of the network, a discharge correction is applied to yield a zero net head loss around the loop. For example, consider the isolated loop in Fig. 10.19. In this loop, the loss of head in the clockwise sense will be given by

$$\begin{aligned} \sum h_{L_c} &= h_{L_{AB}} + h_{L_{BC}} \\ &= \sum kQ_c^n \end{aligned} \tag{10.31}$$

The loss of head for the loop in the counterclockwise sense is

$$\sum h_{L_{cc}} = \sum_{cc} kQ_{cc}^n \tag{10.32}$$

For a solution, the clockwise and counterclockwise head losses have to be equal, or

$$\sum h_{L_c} = \sum h_{L_{cc}}$$

$$\sum kQ_c^n = \sum kQ_{cc}^n$$

As we noted, the first guess for flow in the network is undoubtedly in error; therefore, a correction in discharge, ΔQ, will have to be applied to satisfy the head loss requirement. If the clockwise head loss is greater than the counterclockwise head loss, ΔQ will have to be applied in the counterclockwise sense. That is, subtract ΔQ from the clockwise flows and add it to the counterclockwise flows:

$$\sum k(Q_c - \Delta Q)^n = \sum k(Q_{cc} + \Delta Q)^n \tag{10.33}$$

Expanding the summation on either side of Eq. (10.33) and including only two terms of the expansion, we obtain

$$\sum k(Q_c^n - nQ_c^{n-1}\Delta Q) = \sum k(Q_{cc}^n + nQ_{cc}^{n-1}\Delta Q)$$

Then solving for ΔQ, we get

$$\Delta Q = \frac{\sum kQ_c^n - \sum kQ_{cc}^n}{\sum nkQ_c^{n-1} + \sum nkQ_{cc}^{n-1}} \tag{10.34}$$

Thus if ΔQ as computed from Eq. (10.34) is positive, the correction is applied in a counterclockwise sense (add ΔQ to counterclockwise flows and subtract it from clockwise flows).

A different ΔQ is computed for each loop of the network and applied to the pipes. Some pipes will have two ΔQ's applied because they will be common to two loops. The first set of corrections usually will not yield the final desired result because the solution is approached only by successive approximations. Thus the corrections are applied successively until the corrections are negligible. Experience has shown that for most loop configurations, applying ΔQ as computed by Eq. (10.34) produces too large a correction. Fewer trials are required to solve for Q's if approximately 0.6 of the computed ΔQ is used.

More information on methods of solution of pipe networks is available in references (26) and (27). A search of the Internet under "pipe networks" yields information on software available from various sources.

example 10.12

A simple pipe network with water flow consists of three valves and a junction as shown in the figure. The piezometric head at points 1 and 2 is 1 ft and reduces to zero at point 4. There is a wide-open globe valve in line *A*, a gate valve half open in line *B*, and a wide-open angle valve in line *C*. The pipe diameter in all lines is 2 inches. Find the flow rate in each line. Assume that the head loss in each line is due only to the valves.

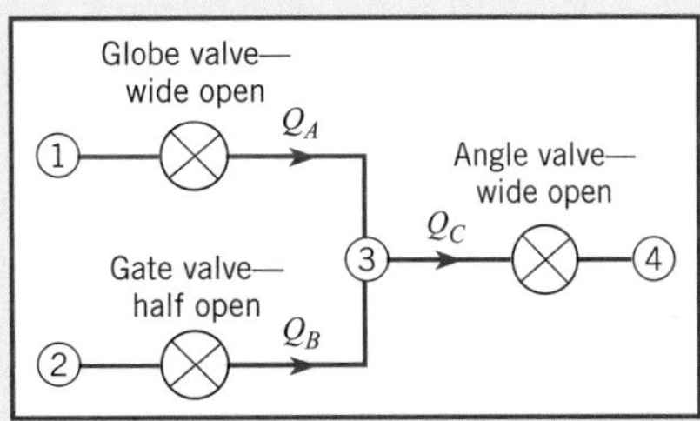

Solution The piezometric heads at points 1 and 2 are equal, so

$$h_{L,1\to3} + h_{L,3\to2} = 0$$

The head loss between points 2 and 4 is 1 ft, so

$$h_{L,2\to3} + h_{L,3\to4} = 1$$

Continuity must be satisfied at point 3, so

$$Q_A + Q_B = Q_C$$

The head loss through a valve is given by

$$h_L = K_V \frac{V^2}{2g}$$
$$= K_V \frac{1}{2g}\left(\frac{Q}{A}\right)^2$$

where K_V is the loss coefficient. For a 2-inch pipe, the head loss becomes

$$h_L = 32.6 K_V Q^2$$

where h_L is in feet and Q is in cfs.

The head loss equation between points 1 and 2 expressed in term of discharge is

$$32.6 K_A Q_A^2 - 32.6 K_B Q_B^2 = 0$$

or

$$K_A Q_A^2 - K_B Q_B^2 = 0$$

where K_A is the loss coefficient for the wide-open globe valve ($K_A = 10$) and K_B is the loss coefficient for the half-open gate valve ($K_B = 5.6$). The head loss equation between points 2 and 4 is

$$32.6 K_B Q_B^2 + 32.6 K_C Q_C^2 = 1$$

or

$$K_B Q_B^2 + K_C Q_C^2 = 0.0307$$

where K_C is the loss coefficient for the wide-open angle valve ($K_C = 5$). The two head loss equations and the continuity equation comprise three equations for Q_A, Q_B, and Q_C. However, the equations are nonlinear and require linearization and solution by iteration (Hardy-Cross approach). The discharge is written as

$$Q = Q_0 + \Delta Q$$

where Q_0 is the starting value and ΔQ is the change. Then

$$Q^2\ .\ Q_0^2 + 2Q_0\Delta Q$$

where the $(\Delta Q)^2$ term is neglected. The equations in terms of ΔQ become

$$\begin{aligned} 2K_A Q_{0,A}\Delta Q_A - 2K_B Q_{0,B}\Delta Q_B &= K_B Q_{0,B}^2 - K_A Q_{0,A}^2 \\ 2K_C Q_{0,C}\Delta Q_C - 2K_B Q_{0,B}\Delta Q_B &= 0.0307 - K_B Q_{0,B}^2 - K_C Q_{0,C}^2 \\ \Delta Q_A + \Delta Q_B &= \Delta Q_C \end{aligned}$$

which can be express in matrix form as

$$\begin{bmatrix} 2K_A Q_{0,A} & -2K_B Q_{0,B} & 0 \\ 0 & 2K_B Q_{0,B} & 2K_C Q_{0,C} \\ 1 & 1 & -1 \end{bmatrix} \begin{Bmatrix} \Delta Q_A \\ \Delta Q_B \\ \Delta Q_C \end{Bmatrix}$$

$$= \begin{bmatrix} K_B Q_{0,B}^2 - K_A Q_{0,A}^2 \\ 0.0307 - K_B Q_{0,B}^2 - K_C Q_{0,C}^2 \\ 0 \end{bmatrix}$$

The procedure begins by selecting values for $Q_{0,A}$, $Q_{0,B}$, and $Q_{0,C}$. Assume $Q_{0,A} = Q_{0,B}$ and $Q_{0,C} = 2Q_{0,A}$. Then from the head loss equation from points 2 to 4

$$\begin{aligned} K_B Q_{0,B}^2 + K_C Q_{0,C}^2 &= 0.0307 \\ (K_B + 4K_C)Q_{0,B}^2 &= 0.0307 \\ (5.6 + 4 \times 5)Q_{0,B}^2 &= 0.0307 \\ Q_{0,B} &= 0.0346 \end{aligned}$$

and $Q_{0,A} = 0.0346$ and $Q_{0,C} = 0.0693$. These values are substituted into the matrix equation to solve for the ΔQ's. The discharges are corrected by $Q_0^{\text{new}} = Q_0^{\text{old}} + \Delta Q$ and substituted into the matrix equation again to yield new ΔQ's. The iterations are continued until sufficient accuracy is obtained. The accuracy is judged by how close the column matrix on the right approaches zero. A table with the results of iterations for this example is shown below.

		Iteration			
	Initial	**1**	**2**	**3**	**4**
Q_A	0.0346	0.0328	0.0305	0.0293	0.0287
Q_B	0.0346	0.0393	0.0384	0.0394	0.0384
Q_C	0.0693	0.0721	0.0689	0.0687	0.0671

This solution technique is called the Newton-Raphson method for nonlinear systems of algebraic equations. It can be implemented easily on a computer. The solution procedure for more complex systems is the same.

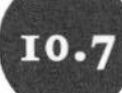

Turbulent Flow in Noncircular Conduits

Basic Development

Earlier in this chapter (Section 10.4), τ_0 was eliminated between Eqs. (10.3) and (10.21) to yield the Darcy-Weisbach equation, Eq. (10.22). It should be noted that Eq. (10.3) was derived by writing the equation of equilibrium in the longitudinal direction for an element of fluid with a circular cross section. If one derives an equation analogous to Eq. (10.3) for flow in a noncircular conduit in which the shear stress acts on the conduit surface having a perimeter P (such as the perimeter of a rectangular conduit) instead of perimeter $2\pi r$, then τ_0 is given by

$$t_0 = \frac{A}{P}\left[-\frac{d}{ds}(p + \gamma z)\right] \tag{10.35}$$

In Eq. (10.3), to which Eq. (10.35) is analogous, the shear stress τ_0 was everywhere constant around the perimeter of the cylindrical element. In Eq. (10.35) for the noncircular conduit, the shear stress is not constant over the perimeter. However, we can still use Eq. (10.21) to relate τ_0 and V, where τ_0 is now the average shear stress on the boundary. Eliminating τ_0 between Eqs. (10.21) and (10.35) and integrating along the pipe yields

$$h_f = \frac{f \cdot L}{4A/P} \frac{V^2}{2g} \tag{10.36}$$

In Eq. (10.36), h_f is the head loss between two points in the conduit, L is the length between the points, and P is the wetted perimeter of the conduit. Thus Eq. (10.36) is the same as the Darcy-Weisbach equation except that D is replaced by $4A/P$. The ratio of the cross-sectional area A to the wetted perimeter P is defined as the *hydraulic radius* R_h. Obviously, for flow of a gas the wetted perimeter P is the perimeter of the duct. Experiments have shown that we can solve flow problems involving noncircular conduits, such as rectangular ducts, if we apply the same methods and equations that we did for pipes but use $4R_h$ in place of D. Consequently, the relative roughness is $k_s/4R_h$, and the Reynolds number is defined as $V(4R_h)/n$.

example 10.13

Air (T = 20°C and p = 101 kPa absolute) flows at a rate of 2.5 m^3/s in a commercial steel rectangular duct 30 cm by 60 cm. What is the pressure drop per 50 m of duct?

Solution First compute Re and $k_s/4R_h$:

$$\text{Re} = \frac{V(4R_h)}{n}$$

Here

$$V = \frac{Q}{A} = \frac{2.5\ \text{m}^3/\text{s}}{0.18\ \text{m}^2} = 13.9\ \text{m/s}$$

The hydraulic radius is given by

$$R_h = \frac{A}{P} = \frac{0.18\ \text{m}^2}{1.8\ \text{m}} = 0.10\ \text{m}$$

$$4R_h = 4(0.10\ \text{m}) = 0.40\ \text{m}$$

$$n = 1.51 \times 10^{-5}\ \text{m}^2/\text{s}$$

Hence

$$\text{Re} = \frac{13.9 \times 0.40}{1.51 \times 10^{-5}} = 3.68 \times 10^5$$

$$\frac{k_s}{4R_h} = \frac{4.6 \times 10^{-5}\ \text{m}}{0.40\ \text{m}} = 1.15 \times 10^{-4}$$

Then from Fig. 10.8, f = 0.015. Thus

$$h_f = \frac{fL}{4R_h} \frac{V^2}{2g}$$

or

$$gh_f = \Delta p_f = \frac{fL}{4R_h} r \frac{V^2}{2}$$

$$r = 1.2 \text{ kg/m}^3$$

Finally,

$$\Delta p_f \text{ per } 50 \text{ m} = \frac{0.015(50 \text{ m})}{0.40 \text{ m}} (1.2 \text{ N} \cdot \text{s}^2/\text{m}^4) \frac{13.9^2}{2} \text{m}^2/\text{s}^2 = 217 \text{ Pa} \quad \triangleleft$$

Uniform Free-Surface Flows

A free-surface flow signifies the flow in a channel or duct with the surface open to the atmosphere. These flows are also known as open-channel flows. Such flows are encountered in culverts and irrigation canals. In this case there is no pressure gradient in the flow direction, so the change in piezometric pressure or head is due solely to elevation changes (gravity effects). Also, the hydraulic radius is based on the wetted area, so the portion of the perimeter on the free surface is not included. A uniform flow requires that the velocity be constant in the flow direction, so the shape of the channel and the depth of fluid will be the same from section to section. Figure 10.20 is an example of uniform free-surface flow in a channel with a rectangular cross section. Note here that the velocity varies across the section but does not change in the flow direction. A more extensive discussion of open channels with nonuniform flows is provided in Chapter 15.

The criterion for determining whether the flow in open channels will be laminar or turbulent is similar to that for flow in pipes. Recall that in pipe flow, if the Reynolds number ($VDr/m = VD/n$) is less than 2000, the flow will be laminar, and if it is greater than about 3000, one can expect the flow to be turbulent. The Reynolds number criterion for open-channel flow is the same if we replace D in the Reynolds number by $4R_h$, as we did in the Darcy-Weisbach equation in the preceding subsection. Thus, we can expect laminar flow to occur in open channels if $V(4R_h)/n < 2000$. The Reynolds number for open channels is usually defined as VR_h/n. Therefore, in open channels, if the Reynolds number is less than 500, we will have laminar flow, and if it is greater than about 750, we can expect to have turbulent flow. A brief analysis of this turbulent criterion will show that water flow in channels will usually be turbulent unless the velocity and/or the depth is very small. Example 10.14 illustrates this point.

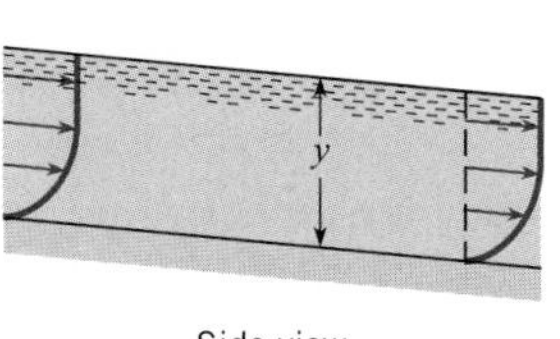

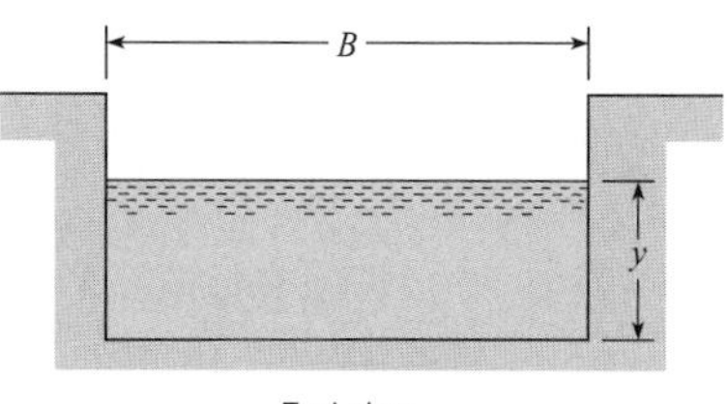

FIGURE 10.20

Open-channel relations.

It should be noted that the wetted perimeter used for calculating the hydraulic radius is the perimeter of the channel that is actually in contact with the flowing liquid. For example, in Fig. 10.20 the hydraulic radius of this channel of rectangular cross section is

$$R_h = \frac{A}{P} = \frac{By}{B+2y} \tag{10.37}$$

One can see that for very wide, shallow channels the hydraulic radius approaches the depth y.

example 10.14

Water (60°F) flows in a 10-ft-wide rectangular channel at a depth of 6 ft. What is the Reynolds number if the mean velocity is 0.1 ft/s? With this velocity, at what maximum depth can we be assured of having laminar flow?

Solution $\mathrm{Re} = VR_h/\nu$

where

$$\begin{aligned} V &= 0.1 \text{ ft/s} \\ R_h &= A/P = By/(B+2y) \\ &= (10 \times 6)/(10 + 2 \times 6) \\ &= 2.73 \text{ ft} \\ \nu &= 1.22 \times 10^{-5} \text{ ft}^2/\text{s (from Table A.5)} \end{aligned}$$

then

$$\mathrm{Re} = (0.1 \text{ ft/s})(2.73 \text{ ft})/(1.22 \times 10^{-5} \text{ ft}^2/\text{s}) = 22{,}377 \quad \triangleleft$$

The maximum Reynolds number at which we can expect to have laminar flow in open channels is 500. Thus, for this limit of Re and for a water velocity of 0.10 ft/s, we can solve for the depth at which this condition will prevail:

$$\mathrm{Re} = VR_h/\nu = (0.10 \text{ ft/s})R_h/(1.22 \times 10^{-5} \text{ ft}^2/\text{s}) = 500$$

Solving for R_h yields

$$R_h = (500)(1.22 \times 10^{-5} \text{ ft}^2/\text{s})/(0.10 \text{ ft/s}) = 0.061 \text{ ft}$$

For rectangular channels

$$R_h = (By)/(B+2y)$$

Thus

$$(By)/(B+2y) = (10y)/(10+2y) = 0.061 \text{ ft}$$

$$y = 0.062 \text{ ft} \quad \triangleleft$$

Example 10.14 shows that indeed the velocity and/or depth must be very small to yield laminar flow of water in a channel. Note also, the depth and hydraulic radius are virtually the same for this case, where the depth is very small relative to the width of the channel.

To determine the head loss for uniform flow in open channels, we utilize Eq. (10.36). That is, the Darcy-Weisbach equation with D replaced by $4R_h$ is used.

example 10.15

Estimate the discharge of water that a concrete channel 10 ft wide can carry if the depth of flow is 6 ft and the slope of the channel is 0.0016.

Solution We use the Darcy-Weisbach equation:

$$h_f = \frac{fL}{4R_h}\frac{V^2}{2g} \quad \text{or} \quad \frac{h_f}{L} = \frac{f}{4R_h}\frac{V^2}{2g}$$

When we have uniform open-channel flow, the slope of the EGL, $S_f = h_f/L$, is the same as the channel slope S_0. Therefore, $h_f/L = S_0$. The foregoing equation then reduces to $V^2/2g = 4R_hS_0/f$, or

$$V = \sqrt{\frac{8g}{f}R_hS_0}$$

Assume $k_s = 0.005$ ft. Then the relative roughness is

$$\frac{k_s}{4R_h} = \frac{0.005 \text{ ft}}{4(60 \text{ ft}^2/22 \text{ ft})} = \frac{0.005 \text{ ft}}{4(2.73 \text{ ft})} = 0.00046$$

Using $k_s/4R_h = 0.00046$ as a guide and referring to Fig. 10.8, we assume that $f = 0.016$. Thus

$$V = \sqrt{\frac{8(32.2 \text{ ft/s}^2)(2.73 \text{ ft})(0.0016)}{0.016}} = \sqrt{70.6 \text{ ft}^2/\text{s}^2} = 8.39 \text{ ft/s}$$

Then

$$\text{Re} = V\frac{4R_h}{\nu} = \frac{8.39 \text{ ft/s}(10.9 \text{ ft})}{1.2(10^{-5} \text{ ft}^2/\text{s})} = 7.62 \times 10^6$$

Using this new value of Re and with $k_s/4R_h = 0.00046$, we read f as 0.016. Our initial guess was good; and now that the velocity is known, we can compute Q:

$$Q = VA = 8.39 \text{ ft/s}(60 \text{ ft}^2) = 503 \text{ cfs}$$ ◁

For rock-bedded channels such as those in some natural streams or unlined canals, the larger rocks produce most of the resistance to flow, and essentially none of this resistance is due to viscous effects. Thus, the friction factor is independent of the Reynolds number. For this type of channel, Limerinos (28) has shown that the resistance coefficient f can be given in terms of the size of rock in the stream bed as

$$f = \frac{1}{\left[1.2 + 2.03\log\left(\frac{R_h}{d_{84}}\right)\right]^2} \tag{10.38}$$

where d_{84} is a measure of the rock size.*

example 10.16

Determine the value of the resistance coefficient, f, for a natural rock-bedded channel that is 100 ft wide and has an average depth of 4.3 ft. The d_{84} size of boulders in the stream bed is 0.72 ft.

Solution This is a fairly wide channel relative to its depth; therefore, R_h is approximated by its depth of 4.3 ft.

$$f = \frac{1}{\left[1.2 + 2.03\log\left(\frac{4.3}{0.72}\right)\right]^2} = 0.130 \qquad \triangleleft$$

The Chezy Equation and Manning Equation

Leaders in open-channel research have recommended the adoption of the methods already presented (involving the Reynolds number and relative roughness) for channel design (29). However, many engineers continue to use the older, more traditional methods for designing open channels. Because of this current use and for historic reasons, the Chezy equation and Manning equation are presented here.

If Eq. (10.36) is rewritten in slightly different form, we have

$$\frac{h_f}{L} = \frac{f}{4R_h}\frac{V^2}{2g} \tag{10.39}$$

For uniform flow, the depth—called *normal depth*—is constant. Consequently, h_f/L is the slope S_0 of the channel, and Eq. (10.39) can be written as

$$R_h S_0 = \frac{f}{8g}V^2$$

or

$$V = C\sqrt{R_h S_0} \tag{10.40}$$

where

$$C = \sqrt{8g/f} \tag{10.41}$$

Since $Q = VA$, we can express the discharge in a channel as

* Most river-worn rocks are somewhat elliptical in shape. Limerinos (28) showed that the intermediate dimension correlates best with f. The d_{84} refers to the size of rock (intermediate dimension) for which 84% of the rocks in the random sample are smaller than the d_{84} size. Details for choosing the sample are given by Wolman (30). The basic procedure entails sampling at least 100 rocks on the channel bottom. For example, a grid with 100 points could be laid out on the channel bottom, and the rock under each grid point would be a rock of the sample from which the d_{84} is determined.

$$Q = CA\sqrt{R_h S_0} \tag{10.42}$$

This equation is known as the *Chezy equation* after a French engineer of that name. For practical application, the coefficient C must be determined, and the usual design formula for C in the SI system of units is given as

$$C = \frac{R_h^{1/6}}{n} \tag{10.43}$$

When we insert this expression for C into Eq. (10.42), we obtain a common form of the discharge equation for uniform flow in open channels for SI units:

$$Q = \frac{1.0}{n} A R_h^{2/3} S_0^{1/2} \tag{10.44}$$

In Eq. (10.44), n is a resistance coefficient called Manning's n, which has different values for different types of boundary roughness. Table 10.4 gives n for various types of boundary surfaces. The major limitation of this approach is that the viscous or relative roughness effects are not present in the design formula. Hence, application outside the range of normal-sized channels carrying water is not recommended.

TABLE 10.4 TYPICAL VALUES OF THE ROUGHNESS COEFFICIENT n

Lined Canals	n
Cement plaster	0.011
Untreated gunite	0.016
Wood, planed	0.012
Wood, unplaned	0.013
Concrete, troweled	0.012
Concrete, wood forms, unfinished	0.015
Rubble in cement	0.020
Asphalt, smooth	0.013
Asphalt, rough	0.016
Corrugated metal	0.024
Unlined Canals	
Earth, straight and uniform	0.023
Earth, winding and weedy banks	0.035
Cut in rock, straight and uniform	0.030
Cut in rock, jagged and irregular	0.045
Natural Channels	
Gravel beds, straight	0.025
Gravel beds plus large boulders	0.040
Earth, straight, with some grass	0.026
Earth, winding, no vegetation	0.030
Earth, winding, weedy banks	0.050
Earth, very weedy and overgrown	0.080

Manning Equation—Traditional System of Units

It can be shown that, in converting from SI to the traditional system of units, one must apply a factor equal to 1.49 if the same value of n is used in the two systems. Thus in the traditional system the discharge equation using Manning's n is given as

$$Q = \frac{1.49}{n} A R_h^{2/3} S_0^{1/2} \tag{10.45}$$

example 10.17

If the channel of Example 10.16 has a slope of 0.0030, what is the discharge in the channel and what is the numerical value of Manning's n for this channel?

Solution From Example 10.16, we have an f value of 0.130, and the approximate value of R_h is 4.3 ft. In Example 10.14 it was shown that

$$V = \sqrt{\frac{8g}{f}}\sqrt{R_h S_0}$$

Therefore

$$V = \left[\sqrt{\frac{(8)(32.2 \text{ ft/s}^2)}{0.130}}\right]\left[\sqrt{(4.3 \text{ ft})(0.0030)}\right] = 5.06 \text{ ft/s}$$

But

$$Q = VA = (5.06)(100 \times 4.3) = 2176 \text{ cfs}$$

We solve for n using Eq. (10.41). [*Note:* We are in the traditional system of units; therefore, we use Eq. (10.45) rather than Eq. (10.44)].

$$n = \frac{1.49}{Q} A R_h^{2/3} S_0^{1/2}$$

$$n = \left(\frac{1.49}{2176 \text{ ft}^3/\text{s}}\right)(100 \times 4.3 \text{ ft}^2)(4.3 \text{ ft})^{2/3}(0.003)^{1/2}$$

$$n = 0.0426$$

◁

example 10.18

Using the Chezy equation with Manning's n, compute the discharge in the channel described in Example 10.15.

Solution

$$Q = \frac{1.49}{n} A R_h^{2/3} S_0^{1/2}$$

From Example 10.15,

$$R_h = \frac{60}{22} = 2.73 \text{ ft} \qquad R_h^{2/3} = 1.95$$

$$S_0^{1/2} = 0.04 \quad \text{and} \quad A = 60 \text{ ft}^2$$

$$n = 0.015 \qquad \text{(Table 10.4)}$$

Then

$$Q = \frac{1.49}{0.015}(60)(1.96)(0.04) = 467 \text{ cfs}$$ ◁

The two results (Examples 10.15 and 10.18) are within expected engineering accuracy for this type of problem. For a more complete discussion of the historical development of Manning's equation and the choice of n values for use in design or analysis, refer to Yen (31) and Chow (32).

Best Hydraulic Section

The quantity $AR^{2/3}$ in Manning's equation [Eqs. (10.44) and (10.45)] is called the section factor, in which $R = A/P$; therefore, the section factor relating to uniform flow is given by $A(A/P)^{2/3}$. Thus, for a channel of given resistance and slope, the discharge will increase with increasing cross-sectional area but decrease with increasing wetted perimeter P. For a given area, A, and a given shape of channel—for example, rectangular cross section—there will be a certain ratio of depth to width (y/B) for which the section factor will be maximum. Such a ratio establishes the *best hydraulic section.* That is, the best hydraulic section is the channel proportion that yields a minimum wetted perimeter for a given cross-sectional area.

example 10.19

Determine the best hydraulic section for a rectangular channel.

Solution For the rectangular channel, $A = By$ and $P = B + 2y$. Let A be constant, then let us minimize P. But

$$P = B + 2y$$

or

$$P = \frac{A}{y} + 2y$$

Thus, we see that the perimeter varies only with y for the given conditions. If we differentiate P with respect to y and set the differential equal to zero, we will have the condition for minimizing P for the given area A:

$$\frac{dP}{dy} = \frac{-A}{y^2} + 2 = 0$$

or

$$\frac{A}{y^2} = 2$$

But $A = By$, so

$$\frac{By}{y^2} = 2 \quad \text{or} \quad y = \frac{1}{2}B \quad \triangleleft$$

Thus, the best hydraulic section for a rectangular channel occurs when the depth is one-half the width of the channel.

It can be shown that the best hydraulic section for a trapezoidal channel is half a hexagon; for the circular section, it is the half circle; and for the triangular section, it is half of a square. Of all the various shapes, the half circle has the best hydraulic section.

The best hydraulic section can be relevant to the cost of the channel. For example, if a trapezoidal channel were to be excavated and if the water surface were to be at adjacent ground level, the minimum amount of excavation (and excavation cost) would result if the channel of best hydraulic section were used.

Uniform Flow in Culverts and Sewers

Sewers are conduits that are used to carry domestic, commercial, or industrial waste (called sewage) from households, businesses, and factories to sewage disposal sites. These conduits are often circular in cross section, but elliptical and rectangular conduits are also used. The volume rate of sewage varies throughout the day and season, but of course sewers are designed to carry the maximum design discharge flowing full or nearly full. At discharges less than the maximum, the sewers will operate as open channels.

Sewage usually consists of about 99% water and 1% solid waste. Because most sewage is so dilute, it is assumed that it has the same physical properties as water for purposes of discharge computations. However, if the velocity in the sewer is too small, the solid particles may settle out and cause blockage of the flow. Therefore, sewers are usually designed to have a minimum velocity of about 2 ft/s (0.60 m/s) at maximum flow condition. This condition is met by choosing a slope on the sewer line to achieve the desired velocity.

example 10.20

A sewer line is to be constructed of concrete pipe and to be laid on a slope of 0.006. If $n = 0.013$ and if the design discharge is 110 cfs, what size pipe (commercially available) would you choose for a full flow condition? What will be the mean velocity in the sewer pipe for these conditions? (It should be noted that concrete pipe is readily available in commercial sizes of 8-in., 10-in., and 12-in. diameter and then in 3-in. increments up to 36-in. diameter. From 36-in. diameter up to 144 in. the sizes are available in 6-in. increments.)

Solution

$$Q = \frac{1.49}{n} A R^{2/3} S_0^{1/2}$$

where $Q = 110\ \text{ft}^3/\text{s}$

$n = 0.013$

$S_0 = 0.006$ (assume atmospheric pressure along the pipe)

Then
$$AR^{2/3} = \frac{(110\ \text{ft}^3/\text{s})(0.013)}{(1.49)(0.006)^{1/2}} = 12.39\ \text{ft}^{8/3}$$

But
$$R = \frac{A}{P} \qquad R^{2/3} = \left(\frac{A}{P}\right)^{2/3}$$

Then
$$AR^{2/3} = \frac{A^{5/3}}{P^{2/3}} = 12.39\ \text{ft}^{8/3}$$

For a pipe flowing full, $A = pD^2/4$ and $P = \pi D$, or

$$\frac{(pD^2/4)^{5/3}}{(pD)^{2/3}} = 12.39\ \text{ft}^{8/3}$$

Solving for diameter yields $D = 3.98$ ft $= 47.8$ in. Use the next commercial size larger, which is $D = 48$ in. ◁

$$A = \frac{pD^2}{4} = 50.3\ \text{ft}^2 \text{ (for pipe flowing full)}$$

Thus
$$V = \frac{Q}{A} = \frac{(110\ \text{ft}^3/\text{s})}{(50.3\ \text{ft}^2)} = 2.19\ \text{ft/s}$$ ◁

A culvert is a conduit placed under a fill such as a highway embankment. It is used to convey streamflow from the uphill side of the fill to the downhill side. Figure 10.21 shows the essential features of a culvert. Culverts are designed to pass the design discharge without adverse effects on the fill. That is, the culvert should be able to convey runoff from a design storm without overtopping the fill and without erosion of the fill at either the upstream or downstream end of the culvert. The design storm, for example, might be the maximum storm that could be expected to occur once in 50 years at the particular site.

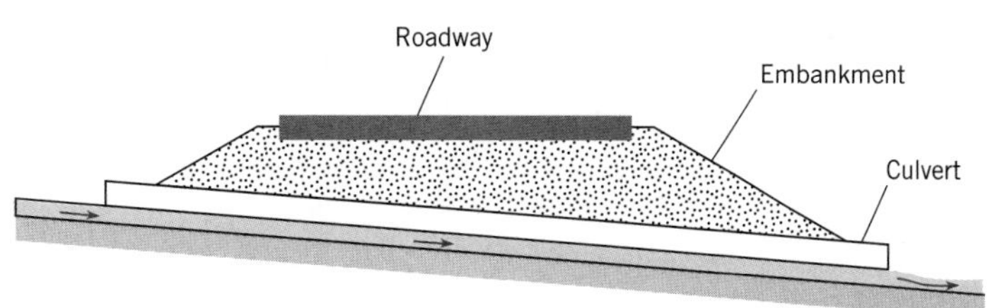

FIGURE 10.21

Culvert under a highway embankment.

The flow in a culvert is a function of many variables, including cross-sectional shape (circular or rectangular), slope, length, roughness, entrance design, and exit design. Flow in a culvert may occur as an open channel throughout its length, it may occur as a completely full pipe, or it may occur as a combination of both. The complete design and analysis of culverts are beyond the scope of this text; therefore, only a simple example is included here. For more extensive treatment of culverts, please refer to Chow (32), Henderson (33), and American Concrete Pipe Assoc. (34).

example 10.21

A 54-in.-diameter culvert laid under a highway embankment has a length of 200 ft and a slope of 0.01. This was designed to pass a 50-year flood flow of 225 cfs under full flow conditions (see the figure below). For these conditions, what head H is required? When the discharge is only 50 cfs, what will be the uniform flow depth in the culvert? Assume $n = 0.012$.

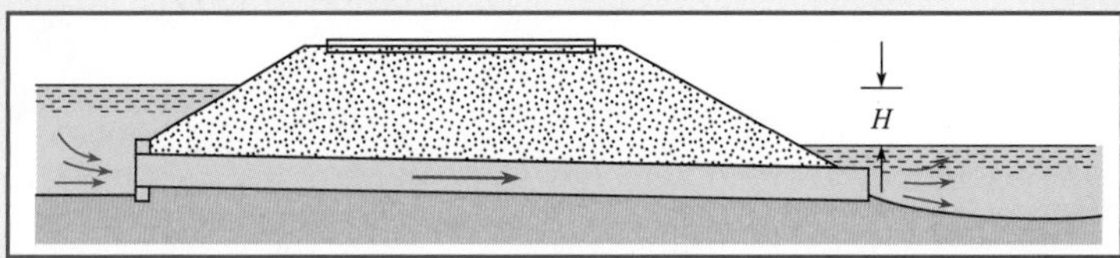

Solution For the flood flow of 225 cfs, one must consider the entrance and exit head losses as well as the head loss in the pipe itself; therefore, use the energy equation to solve this example.

$$\frac{p_1}{g} + \frac{V_1^2}{2g} + z_1 = \frac{p_2}{g} + \frac{V_2^2}{2g} + z_2 + \sum h_L$$

Let points 1 and 2 be at the upstream and downstream water surfaces, respectively.

Thus, $p_1 = p_2 = 0$ gage and $V_1 = V_2 = 0$

Also, $z_1 - z_2 = H$

Then we have

$$H = \sum h_L$$

$$H = \text{pipe head loss} + \text{entrance head loss} + \text{exit head loss}$$

$$H = \frac{V^2}{2g}(K_e + K_E) + \text{pipe head loss}$$

Assume

$$K_e = 0.50 \text{ (from Table 10.3)}$$

$$K_E = 1.00 \text{ (from Table 10.3)}$$

For the pipe head loss, use Eq. (10.45):

$$Q = \frac{1.49}{n} A R_h^{2/3} S_0^{1/2} \tag{10.45}$$

where

$$Q = 225 \text{ ft}^3/\text{s}$$

$$A = \frac{pD^2}{4} = 15.90 \text{ ft}^2$$

$$R_h = \frac{A}{P} = \frac{pD^2/4}{pD} = \frac{D}{4} = 1.125 \text{ ft}$$

$$R_h^{2/3} = (1.125 \text{ ft})^{2/3} = 1.0817 \text{ ft}^{2/3}$$

$$S_0 = \frac{h_f}{L}$$

Then Eq. (10.45) is written as

$$225 = \frac{1.49}{0.012}(15.90 \text{ ft}^2)(1.0817 \text{ ft}^{2/3})\left(\frac{h_f}{200}\right)^{1/2}$$

$$h_f = 2.22 \text{ ft}$$

$$V = \frac{Q}{A} = \frac{225 \text{ ft}^3/\text{s}}{15.90 \text{ ft}^2} = 14.15 \text{ ft/s}$$

Solving for H,

$$H = \frac{14.15^2}{64.4}(0.50 + 1.0) + 2.22$$

$$H = 4.66 \text{ ft} + 2.22 \text{ ft} = 6.88 \text{ ft} \quad \triangleleft$$

For $Q = 50$ cfs, we need to use Eq. (10.45):

$$50 = \frac{1.49}{0.012} A R_h^{2/3} (0.01)^{1/2}$$

However, this culvert will flow only partly full with a Q of 50 cfs. Therefore, the physical relationship will be as shown below.

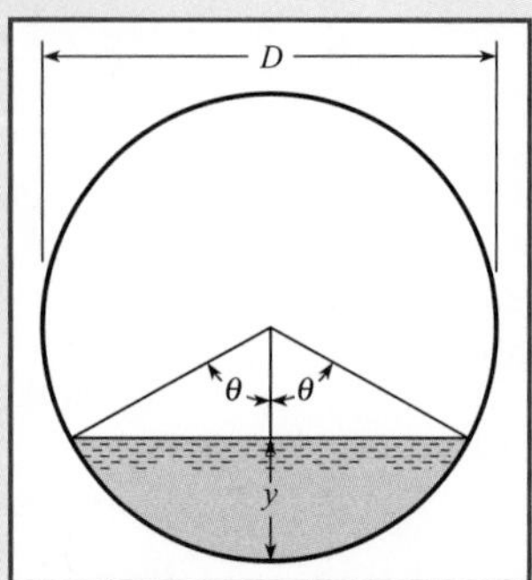

Thus, if the angle θ is given in degrees, the cross-sectional flow area will be given as

$$A = \left[\left(\frac{pD^2}{4}\right)\left(\frac{2\theta}{360^\circ}\right)\right] - \left(\frac{D}{2}\right)^2 (\sin u \ \cos u)$$

The wetted perimeter will be $P = \pi D(u/180^\circ)$, or

$$R_h = \frac{A}{P} = \left(\frac{D}{4}\right)\left[1 - \left(\frac{\sin u \ \cos u}{p(\theta/180^\circ)}\right)\right]$$

Substituting these relations for A and R_h into the discharge equation and solving for θ yields

$$u = 70^\circ$$

Depth of flow:

$$y = \frac{D}{2} - \frac{D}{2}\cos u = \left(\frac{54 \text{ in.}}{2}\right)(1 - 0.342) = 17.8 \text{ in.}$$ ◁

Summary

Flow in conduits is important to a wide variety of industries. As noted in Chapter 7, the analysis of a piping system requires information on the head (or pressure) loss to predict flow rates, power delivery, or power requirements for system operation. The head loss is given by the general equation

$$h_L = \sum f \frac{L}{D} \frac{V^2}{2g} + \sum K \frac{V^2}{2g}$$

where f is the Darcy-Weisbach friction factor, L is the pipe length, D is the diameter, V is the mean velocity and K is a loss coefficient. The first group of terms represents the loss due to friction in straight lengths of pipe. The second group of terms represents the head loss associated with components such as valves, elbows, bends, and transition sections.

The Darcy-Weisbach friction factor is a function of the pipe Reynolds number,

$$\text{Re} = \frac{VD}{n}$$

For Reynolds numbers less than 2000, the pipe flow is laminar. An analytic solution shows the velocity distribution is parabolic (Hagen-Poiseuille flow), and the Darcy-Weisbach friction factor is

$$f = \frac{64}{\text{Re}}$$

For Reynolds numbers greater than 3000, the flow is turbulent and characterized by a near uniform velocity profile with high velocity gradients near the pipe wall. The friction factor depends on the Reynolds number and the relative roughness:

$$f = f\left(\text{Re}, \frac{k_s}{D}\right)$$

where k_s is the equivalent sand grain roughness. The values for friction factor can be obtained from the Moody diagram or empirical equations. For a smooth pipe, the friction factor is independent of the relative roughness and depends only on the Reynolds number. For the fully rough condition, the friction factor is independent of the Reynolds number and depends only on the relative roughness.

The head loss coefficients may be obtained from tables and other sources of information for various flow components.

Noncircular pipes can be analyzed using the hydraulic radius, which is defined as

$$R_h = \frac{A}{P}$$

where A is the cross-sectional area of the conduit and P is the wetted perimeter. To analyze noncircular ducts, the diameter in the equations for circular pipes is replaced by $4R_h$.

The analysis of pipe networks is based on the continuity equation being satisfied at each junction and the head loss between any two junctions being independent of pipe path between the two junctions. A series of equations based on these principles are solved iteratively to obtain the flow rate in each pipe and the pressure at each junction in the network.

In an open-channel flow, the head loss corresponds to the potential energy change associated with the slope of the channel. The discharge in an open channel is given by the Chezy equation:

$$Q = \frac{1}{n} A R_h^{2/3} S_0^{1/2}$$

where A is the flow area, S_0 is the slope of the channel, and n is the resistance coefficient (Manning's n), which has been tabulated for different surfaces.

References

1. Reynolds, O. "An Experimental Investigation of the Circumstances Which Determine Whether the Motion of Water Shall Be Direct or Sinuous and of the Law of Resistance in Parallel Channels." *Phil. Trans. Roy. Soc. London,* 174, part III (1883).

2. Laufer, John. "The Structure of Turbulence in Fully Developed Pipe Flow." *NACA Rept.,* 1174 (1954).

3. Schlichting, Hermann. *Boundary Layer Theory,* 7th ed. McGraw-Hill, New York, 1979.

4. Rouse, Hunter, and Simon Ince. *History of Hydraulics.* Iowa Institute of Hydraulic Research, University of Iowa, 1957.

5. Moody, Lewis F. "Friction Factors for Pipe Flow." *Trans. ASME,* 671 (November 1944).

6. Nikuradse, J. "Strömungsgesetze in rauhen Rohren." *VDI-Forschungsh.,* no. 361 (1933). Also translated in *NACA Tech. Memo,* 1292.

7. Roberson, John A., and C. K. Chen. "Flow in Conduits with Low Roughness Concentration." *J. Hydraulics Div., Am. Soc. Civil Eng.,* 96, no. HY4 (April 1970).

8. Wright, S. J. "A Theory for Predicting the Resistance to Flow in Conduits with Nonuniform Roughness." M. S. thesis, Washington State University, Pullman, WA, 1973.

9. Calhoun, Roger G. "A Statistical Roughness Model for Computation of Large Bed-Element Stream Resistance." M.S. thesis, Washington State University, Pullman, Wash., 1975.

10. Kumar, S. "An Analytical Model for Computation of Rough Pipe Resistance." Ph.D. dissertation, Washington State University, Pullman, WA, 1979.

11. Eldridge, J. R. "An Analytical Method for Predicting Resistance to Flow in Rough Pipes and Open Channels." M.S. thesis, Washington State University, Pullman, WA, 1983.

12. Kumar, S. "An Analytical Method for Computation of Rough Boundary Resistance," Paper No. 5 in *Channel Flow Resistance.* Water Resources Publications, Littleton, CO, 1992.

13. White, F. M. *Viscous Fluid Flow.* McGraw-Hill, New York, 1991.

14. Colebrook, F. "Turbulent Flow in Pipes with Particular Reference to the Transition Region between the Smooth and Rough Pipe Laws." *J. Inst. Civ. Eng.,* vol. 11, 133–156 (1939).

15. Swamee, P. K., and A. K. Jain. "Explicit Equations for Pipe-Flow Problems." *J. Hydraulic Division of the ASCE,* vol. 102, no. HY5 (May 1976).

16. Streeter, V. L., and E. B. Wylie. *Fluid Mechanics,* 7th ed. McGraw-Hill, New York, 1979.

17. Barbin, A. R., and J. B. Jones. "Turbulent Flow in the Inlet Region of a Smooth Pipe." *Trans. ASME, Ser. D: J. Basic Eng.,* vol. 85, no. 1 (March 1963).

18. ASHRAE. *ASHRAE Handbook—1977 Fundamentals.* Am. Soc. of Heating, Refrigerating and Air Conditioning Engineers, Inc., New York, 1977.

19. Crane Co. "Flow of Fluids Through Valves, Fittings and Pipe." Technical Paper No. 410, Crane Co. (1988), 104 N. Chicago St., Joliet, IL 60434.

20. Fried, Irwin, and I. E. Idelchik. *Flow Resistance: A Design Guide for Engineers.* Hemisphere, New York, 1989.

21. Hydraulic Institute. *Engineering Data Book,* 2nd ed., Hydraulic Institute, 30200 Detroit Road, Cleveland, OH 44145.

22. Miller, D. S. *Internal Flow—A Guide to Losses in Pipe and Duct Systems.* British Hydrodynamic and Research Association (BHRA), Cranfield, England (1971).

23. Streeter, V. L. (ed.) *Handbook of Fluid Dynamics.* McGraw-Hill, New York, 1961.

24. Beij, K. H. "Pressure Losses for Fluid Flow in 90% Pipe Bends." *J. Res. Nat. Bur. Std.,* 21 (1938). Information cited in Streeter (28).

25. Cross, Hardy. "Analysis of Flow in Networks of Conduits or Conductors." *Univ. Illinois Bull.,* 286 (November 1936).

26. Hoag, Lyle N., and Gerald Weinberg. "Pipeline Network Analysis by Digital Computers." *J. Am. Water Works Assoc.,* 49 (1957).

27. Jeppson, Roland W. *Analysis of Flow in Pipe Networks.* Ann Arbor Science Publishers, Ann Arbor, MI, 1976.

28. Limerinos, J. T. "Determination of the Manning Coefficient from Measured Bed Roughness in Natural Channels." Water Supply Paper 1898-B, U.S. Geological Survey, Washington, DC, 1970.

29. Committee on Hydromechanics of the Hydraulics Division of American Society of Civil Engineers. "Friction Factors in Open Channels." *J. Hydraulics Div., Am. Soc. Civil Eng.* (March 1963).

30. Wolman, M. G. "The Natural Channel of Brandywine Creek, Pennsylvania." Prof. Paper 271, U.S. Geological Survey, Washington D.C., 1954.

31. Yen, B. C. (ed.) *Channel Flow Resistance: Centennial Of Manning's Formula.* Water Resources Publications, Littleton, CO, 1992.

32. Chow, Ven Te. *Open Channel Hydraulics.* McGraw-Hill, New York, 1959.

33. Henderson, F. M. *Open Channel Flow.* Macmillan, New York, 1966.

34. American Concrete Pipe Assoc. *Concrete Pipe Design Manual.* American Concrete Pipe Assoc., Vienna, Va., 1980.

Problems

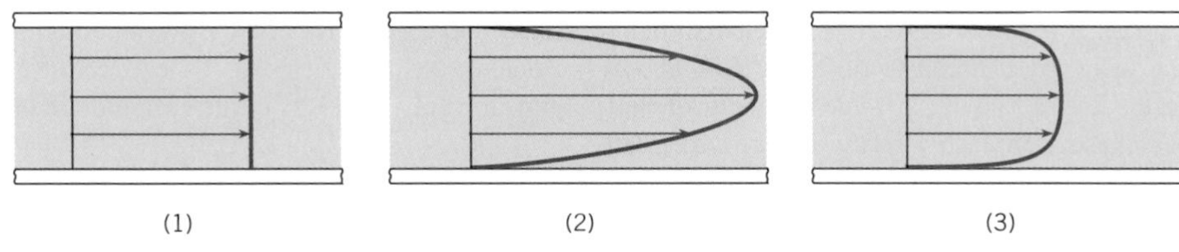

PROBLEM 10.1

10.1 Consider the mean-velocity profiles for flow in the pipes shown. Match the profiles with the following: (a) turbulent flow, (b) obviously a case of hypothetical flow (zero viscosity), (c) laminar case, (d) $\alpha = 1.0$, (e) $\alpha = 1.05$, (f) $\alpha = 2.00$.

10.2 Liquid in the pipe shown in the figure has a specific weight of 8 kN/m^3. The acceleration of the liquid is zero. Is the liquid stationary, moving upward, or moving downward in the pipe? If the pipe diameter is 1 cm and the liquid viscosity is 3.0×10^{-3} $\text{N} \cdot \text{s/m}^2$, what is the magnitude of the mean velocity in the pipe?

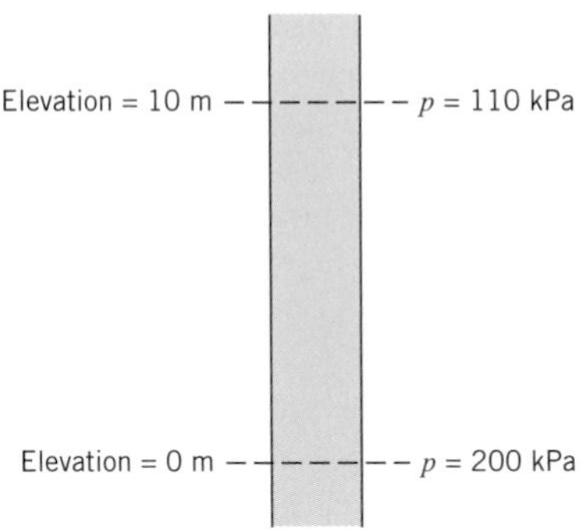

PROBLEM 10.2

10.3 A viscous oil is contained in this cylinder/nozzle system that has a vertical orientation. A valve is instantaneously opened to let the oil drain out of the cylinder. Below are listed words that might characterize the flow at point *A*. Which ones are valid characterizations at the time when the oil surface reaches the level of section 2? (a) unsteady, (b) steady, (c) irrotational, (d) rotational, (e) nonuniform, (f) uniform.

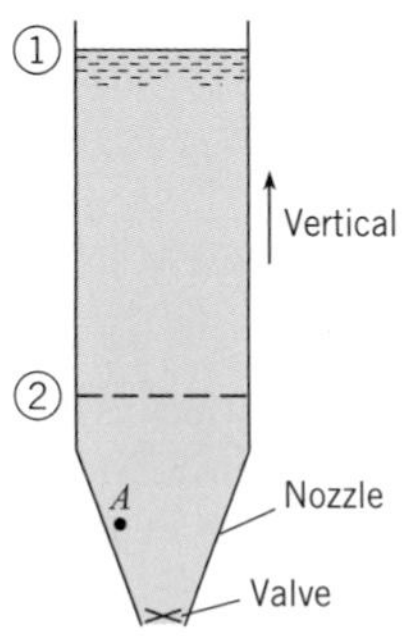

PROBLEM 10.3

10.4 Oil ($S = 0.97$, $m = 10^{-2}$ lbf-s/ft^2) is pumped through a 1-in. pipe at the rate of 0.05 cfs. What is the pressure drop per 100 ft of level pipe?

10.5 Liquid flows downward in a 1-cm, vertical, smooth pipe with a mean velocity of 2.0 m/s. The liquid has a density of 1000 kg/m^3 and a viscosity of 0.06 $\text{N} \cdot \text{s/m}^2$. If the pressure at a given section is 600 kPa, what will be the pressure at a section 10 m below that section?

10.6 A liquid ($\rho = 1000$ kg/m3, $\mu = 10^{-1}$ N · s/2 m^2, $v = 10^{-4}$ m^2/s) flows uniformly with a mean velocity of 1 m/s in a pipe with a diameter of 8 mm. For this condition, will the velocity distribution be logarithmic or parabolic? What will be the ratio of the shear stress at 1 mm from the wall to the shear stress on the wall?

10.7 Glycerine at a temperature of 30°C flows at a rate of 8×10^{-6} $\text{m}^3\text{/s}$ through a horizontal tube with a 30-mm diameter. What is the pressure drop in pascals per 10 m?

10.8 Kerosene ($S = 0.80$ and $T = 68°\text{F}$) flows from the tank shown and through the 1/4-in.-diameter (ID) tube. Determine the mean velocity in the tube and the discharge. Assume the only head loss is in the tube.

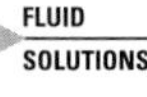

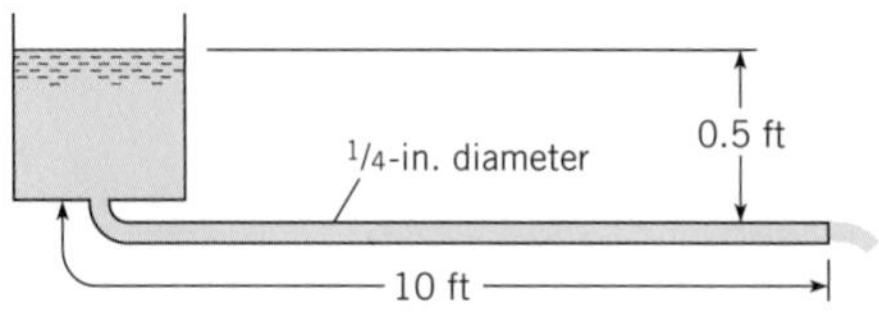

PROBLEM 10.8

10.9 Oil (S = 0.94, m = 0.048 N · s/m^2) is pumped through a horizontal 5-cm pipe. Mean velocity is 0.7 m/s. What is the pressure drop per 10 m of pipe?

FLUID SOLUTIONS **10.10** SAE 10W-30 oil is pumped through an 8-m length of 1-cm-diameter drawn tubing at a discharge of 7.85×10^{-4} m^3/s. There is a pump in the line as shown. The pipe is horizontal, and the pressures at points 1 and 2 are equal. Find the power necessary to operate the pump, assuming the pump has an efficiency of 100%. Properties of SAE 10W-30 oil: kinematic viscosity = 7.6×10^{-5} m^2/s, specific weight = 8630 N/m^3.

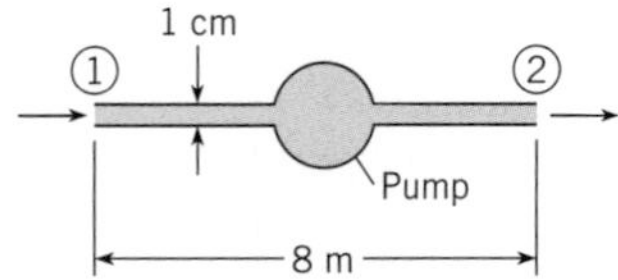

PROBLEM 10.10

10.11 Oil (S = 0.9; $\mu = 10^{-2}$ lbf-s/ft^2; ν = 0.0057 ft^2/s) flows downward in the pipe, which is 0.10 ft in diameter and has a slope of 30° with the horizontal. Mean velocity is 2 ft/s. What is the pressure gradient (dp/ds) along the pipe?

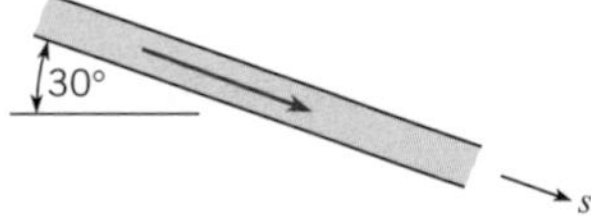

PROBLEM 10.11

10.12 A fluid ($\mu = 10^{-2}$ N · s/m^2; ρ = 800 kg/m^3) flows with a mean velocity of 5 cm/s in a 10-cm smooth pipe. Answer the following questions relating to the given flow conditions.

a. What is the magnitude of the maximum velocity in the pipe?
b. What is the magnitude of the resistance coefficient f?
c. What is the shear velocity for these flow conditions?
d. What is the shear stress at a radial distance of 25 mm from the center of the pipe?

10.13 Kerosene (20°C) flows at a rate of 0.04 m^3/s in a 25-cm pipe. Would you expect the flow to be laminar or turbulent?

10.14 In the pipe system for a given discharge, the ratio of the head loss in a given length of the 1-m pipe to the head loss in the same length of the 2-m pipe is (a) 2, (b) 4, (c) 16, (d) 32.

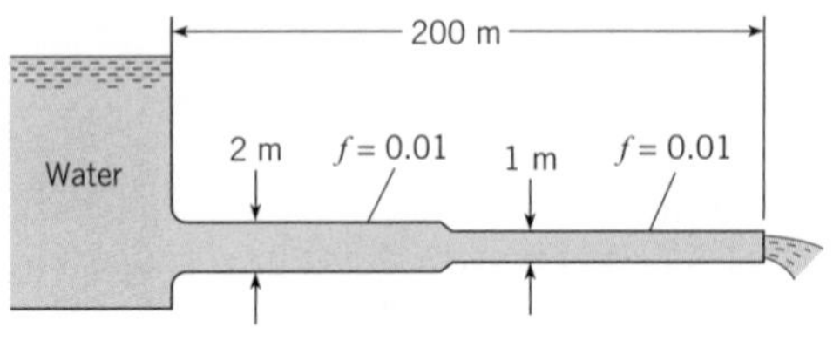

PROBLEM 10.14

10.15 Glycerine (T = 68°F) flows in a pipe with a 1/2-ft diameter at a mean velocity of 2 ft/s Is the flow laminar or turbulent? Plot the velocity distribution across the flow section.

10.16 Glycerine (T = 20°C) flows through a funnel as shown. Calculate the mean velocity of the glycerine exiting the tube. Assume the only head loss is due to friction in the tube.

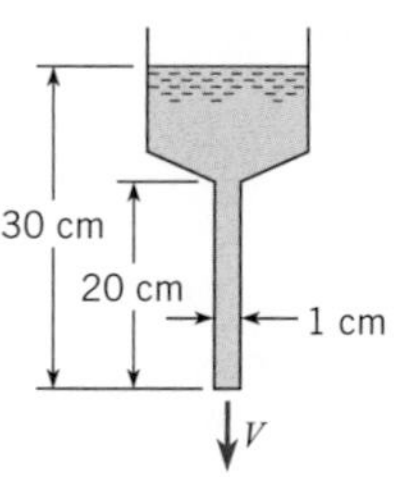

PROBLEM 10.16

10.17 What size of steel pipe should be used to carry 0.2 cfs of castor oil at 90°F a distance of 0.5 mi with an allowable pressure drop of 10 psi (μ = 0.085 lbf-s/ft^2)? Assume S = 0.85.

10.18 Mercury at 20°C flows downward in a long circular tube that is open to the atmosphere at the top and bottom. The tube is vertically oriented. Find the tube diameter for which the flow would just become turbulent (Re = 2000).

10.19 Glycerine (20°C) flows in a 4-cm steel tube with a mean velocity of 40 cm/s. Is the flow laminar or turbulent? What is the shear stress at the center of the tube and at the wall? If the tube is vertical and the flow is downward, will the pressure increase or decrease in the direction of flow? At what rate?

10.20 A small tank with a tube connected to it is to be used as a viscometer for liquids. Design a viscometer utilizing such equipment. State all assumptions and describe the procedure for the viscosity measurements. Assume that liquids to be measured will range from kerosene to glycerine.

10.21 Velocity measurements are made across a 1-ft pipe. The velocity at the center is found to be 2 fps, and the velocity distribution is seen to be parabolic. If the pressure drop is found to be 15 psf per 100 ft of pipe, what is the kinematic viscosity ν of the fluid? Assume that the fluid's specific gravity is 0.90.

10.22 Velocity measurements are made in a 30-cm pipe. The velocity at the center is found to be 1.5 m/s, and the velocity distribution is observed to be parabolic. If the pressure drop is found to be 1.9 kPa per 100 m of pipe, what is the kinematic viscosity ν of the fluid? Assume that the fluid's specific gravity is 0.80.

10.23 Water is pumped through a heat exchanger consisting of tubes 5 mm in diameter and 5 m long. The velocity in each tube is 12 cm/s. The water temperature increases from 20°C at the entrance to 30°C at the exit. Calculate the pressure difference across the heat exchanger, neglecting entrance losses but accounting for the effect of temperature change by using properties at average temperatures.

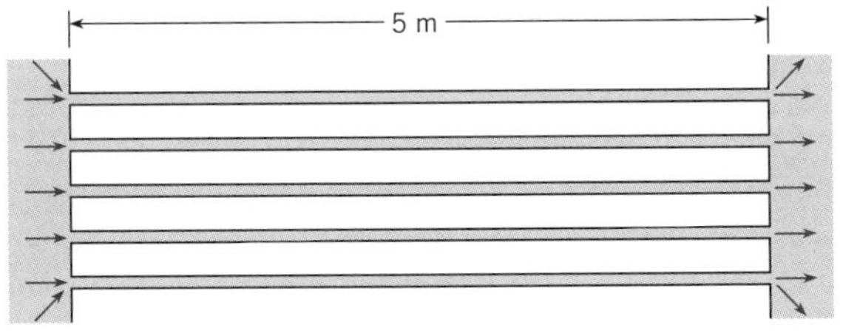

PROBLEM 10.23

10.24 The velocity of oil (S = 0.8) through the 2-in. smooth pipe is 5 ft/s. Here $L = 30$ ft, $z_1 = 2$ ft, $z_2 = 4$ ft, and the manometer deflection is 4 in. Determine the flow direction, the resistance coefficient f, whether the flow is laminar or turbulent, and the viscosity of the oil.

10.25 The velocity of oil (S = 0.8) through the 5-cm smooth pipe is 1.2 m/s. Here $L = 12$ m, $z_1 = 1$ m, $z_2 = 2$ m, and the manometer deflection is 10 cm. Determine the flow direction, the resistance coefficient f, whether the flow is laminar or turbulent, and the viscosity of the oil.

10.26 Flow of a liquid in a smooth 3-cm pipe is thought to yield a head loss of 2 m per meter of pipe when the mean velocity is 1 m/s. If the rate of flow was doubled, would the head loss also be doubled? Explain.

10.27 In a 12-in. smooth pipe, f is 0.017 when oil having a specific gravity of 0.82 flows with a mean velocity of 6 ft/s. What is the viscous shear stress on the wall?

10.28 Consider the flow of oil (ρ = 900 kg/m3; r μ = 10^{-1} N · s/m^2) in a 10-cm smooth pipe and the flow of a gas (r = 1.0 kg/m^3; m = 10^{-5} N · s/m^2) in a 10-cm smooth pipe. Both the oil and the gas flow with a mean velocity of 1 m/s. Will the ratio of the maximum velocity in the oil to the maximum velocity in the gas ($V_{\text{max,oil}}/V_{\text{max,gas}}$) be (a) greater than 1, (b) equal to 1, or (c) less than 1?

10.29 Water (50°F) flows with a speed of 5 ft/s through a horizontal run of PVC pipe. The length of the pipe is 100 ft, and the pipe is schedule 40 with a nominal diameter of 2.5 inches. Calculate (a) the pressure drop in psi, (b) the head loss in feet, and (c) the power in horsepower needed to overcome the head loss. (*Note:* PVC is a type of plastic. A 2.5-in. schedule 40 pipe has an inside diameter of 2.45 in.)

10.30 Water (10°C) flows with a speed of 2 m/s through a horizontal run of PVC pipe. The length of the pipe is 50 m and the pipe is schedule 40 with a nominal diameter of 2.5 inches. Calculate (a) the pressure drop in kilopascals, (b) the head loss in meters, and (c) the power in watts needed to overcome the head loss. (*Note:* PVC is a type of plastic. A 2.5-in. schedule 40 pipe has an inside diameter of 62.2 mm.)

10.31 Water (70°F) flows through a 6-in. smooth pipe at the rate of 2 cfs. What is the resistance coefficient f?

10.32 Water (10°C) flows through a 25-cm smooth pipe at a rate of 0.06 m^3/s. What is the resistance coefficient f?

10.33 Air flows in a 3-cm smooth tube at a rate of 0.015 m^3/s. If T = 20°C and p = 110 kPa absolute, what is the pressure drop per meter of length of tube? FLUID SOLUTIONS

10.34 Glycerine at 20°C flows at 0.6 m/s in the 2-cm commercial steel pipe. Two piezometers are used as shown to measure the piezometric head. The distance along the pipe between the standpipes is 1 m. The inclination of the pipe is 20°. What is the height difference Δh between the glycerine in the two standpipes?

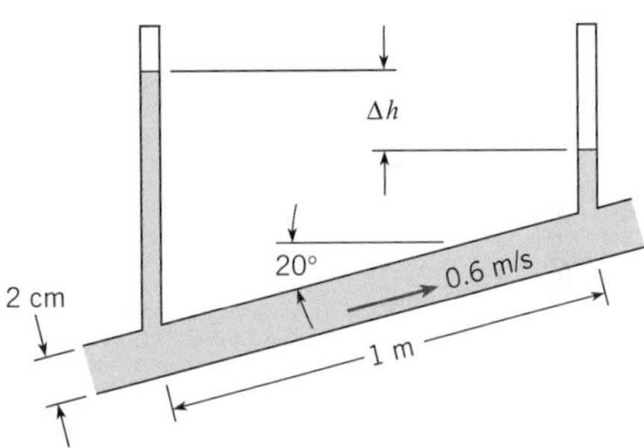

PROBLEM 10.34

10.35 Air flows in a 1-in. smooth tube at a rate of 30 cfm. If T = 80°F and p = 15 psia, what is the pressure drop per foot of length of tube?

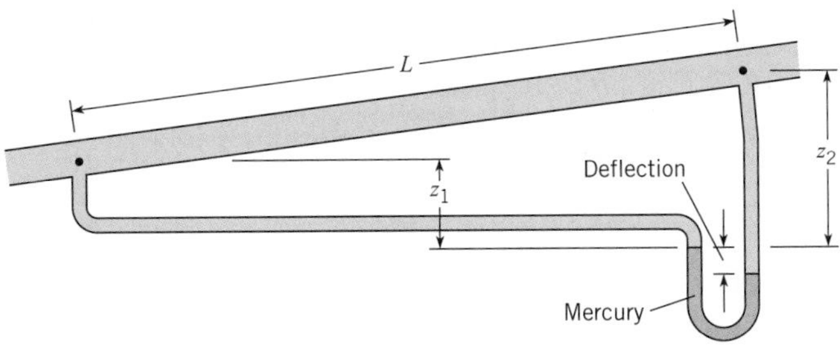

PROBLEMS 10.24, 10.25

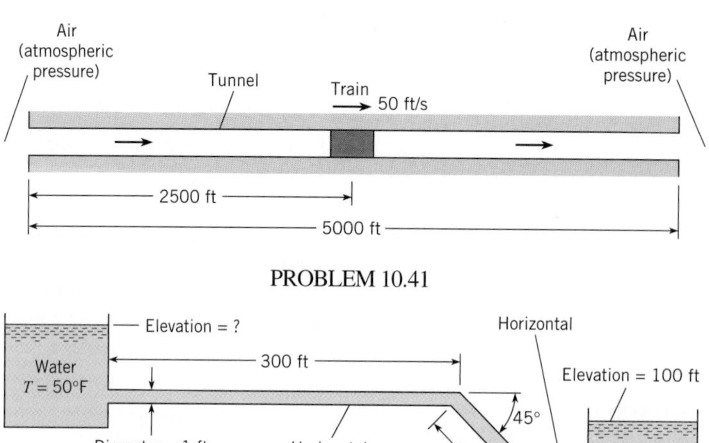

PROBLEM 10.41

PROBLEM 10.43

10.36 A pipe can be used to measure the viscosity of a fluid. A liquid flows in a 1-cm smooth pipe 1 m long with an average velocity of 3 m/s. A head loss of 50 cm is measured. Estimate the kinematic viscosity.

FLUID SOLUTIONS **10.37** Water flows in the pipe shown, and the manometer deflects 90 cm. What is f for the pipe if $V = 3$ m/s?

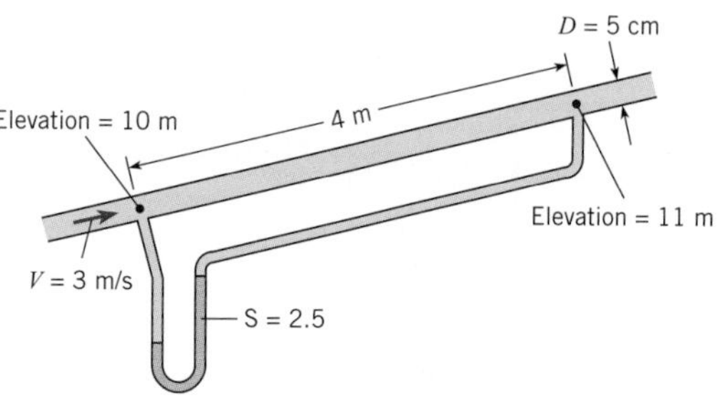

PROBLEM 10.37

10.38 What is f for the flow of water at 10°C through a 10-cm cast-iron pipe with a mean velocity of 4 m/s? Plot the velocity distribution for this flow.

10.39 Consider flow in a long, uniform-diameter pipe, $k_s = 10^{-4}$ ft. If the mean velocity is equal to 1 ft/s, if $d = 0.10$ ft, and if $\nu = 10^{-4}$ ft^2/s, which of the following values is the correct resistance coefficient f? (a) 0.064, (b) 0.044, (c) 0.034, (d) 0.020.

10.40 Water at 20°C flows through a 3-cm smooth brass tube at a rate of 0.002 m^3/s. What is f for this flow?

10.41 A train travels through a tunnel as shown. The train and tunnel are circular in cross section. Clearance is small, causing all air (60°F) to be pushed from the front of the train and discharged from the tunnel. The tunnel is 10 ft in diameter and is concrete. The train speed is 50 fps. Assume the concrete is very rough ($k_s = 0.05$ ft).

a. Determine the change in pressure between the front and rear of the train that is due to *pipe friction* effects.

b. Sketch the energy and hydraulic grade lines for the train position shown.

c. What power is required to produce the air flow in the tunnel?

10.42 Water will be siphoned through a 3/16-in.-diameter 50-in.-long Tygon tube from a jug on an upside-down wastebasket into a graduated cylinder as shown. The initial level of the water in the jug is 21 in. above the table top. The graduated cylinder is a 500-ml cylinder, and the water surface in the cylinder is 12 in. above the table top when the cylinder is full. The bottom of the cylinder is 1/2 in. above the table. The inside diameter of the jug is 7 in. Calculate the time it will take to fill the cylinder from an initial depth of 2 in. of water in the cylinder.

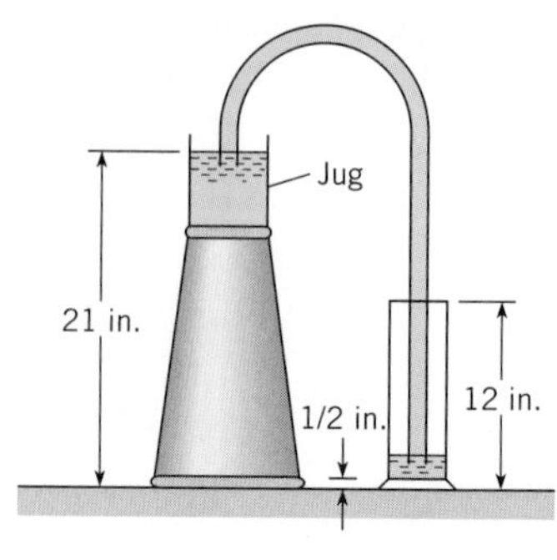

PROBLEM 10.42

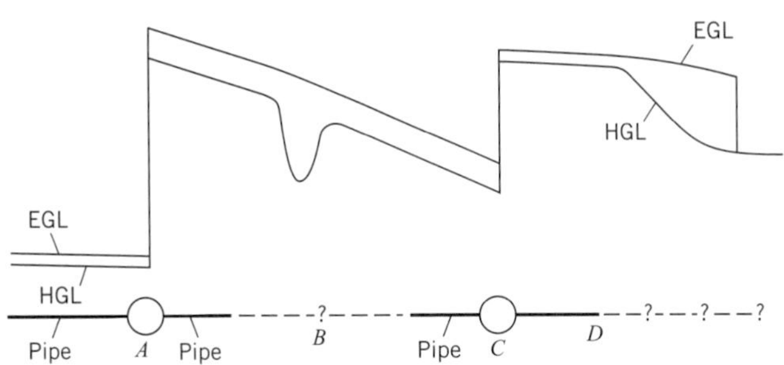

PROBLEM 10.47

10.43 Estimate the elevation required in the upper reservoir to produce a water discharge of 10 cfs in the system. Carefully draw the HGL and the EGL for the system. Where is the point of minimum pressure in the pipe, and what is the magnitude of the pressure at that point? What kind of pipe do you think this is?

10.44 A water turbine is connected to a reservoir as shown. The flow rate in this system is 5 cfs. What power can be delivered by the turbine if its efficiency is 80%? Assume a temperature of 70°F.

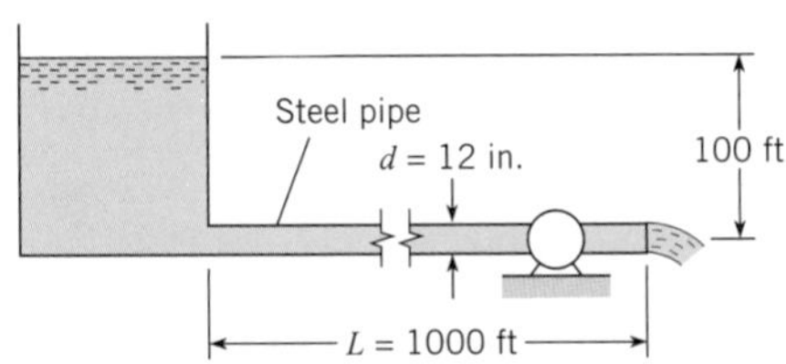

PROBLEM 10.44

10.45 A fluid ($\mu = 10^{-2}$ N · s/m^2; $\rho = 800$ kg/m^3) flows with a mean velocity of 500 mm/s in a 100-mm diameter smooth pipe. Answer the following questions relating to the given flow conditions.

a. What is the magnitude of the maximum velocity in the pipe?
b. What is the magnitude of the resistance coefficient f?
c. What is the shear velocity?
d. What is the shear stress at a radial distance of 25 mm from the center of the pipe?
e. If the discharge is doubled, will the head loss per length of pipe also be doubled?

10.46 Which of the following statements is true of the energy grade line for steady flow in pipes where no pumps or turbines are present?

a. The EGL always slopes downward in the direction of flow.
b. The EGL is steeper for a rough pipe than for a smooth pipe if the pipe diameter and discharge are equal.
c. In a given smooth pipe ($D = 1$ m and $\nu = 10^{-6}$ m^2/s), the slope of the EGL is doubled when the velocity is changed from 1 m/s to 2 m/s.
d. In a given smooth pipe ($D = 1$ m and $\nu = 10^{-2}$ m^2/s), the slope of the EGL is doubled when the velocity is changed from 1 m/s to 2 m/s.
e. In a given rough pipe ($D = 1$ m, $\nu = 10^{-6}$ m^2/s, $k_s = 1$ cm), the slope of the EGL is doubled when the velocity is changed from 1 m/s to 2 m/s.

10.47 Referring to the figure, what do you think are at A and C? What do you think is at B? Beyond D, complete the physical setup that could yield the EGL and HGL shown. What other information is indirectly revealed by the EGL and HGL?

10.48 Water (20°C) flows in a 15-cm cast-iron pipe at a rate of 0.05 m^3/s. For these conditions, determine or estimate the following:

a. Shear stress at the wall, τ_0
b. Shear stress 1 cm from the wall
c. Velocity 1 cm from the wall

10.49 Water flows from reservoir A to reservoir B. Reservoir A has a water-surface elevation of 100 m, and reservoir B has a water-surface elevation of 70 m. The two reservoirs are 100 m apart. Devise a conduit system to transport water from A to B in such a way that part of the conduit will have subatmospheric pressure in it. Assume the system contains neither pumps nor valves. Also, draw the EGL and the HGL for the system, and determine the magnitude of the minimum pressure.

10.50 Water is pumped through a vertical 10-cm new steel pipe to an elevated tank on the roof of a building. The pressure on the discharge side of the pump is 1.6 MPa. What pressure can be expected at a point in the pipe 80 m above the pump when the flow is 0.02 m^3/s? Assume $T = 20$°C.

10.51 Water is draining from a tank through a 1-in. pipe 10 ft long as shown. The pipe is galvanized iron, and the entrance is sharp edged. The distance from the water surface to the end of the pipe is 14 ft. The kinematic viscosity of the water is 1.22×10^{-5} ft^2/s. Calculate the velocity in the pipe.

10.52 A tank and piping system is shown. The galvanized pipe diameter is 1.5 cm and the total length of pipe is 10 m. The two 90° elbows are threaded fittings. The vertical distance from the water surface to the pipe outlet is 5 m. The velocity of the water

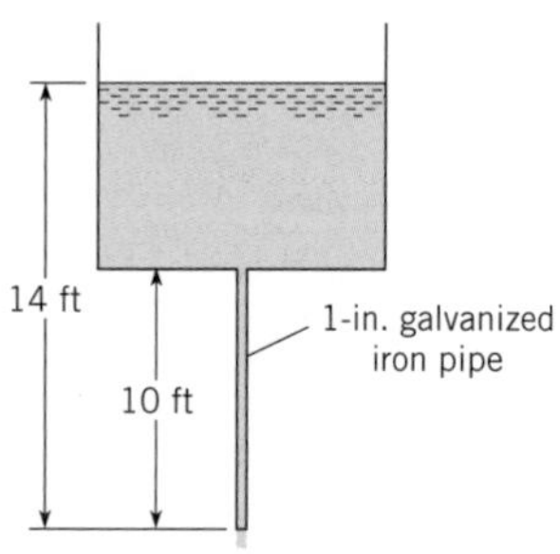

PROBLEM 10.51

in the tank is negligible. Find (a) the exit velocity of the water and (b) the height (h) the water jet would rise on exiting the pipe. The water temperature is 20°C.

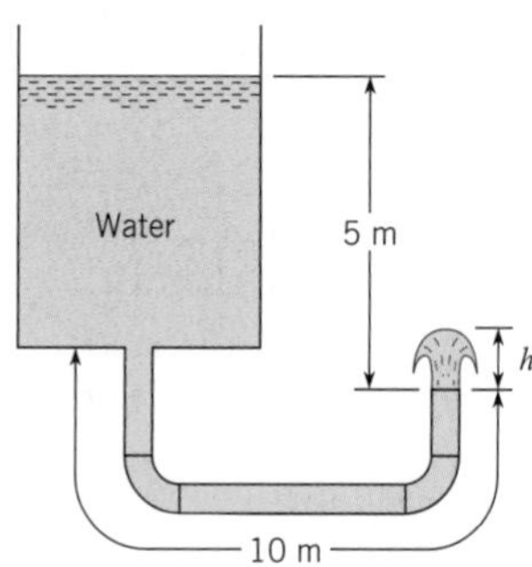

PROBLEM 10.52

10.53 Water (60°F) is pumped from a reservoir to a large, pressurized tank as shown. The steel pipe is 4 in. in diameter and 300 ft long. The discharge is 1 cfs. The initial water levels in the tanks are the same, but the pressure in tank B is 10 psig and tank A is open to the atmosphere. The pump efficiency is 90%. Find the power necessary to operate the pump for the given conditions. Draw the HGL and EGL for the system.

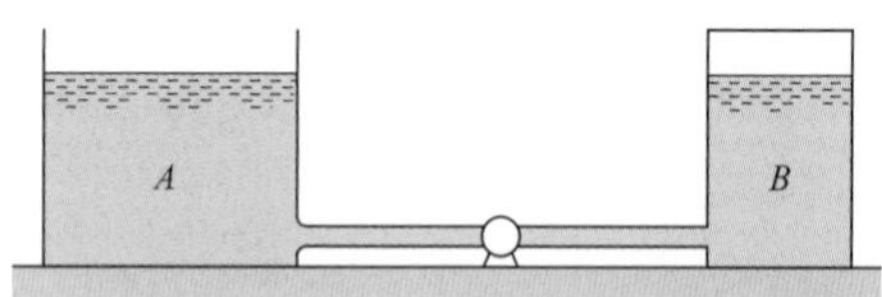

PROBLEM 10.53

10.54 A pump is used to fill a tank from a reservoir as shown. The head provided by the pump is given by $h_p = h_0(1 - (Q^2/Q_{\max}^2))$ where h_0 is 50 meters, Q is the discharge through the pump, and $Q_{\max}$ is 2 m^3/s. Assume $f = 0.018$ and the pipe diameter is 90 cm. Initially the water level in the tank is the same as the level in the reservoir. The cross-sectional area of the tank is 100 m^2. How long will it take to fill the tank to a height, h, of 40 m?

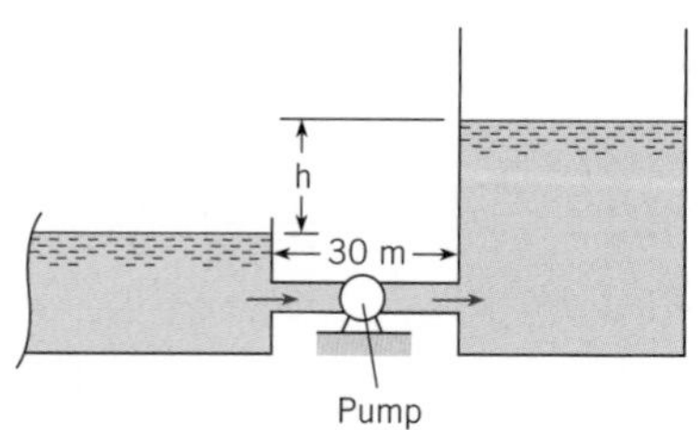

PROBLEM 10.54

10.55 Suppose that it is possible to prevent turbulence from developing in the flow in a pipe. What would be the ratio of the head loss for the laminar flow to that for the turbulent flow when kerosene ($S = 0.82$, $\nu = 2 \times 10^{-6}$ m^2/s) is pumped through a 3-cm smooth pipe with an average velocity of 4 m/s?

10.56 In a 4-in. uncoated cast-iron pipe, 0.02 cfs of water flows at 60°F. Determine f from Fig. 10.8.

10.57 Determine the head loss in 900 ft of a concrete pipe with a 6-in. diameter ($k_s = 0.0002$ ft) carrying 3.0 cfs of fluid. The properties of the fluid are $\nu = 3.33 \times 10^{-3}$ ft^2/s and $\rho = 1.5$ slug/ft^3.

10.58 Points A and B are 1 km apart along a 15-cm new steel pipe. Point B is 20 m higher than A. With a flow from A to B of 0.03 m^3/s of crude oil ($S = 0.82$) at 10°C ($\mu = 10^{-2}$ N · s/m^2), what pressure must be maintained at A if the pressure at B is to be 300 kPa?

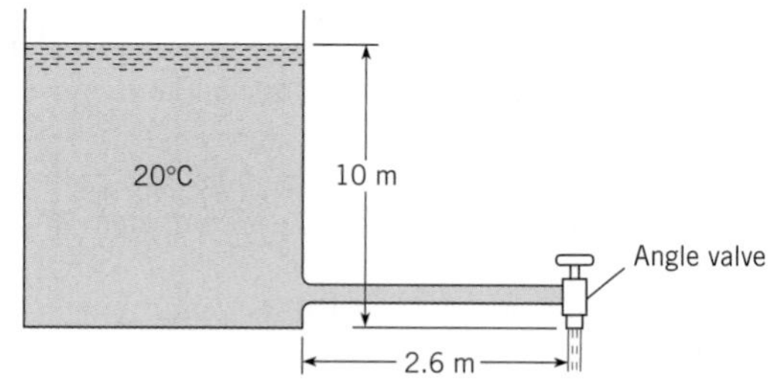

PROBLEM 10.59

10.59 Water flows from a tank through a 2.6-m length of galvanized-iron pipe 26 mm in diameter. At the end of the pipe is an angle valve that is wide open. The height of the water level in the tank above the valve exit is 10 m.

The tank is 2 m in diameter. Calculate the time required to reduce the level in the tank from 10 m to 2 m. Use Eq. (10.26) for f. Write the energy equation and use some numerical scheme or software to solve for the discharge as a function of water level in the tank. Then use the equation

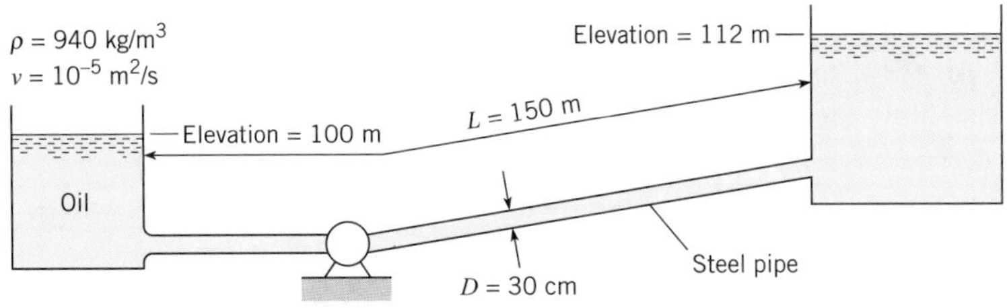

PROBLEM 10.64

$$\frac{dh}{dt} = -\frac{Q}{A}$$

where A is the cross-sectional area of the tank, to relate the discharge to the rate of decrease of the water level in the tank.

10.60 Points A and B are 3 mi apart along a 24-in. new cast-iron pipe carrying water ($T = 50°\text{F}$). Point A is 30 ft higher than B. The pressure at B is 20 psi greater than that at A. Determine the direction and rate of flow.

10.61 If the flow of $0.10\ \text{m}^3/\text{s}$ of water is to be maintained in the system shown, what power must be added to the water by the pump? The pipe is made of steel and is 15 cm in diameter. Draw the EGL and the HGL for the system.

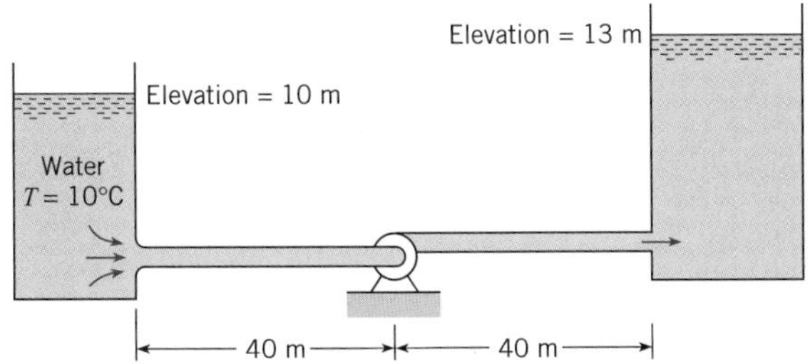

PROBLEM 10.61

10.62 Water at 10°C flows from a reservoir (water surface at 120 m elevation) through 10 km of concrete pipe of 1.0 m diameter and discharges into another reservoir (water surface at 20 m elevation). Neglecting inlet and exit losses, what is the discharge? If riveted steel pipe of the same diameter were used, what would be the discharge? What pump power (in megawatts) would be required to produce a discharge of $2.8\ \text{m}^3/\text{s}$ in the reverse direction (uphill) in the concrete pipe?

10.63 A fluid with $\nu = 10^{-6}\ \text{m}^2/\text{s}$ and $\rho = 800\ \text{kg/m}^3$ flows through the 8-cm galvanized-iron pipe. Estimate the flow rate for the conditions shown in the figure.

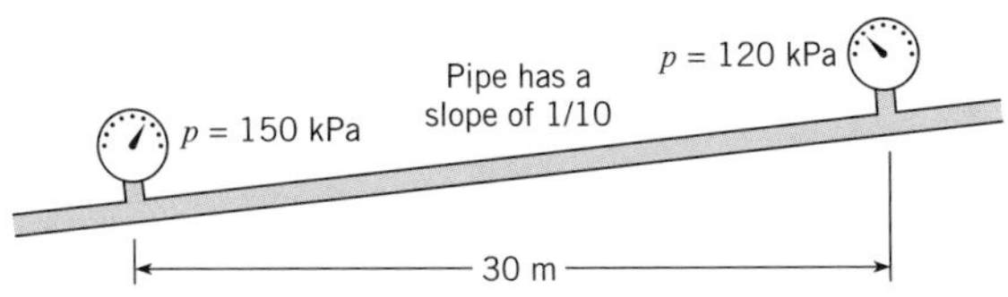

PROBLEM 10.63

10.64 What power must the pump supply to the system to pump the oil from the lower reservoir to the upper reservoir at a rate of $0.20\ \text{m}^3/\text{s}$? Sketch the HGL and the EGL for the system. FLUID SOLUTIONS

10.65 For a 40-cm pipe the resistance coefficient f was found to be 0.06 when the mean velocity was 3 m/s and the kinematic viscosity was $10^{-5}\ \text{m}^2/\text{s}$. If the velocity were doubled, would you expect the head loss per meter of length of pipe to double, triple, or quadruple?

10.66 A cast-iron pipe 1.0 ft in diameter and 200 ft long joins two water (60°F) reservoirs. The upper reservoir has a water-surface elevation of 100 ft, and the lower one has a water-surface elevation of 40 ft. The pipe exits from the side of the upper reservoir at an elevation of 70 ft and enters the lower reservoir at an elevation of 30 ft. There are two wide-open gate valves in the pipe. Draw the EGL and the HGL for the system, and determine the discharge in the pipe.

10.67 An engineer is making an estimate of hydroelectric power for a home owner. This owner has a small stream ($Q = 2$ cfs, $T = 40°\text{F}$) that is located at an elevation $H = 34$ ft above the owner's residence. The owner is proposing to divert the stream and operate a water turbine connected to an electric generator to supply electrical power to the residence. The maximum acceptable head loss in the penstock (a penstock is a conduit that supplies a turbine) is 3 ft. The penstock has a length of 87 ft. If the penstock is going to be fabricated from commercial-grade, plastic pipe, find the minimum diameter that can be used. Neglect minor losses associated with flow through the penstock. Assume that pipes are available in even sizes—that is, 2 in., 4 in., 6 in., etc.

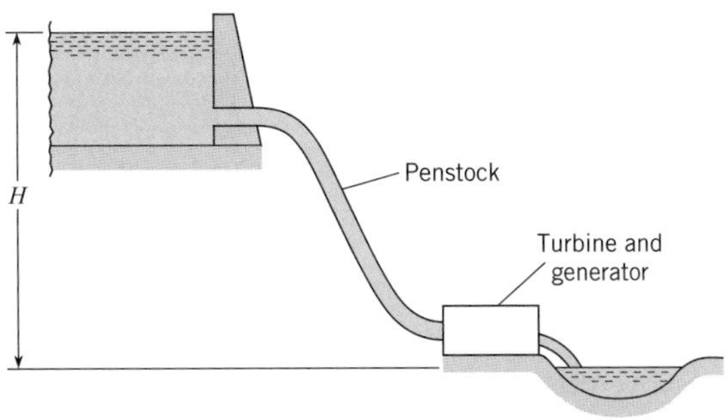

PROBLEM 10.67

10.68 Determine the diameter of commercial steel pipe required to convey 300 cfs of water at 60°F with a head loss of 1 ft per 1000 ft of pipe. Assume pipes are available in the even sizes when the diameters are expressed in inches (that is, 10 in., 12 in., etc.).

10.69 A pipeline is to be designed to carry crude oil (S = 0.93, $\nu = 10^{-5}$ m^2/s) with a discharge of 0.10 m^3/s and a head loss per kilometer of 50 m. What diameter of steel pipe is needed? What power output from a pump is required to maintain this flow? Available pipe diameters are 20, 22, and 24 cm.

10.70 Design a pipe to carry water at a rate of 15 cfs between two reservoirs if the reservoirs are 3 mi apart and the elevation difference between their water surfaces is to be 30 ft. Assume pipes are available in even sizes (that is, diameters of 16 in., 18 in., etc.).

10.71 Utilize the configuration shown in Problem 7.78 to design a cavitation demonstration device. Assume that the conduit with diameter D will be made of clear plastic so that the cavitation phenomenon may be visually observed and assume that the pressure in the water main from which you will draw water for your experiment is 50 psig.

10.72 The water-surface elevation in a reservoir is 120 ft. A straight pipe 100 ft long and 6 in. in diameter conveys water from the reservoir to an open drain. The pipe entrance (it is abrupt) is at elevation 100 ft, and the pipe outlet is at elevation 70 ft. At the outlet the water discharges freely into the air. The water temperature is 50°F. If the pipe is asphalted cast iron, what will be the discharge rate in the pipe? Consider all head losses. Also draw the HGL and the EGL for this system.

FLUID SOLUTIONS **10.73** The water-surface elevation in a reservoir is 600 ft. A straight pipe 1200 ft long and 2 ft in diameter conveys water from the reservoir to an open drain. The pipe entrance (it is rounded with $r/d = 0.1$) is at elevation 570 ft, and the pipe outlet is at elevation 200 ft. At the outlet the water discharges freely into the air. The water temperature is 50°F. The discharge of water in the pipe is 157 cfs. Determine the minimum pressure in the pipe and the equivalent sand roughness, k_s, of the pipe.

10.74 A heat exchanger is being designed as a component of a geothermal power system in which heat is transferred from the geothermal brine to a "clean" fluid in a closed-loop power cycle. The heat exchanger, a shell-and-tube type, consists of 100 galvanized-iron tubes 2 cm in diameter and 5 m long, as shown. The temperature of the fluid is 200°C, the density is 860 kg/m^3, and the viscosity is 1.35×10^{-4} N · s/m^2. The total mass flow rate through the exchanger is 50 kg/s.

a. Calculate the power required to operate the heat exchanger, neglecting entrance and outlet losses.

b. After continued use, 2 mm of scale develops on the inside surfaces of the tubes. This scale has an equivalent roughness of 0.5 mm. Calculate the power required under these conditions.

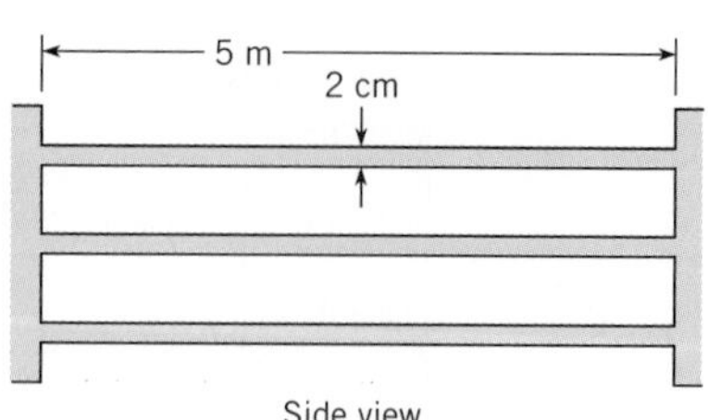

PROBLEM 10.74

10.75 The heat exchanger shown consists of 20 m of drawn tubing 2 cm in diameter with 19 return bends. Water is pumped through the system at 3×10^{-4} m^3/s, entering at 20°C and exiting at 80°C. The elevation difference between the entrance and the exit is 0.8 m. Calculate the pump power required to operate the heat exchanger if the pressure at 1 equals the pressure at 2. Use the viscosity corresponding to the average temperature in the heat exchanger.

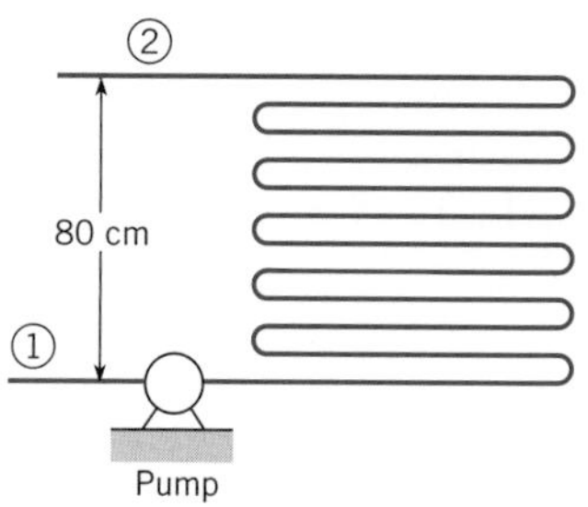

PROBLEM 10.75

10.76 A heat exchanger consists of a closed system with a series of parallel tubes connected by 180° elbows as shown in the figure. There are a total of 14 return elbows. The pipe diameter is 2 cm and the total pipe length is 10 m. The head loss coefficient for each return elbow is 2.2. The tube is copper. Water with an average temperature of 40°C flows through the system with a mean velocity of 10 m/s. Find the power required to operate the pump if the pump is 80% efficient.

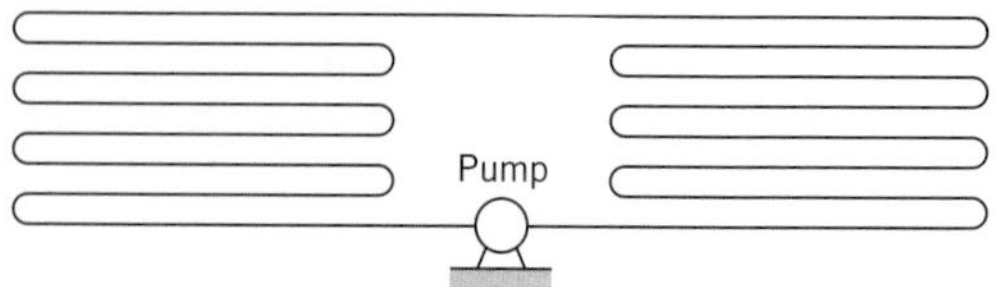

PROBLEMS 10.76, 10.77

10.77 A heat exchanger consists of 15 m of copper tubing with an internal diameter of 15 mm. There are 14 return elbows in the system with a loss coefficient of 2.2 for each elbow. The pump in the system has a pump curve given by

$$h_p = h_{p0}\left[1 - \left(\frac{Q}{Q_{max}}\right)^3\right]$$

where h_{p0} is head provided by the pump at zero discharge and Q_{max} is 10^{-3} m^3/s. Water at 40°C flows through the system. Find the system operating point for values of h_{p0} of 2, 10, and 20 m. Use Eq. (10.26) for the Darcy-Weisbach friction factor. Set up the equations to solve using some available software or root-finding scheme.

10.78 Determine the discharge of water through the system shown. In addition,

a. Draw the HGL and the EGL for the system.
b. Locate the point of maximum pressure.
c. Locate the point of minimum pressure.
d. Calculate the maximum and minimum pressures in the system.

10.79 Gasoline (T = 50°F) is pumped from the gas tank of an automobile to the carburetor through a 1/4-in. fuel line of drawn tubing 10 ft long. The line has five 90° smooth bends with an r/d of 6. The gasoline discharges through a 1/32-in. jet in the carburetor to a pressure of 14 psia. The pressure in the tank is 14.7 psia. The pump is 80% efficient. What power must be supplied to the pump if the automobile is consuming fuel at the rate of 0.12 gpm? Obtain gasoline properties from Figs. A.2 and A.3 in the Appendix.

10.80 Find the loss coefficient K_v of the partially closed valve that is required to reduce the discharge to 50% of the flow with the valve wide open as shown.

10.81 The pressure at a water main is 300 kPa gage. What size of pipe is needed to carry water from the main at a rate of 0.025 m^3/s to a factory that is 160 m from the main? Assume that galvanized-steel pipe is to be used and that the pressure required at the factory is 60 kPa gage at a point 10 m above the main connection.

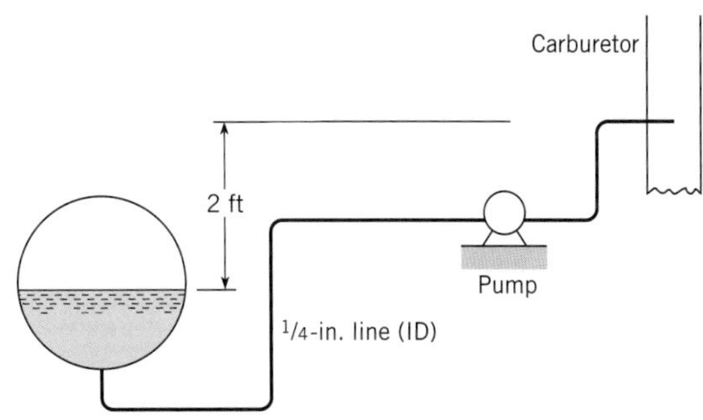

PROBLEM 10.79

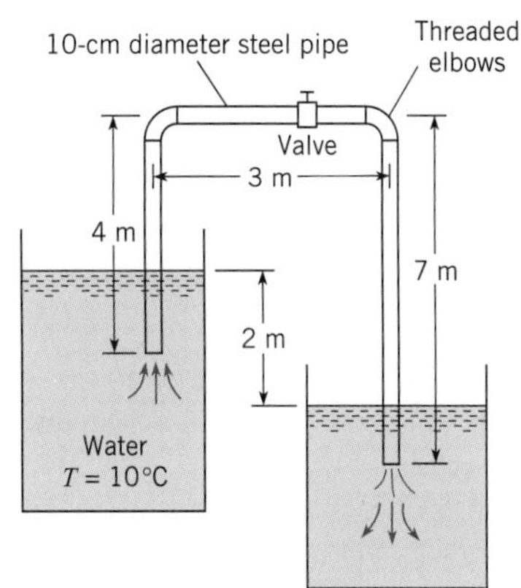

PROBLEM 10.80

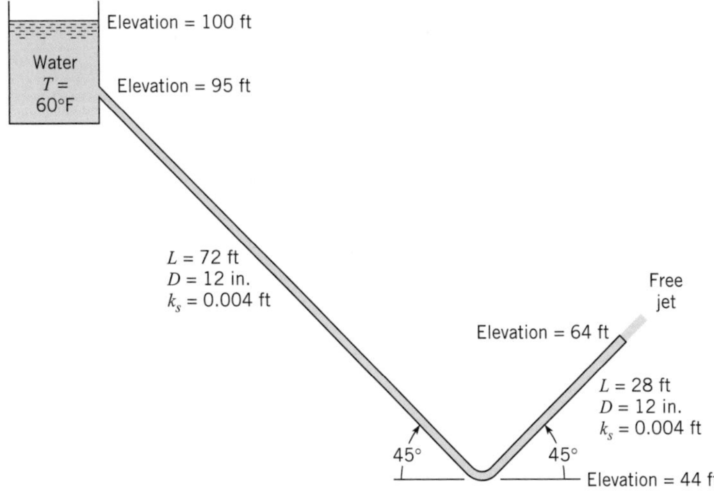

PROBLEM 10.78

10.82 Two reservoirs with a difference in water-surface elevation of 11 ft are joined by 50 ft of 1-ft steel pipe and 30 ft of 6-in. steel pipe in series. The 1-ft line contains three bends $(r/d = 1)$, and the 6-in. line contains two bends $(r/d = 4)$. If the 1-ft and 6-in. lines are joined by an abrupt contraction, determine the discharge. Assume $T = 60°F$.

10.83 The 10-cm galvanized-steel pipe is 1000 m long and discharges water into the atmosphere. The pipeline has an open globe valve and four threaded elbows; $h_1 = 3$ m and $h_2 = 15$ m. What is the discharge, and what is the pressure at A, the midpoint of the line?

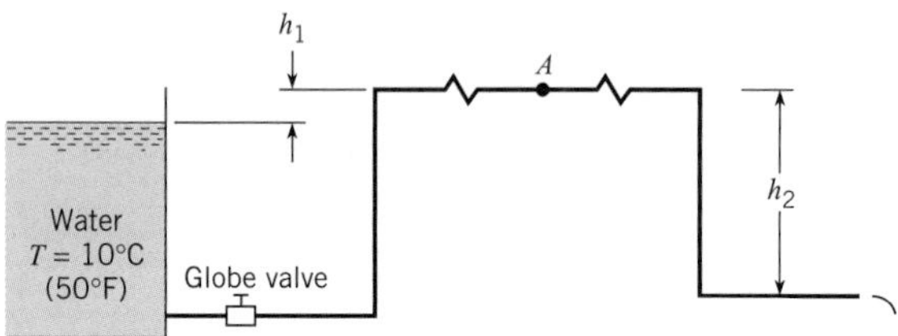

PROBLEM 10.83

10.84 Air (20°C) flows with a speed of 10 m/s through a horizontal rectangular air-conditioning duct. The duct is 20 m long and has a cross section of 4 by 10 in. (102 by 254 mm). Calculate (a) the pressure drop in inches of water and (b) the power in watts needed to overcome head loss. Assume the roughness of the duct is $k_s = 0.004$ mm. Neglect minor losses.

FLUID SOLUTIONS **10.85** The sketch shows a test of an electrostatic air filter. The pressure drop for the filter is 3 inches of water when the air speed is 10 m/s. What is the minor loss coefficient for the filter? Assume air properties at 20°C.

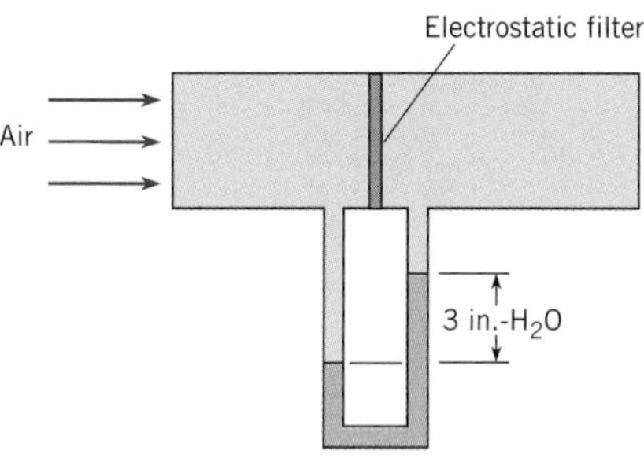

PROBLEM 10.85

10.86 What power must be supplied by the pump to the flow if water ($T = 20°C$) is pumped through the 300-mm steel pipe from the lower tank to the upper one at a rate of 0.314 m^3/s?

10.87 If the pump for Fig. 10.16 is installed in the system of Prob. 10.86, what will be the rate of discharge of water from the lower tank to the upper one?

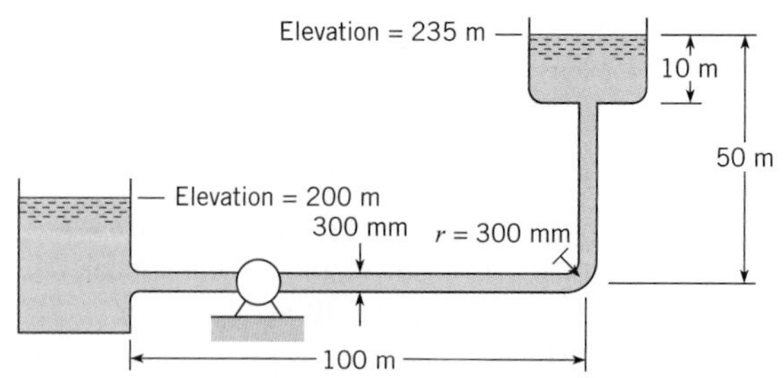

PROBLEM 10.86

10.88 The 3-in.-diameter injector pipe is intended to make the system operate like a jet pump. If water velocity in the injector pipe is 60 ft/s, will the system operate as a pump? In other words, will water be drawn from the lower reservoir and discharged into the upper reservoir? Give computations and/or statements to support your conclusion.

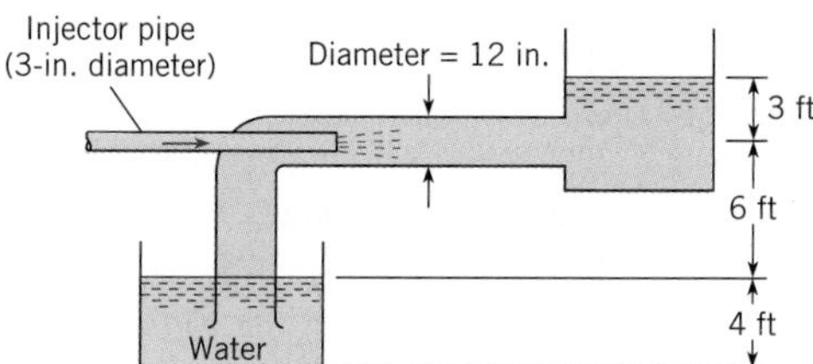

PROBLEM 10.88

10.89 A pump that has the characteristic curve shown in the accompanying graph is to be installed as shown. What will be the discharge of water in the system?

10.90 If the liquid of Prob. 10.89 is a superliquid (zero head loss occurs with the flow of this liquid), then what will be the pumping rate, assuming that the pump curve is the same?

10.91 Water is pumped at a rate of 25 m^3/s from the reservoir and out through the pipe, which has a diameter of 1.50 m. What power must be supplied to the water to effect this discharge?

10.92 Both pipes shown have an equivalent sand roughness k_s of 0.10 mm and a discharge of 0.1 m^3/s. Also, $D_1 = 15$ cm, $L_1 = 50$ m, $D_2 = 30$ cm, and $L_2 = 160$ m. Determine the difference in the water-surface elevation between the two reservoirs.

10.93 Both pipes shown have an equivalent sand roughness k_s of 4×10^{-4} ft and a discharge of 5 cfs. Also, $D_1 = 6$ in., $L_1 = 150$ ft, $D_2 = 12$ in., and $L_2 = 500$ ft. Determine the difference in water-surface elevation between the two reservoirs.

10.94 Estimate the discharge of oil through the pipe shown. Also, draw the HGL and the EGL for the system.

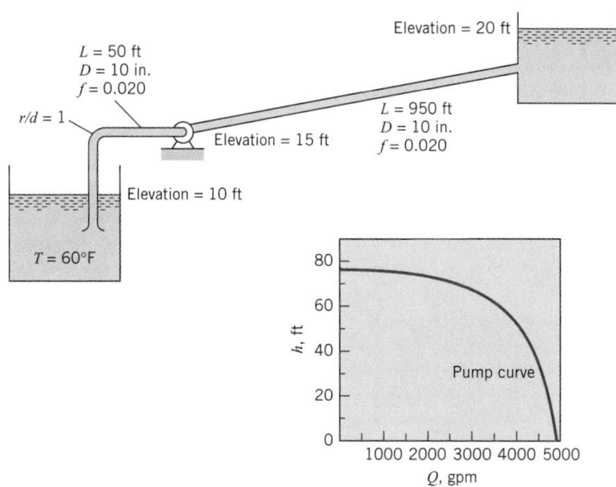

PROBLEMS 10.89, 10.90

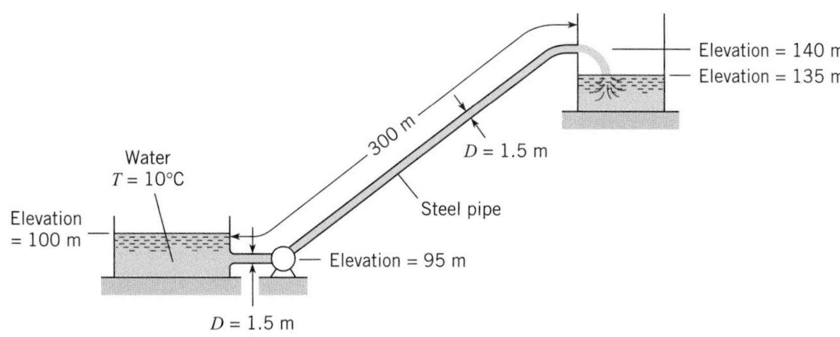

PROBLEM 10.91

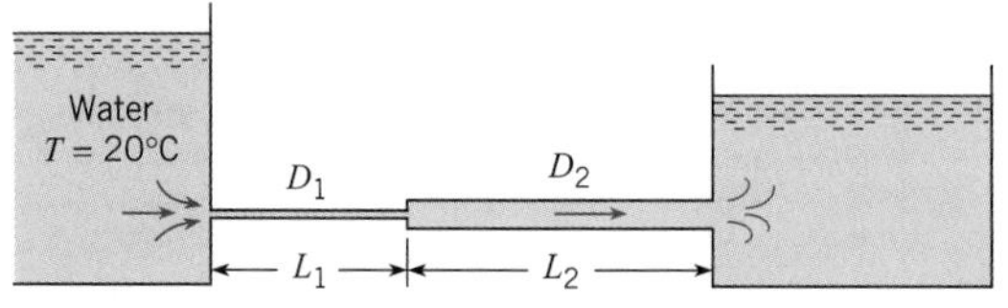

PROBLEMS 10.92, 10.93

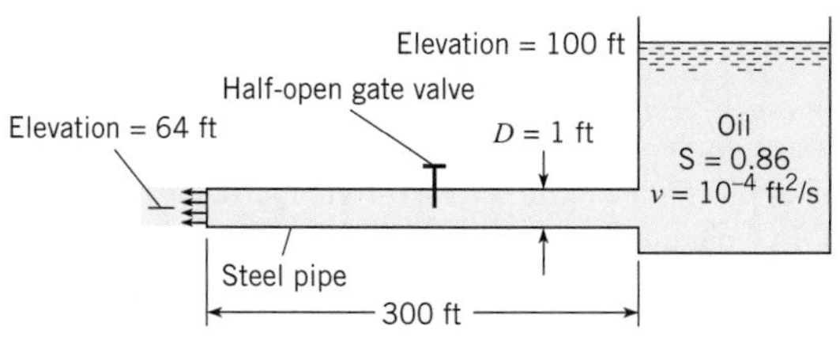

PROBLEM 10.94

10.95 If the deluge through the system shown is 2.0 cfs, what horsepower is the pump supplying to the water? Draw the HGL and the EGL for the system, and determine the water pressure at the midpoint of the long pipe. The four bends have a radius of 12 in., and the 6-in. pipe is smooth.

10.96 If the pump efficiency is 80%, what power must be supplied to the pump in order to pump fuel oil (S = 0.94) at a rate of 1.2 m^3/s up to the high reservoir? Assume that the conduit is a steel pipe and $\nu = 5 \times 10^{-5}$ m^2/s.

10.97 Determine the elevation of the water surface in the upstream reservoir if the discharge in the system is 0.15 m^3/s. Carefully sketch the HGL and the EGL, showing relative magnitudes and slopes. Label $V^2/2g$, ρ/γ, and z at section A-A.

10.98 Liquid discharges from a tank through the piping system shown. There is a venturi section at A and a sudden contraction at B. The liquid discharges to the atmosphere. Sketch the energy and hydraulic gradelines. Where might cavitation occur?

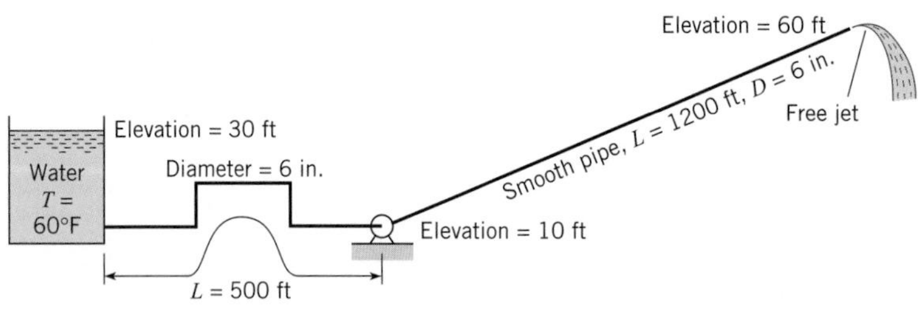

PROBLEM 10.95

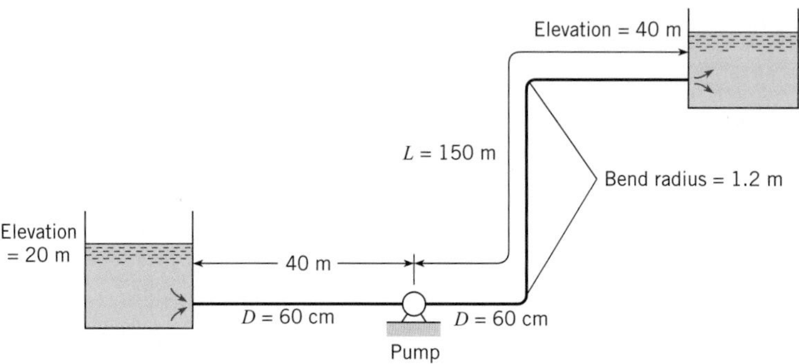

PROBLEM 10.96

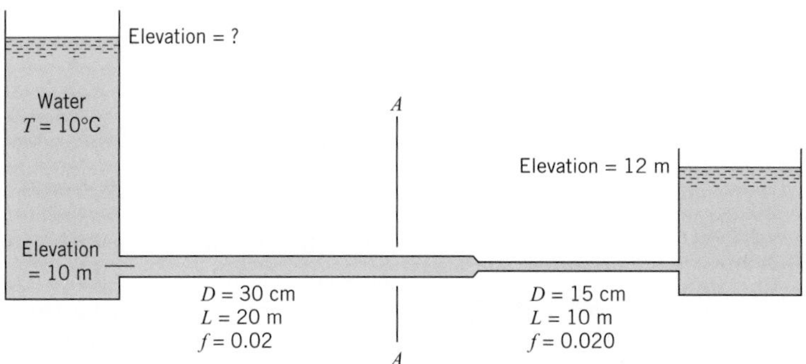

PROBLEM 10.97

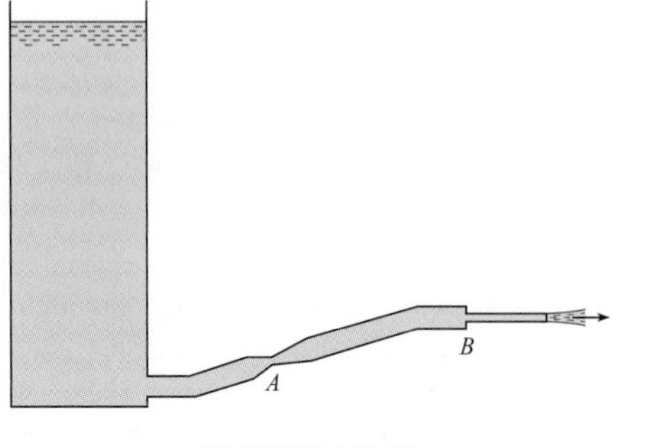

PROBLEM 10.98

10.99 The steel pipe shown carries water from the main pipe *A* to the reservoir and is 2 in. in diameter and 240 ft long. What must be the pressure in pipe *A* to provide a flow of 50 gpm?

10.100 If the water surface elevation in reservoir *B* is 110 m, what must be the water surface elevation in reservoir *A* if a flow of 0.03 m^3/s is to occur in the cast-iron pipe? Draw the HGL and the EGL, including relative slopes and changes in slope.

10.101 Design a pipe system to supply water flow from the elevated tank to the reservoir at a discharge of 2.5 m^3/s. The water surface in the tank is 50 m above the water surface in the reservoir. Write a computer program to solve this type of problem, and then run it to get the desired solution.

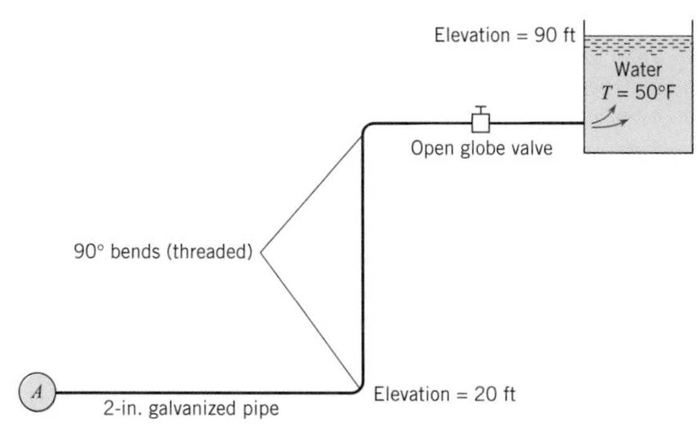

PROBLEM 10.99

10.102 Design a system to supply water flow from the reservoir to the elevated tank of Prob. 10.101 at a discharge of 1.0 m^3/s.

10.103 Design laboratory equipment to illustrate the phenomenon of cavitation utilizing the concept shown in Fig. 5.11. Assume that you have a water source from a main pipe with a pressure of 50 psig.

10.104 Design equipment and an experimental procedure to verify the momentum principle. The guidelines are:

a. A laboratory water source with a pressure of 50 psig in the main pipe is available to you.
b. A jet of water is to be involved in your experiment.
c. A force produced by the momentum effect is to be measured.

10.105 A pipe system consists of a gate valve, wide open ($K_v = 0.2$), in line A and a globe valve, wide open ($K_v = 10$), in line B. The cross-sectional area of pipe A is half of the cross-sectional area of pipe B. The head loss due to the junction, elbows, and pipe friction are negligible compared with the head loss through the valves. Find the ratio of the discharge in line B to that in line A.

10.106 A flow is divided into two branches as shown. A gate valve, half open, is installed in line A, and a globe valve, fully open, is installed in line B. The head loss due to friction in each branch is negligible compared with the head loss across the valves. Find the ratio of the velocity in line A to that in line B (include elbow losses for threaded pipe fittings).

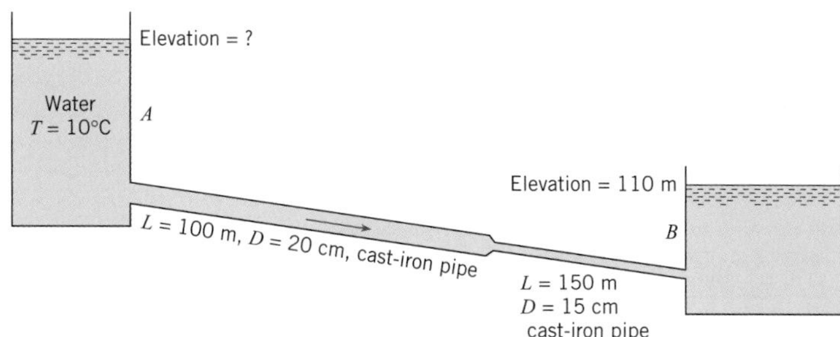

PROBLEM 10.100

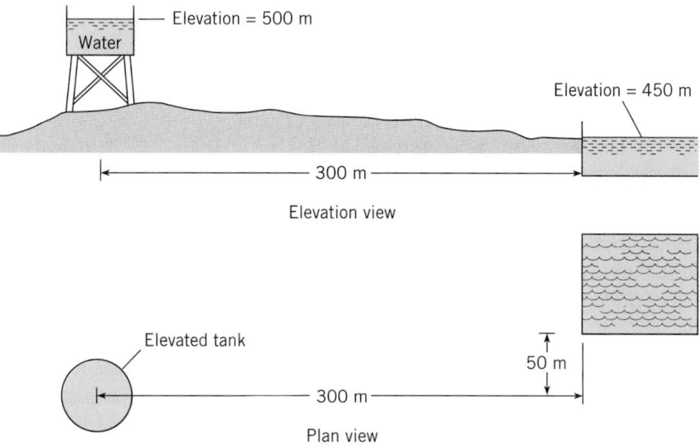

PROBLEMS 10.101, 10.102

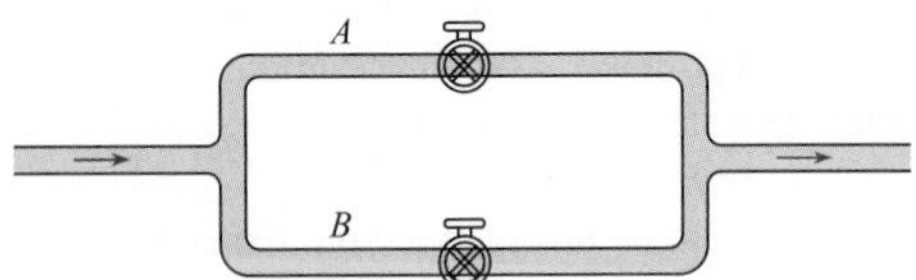

PROBLEMS 10.105, 10.106

FLUID SOLUTIONS **10.107** In the parallel system shown, pipe 1 has a length of 1000 m and is 50 cm in diameter. Pipe 2 is 1500 m long and 40 cm in diameter. The pipe is commercial steel. What is the division of the flow of water at 10°C if the total discharge is to be 1.2 m^3/s?

10.108 Pipes 1 and 2 are the same kind (cast-iron pipe), but pipe 2 is four times as long as pipe 1. They are the same diameter (1 ft). If the discharge of water in pipe 2 is 1 cfs, then what will be the discharge in pipe 1? Assume the same value of f in both pipes.

10.109 Water flows from left to right in this parallel pipe system. The pipe having the greatest velocity is (a) pipe A, (b) pipe B, (c) pipe C.

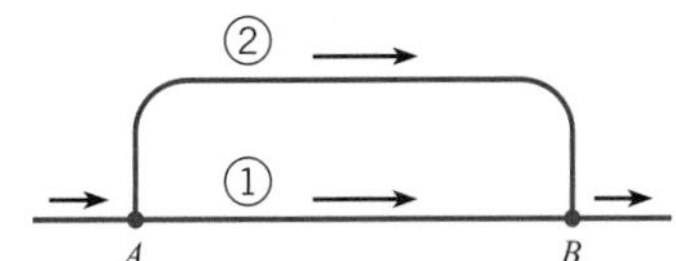

PROBLEMS 10.107, 10.108

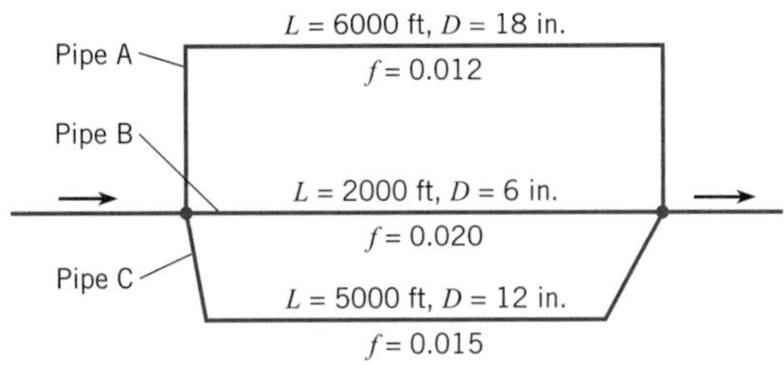

PROBLEM 10.109

10.110 Two pipes are connected in parallel. One pipe is twice the diameter of the other and three times as long. Assume that f in the larger pipe is 0.010 and f in the smaller one is 0.014. Determine the ratio of the discharges in the two pipes.

10.111 With a total flow of 14 cfs, determine the division of flow and the head loss from A to B.

10.112 The pipes shown in the system are all concrete. With a flow of 25 cfs of water, find the head loss and the division of flow in the pipes from A to B. Assume $f = 0.030$ for all pipes.

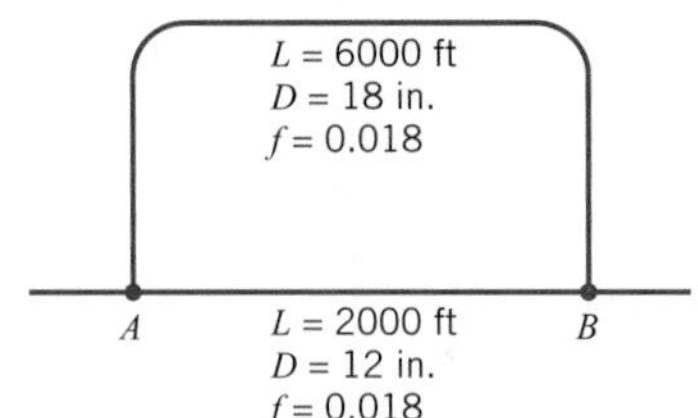

PROBLEM 10.111

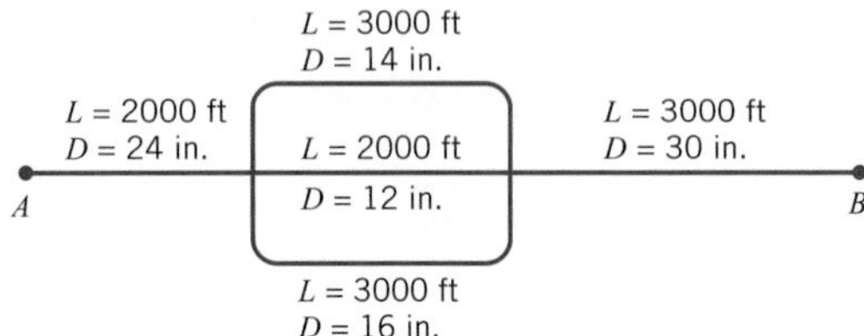

PROBLEM 10.112

10.113 A parallel pipe system is set up as shown. Flow occurs from A to B. To augment the flow, a pump having the characteristics shown in Fig. 10.16 is installed at point C. For a total discharge of 0.60 m^3/s, what will be the division of flow between the pipes and what will be the head loss between A and B? Assume commercial steel pipe.

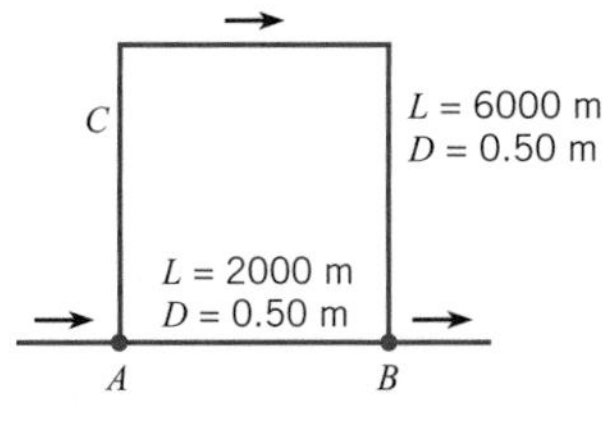

PROBLEM 10.113

10.114 Frequently in the design of pump systems, a bypass line will be installed in parallel to the pump so that some of the fluid can recirculate as shown. The bypass valve then controls the flow rate in the system. Assume that the head-versus-discharge curve for the pump is given by $h_p = 100 - 100Q$, where h_p is in meters and Q is in m^3/s. The bypass line is 10 cm in diameter. Assume the only head loss is that due to the valve, which has a head-loss coefficient of 0.2. The discharge leaving the system is 0.2 m^3/s. Find the discharge through the pump and bypass line.

10.115 Consider air flow in a square duct (4 ft × 4 ft in cross section) and water flow in an open channel (4 ft wide and with a depth of flow of 2 ft). Let the hydraulic radius for the air flow be R_A, and let the hydraulic radius for the water flow be R_W. For these conditions, indicate which one of the following statements

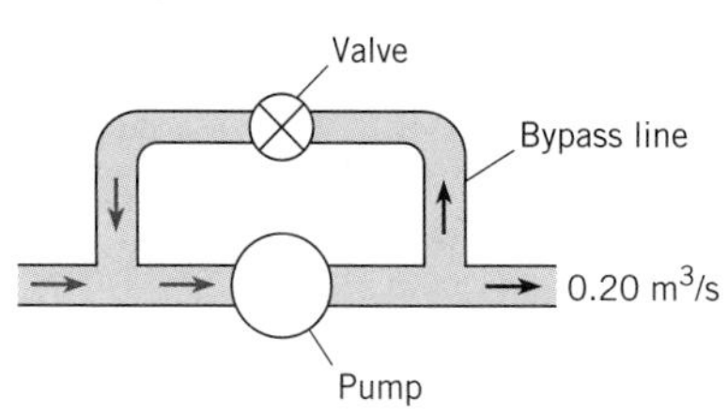

PROBLEM 10.114

relating to hydraulic radius is true: (a) $R_A = R_W$, (b) $R_A = 1.5R_W$, (c) $R_A = 0.67R_W$, (d) $R_A = 0.50R_W$.

10.116 Air at 60°F and atmospheric pressure flows in a horizontal duct with a cross section corresponding to an equilateral triangle (all sides equal). The duct is 100 ft long, and the dimension of a side is 6 in. The duct is constructed of galvanized iron ($k_s = 0.0005$ ft). The mean velocity in the duct is 12 ft/s. What is the pressure drop over the 100-ft length?

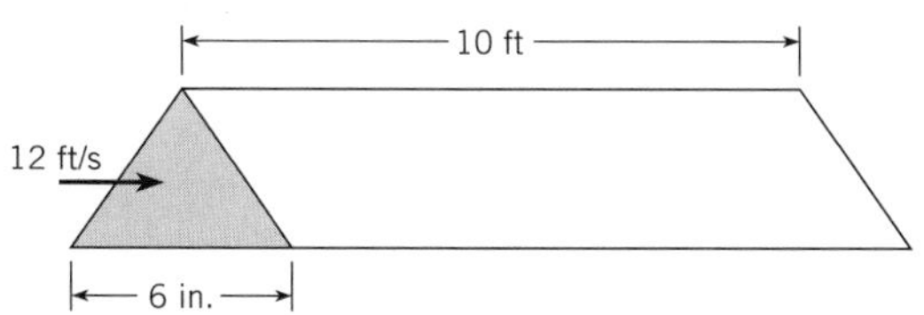

PROBLEM 10.116

10.117 Consider uniform flow of water in these two channels. They both have the same slope, the same wall roughness, and the same cross-sectional area. Then one can conclude that (a) $Q_A = Q_B$, (b) $Q_A < Q_B$, (c) $Q_A > Q_B$.

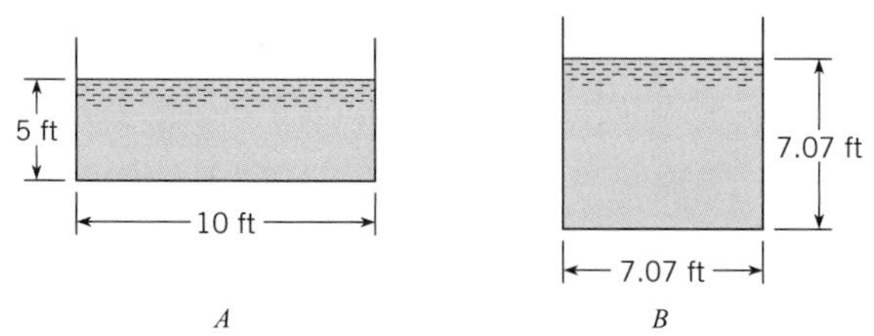

PROBLEM 10.117

10.118 A cold-air duct 100 cm by 15 cm in cross section is 100 m long and made of galvanized iron. This duct is to carry air at a rate of 6 m^3/s at a temperature of 15°C and atmospheric pressure. What is the power loss in the duct?

10.119 An air-conditioning system is designed to have a duct with a rectangular cross section 1 ft by 2 ft, as shown. During construction, a truck driver backed into the duct and made it a trapezoidal section, as shown. The contractor, behind schedule, installed it anyway. For the same pressure drop along the pipe, what will be the ratio of the velocity in the trapezoidal duct to that in the rectangular duct? Assume the Darcy-Weisbach resistance coefficient is the same for both ducts.

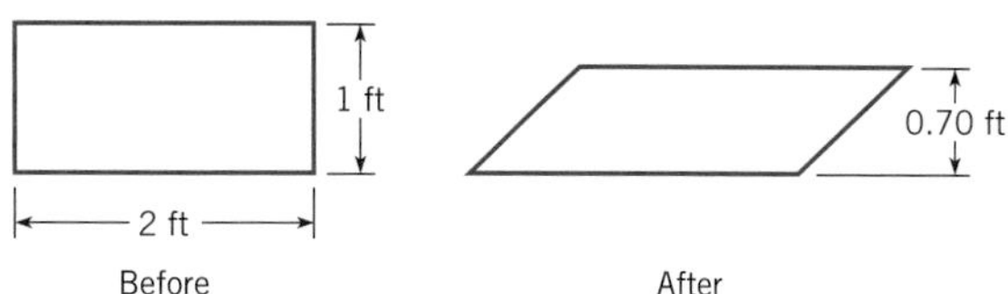

PROBLEM 10.119

10.120 Water ($\nu = 10^{-6}$ m^2/s) flows fully through a concrete ($k_s = 10^{-3}$ m) duct (no free surface) that has a rectangular cross section (70 cm by 100 cm). If the velocity of flow is 10 m/s, the resistance coefficient f will be about (a) 0.01, (b) 0.02, (c) 0.03, (d) 0.04.

10.121 This wood flume has a slope of 0.0015. What will be the discharge of water in it for a depth of 1 m?

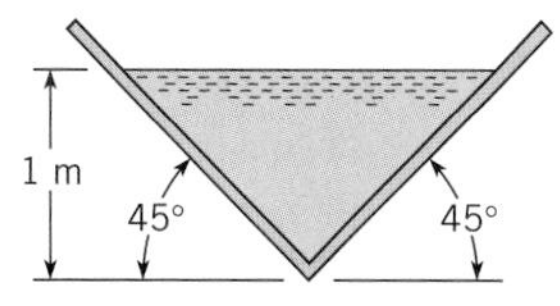

PROBLEM 10.121

10.122 Estimate the discharge in a rock-bedded stream ($d_{84} = 30$ cm) that has an average depth of 2.21 m, a slope of 0.0037, and a width of 48 m. Assume $k_s = d_{84}$.

10.123 Estimate the discharge of water ($T = 10$°C) that flows 1.5 m deep in a long rectangular concrete channel that is 3 m wide and is on a slope of 0.001.

10.124 A rectangular concrete channel is 12 ft wide and has uniform water flow. If the channel drops 5 ft in a length of 8000 ft, what is the discharge? Assume $T = 60$°F. The depth of flow is 4 ft.

10.125 Consider channels of rectangular cross section carrying 100 cfs of water flow. The channels have a slope of 0.001. Determine the cross-sectional areas required for widths of 2 ft, 4 ft, 6 ft, 8 ft, 10 ft, and 15 ft. Plot A versus y/b, and see how the results compare with the accepted result for the best hydraulic section.

10.126 A concrete sewer pipe 3 ft in diameter is laid so it has a drop in elevation of 1.0 ft per 1000 ft of length. If sewage (assume the properties are the same as those of water) flows at a depth of 1.5 ft in the pipe, what will be the discharge?

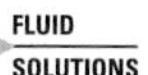

10.127 Determine the discharge in a 5-ft-diameter concrete sewer pipe on a slope of 0.001 that is carrying water at a depth of 4 ft.

10.128 Water flows at a depth of 6 ft in the trapezoidal, concrete-lined channel shown. If the channel slope is 1 ft in 2000 ft, what is the average velocity and what is the discharge?

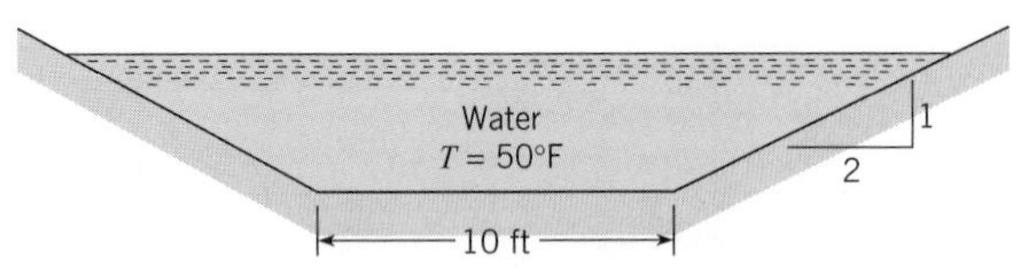

PROBLEM 10.128

10.129 What will be the depth of flow in a trapezoidal concrete-lined channel that has a water discharge of 1000 cfs? The channel has a slope of 1 ft in 500 ft. The bottom width of the channel is 10 ft, and the side slopes are 1 vertical to 1 horizontal.

10.130 What discharge of water will occur in a trapezoidal channel that has a bottom width of 10 ft and side slopes of 1 vertical to 1 horizontal if the slope of the channel is 4 ft/mi and the depth is to be 5 ft? The channel will be lined with concrete.

10.131 A rectangular concrete channel 4 m wide on a slope of 0.004 is designed to carry a water (T = 10°C) discharge of 25 m^3/s. Estimate the uniform flow depth for these conditions. The channel has a rectangular cross section.

10.132 A rectangular troweled concrete channel 12 ft wide with a slope of 10 ft in 8000 ft is designed for a discharge of 500 cfs. For a water temperature of 40°F, estimate the depth of flow.

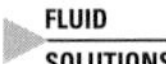

10.133 A concrete-lined trapezoidal channel having a bottom width of 10 ft and side slopes of 1 vertical to 2 horizontal is designed to carry a flow of 3000 cfs. If the slope of the channel is 0.001, what will be the depth of flow in the channel?

10.134 Design a canal having a trapezoidal cross section to carry a design discharge of irrigation water of 900 cfs. The slope of the canal is to be 0.002. The canal is to be lined with concrete, and it is to have the best hydraulic section for the design flow.

10.135 For the given source and loads shown, how will the flow be distributed in the simple network, and what will be the pressures at the load points if the pressure at the source is 60 psi? Assume horizontal pipes and $f = 0.012$ for all pipes.

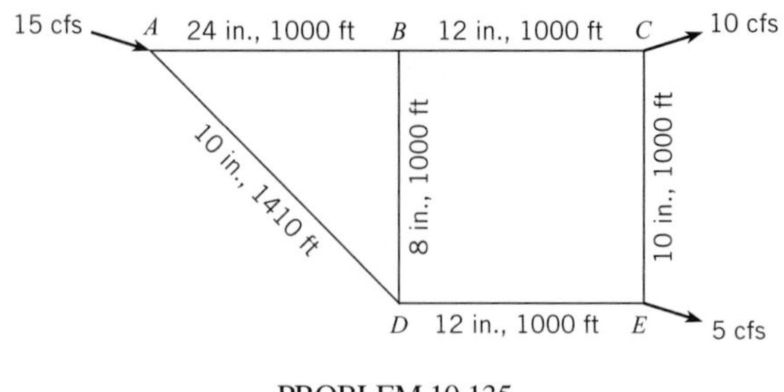

PROBLEM 10.135

10.136 A platform is to be designed to move equipment from place to place on a factory floor. When the equipment is to be moved, the platform on which the equipment is placed will be lifted free of the floor surface by bearing(s) in the platform similar to that described in Prob. 9.33. In this case, however, the fluid is to be air. Scope the design of such a system and make enough calculations to justify the feasibility of such a system. You should make assumptions about size and weight of the platform, etc.

10.137 Design a system for measuring the viscosity of a gas based on the flow in a conduit. The system should be capable of measuring viscosity over a range from 10^{-5} to 5×10^{-5} N · s/m^2.

10.138 One of the problems associated with using a pressure tap to measure the pressure in a conduit carrying sand and other debris is the tendency for the debris to get in the opening of the pressure tap and invalidate the measurement. Develop some ideas for schemes to avert this problem.

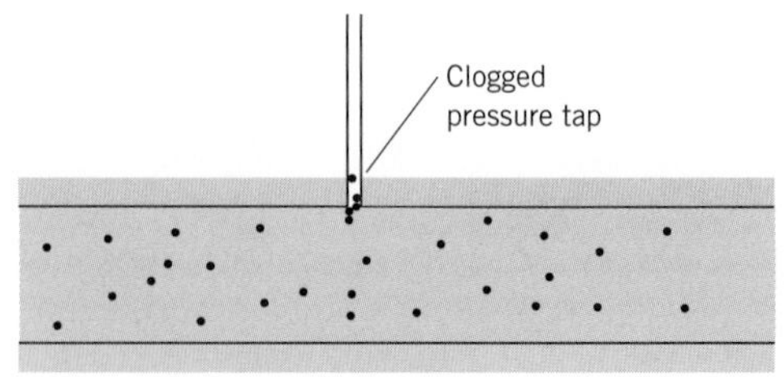

PROBLEM 10.138

Gliders use long slender wings to reduce drag and enhance lift. (Adam Woolfitt/CORBIS.)

C H A P T E R 11

Drag and Lift

A body immersed in a flowing fluid is acted on by both pressure and viscous forces from the flow. The sum of the forces (pressure, viscous, or both) that acts normal to the free-stream direction is the *lift*, and the sum that acts parallel to the free-stream direction is the *drag*. Buoyant or weight forces may also act on the body; however, lift and drag forces are limited by definition to those forces produced by the dynamic action of the flowing fluid. We will first consider the mathematical formulation for lift and drag in terms of the viscous stresses and pressure; then we will consider each more intensively in subsequent sections.

Basic Considerations

Consider the forces acting on the airfoil in Fig. 11.1. The vectors normal to the surface of the airfoil are normal forces per unit area, referred to simply as pressure. As shown here, the pressure is referenced to the free-stream pressure. Because the velocity of the flow over the top of the airfoil is greater than the free-stream velocity, the pressure over the top is negative, or less than the free-stream pressure. This follows directly from application of the Bernoulli equation. Because the velocity along the underside of the wing is less than the free-stream velocity, the pressure there is positive, or greater than the free-stream pressure. Hence both the negative pressure over the top and the positive pressure along the bottom contribute to the lift.

The vectors in Fig. 11.1 that are parallel to the surface of the airfoil represent the shear forces per unit area, referred to simply as shear stress. Except on the front of the airfoil, shear stress acts essentially parallel to the free-stream direction. Hence the shear stress contributes largely to the drag of the airfoil.

The mathematical formulation for lift and drag in terms of the pressure and shear stress can be derived with the aid of Fig. 11.2. Here the pressure and viscous forces acting on a differential area of the surface of the airfoil are shown. The magnitude of the pressure force is $dF_p = p\,dA$, and the magnitude of the viscous force is $dF_V = \tau\,dA$. However, we are interested in separating the forces into components that are normal and parallel to the free-stream direction to determine lift and drag, respectively. Hence the differential lift force is*

* The sign convention on τ is such that a clockwise sense of $\tau\,dA$ on the surface of the foil signifies a positive sign for τ.

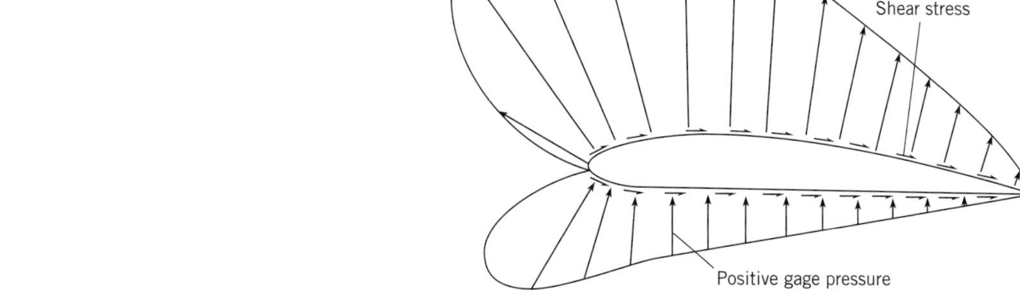

FIGURE 11.1

Pressure and shear stress acting on an airfoil.

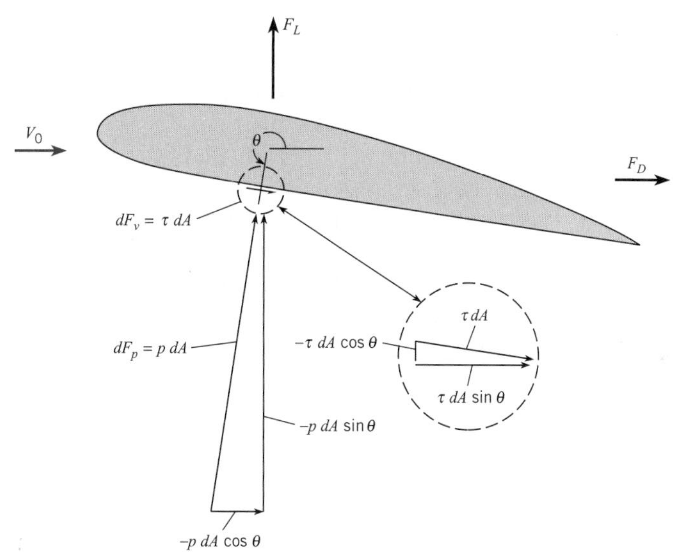

FIGURE 11.2

Pressure and viscous forces acting on a differential element of area.

$$dF_L = -p\ dA\sin\theta - \tau\ dA\cos\theta$$

and the differential drag is

$$dF_D = -p\ dA\cos\theta + \tau\ dA\sin\theta$$

Then the total lift and drag on the airfoil are obtained by integration of the respective differential forces over the entire surface of the airfoil:

$$F_L = \int(-p\sin\theta - \tau\cos\theta)dA \tag{11.1}$$

$$F_D = \int(-p\cos\theta + \tau\sin\theta)dA \tag{11.2}$$

Equations (11.1) and (11.2) are for a two-dimensional flow; that is, there is no velocity component in the direction normal to the page, so the shear stress and pressure force vectors lie in the plane of the page. The same basic principle (separation of forces into directions parallel and normal to the free-stream direction) can easily be extended to three-dimensional flows.

In this chapter we shall refer to two-dimensional and three-dimensional bodies. By a two-dimensional body, we mean a body over which the flow is two-dimensional. For example, a very long cylinder that has flow approaching it from a normal direction is classified as a two-dimensional body because the flow around the ends does not affect the flow pattern and pressure distribution over the central part of the body. However, a short cylinder is classified as a three-dimensional body because the end effects are significant. For a two-dimensional body, the aerodynamic forces and representative areas are sometimes based on a unit length of the body. The two-dimensional body is identified by dark shading of the figures.

Still another classification is the axisymmetric body. Here, if the approach flow is uniform and parallel to the axis of symmetry, then in effect the resulting flow is two-dimensional. That is, for an x-r coordinate system, where x is measured along the axis of symmetry and r is the normal radial distance from the axis, the velocity components exist only in the x and r directions.

Equations (11.1) and (11.2) can be used to evaluate F_L and F_D when the pressure and shear stresses are obtained either analytically or experimentally. However, it is also common to obtain the overall drag and lift by force dynamometer measurements in a wind tunnel. The next section considers the direct application of the pressure and shear stress variation over the surface of a plate to determine the drag on the plate.

Drag of Two-Dimensional Bodies

Drag of a Thin Plate

To illustrate the relative effect of pressure and viscous forces on drag, we shall consider the drag of a plate first oriented parallel to the flow and then oriented normal to the flow. In the parallel position, the only force acting is viscous shear in the direction of the flow. Hence, from our considerations of surface resistance in Chapter 9, the drag for both sides of the plate is given as

$$F_D = 2C_f b\ell\rho\frac{V_0^2}{2}$$

When the plate is turned normal to the flow, as in Fig. 11.3, both pressure and viscous forces act on the plate. However, the viscous forces act only in the transverse direction and, in addition, are symmetrical about the midpoint of the plate. Consequently, the viscous forces do not directly contribute to the lift or drag of the plate. Because the pressure on the plate acts to produce a force only in a direction parallel to the flow, the pressure force contributes totally to the drag of the body. Hence Eq. (11.2) applied to the plate reduces to

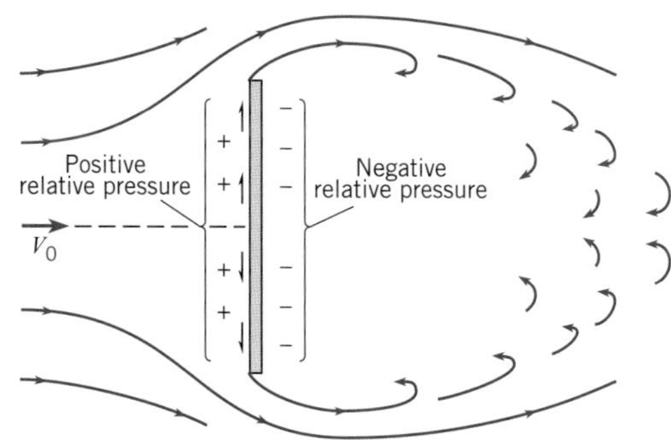

FIGURE 11.3

Flow past a flat plate.

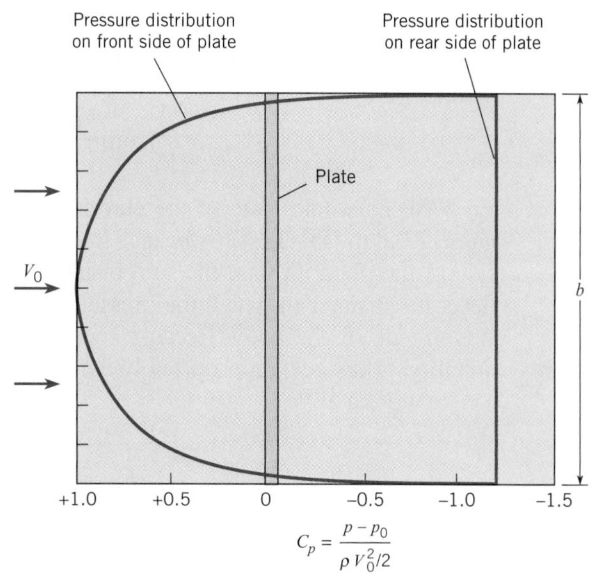

FIGURE 11.4

Pressure distribution on a plate normal to the approach flow for Re > 10^4.

$$F_D = \int (-p\cos\theta)dA$$

The pressures on the upstream and downstream sides of the plate can be obtained experimentally and are usually given in terms of C_p, as shown in Fig. 11.4 for flow with a relatively high value of the Reynolds number ($V_0 b/\nu > 10^4$).

Since the pressure on the downstream side is essentially constant,

$$p = p_0 - 1.2\rho\frac{V_0^2}{2}$$

and since $\theta = 0$, the contribution to drag for the downstream side is

$$F_{D,\text{downstream}} = -\left(p_0 - 1.2\rho\frac{V_0^2}{2}\right)b\ell = -p_0 b\ell + 1.2\rho\frac{V_0^2}{2}b\ell$$

where ℓ is the length of the plate normal to the plane of the paper, and by definition of a two-dimensional body, $\ell \gg b$. For the front side, $\theta = \pi$. Hence $\cos\theta = -1$, and the contribution to drag due to pressure on the upstream side is

$$F_{D,\text{upstream}} = \int_{-b/2}^{b/2} \left(p_0 + C_p \rho \frac{V_0^2}{2}\right) \ell \, dy = p_0 b\ell + \rho \frac{V_0^2}{2} \ell \int_{-b/2}^{b/2} C_p \, dy$$

Then the total drag on the plate is given by

$$\begin{aligned} F_D &= F_{D,\text{upstream}} + F_{D,\text{downstream}} \\ &= \rho \frac{V_0^2}{2} \ell \left(\int_{-b/2}^{b/2} C_p \, dy + 1.2b \right) \end{aligned} \tag{11.3}$$

Evaluation of the first term inside the parentheses on the right-hand side of Eq. (11.3) yields a magnitude of approximately 0.8*b*. Thus the drag of this plate is given as

$$F_D = \rho \frac{V_0^2}{2} b\ell (0.8 + 1.2) \tag{11.4}$$

We must take note of the pure numbers inside the parentheses of Eq. (11.4). The number 0.8 in Eq. (11.4) represents the average pressure coefficient C_p over the upstream side of the plate. In fact, the sum inside the parentheses (0.8 + 1.2) of this equation reflects the manner in which the pressure is distributed over the upstream side and downstream side of the body. Because the drag varies directly with the magnitude of this quantity, it has been appropriately defined as the *coefficient of drag* C_D. Thus Eq. (11.4) can be written as

$$F_D = C_D A_p \rho \frac{V_0^2}{2} \tag{11.5}$$

where C_D is the coefficient of drag, A_p is the projected area of the body, ρ is the fluid density, and V_0 is the free-stream velocity. The projected area A_p is the silhouetted area that would be seen by a person looking at the body from the direction of flow. For example, the projected area of the above plate normal to the flow is $b\ell$, and the projected area of a cylinder with its axis normal to the flow is $d\ell$. In Chapter 8 we saw that C_p is a function of the Reynolds number Re; and because $C_D = f(C_p)$, C_D is also a function of Re. When the drag of bodies is due solely to the shear stress on the body, C_D is still a function of Re because τ is also a function of Re.

Coefficients of Drag for Various Two-Dimensional Bodies

We have already seen that C_D can be determined if the pressure and shear stress distribution around a body are known. The coefficient of drag can also be calculated if the total drag is measured, for example, by means of a force balance in a wind tunnel. Then C_D is calculated using Eq. (11.5) written as follows:

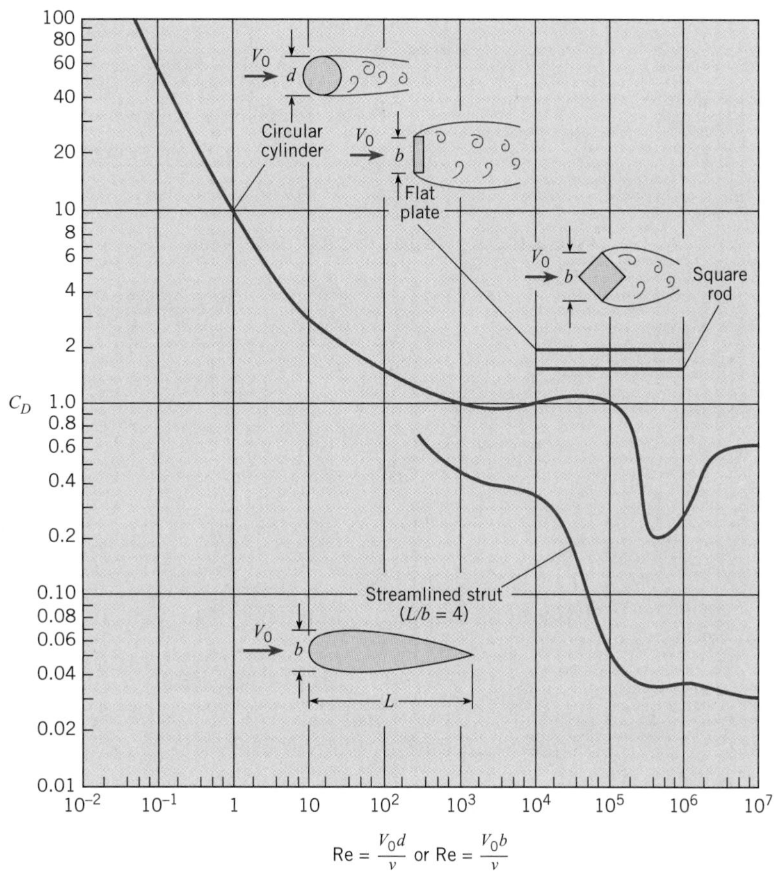

FIGURE 11.5

Coefficient of drag versus Reynolds number for two-dimensional bodies. [Data sources: Bullivant (1), Defoe (2), Goett and Bullivant (3), Jacobs (4), Jones (4), and Lindsey (6)]

$$C_D = \frac{F_D}{A_p \rho V_0^2/2} \tag{11.6}$$

Much of the data (C_D versus Re) found in the literature is obtained in this manner.

The coefficient of drag for the flat plate normal to the free stream and for other two-dimensional bodies, for a wide range of Reynolds numbers, is given in Fig. 11.5. In general, the total drag of a blunt body is partly due to viscous resistance and partly due to pressure variation. The pressure drag is largely a function of the form or shape of the body; hence it is called *form drag*. The viscous drag is often called *skin friction drag*.

example 11.1

A television transmitting antenna is on top of a pipe 30 m high (98.4 ft) and 30 cm (11.8 in.) in diameter, which is on top of a tall building. What will be the total drag of the pipe and the bending moment at the base of the pipe in a 35-m/s (115-ft/s) wind at normal atmospheric pressure and a temperature of 20°C (68°F)?

Solution For the conditions given, the viscosity and density of the air are obtained from the Appendix:

$$\mu = 1.81 \times 10^{-5}\,\text{N}\cdot\text{s/m}^2 \quad (3.78 \times 10^{-7}\ \text{lbf-s/ft}^2)$$

$$\rho = 1.20\ \text{kg/m}^3 \quad (0.00234\ \text{slugs/ft}^3)$$

Next the Reynolds number is calculated:

$$\text{Re} = \frac{V_0 d\rho}{\mu} = \frac{35\ \text{m/s} \times 0.30\ \text{m} \times 1.20\ \text{kg/m}^3}{(1.81 \times 10^{-5}\ \text{N}\cdot\text{s/m}^2)} = 7.0 \times 10^5$$

Then, from Fig. 11.5, $C_D = 0.20$. Now we compute the total drag:

$$F_D = \frac{C_D A_p \rho V_0^2}{2}$$

$$= \frac{(0.2)(30\ \text{m})(0.3\ \text{m})(1.20\ \text{kg/m}^3)(35^2\text{m}^2/\text{s}^2)}{2} = 1323\ \text{N} \quad \triangleleft$$

Assuming that the resultant drag force acts midway up the pole, the moment is

$$M = F_D\left(\frac{L}{2}\right) = (1323\ \text{N})\left(\frac{30}{2}\ \text{m}\right) = 19{,}845\ \text{N}\cdot\text{m} \quad \triangleleft$$

Traditional units:

$$F_D = 0.2(98.4\ \text{ft})\left(\frac{11.8}{12}\ \text{ft}\right)(0.00234\ \text{slugs/ft}^3)\left(\frac{115^2\text{ft}^2/\text{s}^2}{2}\right) = 299\ \text{lbf} \quad \triangleleft$$

$$M = (299\ \text{lbf})\left(\frac{98.4}{2}\ \text{ft}\right) = 14{,}700\ \text{ft-lbf} \quad \triangleleft$$

Discussion of C_D for Two-Dimensional Bodies

At low Reynolds numbers, C_D changes with the Reynolds number. The change is due to the relative change in viscous resistance, which has already been mentioned in Chapter 8. Above $\text{Re} = 10^4$, the flow pattern remains virtually unchanged, thereby producing constant values of C_p over the body. Constancy of C_p at high Reynolds numbers is reflected in the constancy of C_D. This characteristic, the constancy of C_D at high values of Re, is representative of most bodies that have angular form. However, certain bodies with rounded form, such as circular cylinders, show a remarkable decrease in C_D with an increase in Re from about 10^5 to 5×10^5.

This reduction in C_D at a Reynolds number of approximately 10^5 is due to a change in the flow pattern triggered by a change in the character of the boundary layer. For Reynolds numbers less than 10^5, the boundary layer is laminar, and separation occurs about midway between the upstream side and downstream side of the

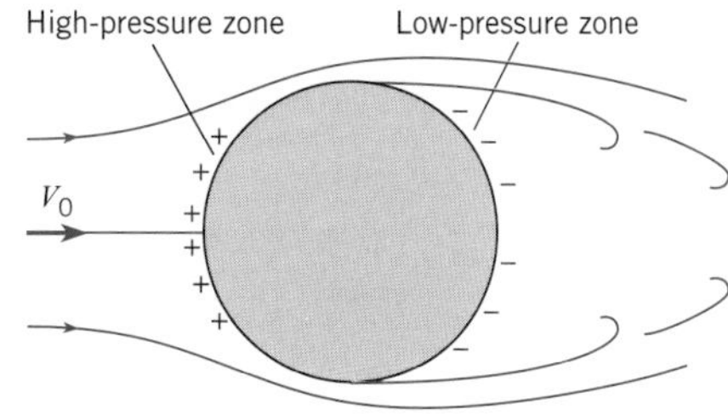

FIGURE 11.6

Flow pattern around a cylinder for $10^3 < Re < 10^5$.

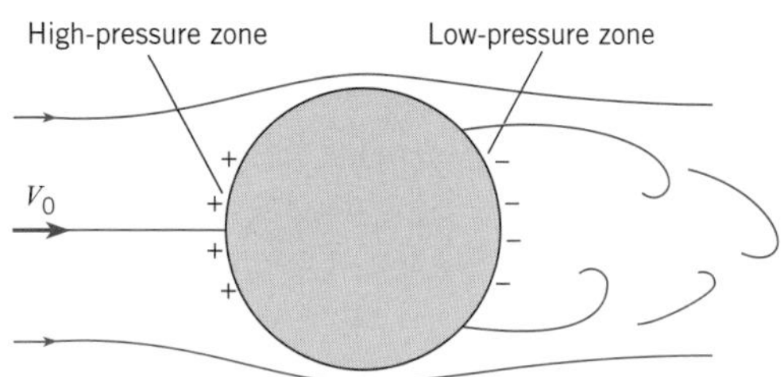

FIGURE 11.7

Flow pattern around a cylinder for $Re > 5 \times 10^5$.

cylinder (Fig. 11.6). Hence the entire downstream half of the cylinder is exposed to a relatively low pressure, which in turn produces a relatively high value for C_D. When the Reynolds number is increased to about 10^5, the boundary layer on the surface of the cylinder becomes turbulent, which causes higher-velocity fluid to be mixed into the region close to the wall of the cylinder. As a consequence of the presence of this high-velocity, high-momentum fluid in the boundary layer, the flow proceeds farther downstream along the surface of the cylinder against the adverse pressure before separation occurs (Fig. 11.7). Hence the flow pattern causes C_D to be reduced for the following reason: With the turbulent boundary layer, the streamlines downstream of the cylinder midsection diverge somewhat before separation, and hence a decrease in velocity occurs before separation. According to the Bernoulli equation, the decrease in velocity produces a pressure at the point of separation that is greater than the pressure at the midsection. Thus the pressure at the point of separation, and also in the zone of separation, is significantly greater under these conditions than when separation occurs farther upstream. Therefore, the pressure difference between the upstream and downstream surfaces of the cylinder is less at high values of Re, yielding a lower drag and a lower C_D.

Because the boundary layer is so thin, it is also very sensitive to other conditions. For example, if the surface of the cylinder is slightly roughened upstream of the midsection, the boundary layer will be forced to become turbulent at lower Reynolds numbers than those for a smooth cylinder surface. The same trend can also be produced by creating abnormal turbulence in the approach flow. The effects of roughness are shown in Fig. 11.8 for cylinders that were roughened with sand grains of size k. A small to medium size of roughness $(10^{-3} < k/d < 10^{-2})$ on a cylinder triggers an early onset of reduction of C_D. However, when the relative roughness is quite large $(10^{-2} < k/d)$, the characteristic dip in C_D is absent.

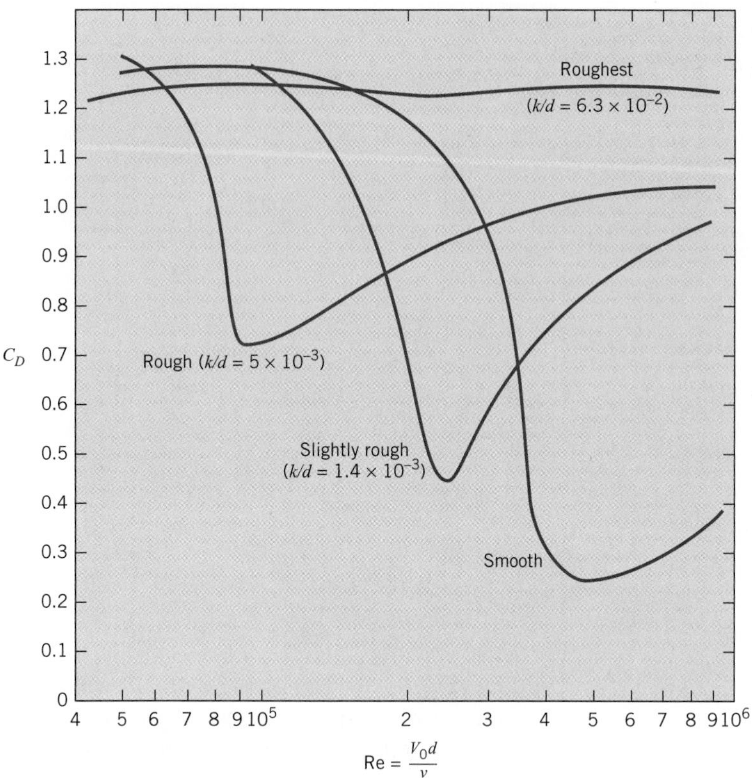

FIGURE 11.8

Effects of roughness on C_D for a cylinder. [After Miller et al. (7)]

11.3 Vortex Shedding from Cylindrical Bodies

Figures 11.6 and 11.7 show the average (temporal mean) flow pattern around a cylinder. The phenomenon becomes more complex, however, when we observe the detailed flow pattern as time passes. Observations show that above $\mathrm{Re} \approx 50$, vortices are formed and shed periodically downstream of the cylinder. Hence, at a given time, the detailed flow pattern might appear as in Fig. 11.9. In this figure a vortex is in the process of formation near the top of the cylinder. Below and to the right of the first vortex is another vortex, which was formed and shed a short time before. Thus the flow process in the wake of a cylinder involves the formation and shedding of vortices alternately from one side and then the other. This phenomenon is of major importance in engineering design, because the alternate formation and shedding of vortices also creates a regular change in pressure with consequent periodicity in side thrust on the cylinder. Vortex shedding was the primary cause of failure of the Tacoma Narrows suspension bridge in the state of Washington in 1940. Another, more commonplace effect of vortex shedding is the "singing" of wires in the wind.

FIGURE 11.9
Formation of a vortex behind a cylinder.

If the frequency of the vortex shedding is in resonance with the natural frequency of the member that produces it, large amplitudes of vibration with resulting large stresses can develop. Experiments show that the frequency of shedding is given in terms of the Strouhal number St, and this in turn is a function of the Reynolds number. Here the Strouhal number is defined as

$$\mathrm{St} = \frac{nd}{V_0} \tag{11.7}$$

where n is the frequency of shedding of vortices from one side of the cylinder, in Hz, d is the diameter of the cylinder, and V_0 is the free-stream velocity.

The relationship between the Strouhal number and the Reynolds number for vortex shedding from a circular cylinder is given in Fig. 11.10.

Other cylindrical and two-dimensional bodies also shed vortices. Consequently, the engineer should always be alert to vibration problems when designing structures that are exposed to wind or water flow.

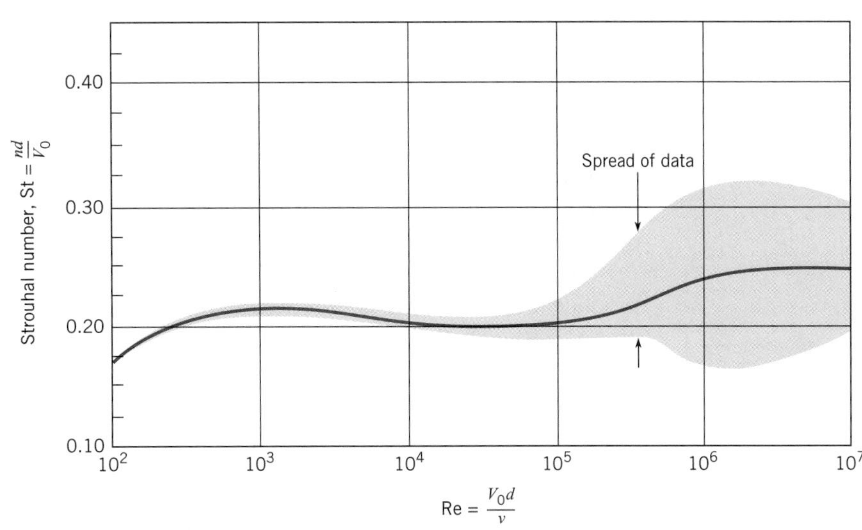

FIGURE 11.10
Strouhal number versus Reynolds number for flow past a circular cylinder. [After Jones (5) and Roshko (8)]

example 11.2

For the cylinder and conditions of Example 11.1, at what frequency will the vortices be shed?

Solution We compute the frequency n from the Strouhal number, which is given in Fig. 11.10 as a function of the Reynolds number. Thus with a Reynolds number of 7.0×10^5 from Example 11.1, we read a Strouhal number of 0.23. But

$$\mathrm{St} = \frac{nd}{V_0}$$

so

$$n = \frac{SV_0}{d} = \frac{0.23 \times 35 \text{ m/s}}{0.30 \text{ m}} = 27 \text{ Hz} \quad \triangleleft$$

11.4 Effect of Streamlining

For Reynolds numbers greater than 10^3, the drag of a cylinder is predominantly due to the pressure variation around the cylinder. The pressure difference between the upstream and downstream sides of the cylinder is the primary cause of drag, and this pressure difference is due largely to separation. Hence, if the separation can be eliminated, the drag will be reduced. That is exactly what streamlining does. Streamlining reduces the extreme curvature on the downstream side of the body, and this process reduces or eliminates separation. Therefore, the coefficient of drag is greatly reduced, as seen in Fig. 11.5 (C_D for the streamlined shape is only about 10% of C_D for the circular cylinder when $\mathrm{Re} \approx 5 \times 10^5$).

When a body is streamlined by elongating it and reducing its curvature, the pressure drag is reduced. However, the viscous drag is increased because there is a greater amount of surface on the streamlined body than on the nonstreamlined body. Consequently, when a body is streamlined to produce minimum drag, there is an optimum condition to be sought. The optimum condition results when the sum of surface drag and pressure drag is minimum.

In this discussion of streamlining, it is interesting to note that streamlining to produce minimum drag at high Reynolds numbers will probably not produce minimum drag at very low Reynolds numbers. For Reynolds numbers less than unity, the majority of the drag of a cylinder is due to the viscous shear stress on the wall of the cylinder. Hence, if the cylinder is streamlined, the viscous shear stress is simply magnified and C_D may actually increase for this range of Re where the viscous resistance is predominant.

At high values of the Reynolds number, another advantage of streamlining is that the periodic formation of vortices is eliminated.

example 11.3

Compare the drag of the cylinder of Example 11.1. with the drag of the streamlined shape shown in Fig. 11.5. Assume that they both have the same projected area and that the streamlined shape is oriented for minimum drag.

Solution Because $F_D = C_D A_p \rho V_0^2/2$, the drag of the streamlined shape will be

$$F_{DS} = F_{DC}\left(\frac{C_{DS}}{C_{DC}}\right)$$

where F_{DS} is the drag of the streamlined shape, F_{DC} is the drag of the cylinder, C_{DS} is the coefficient of drag of the streamlined shape, and C_{DC} is the coefficient of drag of the cylinder. But $C_{DS} = 0.034$, from Fig. 11.5 with Re $= 7.0 \times 10^5$. Then

$$F_{DS} = F_{DC}\left(\frac{0.034}{0.20}\right) = (1323\text{ N})\left(\frac{0.034}{0.20}\right)$$

$$F_{DS} = 225\text{ N} \quad \triangleleft$$

$$\frac{F_{DS}}{F_{DC}} = 0.17 \quad \triangleleft$$

11.5 Drag of Axisymmetric and Three-Dimensional Bodies

The same principles that apply to the drag of two-dimensional bodies also apply to that of axisymmetric and three-dimensional bodies. That is, at very low values of the Reynolds number, the coefficient of drag is given by exact equations relating C_D and Re. At high values of Re, the coefficient of drag becomes constant for angular bodies, whereas rather abrupt changes in C_D occur for rounded bodies. All of these characteristics can be seen in Fig. 11.11, where C_D is plotted against Re for several axisymmetric bodies.

The drag coefficient of a sphere is of special interest because many applications involve the drag of spherical or near-spherical objects such as particles and droplets. Also, the drag of a sphere is often used as a standard of comparison for other shapes. For Reynolds numbers less than 0.5, the flow around the sphere is laminar and amenable to analytical solutions. An exact solution by Stokes yielded the following equation, which is called *Stokes' law,* for the drag of a sphere:

$$F_D = 3\pi\mu V_0 d \tag{11.8}$$

Note that the drag for this laminar flow condition varies directly with the first power of V_0. This is characteristic of all laminar flow processes. For completely turbulent flow, the drag is a function of the velocity to the second power. When Eqs. (11.8) and (11.6) are solved simultaneously, we get the coefficient of drag corresponding to Stokes' law:

$$C_D = \frac{24}{\text{Re}} \tag{11.9}$$

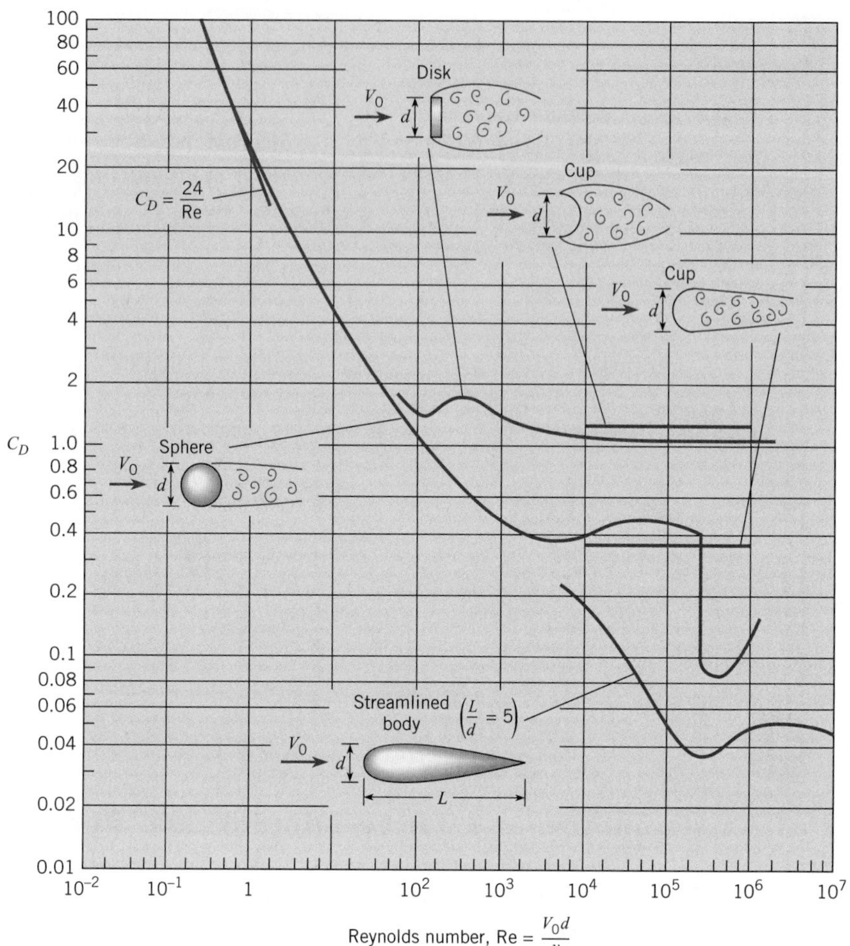

FIGURE 11.11

Coefficient of drag versus Reynolds number for axisymmetric bodies. [Data sources: Abbott (9), Brevoort and Joyner (10), Freeman (11), and Rouse (12)]

Thus for flow past a sphere, when Re ≤ 0.5, one may use the direct relation for C_D given above.

Several correlations for the drag coefficient of a sphere are available (13). One such correlation has been proposed by Clift and Gauvin (14):

$$C_D = \frac{24}{\text{Re}}(1 + 0.15\text{Re}^{0.687}) + \frac{0.42}{1 + 4.25 \times 10^4 \text{Re}^{-1.16}} \quad (11.10)$$

which deviates from the *standard drag curve** by –4% to 6% for Reynolds numbers up to 3×10^5. Note that as the Reynolds number approaches zero, this correlation reduces to the equation for Stokes flow.

Values for C_D for other axisymmetric and three-dimensional bodies at high Reynolds numbers (Re $> 10^4$) are given in Table 11.1. Extensive data on the drag of various shapes available in Hoerner (15).

* The *standard drag curve* represents the best fit of the cumulative data that have been obtained for drag coefficient of a sphere.

TABLE 11.1 APPROXIMATE C_D VALUES FOR VARIOUS BODIES

Type of Body	Length Ratio	Re	C_D
Rectangular plate	$l/b = 1$	$>10^4$	1.18
	$l/b = 5$	$>10^4$	1.20
	$l/b = 10$	$>10^4$	1.30
	$l/b = 20$	$>10^4$	1.50
	$l/b = \infty$	$>10^4$	1.98
Circular cylinder with axis parallel to flow	$l/d = 0$ (disk)	$>10^4$	1.17
	$l/d = 0.5$	$>10^4$	1.15
	$l/d = 1$	$>10^4$	0.90
	$l/d = 2$	$>10^4$	0.85
	$l/d = 4$	$>10^4$	0.87
	$l/d = 8$	$>10^4$	0.99
Square rod	∞	$>10^4$	2.00
Square rod	∞	$>10^4$	1.50
Triangular cylinder (60°)	∞	$>10^4$	1.39
Semicircular shell	∞	$>10^4$	1.20
Semicircular shell	∞	$>10^4$	2.30
Hemispherical shell		$>10^4$	0.39
Hemispherical shell		$>10^4$	1.40
Cube		$>10^4$	1.10
Cube		$>10^4$	0.81
Cone—60° vertex		$>10^4$	0.49
Parachute		$\approx 3 \times 10^7$	1.20

SOURCES: Brevoort and Joyner (10), Lindsey (6), Morrison (16), Roberson et al. (17), Rouse (12), and Scher and Gale (18).

example 11.4

What is the drag of a 12-mm sphere that drops at a rate of 8 cm/s in oil ($\mu = 10^{-1}\ \text{N}\cdot\text{s/m}^2$, S = 0.85)?

Solution

$$\rho = 0.85 \times 1000\ \text{kg/m}^3$$

THEN
$$\text{Re} = \frac{V d\rho}{\mu} = \frac{(0.08\ \text{m/s})(0.012\ \text{m})(850\ \text{kg/m}^3)}{10^{-1}\ \text{N}\cdot\text{s/m}^2} = 8.16$$

Next, from Fig. 11.11,

$$C_D = 5.3$$

$$F_D = \frac{C_D A_p \rho V_0^2}{2}$$

Hence
$$F_D = \frac{(5.3)(\pi/4)(0.012^2\ \text{m}^2)(850\ \text{kg/m}^3)(0.08^2\ \text{m}^2/\text{s}^2)}{2}$$

$$= 1.63 \times 10^{-3}\ \text{N}$$ ◁

example 11.5

A bicyclist travels on a level path. The frontal area of the cyclist and bicycle is 2 ft^2 and the drag coefficient is 0.8. The density of the air is 0.075 lbm/ft^3. The physiological limit of power for this well-trained cyclist is 0.5 hp. Neglecting rolling friction, find the maximum speed of the cyclist.

Solution The power is the product of the drag force and the velocity.

$$P = F_D V = (1/2) C_D A_p \rho V^3$$

Solving for the velocity, we get

$$V = (2P/\rho C_D A_p)^{1/3}$$

$$= \left[\frac{2 \times 0.5\ \text{hp} \times 550\ \text{ft-lbf/s hp}}{0.075\ \text{lbm/ft}^3 \times (1/32.2)\ \text{slugs/lbm} \times 0.8 \times 2\ \text{ft}^2}\right]^{1/3}$$

$$= 52.8\ \text{ft/s} = 36.0\ \text{mph}$$ ◁

Terminal Velocity

The engineer is often required to use drag data in the computation of the *terminal velocity* of a body. When a body is first dropped in the atmosphere or in water, it accelerates under the action of its weight. Then, as the speed of the body increases, the drag increases. Finally, the drag reaches a magnitude such that the sum of all external forces on the body is zero. Hence acceleration ceases, and the body has attained its terminal velocity. Thus the terminal velocity is the maximum velocity attained by a falling body. This assumes that the fluid through which it falls has constant properties over the path of descent. The following two examples illustrate the method of computing the terminal velocity of a body—first in air, then in water.

example 11.6

A 20-mm solid plastic sphere (S = 1.3) is dropped from an airplane. Assuming standard atmospheric conditions (T = 20°C, p_{atm} = 101 kPa absolute), what will be the terminal velocity of the sphere?

Solution The free-body diagram of the sphere is shown in the accompanying figure, where F_B is the buoyant force, F_D is the drag, and W is the weight.

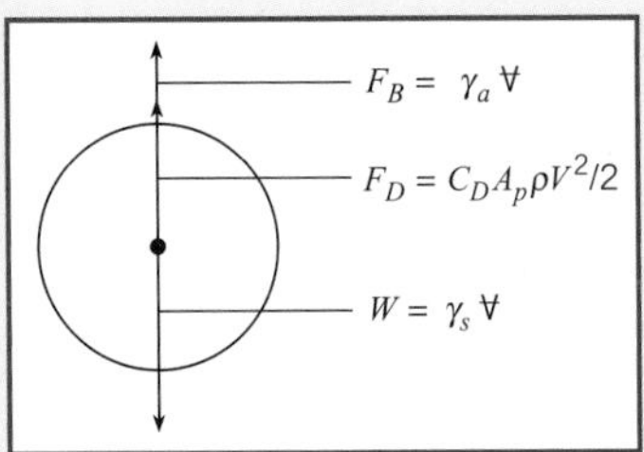

With no acceleration, the drag force plus the buoyant force must equal the weight of the sphere.

$$F_D + F_B = W$$

or

$$\tfrac{1}{2}\rho V_0^2 C_D A_p = (\gamma_s - \gamma_a)\forall$$

where γ_s is the specific weight of the sphere and γ_a is the specific weight of air.

For the sphere

$$A_p = \frac{\pi}{4}d^2 \quad \text{and} \quad \forall = \frac{\pi}{6}d^3$$

The specific weight of the sphere, γ_s, is the specific gravity of the sphere times γ_{water}, or $\gamma_s = 1.3 \times 9.81\ \text{kN/m}^3 = 12.7\ \text{kN/m}^3$. Also, $\rho_{\text{air}} = 1.20\ \text{kg/m}^3$ and $\gamma_a = 1.20 \times 9.81\ \text{N/m}^3$. Thus $\gamma_a \ll \gamma_s$. Hence

$$V_0 = \left(\frac{2\gamma_s(1/6)\pi d^3}{C_D(\pi/4)d^2\rho_{\text{air}}}\right)^{1/2} = \left(\frac{\gamma_s(4/3)d}{C_D\rho_{\text{air}}}\right)^{1/2}$$

$$= \left(\frac{12.7(10^3\ \text{N/m}^3)(4/3)(0.02\ \text{m})}{C_D(1.20\ \text{kg/m}^3)}\right)^{1/2} = \left(\frac{282\ \text{m}^2/\text{s}^2}{C_D}\right)^{1/2} = \frac{16.8\ \text{m/s}}{(C_D)^{1/2}}$$

Obviously, the sphere will drop with a high velocity. Therefore we make a first guess at C_D by choosing a value from the high range of Re. Assume $C_D = 0.40$. We get a velocity of

$$V_0 = \frac{16.8\ \text{m/s}}{(0.40)^{1/2}} = 26.6\ \text{m/s}$$

Now we compute Re based on this velocity of 26.6 m/s, so that we can get a better value of C_D.

$$\text{Re} = \frac{V\,d\rho}{\mu}$$

$$= \frac{(26.6\ \text{m/s})(0.02\ \text{m})(1.20\ \text{kg/m}^3)}{1.81 \times 10^{-5}\ \text{N}\cdot\text{s/m}^2} = 3.5 \times 10^4$$

Then a better estimate of C_D based on $\text{Re} = 3.5 \times 10^4$ is $C_D = 0.48$. Our first guess for C_D was too low. Hence we make another calculation of V_0 based on the more precise value of C_D:

$$V_0 = \frac{16.8\ \text{m/s}}{(0.48)^{1/2}} = 24.2\ \text{m/s}$$

Another calculation based on the new velocity of 24.2 m/s yields a C_D of 0.48. Because the drag coefficient is unchanged from the previous iteration, no further iterations are necessary, so the terminal velocity is

$$V_0 = 24.2\ \text{m/s}$$

This example can also be carried out using a programmable calculator and Eq. (11.10) for the drag coefficient. First a value is assumed for V_0, then the Reynolds number is calculated using

$$\text{Re} = 1326 V_0$$

Equation (11.10) is used to find C_D and a new velocity is calculated using the same equation as above. This velocity is then used to obtain a new Reynolds number and the cycle is repeated until V_0 no longer changes. For the above example, starting off with an initial velocity of 10 m/s, the following iterations result.

Iteration no.	V_0	C_D
1	25.28	0.433
2	24.30	0.478
3	24.34	0.476

This answer is essentially the same as obtained above. The difference is attributed to the accuracy with which we can read the values from Fig. 11.11 and the deviation of Eq. (11.10) from the standard drag curve.

example 11.7

A 20-mm plastic sphere (S = 1.3) is dropped in water. Determine its terminal velocity. Assume $T = 20°C$.

Solution We approach this problem in the same manner as Example 11.6:

$$F_D = (\gamma_s - \gamma_w)(1/6)\pi d^3$$

Then

$$V_0 = \left[\frac{(\gamma_s - \gamma_w)(4/3)d}{C_D\rho_w}\right]^{1/2}$$

$$= \left[\frac{(12.7 - 9.8)(10^3\ \text{N/m}^3)(4/3)(0.02\ \text{m})}{C_D \times 10^3\ \text{kg/m}^3}\right]^{1/2}$$

$$= \left(\frac{0.077}{C_D}\right)^{1/2} = \frac{0.28}{C_D^{1/2}}\ \text{m/s}$$

Now, the velocity of fall will not be as great as in air, so a reasonable first guess at C_D is 0.5. With this value of C_D, the velocity is

$$V_0 = \frac{0.28\ \text{m/s}}{0.5^{1/2}} = 0.40\ \text{m/s}$$

Then $\text{Re} = V\,d\rho/\mu$, where $\rho = 10^3\ \text{kg/m}^3$ and $\mu = 1.00 \times 10^{-3}\ \text{N}\cdot\text{s/m}^2$.

$$\text{Re} = \frac{(0.40\ \text{m/s})(0.02\ \text{m})(10^3\ \text{kg/m}^3)}{1.00(10^{-3}\ \text{N}\cdot\text{s/m}^2)} = 8 \times 10^3$$

From Fig. 11.11, C_D at this Reynolds number is 0.38, which gives a velocity of 0.45 m/s. This velocity, in turn, corresponds to a Reynolds number of 9×10^3 and a drag coefficient of 0.39. Substituting this value into the equation for velocity yields 0.45 m/s. No further iterations are necessary, so the terminal velocity is

$$V_0 = 0.45\ \text{m/s}$$

This same example can be done using a programmable calculator. Starting with an arbitrary velocity, calculate the Reynolds number and then use Eq. (11.10) for the drag coefficient. Calculate a new velocity using the drag coefficient and then use this velocity to repeat the cycle. Starting with an initial velocity of 5 m/s, the procedure gives the following results.

Iteration no.	V_0	C_D
1	0.399	0.491
2	0.401	0.405
3	0.437	0.409

The terminal velocity is 0.437 m/s, which agrees reasonably well with the previous result. The difference is due to the inaccuracy in reading values for the drag coefficient from Fig. 11.11 and the small error in the drag coefficient resulting from using Eq. (11.10).

Effect of Compressibility on Drag

The variation of drag coefficient with Mach number for three axisymmetric bodies is shown in Fig. 11.12. In each case, the drag coefficient increases only slightly with the Mach number at low Mach numbers and then increases sharply as transonic flow ($\text{M} \approx 1$) is approached. Note that the rapid increase in drag coefficient occurs at a higher Mach number (closer to unity) if the body is slender with a pointed nose. The drag coefficient reaches a maximum at a Mach number somewhat larger than unity and then decreases as the Mach number is further increased.

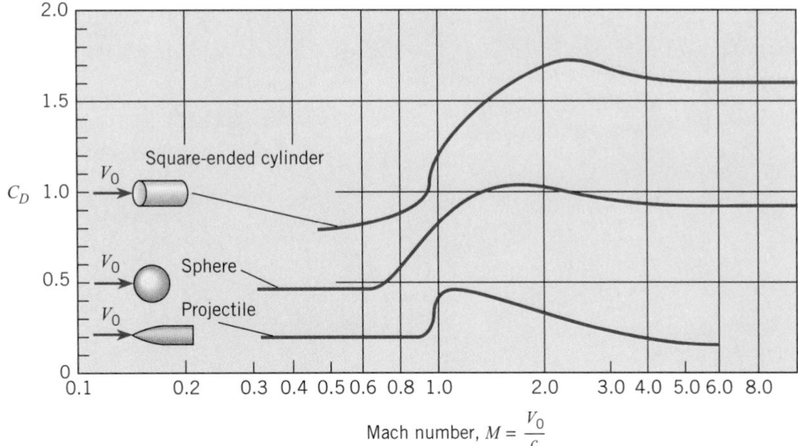

FIGURE 11.12

Drag characteristics of projectile, sphere, and cylinder with compressibility effects. [After Rouse (12)]

The slight increase in drag coefficient with low Mach numbers is attributed to an increase in form drag due to compressibility effects on the pressure distribution. However, as the flow velocity is increased, the maximum velocity on the body finally becomes sonic. The Mach number of the free-stream flow at which sonic flow first appears on the body is called the *critical Mach number.* Further increases in flow velocity result in local regions of supersonic flow ($M > 1$), which lead to wave drag due to shock wave formation and an appreciable increase in drag coefficient.

The critical Mach number for a sphere is approximately 0.6. One notes in Fig. 11.12 that the drag coefficient begins to rise sharply at about this Mach number. The critical Mach number for the pointed body is larger, and, correspondingly, the rise in drag coefficient occurs at a Mach number closer to unity.

The drag coefficient data for the sphere shown in Fig. 11.12 are for a Reynolds number of the order of 10^4. The data for the sphere shown in Fig. 11.11, on the other hand, are for very low Mach numbers. The question then arises about the general variation of the drag coefficient of a sphere with both Mach number and Reynolds number. Information of this nature is often needed to predict the trajectory of a body through the upper atmosphere, and it is sometimes needed to analyze the flows transporting solid-phase particles.

A contour plot of the drag coefficient of a sphere versus both Reynolds and Mach numbers based on available data (19) is shown in Fig. 11.13. One notices the C_D-versus-Re curve from Fig. 11.11 in the $M = 0$ plane. Correspondingly, one sees the C_D-versus-M curve from Fig. 11.12 in the $Re = 10^4$ plane. We see, then, that at low Reynolds numbers C_D decreases with increasing Mach number, whereas at high Reynolds numbers the opposite trend is observed. Using this figure, the engineer can determine the drag coefficient of a sphere at any combination of Re and M. Of course, corresponding C_D contour plots can be generated for any body, provided the data are available.

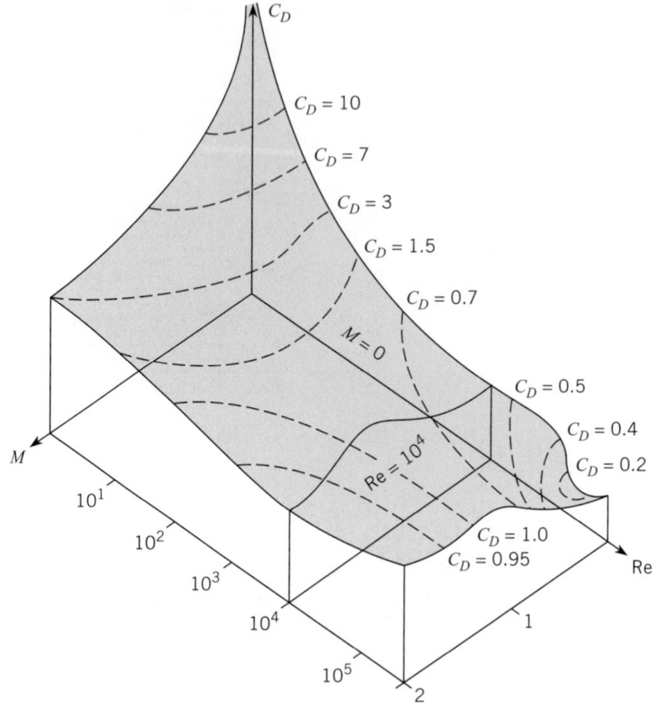

FIGURE 11.13

Contour plot of the drag coefficient of the sphere versus Reynolds and Mach numbers.

11.8 Lift

In Section 11.1 it was shown that a differential pressure between the top and bottom of a body causes a lateral force or lift to be imposed on the body. However, no explanation was given for the cause of the differences in velocity that produce such a pressure distribution. In this section we will consider circulation, which is the basic cause of lift. Then we will consider the lift and drag characteristics of typical airfoils.

Circulation

Consider a closed path such as is shown in Fig. 11.14. Along any differential segment of the path, the velocity can be resolved into components that are tangent and normal to the path. Let us signify the tangential component of velocity as V_L. If we integrate $V_L\, dL$ around the curve, the resulting quantity is called *circulation,* which is represented by the Greek letter Γ (capital gamma). Hence we have

$$\Gamma = \oint V_L\, dL \tag{11.11}$$

Sign convention dictates that in applying Eq. (11.11), we take tangential velocity vectors that have a counterclockwise sense around the curve as negative and take those that have

FIGURE 11.14

Concept of circulation.

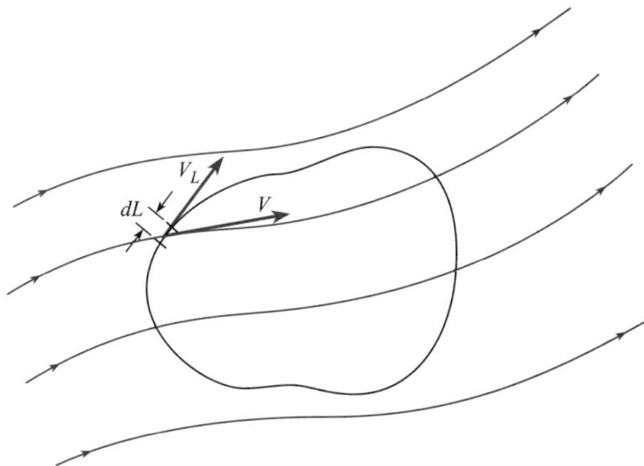

a clockwise direction as having a positive contribution.* The circulation for an irrotational vortex is determined as follows. The tangential velocity at any radius is C/r, where a positive C means a clockwise rotation. Therefore, if we evaluate the circulation about a curve with radius r, the differential circulation is

$$d\Gamma = V_L\, dL = \frac{C}{r_1} r_1\, d\theta = C\, d\theta \tag{11.12}$$

Then, when we integrate this around the entire circle, we obtain

$$\Gamma = \int_0^{2\pi} C\, d\theta = 2\pi C \tag{11.13}$$

One way to induce circulation physically is to rotate a cylinder about its axis. In Fig. 11.15*a* we see the flow pattern produced by such action. The velocity of the fluid next to the surface of the cylinder is equal to the velocity of the cylinder surface itself because of the nonslip condition that must prevail between the fluid and solid. At some distance from the cylinder, however, the velocity decreases with r, much like it does for the irrotational vortex. In the next section we will see how circulation produces lift.

Combination of Circulation and Uniform Flow around a Cylinder

If we now superpose the velocity field produced for uniform flow around a cylinder, Fig. 11.15*b*, onto a velocity field with circulation around a cylinder, Fig. 11.15*a*, we see that the velocity is reinforced on the top side of the cylinder and reduced on the other side (Fig. 11.15*c*). We also observe that the stagnation points have both moved toward the low-velocity side of the cylinder. Consistent with the Bernoulli principle (assuming irrotational flow throughout), we find that the pressure on the high-velocity side is lower than the pressure on the low-velocity side. Hence a pressure differential exists that causes a side thrust, or lift, on the cylinder. According to ideal-flow theory, the lift per unit length of an infinitely long cylinder is given by $F_L/\ell = \rho V_0 \Gamma$, where F_L is the lift on the segment of length ℓ. For this ideal irrotational flow there is no drag on the cylinder. For the

* The sign convention is the opposite of that for the mathematical definition of a line integral.

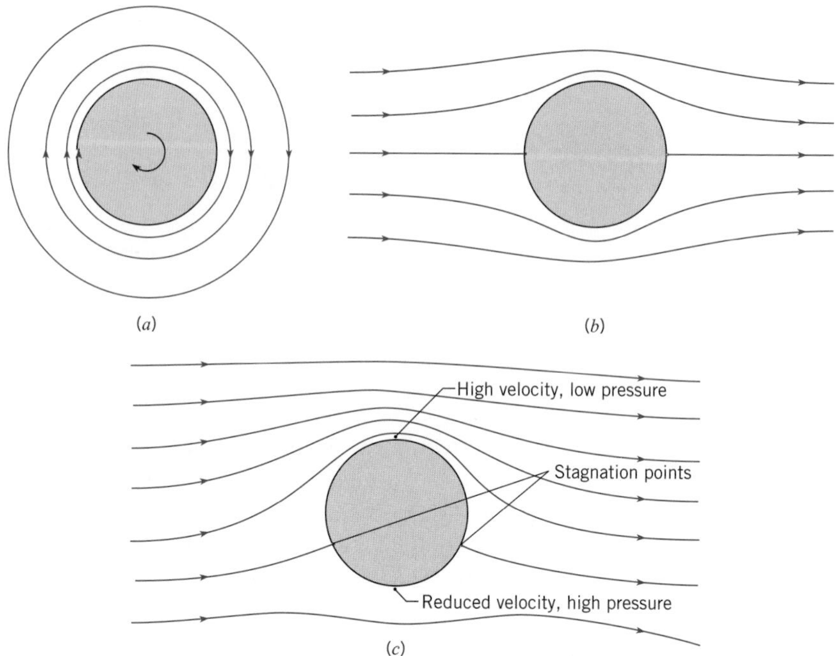

FIGURE 11.15

Ideal flow around a cylinder. (a) Circulation. (b) Uniform flow. (c) Combination of circulation and uniform flow.

real-flow case, separation and viscous stresses do produce drag, and the same viscous effects will reduce the lift somewhat. Even so, the lift is significant when flow occurs past a rotating body or when a body is translating and rotating through a fluid. Hence we have the reason for the "curve" on a pitched baseball or the "drop" on a Ping Pong ball that is given a fore spin. This phenomenon of lift produced by rotation of a solid body is called the *Magnus effect* after a nineteenth-century German scientist who made early studies of the lift on rotating bodies. A paper by Mehta (23) offers an interesting account of the motion of rotating sports balls.

Coefficients of lift and drag for the rotating cylinder with end plates are shown in Fig. 11.16. In this figure, the parameter $r\omega/V_0$ is the ratio of cylinder surface speed to the free-stream velocity, where r is the radius of the cylinder and ω is the angular speed in radians per second. The corresponding curves for the rotating sphere are given in Fig. 11.17. The *lift coefficient* is defined as

$$C_L = \frac{F_L}{A_p \rho V_0^2/2}$$

where A_p is the projected area.

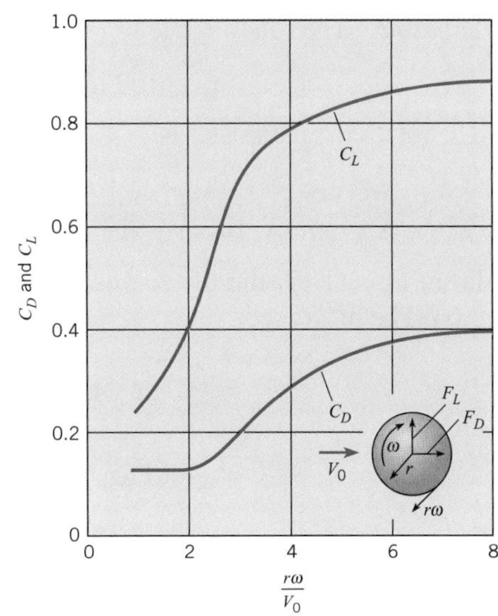

FIGURE 11.16

Coefficients of lift and drag as functions of $r\omega/V_0$ for a rotating cylinder. [After Rouse (12)]

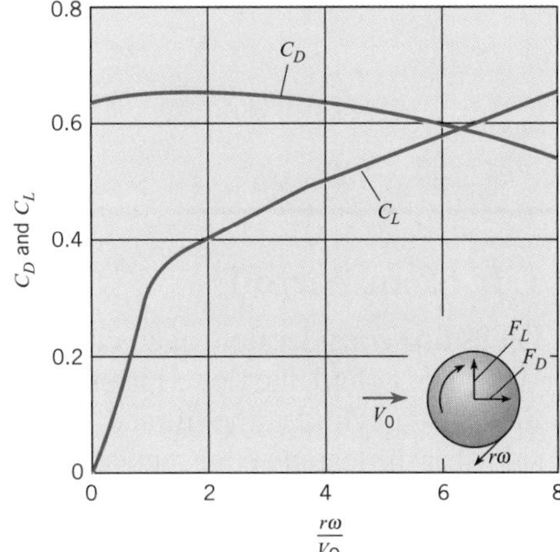

FIGURE 11.17

Coefficients of lift and drag for a rotating sphere. [After Barkla et al. (20). Reprinted with the permission of Cambridge University Press.]

example 11.8

A Ping Pong ball is moving at 10 m/s in air and is spinning at 100 revolutions per second in the clockwise direction. The diameter of the ball is 3 cm. Calculate the lift and drag force and indicate the direction of the lift (up or down). The density of air is 1.2 kg/m^3.

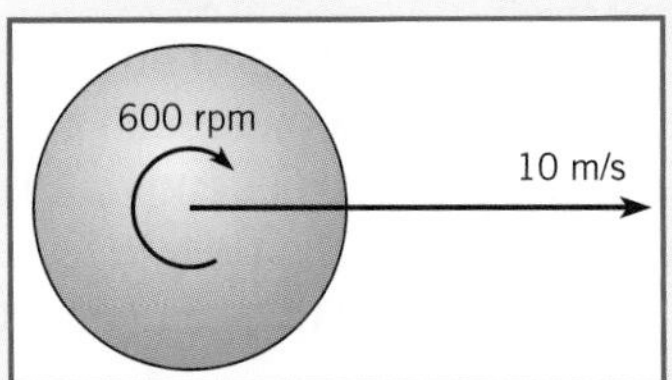

Solution The rotation rate in rad/s is

$$\omega = (100 \text{ rev/s})(2\pi \text{ rad/rev}) = 628 \text{ rad/s}$$

The rotational parameter is

$$\frac{\omega r}{V_0} = \frac{(628 \text{ rad/s})(0.015 \text{ m})}{10 \text{ m/s}} = 0.942$$

From Fig. 11.17, the lift coefficient is approximately 0.26 and the drag coefficient is 0.64. The lift force is

$$F_L = \frac{1}{2}\rho V_0^2 C_L A_p$$

$$= \frac{1}{2}(1.2 \text{ kg/m}^3)(10^2 \text{ m}^2/\text{s}^2)(0.26)\frac{\pi}{4}(0.03^2 \text{ m}^2)$$

$$= 1.10 \times 10^{-2} \text{ N}$$

The force is downward.

The drag force is

$$F_D = \frac{1}{2}\rho V_0^2 C_D A_p$$

$$= 27.1 \times 10^{-3} \text{ N}$$

Lift of an Airfoil

Let us first consider flow of an ideal fluid (nonviscous and incompressible) past an airfoil. With such a fluid, the flow is irrotational, as shown in Fig. 11.18*a*. Here, as for irrotational flow past a cylinder, the lift and drag are zero. There is a stagnation point on the bottom side near the leading edge, and another on the top side near the trailing edge of the foil. In the real flow (viscous fluid) case, the flow pattern around the upstream half of the foil is plausible. However, the flow pattern in the region of the trailing edge, as shown in Fig. 11.18*a*, cannot occur. A stagnation point on the upper side of the foil indicates that fluid must flow from the lower side around the trailing edge and then toward the stagnation point. Such a flow pattern implies an infinite acceleration of the fluid particles as they turn the corner around the trailing edge of the wing. This is a physical impossibility, and as we have seen in previous sections of the text, separation occurs at the sharp edge. As a consequence of the separation, the upstream stagnation point moves to the trailing edge. Flow from both the top and bottom sides of the airfoil in the vicinity of the trailing edge then leaves the airfoil smoothly and essentially parallel to these surfaces at the trailing edge (Fig. 11.18*b*).

To bring theory into line with the physically observed phenomenon, it was hypothesized that a circulation around the airfoil must be induced in just the right amount so that the downstream stagnation point is moved all the way back to the trailing edge of the airfoil, thus allowing the flow to leave the airfoil smoothly at the trailing edge. This is called the *Kutta condition* (21), named after a pioneer in aerodynamic theory. When analyses are

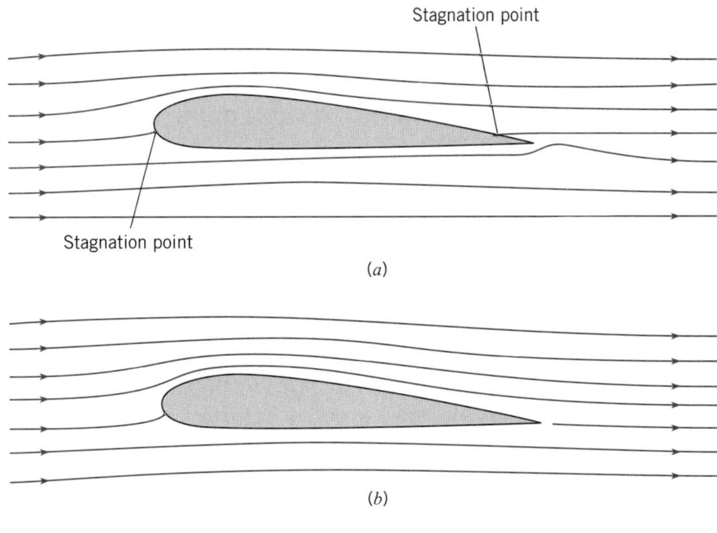

FIGURE 11.18

Patterns of flow around an airfoil. (a) Ideal flow—no circulation. (b) Real flow—circulation.

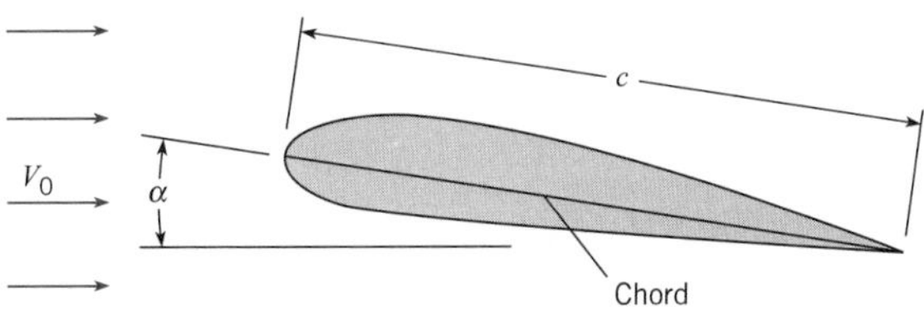

FIGURE 11.19

Definition sketch for an airfoil section.

made with this simple assumption concerning the magnitude of the circulation, very good agreement occurs between theory and experiment for the flow pattern and the pressure distribution, as well as for the lift on a two-dimensional airfoil section (no end effects). Ideal-flow theory then shows that the magnitude of the circulation required to maintain the rear stagnation point at the trailing edge (the Kutta condition) of a symmetric airfoil with a small angle of attack is given by

$$\Gamma = \pi c V_0 \alpha \tag{11.14}$$

where Γ is the circulation, c is the chord length of the airfoil, and α is the angle of attack of the chord of the airfoil with the free-stream direction (see Fig. 11.19 for a definition sketch).

Like that for the cylinder, the lift per unit length for an infinitely long wing is

$$F_L/\ell = \rho V_0 \Gamma$$

The planform area for the length segment ℓ is ℓc. Hence the lift on segment ℓ is

$$F_L = \rho V_0^2 \pi c \ell \alpha \tag{11.15}$$

For an airfoil we define the coefficient of lift as

$$C_L = \frac{F_L}{S\rho V_0^2/2} \tag{11.16}$$

where S is the planform area of the wing—that is, the area seen from the plan view. On combining Eqs. (11.14) and (11.15) and identifying S as the area associated with length segment ℓ, we find that C_L for irrotational flow past a two-dimensional airfoil is given by

$$C_L = 2\pi\alpha \tag{11.17}$$

Equations (11.15) and (11.17) are the theoretical lift equations for an infinitely long airfoil at a small angle of attack. Flow separation near the leading edge of the airfoil produces deviations (high drag and low lift) from the ideal-flow predictions at high angles of attack. Hence experimental wind tunnel tests are always made to evaluate the performance of a given type of airfoil section. For example, the experimentally determined values of lift coefficient versus α for two NACA airfoils are shown in Fig. 11.20. Note in this figure that the coefficient of lift increases with the angle of attack, α, to a maximum value and then decreases with further increase in α. This condition, where C_L starts to decrease with a further increase in α, is called *stall*. Stall occurs because of the onset of separation over the top of the airfoil, which changes the pressure distribution in such a way as not only to decrease lift but also to increase drag. Data for many other airfoil sections are given by Abbott and Von Doenhoff (22).

Airfoils of Finite Length—Effect on Drag and Lift

The drag of a two-dimensional foil at a low angle of attack (no end effects) is primarily viscous drag. However, wings of finite length alsohave an added drag and a reduced lift associated with vortices generated at the wing tips. These vortices occur because the high pressure below the wing and the low pressure on top cause fluid to circulate around the

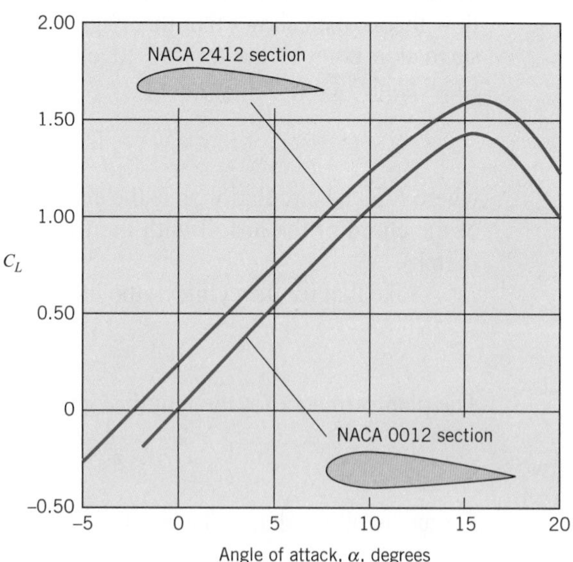

FIGURE 11.20

Values of C_L for two NACA airfoil sections. [After Abbott and Van Doenhoff (22)]

end of the wing from the high- to the low-pressure zone, as shown in Fig. 11.21. This induced flow has the effect of adding a downward component of velocity, w, to the approach velocity V_0. Hence, the "effective" free-stream velocity is now at an angle ($\phi \approx w/V_0$) to the direction of the original free-stream velocity, and the resultant force is tilted back as shown in Fig. 11.22. Thus the effective lift is smaller than the lift for the infinitely long wing because the effective angle of incidence is smaller. This resultant force has a component parallel to V_0 that is called the *induced drag* and is given by $F_L\phi$. Prandtl (25) showed that the induced velocity w for an elliptical spanwise lift distribution is given by the following equation:

$$w = \frac{2F_L}{\pi\rho V_0 b^2} \tag{11.18}$$

where b is the total length (or span) of the finite wing. Hence

$$F_{Di} = F_L\phi = \frac{2F_L^2}{\pi\rho V_0^2 b^2} = \frac{C_L^2}{\pi}\frac{S^2}{b^2}\frac{\rho V_0^2}{2} \tag{11.19}$$

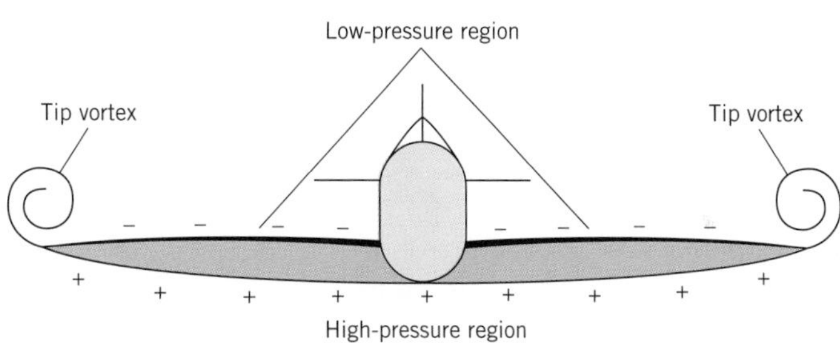

FIGURE 11.21

Formation of tip vortices.

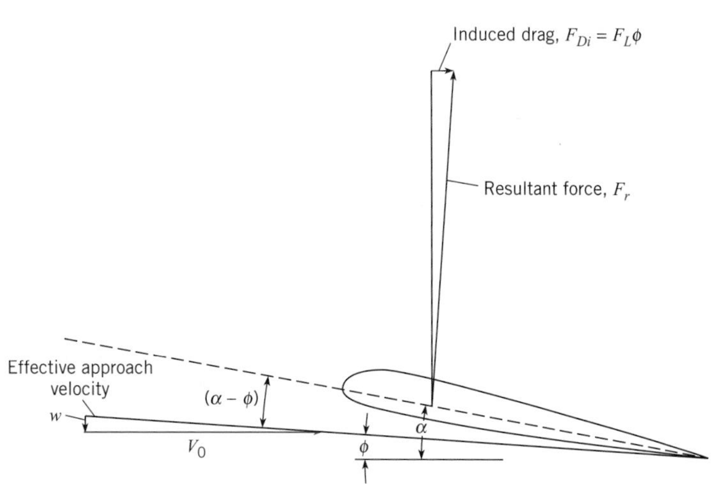

FIGURE 11.22

Definition sketch for induced-drag relations.

From Eq. (11.19) it can be easily shown that the coefficient of induced drag, C_{Di}, is given by

$$C_{Di} = \frac{C_L^{\,2}}{\pi(b^2/S)} \tag{11.20}$$

which happens to represent the minimum induced drag for any wing planform. Here the ratio b^2/S is called the aspect ratio Λ of the wing, and S is the planform area of the wing. Thus, for a given wing section (constant C_L and constant chord c), longer wings (larger aspect ratios) have smaller induced-drag coefficients. The induced drag is a significant portion of the total drag of an airplane at low velocities and must be given careful consideration in airplane design. Aircraft (such as gliders) and even birds (such as the albatross and gull) that are required to be airborne for long periods of time with minimum energy expenditure are noted for their long, slender wings. Such a wing is more efficient because the induced drag is small. To illustrate the effect of finite span, we turn to Fig. 11.23, which shows C_L and C_D versus α for wings with several aspect ratios.

The total drag of a rectangular wing is computed by

$$F_D = (C_{D0} + C_{Di})\frac{bc\rho V_0^2}{2}$$

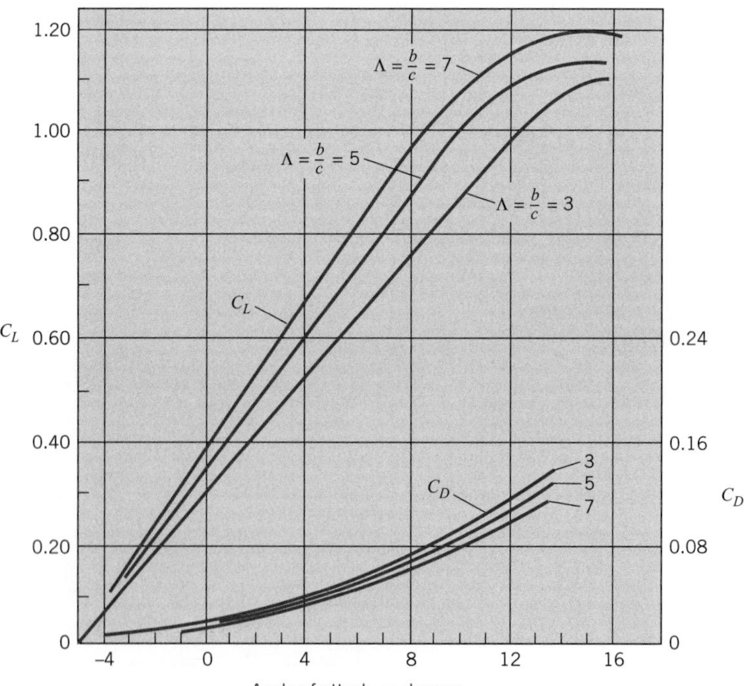

FIGURE 11.23

Coefficients of lift and drag for three wings with aspect ratios of 3, 5, and 7. [After Prandtl (23)]

example 11.9

An airplane with a weight of 10,000 lbf is flying at 600 ft/s at 36,000 ft, where the pressure is 3.3 psia and the temperature is –67°F. The lift coefficient is 0.2. The span of the wing is 54 ft. Calculate the wing area (in sq. ft.) and the minimum induced drag.

Solution The air density can be calculated using the equation of state.

$$\begin{aligned}\rho &= \frac{p}{RT}\\ &= \frac{(3.3\ \text{lbf/in}^2)(144\ \text{in}^2/\text{ft}^2)}{(1716\ \text{ft-lbf/slug-°R})(-67+460\text{°R})}\\ &= 0.000705\ \text{slug/ft}^3\end{aligned}$$

For steady flight, the lift force is equal to the weight,

$$W = F_L = \frac{1}{2}\rho V_0^2 C_L S$$

so

$$\begin{aligned}S &= \frac{2W}{\rho V_0^2 C_L}\\ &= \frac{2\times 10{,}000\ \text{lbf}}{(0.000705\ \text{slug/ft}^3)(600^2\ \text{ft}^2/\text{s}^2)(0.2)}\\ &= 394\ \text{ft}^2\end{aligned}$$

The minimum induced drag coefficient is

$$\begin{aligned}C_{Di} &= \frac{C_L^2}{\pi\left(\frac{b^2}{S}\right)}\\ &= \frac{0.2^2}{\pi\left(\frac{54^2}{394}\right)}\\ &= 0.00172\end{aligned}$$

The induced drag is

$$D_i = \frac{1}{2}\rho V_0^2 C_{Di} S$$

$$= \frac{1}{2}(0.000705 \text{ slug/ft}^3)(600^2 \text{ ft}^2/\text{s}^2)(0.00172)(394 \text{ ft}^2)$$

$$= 86.0 \text{ lbf}$$

where C_{D0} is the coefficient of form drag of the wing section and C_{Di} is the coefficient of induced drag.

A graph showing C_L and C_D versus α is given in Fig. 11.24. Note in this graph that C_D is separated into the induced-drag coefficient C_{Di} and the form drag coefficient C_{D0}.

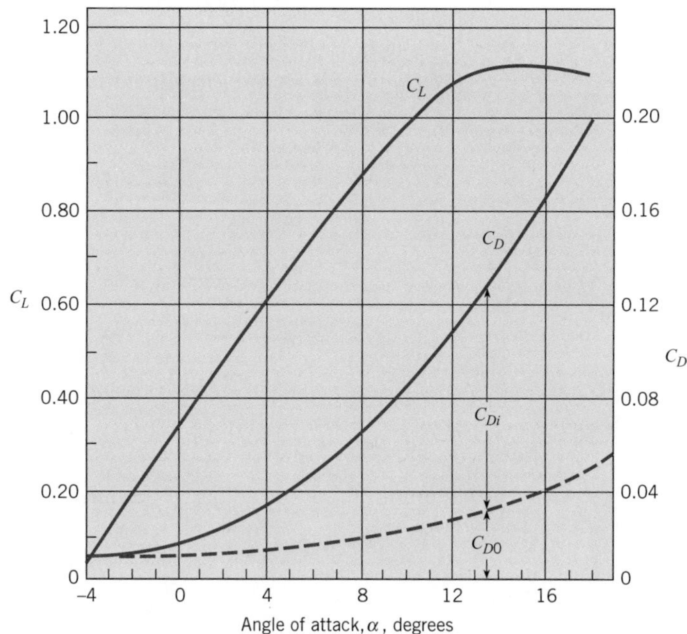

FIGURE 11.24

Coefficients of lift and drag for a wing with an aspect ratio of 5. [After Prandtl (23)]

example 11.10

A light plane (weight = 10 kN) has a wingspan of 10 m and a chord length of 1.5 m. If the lift characteristics of the wing are like those given in Fig. 11.23, what must be the angle of attack for a takeoff speed of 140 km/h? What is the stall speed? Assume two passengers at 800 N each and standard atmospheric conditions.

Solution The lift must be 10 kN + 1.6 kN = 11.6 kN. Hence

$$C_L = \frac{F_L}{S\rho V_0^2/2} = \frac{11{,}600 \text{ N}}{(15 \text{ m}^2)(1.2 \text{ kg/m}^3)[(140{,}000/3600)^2 \text{ m}^2/\text{s}^2]/2} = 0.852$$

Also $$\frac{b}{c} = \frac{10}{1.5} = 6.67$$

Hence from the curves of Fig. 11.23, we get

$$\alpha = 7° \quad \triangleleft$$

Stall will occur at $$C_L = 1.18$$

Then $$11{,}600 = 1.18(15)\left(\frac{1.2}{2}\right)(V_{\text{stall}})^2$$

$$V_{\text{stall}} = 33.0 \text{ m/s} = 119 \text{ km/h} \quad \triangleleft$$

Drag and Lift on Road Vehicles

Early in the development of cars, aerodynamic drag was a minor factor in performance because normal highway speeds were quite low. Thus in the 1920s, coefficients of drag for cars were around 0.80. As highway speeds increased and the science of metal forming became more advanced, cars took on a less angular shape, so that by the 1940s drag coefficients were 0.70 and lower. In the 1970s the average C_D for U.S. cars was approximately 0.55. In the early 1980s the average C_D for American cars dropped to 0.45, and currently auto manufacturers are giving even more attention to reducing drag in designing their cars. All major U.S., Japanese, and European automobile companies now have models with C_D's of about 0.33, and some companies even report C_D's as low as 0.29 on new models. European manufacturers were the leaders in the streamlining of cars because European gasoline prices (including tax) have been for a number of years about three times those in the United States. Table 11.2 shows the C_D for a 1932 Fiat and for other, more contemporary car models.

Great strides have been made in reducing the drag coefficients for passenger cars. However, significant future progress will be very hard to achieve. One of the most streamlined cars was the "Bluebird," which set a world land-speed record in 1938. Its C_D was 0.16. The minimum C_D of well-streamlined racing cars is about 0.20. Thus, lowering the C_D for passenger cars below 0.30 will require exceptional design and workmanship. For example, the underside of most cars is aerodynamically very rough (axles, wheels, muffler, fuel tank, shock absorbers, and so on). One way to smooth the underside is to add a panel to the bottom of the car. But then clearance may become a problem, and adequate dissipation of heat from the muffler may be hard to achieve. Other basic features of the automobile that contribute to drag but are not very amenable to drag reduction modifications are interior air-flow systems for engine cooling, wheels, exterior features such as rear-view mirrors and antennas, and other surface protrusions. The reader is directed to two books on road vehicle aerodynamics, (24) and (25), which address all aspects of the drag and lift of road vehicles in considerably more detail than is possible here.

To produce low-drag vehicles, the basic tear-drop shape is an idealized starting point. This shape can be altered to accommodate the necessary functional features of the

TABLE 11.2 COEFFICIENTS OF DRAG FOR CARS

Make and Model	Profile	C_D
1932 Fiat Balillo		0.60
Volkswagen "Bug"		0.46
Plymouth Voyager		0.36
Toyota Paseo		0.31
Dodge Intrepid		0.31
Ford Taurus		0.30
Mercedes-Benz E320		0.29
Ford Probe V (concept car)		0.14
GM Sunraycer (experimental solar vehicle)		0.12

vehicle. For example, the rear end of the tear-drop shape must be lopped off to yield an overall vehicle length that will be manageable in traffic and will fit in our garages. Also, the shape should be more wide than high. Wind-tunnel tests are always helpful in producing the most efficient design. One such test was done on a $3/8$ scale model of a typical notchback sedan. Wind tunnel test results for such a sedan are shown in Fig. 11.25. Here the centerline pressure distribution (distribution of C_P) for the conventional sedan is shown by a solid line and that for a sedan with a 68-mm rear-deck lip is shown by a

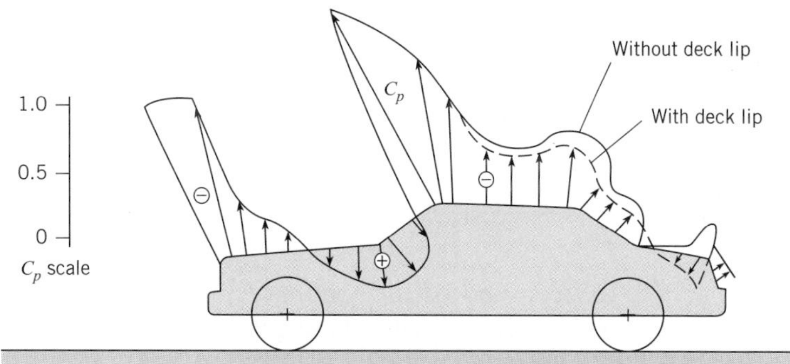

FIGURE 11.25

Effect of rear-deck lip on model surface. Pressure coefficients are plotted normal to the surface. [After Schenkel (25). Reprinted with permission from SAE Paper No. 770389. ©1977 Society of Automotive Engineers, Inc.]

dashed line. Clearly the rear-deck lip causes the pressure on the rear of the car to increase (C_p is less negative), thereby reducing the drag on the car itself. It also decreases the lift, thereby improving traction. Of course, the lip itself produces some drag, and these tests show that the optimum lip height for greatest overall drag reduction is about 20 mm.

Research and development programs to reduce the drag of automobiles continue. As an entry in the PNGV (Partnership for a New Generation of Vehicles), General Motors (26) has exhibited a vehicle with a drag coefficient as low as 0.163, which is approximately one-half that of the typical midsize sedan. These automobiles will have a rear engine to eliminate the exhaust system underneath the vehicle, and allow a flat underbody. Cooling air for the engine is drawn in through inlets on the rear fenders and exhausted out the rear, reducing the drag due to the wake. The protruding rear view mirrors are also removed to reduce the drag. The cumulative effect of these design modifications is a sizable reduction in aerodynamic drag.

The drag of trucks can be reduced by installing vanes near the corners of the truck body to deflect the flow of air more sharply around the corner, thereby reducing the degree of separation. This in turn creates a higher pressure on the rear surfaces of the truck, which reduces the drag of the truck. Problem 11.33 addresses this issue of drag reduction by the use of vanes.

One of the desired features in racing cars is the generation of negative lift to improve the stability and traction at high speeds. One idea (27) is to generate negative gage pressure underneath the car by installing a *ground-effect pod.* This is an airfoil section mounted across the bottom of the car that produces a venturi effect in the channel between the airfoil section and the road surface. The design of ground-effect vehicles involves optimizing design parameters to avoid separation and possible increase in drag. Another scheme to generate negative lift is the use of vanes as shown in Fig. 11.26. Sometimes "gurneys" are mounted on these vanes to reduce separation effects. Gurneys are small ribs mounted on the upper surface of the vanes near the trailing edge to induce local separation, reduce the

FIGURE 11.26

Racing car with negative-lift devices.

separation on the lower surface of the vane, and increase the magnitude of the negative lift. As the speed of racing cars continues to increase, automobile aerodynamics will play an ever-increasing role in traction, stability, and control.

example 11.11

The rear vane installed on the racing car of Fig. 11.26 is at an angle of attack of 8° and has characteristics like those given in Fig. 11.23. Estimate the downward thrust (negative lift) and drag from the vane that is 1.5 m long and has a chord length of 250 mm. Assume the racing car travels at a speed of 270 km/h on a track where normal atmospheric pressure and a temperature of 30°C prevail.

Solution

$$F_L = C_L \ell c \rho V_0^2/2$$

where $\ell = 1.5\text{ m} \quad c = 0.25\text{ m} \quad \ell/c = 6.0$

$$\rho = 1.17\text{ kg/m}^3 \qquad V_0 = 270\text{ km/h} = 75\text{ m/s}$$

$$C_L = 0.93 \qquad \text{(from Fig. 11.23)}$$

Then negative lift is

$$F_L = 0.93 \times 1.5 \times 0.25 \times 1.17 \times (75)^2/2 = 1148\text{ N} \qquad \triangleleft$$

Also $C_D = 0.070$ (from Fig. 11.23)

Therefore $F_D = (0.070/0.93) \times 1148 = 86.4\text{ N}$ ◁

11.9 Summary

A body immersed in a flowing fluid is subjected to pressure and sheer stress forces that produce lift and drag. The drag force, by definition, is parallel to the free-stream velocity and the lift force is perpendicular.

The drag force on a body is evaluated using

$$F_D = C_D \frac{1}{2} \rho V_0^2 A$$

where C_D is the drag coefficient, V_0 is the relative speed between the body and free-stream velocity, and A is the reference area. The drag coefficient is typically a function of Reynolds number based on the relative speed and a characteristic dimension of the body. Values of the drag coefficient for various shapes are determined analytically or experimentally and are published as equations or in tabular or graphical format. The drag force is the combination of two effects, form drag and skin friction drag. Form drag is due to pressure forces acting on the body while skin friction is due to shear stress on the body surface.

The drag coefficient of a sphere for Reynolds numbers less than 0.5 is

$$C_D = \frac{24}{\text{Re}}$$

which is known as *Stokes drag.*

For bluff bodies in high–Reynolds number flows, the drag force is primarily form drag and results from the reduced pressure in the body's wake. For streamlined bodies the form drag is reduced and skin friction drag plays a more important role. The drag coefficients of cylinders and spheres show a marked decrease at Reynolds numbers near 10^5. This effect is attributed to the flow in the boundary layer changing from laminar to turbulent, moving the separation point downstream, reducing the wake region, and decreasing the form drag. The Reynolds number where the drag coefficient decreases is the critical Reynolds number, and the phenomenon is known as the *critical Reynolds number effect.*

Cylinders and bluff bodies in a cross-flow produce vortex shedding in which vortices are released alternately from each side of the body. The frequency of vortex shedding is given by the Strouhal number

$$\text{St} = \frac{nd}{V_0}$$

where n is the rate at which vortices are shed (Hz) from one side and d is the cross-stream dimension of the body.

Increasing the Mach number of flow past a body increases the drag coefficient. The free-stream Mach number where sonic flow first occurs on a body is the critical Mach number.

The lift force on a body is quantified by

$$F_L = C_L \frac{1}{2} \rho V_0^2 S$$

where C_L is the lift coefficient and S is the reference area. The values for the lift coefficient for various bodies are obtained by analysis or experiment.

The lift on an airfoil is due to the circulation produced by the airfoil on the surrounding fluid. This circulatory motion causes a change in the momentum of the fluid and a lift on the airfoil. The lift coefficient for a symmetric two-dimensional wing (no tip effect) is

$$C_L = 2\pi\alpha$$

where α is the angle of attack (expressed in radians) and the reference angle is the product of the chord and a unit length of wing. As the angle of attack increases, the airfoil stalls and the lift coefficient decreases. A wing of finite span produces trailing vortices that reduce the angle of attack and produce an induced drag. The drag coefficient corresponding to the minimum induced drag is

$$C_{Di} = \frac{C_L^2}{\pi(b^2/S)}$$

where b is the wing span and S is the planform area of the wing.

References

1. Bullivant, W. K. "Tests of the NACA 0025 and 0035 Airfoils in the Full Scale Wind Tunnel." *NACA Rept.,* 708 (1941).

2. DeFoe, G. L. "Resistance of Streamline Wires." *NACA Tech. Note,* 279 (March 1928).

3. Goett, H. J., and W. K. Bullivant. "Tests of NACA 0009, 0012, and 0018 Airfoils in the Full Scale Tunnel." *NACA Rept.,* 647 (1938).

4. Jacobs, E. N. "The Drag of Streamline Wires." *NACA Tech. Note,* 480 (December 1933).

5. Jones, G. W., Jr. "Unsteady Lift Forces Generated by Vortex Shedding about a Large, Stationary, and Oscillating Cylinder at High Reynolds Numbers." *Symp. Unsteady Flow, ASME* (1968).

6. Lindsey, W. F. "Drag of Cylinders of Simple Shapes." *NACA Rept.,* 619 (1938).

7. Miller, B. L., J. F. Mayberry, and I. J. Salter. "The Drag of Roughened Cylinders at High Reynolds Numbers." *NPL Rept. MAR Sci.,* R132 (April 1975).

8. Roshko, A. "Turbulent Wakes from Vortex Streets." *NACA Rept.,* 1191 (1954).

9. Abbott, I. H. "The Drag of Two Streamline Bodies as Affected by Protuberances and Appendages." *NACA Rept.,* 451 (1932).

10. Brevoort, M. J., and U. T. Joyner. "Experimental Investigation of the RobinsonType Cup Anemometer." *NACA Rept.,* 513 (1935).

11. Freeman, H. B. "Force Measurements on a 1/40-Scale Model of the U.S. Airship 'Akron.'" *NACA Rept.,* 432 (1932).

12. Rouse, H. *Elementary Mechanics of Fluids.* John Wiley, New York, 1946.

13. Clift, R., J. R. Grace, and M. E. Weber. *Bubbles, Drops and Particles.* Academic Press, San Diego, CA (1978).

14. Clift, R., and W. H. Gauvin. "The Motion of Particles in Turbulent Gas Streams." *Proc. Chemeca '70,* vol. 1, pp. 14–28 (1970).

15. Hoerner, S. F. *Fluid Dynamic Drag.* Published by the author, 1958.

16. Morrison, R. B. (ed). *Design Data for Aeronautics and Astronautics.* John Wiley, New York, 1962.

17. Roberson, J. A., et al. "Turbulence Effects on Drag of Sharp-Edged Bodies." *J. Hydraulics Div., Am. Soc. Civil Eng.* (July 1972).

18. Scher, S. H., and L. J. Gale. "Wind Tunnel Investigation of the Opening Characteristics, Drag, and Stability of Several Hemispherical Parachutes." *NACA Tech. Note,* 1869 (1949).

19. Crowe, C. T., et al. "Drag Coefficient for Particles in Rarefied, Low Mach-Number Flows." In *Progress in Heat and Mass Transfer,* vol. 6, pp. 419–431. Pergamon Press, New York, 1972.

20. Barkla, H. M., et al. "The Magnus or Robins Effect on Rotating Spheres." *J. Fluid Mech.,* vol. 47, part 3 (1971).

21. Kuethe, A. M., and J. D. Schetzer. *Foundations of Aerodynamics.* John Wiley, New York, 1967.

22. Abbott, H., and A. E. Von Doenhoff. *Theory of Wing Sections.* Dover, New York, 1949.

23. Prandtl, L. "Applications of Modern Hydrodynamics to Aeronautics." *NACA Rept.,* 116 (1921).

24. Hucho, Wolf-Heinrich, ed. *Aerodynamics of Road Vehicles.* Butterworth, London, 1987.

25. Schenkel, Franz K. "The Origins of Drag and Lift Reductions on Automobiles with Front and Rear Spoilers." *SAE Paper,* no. 770389 (February 1977).

26. Sharke, P. "Smooth Body." *Mechanical Engineer,* vol. 121, pp. 74–77 (1999).

27. Smith, C. *Engineer to Win.* Motorbooks International, Osceola, WI (1984).

28. Mehta, R. D. "Aerodynamics of Sports Balls." *Annual Review of Fluid Mechanics,* 17, p. 151 (March 1985).

Problems

11.1 A hypothetical pressure coefficient variation over a long (length normal to the page) plate is shown. What is the coefficient of drag for the plate in this orientation and with the given pressure distribution? Assume that the reference area is the surface area (one side) of the plate.

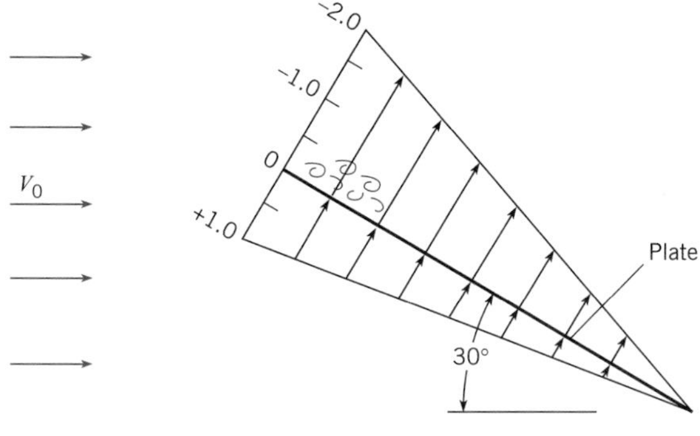

PROBLEM 11.1

11.2 Flow is occurring past the square rod. The pressure coefficient values are as shown. From which direction do you think the flow is coming? (a) SW direction, (b) SE direction, (c) NW direction, (d) NE direction.

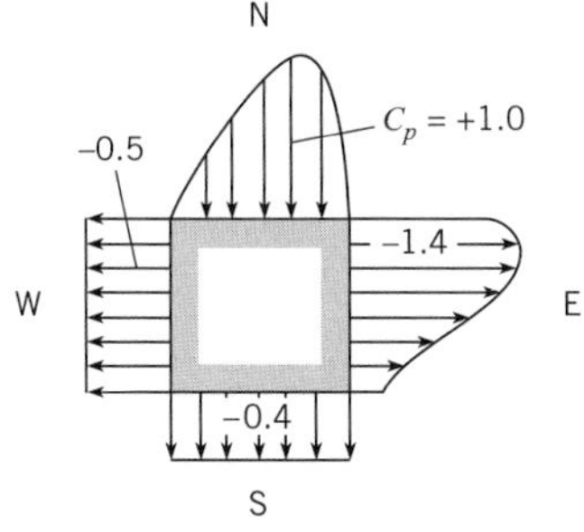

PROBLEM 11.2

11.3 The hypothetical pressure distribution on a rod of triangular (equilateral) cross section is shown, where flow is from left to right. That is, C_p is maximum and equal to +1.0 at the leading edge and decreases linearly to zero at the trailing edges. The pressure coefficient on the downstream face is constant with a value of –0.5. Neglecting skin friction drag, find C_D for the rod.

11.4 The pressure distribution on a rod having a triangular (equilateral) cross-section is shown, where flow is from left to right. What is C_D for the rod?

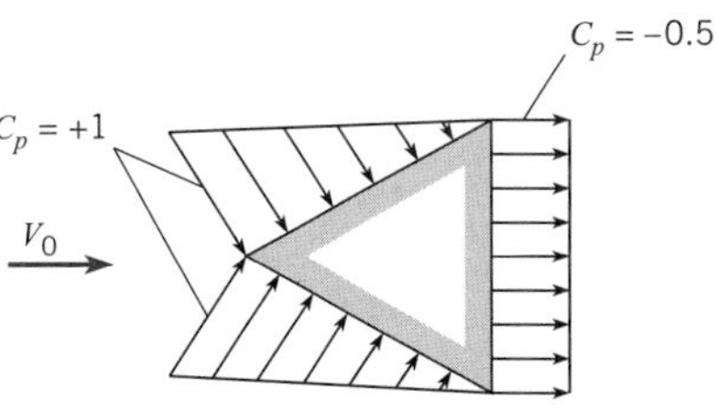

PROBLEM 11.3

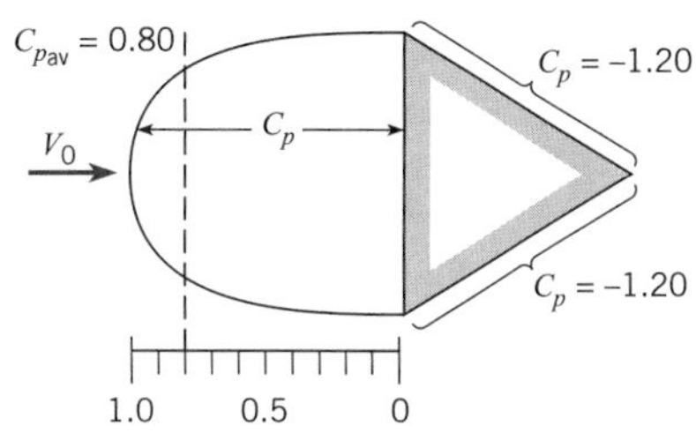

PROBLEM 11.4

11.5 You have access to a wind tunnel that can produce air velocity up to 100 ft/s in a test section that is rectangular (3 ft × 3 ft). Design equipment to determine the coefficient of drag of spheres having different surface roughness. Scope the test procedure in your design.

11.6 Estimate the energy in joules and kcal (food calories) that a runner supplies to overcome aerodynamic drag during a 10-km race. The runner runs a 6:30 pace (i.e., each mile takes 6 minutes and 30 seconds). The product of frontal area and coefficient of drag is $C_DA = 8\ \text{ft}^2$. (One "food calorie" is equivalent to 4186 J.) Assume an air density of 1.22 kg/m^3.

11.7 Compute the overturning moment exerted by a 35-m/s wind on a smokestack that has a diameter of 2.5 m and a height of 75 m. Assume that the air temperature is 20°C and that p_a is 99 kPa absolute. FLUID SOLUTIONS

11.8 What is the moment at the bottom of a flagpole 35 m (115 ft) high and 10 cm in diameter in a 25-m/s wind? The atmospheric pressure is 100 kPa (14.5 psia), and the temperature is 20°C (68°F). FLUID SOLUTIONS

11.9 A 50-cm-diameter pipe is designed to convey air with a range of flow from 2 to 6 m^3/s. Your supervisor asks you to design a flow measuring device that consists of a small cup, disk, or sphere attached to a cantilevered support and placed in the duct. Your supervisor says that by measuring the force acting on the device you can use that to estimate the rate of flow. Design the device and include example computations to show how the device will work. For a more accurate use of the device for indicating flow rate, it would undoubtedly have to be calibrated. Scope the method of calibration.

11.10 A cooling tower, used for cooling recirculating water in a modern steam power plant, is 350 ft high and 250 ft average diameter. Estimate the drag on the cooling tower in a 200 mph wind ($T = 60°\text{F}$).

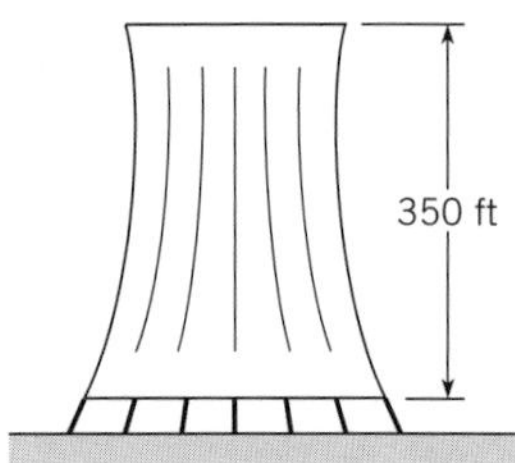

PROBLEM 11.10

11.11 A cylindrical rod of diameter d and length L is rotated in still air about its midpoint in a horizontal plane. Assume the drag force at each section of the rod can be calculated assuming a two-dimensional flow with an oncoming velocity equal to the relative velocity component normal to the rod. Assume C_D is constant along the rod.

a. Derive an expression for the average power needed to rotate the rod.

b. Calculate the power for $\omega = 50$ rad/s, $d = 2$ cm, $L = 1.5$ m, $\rho = 1.2$ kg/m^3, and $C_D = 1.2$.

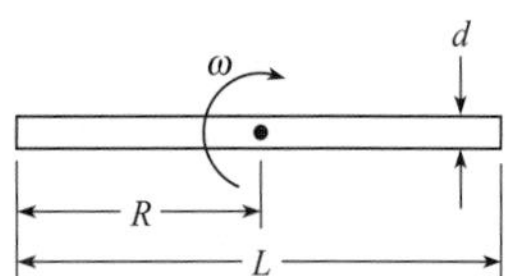

PROBLEM 11.11

FLUID SOLUTIONS **11.12** A Ping Pong ball of mass 2.6 g and diameter 38 mm is supported by an air jet. The air is at a temperature of 18°C and a pressure of 27 in. Hg. What is the minimum speed of the air jet?

11.13 At what frequency will vortices be shed from the flagpole of Prob. 11.8?

11.14 Estimate the wind force on a billboard 10 ft high and 30 ft wide when a 50-mph wind ($T = 60°\text{F}$) is blowing normal to it.

FLUID SOLUTIONS **11.15** Determine the drag of a 8 ft × 8 ft plate held at right angles to a stream of air (60°F and standard atmospheric pressure) having a velocity of 100 fps.

11.16 Estimate the ratio of the drag of a thin square plate (2 m by 2 m) when it is towed through water (10°C) in a normal orientation to that when it is towed edgewise. Assume a towing speed of about 1 m/s.

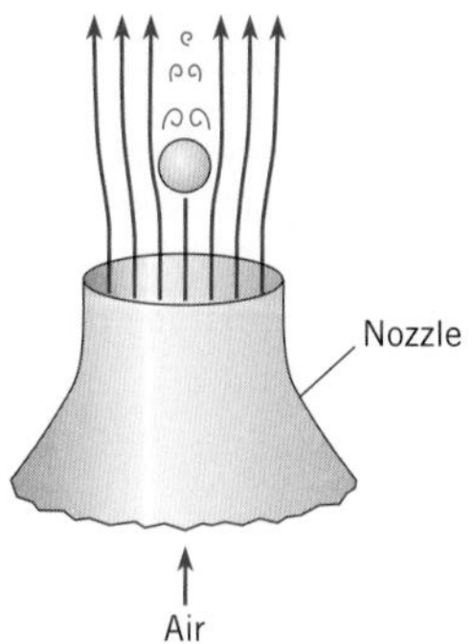

PROBLEM 11.12

11.17 What drag is produced when a disk 0.5 m in diameter is submerged in water at 10°C and towed behind a boat at a speed of 3 m/s? Assume orientation of the disk so that maximum drag is produced.

11.18 A circular billboard having a diameter of 6 m is mounted so as to be freely exposed to the wind. Estimate the total force exerted on the structure by a wind that has a direction normal to the structure and a speed of 30 m/s. Assume $T = 10°\text{C}$ and $p = 101$ kPa absolute.

11.19 Estimate the moment at ground level on a sign-post supporting a sign measuring 2 m by 2 m if the wind is normal to the surface and has a speed of 40 m/s and the center of the sign is 3 m above the ground. Neglect the wind load on the post itself. Assume $T = 10°\text{C}$ and $p = 100$ kPa absolute.

11.20 Estimate the additional power required for the truck when it is carrying the rectangular sign at a speed of 25 m/s over that required when it is traveling at the same speed but is not carrying the sign.

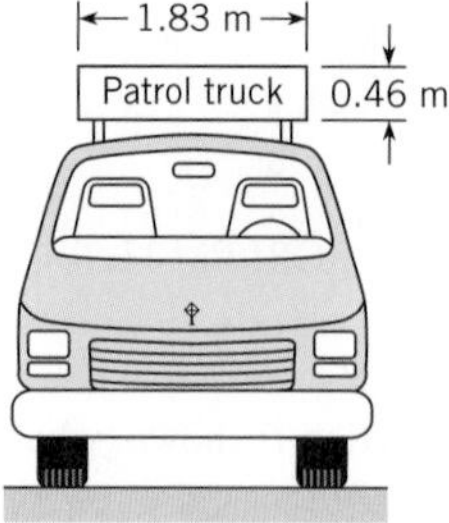

PROBLEM 11.20

11.21 Estimate the added power required for the car when the car-top carrier is used and the car is driven at 80 km/h in a 20 km/h head wind over that required when the carrier is not used in the same conditions.

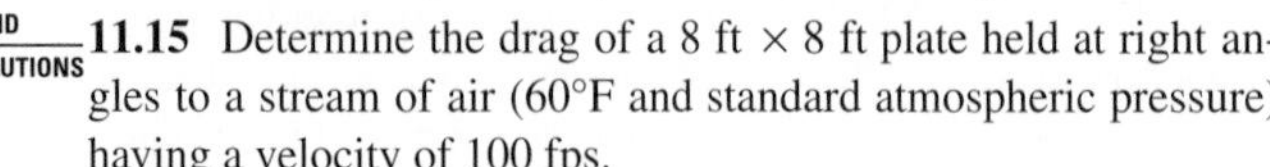

PROBLEM 11.21

11.22 The resistance to motion of an automobile consists of rolling resistance and aerodynamic drag. The rolling resistance is the product of the weight and the coefficient of rolling friction. The weight of an automobile is 3000 lb, and it has a frontal area of 20 ft^2. The drag coefficient is 0.30, and the coefficient of rolling friction is 0.02. Determine the percentage savings in gas mileage that one achieves when one drives at 55 mph instead of 65 mph on a level road. Assume an air temperature of 60°F.

11.23 A car coasts down a very long hill. The weight of the car is 2500 lbf, and the slope of the grade is 6%. The rolling friction coefficient is 0.01. The frontal area of the car is 20 ft^2, and the drag coefficient is 0.32. The density of the air is 0.002 slugs/ft^3. Find the maximum coasting speed of the car in mph.

11.24 An automobile with a mass of 1000 kg is driven up a hill where the slope is 3° (5.2% grade). The automobile is moving at 30 m/s. The coefficient of rolling friction is 0.02, the drag coefficient is 0.4, and the cross-sectional area is 4 m^2. Find the power (in kW) needed for this condition. The air density is 1.2 kg/m^3.

11.25 An automobile is traveling on a level road into a head wind of 15 m/s. The speed of the automobile with respect to the ground is 30 m/s. The drag coefficient is 0.4, and the projected area is 2 m^2. The air density is 1.2 kg/m^3. The car weighs 10,000 N, and the coefficient of rolling friction is 0.02. What power (in kilowatts) is needed to overcome both aerodynamic drag and rolling resistance?

11.26 Estimate the wind force that would act on *you* if you were standing on top of a tower in a 30-m/s (115-ft/s) wind on a day when the temperature was 20°C (68°F) and the atmospheric pressure was 96 kPa (14 psia).

11.27 Windstorms sometimes blow empty boxcars off their tracks. The dimensions of one type of boxcar are shown. What minimum wind velocity normal to the side of the car would be required to blow the car over?

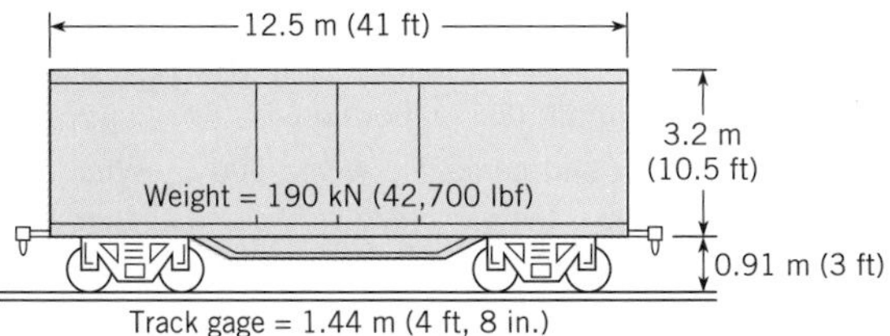

PROBLEM 11.27

11.28 A bicyclist is coasting down a hill with a slope of 8° into a head wind (measured with respect to the ground) of 5 m/s. The mass of the cyclist and bicycle is 80 kg, and the coefficient of rolling friction is 0.02. The drag coefficient is 0.5, and the projected area is 0.5 m^2. The air density is 1.2 kg/m^3. Find the speed of the bicycle in meters per second.

11.29 A bicyclist is capable of delivering 175 W of power to the wheels. The rolling resistance of a bicycle is negligible, so the power equals the aerodynamic drag times the velocity. How fast can the bicyclist travel in a 3-m/s head wind if his or her projected area is 0.5 m^2, the drag coefficient is 0.3, and the air density is 1.2 kg/m^3? FLUID SOLUTIONS

11.30 The drag coefficient of a sports car with a convertible roof is 0.3 when the roof is closed and increases to 0.42 when the roof is open. The resistance to motion consists of rolling resistance and aerodynamic drag. The rolling resistance is the product of the coefficient of rolling friction and the weight. The sports car has a mass of 800 kg, a frontal area of 4 m^2, and a coefficient of rolling friction of 0.05. The maximum power delivered to the wheels is 80 kW. Assuming a temperature of 20°C and a level road, determine the maximum possible speed with the roof closed and with it open.

11.31 A motorist has an automobile that has a drag coefficient of 0.30 and a frontal area (on which the drag coefficient is based) of 2 m^2. The total mass of driver and automobile is 500 kg. The coefficient of rolling friction is 0.1. Traveling into a head wind one day, the driver notes that the gasoline consumption is 20% higher than it is on a still day when she travels at the same speed, 90 km/h. The terrain is flat, and the air density is 1.2 kg/m^3. Find the velocity of the head wind.

11.32 Assume that the horsepower of the engine in the original 1932 Fiat Balillo (see Table 11.2) was 40 bhp (brake horsepower) and that the maximum speed at sea level was 60 mph. Also assume that the projected area of the automobile is 30 ft^2. Assume that the automobile is now fitted with a modern 220-bhp motor with a weight equal to the weight of the original motor; thus the rolling resistance is unchanged. What is the maximum speed of the "souped up" Balillo at sea level?

11.33 One way to reduce the drag of a blunt object is to install vanes to suppress the amount of separation. Such a procedure was used on model trucks in a wind-tunnel study by Kirsch and Bettes (18). For tests on a van-type truck, they noted that without vanes the C_D was 0.78. However, when vanes were installed around the top and side leading edges of the truck body (see the figure), a 25% reduction in C_D was achieved. For a truck with a projected area of 8.36 m^2, what reduction in drag force will be effected by installation of the vanes when the truck travels at 100 km/h? Assume standard atmospheric pressure and a temperature of 20°C.

11.34 A dirigible flies at 25 ft/s at an altitude where the specific weight of the air is 0.07 lbf/ft^3 and the kinematic viscosity is 1.3×10^{-4} ft^2/s. The dirigible has a length-to-diameter ratio

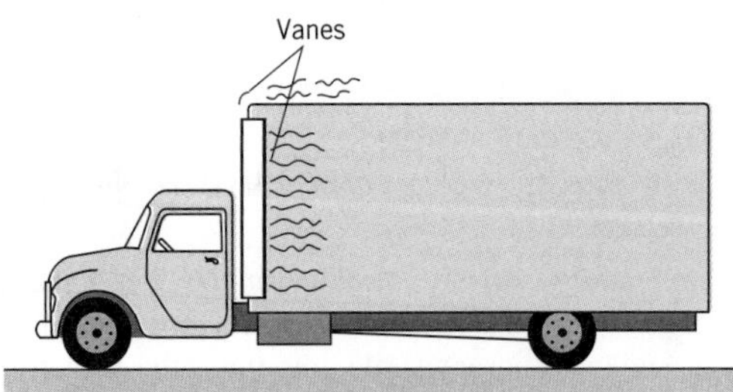

PROBLEM 11.33

of 5 and has a drag coefficient corresponding to the streamlined body in Fig. 11.11. The diameter of the dirigible is 100 ft. What is the power required to propel the dirigible at this speed?

11.35 For the truck of Prob. 11.33, assume that the total resistance is given by $R = F_D + C$, where F_D is the air drag and C is the resistance due to bearing friction, etc. If C is constant at 450 N for the given truck, what fuel savings percentage will be effected by the installation of the vanes when the truck travels at 80 km/h and at 100 km/h?

11.36 If the train of Prob. 9.72 has a form drag coefficient (for the locomotive and irregular undercarriage) of 0.80 and a constant bearing resistance of 3000 N, what will be the total resistance for the train at a speed of 100 km/h and at a speed of 200 km/h? Assume the projected area of 9 m^2 and standard atmospheric conditions prevail. What percentage of the resistance is due to bearing resistance, form drag, and skin friction drag at each of the two speeds?

11.37 Utilizing Stokes' law, design equipment to measure the viscosity of liquids such as water, kerosene, and glycerine. Also write instructions for use of the equipment.

11.38 A sphere 1 ft in diameter weighs 27 lbf in air. If it is released from rest 15 ft below the surface of a tank 30 ft deep that is filled with very viscous oil ($S = 0.85$, $\mu = 1.0$ lbf · s/ft^2), will its terminal velocity be in the laminar flow range? What will be the terminal velocity?

FLUID SOLUTIONS **11.39** A sphere 2 cm in diameter rises in oil at a velocity of 1.5 cm/s. What is the specific weight of the sphere if the oil density is 900 kg/m^3 and the dynamic viscosity is 0.096 N · s/m^2?

11.40 Estimate the terminal velocity of a 1.5-mm plastic sphere in oil. The oil has a specific gravity of 0.95 and a kinematic viscosity of 10^{-4} m^2/s. The plastic has a specific gravity of 1.07. The volume of a sphere is given by $\pi D^3/6$.

11.41 A 2-cm plastic ball with a specific gravity of 1.2 is released from rest in water at 20°C. Find the time and distance needed to achieve 99% of the terminal velocity. Write out the equation of motion by equating the mass times acceleration to the buoyant force, weight, and drag force and solve by developing a computer program, or using available software. Use Eq. (11.10) for the drag coefficient. [*Hint:* The equation of motion can be expressed in the form

$$\frac{dv}{dt} = -\left(\frac{C_D \,\mathrm{Re}}{24}\right)\frac{18\mu}{\rho_b d^2}v + \frac{\rho_b - \rho_w}{\rho_b}g$$

where ρ_b is the density of the ball and ρ_w is the density of the water. This form avoids the problem of the drag coefficient approaching infinity when the velocity approaches zero because $C_D\mathrm{Re}/24$ approaches unity as the Reynolds number approaches zero. An "if-statement" is needed to avoid a singularity in Eq. (11.10) when the Reynolds number is zero.]

11.42 Consider a small air bubble (approximately 4 mm diameter) rising in a very tall column of liquid. Will the bubble accelerate or decelerate as it moves upward in the liquid? Will the drag of the bubble be largely skin friction or form drag? Explain.

11.43 A 120 lbf (534 N) skydiver is free-falling at an altitude of 6500 ft (1980 m). Estimate the terminal velocity in mph for minimum and maximum drag conditions. At maximum drag conditions, the product of frontal area and coefficient of drag is $C_DA = 8$ ft^2 (0.743 m^2). At minimum drag conditions, $C_DA = 1$ ft^2 (0.0929 m^2). Assume the pressure and temperature at sea level are 14.7 psia (101 kPa) and 60°F (15°C). To calculate air properties, use the lapse rate for the U.S. standard atmosphere (see Chapter 3).

PROBLEM 11.43

11.44 If Stokes' law is considered valid below a Reynolds number of 0.5, what is the largest raindrop that will fall in accordance with Stokes' law? FL SO

11.45 What is the terminal velocity of a 0.5-cm hailstone in air that has an atmospheric pressure of 96 kPa absolute and a temperature of 0°C? Assume that the hailstone has a specific weight of 6 kN/m^3.

11.46 A spherical rock weighs 35 N in air and 7 N in water. Estimate its terminal velocity as it falls in water (20°C). FL SO

11.47 A drag chute is used to decelerate an airplane after touchdown. The chute has a diameter of 12 ft and is deployed when the aircraft is moving at 200 ft/s. The mass of the aircraft is 20,000 lbm, and the density of the air is 0.075 lbm/ft^3. Find the initial deceleration of the aircraft due to the chute.

11.48 A paratrooper and parachute weigh 900 N. What rate of descent will they have if the parachute is 7 m in diameter and the air has a density of 1.20 kg/m^3?

11.49 A cylinder of wood 80 cm long and 20 cm in diameter is weighted with lead at one end so that its total weight in air is 200

N. If this cylinder is released at a depth of 100 m in a lake 200 m deep, what will be the terminal velocity of the cylinder? Assume $T = 10°C$.

11.50 This cube is weighted so that it will fall with one edge down as shown. The cube weighs 19.8 N in air. What will be its terminal velocity in water?

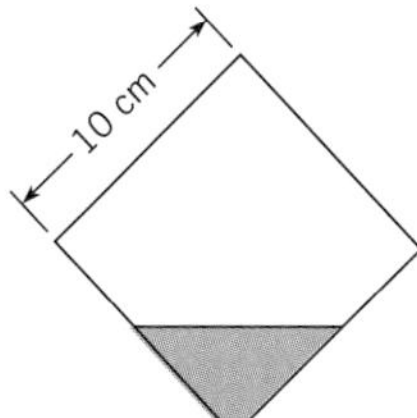

PROBLEM 11.50

11.51 If a balloon weighs 0.15 N (empty) and is inflated with helium to a diameter of 50 cm, what will be its terminal velocity in air (standard atmospheric conditions)? The helium is at standard conditions.

11.52 A balloon is released at sea level in a standard atmosphere (see Chapter 3). The balloon weighs 0.5 N empty and is inflated with helium (at standard conditions) to a diameter of 30 cm. Find the time for the balloon to reach an altitude of 5000 m. Assume the balloon does not change in size and neglect the effects of change in viscosity with temperature. Use Eq. (11.10) for the drag coefficient. [*Hint:* The equation of motion can be expressed in the form

$$\frac{dv}{dt} = -\left(\frac{C_D \,\text{Re}}{24}\right)\frac{18\mu}{\rho_H d^2 F}v + g\left(\frac{\rho_a}{\rho_H F} - 1\right)$$

where

$$F = 1 + \frac{W}{\rho_H g \mathcal{V}}$$

where ρ_a is the air density, ρ_H is the density of the helium at sea level, and $\mathcal{V}$ is the volume of the balloon. This form avoids the problem with drag coefficient as the Reynolds number approaches zero since $C_D \text{Re}/24 \to 1$. However, the singularity in Eq. (11.10) at a Reynolds number of zero has to be circumvented by using an "if" statement.]

11.53 If a balloon weighs 0.01 lbf (empty) and is inflated with helium to a diameter of 12 in., what will be its terminal velocity in air at 60°F and atmospheric pressure? The helium is at the same conditions.

11.54 A cylindrical anchor (vertical axis) made of concrete ($\gamma = 15 \text{ kN/m}^3$) is reeled in at a rate of 1.0 m/s by a man in a boat. If the anchor is 30 cm in diameter and 30 cm long, what tension must be applied to the rope to pull it up at this rate? Neglect the weight of the rope.

11.55 Determine the terminal velocity of a spherical pebble 1/4 in. in diameter that is falling in 75°F water. Assume that the specific gravity of the pebble equals 3.0 and that the kinematic viscosity of the water equals $1 \times 10^{-5} \text{ ft}^2/\text{s}$.

11.56 Determine the terminal velocity in water ($T = 10°C$) of a 10-cm ball that weighs 15 N in air.

11.57 A spherical balloon 2 m in diameter that is used for meteorological observations is filled with helium at standard conditions. The empty weight of the balloon is 3 N. What velocity of ascent will it attain under standard atmospheric conditions?

11.58 A meteor with a density 3000 kg/m^3 enters the atmosphere with a terminal velocity of Mach 1 in air at 20 kPa and −55°C. The meteor is spherical. What is the diameter of the meteor?

11.59 Determine the characteristics of a sphere that is to have a terminal velocity of 0.5 m/s when falling in water (20°C) and a diameter no smaller than 10 cm and no larger than 20 cm.

11.60 A sphere of diameter 0.3 ft rotating at a rate of 50 rad/s is placed in a stream of water (density = 1.94 slugs/ft^3) with velocity 3 ft/s. Determine the lift force on the sphere.

11.61 The baseball is thrown from west to east with a spin about its vertical axis as shown. Under these conditions it will "break" toward the (a) north, (b) south, (c) neither.

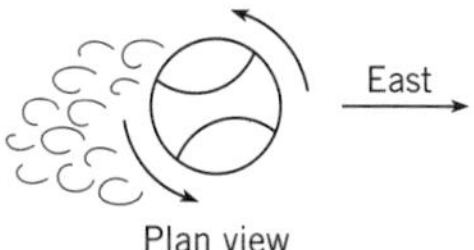

PROBLEM 11.61

11.62 Analyses of pitched baseballs indicate that C_L of a rotating baseball is approximately three times that shown in Fig. 11.17. This greater C_L is due to the added circulation caused by the seams of the ball. What is the lift of a ball pitched at a speed of 85 mph and with a spin rate of 35 rps? Also, how much will the ball be deflected from its original path by the time it gets to the plate as a result of the lift force? *Note:* The mound-to-plate distance is 60 ft, the weight of the baseball is 5 oz, and the circumference is 9 in. Assume standard atmospheric conditions, and assume that the axis of rotation is vertical.

11.63 This circular cylinder is positioned between the vertical walls of a wind tunnel and is rotated as shown. If the wind tunnel has air flow from right to left, choose which force vector (a, b, c, or d) best represents the force that must be applied to the cylinder to hold it in position. Neglect the weight of the cylinder.

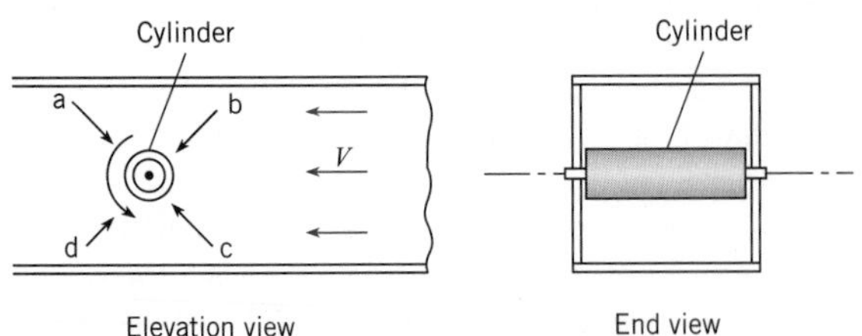

PROBLEM 11.63

11.64 A semiautomatic popcorn popper is shown. After the unpopped corn is placed in screen S, the fan F blows air past the heating coils C and then past the popcorn. When the corn pops, its projected area increases; thus it is blown up and into a container. Unpopped corn has a mass of about 0.15 g per kernel and an average diameter of approximately 6 mm. When the corn pops, its average diameter is about 18 mm. Within what range of airspeeds in the chamber will the device operate properly?

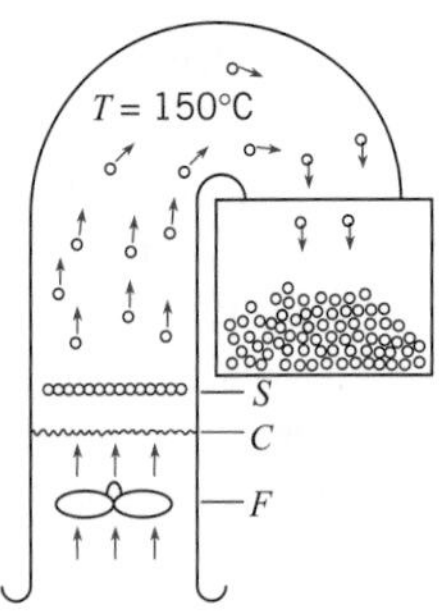

PROBLEM 11.64

11.65 Hoerner (14) presents data that show that fluttering flags of moderate-weight fabric have a drag coefficient (based on the flag area) of about 0.14. Thus the total drag is about 14 times the skin friction drag alone. Design a flagpole that is 100 ft high and is to fly an American flag 6 ft high. Make your own assumptions regarding other required data.

FLUID SOLUTIONS **11.66** For the conditions of Prob. 11.1 what is C_L?

FLUID SOLUTIONS **11.67** An airplane wing having the characteristics shown in Fig. 11.23 is to be designed to lift 2000 lbf when the airplane is cruising at 200 ft/s with an angle of attack of 3°. If the chord length is to be 4 ft, what span of wing is required? Assume $\rho = 0.0024 \text{ slugs/ft}^3$.

11.68 A boat of the hydrofoil type has a lifting vane with an aspect ratio of 4 that has the characteristics shown in Fig. 11.23. If the angle of attack is 4° and the weight of the boat is 5 tons, what foil dimensions are needed to support the boat at a velocity of 60 fps?

11.69 One wing (wing A) is identical (same cross section) to another wing (wing B) except that wing B is twice as long as wing A. Then for a given wind speed past both wings and with the same angle of attack, one would expect the total lift of wing B to be (a) the same as that of wing A, (b) less than that of wing A, (c) double that of wing A, (d) more than double that of wing A.

11.70 What happens to the value of the induced drag coefficient for an aircraft that increases speed in level flight? (a) it increases, (b) it decreases, (c) it does not change.

11.71 The total drag coefficient for an airplane wing is $C_D = C_{D_0} + C_L^2/\pi\Lambda$, where C_{D_0} is the form drag coefficient, C_L is the lift coefficient and Λ is the aspect ratio of the wing. The power is given by $P = F_DV = 1/\ 2C_D\rho V^3S$. For level flight the lift is equal to the weight, so $W/S = 1/2\rho C_LV^2$, where W/S is called the "wing loading." Find an expression for V for which the power is a minimum in terms of $V_{\text{MinPower}} = f(\rho, \Lambda, W/S, C_{D_0})$, and find the V for minimum power when $\rho = 1 \text{ kg/m}^3$, $\Lambda = 10$, $W/S = 600 \text{ N/m}^2$, and $C_D = 0.02$.

11.72 The airstream affected by the wing of an airplane can be considered to be a cylinder (stream tube) with a diameter equal to the wing span, b. Far downstream from the wing, the tube is deflected through an angle θ from the original direction. Apply the momentum equation to the stream tube between sections 1 and 2 and find the lift of the wing as a function of b, ρ, V, and θ. Relating the lift to the lift coefficient, find θ as a function of b, C_L, and wing area, S. Using the relation for induced drag, $F_{Di} = F_L\theta/2$, show that $C_{Di} = C_L^2/\pi\Lambda$, where Λ is the wing aspect ratio.

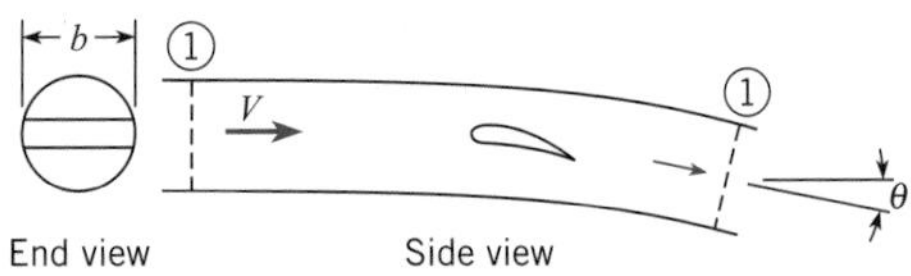

PROBLEM 11.72

11.73 The landing speed of an airplane is 8 m/s faster than its stalling speed. The lift coefficient at landing speed is 1.2, and the maximum lift coefficient (stall condition) is 1.4. Calculate both the landing speed and the stalling speed.

11.74 An airplane has a rectangular-planform wing that has an elliptical spanwise lift distribution. The airplane has a mass of 1200 kg, a wing area of 20 m^2, and a wingspan of 14 m, and it is flying at 60 m/s at 3000-m altitude in a standard atmosphere. If the form drag coefficient is 0.01, calculate the total drag on the wing and the power ($P = F_DV$) necessary to overcome the drag.

11.75 The figure shows the relative pressure distribution for a Göttingen 387-FB lifting vane (19) when the angle of attack is 8°. If such a vane with a 20-cm chord were used as a hydrofoil at a depth of 70 cm, at what speed in 10°C fresh water would cavitation begin? Also, estimate the lift per unit of length of foil at this speed.

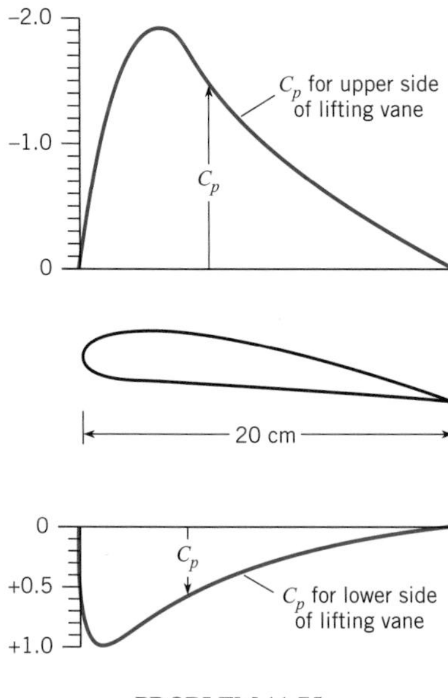

PROBLEM 11.75

11.76 Consider the distribution of C_p as given for the wing section in Prob. 11.75. For this distribution of C_p, the lift coefficient C_L will fall within which range of values: (a) $0 < C_L < 1.0$; (b) $1.01 < C_L < 2.0$; (c) $2.01 < C_L < 3.0$; (d) $3.0 < C_L$?

11.77 The total drag coefficient for a wing with an elliptical lift distribution is $C_D = C_{D_0} + C_L^2/\pi\Lambda$, where Λ is the aspect ratio. Derive an expression for C_L that corresponds to minimum C_D/C_L (maximum C_L/C_D) and the corresponding C_L/C_D.

11.78 A glider at 1000-m altitude has a mass of 200 kg and a wing area of 20 m^2. The glide angle is 1.7°, and the air density is 1.2 kg/m^3. If the lift coefficient of the glider is 0.8, how many minutes will it take to reach sea level on a calm day?

11.79 The wing loading on an airplane is defined as the aircraft weight divided by the wing area. An airplane with a wing loading of 2000 N/m^2 has the aerodynamic characteristics given by Fig. 11.24. Under cruise conditions the lift coefficient is 0.3. If the wing area is 10 m^2, find the drag force.

11.80 An ultralight airplane has a wing with an aspect ratio of 5 and with lift and drag coefficients corresponding to Fig. 11.23. The planform area of the wing is 200 ft^2. The weight of the airplane and pilot is 400 lbf. The airplane flies at 50 ft/s in air with a density of 0.002 slugs/ft^3. Find the angle of attack and the drag force on the wing.

11.81 Your objective is to design a human-powered aircraft using the characteristics of the wing in Fig. 11.23. The pilot weighs 130 pounds and is capable of outputting 1/2 horsepower (225 ft-lbf/s) of continuous power. The aircraft without the wing has a weight of 40 lbf, and the wing can be designed with a weight of 0.12 lbf per square foot of wing area. The drag consists of the drag of the structure plus the drag of the wing. The drag coefficient of the structure, C_{D_0} is 0.05, so that the total drag on the craft will be

$$F_D = (C_{D_0} + C_D)\frac{1}{2}\rho V_0^2 S$$

where C_D is the drag coefficient from Fig. 11.23. The power required is equal to $F_D V_0$. The air density is 0.00238 slugs/ft^3. Assess if the airfoil is adequate and, if it is, find the optimum design (wing area and aspect ratio).

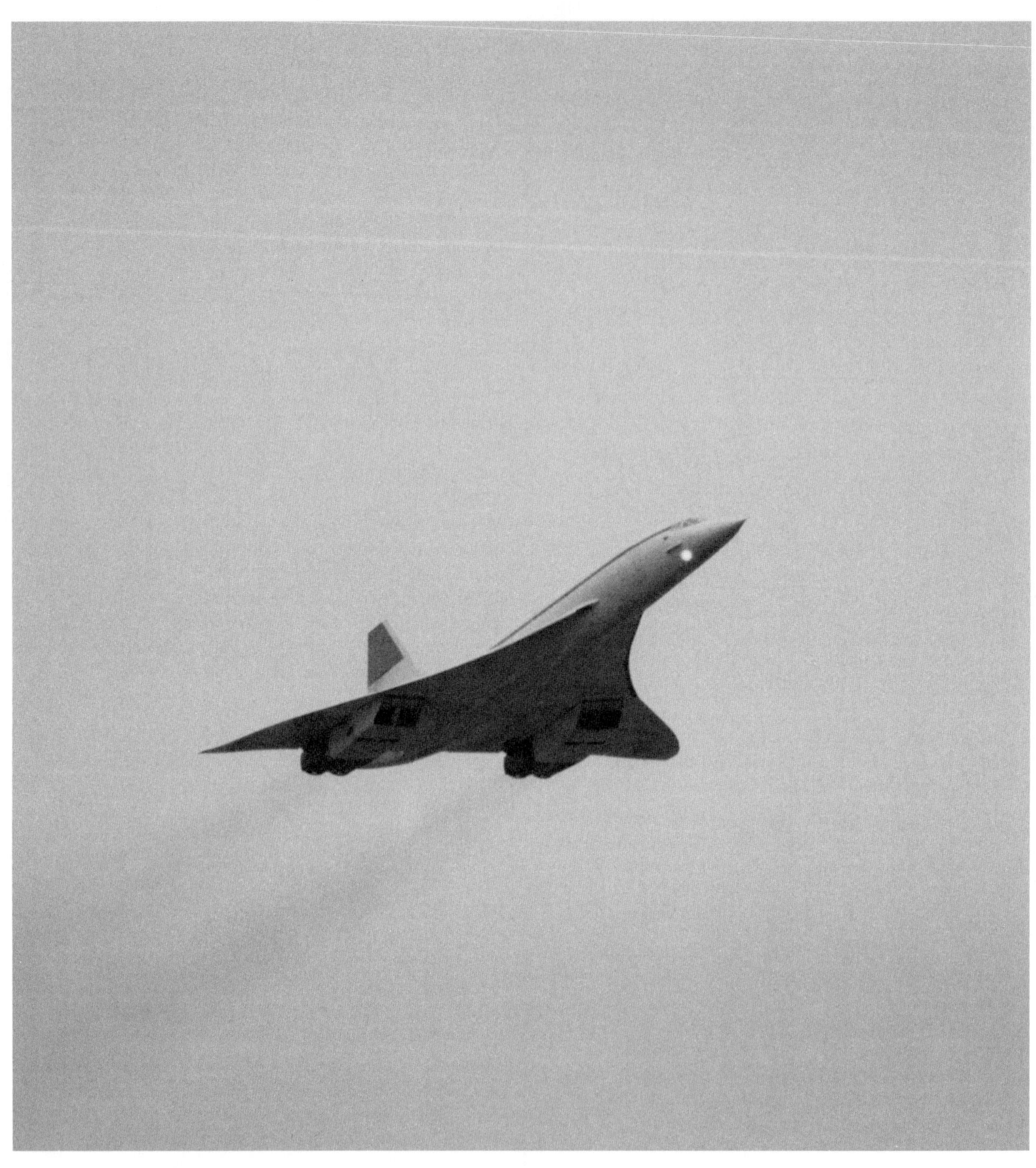

The Concorde was the first commercial supersonic transport. It carried 100 passengers and flew at 1336 mph (mach 2) at an altitude of 55,000 ft. The shortest flight time from New York to London was 2 hours and 54 minutes. The last flight of the Concorde was October 24, 2003. (PhotoDisc, Inc./Getty Images)

CHAPTER 12

Compressible Flow

Up to this point in our study of fluid flow, we have assumed that the density of the fluid is constant. This assumption is well founded for the steady flow of liquids because very large, atypical pressure changes effect only small density variations. For example, an increase in pressure of 20 MPa changes the density of water by less than 1%. On the other hand, pressure changes normally encountered in gas flow problems can cause significant density changes. For example, if the same 20 MPa pressure change were applied to air initially at atmospheric pressure, its density would change by 4370%. A change of this magnitude cannot be regarded as negligible!

The variables that describe the state of a flowing liquid are velocity, pressure, and temperature. The equations available to solve for these variables derive from the conservation of mass, momentum, and energy. If, in addition, density is a variable, one more equation is necessary. This is the equation of state, which for an ideal gas is

$$p = \rho RT \tag{12.1}$$

and which was introduced in Chapter 2. The developments in this chapter will be based on ideal-gas relationships.

12.1 Wave Propagation in Compressible Fluids

The speed of a flowing liquid is typically much less than the speed at which a pressure disturbance is propagated through the liquid. Gas flows, on the other hand, can achieve speeds that are comparable to and even exceed the speed at which pressure disturbances are propagated. In this situation, with compressible fluids, the propagation speed is an important parameter and must be incorporated into the flow analysis. In this section we will show how the speed of an infinitesimal pressure disturbance can be evaluated and what its significance is to flow of a compressible fluid.

Speed of Sound

Everyone has had the experience during a thunderstorm of seeing lightning flash and hearing the accompanying thunder an instant later. Obviously, the sound was produced by the lightning, so the sound wave must have traveled at a finite speed. If the air were totally incompressible (if that were possible), we would hear the thunder and see the flash simultaneously, because all disturbances propagate at infinite speed through incompressible media.* It is analogous to striking one end of a bar of incompressible material and recording instantaneously the response at the other end. Actually, all materials are compressible to some degree and propagate disturbances at finite speeds.

The *speed of sound* is defined as the rate at which an *infinitesimal* disturbance (pressure pulse) propagates in a medium with respect to the frame of reference of that medium. Actual sound waves, comprised of pressure disturbances of finite amplitude, such that the ear can detect them, travel only slightly faster than the "speed of sound." We found in Chapter 6 that the speed at which a pressure wave travels through a fluid depends on the bulk modulus of the fluid and its density. We performed that analysis by considering the unsteady flow within a control volume as the wave passed through the control volume. Here we will derive an equation for the speed of sound assuming the control volume moves with the wave, thereby analyzing a steady-flow problem.

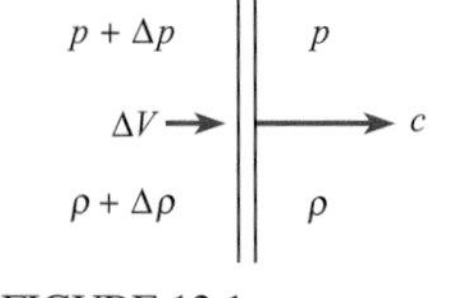

FIGURE 12.1

Section of a sound wave.

Let us consider a small section of a pressure wave as it propagates at velocity c through a medium, as depicted in Fig. 12.1. As the wave travels through the gas at pressure p and density ρ, it produces infinitesimal changes of Δp, $\Delta\rho$, and ΔV. We realize that these changes must be related through the laws of conservation of mass and momentum. Let us draw a control surface around the wave and let the control volume travel with the wave. The velocities, pressures, and densities relative to the control volume (which is assumed to be very thin) are shown in Fig. 12.2. Conservation of mass in a steady flow requires that the net mass flux across the control surface be zero. Thus

$$-\rho c A + (\rho + \Delta\rho)(c - \Delta V)A = 0 \tag{12.2}$$

where A is the cross-sectional area of the control volume. Neglecting products of higher-order terms ($\Delta\rho\Delta V$) and dividing by the area reduces the conservation-of-mass equation to

$$-\rho\Delta V + c\Delta\rho = 0 \tag{12.3}$$

The momentum equation for steady flow,

FIGURE 12.2

Flow relative to the sound wave.

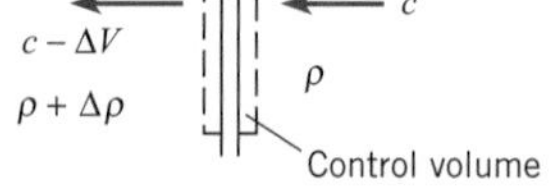

* Actually, the thunder would be heard before the lightning was seen, because light also travels at a finite, though very high, speed! However, this would violate one of the basic tenets of relativity theory. No medium can be completely incompressible and propagate disturbances exceeding the speed of light.

$$\sum \mathbf{F} = \dot{m}_o \mathrm{v}_o - \dot{m}\mathrm{v}_i \tag{12.4}$$

applied to the control volume containing the pressure wave gives

$$(p + \Delta p)A - pA = (-c)(-\rho A c) + (-c + \Delta V)\rho A c \tag{12.5}$$

where the direction to the right is defined as positive. The momentum equation reduces to

$$\Delta p = \rho c \Delta V \tag{12.6}$$

Substituting the expression for ΔV obtained from Eq. (12.3) into Eq. (12.6) gives

$$c^2 = \frac{\Delta p}{\Delta \rho} \tag{12.7}$$

which shows how the speed of propagation is related to the pressure and density change across the wave. We immediately see from this equation that if the flow were ideally incompressible, $\Delta\rho = 0$, the propagation speed would be infinite, which confirms the argument presented earlier.

Equation (12.7) gives us an expression for the speed of a general pressure wave. The sound wave is a special type of pressure wave. By definition, a sound wave produces only infinitesimal changes in pressure and density, so it can be regarded as a reversible process. There is also negligibly small heat transfer, so one can assume the process is *adiabatic*. A reversible, adiabatic process is an *isentropic* process; thus the resulting expression for the speed of sound is

$$c^2 = \left.\frac{\partial p}{\partial \rho}\right|_s \tag{12.8}$$

This equation is valid for the speed of sound in any substance. However, for many substances the relationship between p and ρ at constant entropy is not very well known.

To reiterate, the speed of sound is the speed at which an infinitesimal pressure disturbance travels through a fluid. Waves of finite strength (finite pressure change across the wave) travel faster than sound waves. Sound speed is the *minimum* speed at which a pressure wave can propagate through a fluid.

We shall now determine the speed of sound in an ideal gas. It can be shown from thermodynamics—see Oswatitsch (4), for example—that the following relationship between pressure and density holds for an isentropic process.

$$\frac{p}{\rho^k} = \text{constant} \tag{12.9}$$

Here k is the ratio of specific heats—that is, the ratio of specific heat at constant pressure to that at constant volume.

$$k = \frac{c_p}{c_v} \tag{12.10}$$

The values of k for some commonly used gases are given in Table A.2 in the Appendix. Taking the derivative of Eq. (12.9) to obtain $\partial p/\partial\rho|_s$ results in

$$\left.\frac{\partial p}{\partial \rho}\right|_s = \frac{kp}{\rho} \tag{12.11}$$

However, from the equation of state for an ideal gas,

$$\frac{p}{\rho} = RT$$

so the *speed of sound* is given by

$$c = \sqrt{kRT} \tag{12.12}$$

Thus we find that the speed of sound in an ideal gas varies with the square root of the temperature. Using this equation to predict sound speeds in real gases at standard conditions gives results very near the measured values. Of course, if the state of the gas is far removed from ideal conditions (high pressures, low temperatures), then using Eq. (12.12) is not advisable.

example 12.1

Calculate the speed of sound in air at 15°C.

Solution From Table A.2 we find that $k = 1.4$ and $R = 287$ J/kg K for air. Using the speed-of-sound equation, we calculate

$$c = [(1.4)(287 \text{ J/kg K})(288 \text{ K})]^{1/2} = 340 \text{ m/s} \quad \triangleleft$$

Mach Number

We shall now demonstrate in a very simple way the significance of sound in a compressible flow. Consider the airfoil traveling at speed V in Fig. 12.3. As this airfoil travels through the fluid, the pressure disturbance generated by the airfoil's motion propagates as a wave at sonic speed ahead of the airfoil. These pressure disturbances travel a considerable distance ahead of the airfoil before being attenuated by the viscosity of the fluid, and they "warn" the upstream fluid that the airfoil is coming (the Paul Reveres of fluid flow!). In turn, the fluid particles begin to move apart in such a way that there is a smooth flow over the airfoil by the time it arrives. If a pressure disturbance created by the airfoil is essentially attenuated in time Δt, then the fluid at a distance $\Delta t(c - V)$ ahead is alerted to prepare for the airfoil's impending arrival.

What happens as the speed of the airfoil is increased? Obviously, the relative velocity $c - V$ is reduced, and the upstream fluid has less time to prepare for the airfoil's arrival. The flow field is modified by smaller streamline curvatures, and the form drag on the airfoil is increased. If the airfoil speed increases to the speed of sound or greater, the

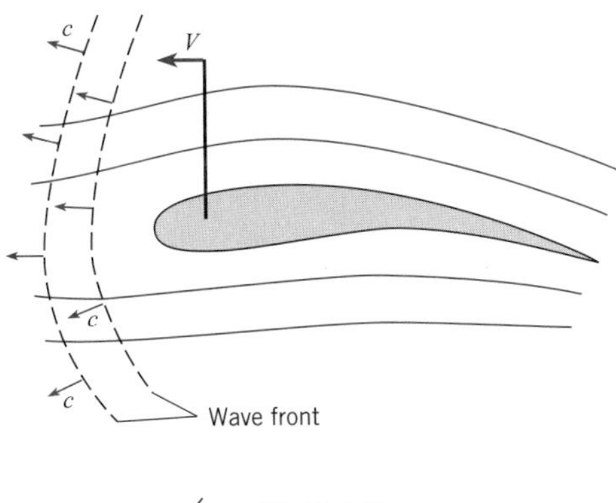

FIGURE 12.3

Propagation of a sound wave by an airfoil.

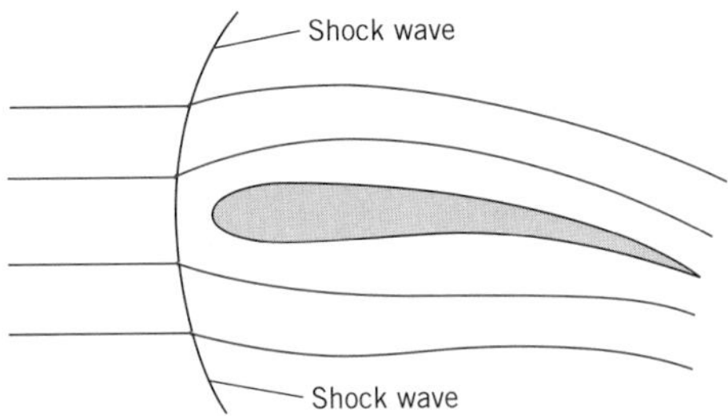

FIGURE 12.4

Standing shock wave in front of an airfoil.

fluid has no warning whatsoever that the airfoil is coming and cannot prepare for its arrival. Nature, at this point, resolves the problem by creating a shock wave that stands off the leading edge, as shown in Fig. 12.4. As the fluid passes through the shock wave near the leading edge, it is decelerated to a speed less than sonic speed and therefore has time to divide and flow around the airfoil. Shock waves will be treated in more detail in Section 12.3.

Another approach to appreciating the significance of sound propagation in a compressible fluid is to consider a point source of sound moving in a quiescent fluid, as shown in Fig. 12.5. The sound source is moving at a speed less than the local sound speed in Fig. 12.5*a* and faster than the local sound speed in Fig. 12.5*b*. At time $t = 0$ a sound pulse is generated and propagates radially outward at the local speed of sound. At time t_1 the sound source has moved a distance Vt_1, and the circle representing the sound wave emitted at $t = 0$ has a radius of ct_1. The sound source emits a new sound wave at t_1 that propagates radially outward. At time t_2 the sound source has moved to Vt_2, and the sound waves have moved outward as shown.

When the sound source moves at a speed less than the speed of sound, the sound waves form a family of nonintersecting eccentric circles, as shown in Fig. 12.5*a*. For an observer stationed at *A* the frequency of the sound pulses would appear higher than the emitted frequency because the sound source is moving toward the observer. In fact, the observer at *A* will detect a frequency of

$$f = f_0/(1 - V/c)$$

where f_0 is the emitting frequency of the moving sound source. This change in frequency is known as the *Doppler effect*.

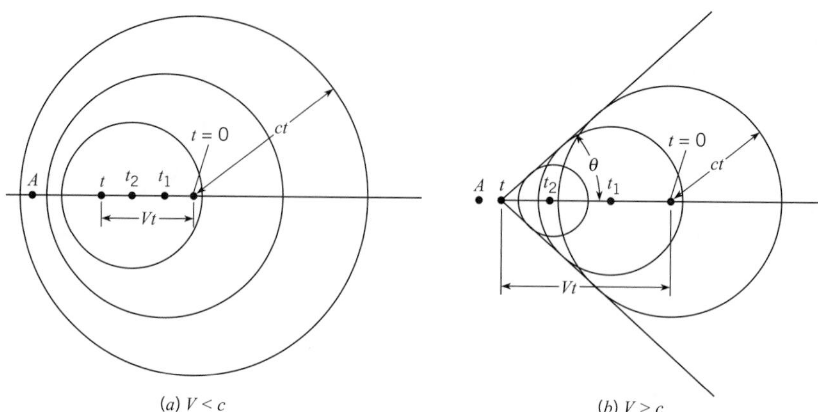

FIGURE 12.5

Sound field generated by a moving point source of sound.

When the sound source moves faster than the local sound speed, the sound waves intersect and form the locus of a cone with a half-angle of

$$\theta = \sin^{-1}(c/V)$$

The observer at A will not detect the sound source until it has passed. In fact, only an observer within the cone is aware of the moving sound source.

In view of the physical arguments given above, it is apparent that an important parameter relating to sound propagation and compressibility effects is the ratio V/c. This parameter, already introduced in Chapter 8, was first proposed by Ernst Mach, an Austrian scientist, and bears his name. The Mach number is defined as

$$\mathrm{M} = \frac{V}{c} \tag{12.13}$$

The conical wave surface depicted in Fig. 12.5*b* is known as a *Mach wave* and the conical half-angle as the *Mach angle.*

Besides the qualitative argument presented above for Mach number, we also recall from Chapter 8 that the Mach number is the ratio of the inertial to elastic forces acting on the fluid. If the Mach number is small, the inertial forces are ineffective in compressing the fluid and the fluid can be regarded as incompressible.

Compressible flows are characterized by their Mach number regimes as follows:

$\mathrm{M} < 1$	Subsonic flow
$\mathrm{M} \approx 1$	Transonic flow
$\mathrm{M} > 1$	Supersonic flow

Flows with Mach numbers exceeding 5 are sometimes referred to as *hypersonic.* Airplanes designed to travel near sonic speeds and faster are equipped with Mach meters because of the significance of the Mach number with respect to aircraft performance.

example 12.2

The Concorde is traveling at 1400 km/h at an altitude of 12,000 m, where the temperature is –56°C. Determine the Mach number at which the airplane is flying and characterize the flow.

Solution First we must determine the speed of sound.

$$c = \sqrt{kRT} = [(1.4)(287\ \text{J/kg K})(217\ \text{K})]^{1/2} = 295\ \text{m/s}$$

The speed of the airplane in meters per second is

$$V = (1400\ \text{km/h})\left(\frac{1}{3600}\ \text{h/s}\right)(1000\ \text{m/km}) = 389\ \text{m/s}$$

The Mach number is therefore

$$\text{M} = \frac{389}{295} = 1.32 \qquad \triangleleft$$

and the flow is supersonic.

12.2

Mach Number Relationships

We are already familiar with the Bernoulli equation and the energy equation and with their utility in determining fluid properties along streamlines in liquid flow. In this section, we shall learn how the Mach number is used to determine fluid properties in compressible flows. Let us consider a control volume bounded by two streamlines in a steady compressible flow, as shown in Fig. 12.6. Applying the energy equation, Eq. (7.12), to this control volume, realizing that the shaft work is zero, gives

$$-\dot{m}_1\left(h_1 + \frac{V_1^2}{2} + gz_1\right) + \dot{m}_2\left(h_2 + \frac{V_2^2}{2} + gz_2\right) = \dot{Q} \tag{12.14}$$

As pointed out in Chapter 7, the elevation terms (z_1 and z_2) can usually be neglected for gaseous flows. If the flow is adiabatic ($\dot{Q} = 0$), the energy equation reduces to

$$\dot{m}_1\left(h_1 + \frac{V_1^2}{2}\right) = \dot{m}_2\left(h_2 + \frac{V_2^2}{2}\right) \tag{12.15}$$

From the principle of continuity, the mass flow rate is constant, $\dot{m}_1 = \dot{m}_2$, so

$$h_1 + \frac{V_1^2}{2} = h_2 + \frac{V_2^2}{2} \tag{12.16}$$

Since positions 1 and 2 are arbitrary points on the same streamline, we say that

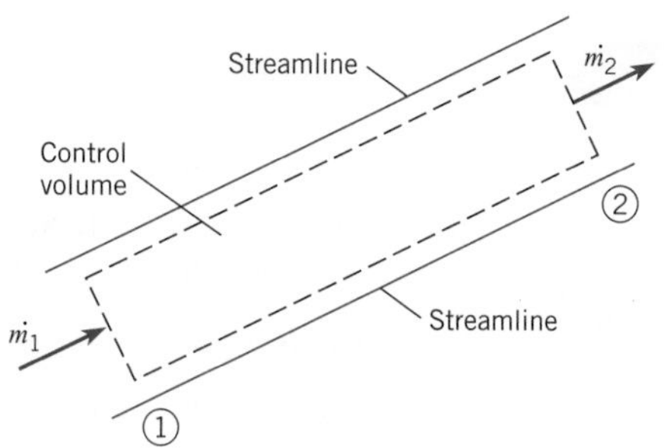

FIGURE 12.6

Control volume enclosed by streamlines.

$$h + \frac{V^2}{2} = \text{constant along a streamline in an adiabatic flow} \tag{12.17}$$

The constant in this expression is called the *total enthalpy,* h_t. It is the enthalpy that would arise if the flow velocity were brought to zero in a reversible, adiabatic process. Thus the energy equation along a streamline under adiabatic conditions is

$$h + \frac{V^2}{2} = h_t \tag{12.18}$$

If h_t is the same for all streamlines, the flow is *homenergic.*

It is instructive at this point to compare Eq. (12.18) with the Bernoulli equation. Expressing the specific enthalpy as the sum of the specific internal energy and p/ρ, Eq. (12.18) becomes

$$u + \frac{p}{\rho} + \frac{V^2}{2} = \text{constant}$$

If the fluid is incompressible and there is no heat transfer, the specific internal energy is constant and the equation reduces to the Bernoulli equation (excluding the hydrostatic pressure terms).

Temperature

The enthalpy of an ideal gas can be written as

$$h = c_p T \tag{12.19}$$

where c_p is the specific heat at constant pressure. Substituting this relation into Eq. (12.18) and dividing by $c_p T$, we obtain

$$1 + \frac{V^2}{2c_p T} = \frac{T_t}{T} \tag{12.20}$$

where T_t is the total temperature. From thermodynamics (4) we know that for an ideal gas

$$c_p - c_v = R \tag{12.21}$$

or

$$k - 1 = \frac{R}{c_v} = \frac{kR}{c_p}$$

Therefore

$$c_p = \frac{kR}{k-1} \tag{12.22}$$

Substituting this expression for c_p back into Eq. (12.20) and realizing that kRT is the speed of sound squared results in the *total temperature* equation

$$T_t = T\left(1 + \frac{k-1}{2}\mathrm{M}^2\right) \tag{12.23}$$

The temperature T is called the *static temperature*—the temperature that would be registered by a thermometer moving with the flowing fluid. Total temperature is analogous to total enthalpy in that it is the temperature that would arise if the velocity were brought to zero isentropically. If the flow is adiabatic, the total temperature is constant along a streamline. If not, the total temperature varies according to the amount of thermal energy transferred.

example 12.3

An aircraft is flying at M = 1.6 at an altitude where the atmospheric temperature is –50°C. The temperature on the aircraft's surface is approximately the total temperature. Estimate the surface temperature, taking $k = 1.4$.

Solution This problem can be visualized as the aircraft being stationary and an airstream with a static temperature of –50°C flowing past the aircraft at a Mach number of 1.6. The static temperature in absolute temperature units is

$$T = 223 \text{ K}$$

Using the equation for the total temperature gives

$$T_t = 223[1 + 0.2(1.6)^2] = 337 \text{ K or } 64°\text{C}$$ ◁

Pressure

If the flow is isentropic, thermodynamics shows that the following relationship for pressure and temperature of an ideal gas between two points on a streamline is valid (1):

$$\frac{p_1}{p_2} = \left(\frac{T_1}{T_2}\right)^{k/(k-1)} \tag{12.24}$$

Isentropic flow means that there is no heat transfer, so the total temperature is constant along the streamline. Therefore

$$T_t = T_1\left(1 + \frac{k-1}{2}\mathrm{M}_1^2\right) = T_2\left(1 + \frac{k-1}{2}\mathrm{M}_2^2\right) \tag{12.25}$$

Solving for the ratio T_1/T_2 and substituting into Eq. (12.24) shows that the pressure variation with the Mach number is given by

$$\frac{p_1}{p_2} = \left\{\frac{1 + [(k-1)/2]\mathrm{M}_2^2}{1 + [(k-1)/2]\mathrm{M}_1^2}\right\}^{k/(k-1)} \tag{12.26}$$

In that the equation of state is used to derive Eq. (12.24), absolute pressures must always be used in calculations with these equations.

The *total pressure* in a compressible flow is defined as

$$p_t = p\left(1 + \frac{k-1}{2}\mathrm{M}^2\right)^{k/(k-1)} \tag{12.27}$$

which is the pressure that would result if the flow were decelerated to zero speed reversibly and adiabatically. Unlike total temperature, total pressure may not be constant along streamlines in adiabatic flows. For example, we will discover that flow through a shock wave, though adiabatic, is not reversible and, therefore, not isentropic. The total pressure variation along a streamline in an adiabatic flow can be obtained by substituting Eqs. (12.27) and (12.25) into Eq. (12.26) to give

$$\frac{p_{t_1}}{p_{t_2}} = \frac{p_1}{p_2}\left\{\frac{1 + [(k-1)/2]\mathrm{M}_1^2}{1 + [(k-1)/2]\mathrm{M}_2^2}\right\}^{k/(k-1)} = \frac{p_1}{p_2}\left(\frac{T_2}{T_1}\right)^{k/(k-1)} \tag{12.28}$$

Unless the flow is also reversible and Eq. (12.24) is applicable, the total pressures at points 1 and 2 will not be equal. However, if the flow is isentropic, total pressure is constant along streamlines.

Density

Analogous to the total pressure, the *total density* in a compressible flow is given by

$$\rho_t = \rho\left(1 + \frac{k-1}{2}\mathrm{M}^2\right)^{1/(k-1)} \tag{12.29}$$

where ρ is the local or static density. If the flow is isentropic, then ρ_t is a constant along streamlines and Eq. (12.29) can be used to determine the variation of gas density with the Mach number.

In literature dealing with compressible flows, one often finds reference to "stagnation" conditions—that is, "stagnation temperature" and "stagnation pressure." By definition, *stagnation* refers to the conditions that exist at a point in the flow where the velocity is zero, regardless of whether or not the zero velocity has been achieved by an adiabatic, or reversible process. For example, if one were to insert a Pitot tube into a compressible flow, strictly speaking one would measure stagnation pressure, not total pressure, since the deceleration of the flow would not be reversible. In most cases, however, the difference between stagnation and total pressure is negligibly small.

Kinetic Pressure

The kinetic pressure, $q = \rho V^2/2$, is often used, as we have seen in Chapter 11, to calculate aerodynamic forces with the use of appropriate coefficients. It can also be related to the Mach number. Using the equation of state for an ideal gas to replace ρ gives

$$q = \frac{1}{2}\frac{pV^2}{RT} \tag{12.30}$$

Then using the equation for the speed of sound, Eq. (12.12), results in

$$q = \frac{k}{2}p\mathrm{M}^2 \tag{12.31}$$

where p must always be an absolute pressure since it derives from the equation of state.

example 12.4

The drag coefficient for a sphere at a Mach number of 0.7 is 0.95. Determine the drag force on a sphere 10 mm in diameter in air if $p = 101$ kPa.

Solution The drag force on a sphere is

$$F_D = \frac{1}{2}\rho V^2 C_D A_p = qC_D A_p$$

where A_p is the projected area. The kinetic pressure is

$$q = \frac{1.4}{2}(101\text{ kPa})(0.7)^2 = 34.6\text{ kPa}$$

The drag force is calculated to be

$$F_D = 0.95\left(34.6 \times 10^3 \frac{\text{N}}{\text{m}^2}\right)\left(\frac{\pi}{4}\right)(10^{-2})^2\text{ m}^2 = 2.6\text{ N} \quad \triangleleft$$

There is one very important fact about compressible flow that must be stressed: The Bernoulli equation is not valid for compressible flows! Let us see what would happen if one decided to measure the Mach number of a high-speed air flow with a Pitot-static tube, assuming that the Bernoulli equation was valid. Let us say a total pressure of 180 kPa and a static pressure of 100 kPa were measured. By the Bernoulli equation the kinetic pressure is equal to the difference between the total and static pressures, so

$$\frac{1}{2}\rho V^2 = p_t - p \qquad \text{or} \qquad \frac{k}{2}p\mathrm{M}^2 = p_t - p$$

Solving for the Mach number,

$$\mathrm{M} = \sqrt{\frac{2}{k}\left(\frac{p_t}{p} - 1\right)}$$

and substituting in the measured values, one obtains

$$\mathrm{M} = 1.07$$

Now, what should have been done? The expression relating the total and static pressures in a compressible flow is Eq. (12.27). Solving that equation for the Mach number gives

$$\mathrm{M} = \left\{\frac{2}{k-1}\left[\left(\frac{p_t}{p}\right)^{(k-1)/k} - 1\right]\right\}^{1/2} \tag{12.32}$$

and substituting in the measured values yields

$$\mathrm{M} = 0.96$$

Thus applying the Bernoulli equation would have led one to say that the flow was supersonic, whereas the flow was actually subsonic. In the limit of low velocities ($p_t/p \longrightarrow 1$), Eq. (12.32) reduces to the expression derived from the Bernoulli equation, which is indeed valid for very low ($\mathrm{M} \ll 1$) Mach numbers.

It is instructive to see how the pressure coefficient at the stagnation (total pressure) condition varies with Mach number. The pressure coefficient is given by

$$C_p = \frac{p_t - p}{\frac{1}{2}\rho V^2}$$

Using Eq. (12.31) for the kinetic pressure enables us to express C_p as a function Mach number and the ratio of specific heats.

$$C_p = \frac{2}{k\mathrm{M}^2}\left[\left(1 + \frac{k-1}{2}\mathrm{M}^2\right)^{k/(k-1)} - 1\right]$$

The variation of C_p with Mach number is shown in Fig. 12.7. At a Mach number of zero, the pressure coefficient is unity, which corresponds to incompressible flow. The pressure coefficient begins to depart significantly from unity at a number of about 0.3. From this observation we infer that compressibility effects in the flow field are unimportant for Mach numbers less than 0.3.

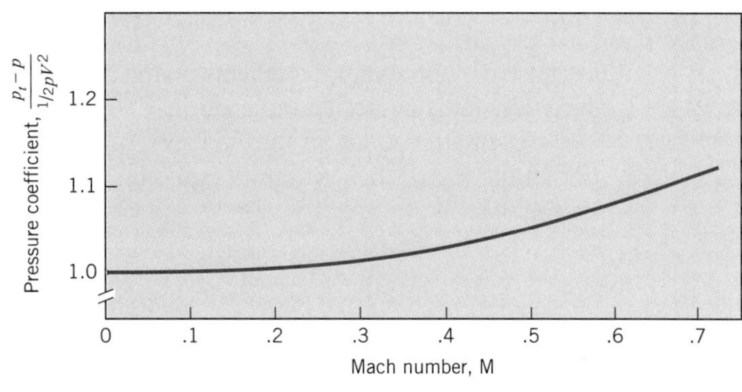

FIGURE 12.7

Variation of the pressure coefficient with Mach number.

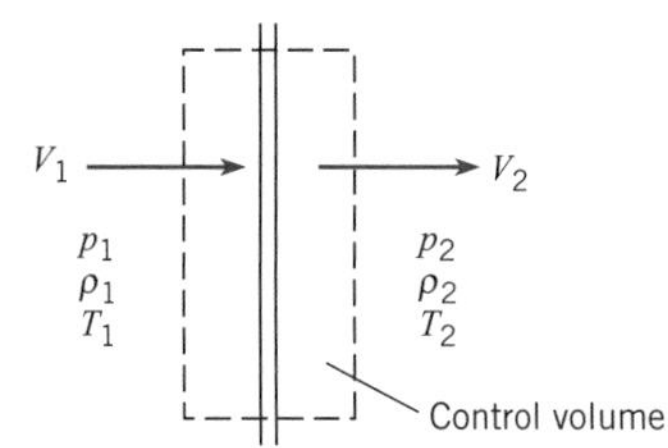

FIGURE 12.8

Control volume enclosing a normal shock wave.

12.3 Normal Shock Waves

Normal shock waves are wave fronts normal to the flow across which a supersonic flow is decelerated to a subsonic flow with an attendant increase in static temperature, pressure, and density. The normal shock wave is analogous to the water hammer introduced in Chapter 6 and somewhat analogous to the hydraulic jump, which will be introduced in Chapter 15.

Change in Flow Properties across a Normal Shock Wave

The most straightforward way to analyze a normal shock wave is to draw a control surface around the wave, as shown in Fig. 12.8, and write down the continuity, momentum, and energy equations. The net mass flux into the control volume is zero because the flow is steady. Therefore

$$-r_1 V_1 A + \rho_2 V_2 A = 0 \tag{12.33}$$

where A is the cross-sectional area of the control volume. Equating the net pressure forces acting on the control surface to the net efflux of momentum from the control volume gives

$$\rho_1 V_1 A(-V_1 + V_2) = (p_1 - p_2)A \tag{12.34}$$

The energy equation can be expressed simply as

$$T_{t_1} = T_{t_2} \tag{12.35}$$

because the temperature gradients on the control surface are assumed negligible and thus heat transfer is neglected (adiabatic).

Using the equation for the speed of sound, Eq. (12.12), and the equation of state for an ideal gas, the continuity equation can be rewritten to include the Mach number as follows:

$$\frac{p_1}{RT_1}\mathrm{M}_1\sqrt{kRT_1} = \frac{p_2}{RT_2}\mathrm{M}_2\sqrt{kRT_2} \tag{12.36}$$

The Mach number can be introduced into the momentum equation in the following way:

$$\rho_2 V_2^2 - \rho_1 V_1^2 = p_1 - p_2$$

$$p_1 + \frac{p_1}{RT_1}V_1^2 = p_2 + \frac{p_2}{RT_2}V_2^2 \tag{12.37}$$

$$p_1(1 + k\mathrm{M}_1^2) = p_2(1 + k\mathrm{M}_2^2)$$

Rearranging Eq. (12.37) for the *static-pressure ratio* across the shock wave results in

$$\frac{p_2}{p_1} = \frac{(1 + k\mathrm{M}_1^2)}{(1 + k\mathrm{M}_2^2)} \tag{12.38}$$

As we shall show later, the Mach number of a normal shock wave is always greater than unity upstream and less than unity downstream, so the static pressure always increases across a shock wave.

We can rewrite the energy equation in terms of the temperature and Mach number, as we did in Eq. (12.23), by utilizing the fact that $T_{t_2}/T_{t_1} = 1$, to obtain the *temperature ratio* across the shock wave.

$$\frac{T_2}{T_1} = \frac{\{1 + [(k-1)/2]\mathrm{M}_1^2\}}{\{1 + [(k-1)/2]\mathrm{M}_2^2\}} \tag{12.39}$$

Substituting Eqs. (12.38) and (12.39) into Eq. (12.36) yields the following relationship for the Mach numbers upstream and downstream of a normal shock wave:

$$\frac{\mathrm{M}_1}{1 + k\mathrm{M}_1^2}\left(1 + \frac{k-1}{2}\mathrm{M}_1^2\right)^{1/2} = \frac{\mathrm{M}_2}{1 + k\mathrm{M}_2^2}\left(1 + \frac{k-1}{2}\mathrm{M}_2^2\right)^{1/2} \tag{12.40}$$

Then, solving this equation for M_2 as a function of M_1, we obtain two solutions. One solution is trivial, $\mathrm{M}_1 = \mathrm{M}_2$, which corresponds to no shock wave in the control volume. The other solution gives the Mach number downstream of the shock wave:

$$\mathrm{M}_2^2 = \frac{(k-1)\mathrm{M}_1^2 + 2}{2k\mathrm{M}_1^2 - (k-1)} \tag{12.41}$$

Note: Because of the symmetry of Eq. (12.40), we can also use Eq. (12.41) to solve for M_1 given M_2, by simply interchanging the subscripts on the Mach numbers.

Setting $M_1 = 1$ in Eq. (12.41) results in M_2 also being equal to unity. Equations (12.38) and (12.39) also show that there would be no pressure or temperature increase across such a wave. In fact, the wave corresponding to $M_1 = 1$ is the sound wave across which, by definition, pressure and temperature changes are infinitesimal. Thus the sound wave represents a degenerate normal shock wave.

example 12.5

A normal shock wave occurs in air flowing at a Mach number of 1.5. The static pressure and temperature of the air upstream of the shock wave are 100 kPa absolute and 15°C. Determine the Mach number, pressure, and temperature downstream of the shock wave.

Solution We use the equation for the Mach number downstream of the shock wave:

$$M_2^2 = \frac{(0.4)(1.5)^2 + 2}{(2.8)(1.5)^2 - 0.4} = 0.49$$

$$M_2 = 0.7 \quad \triangleleft$$

The equations for pressure ratio and temperature ratio provide the downstream pressure and temperature:

$$p_2 = p_1\left(\frac{1 + kM_1^2}{1 + kM_2^2}\right)$$

$$= (100 \text{ kPa})\left[\frac{1 + (1.4)(1.5)^2}{1 + (1.4)(0.7)^2}\right] = 246 \text{ kPa, absolute} \quad \triangleleft$$

$$T_2 = T_1\left\{\frac{1 + [(k-1)/2]M_1^2}{1 + [(k-1)/2]M_2^2}\right\}$$

$$= (288 \text{ K})\left[\frac{1 + (0.2)(2.25)}{1 + (0.2)(0.49)}\right] = 380 \text{ K or } 107°\text{C} \quad \triangleleft$$

Note that absolute pressures and temperatures must always be used in carrying out these calculations. The changes in flow properties across a shock wave are presented in Table A.1 in the Appendix for a gas, such as air, for which $k = 1.4$.

A shock wave is an adiabatic process in which no shaft work is done. Thus for ideal gases the total temperature (and total enthalpy) is unchanged across the wave. The total pressure, however, does change across a shock wave. The total pressure upstream of the wave in Example 12.5 is

$$p_{t_1} = p_1\left(1 + \frac{k-1}{2}M_1^2\right)^{k/(k-1)}$$

$$= 100 \text{ kPa}[1 + (0.2)(2.25)]^{3.5} = 367 \text{ kPa}$$

The total pressure downstream of the same wave is

$$p_{t_2} = p_2\left(1 + \frac{k-1}{2}M_2^2\right)^{k/(k-1)}$$

$$= 246 \text{ kPa}[1 + (0.2)(0.49)]^{3.5} = 341 \text{ kPa}$$

Thus we see that the total pressure decreases through the wave, which, as we will see later, is because the flow through the shock wave is not an isentropic process. Total pressure remains constant along streamlines only in isentropic flow. Values for the ratio of total pressure across a normal shock wave are also provided in Table A.1 in the Appendix.

Existence of Shock Waves Only in Supersonic Flows

Let us look back at Eq. (12.41), which gives the Mach number downstream of a normal shock wave. If one were to substitute a value for M_1 less than unity, it is easy to see that one would obtain a value for M_2 larger than unity. For example, if $M_1 = 0.5$ in air, then

$$M_2^2 = \frac{(0.4)(0.5)^2 + 2}{(2.8)(0.5)^2 - 0.4}$$

$$M_2 = 2.65$$

Is it possible to have a shock wave in a subsonic flow across which the Mach number becomes supersonic? We also find that the total pressure would increase across such a wave; that is,

$$\frac{p_{t_2}}{p_{t_1}} > 1$$

The existence of such a wave would be a significant scientific discovery!

The only way to determine whether such a solution is possible is to invoke the second law of thermodynamics, which states that for any process the entropy of the universe must remain unchanged or increase.

$$\Delta s_{\text{univ}} \geq 0 \qquad (12.42)$$

Because the shock wave is an adiabatic process, there is no change in the entropy of the surroundings; thus the entropy of the system must remain unchanged or increase.

$$\Delta s_{\text{sys}} \geq 0 \qquad (12.43)$$

The entropy change of an ideal gas between pressures p_1 and p_2 and temperatures T_1 and T_2 is given by Van Wylen and Sonntag (1):

$$\Delta s_{1\to 2} = c_p \ln\frac{T_2}{T_1} - R\ln\frac{p_2}{p_1} \tag{12.44}$$

Using the relationship between c_p and R, Eq. (12.22), we can express the entropy change as

$$\Delta s_{1\to 2} = R\ln\left[\frac{p_1}{p_2}\left(\frac{T_2}{T_1}\right)^{k/(k-1)}\right] \tag{12.45}$$

Note that the quantity in the square brackets is simply the total pressure ratio as given by Eq. (12.28). Therefore the entropy change across a shock wave can be rewritten as

$$\Delta s = R\ln\frac{p_{t_1}}{p_{t_2}} \tag{12.46}$$

A shock wave across which the Mach number changes from subsonic to supersonic would give rise to a total pressure ratio less than unity and a corresponding decrease in entropy,

$$\Delta s_{\text{sys}} < 0$$

which violates the second law of thermodynamics. Therefore shock waves can exist only in supersonic flow.

The total pressure ratio approaches unity for sound waves, which conforms with the definition that they are isentropic ($\ln 1 = 0$).

example 12.6

Find the entropy increase across the shock wave considered in Example 12.5.

Solution

$$\Delta s = R\ln\frac{p_{t_1}}{p_{t_2}}$$

$$p_{t_1} = 367 \text{ kPa} \quad \text{and} \quad p_{t_2} = 341 \text{ kPa}$$

$$\Delta s = (287 \text{ J/kg K})\left(\ln\frac{367}{341}\right) = 21 \text{ J/kg K}$$ ◁

More examples of shock waves will be given in the next section. We shall conclude this section by qualitatively discussing other features of shock waves.

Besides the normal shock waves studied here, there are oblique shock waves that are inclined with respect to the flow direction. Let us look once again at the shock wave structure in front of a blunt body, as depicted qualitatively in Fig. 12.9. The portion of the shock wave immediately in front of the body behaves like a normal shock wave. As the shock wave bends in the free-stream direction, oblique shock waves result. The same relationships derived above for the normal shock waves are valid for the

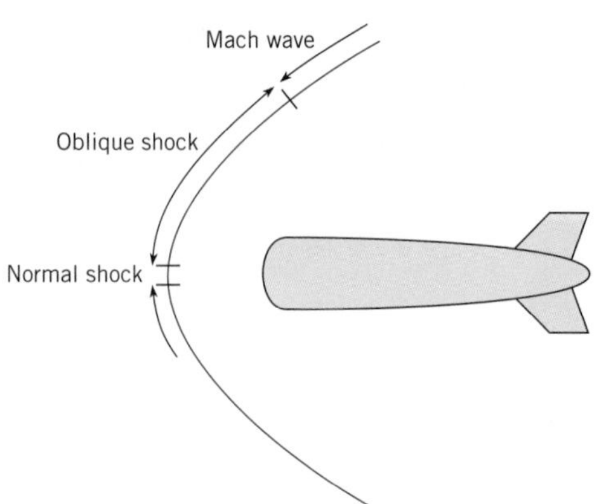

FIGURE 12.9

Shock wave structure in front of a blunt body.

velocity components normal to oblique waves. The oblique shock waves continue to bend in the downstream direction until the Mach number of the velocity component normal to the wave is unity. Then the oblique shock has degenerated into a so-called Mach wave across which changes in flow properties are infinitesimal.

The familiar sonic booms are the result of weak oblique shock waves that reach ground level. One can appreciate the damage that would ensue from stronger oblique shock waves if aircraft were permitted to travel at supersonic speeds near ground level.

12.4 Isentropic Compressible Flow through a Duct with Varying Area

We are already familiar with incompressible flow through ducts of varying cross-sectional area, such as the venturi tube. As the flow approaches the throat (smallest area), the velocity increases and the pressure decreases; then as the area again increases, the velocity decreases. The same velocity–area relationship is not always found for compressible flows.

Dependence of the Mach Number on Area Variation

Consider the duct of varying area shown in Fig. 12.10. It is assumed that the flow is isentropic and that the flow properties at each section are uniform. This type of analysis, in which the flow properties are assumed to be uniform at each section yet in which the cross-sectional area is allowed to vary (nonuniform), is still classified as "one-dimensional."

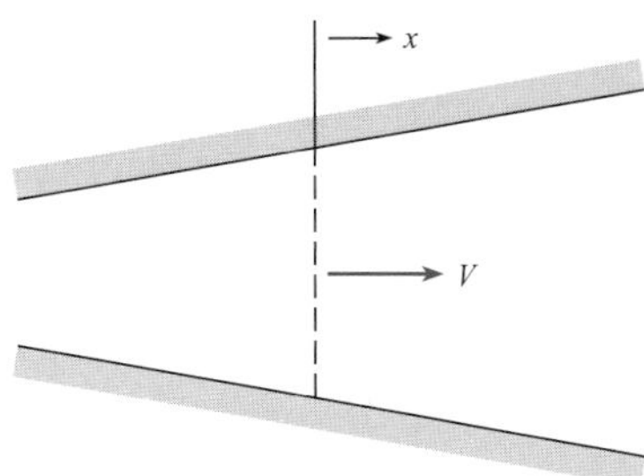

FIGURE 12.10

Duct with variable area.

The mass flow through the duct is given by

$$\dot{m} = \rho A V \tag{12.47}$$

where A is the duct's cross-sectional area. Since the mass flow is constant along the duct, we have

$$\frac{d\dot{m}}{dx} = \frac{d(\rho A V)}{dx} = 0 \tag{12.48}$$

which can be written as*

$$\frac{1}{\rho}\frac{d\rho}{dx} + \frac{1}{A}\frac{dA}{dx} + \frac{1}{V}\frac{dV}{dx} = 0 \tag{12.49}$$

The flow is assumed to be inviscid, so Euler's equation for steady flow is applicable:

$$\rho V \frac{dV}{dx} + \frac{dp}{dx} = 0 \tag{12.50}$$

Making use of Eq. (12.8), which relates $dp/d\rho$ to the speed of sound in an isentropic flow, gives

$$\frac{-V}{c^2}\frac{dV}{dx} = \frac{1}{\rho}\frac{d\rho}{dx} \tag{12.51}$$

This equation is now used to eliminate ρ in Eq. (12.49). The result is

$$\frac{1}{V}\frac{dV}{dx} = \frac{(1/A)(dA/dx)}{\mathrm{M}^2 - 1} \tag{12.52}$$

which, though simple, leads to the following important, far-reaching conclusions.

* This step can easily be seen by first taking the logarithm of Eq. (12.47):

$$\ln(\rho A V) = \ln\rho + \ln A + \ln V$$

and then taking the derivative of each term:

$$\frac{d}{dx}[\ln(\rho A V)] = 0 = \frac{1}{\rho}\frac{d\rho}{dx} + \frac{1}{A}\frac{dA}{dx} + \frac{1}{V}\frac{dV}{dx}$$

Subsonic Flow

For subsonic flow, $M^2 - 1$ is negative, which means that a decreasing area leads to an increasing velocity, and, correspondingly, an increasing area leads to a decreasing velocity. This velocity–area relationship agrees with our experience relating to flow through pipes with section changes.

Supersonic Flow

For supersonic flow, $M^2 - 1$ is positive, so a decreasing area leads to a decreasing velocity, and an increasing area leads to an increasing velocity. Thus the velocity at the minimum area of a duct with supersonic compressible flow is a minimum. This is the principle underlying the operation of diffusers on jet engines for supersonic aircraft, as shown in Fig. 12.11. The purpose of the diffuser is to decelerate the flow so that there is sufficient time for combustion in the chamber. Then the diverging nozzle accelerates the flow again to achieve a larger kinetic energy of the exhaust gases and an increased engine thrust.

Transonic Flow (M ≈ 1)

Stations along a duct corresponding to $dA/dx = 0$ represent either a local minimum or a local maximum in the duct's cross-sectional area, as illustrated in Fig. 12.12. If at these stations the flow were either subsonic ($M < 1$) or supersonic ($M > 1$), then by Eq. (12.52) $dV/dx = 0$, so the flow velocity would have either a maximum or a minimum value. In particular, if the flow were supersonic through the duct of Fig. 12.12*a*, then the velocity would be a minimum at the throat; if subsonic, a maximum.

Now, what happens if the Mach number is unity? Equation (12.52) tells us that if the Mach number is unity and dA/dx is not equal to zero, the velocity gradient dV/dx is infinite—a physically impossible situation. Therefore, dA/dx must be zero where the Mach number is unity in order for a finite, physically reasonable velocity gradient to exist.*

We can argue one step further here to show that sonic flow can occur only at a minimum area. Consider Fig. 12.12*a*. If the flow is initially subsonic, the converging duct accelerates the flow toward a sonic velocity. If the flow is initially supersonic, the converging duct decelerates the flow toward a sonic velocity. Using this same reasoning, one can prove that sonic flow is impossible in the duct depicted in Fig. 12.12*b*. If the

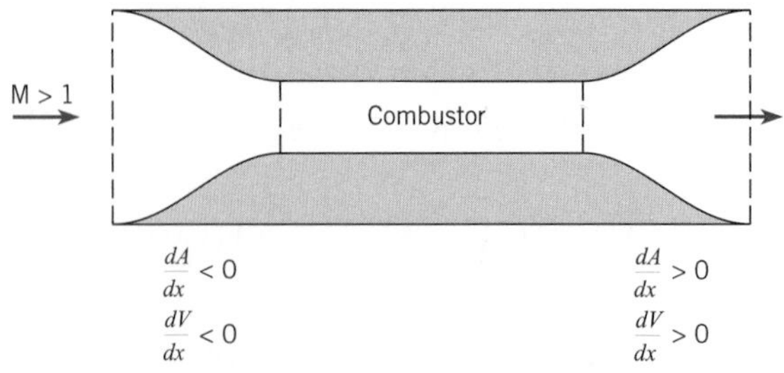

FIGURE 12.11

Engine for supersonic aircraft.

* Actually, the velocity gradient is indeterminate because the numerator and denominator are both zero. It can be shown by application of L'Hôpital's rule, however, that the velocity gradient is finite.

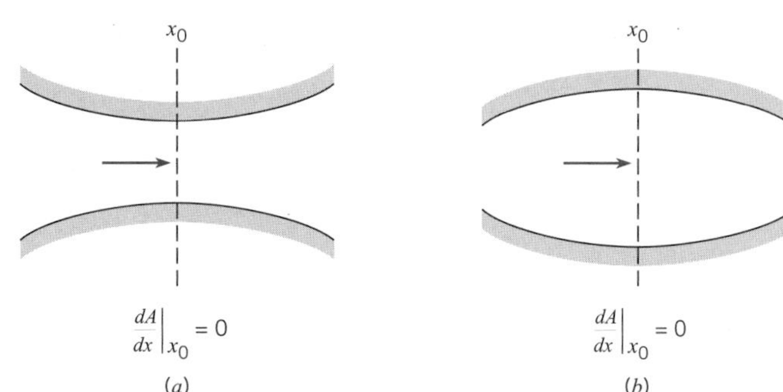

FIGURE 12.12
Duct contours for which dA/dx is zero.

flow is initially supersonic, the diverging duct increases the Mach number even more. If the flow is initially subsonic, the diverging duct decreases the Mach number; thus sonic flow cannot be achieved at a maximum area. Hence the Mach number in a duct of varying cross-sectional area can be unity only at a local area minimum (throat). This does not imply, however, that the Mach number must always be unity at a local area minimum.

Laval Nozzle

The Laval nozzle is a duct of varying area that produces supersonic flow. The nozzle is named after its inventor, de Laval (1845–1913), a Swedish engineer. According to the foregoing discussion, the nozzle must consist of a converging section to accelerate the subsonic flow, a throat section for transonic flow, and a diverging section to further accelerate the supersonic flow. Thus the shape of the Laval nozzle is as shown in Fig. 12.13.

One very important application of the Laval nozzle is the supersonic wind tunnel, which has been an indispensable tool in the development of supersonic aircraft. Basically, the wind tunnel, as illustrated in Fig. 12.14, consists of a high-pressure source of gas, a Laval nozzle to produce supersonic flow, and a test section. The high-pressure source may be from a large pressure tank, which is connected to the Laval nozzle through a regulator valve to maintain a constant upstream pressure, or from a pumping system that provides a continuous high-pressure supply of gas.

The equations relating to the compressible flow through a Laval nozzle have already been developed. Since the mass flow rate is the same at every cross section along the nozzle, we have

$$\rho VA = \text{constant}$$

and the constant is usually evaluated corresponding to those conditions that exist when the Mach number is unity. Thus

$$\rho VA = \rho_* A_* V_* \tag{12.53}$$

where the asterisk signifies conditions wherein the Mach number is equal to unity. Rearranging Eq. (12.53) gives

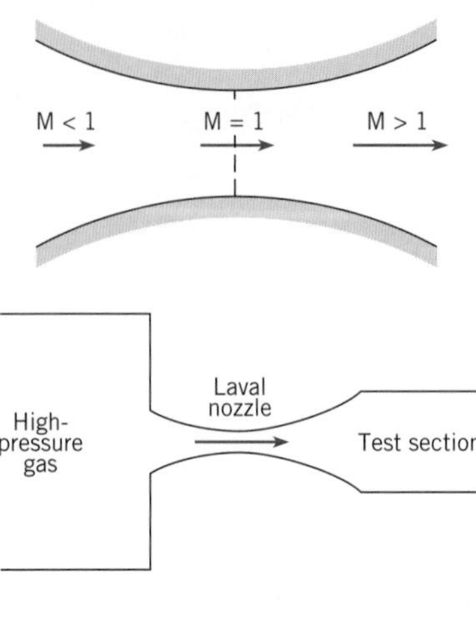

FIGURE 12.13

Laval nozzle.

FIGURE 12.14

Wind tunnel.

$$\frac{A}{A_*} = \frac{\rho_* V_*}{\rho V}$$

However, the velocity is the product of the Mach number and the local speed of sound. Therefore

$$\frac{A}{A_*} = \frac{\rho_*}{\rho} \frac{M_* \sqrt{kRT_*}}{M \sqrt{kRT}} \tag{12.54}$$

By definition $M_* = 1$, so

$$\frac{A}{A_*} = \frac{\rho_*}{\rho} \left(\frac{T_*}{T} \right)^{1/2} \frac{1}{M} \tag{12.55}$$

Because the flow in a Laval nozzle is assumed to be isentropic, the total temperature and total pressure (and total density) are constant throughout the nozzle. From Eq. (12.29), we have

$$\frac{\rho_*}{\rho} = \left\{ \frac{1 + [(k-1)/2]M^2}{(k+1)/2} \right\}^{1/(k-1)}$$

and from Eq. (12.25), the temperature ratio is given by

$$\frac{T_*}{T} = \frac{1 + [(k-1)/2]M^2}{(k+1)/2}$$

Substituting these expressions into Eq. (12.55) yields the following relationship for *area ratio* as a function of Mach number in a Laval nozzle:

$$\frac{A}{A_*} = \frac{1}{M} \left\{ \frac{1 + [(k-1)/2]M^2}{(k+1)/2} \right\}^{(k+1)/2(k-1)} \tag{12.56}$$

This equation is valid, of course, for all Mach numbers—subsonic, transonic, and supersonic. The area ratio A/A_* is the ratio of the area at the station where the Mach number is M to the area where M is equal to unity. Many supersonic wind tunnels are designed to maintain the same test section area and to vary the Mach number by varying the throat area.

example 12.7

Suppose we are designing a supersonic wind tunnel to operate with air at a Mach number of 3. If the throat area is 10 cm^2, what must the cross-sectional area of the test section be?

Solution Putting $k = 1.4$ for air and M = 3 in the area ratio equation gives

$$\frac{A}{A_*} = \frac{1}{3}\left[\frac{1 + (0.2)3^2}{1.2}\right]^3 = 4.23$$

Thus the area of the test section must be 42.3 cm^2. ◁

Example 12.7 demonstrates that it is a straightforward task to calculate the area ratio given the Mach number and ratio of specific heats. However, in practice, one usually knows the area ratio and wishes to determine the Mach number. It is not possible to solve Eq. (12.56) for the Mach number as an explicit function of the area ratio. For this reason, compressible-flow tables have been developed that allow one to obtain the Mach number easily given the area ratio.

Let us look again at Table A.1 in the Appendix. This table has been developed for a gas, such as air, for which $k = 1.4$. The symbols that head each column are defined at the beginning of the table. Tables for both subsonic and supersonic flow are provided.

example 12.8

A wind tunnel using air has an area ratio of 10. The absolute total pressure and temperature are 4 MPa and 350 K. Find the Mach number, pressure, temperature, and air velocity in the test section.

Solution From the table for supersonic flow,

M	A/A_*
3.5	6.79
4.0	10.72

we find that the Mach number must be between 3.5 and 4.0. Interpolating between the two points gives M = 3.91 at $A/A_* = 10.0$ ◁

The compressible-flow tables can also be used to interpolate for T/T_t and p/p_t, or these ratios can be calculated using the total pressure and total temperature equations. The results are

$$\frac{p}{p_t} = 0.00743 \quad \text{and} \quad \frac{T}{T_t} = 0.246$$

In the test section,

$$p = 29.7 \text{ kPa} \quad \text{and} \quad T = 86 \text{ K} \qquad \triangleleft$$

The velocity in the test section is obtained from

$$V = \text{M}c = \text{M}\sqrt{kRT} = 727 \text{ m/s} \qquad \triangleleft$$

Mass Flow Rate through a Laval Nozzle

An important consideration in the design of a supersonic wind tunnel is size. A large wind tunnel requires a large mass flow rate, which, in turn, requires a large pumping system for a continuous-flow tunnel or a large tank for sufficient run time in an intermittent tunnel. The easiest station at which to calculate the mass flow rate is the throat, because at this station the Mach number is unity.

$$\dot{m} = \rho_* A_* V_* = \rho_* A_* \sqrt{kRT_*}$$

It is more convenient, however, to express the mass flow in terms of total conditions. The local density and static temperature at sonic velocity are related to the total density and temperature by

$$\frac{T_*}{T_t} = \left(\frac{2}{k+1}\right)$$

$$\frac{\rho_*}{\rho_t} = \left(\frac{2}{k+1}\right)^{1/(k-1)}$$

which, when substituted into the foregoing equation, give

$$\dot{m} = \rho_t \sqrt{kRT_t} A_* \left(\frac{2}{k+1}\right)^{(k+1)/2(k-1)} \tag{12.57}$$

Usually, the total pressure and temperature are known. Using the equation of state for an ideal gas to eliminate ρ_t, we have the expression for *critical mass flow rate*

$$\dot{m} = \frac{p_t A_*}{\sqrt{RT_t}} k^{1/2} \left(\frac{2}{k+1}\right)^{(k+1)/2(k-1)} \tag{12.58}$$

For gases with a ratio of specific heats of 1.4,

$$\dot{m} = 0.685\frac{p_t A_*}{\sqrt{RT_t}} \tag{12.59}$$

and for gases with $k = 1.67$,

$$\dot{m} = 0.727\frac{p_t A_*}{\sqrt{RT_t}} \tag{12.60}$$

example 12.9

A supersonic wind tunnel with a square test section 15 cm by 15 cm is being designed to operate at a Mach number of 3 using air. The static temperature and pressure in the test section are –20°C and 50 kPa. Calculate the mass flow rate.

Solution From Example 12.7 the area ratio for a Mach 3 wind tunnel is 4.23. Thus the area of the throat must be

$$A_* = \frac{225}{4.23} = 53.2\ \text{cm}^2 = 0.00532\ \text{m}^2$$

The total pressure is obtained from the total pressure equation.

$$p_t = p\left(1 + \frac{k-1}{2}\text{M}^2\right)^{k/(k-1)} = 50(36.7) = 1836\ \text{kPa} = 1.836\ \text{MPa}$$

The total temperature is

$$T_t = T\left(1 + \frac{k-1}{2}\text{M}^2\right) = 253(2.8) = 708\ \text{K}$$

Finally, the mass flow rate from Eq. (12.59) is

$$\dot{m} = \frac{(0.685)[1.836(10^6\ \text{N/m}^2)](0.00532\ \text{m}^2)}{[(287\ \text{J/kg K})(708\ \text{K})]^{1/2}} = 14.8\ \text{kg/s} \quad \triangleleft$$

A pump capable of moving air at this rate against a 1.8-MPa pressure would require over 6000 kW of power input. Such a system would be large and costly to build and to operate.

Classification of Nozzle Flow by Exit Conditions

Let us now take a qualitative look at the pressure distribution in a Laval nozzle. Consider the Laval nozzle depicted in Fig. 12.15 with the corresponding pressure and Mach number distributions plotted beneath it. The pressure at the nozzle entrance is very near the total pressure, because the Mach number is small. As the area decreases toward the throat, the Mach number increases and the pressure decreases. The static-to-total-pressure ratio at the throat, where conditions are sonic, is called the *critical pressure ratio*. It has a value of

$$\frac{p_*}{p_t} = \left(\frac{2}{k+1}\right)^{k/(k-1)}$$

which for air with $k = 1.4$ is

$$\frac{p_*}{p_t} = 0.528$$

It is called a critical pressure ratio because to achieve sonic flow with air in a nozzle, it is necessary that the exit pressure be at least less than 0.528 of the total pressure. The pressure continues to decrease until it reaches the exit pressure corresponding to the nozzle exit area ratio. Similarly, the Mach number monotonically increases with distance down the nozzle.

Now what happens if the nozzle exit pressure p_e is different from the back pressure (the pressure to which the nozzle exhausts)? If the exit pressure is higher than the back pressure, an expansion wave exists at the nozzle exit, as shown in Fig. 12.16*a*. These waves, which will not be studied here, effect a turning and further acceleration of the flow to achieve the back pressure. As one watches the exhaust of a rocket motor as it rises through the ever-decreasing pressure of higher altitudes, one can see the plume fan out as the flow turns more to achieve the lower pressure. A nozzle for which the exit pressure is larger than the back pressure is called an *underexpanded nozzle* because the flow could have expanded further.

If the exit pressure is less than the back pressure, shock waves occur. If the exit pressure is only slightly less than the back pressure, then pressure equalization can be obtained by oblique shock waves at the nozzle exit, as shown in Fig. 12.16*b*.

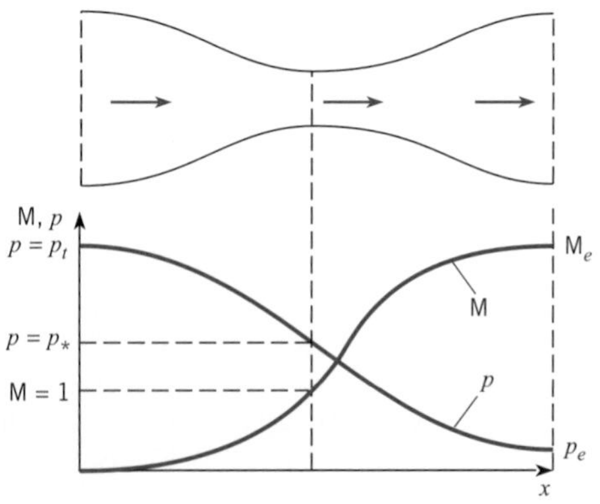

FIGURE 12.15

Distribution of static pressure and Mach number in a Laval nozzle.

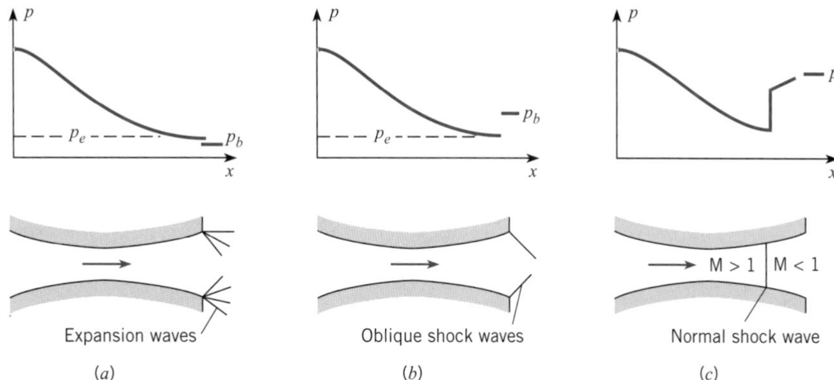

FIGURE 12.16

Conditions at a nozzle exit. (a) Expansion waves. (b) Oblique shock waves. (c) Normal shock wave.

If, however, the difference between back pressure and exit pressure is larger than can be accommodated by oblique shock waves, a normal shock wave will occur in the nozzle, as shown in Fig. 12.16*c*. A pressure jump occurs across the normal shock wave. The flow becomes subsonic and decelerates in the remaining portion of the diverging section in such a way that the exit pressure is equal to the back pressure. As the back pressure is further increased, the shock wave moves toward the throat region until, finally, there is no region of supersonic flow. A nozzle in which the exit pressure corresponding to the exit area ratio of the nozzle is less than the back pressure is called an *overexpanded nozzle*. Any flow that exits from a duct (or pipe) subsonically must always exit at the local back pressure.

A nozzle with supersonic flow in which the exit pressure is equal to the back pressure is *ideally expanded*.

example 12.10

The total pressure in a nozzle with an area ratio (A/A_*) of 4 is 1.3 MPa. Air is flowing through the nozzle. If the back pressure is 100 kPa, is the nozzle overexpanded, ideally expanded, or underexpanded?

Solution Interpolating between two area ratio values in the supersonic flow part of Table A.1 of the Appendix,

M	A/A_*
2.90	3.850
3.0	4.235

gives M = 2.94 at $A/A_* = 4.0$. The corresponding pressure ratio is

$$\frac{p}{p_t} = 0.0298$$

so

$$p = 38.7 \text{ kPa}$$

Therefore the nozzle is overexpanded. ◁

example 12.11

The Laval nozzle shown in the figure has an expansion ratio of 4 (exit area/ throat area). Air flows through the nozzle, and a normal shock wave occurs where the area ratio is 2. The total pressure upstream of the shock is 1 MPa. Determine the static pressure at the exit.

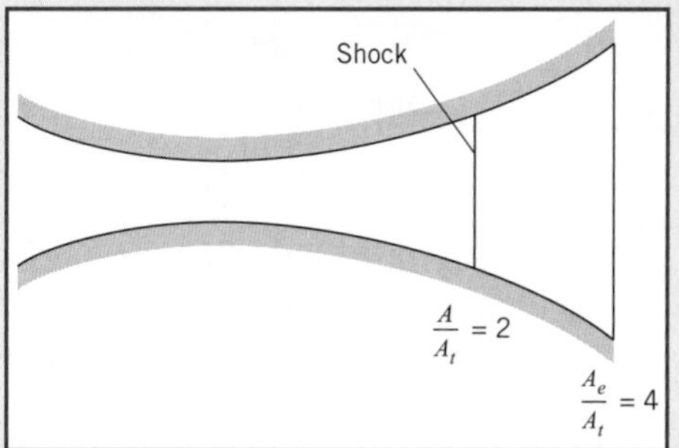

Solution From the supersonic-flow part of Table A.1, we find that the Mach number corresponding to an area ratio of 2 is 2.20. From the same line in that table, we find that the Mach number downstream of the shock is 0.547 and the ratio of total pressures across the shock wave is

$$\frac{p_{t_2}}{p_{t_1}} = 0.6281$$

Thus the total pressure downstream of the wave is

$$p_{t_2} = 0.6281 \times 1 \text{ MPa} = 6.28 \text{ kPa}$$

From the subsonic part of Table A.1, the value for A/A_* corresponding to the Mach number behind the shock wave is 1.26. Thus the ratio of nozzle area to throat area that would be needed to develop sonic flow downstream of the shock wave (if that were to be done) is

$$\frac{A_*}{A_t} = \frac{A/A_t}{A/A_*} = \frac{2}{1.26} = 1.59$$

Hence the ratio of the exit area to the area needed to develop sonic flow downstream of the shock wave is

$$\frac{A_e}{A_*} = \frac{A_e/A_t}{A_*/A_t} = \frac{4}{1.59} = 2.52$$

From Table A.1, the subsonic Mach number corresponding to this area ratio is

$$\mathrm{M}_e = 0.24$$

The static pressure corresponding to this Mach number and a total pressure of 628 kPa is obtained from the total pressure equation:

$$p_e = \frac{628 \text{ kPa}}{[1 + (0.2)(0.24)^2]^{3.5}} = 603 \text{ kPa} \quad \triangleleft$$

Mass Flow through a Truncated Nozzle

The *truncated nozzle* is a Laval nozzle cut off at the throat, as shown in Fig. 12.17. The nozzle exits to a back pressure p_b. This type of nozzle is important to engineers because of its frequent use as a flow-metering device for compressible flows.

To calculate the mass flow, we must first determine whether the flow at the exit is sonic or subsonic. Of course, the flow at the exit could never be supersonic, since the nozzle area does not diverge. First we calculate the value of the *critical pressure ratio*:

$$\frac{p_*}{p_t} = \left(\frac{2}{k+1}\right)^{k/(k-1)}$$

which, for air, is 0.528. We then evaluate the ratio of back pressure to total pressure, p_b/p_t, and compare it with the critical pressure ratio:

1. If $p_b/p_t < p_*/p_t$, the exit pressure is higher than the back pressure, so the exit flow must be sonic. Pressure equilibration is achieved after exit by a series of expansion waves. The mass flow is calculated using Eq. (12.58), where A_* is the area at the truncated station.

2. If $p_b/p_t > p_*/p_t$, the flow exits subsonically. If we were to irrationally assume that the flow exited at the speed of sound, then the exit pressure p_* would be less than p_b. There can be no shock waves in a sonic flow (only sound waves) to raise the exit pressure to the back pressure. Therefore, the flow adjusts itself to the back pressure by exiting subsonically.

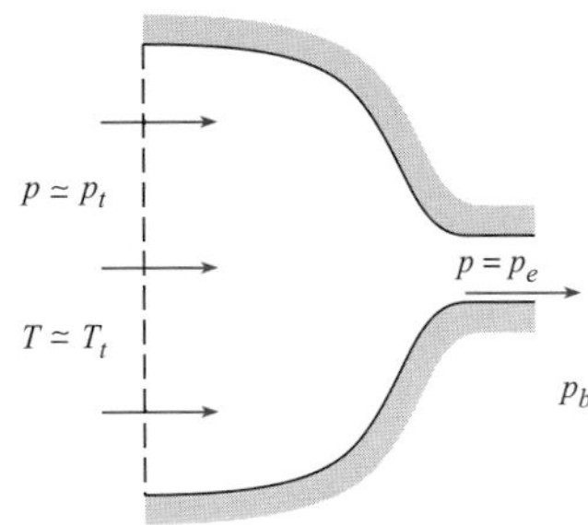

FIGURE 12.17

Truncated nozzle.

In case 2, one must first determine the Mach number at the exit by using Eq. (12.32):

$$\mathrm{M}_e = \sqrt{\frac{2}{k-1}\left[\left(\frac{p_t}{p_b}\right)^{(k-1)/k} - 1\right]}$$

Then, using this value for Mach number, one calculates the static temperature and speed of sound at the exit:

$$T_e = \frac{T_t}{\{1 + [(k-1)/2]\mathrm{M}_e^2\}}$$

$$c_e = \sqrt{kRT_e}$$

The gas density at the nozzle exit is determined by using the exit temperature and back pressure:

$$\rho_e = \frac{p_b}{RT_e}$$

Finally, the mass flow is given by

$$\dot{m} = \rho_e A_e \mathrm{M}_e c_e$$

where A_e is the area at the truncated section.

example 12.12

Air exhausts through a truncated nozzle 3 cm in diameter from a reservoir at a pressure of 160 kPa and a temperature of 80°C. Calculate the mass flow rate if the back pressure is 100 kPa.

Solution First we must determine the nature of the flow at the nozzle exit by evaluating the pressure ratio $p_b/p_t = 100/160 = 0.625$. Because 0.625 is larger than the critical pressure ratio for air (0.528), the flow at the nozzle exit must be subsonic. The Mach number is

$$\mathrm{M}_e^2 = \frac{2}{k-1}\left[\left(\frac{p_t}{p_b}\right)^{(k-1)/k} - 1\right]$$

$$\mathrm{M}_e = 0.85$$

The static temperature at the exit is

$$T_e = \frac{T_t}{\{1 + [(k-1)/2]\mathrm{M}_e^2\}} = 308 \text{ K}$$

Correspondingly, the density at the exit is

$$\rho_e = \frac{p_b}{RT_e} = \frac{100 \times 10^3 \text{ N/m}^2}{(287 \text{ J/kg K})(309 \text{ K})} = 1.13 \text{ kg/m}^3$$

The speed of sound at the exit is

$$c_e = [(1.4)(287 \text{ J/kg K})(309 \text{ K})]^{1/2} = 352 \text{ m/s}$$

Finally, we calculate the mass flow rate to be

$$\dot{m} = (1.13 \text{ kg/m}^3)(0.785)(0.03^2 \text{ m}^2)(0.85)(352 \text{ m/s}) = 0.239 \text{ kg/s} \quad \triangleleft$$

Had p_b/p_t been less than 0.528, then we would have used Eq. (12.58) to calculate the mass flow rate.

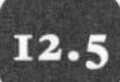

12.5 Compressible Flow in a Pipe with Friction

The flow of liquid through a pipe was studied in Chapter 10. The analysis of compressible flow in a pipe is somewhat more difficult because of the dependence of density and pressure on temperature. The problem is also complicated by the fact that wall friction and heat transfer cannot be simply combined into a single head loss parameter because of their distinct effect on the Mach number distribution along the pipe. In most engineering problems, however, the effect of wall friction is the most significant parameter. Thus it will be studied here. Two problems will be considered: first, flow in an insulated pipe in which the fluid is treated as an adiabatic system, and second, an isothermal flow that approximates flow in long pipelines.

Adiabatic Flow

The conservation-of-mass equation for uniform flow through a constant-area duct is

$$\rho V = \text{constant}$$

which, expressed in differential form, becomes

$$\frac{dV}{V} + \frac{d\rho}{\rho} = 0 \tag{12.61}$$

The conservation-of-energy equation, Eq. (12.17), can be written as

$$h + \frac{V^2}{2} = \text{constant}$$

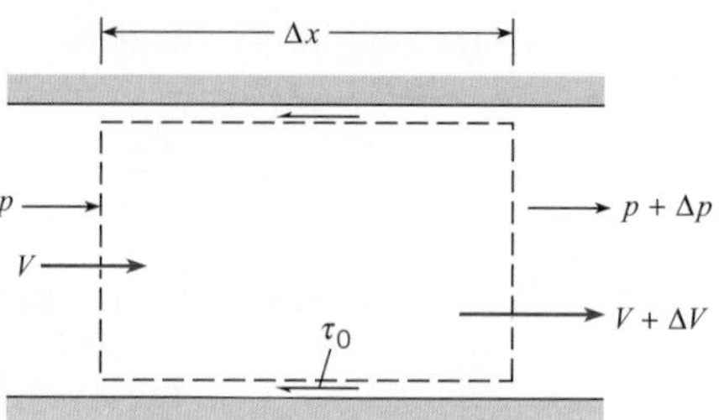

FIGURE 12.18

Control volume in a pipe.

since the flow is adiabatic. Using the relationships between enthalpy and temperature for an ideal gas, Eqs. (12.19) and (12.22), we can rewrite the energy equation in differential form as follows:

$$\frac{kR\,dT}{k-1} + VdV = 0 \tag{12.62}$$

The conservation-of-momentum equation can be obtained by applying the momentum equation to a control volume of length Δx contained in a pipe, as shown in Fig. 12.18. Equating the forces acting on the system to the net efflux of momentum from the control volume results in

$$A[p - (p + \Delta p)] - \tau_0 C\Delta x = \rho VA(-V + V + \Delta V) \tag{12.63}$$

where C is the circumference of the pipe and τ_0 is the shear stress at the wall. Introducing the Darcy-Weisbach resistance coefficient for τ_0, Eq. (10.21),

$$\tau_0 = \frac{f\rho V^2}{8}$$

and simplifying, we can rewrite the momentum equation in differential form as follows:

$$\rho VdV + dp + \frac{f\rho V^2 dx}{2D} = 0 \tag{12.64}$$

where D is the pipe diameter.

Mach Number Distribution along a Pipe

Our goal now is to combine the foregoing conservation equations with the equation of state, Eq. (12.1), to obtain an expression for the Mach number distribution along a pipe. Dividing each term in Eq. (12.64) by the pressure p, and realizing that

$$\frac{p}{\rho} = RT = \frac{c^2}{k}$$

from Eqs. (12.1) and (12.2), we obtain

$$k\mathrm{M}^2\frac{dV}{V} + \frac{dp}{p} + \frac{kf\mathrm{M}^2 dx}{2D} = 0 \tag{12.65}$$

The equation of state can be written in differential form:

$$\frac{dp}{p} = \frac{d\rho}{\rho} + \frac{dT}{T} \tag{12.66}$$

Using the continuity equation, Eq. (12.61), to replace ρ by V and the energy equation, Eq. (12.62), to replace T by V yields

$$\frac{dp}{p} = \frac{-dV}{V} - (k-1)\mathrm{M}^2\frac{dV}{V} \tag{12.67}$$

which when substituted in the momentum equation, Eq. (12.65), results in

$$(\mathrm{M}^2 - 1)\frac{dV}{V} + \frac{kf\mathrm{M}^2}{2}\frac{dx}{D} = 0 \tag{12.68}$$

The Mach number is defined as

$$\mathrm{M} = \frac{V}{(kRT)^{1/2}}$$

which can be written in differential form as follows:

$$\frac{d\mathrm{M}}{\mathrm{M}} = \frac{dV}{V} - \frac{1}{2}\frac{dT}{T} \tag{12.69}$$

Again using Eq. (12.62) to eliminate T yields

$$\frac{d\mathrm{M}}{\mathrm{M}} = \frac{dV}{V}\left[1 + \frac{(k-1)\mathrm{M}^2}{2}\right] \tag{12.70}$$

Using this equation to eliminate V in Eq. (12.68) results in the following differential equation for the Mach number and distance:

$$\frac{(1-\mathrm{M}^2)d\mathrm{M}}{\mathrm{M}^3\{1 + [(k-1)/2]\mathrm{M}^2\}} = \frac{kf\,dx}{2D} \tag{12.71}$$

This equation tells us that if the flow is subsonic, then $d\mathrm{M}/dx > 0$ and the Mach number increases with distance along the pipe. Conversely, if the flow is supersonic, then $d\mathrm{M}/dx < 0$ and the Mach number decreases along the pipe. Thus the effect of wall friction is always to cause the Mach number to approach unity. It is impossible for the Mach number of a compressible flow in a pipe to change from subsonic to supersonic. Consequently, the maximum Mach number that an initially subsonic flow can attain is unity, and this can be reached only at the end of the pipe. Shock waves, of course, can occur in the pipe to change an initially supersonic flow to a subsonic flow.

From Chapter 10 we know that the resistance coefficient f is a function of the Reynolds number and the relative roughness of the pipe. The Reynolds number is constant along the length of a pipe transporting a liquid. The continuity equation requires also that ρV be constant along a pipe transporting a compressible fluid. Temperature, however, may vary by as much as 20% for the subsonic flow of air in a pipe, which corresponds to a viscosity variation of approximately 10%. Referring back to Fig. 10.8 on p. 382, one notes that a 10% change in Reynolds number gives rise to a considerably smaller change in f if the flow is turbulent, which is usually the case. Thus it is reasonable to assume when integrating Eq. (12.71) that f is a constant and equal to the average value, $\bar{f}$, in the pipe.

We are now ready to integrate Eq. (12.71) to determine the variation of the Mach number with distance along the pipe. The left-hand side of the equation can be reduced to a sum of partial fractions to facilitate integration:

$$\left(\frac{1}{\mathrm{M}^3} - \frac{k+1}{2\mathrm{M}} + \frac{(k+1)(k-1)\mathrm{M}}{4\{1+[(k-1)/2]\mathrm{M}^2\}}\right) d\mathrm{M} = \frac{k\bar{f}\,dx}{2D} \tag{12.72}$$

Integrating each side gives

$$\frac{-1}{2\mathrm{M}^2} - \frac{k+1}{2}\ln \mathrm{M} + \frac{k+1}{4}\ln\left(1 + \frac{k-1}{2}\mathrm{M}^2\right) = \frac{k\bar{f}x}{2D} + C \tag{12.73}$$

where C is the integration constant. It is convenient to evaluate C by defining x_* as the distance corresponding to a Mach number of unity.

$$C = \frac{-k\bar{f}x_*}{2D} - \frac{1}{2} + \frac{k+1}{4}\ln\left(\frac{k+1}{2}\right) \tag{12.74}$$

Substituting this expression for C into Eq. (12.73) results in

$$\frac{1-\mathrm{M}^2}{k\mathrm{M}^2} + \frac{k+1}{2k}\ln\left[\frac{(k+1)\mathrm{M}^2}{2+(k-1)\mathrm{M}^2}\right] = \frac{\bar{f}(x_* - x_\mathrm{M})}{D} \tag{12.75}$$

where x_M is the distance corresponding to a Mach number M.

example 12.13

The initial Mach number of the flow of air in a pipe is 0.2. The average value of f is 0.015. Calculate the distance along the pipe (in pipe diameters) required to achieve sonic flow and to achieve a Mach number of 0.8.

Solution Substituting M $= 0.2$ and $k = 1.4$ into Eq. (12.75) gives

$$\frac{\bar{f}(x_* - x_{0.2})}{D} = 14.53$$

Thus the number of pipe diameters to reach sonic flow is

$$\frac{x_* - x_{0.2}}{D} = 969 \quad \triangleleft$$

Substituting M $= 0.8$ into Eq. (12.75) gives

$$\frac{x_* - x_{0.8}}{D} = 0.07 \quad \triangleleft$$

The distance required to increase the Mach number from 0.2 to 0.8 can be obtained by subtraction.

$$\frac{\bar{f}(x_{0.8}-x_{0.2})}{D}=\frac{\bar{f}(x_*-x_{0.2})}{D}-\frac{\bar{f}(x_*-x_{0.8})}{D}=14.53-0.07=14.46$$

$$\frac{x_{0.8}-x_{0.2}}{D}=964 \quad \triangleleft$$

Note that the Mach number increases very rapidly near the end of the pipe.

Thus we see that it is relatively easy to solve for distance along a pipe once the Mach number is known. However, it is more difficult to solve for the change in Mach number given a distance along the pipe. For this reason a plot of Mach number versus $\bar{f}(x_* - x_M)/D$ to use in solving problems of this type is presented in Fig. 12.19.

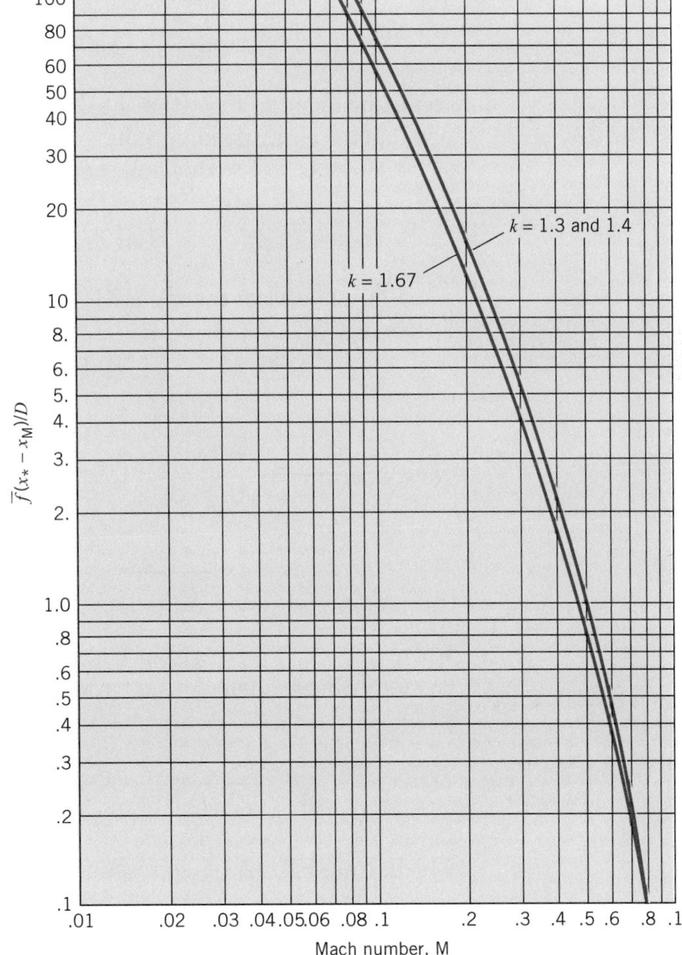

FIGURE 12.19

Variation of $\bar{f}(x_* - x_M)/D$ *with Mach number.*

example 12.14

Air flows at 60 m/s into a commercial steel pipe that has a 5-cm diameter. The pressure and temperature of the air are 1 MPa and 100°C. Determine the Mach number at a distance of 50 m down the pipe.

Solution First, we must evaluate f. Referring to Table A.3 in the Appendix, we find that the dynamic viscosity at 100°C is 2.17×10^{-5} N · s/m². Using the ideal-gas equation, we calculate

$$\rho = \frac{p}{RT} = \frac{10^6 \text{ N/m}^2}{(287 \text{ J/kg K})(373 \text{ K})} = 9.34 \text{ kg/m}^3$$

The Reynolds number, then, is

$$\text{Re} = \frac{(60)(9.34)(0.05)}{2.17(10^{-5})} = 1.29(10^6)$$

Referring again to Fig. 10.8 and Table 10.2, we determine that $k_s/D = 0.001$ and that the corresponding value of f is 0.0195.

The speed of sound at the entrance to the pipe is

$$c = (kRT)^{1/2} = [1.4(287 \text{ J/kg K})(373 \text{ K})]^{1/2} = 387 \text{ m/s}$$

Thus the initial Mach number is M = 60/387 = 0.16, and reference to Fig. 12.19 shows that

$$\frac{\bar{f}(x_* - x_{0.16})}{D} = 24$$

Now we can write

$$\frac{\bar{f}(x_\text{M} - x_{0.16})}{D} = \frac{\bar{f}(x_* - x_{0.16})}{D} - \frac{\bar{f}(x_* - x_\text{M})}{D}$$

where $x_\text{M} - x_{0.16}$ is the distance along the pipe. Substituting the foregoing values for distance, diameter, and average resistance coefficient into this equation yields

$$\frac{\bar{f}(x_* - x_\text{M})}{D} = 24 - \frac{(50)(0.0195) \text{ m}}{0.05 \text{ m}} = 4.5$$

which, from Fig. 12.19, corresponds to a Mach number of

$$\text{M} = 0.32$$ ◁

This calculation was done by taking the initial value of f as the average value. We now ask ourselves how valid this procedure is. The total temperature at the entrance to the tube can be calculated using the total temperature equation:

$$T_t = (373\text{ K})\left[1 + \frac{0.4}{2}(0.16^2)\right] = 375\text{ K}$$

Since it is assumed that the flow is adiabatic, the total temperature does not change along the pipe. Therefore the static temperature 50 m along the pipe is

$$T = \frac{375\text{ K}}{1 + (0.4/2)(0.32^2)} = 367\text{ K}$$

Thus the temperature changes approximately 8 K, and the viscosity change resulting from this temperature change would be approximately 1%. Therefore, the change in Reynolds number is negligible, and the initial friction factor can be used for the average value.

Variation of Pressure with Distance

The differential equation for pressure as a function of velocity and Mach number is Eq. (12.67),

$$\frac{dp}{p} = \frac{-dV}{V}[1 + (k-1)\text{M}^2]$$

Using Eq. (12.70) we can obtain a differential expression for pressure as a function solely of Mach number:

$$\frac{dp}{p} = \frac{-d\text{M}}{\text{M}}\left\{\frac{1 + (k-1)\text{M}^2}{1 + [(k-1)/2]\text{M}^2}\right\} \tag{12.76}$$

From this expression we conclude that $dp/d\text{M} < 0$. This means that pressure decreases with increasing Mach number and increases with decreasing Mach number. Thus for subsonic flow in a pipe, the pressure decreases with distance, the negative pressure gradient providing the force to overcome the wall shear force and accelerate the fluid.

Carrying out the division of the factors contained in the brackets in Eq. (12.76) and dividing by M, we arrive at

$$\frac{dp}{p} = \left\{-\frac{1}{\text{M}} - \frac{[(k-1)/2]\text{M}}{1 + [(k-1)/2]\text{M}^2}\right\}d\text{M} \tag{12.77}$$

Integrating each side, we obtain

$$\ln p = -\ln \text{M} - \frac{1}{2}\ln\left(1 + \frac{k-1}{2}\text{M}^2\right) + C \tag{12.78}$$

We can evaluate the constant of integration C by setting $p = p_*$ at $\text{M} = 1$, so we have

$$\ln p_* = -\frac{1}{2}\ln\left(\frac{k+1}{2}\right) + C$$

which, when substituted back into Eq. (12.78), gives

$$\frac{p_M}{p_*} = \frac{1}{M}\left[\frac{k+1}{2+(k-1)M^2}\right]^{1/2} \tag{12.79}$$

where p_M is the pressure at a Mach number M. The variation of M with p_M/p_* is plotted in Fig. 12.20.

example 12.15

Calculate the pressure at 50 m down the pipe of Example 12.14.

Solution The pressure ratio corresponding to the initial and final Mach numbers can be written as

$$\frac{p_{0.32}}{p_{0.16}} = \frac{p_{0.32}}{p_*}\frac{p_*}{p_{0.16}}$$

Using Fig. 12.20, we find

$$\frac{p_{0.32}}{p_{0.16}} = \frac{3.4}{6.8} = 0.50$$

Therefore the pressure at the 50-m distance is

$$p = 0.50(10^6)\text{ Pa} = 500\text{ kPa}$$

◁

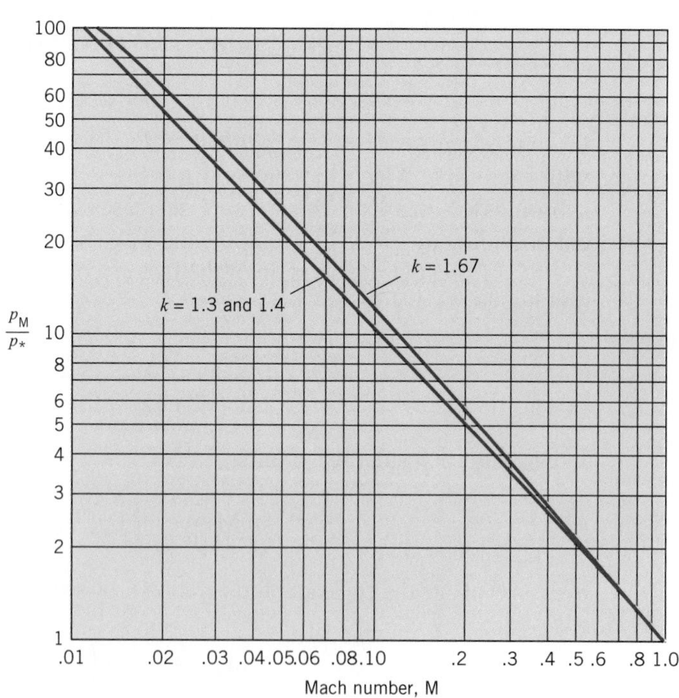

FIGURE 12.20

Variation of p_M/p_* with Mach number for adiabatic viscous flow in a constant-area duct.

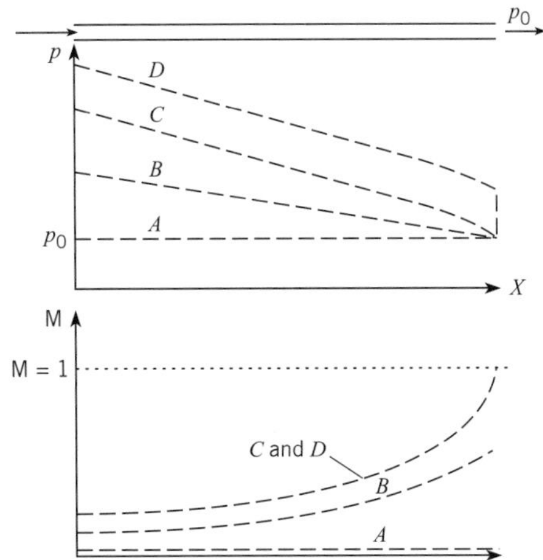

FIGURE 12.21

Distribution of static pressure and Mach number along a pipe.

Let us now consider how the pressure and Mach number vary along a pipe that discharges to the atmosphere as the upstream static pressure is increased. Consider the qualitative distributions of static pressure and Mach number along the pipe shown in Fig. 12.21. The pipe discharges to a back pressure p_b. Cases *A* through *D* represent a continuous increase in the upstream static pressure:

Case A The static pressure is uniform along the pipe and equal to p_b. Thus there is no flow in the pipe, and the Mach number is everywhere zero.

Case B The static pressure uniformly decreases to p_b while the Mach number increases along the pipe.

Case C The static pressure decreases more rapidly and causes the flow to accelerate to a Mach number of unity at the exit. In this case $p_b = p_*$.

Case D The static-pressure distribution is nearly identical to that in case *C* but shifted to a higher value. Thus the pipe discharges at a pressure higher than p_b, and pressure equilibration is achieved by a series of expansion waves. The Mach number distribution differs little from that for case *C*, the flow continuing to discharge at sonic speed.

example 12.16

A brass tube is 3 cm in diameter and 8 m long. The total temperature of the air flow in the tube is 300 K. The pipe discharges to the atmosphere, where the pressure is 100 kPa. Calculate the mass flow in the tube (a) when the inlet static pressure is 120 kPa and (b) when it is 400 kPa.

Solution The approach used to solve this problem depends on whether the flow exits subsonically (case *B*) or sonically (case *D*). If the flow exits subsonically, then the mass flow rate must be such that the exit pressure is equal to the atmospheric pressure. On the other hand, if the flow exits sonically, the mass flow rate is determined by finding the pressure level that gives the correct resistance coefficient *f* in the pipe.

In order to establish which approach to use, we first determine the upstream pressure corresponding to case C, for which the flow exits sonically at atmospheric pressure. The static temperature and speed of sound at the exit for case C are

$$T_e = (300\ \text{K})\frac{2}{k+1} = 250\ \text{K}$$

$$c_e = [(1.4)(287\ \text{J/kg K})(250\ \text{K})]^{1/2} = 317\ \text{m/s}$$

The gas density at the exit is found to be

$$\rho_e = \frac{100 \times 10^3\ \text{N/m}^2}{(287\ \text{J/kg K})(250\ \text{K})} = 1.39\ \text{kg/m}^3$$

Thus the Reynolds number at the exit for case C has a value of

$$\text{Re} = \frac{(317\ \text{m/s})(1.39\ \text{kg/m}^3)(0.03\ \text{m})}{1.5(10^{-5}\ \text{N}\cdot\text{s/m}^2)} = 8.85 \times 10^5$$

Referring to Table 10.2, we find that the relative roughness for the 3-cm brass tube is 0.00005. The corresponding value of the resistance coefficient is found in Fig. 10.8 to be 0.013. Evaluating the parameter $\bar{f}(x_* - x_M)/D$ and entering Fig. 12.19, we find that the initial Mach number would be 0.35. Finally, Fig 12.20 tells us that the pressure ratio corresponding to this Mach number is 3.1. Thus for case C, the static pressure 8 m upstream would be 310 kPa. From this we conclude that part (a) corresponds to case B and part (b) to case D.

a. This problem must be solved using an iterative approach. We shall assume an initial Mach number, calculate the exit Mach number as done in Example 12.14, and then determine the exit pressure as done in Example 12.15. The initial Mach number is varied until the exit pressure matches the atmospheric pressure. The most straightforward approach is to generate a table such as that shown, where M_e is the exit Mach number.

M	***T*, K**	**Re**	$\bar{f}$	$\mathbf{M}_e$	p/p_e	p_e, **kPa**
0.1	299	5.4×10^4	0.020	0.104	1.04	115
0.2	298	1.08×10^5	0.0185	0.24	1.21	99
0.19	298	1.02×10^5	0.0190	0.23	1.19	101

By interpolation we find that p_e = 100 kPa when M = 0.195. The temperature, density, and speed of sound at this Mach number are

$$T = 298\ \text{K} \qquad \rho = 1.40\ \text{kg/m}^3 \qquad c = 346\ \text{m/s}$$

Thus the mass flow rate is

$$\dot{m} = (0.195)(346\ \text{m/s})(1.40\ \text{kg/m}^3)(0.785)(0.03^2\ \text{m}^2) = 0.067\ \text{kg/s}$$

◁

b. The way to solve this part is to use the iterative approach again but on $\bar{f}$. We begin by assuming an $\bar{f}$, calculating the upstream Mach number and Reynolds number, and finding a new $\bar{f}$, which is used as the assumed value for the next iteration. The solution is the value at which $\bar{f}$ no longer changes. A check is always made to be sure that p_* exceeds p_b. The iteration using $\bar{f} = 0.02$ as the initial assumed value is demonstrated in the accompanying table.

f	$\bar{f}(x_* - x_M)/D$	M	Re	$\bar{f}$	p/p_*	p/p_b
0.2	5.33	0.3	5.4×10^5	0.0138	3.6	1.11
0.0138	3.67	0.34	6.1×10^5	0.0135	3.25	1.23
0.0135	3.59	0.35	6.3×10^5	0.0135	3.15	1.26

Thus the initial Mach number is 0.35, the exit Mach number is unity, and the exit pressure exceeds the atmospheric pressure. The temperature, density, and speed of sound at M = 0.35 are

$$T = 293 \text{ K} \qquad \rho = 4.77 \text{ kg/m}^3 \qquad c = 342 \text{ m/s}$$

The mass flow rate is

$$\dot{m} = (0.35)(342 \text{ m/s})(4.77 \text{ kg/m}^3)(0.785)(0.03^2 \text{ m}^2) = 0.403 \text{ kg/s} \quad \triangleleft$$

Isothermal Flow

The analysis of isothermal flow in a constant-area duct is simplified by the fact that the energy equation is

$$T = \text{constant}$$

This also means that the speed of sound in the duct is constant.

The momentum equation for flow in the duct is given by Eq. (12.65):

$$k\text{M}^2 \frac{dV}{V} + \frac{dp}{p} + \frac{kf\text{M}^2 dx}{2D} = 0$$

Because the temperature is constant, the pressure is proportional to the density [see Eq. (12.66)]. Thus

$$\frac{dp}{p} = \frac{d\rho}{\rho} \tag{12.80}$$

Using the continuity equation, Eq. (12.16), to relate ρ and V gives

$$\frac{d\rho}{\rho} = -\frac{dV}{V} \tag{12.81}$$

which, substituted back into Eq. (12.65), yields

$$\frac{d\text{M}}{dx} = \frac{f}{2D} \frac{k\text{M}^3}{1 - k\text{M}^2} \tag{12.82}$$

One notes that if the Mach number is less than $1/\sqrt{k}$, then it increases with distance, whereas the opposite trend is noted for Mach numbers exceeding $1/\sqrt{k}$. Thus the Mach number must always approach $1/\sqrt{k}$ for isothermal flows, compared with unity for adiabatic flows.

Mach Number Distribution along a Constant-Area Duct

The Mach number distribution along a constant-area duct is determined by integrating Eq. (12.82) for Mach number as a function of distance. Rewriting Eq. (12.82) as

$$d\mathrm{M}\,\frac{(1-k\mathrm{M}^2)}{k\mathrm{M}^3}=\frac{f\,dx}{2D} \tag{12.83}$$

or

$$\frac{d\mathrm{M}}{k\mathrm{M}^3}-\frac{d\mathrm{M}}{\mathrm{M}}=\frac{f\,dx}{2D} \tag{12.84}$$

we can integrate each side to obtain

$$\frac{1}{-2k\mathrm{M}^2}-\ln\mathrm{M}=\frac{fx}{2D}+C \tag{12.85}$$

When we set x_T as the maximum length at which $\mathrm{M}=1/\sqrt{k}$, the constant of integration becomes

$$-\frac{1}{2}-\ln\left(\frac{1}{\sqrt{k}}\right)-\frac{fx_T}{2D}=C \tag{12.86}$$

and substituting this constant in Eq. (12.85) gives

$$\frac{f(x_T-x_\mathrm{M})}{D}=\ln(D\mathrm{M}^2)+\frac{(1-k\mathrm{M}^2)}{k\mathrm{M}^2} \tag{12.87}$$

The variation of $f(x_T-x_\mathrm{M})/D$ with $k\mathrm{M}^2$ is plotted in Fig. 12.22.

Pressure Variation

In isothermal flow the speed of sound is constant, so velocity is directly proportional to the Mach number. Also, the pressure is directly proportional to the density. Substituting these relationships into Eq. (12.81), we obtain

$$\frac{dp}{p}=-\frac{d\mathrm{M}}{\mathrm{M}} \tag{12.88}$$

Integrating Eq. (12.88) yields

$$\ln p=-\ln\mathrm{M}+C \tag{12.89}$$

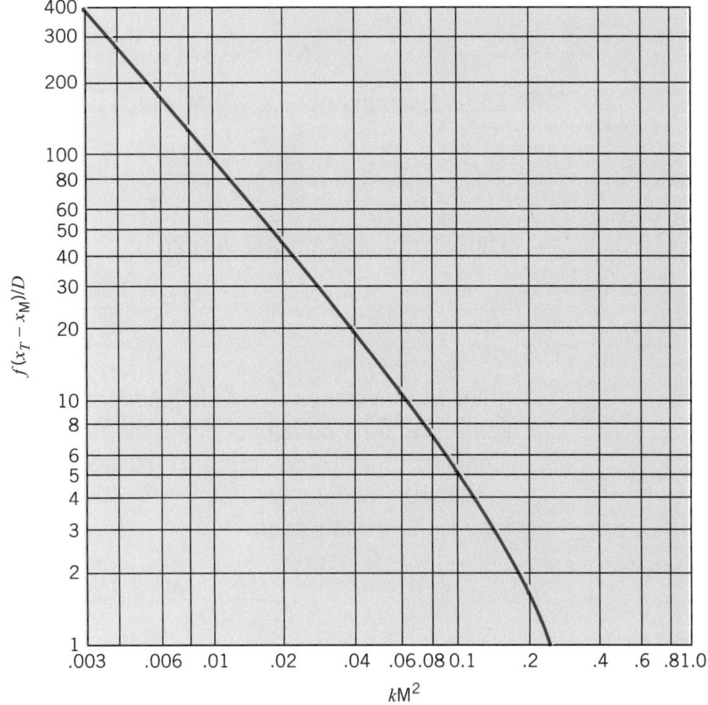

FIGURE 12.22

Variation of $f(x_T - x_M)/D$ with kM^2 for isothermal viscous flow in a constant-area duct.

Letting p_T be the pressure corresponding to the maximum distance, we find the constant of integration and substitute it back into Eq. (12.89) to give

$$\frac{p_M}{p_T} = \frac{1}{\sqrt{k}M} \tag{12.90}$$

Similar expressions can be found for the variation of density, total pressure, and total temperature along the duct.

example 12.17

Methane is to be transported at 15°C in 50-cm commercial steel pipe that is 1 km long. The pressure at the pipe exit is 100 kPa. Determine the maximum flow rate through the pipe and the pressure at the pipe entrance.

Solution From Table A.2 in the Appendix, the value of k for methane is 1.31, the gas constant is 518 J/kg K, and the kinematic viscosity at 15°C is 1.59×10^{-5} m^2/s.

The Mach number at the pipe exit is

$$M = \frac{1}{\sqrt{k}} = \frac{1}{\sqrt{1.31}} = 0.874$$

The corresponding velocity is

$$V = \mathrm{M}c = 0.874 \times \sqrt{1.31 \times 518 \times 288} = 386 \text{ m/s}$$

The corresponding mass flow, which is the maximum flow rate, is

$$\dot{m} = \rho A V = \left(\frac{10^5}{518 \times 288}\right)(0.5)^2\left(\frac{\pi}{4}\right)(386) = 50.8 \text{ kg/s} \quad \triangleleft$$

The Reynolds number at the exit is

$$\mathrm{Re} = \frac{VD}{\nu} = \frac{(386)(0.5)}{1.59 \times 10^{-5}} = 1.2 \times 10^7$$

Since ρV is constant along the pipe (continuity equation) and μ is constant (depends primarily on temperature), the Reynolds number is the same everywhere along the pipe. Therefore, the resistance coefficient does not change. Referring to Fig. 10.8 and Table 10.2, we find that f is 0.012. Thus the left-hand side of Eq. (12.87) has the value

$$\frac{f(x_T - x_\mathrm{M})}{D} = \frac{(0.012)(10^3)}{0.5} = 24$$

Using Fig. 12.22, we find

$$k\mathrm{M}^2 = 0.035$$

The pressure at the entrance is found using Eq. (12.90):

$$p_\mathrm{M} = \frac{100 \text{ kPa}}{\sqrt{0.035}} = 535 \text{ kPa} \quad \triangleleft$$

12.6 Summary

In this chapter we have treated the adiabatic and isothermal flow of a viscous compressible gas in a constant-area duct. The reader is referred to more specialized texts on compressible flow, such as that by Owczarek (2) and Anderson (3), for studies of other types of flows, such as the effect of heat addition due to chemical reaction.

The speed of sound is the speed at which an infinitesimal pressure disturbance travels through a fluid. The speed of sound in an ideal gas is

$$c = \sqrt{kRT}$$

where k is the ratio of specific heats, R is the gas constant, and T is the absolute temperature. The Mach number is defined as

$$\mathrm{M} = \frac{V}{c}$$

Compressible flows are classified as

$\mathrm{M} < 1$	subsonic
$\mathrm{M} \simeq 1$	transonic
$\mathrm{M} > 1$	supersonic

In general, if the Mach number is less than 0.3, a steady flow can be regarded as incompressible.

For an adiabatic flow (no heat transfer), the temperature varies along a streamline according to

$$T = T_t\left(1 + \frac{k-1}{2}\mathrm{M}^2\right)^{-1}$$

where T_t, the total temperature, is the temperature attained if the flow is decelerated to zero velocity. If the flow is isentropic, the pressure varies along a streamline as

$$p = p_t\left(1 + \frac{k-1}{2}\mathrm{M}^2\right)^{k/(1-k)}$$

where p_t is the total pressure, the pressure achieved if the flow is decelerated to zero velocity isentropically.

A normal shock wave is a narrow region where a supersonic flow is decelerated to a subsonic flow with an attendant rise in pressure, temperature, and density. The total temperature does not change through a shock wave but the total pressure decreases. The shock wave is a nonisentropic process and can only occur in supersonic flows.

A Laval nozzle is a duct with a converging and expanding area that is used to accelerate a compressible fluid to supersonic speeds. Sonic flow can occur only at the nozzle throat (minimum area). The ratio of the area at a location in the nozzle to the throat area, A/A_*, is a function of the Mach number at the location and the ratio of specific heats. The flow rate through a Laval nozzle is given by

$$\dot{m} = 0.685\frac{p_t A_*}{\sqrt{RT_t}}$$

A Laval nozzle is classified by comparing the pressure at the exit, p_e, for supersonic flow in the nozzle with the back (ambient) pressure, p_b.

$p_e/p_b > 1$	underexpanded
$p_e/p_b = 1$	ideally expanded
$p_e/p_b < 1$	overexpanded

Shock waves occur in overexpanded nozzles, yielding a subsonic flow at the exit.

The Mach number of a compressible, adiabatic flow in a straight pipe with constant diameter increases with distance along the pipe. The maximum Mach number achievable is unity at the end of the pipe. A flow cannot change from subsonic to supersonic flow in a pipe but can change from supersonic to subsonic through a shock wave. The pressure at the end of a pipe is equal to the back pressure if the exit Mach number is subsonic. The exit pressure can be equal to or exceed the back pressure if the exit Mach number is unity.

References

1. Van Wylen, G. J., and R. E. Sonntag. *Fundamentals of Classical Thermodynamics.* John Wiley, New York, *1965.*
2. Owczarek, J. A. *Fundamentals of Gas Dynamics.* McGraw-Hill, New York, 1971.
3. Anderson, J. D., Jr. *Modern Compressible Flow with Historical Perspective.* McGraw-Hill, New York, 1982.

Problems

12.1 How fast (in meters per second) will a sound wave travel in methane at 0°C?

FLUID SOLUTIONS **12.2** Calculate the speed of sound in helium at 50°C.

12.3 Calculate the speed of sound in hydrogen at 68°F.

FLUID SOLUTIONS **12.4** How much faster will a sound wave propagate in helium than in nitrogen if the temperature of both gases is 20°C?

12.5 Determine what the equation for the speed of sound in an ideal gas would be if the sound wave were an isothermal process.

12.6 The relationship between pressure and density for the propagation of a sound wave through a fluid is

$$p - p_0 = E_v \ln(\rho/\rho_0)$$

where p_0 and ρ_0 are the reference pressure and density (constants) and E_v is the bulk modulus of elasticity. Determine the equation for the speed of a sound wave in terms of E_v and ρ. Calculate the sound speed for water with $\rho = 1000\ \text{kg/m}^3$ and $E_v = 2.20\ \text{GN/m}^2$.

FLUID SOLUTIONS **12.7** A supersonic aircraft is flying at Mach 1.5 through air at –30°C. What temperature could be expected on exposed aircraft surfaces? What is the airspeed behind the shock?

12.8 What is the temperature on the nose of a supersonic fighter flying at Mach 2 through air at 273 K?

12.9 A high-performance aircraft is flying at a Mach number of 1.8 at an altitude of 10,000 m, where the temperature is –44°C and the pressure is 30.5 kPa.

a. How fast is the aircraft traveling in kilometers per hour?

b. The total temperature is an estimate of surface temperature on the aircraft. What is the total temperature under these conditions?

c. Calculate the total pressure under these conditions.

d. If the aircraft slows down, at what speed (kilometers per hour) will the Mach number be unity?

12.10 An airplane travels at 800 km/h at sea level where the temperature is 15°C. How fast would the airplane be flying at the same Mach number at an altitude where the temperature was –40°C?

12.11 An airplane flies at a Mach number of 0.95 at a 10,000-m altitude, where the static temperature is –44°C and the pressure is 30 kPa absolute. The lift coefficient of the wing is 0.05. Determine the wing loading (lift force / wing area).

12.12 An object is immersed in an airflow with a static pressure of 200 kPa absolute, a static temperature of 20°C, and a velocity of 250 m/s. What are the pressure and temperature at the stagnation point?

12.13 An airflow at M $= 0.75$ passes through a conduit with a cross-sectional area of 50 cm^2. The total absolute pressure is 360 kPa, and the total temperature is 10°C. Calculate the mass flow rate through the conduit.

12.14 Oxygen flows from a reservoir in which the temperature is 200°C and the pressure is 300 kPa absolute. Assuming isentropic flow, calculate the velocity, pressure, and temperature when the Mach number is 0.9.

12.15 One problem in creating high-Mach-number flows is condensation of the oxygen component in the air when the temperature reaches 50 K. If the temperature of the reservoir is 300 K and the flow is isentropic, at what Mach number will condensation of oxygen occur?

12.16 Hydrogen flows from a reservoir where the temperature is 20°C and the pressure is 500 kPa absolute to a section 2 cm in diameter where the velocity is 250 m/s. Assuming isentropic flow, calculate the temperature, pressure, Mach number, and mass flow rate at the 2-cm section.

12.17 The total pressure in a Mach-2.5 wind tunnel operating with air is 600 kPa absolute. A sphere 2 cm in diameter, positioned in the wind tunnel, has a drag coefficient of 0.95. Calculate the drag of the sphere.

12.18 Using Eq. (12.27), develop an expression for the pressure coefficient at stagnation conditions—that is, $C_p = (p_t - p)/[(1/2)\rho V^2]$—in terms of Mach number and ratio of specific heats, $C_p = f(k, \mathrm{M})$. Evaluate C_p at M $= 0$, 2, and 4 for $k = 1.4$. What would its value be for incompressible flow?

12.19 For low velocities, the total pressure is only slightly larger than the static pressure. Thus one can write $p_t/p = 1 + \epsilon$, where ϵ is a small positive number ($\epsilon \ll 1$). Using this approximation, show that, as $\epsilon \to 0$ ($\mathrm{M} \to 0$), Eq. (12.32) reduces to

$$\mathrm{M} = \left[\frac{2(p_t/p - 1)}{k}\right]^{1/2}$$

12.20 A normal shock wave exists in a 500-m/s stream of nitrogen having a static temperature of –50°C and a static pressure of 70 kPa. Calculate the Mach number, pressure, and temperature downstream of the wave and the entropy increase across the wave.

12.21 A normal shock wave exists in a Mach 2 stream of air having a static temperature and pressure of 45°F and 30 psia. Calculate the Mach number, pressure, and temperature downstream of the shock wave.

12.22 A Pitot-static tube is used to measure the Mach number on a supersonic aircraft. The tube, because of its bluntness, creates a normal shock wave as shown. The absolute total pressure downstream of the shock wave (p_{t_2}) is 150 kPa. The static pressure of the free stream ahead of the shock wave (p_1) is 40 kPa and is sensed by the static pressure tap on the probe. Determine the Mach number (M_1) graphically.

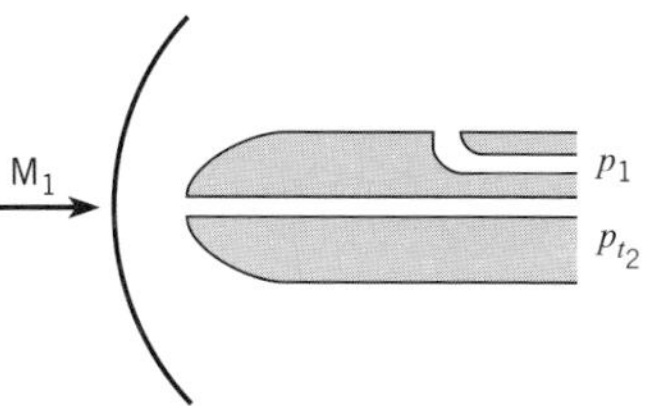

PROBLEM 12.22

12.23 A shock wave occurs in a methane stream in which the Mach number is 3, the static pressure is 100 kPa absolute, and the static temperature is 20°C. Determine the downstream Mach number, static pressure, static temperature, and density.

12.24 The Mach number downstream of a shock wave in helium is 0.9, and the static temperature is 100°C. Calculate the velocity upstream of the wave.

12.25 Show that the lowest Mach number possible downstream of a normal shock wave is

$$\mathrm{M}_2 = \sqrt{\frac{k-1}{2k}}$$

and that the largest density ratio possible is

$$\frac{\rho_2}{\rho_1} = \frac{k+1}{k-1}$$

What are the limiting values of M_2 and ρ_2/ρ_1 for air?

12.26 Show that the Mach number downstream of a weak wave ($\mathrm{M} \simeq 1$) is approximated by

$$\mathrm{M}_2^2 = 2 - \mathrm{M}_1^2$$

[*Hint:* Let $\mathrm{M}_1^2 = 1 + \epsilon$, where $\epsilon \ll 1$, and expand Eq. (12.41) in terms of ϵ.] Compare values for M_2 obtained using this equation with values for M_2 from Table A.1 for $\mathrm{M}_1 = 1$, 1.05, 1.1, and 1.2.

12.27 Develop a computer program for calculating the mass flow in a truncated nozzle. The input to the program would be total pressure, total temperature, back pressure, ratio of specific heats, gas constant, and nozzle diameter. Run the program and compare the results with Example 12.12. Run the program for back pressures of 80, 90, 110, 120, and 130 kPa and make a table for the variation of mass flow rate with back pressure. What trends do you observe?

This program will be useful for Probs. 12.28, 12.29, 12.31, and 12.32.

12.28 The truncated nozzle shown in the figure is used to meter the mass flow of air in a pipe. The area of the nozzle is 3 cm^2. The total pressure and total temperature measured upstream of the nozzle in the pipe are 300 kPa absolute and 20°C. The pressure downstream of the nozzle (back pressure) is 90 kPa absolute. Calculate the mass-flow rate.

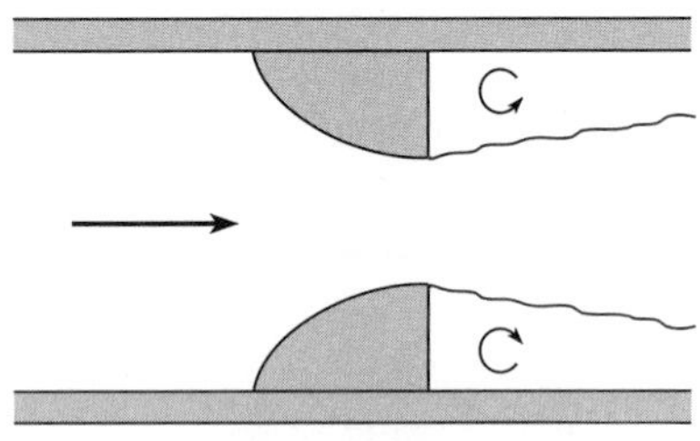

PROBLEM 12.28

FLUID SOLUTIONS **12.29** The truncated nozzle shown in Prob. 12.28 is used to monitor the mass flow rate of methane. The area of the nozzle is 3 cm^2, and the area of the pipe is 12 cm^2. The upstream total pressure and total temperature are 150 kPa absolute and 30°C. The back pressure is 100 kPa.

a. Calculate the mass flow rate of methane.

b. Calculate the mass flow rate assuming the Bernoulli equation is valid, the density being the density of the gas at the nozzle exit.

12.30 A truncated nozzle with an exit area of 4 cm^2 is used to measure a mass flow of air of 0.30 kg/s. The static temperature of the air at the exit is 10°C, and the back pressure is 100 kPa. Determine the total pressure.

12.31 A truncated nozzle with a 12-cm^2 exit area is supplied from a helium reservoir in which the absolute pressure is first 130 kPa and then 350 kPa. The temperature in the reservoir is 28°C, and the back pressure is 100 kPa. Calculate the mass flow rate of helium for the two reservoir pressures.

12.32 A sampling probe is used to draw gas samples from a gas stream for analysis. In sampling, it is important that the velocity entering the probe equal the velocity of the gas stream (isokinetic condition). Consider the sampling probe shown, which has a truncated nozzle inside it to control the mass-flow rate. The probe has an inlet diameter of 4 mm and a truncated nozzle diameter of 2 mm. The probe is in a hot-air stream with a static temperature of 600°C, a static pressure of 100 kPa absolute, and a velocity of 60 m/s. Calculate the pressure required in the probe (back pressure) to maintain the isokinetic sampling condition.

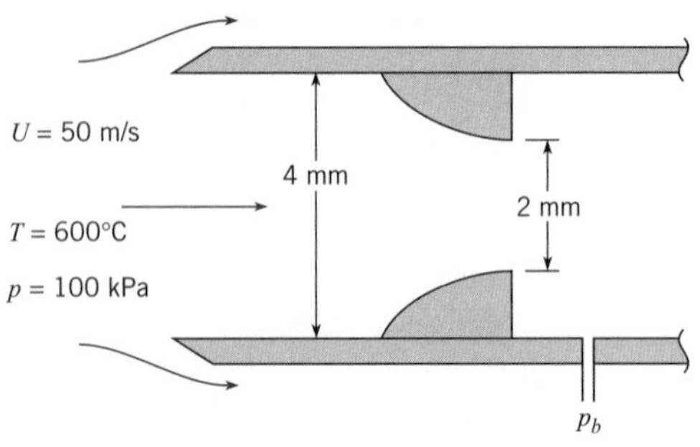

PROBLEM 12.32

12.33 Develop a computer program that requires the Mach number and ratio of specific heats as input and prints out the area ratio, the ratio of static to total pressure, the ratio of static to total temperature, the ratio of density to total density, and the ratio of pressure before and after a shock wave. Run the program for a Mach number of 2 and a ratio of specific heats of 1.4, and compare with results in Table A.1. Then run the program for the same Mach number with ratios of specific heats of 1.3 (carbon dioxide) and 1.67 (helium).

This program will be useful for Probs. 12.35, 12.36, 12.39, 12.43, 12.44, 12.46, and 12.47.

12.34 Develop a computer program that, given the area ratio, ratio of specific heats, and flow condition (subsonic or supersonic) as input, provides the Mach number. This will require some numerical root-finding scheme. Run the program for an area ratio of 5 and ratio of specific heats of 1.4. Compare the results with those in Table A.1. Then run the program for the same area ratio but with the ratios of specific heats of 1.67 (helium) and 1.31 (methane).

This program will be useful for Probs. 12.37, 12.38 and 12.41–12.47.

12.35 A wind tunnel is designed to have a Mach number of 3, a static pressure of 1.5 psia, and a static temperature of –10°F in the test section. Determine the area ratio of the nozzle required and the reservoir conditions that must be maintained if air is to be used.

12.36 A Laval nozzle is to be designed to operate supersonically and expand ideally to an absolute pressure of 30 kPa. If the stagnation pressure in the nozzle is 1 MPa, calculate the nozzle area ratio required. Determine the nozzle throat area for a mass flow of 5 kg/s and a stagnation temperature of 550 K. Assume that the gas is nitrogen.

12.37 A rocket nozzle with an area ratio of 4 is operating at a total absolute pressure of 1.3 MPa and exhausting to an atmosphere with an absolute pressure of 30 kPa. Determine whether the nozzle is overexpanded, underexpanded, or ideally expanded. Assume $k = 1.4$.

12.38 Repeat Prob. 12.37 for a nozzle with the same area ratio but with a ratio of specific heats of 1.2. Classify the nozzle flow.

12.39 A Laval nozzle with an exit area ratio of 1.688 exhausts air from a large reservoir into ambient conditions at $p = 100$ kPa.

a. Show that the reservoir pressure must be 782.5 kPa to achieve ideally expanded exit conditions at $\mathrm{M} = 2$.

b. What are the static temperature and pressure at the throat if the reservoir temperature is 17°C with the pressure as in (a)?

c. If the reservoir pressure were lowered to 700 kPa, what would be the exit condition (overexpanded, ideally expanded, underexpanded, subsonic flow in entire nozzle)?

d. What reservoir pressure would cause a normal shock to form at the exit?

12.40 Determine the Mach number and area ratio at which the dynamic pressure is maximized in a Laval nozzle with air. [*Hint:* Express q in terms of p and M, and use Eq. (12.27) for p. Differentiate with respect to M and equate to zero.]

12.41 A rocket motor operates at an altitude where the atmospheric pressure is 25 kPa. The expansion ratio of the nozzle is 4 (exit area / throat area). The chamber pressure of the motor (total pressure) is 1.2 MPa, and the chamber temperature (total temperature) is 3000°C. The ratio of the specific heats of the exhaust gas is 1.2, and the gas constant is 400 J / kg K. The throat area of the rocket nozzle is 100 cm^2.

a. Determine the Mach number, density, pressure, and velocity at the nozzle exit.

b. Determine the mass flow rate.

c. Calculate the thrust of the rocket using (see Example 6.10)

$$T = \dot{m}V_e + (p_e - p_0)A_e$$

d. What would the chamber pressure of the rocket have to be to have an ideally expanded nozzle? Calculate the rocket thrust under this condition.

12.42 A rocket motor is being designed to operate at sea level, where the pressure is 100 kPa absolute. The chamber pressure (total pressure) is 2.0 MPa, and the chamber temperature (total temperature) is 3300 K. The throat area of the nozzle is 10 cm^2. The ratio of the specific heats (k) of the exhaust gas is 1.2, and the gas constant is 400 J / kg K.

a. Determine the nozzle expansion ratio that is required to achieve an ideally expanded nozzle, and determine the nozzle thrust under these conditions (see Prob. 12.41 for the thrust equation).

b. Determine the thrust that would be obtained if the expansion ratio were reduced by 10% to achieve an underexpanded nozzle.

12.43 Air flows through a Laval nozzle with an expansion ratio of 4. The total pressure of the air entering the nozzle is 200 kPa, and the back pressure is 100 kPa. Determine the area ratio at which the shock wave occurs in the expansion section of the nozzle. (*Hint:* This problem can be solved graphically by calculating the exit pressure corresponding to different shock wave locations and finding the location where the exit pressure is equal to the back pressure.)

12.44 A rocket nozzle has the configuration shown. The diameter of the throat is 4 cm, and the exit diameter is 8 cm. The half-angle of the expansion cone is 15°. Gases with a specific heat of 1.2 flow into the nozzle with a total pressure of 250 kPa. The back pressure is 100 kPa. First, using an iterative or graphical method, determine the area ratio at which the shock occurs. Then determine the shock wave's distance from the throat in centimeters.

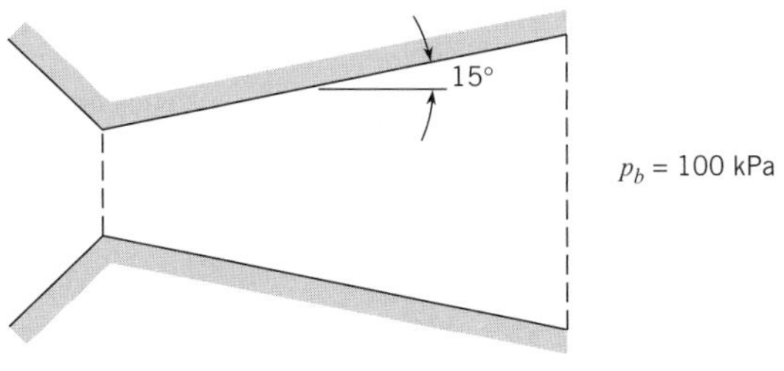

PROBLEM 12.44

12.45 A normal shock wave occurs in a nozzle at an area ratio of 4. Determine the entropy increase if the gas is hydrogen.

12.46 Consider air flow in the variable-area channel shown. Determine the Mach number, static pressure, and stagnation pressure at station 3. Assume isentropic flow except for normal shock waves. FLUID SOLUTIONS

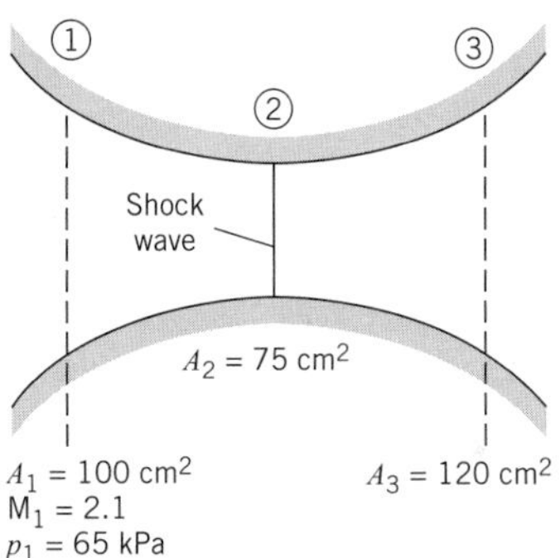

PROBLEM 12.46

12.47 Determine the atmospheric pressure necessary for the shock wave to position itself as shown. The fluid is air.

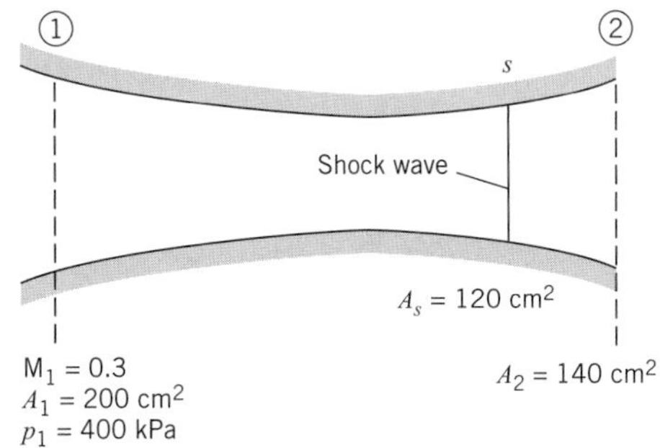

PROBLEM 12.47

12.48 Develop a computer program that, given $f(x-x_*)/D$ for compressible, adiabatic flow in a pipe, returns the initial Mach number and the ratio of the pressure to the pressure at sonic conditions, p_M/p_*. This will require some numerical root-finding scheme. Run the program for $f(x-x_*)/D$ equal to 1, 10, and 100 with $k = 1.4$ and compare the results with those in Fig. 12.19. Then run the program for the same values but with $k = 1.31$ (natural gas).

This program will be useful for Probs. 12.49, 12.50, and 12.53–12.56.

FLUID SOLUTIONS **12.49** An engineer is designing a piping system for air flow. The pipe must be 10 m long, must be fabricated from brass, and must carry a mass flow of 0.2 kg/s. The total temperature of the flow in the pipe is 373 K, and the tolerable pressure loss is 140 kPa. The pipe discharges to a pressure of 100 kPa. Determine the diameter of the pipe.

12.50 The velocity of air entering a commercial steel pipe with a 1-in. diameter is 120 ft/s. The static temperature and pressure are 67°F and 30 psia. Calculate the length of pipe needed to achieve sonic flow; also calculate the pressure at the end of the pipe. Assume the viscosity of the air is independent of pressure.

FLUID SOLUTIONS **12.51** The Mach number of air flowing out the exit of a 3-cm brass tube is 0.9. The total temperature of the air flow is 373 K, and the atmospheric pressure is 100 kPa. How far upstream is the Mach number equal to 0.2?

12.52 The Mach numbers at the inlet and exit of a pipe 0.5 in. in diameter and 20 ft long are 0.2 and 0.7, respectively. Calculate $\bar{f}$ in the pipe for $k = 1.4$.

12.53 Oxygen flows through a wrought-iron pipe 2.5 cm in diameter and 10 m long. It discharges to the atmosphere, where the pressure is 100 kPa. If the absolute static pressure at the beginning of the pipe is 300 kPa and the total temperature is 293 K, calculate the mass flow through the pipe.

12.54 Repeat Prob. 12.53, using an absolute static pressure of 500 kPa at the beginning of the pipe.

FLUID SOLUTIONS **12.55** A pressure hose 10 ft long is to be connected to the outlet of the regulator valve on a nitrogen bottle. The hose must deliver 0.06 lbm/s when the regulator pressure is 45 psia. The hose exhausts to a back pressure of 7 psia. The total temperature is 100°F. Assume the hose has a roughness equivalent to that of galvanized iron, and calculate the required hose diameter.

12.56 An air blower and pipe system is to be designed to convey agricultural products through a 20-cm steel pipe. The pipe is to be 120 m long, and the outlet velocity is to be 50 m/s. If the pipe discharges into the atmosphere (100 kPa, 15°C), what will be the pressure, velocity, and density of the air at the inlet end of the pipe? Assume that the ratio of specific heats for the particle-laden flow is 1.4.

12.57 Methane is pumped into a 15-cm steel pipe at a pressure of 1.2 MPa and a temperature of 320 K ($\mu = 1.5 \times 10^{-5}$ N · s/m^2) and with a velocity of 20 m/s. What is the pressure 3000 m downstream?

12.58 Hydrogen is transported in an underground pipeline. The pipe is 50 m long and 10 cm in diameter; it is maintained at a temperature of 15°C. The initial pressure and velocity are 250 kPa and 200 m/s. Determine the pressure drop in the pipe.

12.59 Helium flows in a 5-cm brass tube 100 m long that is maintained at a temperature of 15°C. The entrance pressure is 120 kPa, and the exit pressure is 100 kPa. Determine the mass flow rate in the pipe. (This problem requires an iterative solution.)

12.60 Design a supersonic wind tunnel that achieves a Mach number of 1.5 in a test section 5 cm by 5 cm. The tunnel is to be attached to a vacuum tank as shown. After the tank is evacuated, the valve is opened and atmospheric air is drawn through the tunnel into the tank. The tunnel should operate for 30 seconds before the pressure rises to the point in the tank that supersonic flow is no longer achievable. Do a preliminary design of this system including details such as nozzle dimensions, configuration, and tank size.

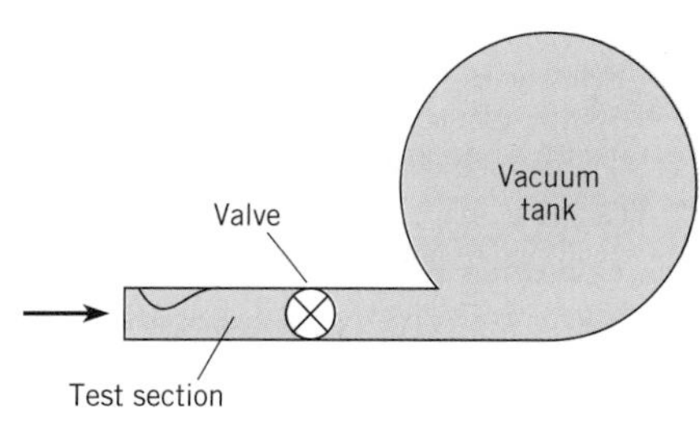

PROBLEM 12.60

12.61 Truncated nozzles are often used for flowmetering devices. Assume that you have to design a truncated nozzle, or a series of truncated nozzles, to measure the performance of an air compressor. The compressor is rated at 100 scfm (standard cubic feet per minute) at 120 psig. (A standard cubic foot is the volume the air would occupy at atmospheric conditions.) A performance curve for the compressor would be a plot of flow rate versus supply pressure. Explain how you would carry out the test program.

Very small solid particles are often entrained in fluids such as air or water to enhance the visualization of the flow. The pattern of flow that occurs when an air jet is discharged into quiescent air is revealed in this photo. This image is produced by a laser light shining on micron-sized particles of titanium oxide. (Courtesy Cecilia D. Richards.)

CHAPTER 13

Flow Measurements

The most common flow measurements are pressure, rate of flow, and velocity. In Chapter 3 several methods were presented for measuring pressure, and in Chapter 5 the basic theory of stagnation and Pitot tubes for the measurement of velocity was given. In this chapter we will consider the Pitot, static, and stagnation tubes in more detail, and we will describe other ways of measuring velocity. Next, we will describe measurement of flow rate, followed by a discussion of measurements in compressible flow. The chapter concludes with a discussion of experimental accuracy.

Instruments for the Measurement of Velocity and Pressure

Stagnation Tube

In Chapter 5, where the stagnation tube was introduced, it was assumed that viscous effects were negligible. This assumption is valid for tubes of normal size when we are measuring moderately high-velocity air or water. However, if the velocity to be measured is very low or the tube is very small, viscous effects become significant, and a correction must be applied to the basic equation. That is, as the Reynolds number decreases, the viscous effects become significant. This influence is shown in Fig. 13.1, where the pressure coefficient $C_p = (p - p_0)/(rV_0^2/2)$ is plotted as a function of the Reynolds number.

In Fig. 13.1 it is seen that when the Reynolds number for the circular stagnation tube is greater than 60, the error in measured velocity is less than 1%. For boundary layer measurements a stagnation tube with a flattened end can be used. By flattening the end of the tube, the velocity measurement can be taken nearer the boundary than if a circular tube were used. For these flattened tubes, the pressure coefficient remains near unity for a Reynolds number as low as 30. See reference (15) for more information on flattened-end stagnation tubes.

Flow Direction with Stagnation-Type Tubes

When the direction of flow is not known, simple pressure-type *yaw meters* can be used to sense the flow direction. Three types of yaw meters are shown in Fig. 13.2. The first two can be used for two-dimensional flow, where flow direction in only one plane needs to be found, and the third is used for determining flow direction in three dimensions. In all these devices, the tube is turned until the pressure on symmetrically opposite openings

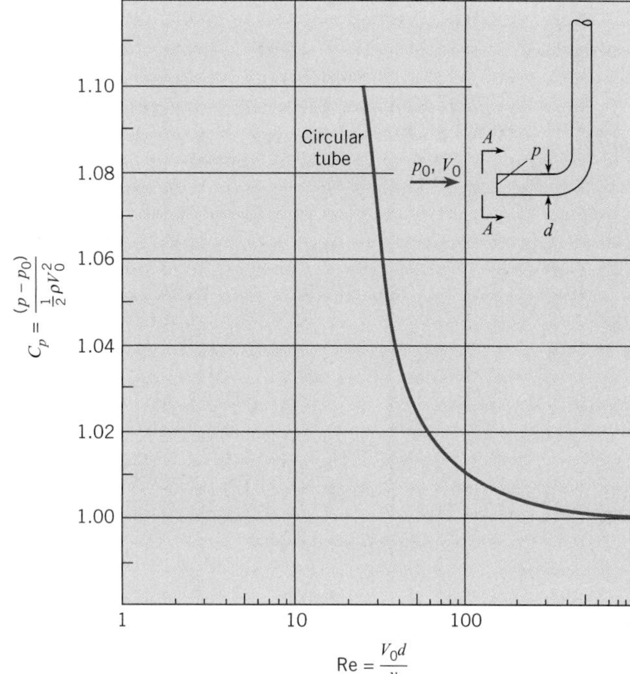

FIGURE 13.1

Viscous effects on C_p. [After Hurd, Chesky, and Shapiro (1). Used with permission from ASME.]

is equal. This pressure is sensed by a differential pressure gage or manometer connected to the openings in the yaw meter. The flow direction is sensed when a null reading is indicated on the differential gage.

Static Tube

At times one wants to measure the static pressure in a flowing fluid. This is accomplished by sensing the pressure at a point along the static tube where the pressure is the same as the free-stream pressure. Such a tube is shown in Fig. 13.3. This is like the Pitot tube except that the stagnation pressure tap is omitted. The placement of the holes along the probe is important for sensing the static pressure, because the rounded nose on the tube causes some decrease of pressure along the tube and the downstream stem causes an increase in pressure in front of it. Hence the location for sensing the static pressure must be at the point where these two effects cancel each other. Experiments reveal that the optimum location is at a point approximately six diameters downstream of the front of the tube and eight diameters upstream from the stem.

The Vane or Propeller Anemometer

The basic vane or propeller anemometer has been used in various applications for a number of years to measure the velocity of either gases or liquids. Basically, the type used for air flows, Fig. 13.4, consists of vanes (propeller blades) attached to a rotor that drives a low-friction gear train that, in turn, drives a pointer that indicates feet on a dial. Thus if the

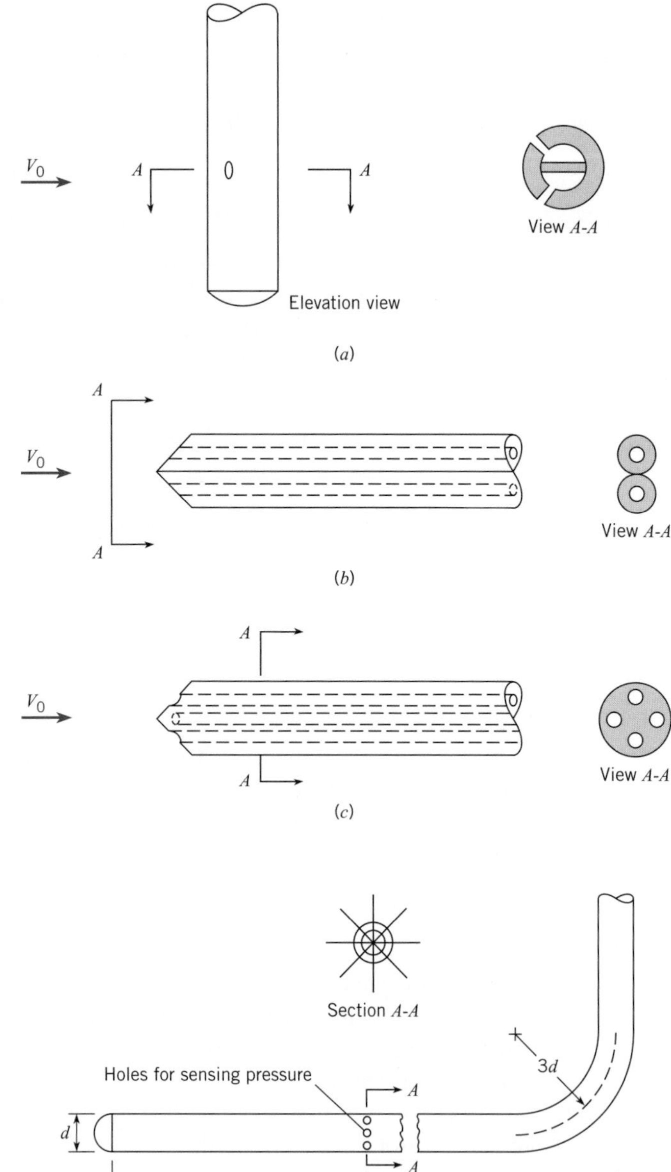

FIGURE 13.2

Various types of yaw meters. (a) Cylindrical-tube yaw meter. (b) Two-tube yaw meter. (c) Three-dimensional yaw meter.

FIGURE 13.3

Static tube.

anemometer is held in an airstream for 1 min and the pointer indicates a 300-ft change on the scale, the average airspeed is 300 ft/min.

Another type of vane anemometer is used for measuring the velocity of flowing water. The usual commercial anemometer of this type is connected to an electrical circuit in such a manner that an electrical signal is triggered for a given number of revolutions. Thus the frequency of rotation of the vane is obtained, which is directly converted to velocity by a calibration curve.

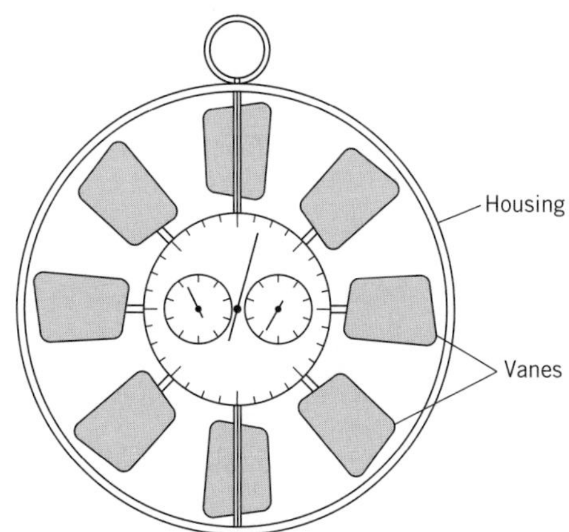

FIGURE 13.4

Vane anemometer for measuring velocity of air flows.

Vane anemometers are also used inside pipes, where they are calibrated to indicate the flow rate directly in cubic meters per second, or in any other appropriate units desired.

Another common anemometer used in meteorological measurements is an ordinary propeller attached to the forward part of a wind vane, all of which is mounted on a mast. The propeller drives an electromagnetic generator, the voltage from which is proportional to the wind speed.

Cup Anemometer

The most common type of cup anemometer is that used by meteorologists to measure wind velocity (Fig. 13.5). However, hydraulic engineers also use a cup anemometer to measure the velocity of flow in streams and rivers. Again, the frequency of rotation is directly related to the velocity of flow by appropriate calibration data.

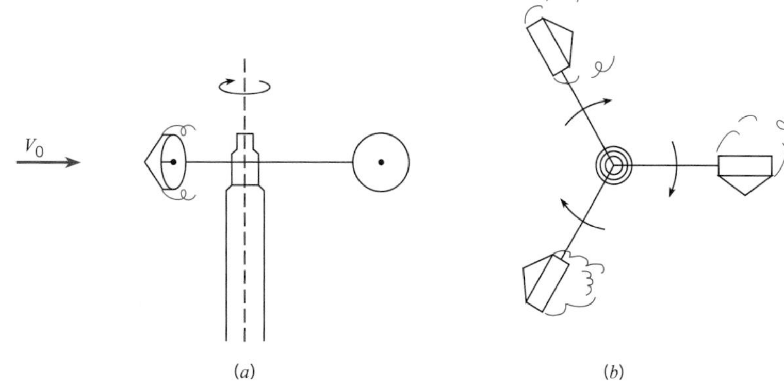

FIGURE 13.5

Cup anemometer.
(a) Elevation view.
(b) Top view.

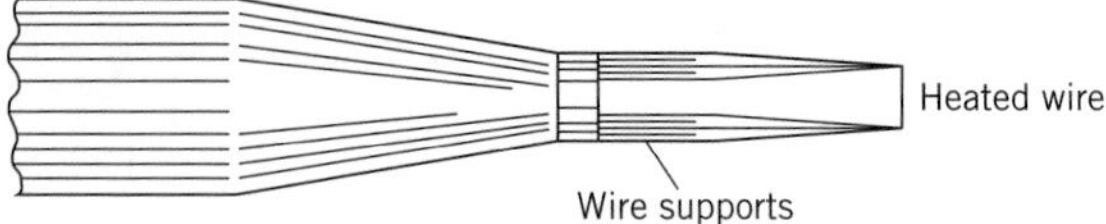

FIGURE 13.6

Probe for hot-wire anemometer (enlarged).

Hot-Wire and Hot-Film Anemometers

The velocity-measuring devices we have already considered are suitable for measuring velocity that either is steady or changes slowly with time. However, if we want to measure the velocity fluctuations due to the eddies in turbulent flow, the response of the aforementioned instruments is too slow to record the rapid changes in velocity. These instruments are also too large for local velocity measurements, which one might wish to make in a thin boundary layer. The hot-wire anemometer is an instrument that is very sensitive to rapid fluctuations in velocity, and its sensing element is small so that its presence does not seriously disturb the nature of the surrounding flow. Another advantage of the hot-wire anemometer is that it is sensitive to low-velocity flows, a characteristic lacking in the Pitot tube. The main disadvantages of the hot-wire anemometer are its delicate nature (the sensor wire is easily broken) and its relatively high cost.

The basic principle of the hot-wire anemometer is described as follows: A wire of very small diameter—the sensing element of the hot-wire anemometer—is welded to supports as shown in Fig. 13.6. In operation the wire either is heated by a fixed flow of electric current (the constant-current anemometer) or is maintained at a constant temperature by adjusting the current (the constant-temperature anemometer).

A flow of fluid past the hot wire causes the wire to cool because of convective heat transfer. In the constant-current anemometer, the cooling of the wire causes its resistance to change, and a corresponding voltage change occurs across the wire. Because the rate of cooling is a function of the speed of flow past the heated wire, the voltage across the wire is correlated with the flow velocity. The more popular type of anemometer, the constant-temperature anemometer, operates by varying the current in such a manner as to keep the resistance (and temperature) constant. The flow of current is correlated with the speed of the flow: the higher the speed, the greater the current needed to maintain a constant temperature. Typically, the wires are 1 to 2 mm in length and heated to 150°C. The wires may be 10 μm or less in diameter; the time response improves with the smaller wire. The lag of the wire's response to a change in velocity (thermal inertia) can be compensated for more easily, using modern electronic circuitry, in constant-temperature anemometers than in constant-current anemometers. The signal from the hot wire is processed electronically to give the desired information, such as mean velocity or the root-mean-square of the velocity fluctuation.

To illustrate the versatility of these instruments, note that the hot-wire anemometer can measure accurately gas flow velocities from 30 cm/s to 150 m/s; it can measure fluctuating velocities with frequencies up to 100,000 Hz; and it has been used satisfactorily for both gases and liquids.

The single hot wire mounted normal to the mean flow direction measures the fluctuating component of velocity in the mean flow direction. Other probe configurations and electronic circuitry can be used to measure other components of velocity.

For velocity measurements in liquids or dusty gases, where wire breakage is a problem, the hot-film anemometer is more suitable. This anemometer consists of a thin conducting metal film (less than 0.1 μm thick) mounted on a ceramic support, which may be 50 μm in diameter. The hot film operates in the same fashion as the hot wire. Recently, the split film has been introduced. It consists of two semicylindrical films mounted on the same cylindrical support and electrically insulated from each other. The split film provides both speed and directional information.

For more detailed information on the hot-wire and hot-film anemometers, see King (2) and Lomas (3).

Laser-Doppler Anemometer

The laser-Doppler anemometer (LDA) is a relatively new instrument for taking fluid velocity measurements. The major advantage of the LDA over the Pitot tube and hot-wire anemometer is that the flow field is not disturbed by the presence of a probe or wire support. A further advantage is that the velocity is measured in a very small flow volume, providing excellent spatial resolution.

There are several different configurations for the LDA, depending on the properties and accessibility of the flow. One configuration, known as the dual-beam mode, is shown in Fig. 13.7. The laser beam, which is a highly coherent, monochromatic light source, is first split into two parallel beams and then passed through a converging lens. The point where the two beams cross is the measuring volume, which might best be described as an ellipsoid that is typically 0.3 mm in diameter and 2 mm long, illustrating the excellent spatial resolution achievable. The interference of the two beams generates a series of light and dark fringes in the measuring volume perpendicular to the plane of the two beams. As a particle passes through the fringe pattern, light is scattered and a portion of the scattered light passes through the collecting lens toward the photodetector. A typical signal obtained from the photodetector is shown in the figure.

It can be shown from optics theory that the spacing between the fringes is given by

$$\Delta x = \frac{l}{2\sin f}$$

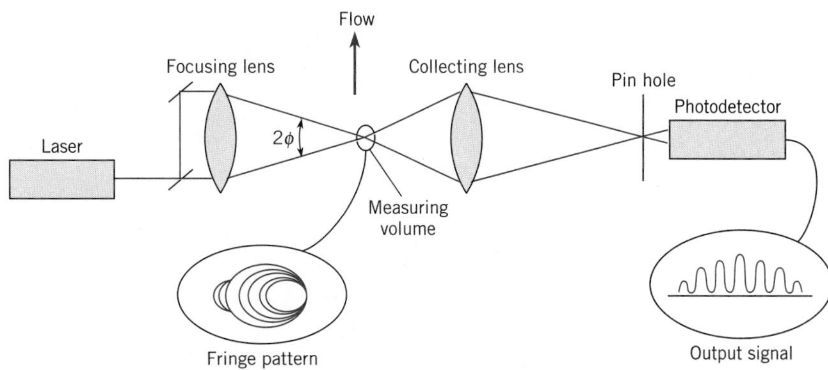

FIGURE 13.7

Dual-beam laser-Doppler anemometer.

where λ is the wavelength of the laser beam and ϕ is the half-angle between the crossing beams. By suitable electronic circuitry, the frequency of the signal (f) is measured, so the velocity is given by

$$U = \frac{\Delta x}{\Delta t} = \frac{lf}{2\sin f}$$

The operation of the laser-Doppler anemometer depends on the presence of particles in the flow to scatter the light, particles sufficiently small that they always move at the fluid velocity. In liquid flows, the impurities of the fluid typically serve as scattering centers. In gaseous flows, however, it is sometimes necessary to "seed" the flow with small particles. Smoke is often used for this seeding.

The laser-Doppler anemometer has specific application in flow measurements requiring fine spatial resolution and minimum disturbance of the flow. Currently, many industrial applications of the LDA are being found to measure velocities of materials such as melted glass, in which it is not possible to mount a mechanical probe.

Laser-Doppler anemometers that provide two or three velocity components of a particle traveling through the measuring volume are now available. This is accomplished by using laser-beam pairs of different colors (wavelengths). The measuring volumes for each color are positioned at the same physical location but oriented differently to measure a different component. The signal-processing system can discriminate the signals from each color and thereby provide component velocities.

Another recent technological advance in laser-Doppler anemometry is the use of fiber optics. The fiber optics transmit the laser beams from the laser to a probe that contains optical elements to cross the beams and generate a measuring volume. Thus measurements at different locations can be made by moving the probe and without moving the laser. For more applications of the laser-Doppler technique see Durst (4).

Marker Methods

The procedure for the marker method for determining velocity stems from the basic concept of the pathline (see Chapter 4). Identifiable particles are placed in the stream. Then,

FIGURE 13.8

Combined time-streak markers (hydrogen bubbles); flow is from left to right. [After Kline (5) Courtesy of Education Development Center, Inc., Newton, MA.]

by analyzing the motion of these particles, one can deduce the velocity of the flow itself. Of course, this requires that the markers follow virtually the same path as the surrounding fluid elements. It means, then, that the marker must have nearly the same density as the fluid or that it must be so small that its motion relative to the fluid is negligible. Thus for water flow it is common to use colored droplets from a liquid mixture that has nearly the same density as the water. For example, Macagno (6) used a mixture of *n*-butyl phthalate and xylene with a bit of white paint to yield a mixture that had the same density as water and could be photographed effectively. Solid particles, such as plastic beads, that have densities near that of the liquid being studied can also be used as markers.

Hydrogen bubbles have also been used for markers in water flow. Here an electrode placed in flowing water causes small bubbles to be formed and swept downstream, thus revealing the motion of the fluid. The wire must be very small so that the resulting bubbles do not have a significant rise velocity with respect to the water. By pulsing the current through the electrode, it is possible to add a time frame to the visualization technique, thus making it a useful tool for velocity measurements. Figure 13.8 shows patches of tiny hydrogen bubbles that were released with a pulsing action from noninsulated segments of a wire located to the left of the picture. Flow is from left to right, and the necked-down section of the flow passage has higher water velocity. Therefore, the patches are longer in that region. Next to the walls the patches of bubbles are shorter, indicating less distance traveled per unit of time. Other details concerning the marker methods of flow visualization are described by Macagno (6).

A relatively new marker method is particle image velocimetry (PIV), which provides a measurement of the velocity field. In PIV, the marker or seeding particles may be minuscule spheres of aluminum, glass, or polystyrene. Or they may be oil droplets, oxygen bubbles (liquids only) or smoke particles (gases only). The seeding particles are illuminated in order to produce a photographic record of their motion. In particular, a

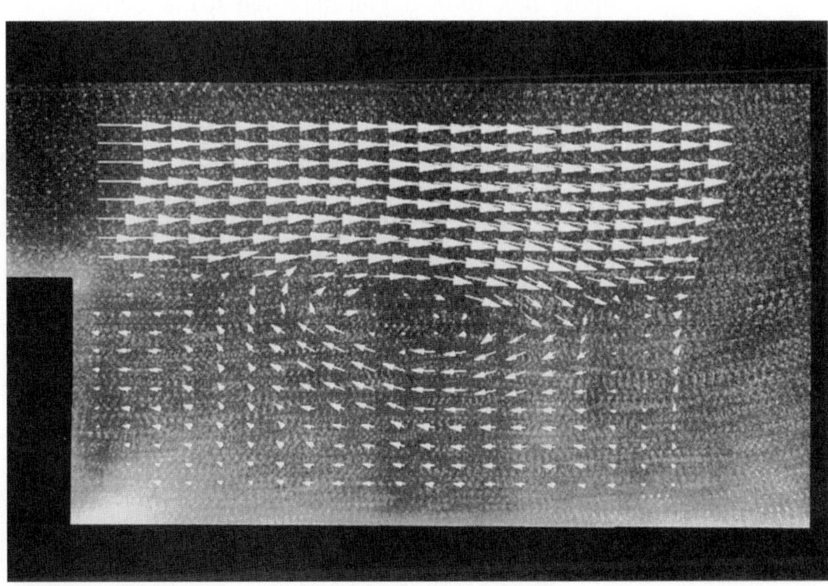

FIGURE 13.9

Velocity vectors from PIV measurements. (Courtesy TSI, Inc., and Florida State University.)

sheet of light passing through a cross section of the flow is pulsed on twice, and the scattered light from the particles is recorded by a camera. The first pulse of light records the position of each particle at time t, and the second pulse of light records the position at time $t + \Delta t$. Thus, the displacement $\mathbf{\Delta r}$ of each particle is recorded on the photograph. Dividing $\mathbf{\Delta r}$ by Δt yields the velocity of each particle. Because PIV uses a sheet of light, the method provides a simultaneous measurement of velocity at locations throughout a cross section of the flow. Hence, PIV is identified as a whole-field technique. Other velocity measurements, the LDV method, for example, are limited to measurements at one location.

PIV measurement of the velocity field for flow over a backward-facing step is shown in Fig. 13.9. This experiment was carried out in water using 15-μm-diameter, silver-coated hollow spheres as seeding particles. Notice that the PIV method provided data over the cross section of the flow. While the data shown in Fig. 13.9 are qualitative, numerical values of the velocity at each location are also available.

The PIV method is typically performed using digital hardware and computers. For example, images may be recorded with a digital camera. Each resulting digital image is evaluated with software that calculates the velocity at points throughout the image. This evaluation proceeds by dividing the image into small sub-areas called "interrogation areas." Within a given interrogation area, the displacement vector ($\mathbf{\Delta r}$) of each particle is found by using statistical techniques (auto- and cross-correlation). After processing, the PIV data are typically available on a computer screen. Additional information on PIV systems is provided by Raffel et al. (7).

Smoke is often used as a marker in gaseous flows. One technique is to suspend a wire vertically across the flow field and allow oil to flow down the wire. The oil tends to accumulate in droplets along the wire. Applying a voltage to the wire vaporizes the oil, creating streaks from the droplets. Figure 13.10 is an example of a flow pattern revealed by such a method. Smoke generators that provide smoke by heating oils are also commercially available. It is also possible to position a thin sheet of laser light through the smoke field to obtain an improved spatial definition of the flow field indicated by the smoke. Another technique is to introduce titanium tetrachloride ($TiCl_4$) in a dried air

FIGURE 13.10

Flow pattern in the wake of a flat plate.

flow, which reacts with the water vapor in the ambient air to produce micron-sized titanium oxide particles, which serve as tracers. The flow pattern obtained for an upward-flowing air jet using this technique in conjunction with a laser light sheet is shown in the photograph at the beginning of this chapter (p. 531). This jet is subjected to an acoustic field, which enhances the vortex shedding pattern observed in the jet.

13.2 Instruments and Procedures for Measurement of Flow Rate

The methods of flow measurement can be classified in a broad sense as either direct or indirect. Direct methods involve the actual measurement of the quantity of flow (volume or weight) for a given time interval. Indirect methods involve the measurement of a pressure change (or some other variable), which in turn is directly related to the rate of flow. *Venturi meters, orifices,* and *flow nozzles* are all devices that apply indirect methods to measure the rate of flow in closed conduits. *Weirs* are devices that employ indirect means to obtain flow rates in open channels. Still another indirect device is the *electromagnetic flow meter,* which operates on the principle that a voltage is generated when a conductor moves in a magnetic field. All of these methods and several others, as well as the *velocity–area integration* of flow measurement, will be discussed in this section.

Direct Volume or Weight Measurements

One of the most accurate methods of obtaining flow rates of liquids is to collect a sample of the flowing fluid over a given period of time t. Then the sample is weighed, and the average weight rate of flow is W/t, where W is the weight of the sample. The volume of a sample can also be measured (usually in a calibrated tank), and from this the average volume rate of flow is calculated as $\forall/t$, where $\forall$ is the volume of the sample.

Velocity–Area Integration

If the velocity of flow in a pipe is symmetrical, the distribution of the velocity along a radial line can be used to determine the volume rate of flow (discharge) in the pipe. The discharge is ob-

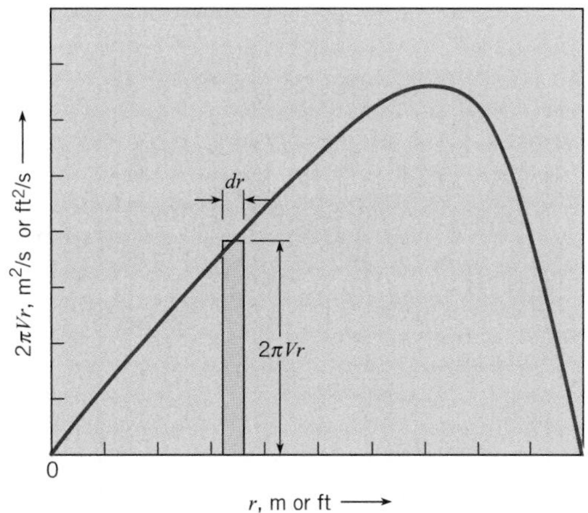

FIGURE 13.11 *Graphical integration in a pipe.*

tained by numerically or graphically integrating over the cross-sectional area of the pipe. Thus a Pitot tube or hot-wire anemometer velocity traverse across the flow section provides the primary data from which the discharge is evaluated. One procedure for evaluating this discharge is given in the next paragraph.

From test data of V versus r, one computes $2\pi Vr$ for various values of r. Then when $2\pi Vr$ versus r is plotted, the area under the resulting curve, Fig. 13.11, is equal to the discharge. This is so because $dQ = V\,dA = V(2\pi r\,dr)$, which is given by an elemental strip of area in Fig. 13.11. Hence the total area is equivalent to the total discharge.

example 13.1

The data given in the table are for a velocity traverse of air flow in a pipe 100 cm in diameter. What is the volume rate of flow in cubic meters per second?

r, cm	V, m/s
0.00	50.0
5.00	49.5
10.00	49.0
15.00	48.0
20.00	46.5
25.00	45.0
30.00	43.0
35.00	40.5
40.00	37.5
45.00	34.0
47.50	25.0
50.00	0.0

Solution First make a table of $2\pi Vr$ versus r as shown below. Now plot $2\pi Vr$ versus r. This plot is shown in the figure below the table. The area under the curve is 29.3 m^3/s. Hence

$$Q = 29.3\ \mathrm{m^3/s}$$ ◁

r, cm	$2\pi Vr$, m^2 s
0.00	0.0
5.00	15.6
10.00	30.8
15.00	45.2
20.00	58.4
25.00	70.7
30.00	81.1
35.00	89.1
40.00	94.2
45.00	96.1
47.50	74.6
50.00	0.0

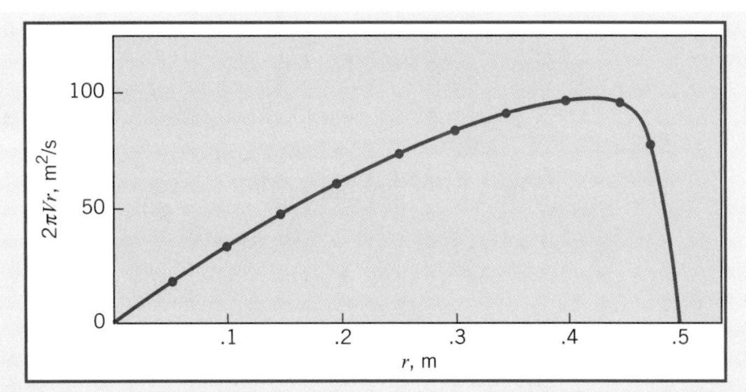

The foregoing procedure involving velocity–area integration is applicable to pipes where the velocity distribution is symmetrical with the axis of the pipe. However, even for flows that are unsymmetrical, it should be obvious that by summing $V\Delta A$ over a flow section one can obtain the total flow rate. Such a procedure is commonly used to obtain the discharge in streams and rivers.

Orifices

A restricted opening through which fluid flows is an *orifice*. If the geometric characteristics of the orifice plus the properties of the fluid are known, then the orifice can be used to measure flow rates. Consider flow through the sharp-edged pipe orifice shown in Fig. 13.12. Note that the streamlines continue to converge a short distance downstream of the plane of the orifice. Hence the minimum-flow area is actually smaller than the area of the orifice. To relate the minimum-flow area, often called the contracted area of the jet, or *vena contracta*, to the area of the orifice A_o, we use the contraction coefficient, which is defined as

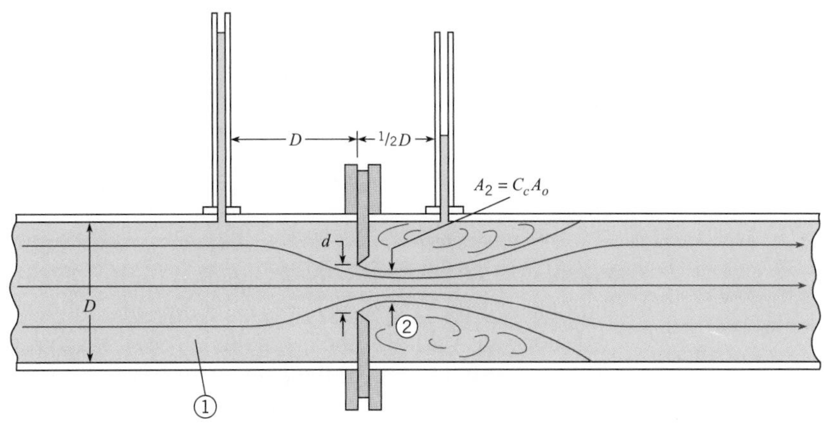

FIGURE 13.12

Flow through a sharp-edged pipe orifice.

$$A_j = C_c A_o$$

$$C_c = \frac{A_j}{A_o}$$

Then, for a circular orifice,

$$C_c = \frac{(p/4)d_j^2}{(p/4)d^2} = \left(\frac{d_j}{d}\right)^2$$

Because d_j and d_2 are identical, we also have $C_c = (d_2/d)^2$. At low values of the Reynolds number, C_c is a function of the Reynolds number. However, at high values of the Reynolds number, C_c is only a function of the geometry of the orifice. For d/D ratios less than 0.3, C_c has a value of approximately 0.62. However, as d/D is increased to 0.8, C_c increases to a value of 0.72.

We begin the derivation of the discharge equation for the orifice by writing the Bernoulli equation between section 1 and section 2 in Fig. 13.12:

$$\frac{p_1}{g} + \frac{V_1^2}{2g} + z_1 = \frac{p_2}{g} + \frac{V_2^2}{2g} + z_2$$

V_1 is eliminated by means of the continuity equation $V_1A_1 = V_2A_2$. Then solving for V_2 gives

$$V_2 = \left\{\frac{2g[(p_1/g + z_1) - (p_2/g + z_2)]}{1 - (A_2/A_1)^2}\right\}^{1/2} \tag{13.1a}$$

However, $A_2 = C_cA_0$ and $h = p/\gamma + z$, so Eq. (13.1*a*) reduces to

$$V_2 = \sqrt{\frac{2g(h_1 - h_2)}{1 - C_c^2A_o^2/A_1^2}} \tag{13.1b}$$

Our primary objective is to obtain an expression for discharge in terms of h_1, h_2, and the geometric characteristics of the orifice. The discharge is given by V_2A_2. Hence, when we multiply both sides of Eq. (13.1*b*) by $A_2 = C_cA_o$, we obtain the desired result:

$$Q = \frac{C_cA_o}{\sqrt{1 - C_c^2A_o^2/A_1^2}}\sqrt{2g(h_1 - h_2)} \tag{13.2}$$

Equation (13.2) is the discharge equation for the flow of an incompressible inviscid fluid through an orifice. However, it is valid only at relatively high Reynolds numbers. For low and moderate values of the Reynolds number, viscous effects are significant, and an additional coefficient called the *coefficient of velocity* C_v must be applied to the discharge equation to relate the ideal to the actual flow.* Thus for viscous flow through an orifice, we have the following discharge equation:

$$Q = \frac{C_v C_c A_o}{\sqrt{1 - C_c^2 A_o^2 / A_1^2}} \sqrt{2g(h_1 - h_2)}$$

The product $C_v C_c$ is called the *discharge coefficient* C_d, and the combination $C_v C_c / (1 - C_c^2 A_o^2 / A_1^2)^{1/2}$ is called the *flow coefficient* K. Thus we have $Q = KA_o\sqrt{2g(h_1 - h_2)}$, where

$$K = \frac{C_d}{\sqrt{1 - C_c^2 A_o^2 / A_1^2}}$$

If Δh is defined as $h_1 - h_2$, then the final form of the *orifice equation* reduces to

$$Q = KA_o\sqrt{2g\Delta h} \tag{13.3a}$$

If a differential pressure transducer is connected across the orifice, it will sense a piezometric pressure change that is equivalent to $\gamma\Delta h$, so the orifice equation becomes

$$Q = KA_o\sqrt{2\frac{\Delta p_z}{r}} \tag{13.3b}$$

Experimentally determined values of K as a function of d/D and Reynolds number based on orifice size are given in Fig. 13.13. If Q is given, Re_d is equal to $4Q/\pi d\nu$. Then K is obtained from Fig. 13.13 (using the vertical lines and the bottom scale), and Δh is computed from Eq. (13.3*a*) or Δp_z can be computed from Eq. (13.3*b*). However, we are often confronted with the problem of determining the discharge Q when a certain value of Δh or a certain value of Δp_z is given. When Q is to be determined, we do not have a direct way to obtain K by entering Fig. 13.13 with Re because Re is a function of the flow rate, which is still unknown. Hence another scale, which does not involve Q, is constructed on the graph of Fig. 13.13. The variables for this scale are obtained in the following manner: Because $\text{Re}_d = 4Q/\pi d\nu$ and $Q = K\pi d^2/4\sqrt{2g\Delta h}$, we can write Re_d in terms of Δh as

$$\text{Re}_d = K\sqrt{2g\Delta h}\frac{d}{\nu}$$

or

$$\frac{\text{Re}_d}{K} = \sqrt{2g\Delta h}\frac{d}{\nu}$$

Thus the slanted dashed lines and the top scale are used in Fig. 13.13 when Δh is known and the flow rate is to be determined. If a certain value of Δp is given, one can enter Fig. 13.13 by using $\Delta p_z/\rho$ in place of $g\Delta h$ in the parameter at the top of Fig. 13.13.

The literature on orifice flow contains numerous discussions concerning the optimum placement of pressure taps on both the upstream side and the downstream side of an orifice. The data given in Fig. 13.13 are for "corner taps." That is, on the upstream side the pressure readings were taken immediately upstream of the orifice plate (at the corner of the orifice

* At low Reynolds numbers the coefficient of velocity may be quite small; however, at Reynolds numbers above 10^5, C_v typically has a value close to 0.98. See Lienhard (8) for C_v analyses.

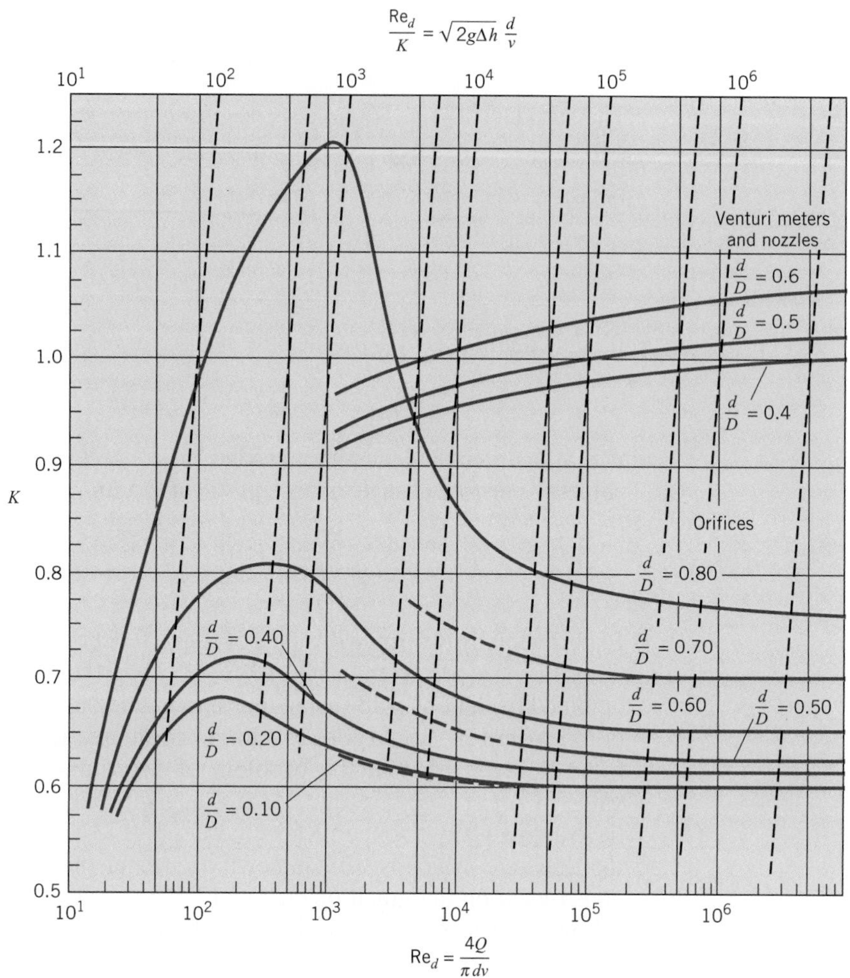

FIGURE 13.13

Flow coefficient K and Re_d/K versus the Reynolds number for orifices, nozzles, and venturi meters. [After Tuve and Sprenkle (9) and ASME (10). Permission to use Tuve granted by Instrumentation & Control Systems *magazine, formerly* Instruments *magazine.]*

plate and the pipe wall), and the downstream tap was at a similar downstream location. However, pressure data from flange taps (1 in. upstream and 1 in. downstream) and from the taps shown in Fig. 13.12 all yield virtually the same values for K—the differences are no greater than the deviations involved in reading Fig. 13.13. For more precise values of K with specific types of taps, see the ASME report on fluid meters (1).

Head Loss for Orifices

Some head loss occurs between the upstream side of the orifice and the vena contracta. However, this head loss is very small compared with the head loss that occurs downstream of the vena contracta. This downstream portion of the head loss is like that for an abrupt expansion. If we neglect all head loss except that due to the expansion of the flow, we have

$$h_L = \frac{(V_2 - V_1)^2}{2g}$$

where V_2 is the velocity at the vena contracta and V_1 is the velocity in the pipe. It can be shown that the ratio of this expansion loss, h_L, to the change in head across the orifice, Δh, is given as

$$\frac{h_L}{\Delta h} = \frac{\frac{V_2}{V_1} - 1}{\frac{V_2}{V_1} + 1} \tag{13.4}$$

Table 13.1 shows how the ratio increases with increasing values of V_2/V_1. It is obvious that an orifice is very inefficient from the standpoint of energy conservation.

TABLE 13.1 RELATIVE HEAD LOSS FOR ORIFICES

$V_2/V_1 \rightarrow$	1	2	4	6	8	10
$h_L/\Delta h \rightarrow$	0	0.33	0.60	0.71	0.78	0.82

example 13.2

A 15-cm orifice is located in a horizontal 24-cm water pipe, and a water-mercury manometer is connected to either side of the orifice. When the deflection on the manometer is 25 cm, what is the discharge in the system, and what head loss is produced by the orifice? Assume the water temperature is 20°C.

Solution The discharge is given by the orifice equation: $Q = KA_o\sqrt{2g\Delta h}$. To either enter Fig. 13.13 or use the orifice equation, we need first to evaluate Δh, the change in piezometric head in meters of fluid that is flowing. This is obtained by applying the manometer equation to the manometer shown in the figure.

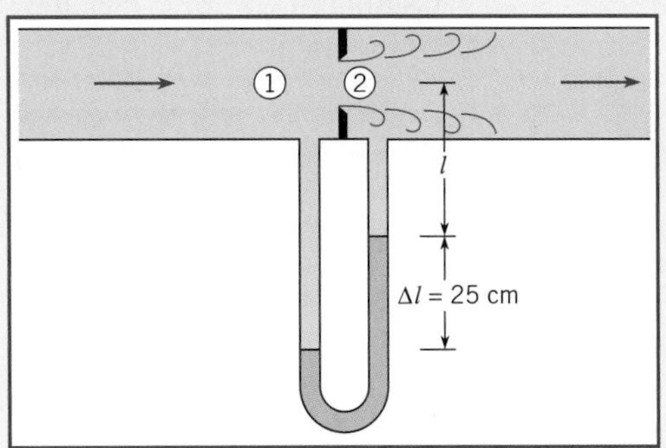

Writing the manometer equation from point 1 to point 2, we have

$$p_1 + g_w(l + \Delta l) - g_{Hg}\Delta l - g_w l = p_2$$

Then
$$\Delta h = \frac{p_1 - p_2}{g_w} = \Delta l \frac{g_{Hg} - g_w}{g_w} = \Delta l\left(\frac{g_{Hg}}{g_w} - 1\right)$$

For this example,

$$\Delta h = (0.25 \text{ m})(13.6 - 1) = 3.15 \text{ m of water}$$

The kinematic viscosity of water at 20°C is 1.0×10^{-6} m^2/s. We now can compute $d\sqrt{2g\Delta h}/n$, the parameter that we need to enter Fig. 13.13:

$$\frac{d\sqrt{2g\Delta h}}{n} = \frac{0.15 \text{ m}\sqrt{2(9.81 \text{ m/s}^2)(3.15 \text{ m})}}{1.0 \times 10^{-6} \text{ m}^2/\text{s}} = 1.2 \times 10^6$$

From Fig. 13.13 with $d/D = 0.625$, we read K to be 0.66 (interpolated). Hence

$$Q = 0.66A_o\sqrt{2g\Delta h}$$
$$= 0.66\frac{p}{4}d^2\sqrt{2(9.81 \text{ m/s}^2)(3.15 \text{ m})}$$
$$= 0.66(0.785)(0.15^2 \text{ m}^2)(7.86 \text{ m/s}) = 0.092 \text{ m}^3/\text{s} \qquad \triangleleft$$

The head loss is given by $h_L = (V_2 - V_1)^2/2g$. We can solve for V_2 if we know the coefficient of contraction $[V_2 = Q/(C_cA_o)]$. We can solve for the coefficient of contraction C_c from the definition of the flow coefficient K:

$$K = \frac{C_d}{1 - C_c^2A_o^2/A_1^2}$$

where for this example $K = 0.66$. The ratio $(A_o/A_1)^2 = (0.625)^4 = 0.1526$ and $C_d = C_vC_c$.

Assuming $C_v = 0.98$ (see the footnote on page 545) and solving for C_c, we obtain $C_c = 0.633$. Then

$$V_2 = Q/(C_cA_o) = (0.092 \text{ m}^3/\text{s})/[(0.633)(p/4)(0.15^2 \text{ m}^2)] = 8.23 \text{ m/s}$$
$$V_1 = Q/A_{pipe} = (0.092 \text{ m}^3/\text{s})/[(p/4)(0.24^2 \text{ m}^2)] = 2.03 \text{ m/s}$$

and

$$h_L = (V_2 - V_1)^2/2g = (8.23 - 2.03)^2/(2 \times 9.81)$$
$$= 1.96 \text{ m} \qquad \triangleleft$$

example 13.3

An air-water manometer is connected to either side of an 8-in. orifice in a 12-in. water pipe. If the maximum flow rate is 2 cfs, what is the deflection on the manometer? The water temperature is 60°F.

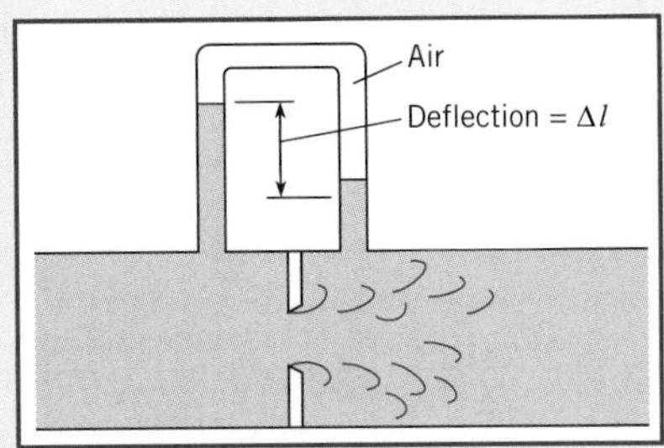

Solution We first compute the Reynolds number $\mathrm{Re}_d = 4Q/\pi d\nu$ so that we can enter Fig. 13.13 to obtain the flow coefficient K. Then K will be used in the orifice equation to compute the deflection Δl.

$$\mathrm{Re}_d = \frac{4Q}{p\ dn} = \frac{(4)(2)}{3.14(8/12)(1.22)(10^{-5})} = 3.1 \times 10^5$$

From Fig. 13.13, by interpolating between curves of $d/D = 0.6$ and $d/D = 0.7$ for $d/D = 8/12 = 0.667$, we read K to be approximately 0.68. Then from $Q = KA_o\sqrt{2g\Delta h}$ we obtain

$$\Delta h = \frac{Q^2}{2gK^2A_o^2} = \frac{4}{64.4(0.68^2)[(p/4)(8/12)^2]^2} = 1.1 \text{ ft}$$

The deflection is related to Δh by

$$\Delta h = \Delta l\left(\frac{g_w - g_{\text{air}}}{g_w}\right)$$

But $\gamma_w \gg \gamma_{\text{air}}$, so

$$\Delta l = \Delta h = 1.1 \text{ ft}$$ ◁

The sharp-edged orifice can also be used to measure the mass flow rate of gases. The discharge equation [Eq. (13.3*b*)] is multiplied by the upstream gas density and an empirical factor to account for compressibility effects (1). The resulting equation is

$$\dot{m} = YA_oK\sqrt{2r_1(p_1 - p_2)}$$

where K, the flow coefficient, is found using Fig. 13.13 and Y is the compressibility factor given by the empirical equation

$$Y = 1 - \left\{ \frac{1}{k}\left(1 - \frac{p_2}{p_1}\right)\left[0.41 + 0.35\left(\frac{A_o}{A_1}\right)^2\right]\right\}$$

In this case both the pressure difference across the orifice and the absolute pressure of the gas are needed. One must remember when using the equation for the compressibility factor that the absolute pressure must be used.

example 13.4

The mass flow rate of natural gas is to be measured using a sharp-edged orifice. The upstream pressure of the gas is 101 kPa absolute, and the pressure difference across the orifice is 10 kPa. The upstream temperature of the methane is 15°C. The pipe diameter is 10 cm, and the orifice diameter is 7 cm. What is the mass flow rate?

Solution We first find the flow coefficient K. The density and kinematic viscosity at the upstream conditions are 0.678 kg/m^3 and 1.59×10^{-5} m^2/s, respectively. The parameter to be used to find the flow coefficient from Fig. 13.13 is

$$\frac{d}{\nu}\sqrt{2\frac{\Delta p}{r_1}} = \frac{0.07}{1.59(10^{-5})}\sqrt{2\frac{10^4}{0.678}} = 7.56 \times 10^5$$

From Fig. 13.13, the flow coefficient is 0.7. The compressibility factor is

$$Y = 1 - \left\{\frac{1}{1.31}\left(1 - \frac{91}{100}\right)(0.41 + 0.35 \times 0.7^4)\right\} = 0.962$$

The mass flow rate of methane is

$$\dot{m} = 0.962 \times \frac{P}{4} \times 0.07^2 \times 0.7\sqrt{2 \times 0.678 \times 10^4} = 0.302 \text{ kg/s}$$ ◁

The foregoing examples involved the determination of either Q or Δh for a given size of orifice. Another type of problem is determination of the diameter of the orifice for a given Q and Δh. For this type of problem a trial-and-error procedure is required. Because one knows the approximate value of K, that is guessed first. Then the diameter is solved for, after which a better value of K can be determined, and so on.

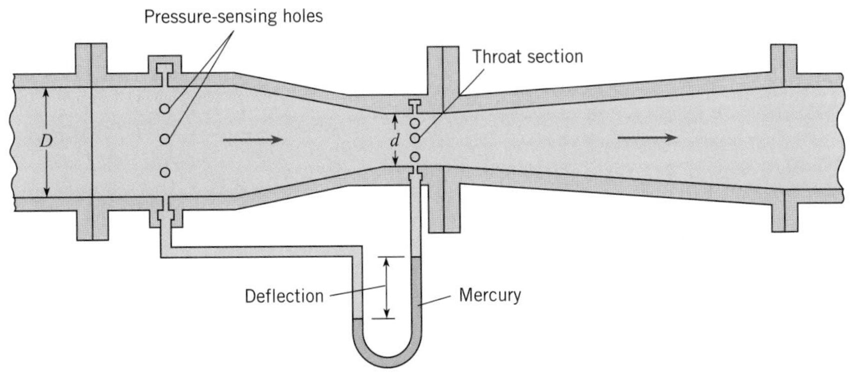

FIGURE 13.14

Typical venturi meter.

Venturi Meter

The orifice is a simple and accurate device for the measurement of flow; however, the head loss for an orifice is quite large. A device that operates on the same principle as the orifice but with a much smaller head loss is the *venturi meter.* The lower head loss results from streamlining the flow passage, as shown in Fig. 13.14. Such streamlining eliminates any jet contraction beyond the smallest flow section. Consequently, the coefficient of contraction has a value of unity, and the basic *venturi equation* is

$$Q = \frac{A_t C_d}{\sqrt{1-(A_t/A_p)^2}}\sqrt{2g(h_p - h_t)} \qquad (13.5)$$

$$Q = KA_t\sqrt{2g\Delta h} \qquad (13.6)$$

where A_t is the throat area and Δh is the difference in piezometric head between the venturi entrance (pipe) and the throat. Note that the venturi equation is the same as the orifice equation. However, K for the venturi meter approaches unity at high values of the Reynolds number and small d/D ratios. This trend can be seen in Fig. 13.13, where values of K for the venturi meter are plotted along with similar data for the orifice.

example 13.5

The pressure difference between the taps of a horizontal venturi meter carrying water is 35 kPa. If $d = 20$ cm and $D = 40$ cm, what is the discharge of water at 10°C?

Solution First we compute Δh. No elevation change occurs between the upstream section and the downstream section. Therefore Δh is

$$\Delta h = \frac{\Delta p}{g_{\text{water}}} = \frac{35{,}000\ \text{N/m}^2}{9810\ \text{N/m}^3} = 3.57\ \text{m of water}$$

From Table A.5 in the Appendix we get $\nu = 1.31 \times 10^{-6}\ \text{m}^2/\text{s}$. Then we compute

$$\frac{d\sqrt{2g\Delta h}}{n} = \frac{0.20\sqrt{2(9.81)(3.57)}}{1.31(10^{-6})} = 1.28 \times 10^6$$

Then

$$K = 1.02$$

Now we compute the discharge:

$$Q = 1.02A_2\sqrt{2g\Delta h}$$
$$= 1.02(0.785)(0.20^2)\sqrt{2(9.81)(3.57)} = 0.268\ \text{m}^3/\text{s}$$

◁

Flow Nozzles

The plate orifice must have a sharp upstream edge to yield the flow coefficients given in Fig. 13.13. If the upstream edge becomes rounded because of chemical or mechanical action, the orifice will not be a reliable flow-measuring device (C_d will increase). The *flow nozzle,* Fig. 13.15, has the advantage that it is less susceptible to wear. The flow nozzle has approximately the same flow coefficients as the venturi meter when the pressure taps are connected as in Fig. 13.15. However, the overall head loss across a flow nozzle is like that for the sharp-edged orifice. That is, the overall head loss is that of an abrupt expansion.

Electromagnetic Flow Meter

All of the flow meters described so far require that some sort of obstruction be placed in the flow. The obstruction may be the rotor of a vane anemometer or the reduced cross section of an orifice or venturi meter. A meter that neither obstructs the flow nor requires pressure taps, which are subject to clogging, is the *electromagnetic flow meter.* Its basic

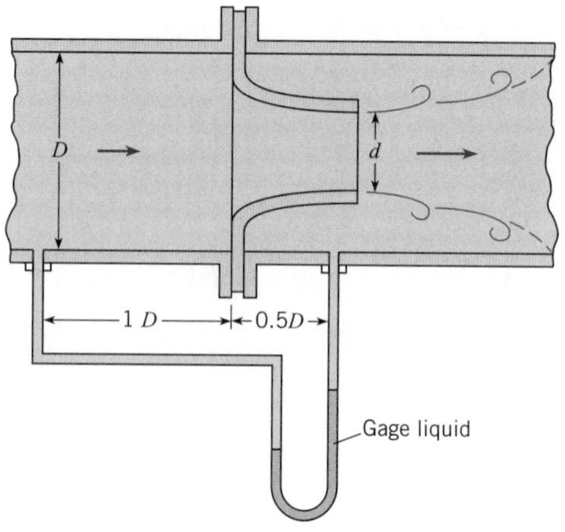

FIGURE 13.15

Typical flow nozzle.

principle is that a conductor that moves in a magnetic field produces an electromotive force. Hence liquids having a degree of conductivity generate a voltage between the electrodes, as in Fig. 13.16, and this voltage is proportional to the velocity of flow in the conduit. It is interesting to note that the basic principle of the electromagnetic flow meter was investigated by Faraday in 1832. However, practical application of the principle was not made until approximately a century later, when it was used to measure blood flow. Recently, with the need for a meter to measure the flow of liquid metal in nuclear reactors and with the advent of sophisticated electronic signal detection, this type of meter has found extensive commercial use.

The main advantages of the electromagnetic flow meter are that the output signal varies linearly with the flow rate and that the meter causes no resistance to the flow. The major disadvantages are its high cost and its unsuitability for measuring gas flow.

For a summary of the theory and application of the electromagnetic flow meter, the reader is referred to Shercliff (11). This reference also includes a comprehensive bibliography on the subject.

Ultrasonic Flow Meter

Another form of nonintrusive flow meter that is used in diverse applications ranging from blood flow measurement to open-channel flow is the *ultrasonic flow meter*. Basically, there are two different modes of operation for ultrasonic flow meters. One mode involves measuring the difference in travel time for a sound wave traveling upstream and downstream between two measuring stations. The difference in travel time is proportional to flow velocity. The second mode of operation is based on the Doppler effect. When an ultrasonic beam is projected into an inhomogeneous fluid, some acoustic energy is scattered back to the transmitter at a different frequency (Doppler shift). The measured frequency difference is related directly to the flow velocity.

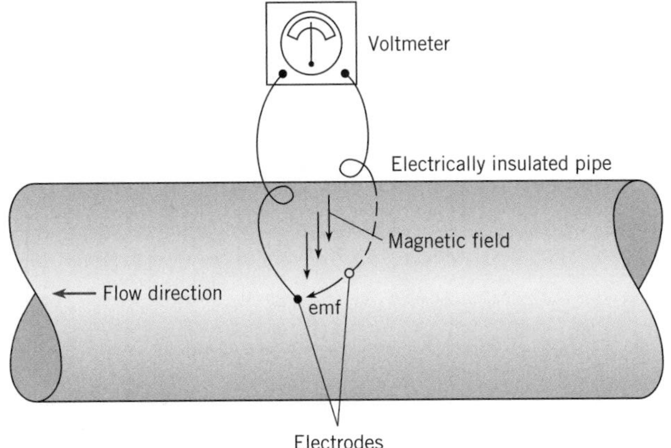

FIGURE 13.16

Electromagnetic flow meter.

Turbine Flow Meter

The *turbine flow meter* consists of a wheel with a set of curved vanes (blades) mounted inside a duct. The volume rate of flow through the meter is related to the rotational speed of the wheel. This rotational rate is generally measured by a blade passing an electromagnetic pickup mounted in the casing. The meter must be calibrated for the given flow conditions. The turbine meter is versatile in that it can be used for either liquids or gases. It has an accuracy of better than 1% over a wide range of flow rates, and it operates with small head loss. The turbine flow meter is used extensively in monitoring flow rates in fuel-supply systems.

Vortex Flow Meter

The *vortex flow meter* consists of a cylinder mounted across a duct. This cylinder sheds vortices and gives rise to an oscillatory flow field. Proper design of the cylindrical element yields a constant Strouhal number for vortex shedding for a range of Reynolds numbers from 10^4 to 10^6. Over this flow range the fluid velocity and volume flow rate are directly proportional to the frequency of oscillation, which can be measured by several different methods. An advantage of this meter is that it has no moving parts (reliability), but it does give rise to a head loss comparable to that from other obstruction-type meters.

Rotameter

The *rotameter* consists of a vertical tapered tube through which the fluid flows upward and inside of which is located the rotor or active element of the meter (Fig. 13.17). Vanes

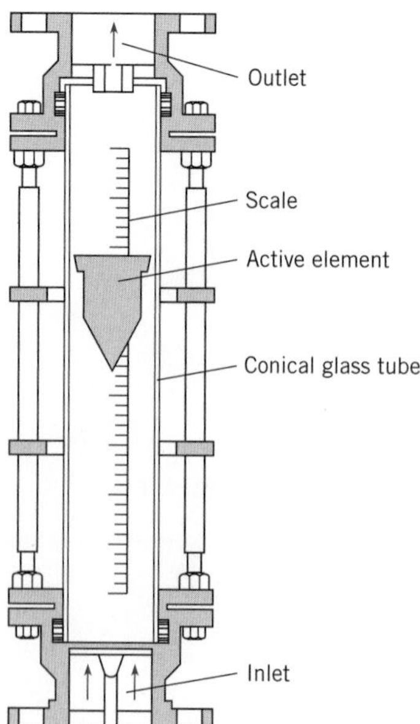

FIGURE 13.17

Rotameter.

cause the rotor to rotate slowly about the axis of the tube, thus keeping it centered within the tube. Because the velocity is lower at the top of the tube (greater flow section there) than at the bottom, the rotor seeks a neutral position where the drag on it just balances its weight. Thus the rotor "rides" higher or lower in the tube, depending on the rate of flow. A calibrated scale on the side of the tube indicates the rate of flow. Although venturi and orifice meters have better accuracy (approximately 1% of full scale) than the rotameter (approximately 5% of full scale), the rotameter offers other advantages, such as simplicity of design.

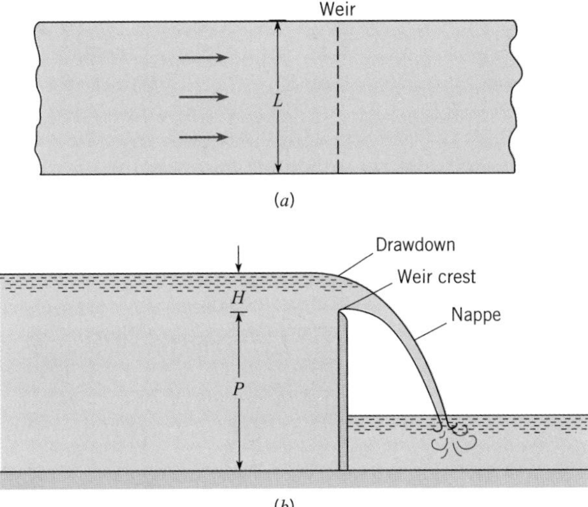

FIGURE 13.18

Definition sketch for sharp-crested weir.
(a) Plan view.
(b) Elevation view.

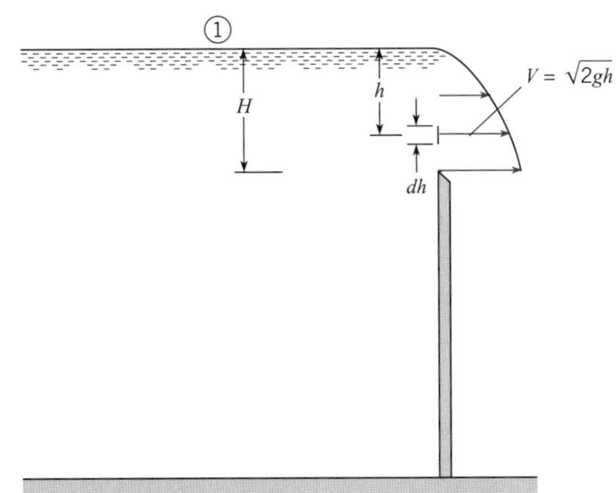

FIGURE 13.19

Theoretical velocity distribution over a weir.

Rectangular Weir

A *weir* is an obstruction in an open channel over which liquid flows. The discharge over the weir is a function of the weir geometry and of the *head* on the weir. Consider flow over the weir in a rectangular channel, shown in Fig. 13.18. The head H on the weir is defined as the vertical distance between the weir crest and the liquid surface taken far enough upstream of the weir to avoid local free-surface curvature (see Fig. 13.18).

The basic discharge equation for the weir is derived by integrating $V\,dA = VL\,dh$ over the total head on the weir. Here L is the length of the weir and V is the velocity at any given distance h below the free surface. Neglecting streamline curvature and assuming negligible velocity of approach upstream of the weir, we obtain an expression for V by writing the Bernoulli equation between a point upstream of the weir and a point in the plane of the weir (see Fig. 13.19). This equation is

$$\frac{p_1}{g} + H = (H - h) + \frac{V^2}{2g} \tag{13.7}$$

Here the reference elevation is the elevation of the crest of the weir, and the reference pressure is atmospheric pressure. Therefore $p_1 = 0$, and Eq. (13.7) reduces to

$$V = \sqrt{2gh}$$

Then $dQ = \sqrt{2ghL}\,dh$, and the discharge equation becomes

$$\begin{aligned} Q &= \int_0^H \sqrt{2ghL}\,dh \\ &= \tfrac{2}{3}L\sqrt{2g}H^{3/2} \end{aligned} \tag{13.8}$$

In the case of actual flow over a weir, the streamlines converge downstream of the plane of the weir, and viscous effects are not entirely absent. Consequently, a discharge coefficient C_d must be applied to the basic expression on the right-hand side of Eq. (13.8) to bring the theory in line with the actual flow rate. Thus the *rectangular weir equation* is

$$\begin{aligned} Q &= \tfrac{2}{3}C_d\sqrt{2g}LH^{3/2} \\ &= K\sqrt{2g}LH^{3/2} \end{aligned} \tag{13.9}$$

For low-viscosity liquids, the flow coefficient K is primarily a function of the relative head on the weir, H/P. An empirically determined equation for K adapted from Kindsvater and Carter (12) is

$$K = 0.40 + 0.05\frac{H}{P} \tag{13.10}$$

This is valid up to an H/P value of 10 as long as the weir is well ventilated so that atmospheric pressure prevails on both the top and the bottom of the weir nappe.

example 13.6

The head on a rectangular weir that is 60 cm high in a rectangular channel that is 1.3 m wide is measured to be 21 cm. What is the discharge of water over the weir?

Solution

$$Q = K\sqrt{2g}LH^{3/2} \qquad \text{where } K = 0.40 + 0.05\frac{H}{P}$$

Hence $$K = 0.40 + 0.05\left(\frac{21}{60}\right) = 0.417$$

Then $$Q = 0.417\sqrt{2(9.81)}(1.3)(0.21^{3/2}) = 0.23 \text{ m}^3/\text{s}$$

When the rectangular weir does not extend the entire distance across the channel, as in Fig. 13.20, additional end contractions occur. Therefore, K will be smaller than for the weir without end contractions. The reader is referred to King and Brater (13) for additional information on flow coefficients for weirs.

Triangular Weir

A definition sketch for the triangular weir is shown in Fig. 13.21. The primary advantage of the triangular weir is that it has a higher degree of accuracy over a much wider range of flow than does the rectangular weir, because the average width of the flow section increases as the head increases.

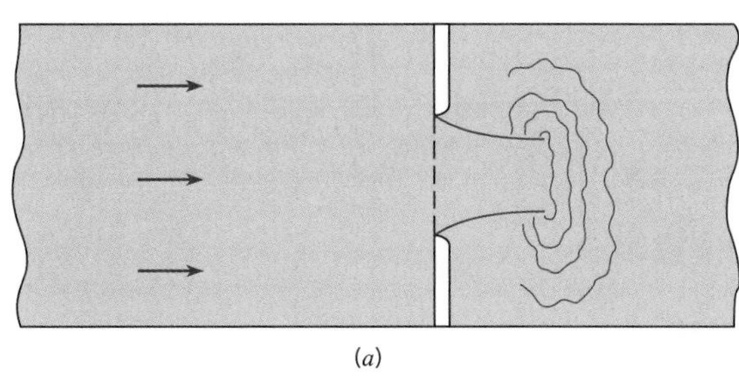

(a)

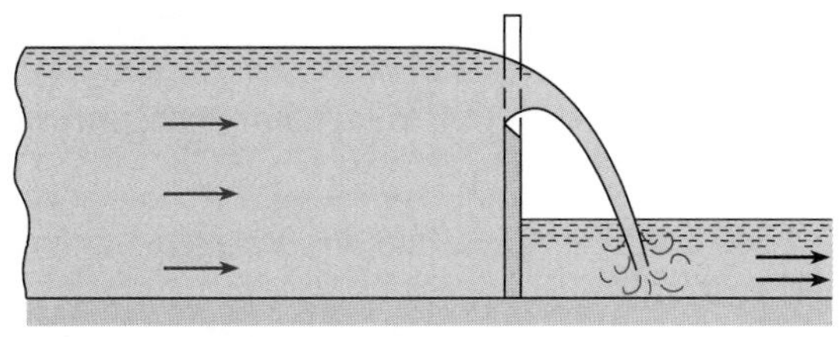

(b)

FIGURE 13.20

Rectangular weir with end contractions.
(a) Plan view.
(b) Elevation view.

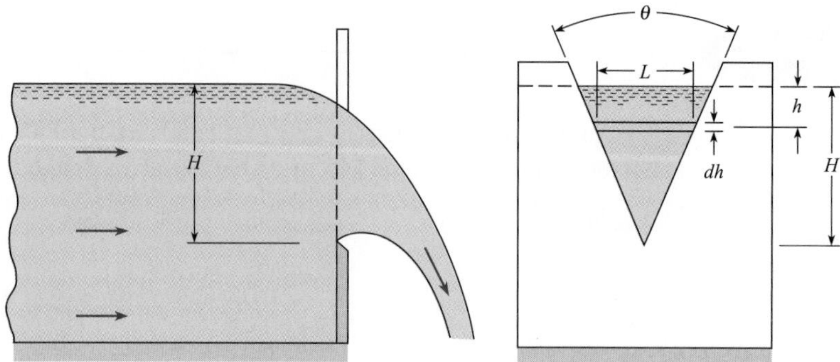

FIGURE 13.21
Definition sketch for the triangular weir.

The basic discharge equation for the triangular weir is derived in the same manner as that for the rectangular weir. The differential discharge $dQ = V\,dA = VL\,dh$ is integrated over the total head on the weir. Thus we have

$$Q = \int_0^H \sqrt{2gh}\,(H-h)2\tan\left(\frac{u}{2}\right)dh$$

which integrates to

$$Q = \frac{8}{15}\sqrt{2g}\tan\left(\frac{u}{2}\right)H^{5/2}$$

However, a coefficient of discharge must still be used with the basic equation. Hence we have

$$Q = \frac{8}{15}C_d\sqrt{2g}\tan\left(\frac{u}{2}\right)H^{5/2} \tag{13.11}$$

Experimental results with water flow over weirs with $\theta = 60°$ and $H > 2$ cm indicate that C_d has a value of 0.58. Hence the *triangular weir equation* with these limitations is

$$Q = 0.179\sqrt{2g}H^{5/2} \tag{13.12}$$

example 13.7

The head on a 60° triangular weir is measured to be 43 cm. What is the flow of water over the weir?

Solution We use the discharge equation $Q = 0.179\sqrt{2g}H^{5/2}$. Hence

$$Q = 0.179 \times \sqrt{2 \times 9.81} \times (0.43)^{5/2} = 0.096\ \text{m}^3/\text{s}$$

◁

More details about flow-measuring devices for incompressible flow can be found in references (14) and (15).

Measurement in Compressible Flow

Many of the measuring techniques used to determine the velocities and flow rates of liquids can be applied to compressible flows. However, since the Bernoulli equation is invalid for compressible flow, alternative relations must be used to correlate velocity and discharge with pressure difference.

Pressure Measurements

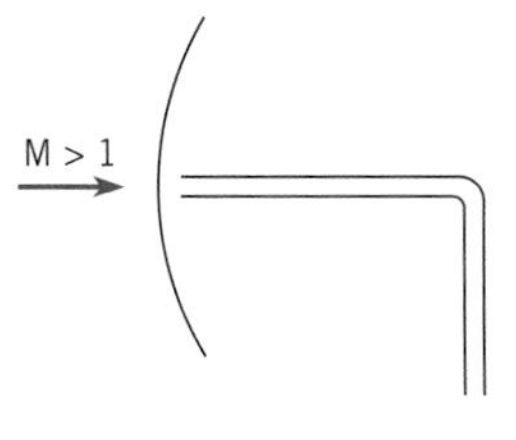

FIGURE 13.22

Stagnation tube in supersonic flow.

Static-pressure measurements can be made using the conventional static-pressure taps of a probe. However, if the boundary layer is disturbed by the presence of a shock wave in the vicinity of the pressure tap, the reading may not give the correct static pressure. The effect of the shock wave on the boundary layer is smaller if the boundary layer is turbulent. Therefore an effort is sometimes made to trip the boundary layer and ensure a turbulent boundary layer in the region of the pressure tap.

The stagnation pressure can be measured with a stagnation tube aligned with the local velocity vector. If the flow is supersonic, however, a shock wave forms around the tip of the probe, as shown in Fig. 13.22, and the stagnation pressure measured is that downstream of the shock wave and not that of the free stream. The stagnation pressure in the free stream can be calculated using the normal shock relationships, provided the free-stream Mach number is known.

Mach Number and Velocity Measurements

A Pitot-static tube can be used to measure Mach numbers in compressible flows. Taking the measured stagnation pressure as the total pressure, one can calculate the Mach number in subsonic flows from the total-to-static-pressure ratio according to Eq. (12.32):

$$\mathrm{M} = \left\{ \frac{2}{k-1} \left[\left(\frac{p_t}{p} \right)^{(k-1)/k} - 1 \right] \right\}^{1/2}$$

It is interesting to note here that one must measure the stagnation and static pressures separately to determine the pressure ratio, whereas one needs only the pressure difference to calculate the velocity of a liquid flow.

If the flow is supersonic, then the indicated stagnation pressure is the pressure behind the shock wave standing off the tip of the tube. By taking this pressure as the total pressure downstream of a normal shock wave and the measured static pressure as the static pressure upstream of the shock wave, one can determine the Mach number of the free stream (M_1) from the static-to-total-pressure ratio (p_1/p_{t_2}) according to the expression

$$\frac{p_1}{p_{t_2}} = \frac{\{[2k/(k+1)]M_1^2 - [(k-1)/(k+1)]\}^{1/(k-1)}}{\{[(k+1)/2]M_1^2\}^{k/(k-1)}} \tag{13.13}$$

which is called the *Rayleigh supersonic Pitot formula.* Note, however, that M_1 is an implicit function of the pressure ratio and must be determined graphically or by some numerical procedure. Many normal-shock tables, such as those in reference (16), have p_1/p_{t_2} tabulated versus M_1, which enables one to find M_1 quite easily by interpolation.

Once the Mach number is determined, more information is needed to evaluate the velocity—namely, the local speed of sound. This can be done by inserting a probe into the flow to measure total temperature and then calculating the static temperature using Eq. (12.23):

$$T = \frac{T_t}{1 + [(k-1)/2]M_1^2}$$

The local speed of sound is then determined by Eq. (12.12):

$$c = \sqrt{kRT}$$

and the velocity is calculated from

$$V = M_1 c$$

The hot-wire anemometer can also be used to measure velocity in compressible flows, provided it is calibrated to account for Mach number effects.

Mass Flow Measurement

Measuring the flow rate of a compressible fluid using a truncated nozzle was discussed in some detail in Chapter 12. Basically, the flow nozzle is a truncated nozzle located in a pipe, so the equations developed in Chapter 12 can be used to determine the flow rate through the flow nozzle. Strictly speaking, the flow rate so calculated should be multiplied by the discharge coefficient. For the high Reynolds numbers characteristic of compressible flows, however, the discharge coefficient can be taken as unity. If the flow at the throat of the flow nozzle is sonic, it is conceivable that the complex flow field existing

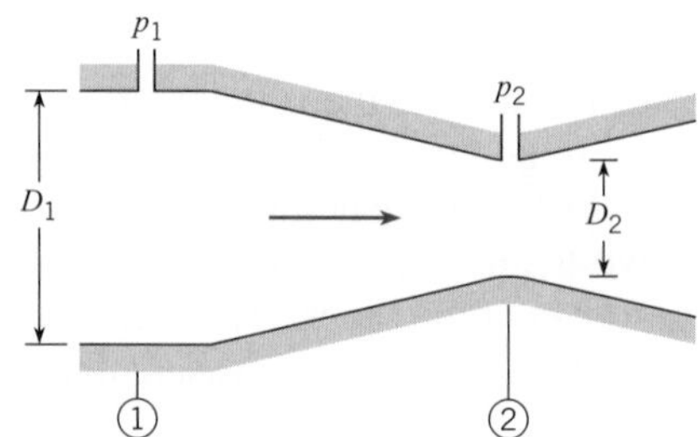

FIGURE 13.23

Venturi meter.

downstream of the nozzle will make the reading from the downstream pressure tap difficult to interpret. That is, there can be no assurance that the measured pressure is the true back pressure. In such a case, it is advisable to use a venturi meter because the pressure is measured directly at the throat.

The mass flow rate of a compressible fluid through a venturi meter can easily be analyzed using the equations developed in Chapter 12. Consider the venturi meter shown in Fig. 13.23. Writing the energy equation, Eq. (12.16), for the flow of an ideal gas between stations 1 and 2 gives

$$\frac{V_1^2}{2} + \frac{kRT_1}{k-1} = \frac{V_2^2}{2} + \frac{kRT_2}{k-1} \tag{13.14}$$

By conservation of mass, the velocity V_1 can be expressed as

$$V_1 = \frac{r_2 A_2 V_2}{r_1 A_1}$$

Substituting this result into Eq. (13.14), using the ideal-gas law to eliminate temperature, and solving for V_2 gives

$$V_2 = \left\{ \frac{[2k/(k-1)][(p_1/r_1) - (p_2/r_2)]}{1 - (r_2 A_2 / r_1 A_1)^2} \right\}^{1/2} \tag{13.15}$$

Assuming that the flow is isentropic,

$$\frac{p_1}{p_2} = \left(\frac{r_1}{r_2}\right)^k$$

the equation for the velocity at the throat can be rewritten as

$$V_2 = \left\{ \frac{[2k/(k-1)](p_1/r_1)[1 - (p_2/p_1)^{(k-1)/k}]}{1 - (p_2/p_1)^{2/k}(D_2/D_1)^4} \right\}^{1/2} \tag{13.16}$$

The mass flow is obtained by multiplying V_2 by $\rho_2 A_2$. This analysis, however, has been based on a one-dimensional flow, and two-dimensional effects can be accounted for by the discharge coefficient C_d. Thus we finally have

$$\dot{m} = C_d r_2 A_2 V_2 = C_d A_2 \left(\frac{p_2}{p_1}\right)^{1/k} \left\{ \frac{[2k/(k-1)]p_1 r_1 [1 - (p_2/p_1)^{(k-1)/k}]}{1 - (p_2/p_1)^{2/k}(D_2/D_1)^4} \right\}^{1/2} \tag{13.17}$$

This equation is valid for all flow conditions, subsonic or supersonic, provided no shock waves occur between station 1 and station 2. It is good design practice to avoid supersonic flows in the venturi meter in order to prevent the formation of shock waves and the attendant total pressure losses. Also, the discharge coefficient can generally be taken as unity if no shock waves occur between 1 and 2.

example 13.8

Calculate the mass flow rate of air through a venturi meter with a throat having a 1-cm diameter (D_2) in a pipe having a 3-cm diameter (D_1). The upstream static pressure is 150 kPa, and the throat pressure is 100 kPa. The static temperature of the air in the pipe is 27°C.

Solution From Table A.2 in the Appendix, we find that $k = 1.4$ and $R = 287$ J/kg K. The gas density in the pipe is

$$r_1 = \frac{p_1}{RT_1} = \frac{150 \times 10^3 \text{ N/m}^2}{(287 \text{ J/kg K})(300 \text{ K})} = 1.74 \text{ kg/m}^3$$

Substituting the appropriate values in Eq. (13.17), we find that

$$\dot{m} = 1 \times 0.785 \times 10^{-4} \text{ m}^2 \left(\frac{1}{1.5}\right)^{0.714}$$
$$\times \left\{\frac{7 \times 150 \times 10^3 \text{ N/m}^2 \times 1.74 \text{ kg/m}^3 [1 - (1/1.5)^{0.286}]}{[1 - (1/1.5)^{1.43}(1/3)^4]}\right\}^{1/2}$$
$$= 0.0264 \text{ kg/s}$$

◁

Shock Wave Visualization

When studying the qualitative features of a supersonic flow in a wind tunnel, it is important to be able to locate and identify the shock wave pattern. Unfortunately, shock waves cannot be seen with the naked eye, so the application of some type of optical technique is necessary. There are three techniques by which shock waves can be seen: the shadowgraph, the interferometer, and the schlieren system. Each technique has its special application related to the type of information on density variation that is desired. The schlieren technique, however, finds frequent use in shock wave visualization.

An illustration of the essential features of the schlieren system is given in Fig. 13.24. Light from the source s is collimated by lens L_1 to produce a parallel-light beam.

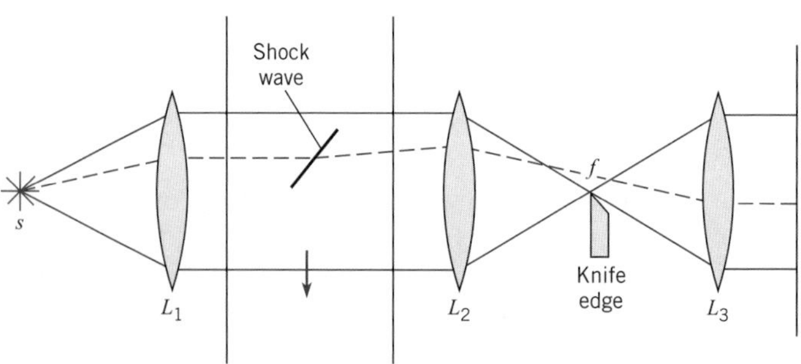

FIGURE 13.24

Schlieren system.

The light then passes through a second lens L_2 and produces an image of the source at plane *f.* A third lens L_3 focuses the image on the display screen. A sharp edge, usually called the *knife edge,* is positioned at plane *f* so as to block out a portion of the light.

If a shock wave occurs in the test section, the light is refracted by the density change across the wave. As illustrated by the dashed line in Fig. 13.24, the refracted ray escapes the blocking effect of the knife edge and the shock wave appears as a lighter region on the screen. Of course, if the beam is refracted in the other direction, the knife edge blocks out more light, and the shock wave appears as a darker region. The contrast can be increased by intercepting more light with the knife edge.

Interferometry

The *interferometer* allows one to map contours of constant density and to measure the density changes in the flow field. The underlying principle is the phase shift of a light beam on passing through media of different densities. The system now employed almost universally is the Mach-Zender interferometer, shown in Fig. 13.25. Light from a common source is split into two beams as it passes through the first half-silvered mirror. One beam passes through the test section, the other through the reference section. The two beams are then recombined and projected onto a screen or photographic plate. If the density in the test section and that in the reference section are the same, there is no phase shift between the two beams, and the screen is uniformly bright. However, a change of density in the test section changes the light speed of the test section beam, and a phase shift is generated between the two beams. Upon recombination of the beams, this phase shift gives rise to a series of dark and light bands on the screen. Each band represents a uniform-density contour, and the change in density across each band can be determined for a given system.

FIGURE 13.25

Schematic diagram of a Mach-Zender interferometer.

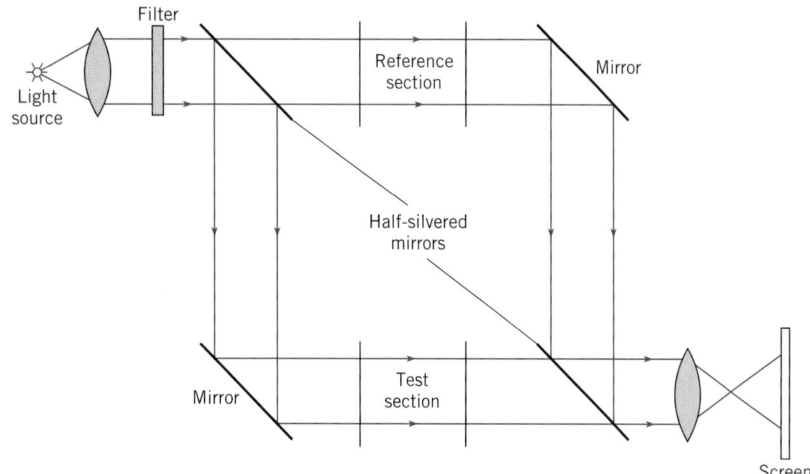

13.4

Accuracy of Measurements

When a parameter is measured, it is important to assess the accuracy of the measurement. The resulting analysis, called an *uncertainty analysis,* provides an estimate of the upper and lower bounds of the parameter. For example, if Q is a measured value of discharge, uncertainty analysis provides an estimate of the uncertainty U_Q in this measurement. The measurement would then be reported as $Q \pm U_Q$.

Commonly, a parameter of interest is not directly measured but is calculated from other variables. For example, discharge for an orifice meter is calculated using Eq. (13.3). Such an equation is called a data reduction equation. Consider a data reduction equation of the form

$$x = f(y_1, y_2, \ldots, y_n)$$

where x is the parameter of interest, and y_1 through y_n are the independent variables. Then, the uncertainty in x, which is written as U_x, is given by

$$U_x = \left[\left(\frac{\partial x}{\partial y_1} U_{y_1}\right)^2 + \left(\frac{\partial x}{\partial y_2} U_{y_2}\right)^2 + \cdots + \left(\frac{\partial x}{\partial y_n} U_{y_n}\right)^2\right]^{0.5} \tag{13.18}$$

where U_{y_i} is the uncertainty in variable y_i. Equation (13.18), known as the *uncertainty equation,* is very useful for quantifying the accuracy of an experimental measurement, and for planning experiments. Additional information about uncertainty analysis is provided by Coleman and Steele (17).

example 13.9

For the orifice meter described in Example 13.2, estimate the uncertainty of the calculated discharge. Assume that uncertainty in K is 0.03, the uncertainty in diameter is 0.15 mm, and the uncertainty in measured head is 10 mm Hg.

Solution The data reduction equation for the orifice meter (13.3*a*) is

$$Q = K(pd^2/4)\sqrt{2gh}$$

where we have substituted $\pi d^2/4$ for area and used the symbol h in place of Δh to simplify subsequent notation. The uncertainties in Q arise because of uncertainties in K, d, g, and h. Neglecting the uncertainties in g, the data reduction equation has the functional form $Q = f(K, d, h)$. Applying the uncertainty equation (13.18) to this problem gives

$$U_Q^2 = \left(\frac{\partial Q}{\partial K} U_K\right)^2 + \left(\frac{\partial Q}{\partial d} U_d\right)^2 + \left(\frac{\partial Q}{\partial h} U_h\right)^2$$

Evaluating each partial derivative and then dividing both sides of this equation by Q^2 results in

$$\left(\frac{U_Q}{Q}\right)^2 = \left(\frac{U_K}{K}\right)^2 + \left(\frac{2U_d}{d}\right)^2 + \left(\frac{U_h}{2h}\right)^2$$

Using values from Example 13.2,

$$\left(\frac{U_Q}{Q}\right)^2 = \left(\frac{0.03}{0.66}\right)^2 + \left(\frac{2 \times 0.15}{150}\right)^2 + \left(\frac{10}{2 \times 250}\right)^2$$

$$\left(\frac{U_Q}{Q}\right)^2 = (20.7 \times 10^{-4}) + (0.04 \times 10^{-4}) + (4 \times 10^{-4})$$

$$U_Q = 0.0497Q = 0.0497 \times (0.092 \text{ m}^3/\text{s})$$

$$= 0.0046 \text{ m}^3/\text{s} \quad \triangleleft$$

Thus the discharge is $Q \pm U_Q = (0.092 \pm 0.0046)$ m³/s. The above calculations reveal that the primary source of uncertainty in the discharge is due to U_K. The term U_h has a small effect, and U_d has a negligible effect.

13.5

Summary

There are many methods and instruments for measuring velocity, pressure and flow rate. These include

For velocity measurement: stagnation tube, Pitot tube, yaw meter, vane and cup anemometers, hot-wire and hot-film anemometers, laser-Doppler anemometer, and particle image velocimetry

For pressure measurement: static tube, piezometer, differential manometer, Bourdon-tube gage, and several types of pressure transducers

For flow rate measurement: direct volume or weight measurement, velocity–area integration, orifice, flow nozzle, venturi, electromagnetic flow meter, ultrasonic flow meter, turbine flow meter, vortex flow meter, rotameter, and weirs

Flow rate or discharge for a flowmeter that uses a restricted opening (i.e., an orifice, flow nozzle, or venturi) is calculated using

$$Q = KA_o\sqrt{2g\Delta h} = KA_o\sqrt{2\Delta p_z/r}$$

where K is a flow coefficient that depends on Reynolds number and the type of flow meter, A_o is the area of the opening, Δh is the change in piezometric head across the flowmeter, and Δp_z is drop in piezometric pressure across the flowmeter.

Discharge for a rectangular weir of length L is given by

$$Q = K\sqrt{2g}LH^{3/2}$$

where K is the flow coefficient that depends on H/P. The term H is the height of the water above the crest of the weir, as measured upstream of the weir, and P is the height of the weir. Discharge for a 60° triangular weir with $H > 2$ cm is given by

$$Q = 0.179\sqrt{2g}H^{5/2}$$

When flow is compressible, instruments such as the stagnation tube, hot- wire anemometer, Pitot tube, and flow nozzle may be used. However, equations correlating velocity and discharge need to be altered to account for the effects of compressibility. To observe shock waves in compressible flow, a schlieren technique or an interferometer may be used.

Uncertainty analysis provides a way to quantify the accuracy of a measurement. When a parameter of interest x is evaluated using an equation of the form $x = f(y_1, y_2, \ldots, y_n)$, where y_1 through y_n are the independent variables, the uncertainty in x (U_x) is given by

$$U_x = \left[\left(\frac{\partial x}{\partial y_1}U_{y_1}\right)^2 + \left(\frac{\partial x}{\partial y_2}U_{y_2}\right)^2 + \cdots + \left(\frac{\partial x}{\partial y_n}U_{y_n}\right)^2\right]^{0.5}$$

where U_{y_i} is the uncertainty in variable y_i. This equation, known as the uncertainty equation, is very useful for estimating uncertainty and for planning experiments.

References

1. Hurd, C. W., K. P. Chesky, and A. H. Shapiro. "Influence of Viscous Effects on Impact Tubes." *Trans. ASME J. Applied Mechanics,* vol. 75 (June 1953).

2. King, H. W., and E. F. Brater. *Handbook of Hydraulics.* McGraw-Hill, New York, 1963.

3. Lomas, Charles C. *Fundamentals of Hot Wire Anemometry.* Cambridge University Press, 1986.

4. Durst, Franz. *Principles and Practice of Laser-Doppler Anemometry.* Academic Press, New York 1981.

5. Kline, J. J. "Flow Visualization." In *Illustrated Experiments in Fluid Mechanics: The NCFMF Book of Film Notes.* Educational Development Center, 1972.

6. Macagno, Enzo O. "Flow Visualization in Liquids." *Iowa Inst. Hydraulic Res. Rept.*, 114 (1969).

7. Raffel, M., C. Wilbert, and J. Kompenhans. *Particle Image Velocimetry.* Springer, New York, 1998.

8. Lienhard, J. H., V, and J. H. Lienhard, IV. "Velocity Coefficients for Free Jets from Sharp-Edged Orifices." *Trans. ASME J. Fluids Engineering*, 106, (March 1984).

9. Tuve, G. L., and R. E. Sprenkle. "Orifice Discharge Coefficients for Viscous Liquids." *Instruments*, vol. 8 (1935).

10. ASME. *Fluid Meters*, 6th ed. ASME, 1971.

11. Shercliff, J. A. *Electromagnetic Flow-Measurement.* Cambridge University Press, New York, 1962.

12. Kindsvater, Carl E., and R. W. Carter. "Discharge Characteristics of Rectangular Thin-Plate Weirs." *Trans. Am. Soc. Civil Eng.*, 124 (1959), 772–822.

13. King, L. V. *Phil. Trans. Roy. Soc. London, Ser. A*, 14 (1914), 214.

14. Miller, R. W. *Flow Measurement Engineering Handbook*. McGraw-Hill, New York, 1983.

15. Scott, R. W. W., ed. *Developments in Flow Measurement–1*. Applied Science, Englewood, NJ, 1982.

16. NACA. "Equations, Tables, and Charts for Compressible Flow." TR 1135 (1953).

17. Coleman, Hugh W., and W. Glenn Steele, *Experimentation and Uncertainty Analysis for Engineers*. Wiley, New York, 1989.

Problems

13.1 Without exceeding an error of 2.5%, what is the minimum air velocity that can be obtained using a 1-mm circular stagnation tube if the formula

$$V = \sqrt{2\Delta p_{\text{stag}}/r} = \sqrt{2gh_{\text{stag}}}$$

is used for computing the velocity? Assume standard atmospheric conditions.

13.2 Without exceeding an error of 1%, what is the minimum water velocity that can be obtained using a 1-mm circular stagnation tube if the formula

$$V = \sqrt{2\Delta p_{\text{stag}}/r} = \sqrt{2gh_{\text{stag}}}$$

is used for computing the velocity? Assume the water temperature is 20°C.

13.3 A stagnation tube 2 mm in diameter is used to measure the velocity in a stream of air as shown. What is the air velocity if the deflection on the air-water manometer is 1.0 mm? Air temperature = 10°C, and p = 100 kPa.

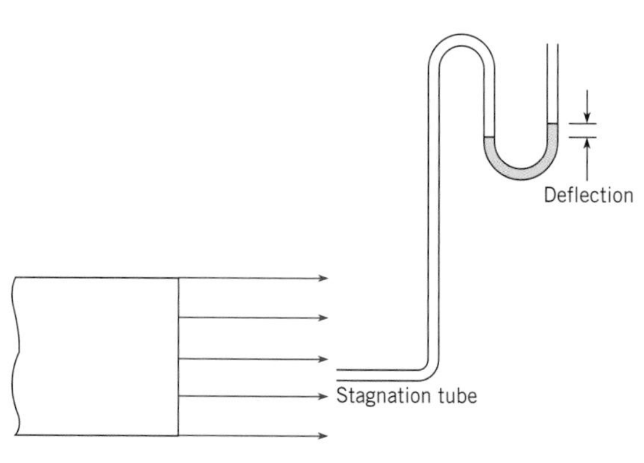

PROBLEMS 13.3, 13.4

13.4 If the velocity in an airstream (p_a = 98 kPa, T = 10°C) is 12 m/s, what deflection will be produced in an air-water manometer if the stagnation tube is 2 mm in diameter?

13.5 What would be the error in velocity determination if one used a C_p value of 1.00 for a circular stagnation tube instead of the true value? Assume the measurement is made with a stagnation tube 2 mm in diameter that is measuring air (T = 25°C, p = 100 kPa absolute) velocity for which the stagnation pressure reading is 5.00 Pa. FLUID SOLUTIONS

13.6 A velocity-measuring probe used frequently for measuring stack gas velocities is shown. The probe consists of two tubes bent away from and toward the flow direction and cut off on a plane normal to the flow direction, as shown. Assume the pressure coefficient is 1.0 at A and −0.4 at B. The probe is inserted in a stack where the temperature is 300°C and the pressure is 100 kPa absolute. The gas constant of the stack gases is 410 J/kg K. The probe is connected to a water manometer, and a 1.0-cm deflection is measured. Calculate the stack gas velocity.

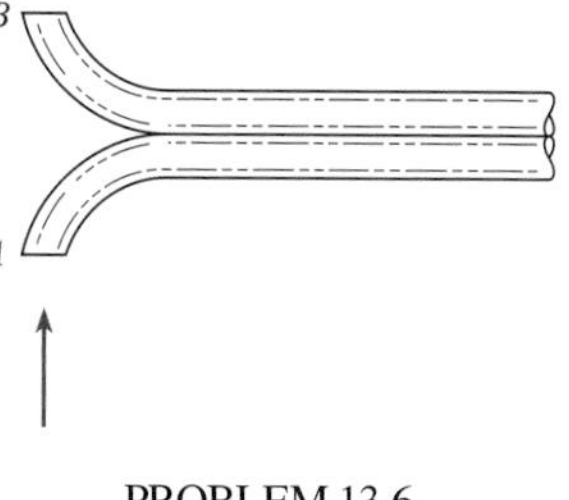

PROBLEM 13.6

13.7 Water from a pipe is diverted into a tank for 3.5 min. If the weight of diverted water is measured to be 14 kN, what is the discharge in cubic meters per second? Assume the water temperature is 20°C.

13.8 Water from a test apparatus is diverted into a calibrated volumetric tank for 5 min. If the volume of diverted water is measured to be 80 m^3, what is the discharge in cubic meters per second, gallons per minute, and cubic feet per second?

13.9 A velocity traverse in a 24-cm oil pipe yields the data in the table. What are the discharge, mean velocity, and ratio of maximum to mean velocity? Does the flow appear to be laminar or turbulent?

r, cm	*V*, m/s	*r*, cm	*V*, m/s
0	8.7	7	5.8
1	8.6	8	4.9
2	8.4	9	3.8
3	8.2	10	2.5
4	7.7	10.5	1.9
5	7.2	11.0	1.4
6	6.5	11.5	0.7

13.10 A velocity traverse inside a 16-in. circular air duct yields the data in the table. What is the rate of flow in cubic feet per second and cubic feet per minute? What is the ratio of V_{max} to V_{mean}? Does it appear that the flow is laminar or turbulent? If $p = 14.3$ psia and $T = 70°F$, what is the mass-flow rate?

*y**	*V*, ft/s	*y**	*V*, ft/s
0.0	0	2.0	110
0.1	72	3.0	117
0.2	79	4.0	122
0.4	88	5.0	126
0.6	93	6.0	129
1.0	100	7.0	132
1.5	106	8.0	135

*Distance from pipe wall, in.

13.11 The asymmetry of the flow in stacks means that flow velocity must be measured at several locations on the cross-flow plane. Consider the cross section of the cylindrical stack shown. The two access holes through which probes can be inserted are separated by 90°. Velocities can be measured at the five points shown (five-point method).

a. Determine the ratio r_m/D such that the areas of the five measuring segments are equal.

b. Determine the ratio r/D (probe location) that corresponds to the centroid of the segment.

c. The data in the table are taken for a stack 2 m in diameter in which the gas temperature is 300°C, the pressure is 110 kPa absolute, and the gas constant is 400 J/kg K. The data represent the deflection on a water manometer connected to a conventional Pitot tube located at the measuring stations. Calculate the mass flow rate.

Station	Δh, cm
1	1.2
2	1.1
3	1.1
4	0.9
5	1.05

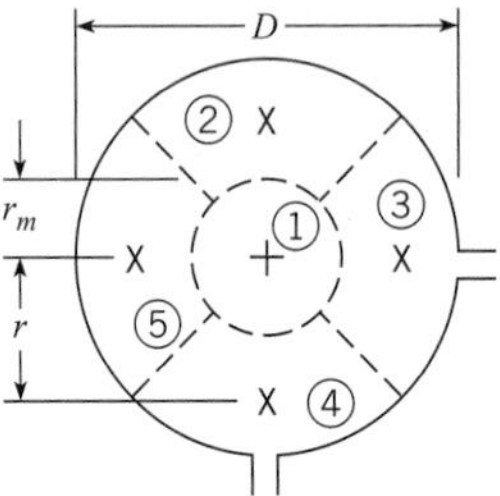

PROBLEM 13.11

13.12 Repeat Prob. 13.11 for the case in which three access holes are separated by 60° and seven measuring points are used. The diameter of the stack is 1.5 m, the gas temperature is 250°C, the pressure is 115 kPa absolute, and the gas constant is 420 J/kg K. The data in the following table represent the deflection of a water manometer connected to a conventional Pitot tube at the measuring stations. Calculate the mass flow rate.

Station	Δh, mm
1	8.2
2	8.6
3	8.2
4	8.9
5	8.0
6	8.5
7	8.4

13.13 Theory and experimental verification indicate that the mean velocity along a vertical line in a wide stream is closely approximated by the velocity at 0.6 depth. If the indicated velocities at 0.6 depth in a river cross section are measured, what is the discharge in the river?

13.14 A laser-Doppler anemometer (LDA) system is being used to measure the velocity of air in a tube. The laser is an argon-ion laser with a wavelength of 4880 angstroms. The angle between the laser beams is 15°. The time interval is determined by measuring the time between five spikes, as shown, on the signal from the photodetector. The time interval between the five spikes is 12 microseconds. Find the velocity.

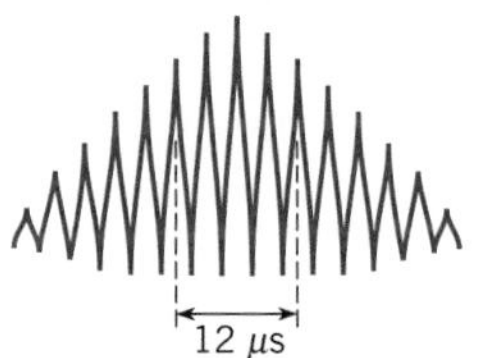

PROBLEM 13.14

13.15 For the jet and orifice shown, determine C_v, C_c, and C_d.

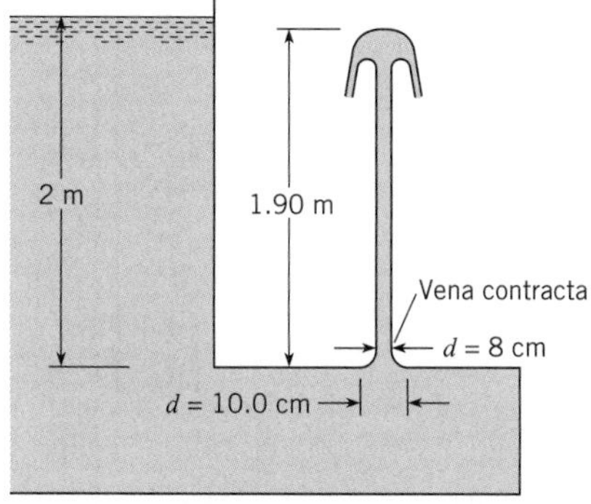

PROBLEM 13.15

13.16 A fluid jet discharging from a 3-cm orifice has a diameter of 2.6 cm at its vena contracta. What is the coefficient of contraction?

13.17 Figure 13.12 is of a sharp-edged orifice. Note that the metal surface immediately downstream of the leading edge makes an acute angle with the metal of the upstream face of the orifice. Do you think the orifice would operate the same (have the same flow coefficient, K) if that angle were 90°? Explain how you came to your conclusion.

13.18 New orifices such as that shown in Fig. 13.12 will have definite flow coefficients as given in Fig. 13.13. With age, however, physical changes could occur to the orifice. Explain what changes these might be and how (if at all) these physical changes might affect the flow coefficients.

13.19 A 5-in. orifice is placed in a 10-in. pipe, and a mercury manometer is connected to either side of the orifice. If the flow rate of water (60°F) through this orifice is 3.0 cfs, what will be the manometer deflection? FLUID SOLUTIONS

13.20 Determine the discharge of water through this 6-in. orifice that is installed in a 12-in. pipe.

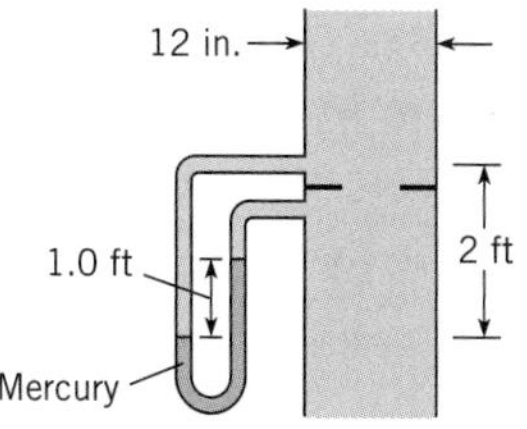

PROBLEM 13.20

13.21 The flow coefficient values for orifices given in Fig. 13.13 were obtained by testing orifices in relatively smooth pipes. If an orifice were used in a pipe that was very rough, do you think you would get a valid indication of discharge by using the flow coefficient of Fig. 13.13? Justify your conclusion.

13.22 Determine the discharge of water ($T = 60°F$) through the orifice shown if $h = 4$ ft, $D = 5$ in., and $d = 2.5$ in.

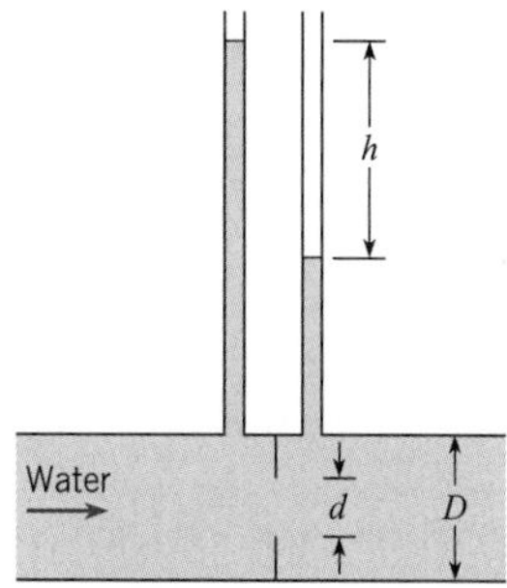

PROBLEM 13.22

13.23 A pressure transducer is connected across an orifice to measure the flow rate of kerosene at 20°C. The pipe diameter is 3 cm, and the ratio of orifice diameter to pipe diameter is 0.6. The pressure differential as indicated by the transducer is 15 kPa. What is the mean velocity of the kerosene in the pipe?

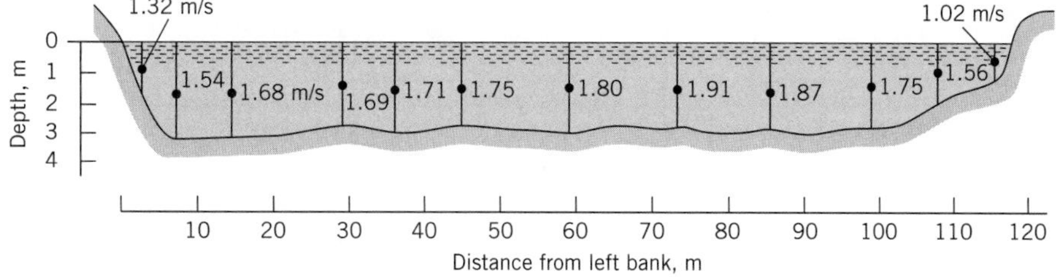

PROBLEM 13.13

13.24 The 10-cm orifice in the horizontal 30-cm pipe shown is the same size as the orifice in the vertical pipe. The manometers are mercury-water manometers, and water ($T = 20°\text{C}$) is flowing in the system. The gages are Bourdon tube gages. The flow, at a rate of 0.1 m^3/s, is to the right in the horizontal pipe and therefore downward in the vertical pipe. Is Δp as indicated by gages A and B the same as Δp as indicated by gages D and E? Determine their values. Is the deflection on manometer C the same as the deflection on manometer F? Determine the deflections.

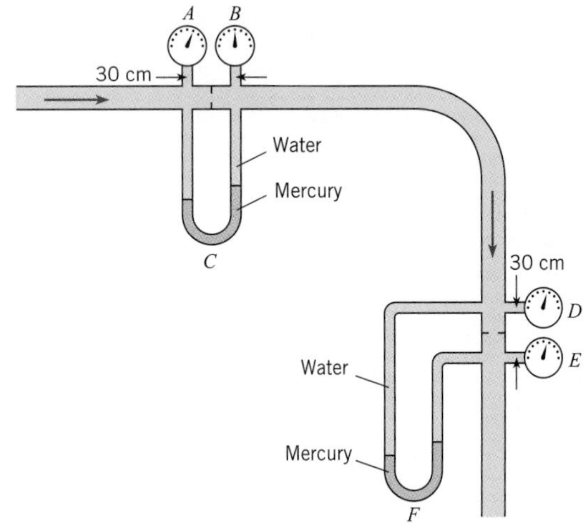

PROBLEM 13.24

13.25 A 15-cm plate orifice at the end of a 30-cm pipe is enlarged to 20 cm. With the same pressure drop across the orifice (approximately 50 kPa), what will be the percentage increase in discharge?

13.26 If water (20°C) is flowing through this 10-cm orifice, estimate the rate of flow.

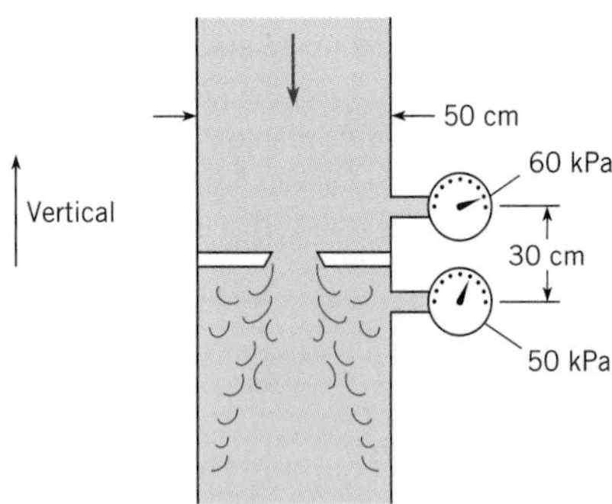

PROBLEM 13.26

13.27 A pressure transducer is connected across an orifice as shown. The pressure at the upstream pressure tap is p_1 and the pressure at the downstream tap is p_2. The pressure at the transducer connected to the upstream tap is $p_{T,1}$ and to the downstream pressure tap, $p_{T,2}$. Show that the difference in piezometric pressure defined as $(p_1 + \gamma z_1) - (p_2 + \gamma z_2)$ is equal to the pressure difference across the transducer, $p_{T,1} - p_{T,2}$.

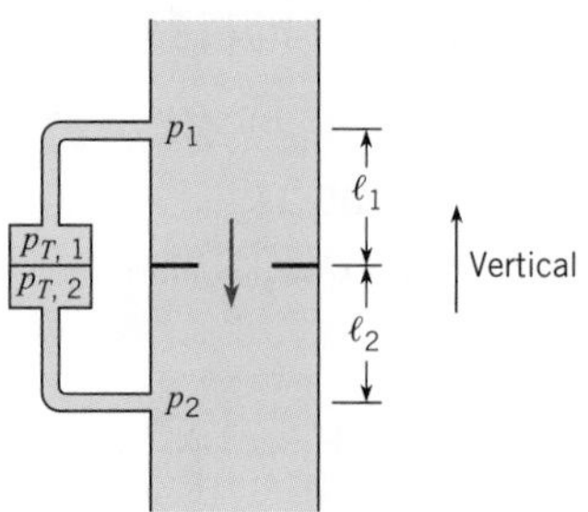

PROBLEM 13.27

13.28 Water ($T = 50°\text{F}$) is pumped at a rate of 20 cfs through the system shown in the figure. What differential pressure will occur across the orifice? What power must the pump supply to the flow for the given conditions? Also, draw the HGL and the EGL for the system. Assume $f = 0.015$ for the pipe.

13.29 Determine the size of orifice required in a 15-cm pipe to measure 0.03 m^3/s of water with a deflection of 1 m on a mercury-water manometer.

13.30 What is the discharge of gasoline ($S = 0.68$) in a 10-cm horizontal pipe if the differential pressure across a 6-cm orifice in the pipe is 50 kPa?

13.31 What size orifice is required to produce a change in head of 6 m for a discharge of 2 m^3/s of water in a pipe 1 m in diameter?

13.32 An orifice is to be designed to have a change in pressure of 50 kPa across it (measured with a differential-pressure transducer) for a discharge of 3.0 m^3/s of water in a pipe 1.2 m in diameter. What diameter should the orifice have to yield the desired results?

13.33 Hemicircular orifices such as the one shown are sometimes used to measure the flow rate of liquids that also transport sediments. The opening at the bottom of the pipe allows free passage of the sediment. Derive a formula for Q as a function of Δp, D, and other relevant variables associated with the problem. Then, using that formula and guessing any unknown data, estimate the water discharge through such an orifice when Δp is read as 80 kPa and flow is in a 30-cm pipe.

13.34 Water flows through a venturi meter that has a 30-cm throat. The venturi meter is in a 60-cm pipe. What deflection will occur on a mercury-water manometer connected between the upstream and throat sections if the discharge is 0.75 m^3/s? Assume $T = 20°\text{C}$.

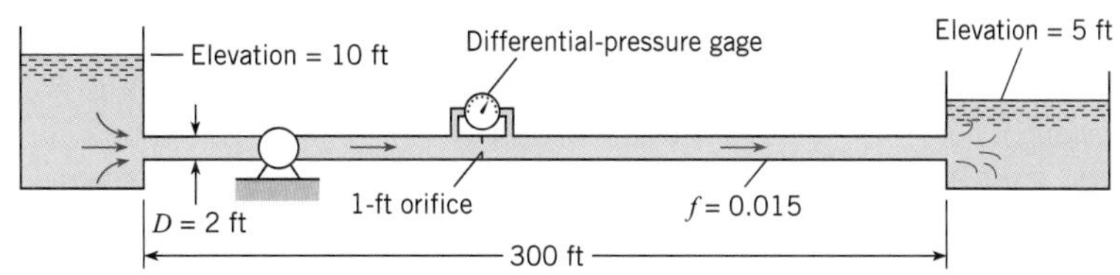

PROBLEM 13.28

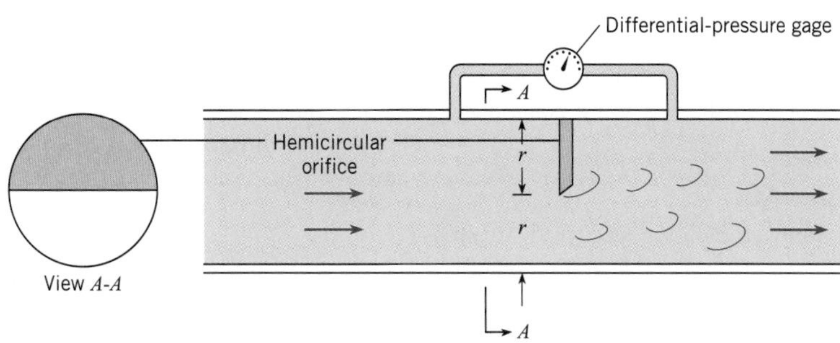

PROBLEM 13.33

13.35 What is the throat diameter required for a venturi meter in a 200-cm horizontal pipe carrying water with a discharge of 10 m^3/s if the differential pressure between the throat and the upstream section is to be limited to 200 kPa at this discharge?

13.36 Estimate the rate of flow of water through the venturi meter shown.

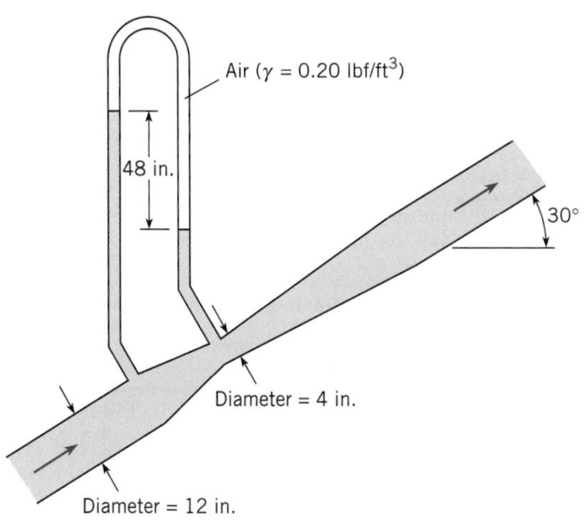

PROBLEM 13.36

13.37 When no flow occurs through the venturi meter, the indicator on the differential-pressure gage is straight up and indicates a Δp of zero. When 5 cfs of water flow to the right, the differential-pressure gage indicates $\Delta p = +10$ psi. If the flow is now reversed and 5 cfs flow to the left through the venturi meter, in which range would Δp fall? (a) $\Delta p < -10$ psi, (b) $-10 \text{ psi} < \Delta p < 0$, (c) $0 < \Delta p < 10$ psi, (d) $\Delta p = 10$ psi.

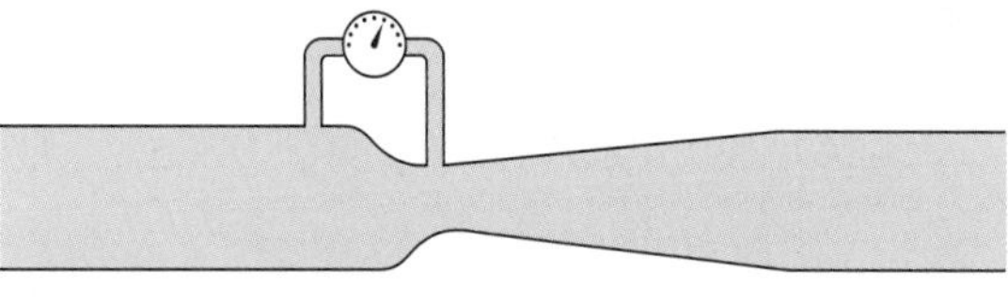

PROBLEM 13.37

13.38 The pressure differential across this venturi meter is 100 kPa. What is the discharge of water through it?

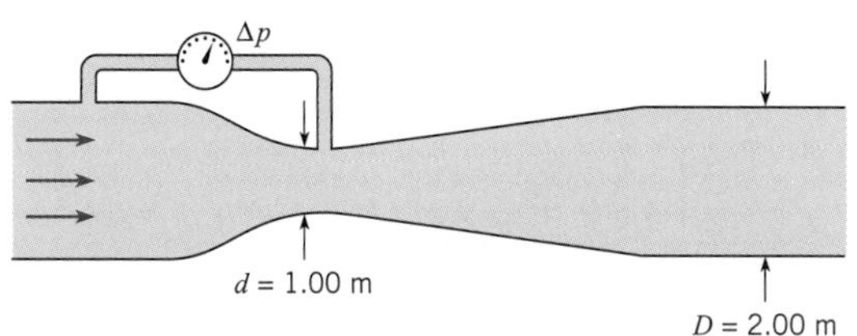

PROBLEM 13.38

13.39 Engineers are calibrating a poorly designed venturi meter for the flow of an incompressible liquid by relating the pressure difference between taps 1 and 2 to the discharge. By applying the Bernoulli equation and assuming a quasi–one-dimensional flow (velocity uniform across every cross section), the engineers find that

$$Q_0 = A_2[2(p_1 - p_2)/r]^{0.5}[1-(d/D)^4]^{-0.5}$$

where D and d are the duct diameters at stations 1 and 2. However, they realize that the flow is not quasi–one-dimensional and that the pressure at tap 2 is not equal to the average pressure in the throat because of streamline curvature. Thus the engineers introduce a correction factor K into the foregoing equation to yield

$$Q = KQ_0$$

Use your knowledge of pressure variation across curved streamlines to decide whether K is larger or smaller than unity, and support your conclusion by presenting a rational argument.

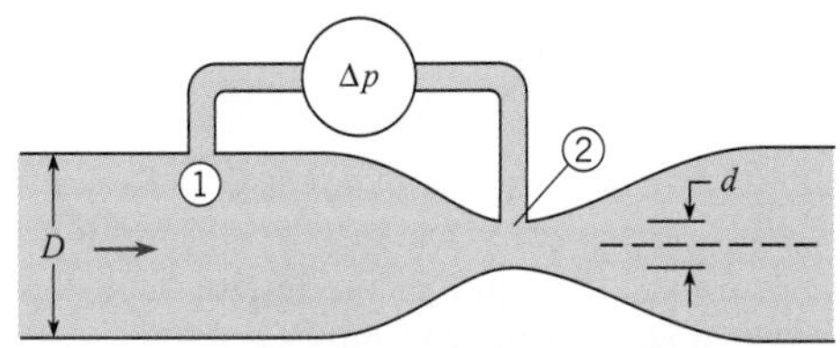

PROBLEM 13.39

13.40 The differential-pressure gage on the venturi meter shown reads 6.2 psi, h = 25 in., d = 6 in., D = 12 in. What is the discharge of water in the system? Assume T = 50°F.

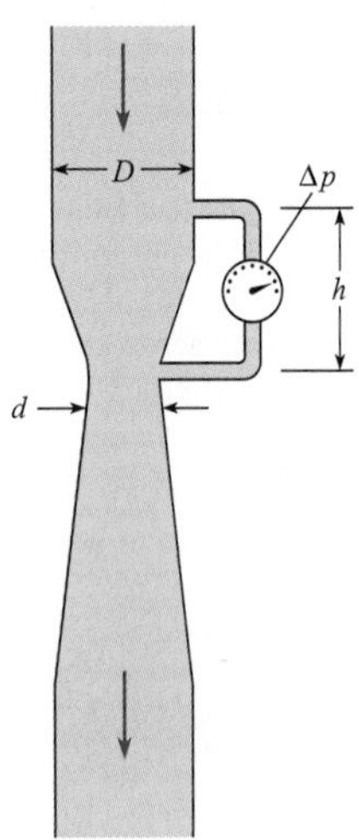

PROBLEMS 13.40, 13.41

13.41 The differential-pressure gage on the venturi meter reads 45 kPa, d = 20 cm, D = 40 cm, and h = 80 cm. What is the discharge of gasoline (S = 0.69, $\mu = 3 \times 10^{-4}$ N · s/m^2) in the system?

13.42 A flow nozzle has a throat diameter of 2 cm and a beta ratio (d/D) of 0.5. Water flows through the nozzle, creating a pressure difference across the nozzle of 8 kPa. The viscosity of the water is 10^{-6} m^2/s, and the density is 1000 kg/m^3. Find the discharge.

13.43 Water flows through an annular venturi consisting of a body of revolution mounted inside a pipe. The pressure is measured at the minimum area and upstream of the body. The pipe is 5 cm in diameter, and the body of revolution is 2.5 cm in diameter. A head difference of 1 m is measured across the pressure taps. Find the discharge in cubic meters per second.

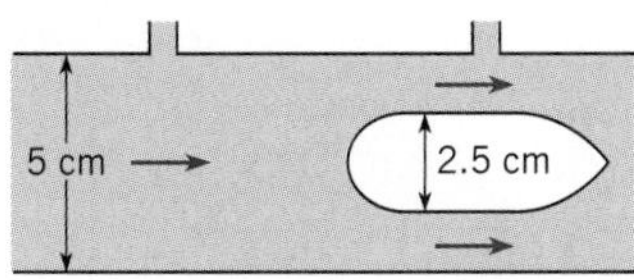

PROBLEM 13.43

13.44 What is the head loss in terms of $V_0^2/2g$ for the flow nozzle shown?

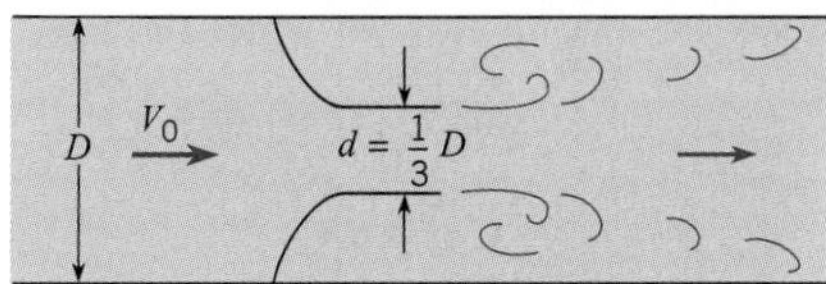

PROBLEM 13.44

13.45 A vortex flow meter is used to measure the discharge in a duct 5 cm in diameter. The diameter of the shedding element is 1 cm. The Strouhal number based on the shedding frequency from one side of the element is 0.2. A signal frequency of 50 Hz is measured by a pressure transducer mounted downstream of the element. What is the discharge in the duct?

13.46 A rotameter operates by aerodynamic suspension of a weight in a tapered tube. The scale on the side of the rotameter is calibrated in scfm of air—that is, cubic feet per minute at standard conditions (p = 1 atm and T = 68°F). By considering the balance of weight and aerodynamic force on the weight inside the tube, determine how the readings would be corrected for nonstandard conditions. In other words, how would the actual cubic feet per minute be calculated from the reading on the scale, given the pressure, temperature, and gas constant of the gas entering the rotameter?

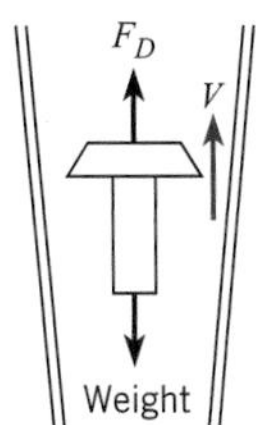

PROBLEM 13.46

13.47 A rotameter is used to measure the flow rate of a gas with a density of 1.1 kg/m^3. The scale on the rotameter indicates 5 liters/s. However, the rotameter is calibrated for a gas with a density of 1.2 kg/m^3. What is the actual flow rate of the gas (in liters per second)?

13.48 One mode of operation of ultrasonic flow meters is to measure the travel times between two stations for a sound wave traveling upstream and then downstream with the flow. The downstream propagation speed with respect to the measuring stations is $c + V$, where c is the sound speed and V is the flow velocity. Correspondingly, the upstream propagation speed is $c - V$.

a. Derive an expression for the flow velocity in terms of the distance between the two stations, L; the difference in travel times, Δt; and the sound speed.

b. The sound speed is typically much larger than V ($c \gg V$). With this approximation, express V in terms of L, c, and Δt.

c. A 10-ms time difference is measured for waves traveling 20 m in a gas where the speed of sound is 300 m/s. Calculate the flow velocity.

13.49 Water flows over a rectangular weir that is 4 m wide and 25 cm high. If the head on the weir is 20 cm, what is the discharge in cubic meters per second?

13.50 The head on a 60° triangular weir is 35 cm. What is the discharge over the weir in cubic meters per second?

13.51 Water flows over two rectangular weirs. Weir A is 5 ft long in a channel 10 ft wide; weir B is 5 ft long in a channel 5 ft wide. Both weirs are 2 ft high. If the head on both weirs is 1.00 ft, then one can conclude that (a) $Q_A = Q_B$, (b) $Q_A > Q_B$, (c) $Q_A < Q_B$.

13.52 A 1-ft-high rectangular weir (weir 1) is installed in a 2-ft-wide rectangular channel, and the head on the weir is observed for a discharge of 10 cfs. Then the 1-ft weir is replaced by a 2-ft-high rectangular weir (weir 2), and the head on the weir is observed for a discharge of 10 cfs. The ratio H_1/H_2 should be (a) equal to 1.00, (b) less than 1.00, (c) greater than 1.00.

13.53 A 3-m-long rectangular weir is to be constructed in a 3-m-wide rectangular channel, as shown (*a*). The maximum flow in the channel will be 4 m^3/s. What should be the height P of the weir to yield a depth of water of 2 m in the channel upstream of the weir?

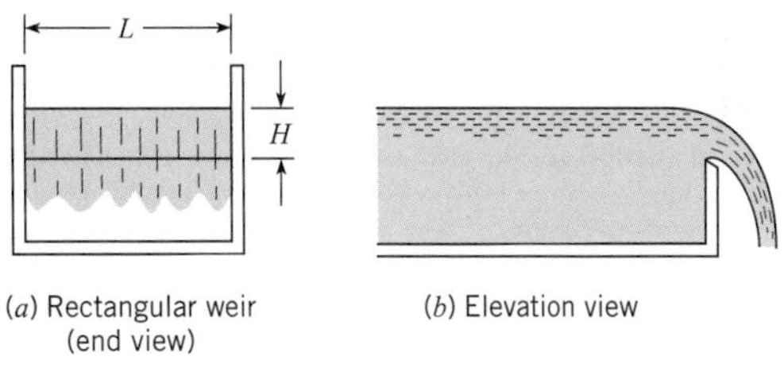

PROBLEMS 13.53, 13.54, 13.55, 13.56

13.54 Consider the rectangular weir described in Prob. 13.53. When the head is doubled, the discharge is (a) doubled, (b) less than doubled, (c) more than doubled.

13.55 A basin is 50 ft long, 2 ft wide, and 4 ft deep. A sharp-crested rectangular weir is located at one end of the basin, and it spans the width of the basin (the weir is 2 ft long). The crest of the weir is 2 ft above the bottom of the basin. At a given instant water in the basin is 3 ft deep; thus water is flowing over the weir and out of the basin. Estimate the time it will take for the water in the basin to go from the 3-ft depth to a depth of 2 ft 2 in.

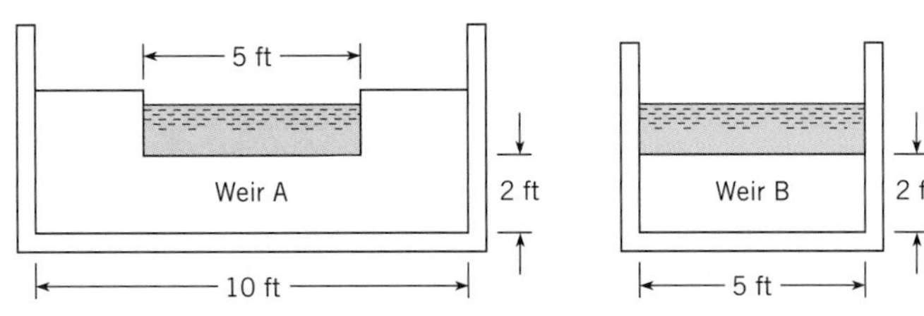

PROBLEM 13.51

13.56 Water at 50°F is piped from a reservoir to a channel like that shown. The pipe from the reservoir to the channel is a 4-in. steel pipe 100 ft in total length. There are two 90° bends, $r/D = 1$, in the line, and the entrance and exit are sharp edged. The weir is 2 ft long. The elevation of the water surface in the reservoir is 100 ft, and the elevation of the bottom of the channel is 70 ft. The crest of the weir is 3 ft above the bottom of the channel. For steady flow conditions determine the water surface elevation in the channel and the discharge in the system.

FLUID SOLUTIONS **13.57** At one end of a rectangular tank 1 m wide is a sharp-crested rectangular weir 1 m high. In the bottom of the tank is a 10-cm sharp-edged orifice. If 0.10 m^3/s of water flows into the tank and leaves the tank both through the orifice and over the weir, what depth will the water in the tank attain?

13.58 The weirs shown in Figs. 13.18, 13.19, 13.20, and 13.21 all have sharp edges. Do you think the weir would operate differently if the edges were not sharp? Justify your answer. For the weir of Fig. 13.18, it is assumed that atmospheric pressure exists both above and below the nappe. To ensure that such prevails, a hole can be drilled in the side of the channel (downstream of the weir) to let this space be at atmospheric pressure. What if such a hole or vent were not there? Explain in detail what might happen without such a vent and how (if at all) it would affect the flow and flow coefficient.

13.59 What is the water discharge over a rectangular weir 3 ft high and 10 ft long in a rectangular channel 10 ft wide if the head on the weir is 1.8 ft?

13.60 A reservoir is supplied with water at 60°F by a pipe with a venturi meter as shown. The water leaves the reservoir through a triangular weir with an included angle of 60°. The flow coefficient of the venturi is unity, the area of the venturi throat is 12 in.^{2}, and the measured Δp is 10 psi. Find the head, H, of the triangular weir.

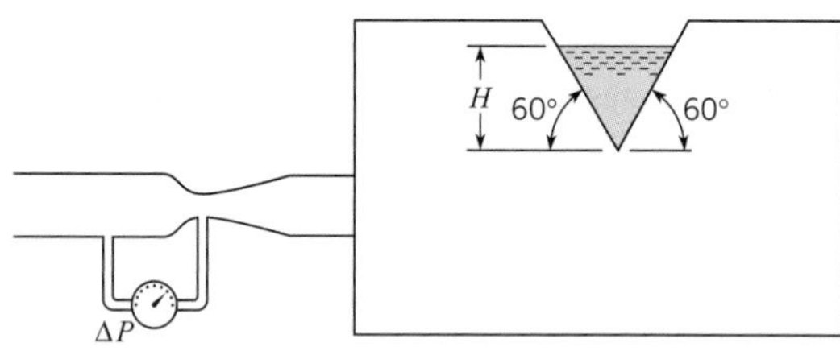

PROBLEM 13.60

FLUID SOLUTIONS **13.61** At a particular instant water flows into the tank shown through pipes A and B, and it flows out of the tank over the rectangular weir at C. The tank width and weir length (dimensions normal to page) are 2 ft. Then, for the given conditions, is the water level in the tank rising or falling?

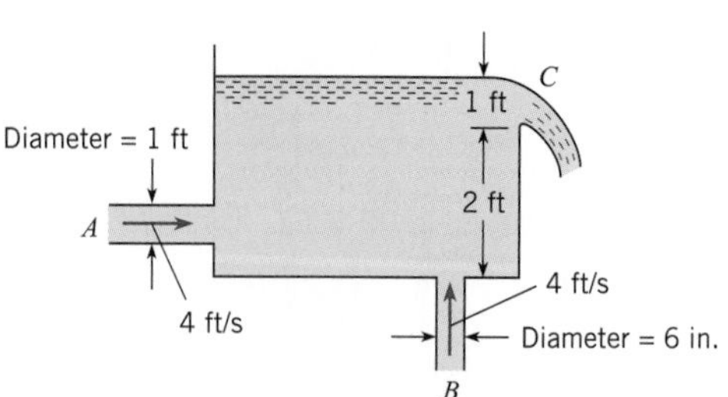

PROBLEM 13.61

13.62 Water flows from the first reservoir to the second over a rectangular weir with a width-to-head ratio of 3. The height P of the weir is twice the head. The water from the second reservoir flows over a 60° triangular weir to a third reservoir. The discharge across both weirs is the same. Find the ratio of the head on the rectangular weir to the head on the triangular weir.

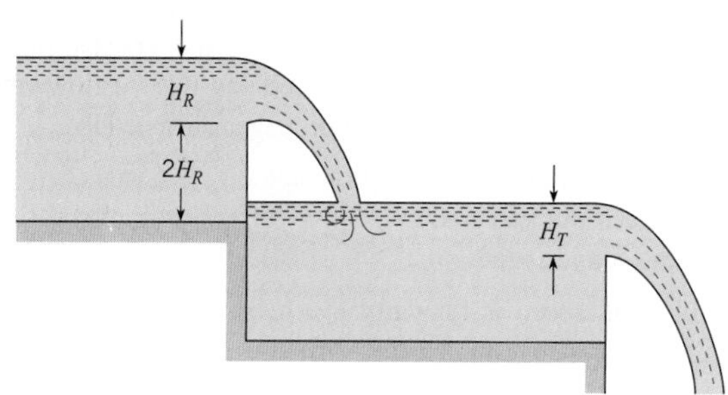

PROBLEMS 13.62, 13.63

13.63 Given the initial conditions of Prob. 13.62, tell, qualitatively and quantitatively, what will happen if the flow entering the first reservoir is increased 50%.

13.64 A rectangular irrigation canal 3 m wide carries water with a discharge of 6 m^3/s. What height of rectangular weir installed across the canal will raise the water surface to a level 2 m above the canal floor?

13.65 The head on a 60° triangular weir is 1.2 ft. What is the discharge of water over the weir?

13.66 An engineer is designing a triangular weir for measuring the flow rate of a stream of water that has a discharge of 10 cfm. The weir has an included angle of 45° and a coefficient of discharge of 0.6. Find the head on the weir.

13.67 A pump is used to deliver water at 10°C from a well to a tank. The bottom of the tank is 2 m above the water surface in the well. The pipe is commercial steel 2.5 m long with a diameter of 5 cm. The pump develops a head of 20 m. A triangular weir with an included angle of 60° is located in a wall of the tank with the bottom of the weir 1 m above the tank floor. Find the level of the water in the tank above the floor of the tank.

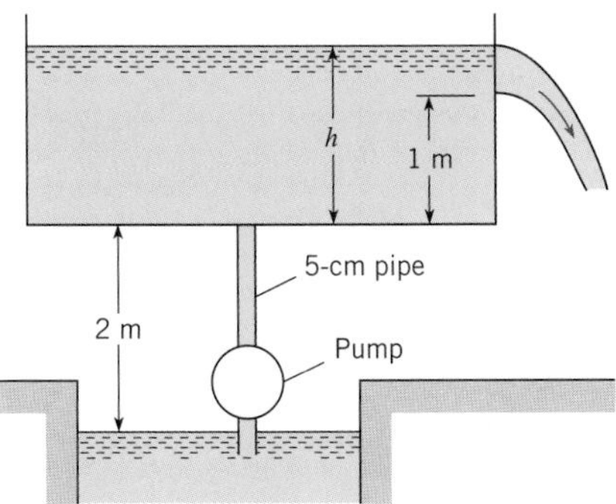

PROBLEM 13.67

13.68 A Pitot tube is used to measure the Mach number in a compressible subsonic flow of air. The stagnation pressure is 140 kPa, and the static pressure is 100 kPa. The total temperature of the flow is 300 K. Determine the Mach number and the flow velocity.

13.69 Use the normal shock wave relationships developed in Chapter 12 to derive the Rayleigh supersonic Pitot formula.

13.70 The static and stagnation pressures measured by a Pitot tube in a supersonic air flow are 54 kPa and 200 kPa, respectively. The total temperature is 350 K. Determine the Mach number and the velocity of the free stream.

13.71 A venturi meter is used to measure the flow of helium in a pipe. The pipe is 1 cm in diameter, and the throat diameter is 0.5 cm. The measured upstream and throat pressures are 120 kPa and 80 kPa, respectively. The static temperature of the helium in the pipe is 17°C. Determine the mass flow rate.

13.72 The mass flow rate of methane is measured with a square-edged orifice. The pipe diameter is 2 cm, and the orifice diameter is 0.8 cm. The upstream and downstream pressure taps measure 150 kPa and 120 kPa absolute, respectively. The static temperature in the pipe is 300 K. Determine the mass flow rate.

13.73 The mass flow rate of air is measured by an orifice with a diameter of 1 cm in a 2-cm pipe. The upstream pressure is 150 kPa, and the downstream pressure is 100 kPa (absolute). The upstream air density is 1.8 kg/m^3, and the kinematic viscosity is 1.8×10^{-5} m^2/s. The ratio of specific heats, k, is 1.4. Calculate the mass flow rate.

13.74 Hydrogen at atmospheric pressure and 15°C flows through a sharp-edged orifice with a beta ratio, d/D, of 0.5 in a 2-cm pipe. The pipe is horizontal, and the pressure change across the orifice is 1 kPa. The flow coefficient is 0.62. Find the mass flow (in kilograms per second) through the orifice.

13.75 A hole 0.2 in. in diameter is accidentally punctured in a line carrying natural gas (methane). The pressure in the pipe is 50 psig, and the atmospheric pressure is 14 psia. The temperature in the line is 70°F. What is the rate at which the methane leaks through the hole (in lbm/s)? The hole can be treated as a truncated nozzle. FLUID SOLUTIONS

13.76 Weirs are often used in places where they might be subject to considerable physical effects such as abrasion. One example where weirs are used is in the measurement of flow in irrigation canals. Think of and list all of the physical effects not indicated in your text that a weir might be subjected to. For each of these effects explain how (if at all) the effect might influence the flow over the weir. That is, for a given head, would the discharge be greater or less than given by the weir equation?

13.77 Hydraulic laboratories are often designed so that a source of water for experiments includes a constant-head tank located at a higher elevation than the laboratory floor from which water can be drawn. The water surface in the tank is maintained at a fairly constant level even though the discharge to experiments may vary. This is accomplished by supplying water to the constant-head tank at a rate higher than needed by the experiment, and the excess water (water not used by the experiment) is discharged over a weir and sent back to the reservoir through a drain pipe. The weir crest is made long so that only a small variation in head (relative to head available for experimentation) occurs with a variation in flow over the weir. Assume that such a setup is available to you and that the maximum available discharge is 5 cfs. Given the basic water source described above, design a piece of equipment (including associated measuring devices) that could be used to determine the coefficient of contraction for flow through an orifice.

13.78 Given the laboratory setup described in Prob. 13.77, design test equipment to determine the resistance coefficient, f, of a 2-in.-diameter steel pipe.

13.79 Given the laboratory setup described in Prob. 13.77, design test equipment to determine the loss coefficients of 2-in. gate and globe valves.

13.80 Consider the stagnation tube of Prob. 13.3. If the uncertainty in the manometer measurement is 0.1 mm, calculate the velocity and the uncertainty in the velocity. Assume that $C_p = 1.00$, $\rho_{air} = 1.25$ kg/m^3, and the only uncertainty is due to the manometer measurement. FLUID SOLUTIONS

13.81 Consider the orifice meter in Prob. 13.20. Calculate the flow rate and the uncertainty in the flow rate. Assume the following values of uncertainty: 0.03 in flow coefficient, 0.05 in. in orifice diameter, and 0.5 in. in height of mercury.

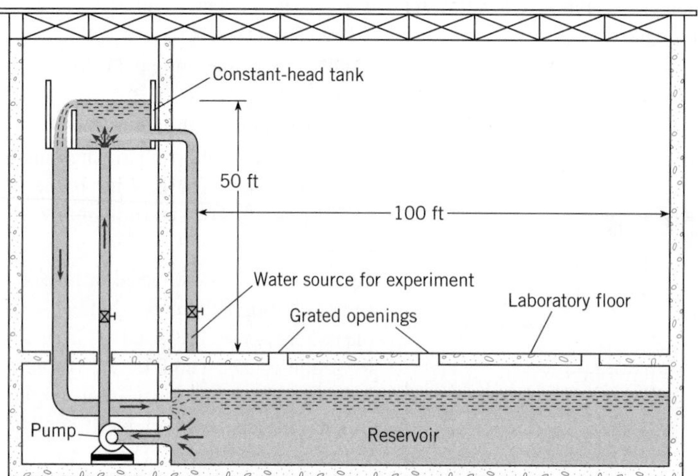

PROBLEMS 13.77, 13.78, 13.79

13.82 Consider the weir in Prob. 13.59. Calculate the discharge and the uncertainty in the discharge. Assume the uncertainty in K is 5%, in H is 3 in., and in L is 1 in.

13.83 Conventional instruments like the Pitot tube cannot be used to measure low-speed velocities of air because the difference between the stagnation and static pressure is too small to measure accurately. Develop some ideas for other schemes to measure air velocities from 1 to 10 ft/s. The device should be reasonably cheap and portable.

13.84 The volume flow rate of gas discharging from a small tube is less than a liter per minute. Devise a scheme to measure the flow rate.

13.85 Design a scheme to measure the density of a flowing fluid by using a combination of flow meters.

Wind turbines are a source of renewable, pollution-free energy. Commercial wind turbines have rotors between 50 and 200 feet in diameter and have a generating capacity from 50 to 1500 kilowatts. European countries are currently developing aggressive wind energy programs. (PhotoDisc, Inc./Getty Images)

CHAPTER 14

Turbomachinery

Machines to move fluids or to extract power from moving fluids have been designed and used by human beings since the beginning of recorded history. Ancient designs included buckets attached to a rope to transport water from a well or river. In the third century B.C., Archimedes invented the screw pump, which the Romans later used in their water supply systems. Water wheels were used in ancient China for grinding grain.

Fluid machines are used everywhere. They are the essential components of the automobiles we drive, of the supply systems for the water we drink, of the power generation plants for the electricity we use, and of the air-conditioning and heating systems for the comfort we enjoy.

Fluid machines are separated into two broad categories: positive displacement machines and turbomachines. Positive displacement machines operate by forcing fluid into or out of a chamber. Examples include the bicycle tire pump, the gear pump, the peristaltic pump, and the human heart. Turbomachines involve the flow of fluid through rotating blades or rotors that remove or add energy to the fluid. Examples include propellers, fans, water pumps, windmills, and compressors. Axial-flow turbomachines operate with the flow entering and leaving the machine in the axial direction. A radial-flow machine can have the flow either entering or leaving the machine in the radial direction.

For turbomachines that add energy to a fluid, the label "pumps" is generally associated with liquids and the labels "fans," "blowers," and "compressors" are associated with gases. Turbines refer to machines that extract energy from flowing fluids such as wind, hot gases, and water. Correspondingly, the machines are labeled wind turbines, gas turbines, and hydraulic turbines.

In this chapter we will address turbomachines. We will first introduce the characteristics of propellers and then extend these concepts to axial-flow pumps and fans. We will then study radial-flow pumps and fans. Finally, we will consider the elementary characteristics of turbines.

Propeller Theory

The design of a propeller is based on the fundamental principles of airfoil theory. For example, if we observe a section of the propeller in Fig. 14.1, we see the analogy between the lifting vane and the propeller. This propeller is rotating at an angular speed ω, and the

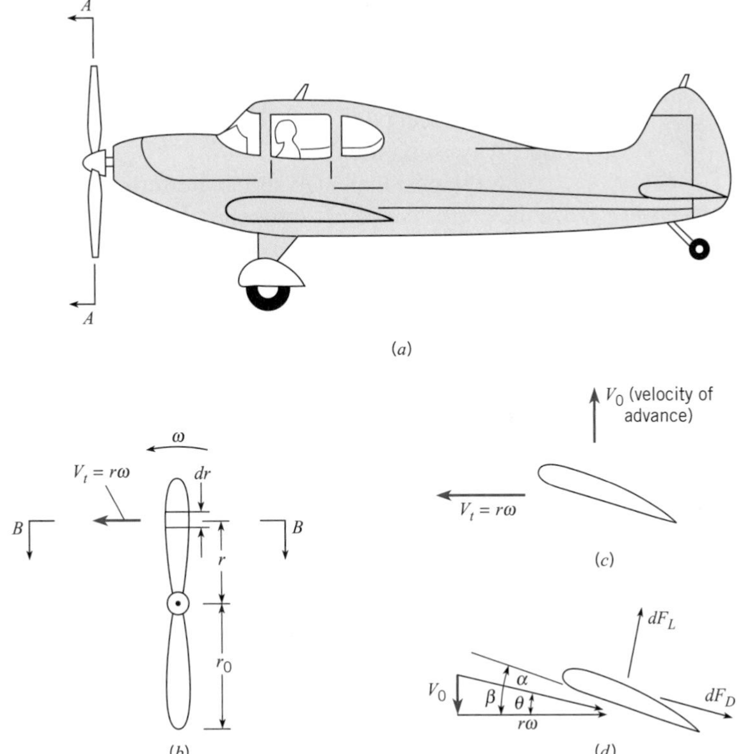

FIGURE 14.1

Propeller motion.
(a) Airplane motion.
(b) View A-A.
(c) View B-B.
(d) Velocity relative to blade element.

speed of advance of the airplane and propeller is V_0. If we focus on an elemental section of the propeller, Fig. 14.1*c*, we note that the given section has a velocity made of components V_0 and V_t. Here V_t is tangential velocity, $V_t = r\omega$, resulting from the rotation of the propeller. Reversing and adding the velocity vectors V_0 and V_t yield the velocity of the air relative to the particular propeller section (Fig. 14.1*d*).

The angle θ is given by

$$\theta = \arctan\left(\frac{V_0}{r\omega}\right) \tag{14.1}$$

For a given forward speed and rotational rate, this angle is a minimum at the propeller tip ($r = r_0$) and increases toward the hub as the radius decreases. The angle β is known as the *pitch angle*. The local angle of attack of the elemental section is

$$\alpha = \beta - \theta \tag{14.2}$$

The propeller can be analyzed as a series of elemental sections producing lift and drag, which provide the propeller thrust and create resistive torque. This torque multiplied by the rotational speed is the power input to the propeller.

The propeller is designed to produce thrust, and since the greatest contribution to thrust comes from the lift force F_L, the goal is to maximize lift and minimize drag, F_D. For a given shape of propeller section, the optimum angle of attack can be determined from data such as are given in Fig. 11.23. Because the angle θ increases with decreasing radius, the local pitch angle has to change to achieve the optimum angle of attack. This is done by twisting the blade.

A dimensional analysis can be performed to determine the π-groups that characterize the performance of a propeller. For a given propeller shape and pitch distribution, the thrust of a propeller, T, will depend on the propeller diameter, D, the rotational speed, n, the forward speed, V_0, the fluid density, ρ, and the fluid viscosity, μ.

$$T = f(D, \omega, V_0, \rho, \mu) \tag{14.3}$$

Performing a dimensional analysis using the methods presented in Chapter 8, we find

$$\frac{T}{\rho n^2 D^4} = f\left(\frac{V_0}{nD}, \frac{\rho D^2 n}{\mu}\right) \tag{14.4}$$

It is conventional practice to express the rotational rate, n, as revolutions per second (rps). The grouping on the left is called the *thrust coefficient*,

$$C_T = \frac{T}{\rho n^2 D^4} \tag{14.5}$$

The first π-group on the right is the *advance ratio*. The second group is a Reynolds number proportional to the tip speed and diameter of the propeller. For most applications, the Reynolds number is high, and experience shows that the thrust coefficient is unaffected by the Reynolds number, so

$$C_T = f\left(\frac{V_0}{nD}\right) \tag{14.6}$$

The angle θ at the propeller tip is related to the advance ratio by

$$\theta = \arctan\left(\frac{V_0}{\omega r_0}\right) = \arctan\left(\frac{1}{\pi}\frac{V_0}{nD}\right) \tag{14.7}$$

As the advance ratio increases and θ increases, the local angle of attack at the blade element decreases, the lift increases, and the thrust coefficient goes down. This trend illustrated in Fig. 14.2, which shows the dimensionless performance curves for a typical propeller. Ultimately, an advance ratio is reached where the thrust coefficient goes to zero.

Performing a dimensional analysis for the power, P, shows

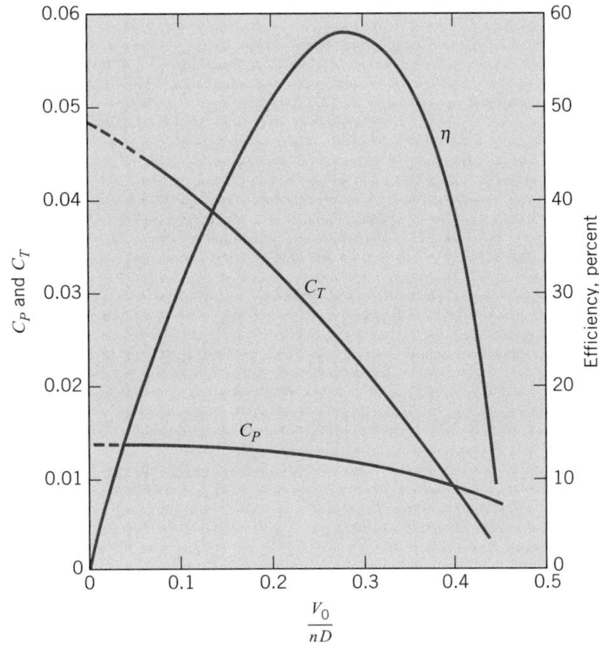

FIGURE 14.2

Dimensionless performance curves for a typical propeller; $D = 2.90$ m, $n = 1400$ rpm. [After Weick (1)]

$$\frac{P}{\rho n^3 D^5} = f\left(\frac{V_0}{nD}, \frac{\rho D^2 n}{\mu}\right) \tag{14.8}$$

The π-group on the left is the *power coefficient*,

$$C_P = \frac{P}{\rho n^3 D^5} \tag{14.9}$$

As with the thrust coefficient, the power coefficient is not significantly influenced by the Reynolds number at high Reynolds numbers, so C_P reduces to a function of the advance ratio only.

$$C_P = f\left(\frac{V_0}{nD}\right) \tag{14.10}$$

The functional relationship between C_P and V_0/nD for an actual propeller is also shown in Fig. 14.2. Even though the thrust coefficient approaches zero for a given advance ratio, the power coefficient shows little decrease because it still takes power to overcome the torque on the propeller blade.

The curves for C_T and C_P are evaluated from performance characteristics of a given propeller operating at different values of V_0 as shown in Fig. 14.3. Even though the data for the curves are obtained for a given propeller, the values for C_T and C_P, as a function

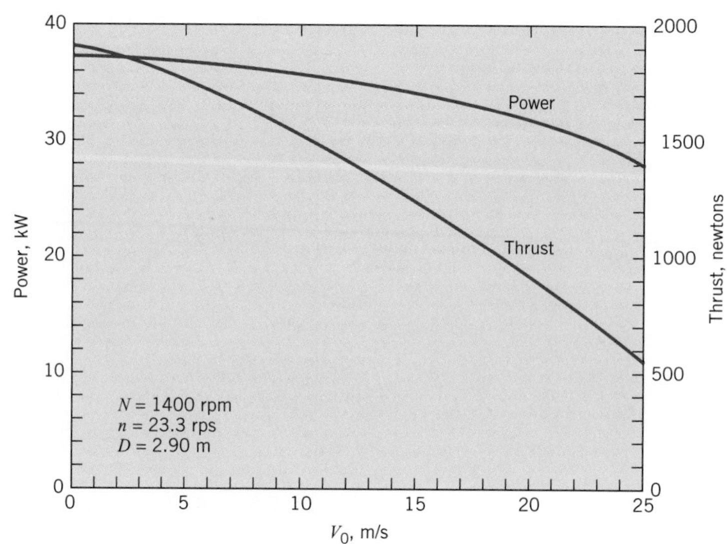

FIGURE 14.3

Power and thrust of a propeller 2.90 m in diameter at a rotational speed of 1400 rpm. [After Weick (2)]

of advance ratio, can be applied to geometrically similar propellers of different sizes and angular speeds.* Example 14.1 illustrates such an application.

example 14.1

A propeller having the characteristics shown in Fig. 14.2 is to be used to drive a swamp boat. If the propeller is to have a diameter of 2 m and a rotational speed of $N = 1200$ rpm, what should be the thrust starting from rest? If the boat resistance (air and water) is given by the empirical equation $F_D = 0.003\rho V_0^2/2$, where V_0 is the boat speed in meters per second, F_D is the drag, and ρ is the mass density of the water, what will be the maximum speed of the boat and what power will be required to drive the propeller? Assume $\rho_{\text{air}} = 1.20\ \text{kg/m}^3$.

Solution First we evaluate the thrust starting from rest:

$$\frac{V_0}{nD} = \frac{(0\ \text{m/s})}{(20\ \text{rps})(2\ \text{m})} = 0$$

Then, referring to Fig. 14.2, we see that $C_T = 0.048$ for $V_0/nD = 0$. Since $C_T = F_T/\rho_a D^4 n^2$, we can compute F_T:

$$F_T = C_T \rho_a D^4 n^2 = 0.048(1.20\ \text{kg/m}^3)(2\ \text{m})^4(20\ \text{rps})^2 = 369\ \text{N} \quad \triangleleft$$

* The speed of sound was not included in the dimensional analysis. However, the propeller performance is reduced as the Mach number based on the propeller tip speed leads to shock waves and other compressible-flow effects.

To obtain the maximum speed, we should recognize that

$$F_T = F_D$$

or

$$C_T \rho_a D^4 n^2 = 0.003 \rho_w \left(\frac{V_0^2}{2} \right)$$

Here ρ_w is the mass density of water, 1000 kg/m^3. To determine the equilibrium condition where $F_T = F_D$, we plot F_T and F_D versus V_0. Where the two curves intersect is the point of equilibrium. The data we need to construct such curves are included in the following table. When we plot F_T and F_D against V_0, we obtain the graph shown below. The curves intersect at $V_0 = 11$ m/s. Hence the boat will attain a maximum speed of 11 m/s. ◁

V_0	V_0/nD	C_T	$F_T = C_T \rho_a D^4 n^2$	$F_D = 0.003 \rho_w V_0^2/2$
5 m/s	0.125	0.040	307 N	37.5 N
10 m/s	0.250	0.027	207 N	150 N
15 m/s	0.375	0.012	90 N	337 N

The input power is $P = C_P \rho_a D^5 n^3$; and since the maximum C_P is 0.014 when $V_0/nD = 0$, we compute the power input as

$$P = 0.014(1.20\ \text{kg/m}^3)(2\ \text{m})^5(20\ \text{rps})^3$$
$$= 4300\ \text{m} \cdot \text{N/s} = 4.30\ \text{kW}$$ ◁

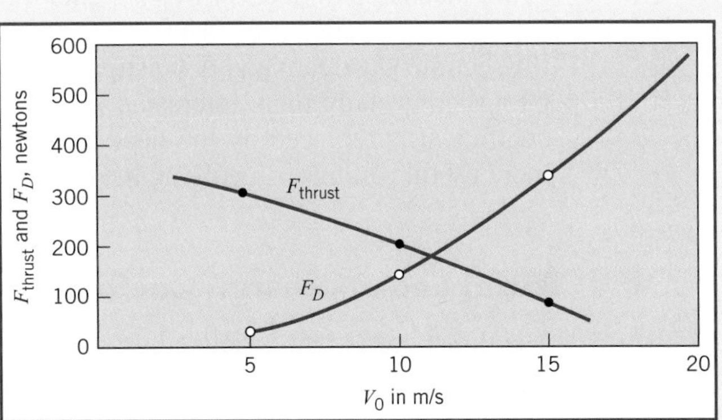

The efficiency of a propeller is defined as the ratio of the power output— that is, thrust times velocity of advance—to the power input. Hence the efficiency η is given as

$$\eta = \frac{F_T V_0}{P} = \frac{C_T \rho D^4 n^2 V_0}{C_P \rho D^5 n^3}$$

or

$$\eta = \frac{C_T}{C_P}\left(\frac{V_0}{nD}\right) \tag{14.11}$$

The variation of efficiency with advance ratio for a typical propeller is also shown in Fig. 14.2. The efficiency can be calculated directly from C_T and C_P performance curves. Note at low advance ratios, the efficiency increases with advance ratio and then reaches a maximum value before the decreasing thrust coefficient causes the efficiency to drop toward zero. The maximum efficiency represents the best operating point for fuel efficiency.

Many propeller systems are designed to have variable pitch; that is, pitch angles can be changed during propeller operation. Different efficiency curves corresponding to varying pitch angle are shown in Fig. 14.4. The envelope for the maximum efficiency is also shown in the figure. During operation of the aircraft, the pitch angle can be controlled to achieve maximum efficiency corresponding to the propeller rpm and forward speed.

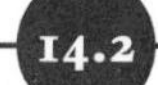

14.2 Axial-Flow Pumps

Pressure Changes

Axial-flow pumps (or blowers) are designed so that the impeller (much like the propeller discussed in Section 14.1) is enclosed within a housing (see Fig. 14.5). The action of the propeller applies a force to the fluid that causes a pressure change between the sections upstream and downstream of the pump. Because engineers working with pumps are usually more interested in the change in head than in the thrust, the thrust coefficient is replaced by the head coefficient, and the ratio V_0/nD is replaced by a more useful parameter involving the discharge.

Head and Discharge Coefficients for Pumps

The thrust coefficient is defined as $F_T/\rho D^4 n^2$, and if the same variables are applied to flow in an axial pump, the thrust can be expressed as $F_T = \Delta p A = \gamma \Delta H A$ or

$$C_T = \frac{\gamma \Delta H A}{\rho D^4 n^2} = \frac{\pi}{4}\frac{\gamma \Delta H D^2}{\rho D^4 n^2} = \frac{\pi}{4}\frac{g \Delta H}{D^2 n^2} \tag{14.12}$$

Now if we define a new parameter, the *head coefficient* C_H, using the variables of Eq. (14.12), we have

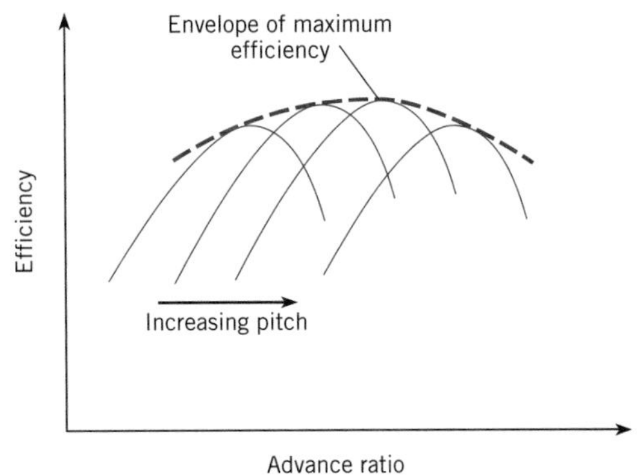

FIGURE 14.4

Efficiency curves for variable pitch propeller.

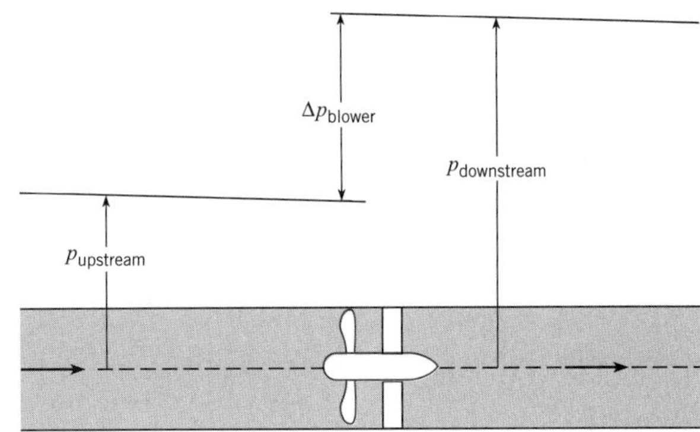

FIGURE 14.5

Axial-flow blower in a duct.

$$C_H = \frac{4}{\pi} C_T = \frac{\Delta H}{D^2 n^2 / g} \tag{14.13}$$

The independent parameter relating to propeller operation is V_0/nD; however, multiplying the numerator and denominator by the diameter squared gives V_0D^2/nD^3, and V_0D^2 is proportional to the discharge. Thus the independent parameter for pump similarity studies is Q/nD^3, and this is termed the *discharge coefficient* C_Q. The power coefficient used for pumps is exactly like the power coefficient used for propellers. Summarizing, the dimensionless parameters used in similarity analyses of pumps are as follows:

$$C_H = \frac{\Delta H}{D^2 n^2 / g} \tag{14.14}$$

$$C_P = \frac{P}{\rho D^5 n^3} \tag{14.15}$$

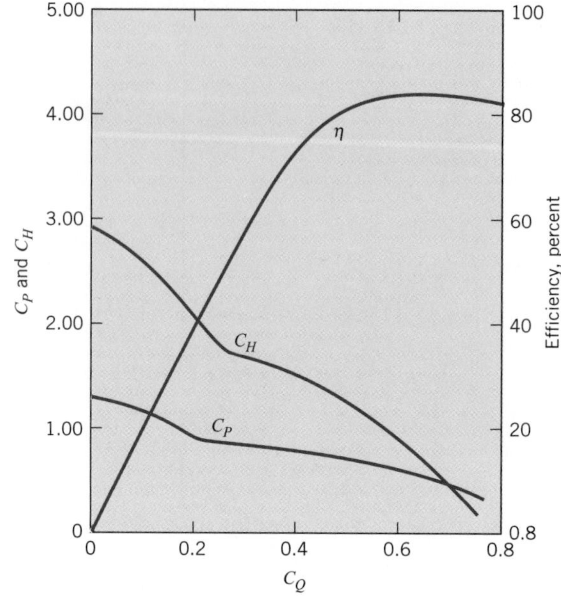

FIGURE 14.6

Dimensionless performance curves for a typical axial-flow pump. [After Stepanoff (3)]

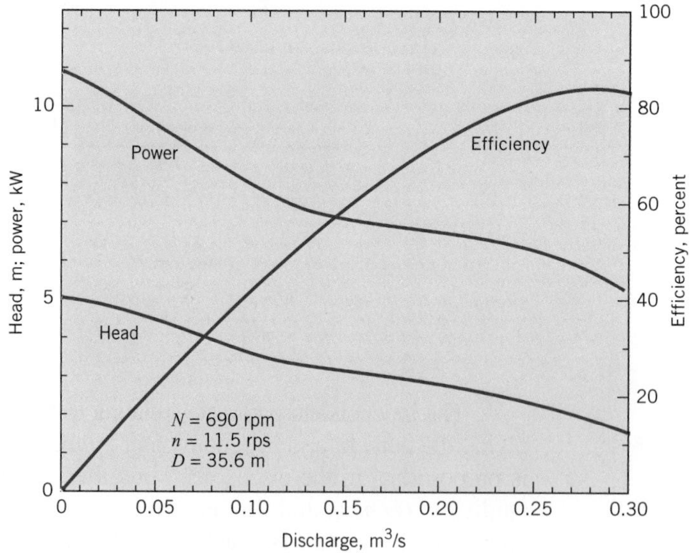

FIGURE 14.7

Performance curves for a typical axial-flow pump. [After Stepanoff (3)]

$$C_Q = \frac{Q}{nD^3} \tag{14.16}$$

where C_H and C_P are functions of C_Q for a given type of pump.

Figure 14.6 is a set of curves of C_H and C_P versus C_Q for a typical axial-flow pump. Also plotted on this graph is the efficiency of the pump as a function of C_Q. The dimensional curves (head and power versus Q for a constant speed of rotation) from which Fig. 14.6 was developed are shown in Fig. 14.7. Because curves like those shown in Fig. 14.6 or Fig. 14.7 characterize pump performance, they are often called *characteristic curves* or *performance curves*. These curves are obtained by experiment.

Performance curves are used to predict prototype operation from model tests or the effect of change of speed of the pump. Following are examples of these applications.

example 14.2

For the pump represented by Figs. 14.6 and 14.7, what discharge of water in cubic meters per second will occur when the pump is operating against a 2-m head and at a speed of 600 rpm? What power in kilowatts is required for these conditions?

Solution First compute C_H. Here

$$D = 35.6 \text{ cm} \qquad \text{and} \qquad n = 10 \text{ rps}$$

Then

$$C_H = \frac{2 \text{ m}}{(0.356 \text{ m})^2(10^2 \text{ s}^{-2})/(9.81 \text{ m/s}^2)} = 1.55$$

Next, using a value of 1.55 for C_H, read a value of 0.40 for C_Q from Fig. 14.6. Hence Q is calculated as follows:

$$C_Q = 0.40 = \frac{Q}{nD^3}$$

or

$$Q = 0.40(10 \text{ s}^{-1})(0.356 \text{ m})^3 = 0.180 \text{ m}^3/\text{s} \quad \triangleleft$$

From Fig. 14.6 the value of C_P is 0.72 for $C_Q = 0.40$. Then

$$\begin{aligned} P &= 0.72\rho D^5 n^3 \\ &= 0.72(10^3 \text{ kg/m}^3)(0.356 \text{ m})^5(10 \text{ s}^{-1})^3 \\ &= 4.12 \text{ km} \cdot \text{N/s} = 4.12 \text{ kJ/s} = 4.12 \text{ kW} \quad \triangleleft \end{aligned}$$

example 14.3

If a 30-cm axial-flow pump having the characteristics shown in Fig. 14.6 is operated at a speed of 800 rpm, what head ΔH will be developed when the water-pumping rate is 0.127 m^3/s? What power is required for this operation?

Solution First compute $C_Q = Q/nD^3$, where

$$Q = 0.127 \text{ m}^3/\text{s}$$

$$n = \frac{800}{60} = 13.3 \text{ rps}$$

$$D = 30 \text{ cm}$$

Then $$C_Q = \frac{0.127\ \text{m}^3/\text{s}}{(13.3\ \text{s}^{-1})(0.30\ \text{m})^3} = 0.354$$

Now enter Fig. 14.6 with a value for C_Q of 0.354 and read a value of 1.70 for C_H and a value of 0.80 for C_P. Then

$$\Delta H = \frac{C_H D^2 n^2}{g} = \frac{1.70(0.30\ \text{m})^2(13.3\ \text{s}^{-1})^2}{(9.81\ \text{m/s}^2)} = 2.76\ \text{m} \quad \triangleleft$$

and $$P = C_P \rho D^5 n^3$$

$$= 0.80(10^3\ \text{kg/m}^3)(0.30\ \text{m})^5(13.3\ \text{s}^{-1})^3 = 4.57\ \text{kW} \quad \triangleleft$$

Range of Application of Axial-Flow Machines

In practical applications, axial-flow machines are best suited for relatively low heads and high rates of flow. Hence pumps used for dewatering lowlands, such as those behind dikes, are almost always of the axial-flow type. Water turbines in low-head dams (less than 30 m) where the flow rate and power production are large are also generally of the axial type. For larger heads, radial- or mixed-flow machines are more efficient. These are discussed in the next section.

Radial-Flow Machines

Centrifugal Pumps

In Fig. 14.8 the type of impeller that is used for many radial-flow pumps is shown. Such pumps are often called *centrifugal pumps*. Fluid from the inlet pipe enters the pump through the eye of the impeller and then travels outward between the vanes of the impeller to its edge, where the fluid enters the casing of the pump and is then conducted to the discharge pipe. The principle of the radial-flow pump is different from that of the axial-flow pump in that the change in pressure results in large part from rotary action (pressure increases outward like that in the rotating tank in Section 4.7) produced by the rotating impeller. Additional pressure increase is produced in the radial-flow pump when the high-velocity flow leaving the impeller is reduced in the expanding section of the casing.

Although the basic designs are different for radial- and axial-flow pumps, it can be shown that the same similarity parameters (C_Q, C_P, and C_H) apply for both types. Thus the methods that have already been discussed for relating size, speed, and discharge in axial-flow machines also apply to radial-flow machines.

The major practical difference between axial- and radial-flow pumps so far as the user is concerned is the difference in the performance characteristics of the two. In Fig. 14.9 the dimensional performance curves for a typical radial-flow pump operating at a constant speed of rotation are shown. In Fig. 14.10 the dimensionless performance curves for the same pump are shown. Note that the power required at shutoff flow is less

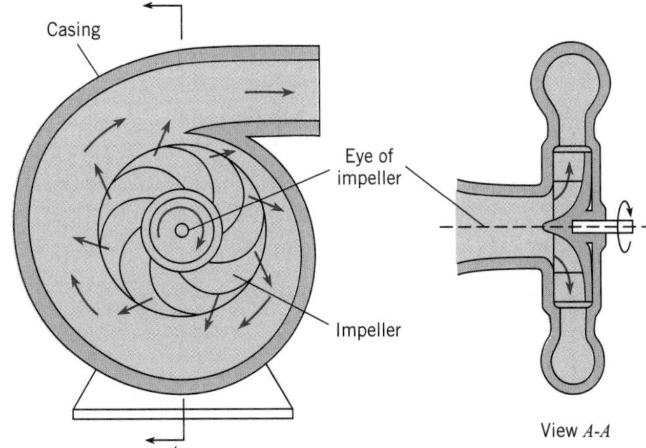

FIGURE 14.8

Centrifugal pump.

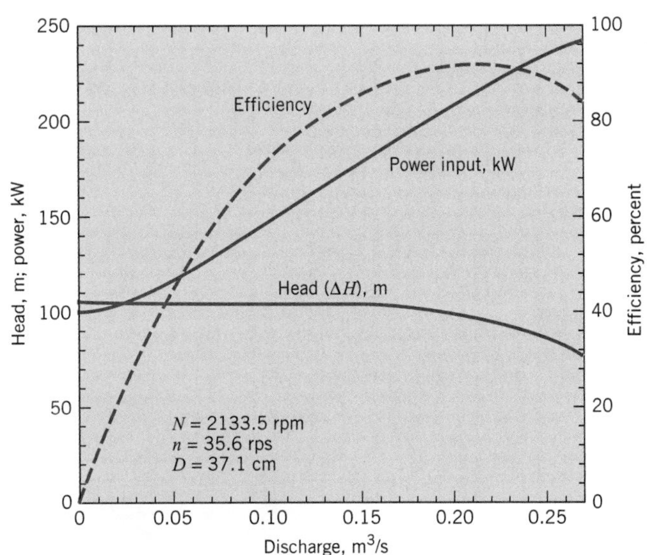

FIGURE 14.9

Performance curves for a typical centrifugal pump; D = 37.1 cm. [After Daugherty and Franzini (4). Used with the permission of the McGraw-Hill Companies.]

than that required for flow at maximum efficiency. Normally, the motor to drive the pump is chosen for conditions of maximum pump efficiency. Hence the flow can be throttled between the limits of shutoff condition and normal operating conditions without any chance of overloading the pump motor. Such is not the case for an axial-flow pump, as seen in Fig. 14.6. In that case, when the pump flow is throttled below maximum-efficiency conditions, the required power increases with decreasing flow, thus leading to the possibility of overloading at low-flow conditions. For very large installations, special operating procedures are followed in order to avoid such overloading. For instance, the valve in the bypass from the pump discharge back to the pump inlet can be adjusted to maintain a constant flow through the pump. However, for small-scale applications, it is often desirable to have complete flexibility in flow control without the complexity of special operating procedures. In this latter case, a radial-flow pump offers a distinct advantage. Radial-flow pumps are manufactured in sizes from 1 hp or less and heads of 50 or 60 ft to thousands of horsepower and heads of several hundred feet. Figure 14.11 shows a cutaway view of a single-suction, single-stage, horizontal-shaft radial pump. Another common design has flow entering the impeller from both sides, as shown in Figs. 14.12 and 14.13. Such a *double-suction impeller* is equivalent to two single-suction impellers placed back to back and made as a single casting. This arrangement gives balanced end thrust on the shaft of the impeller.

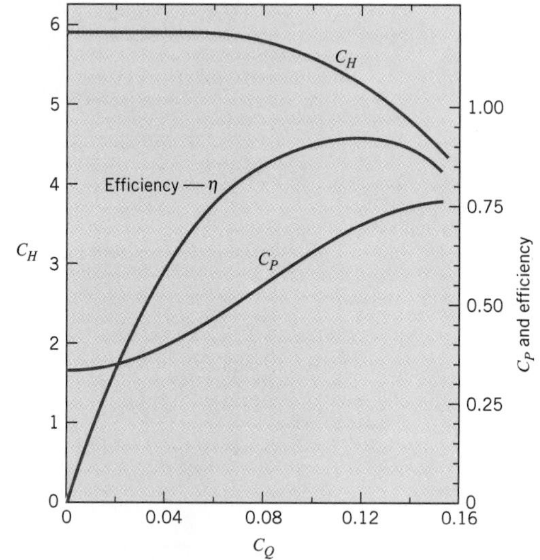

FIGURE 14.10

Dimensionless performance curves for a typical centrifugal pump, from data given in Fig. 14.9. [After Daugherty and Franzini (4)]

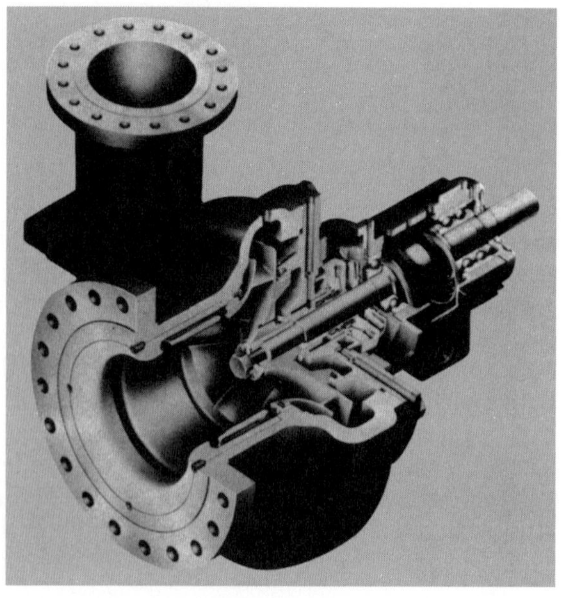

FIGURE 14.11

Cutaway view of a single-suction, single-stage, horizontal-shaft radial pump. (Courtesy of Ingersol Rand Co.)

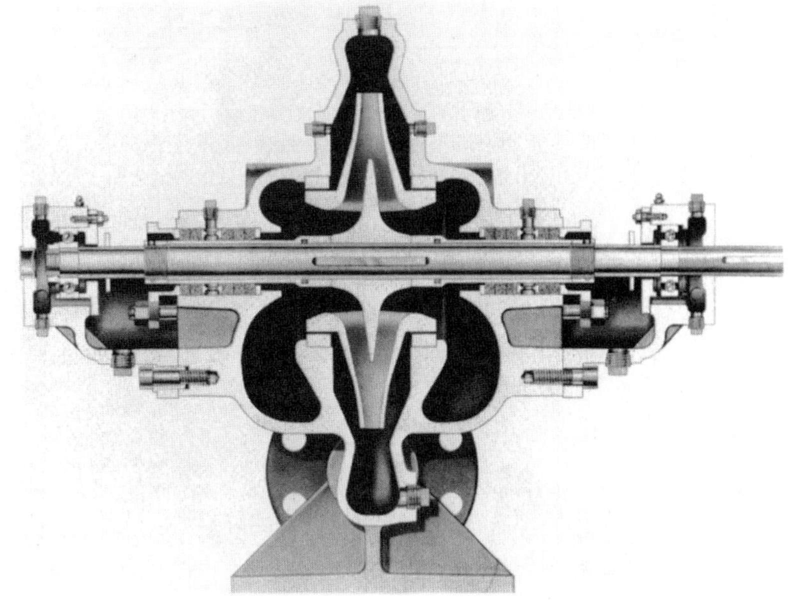

FIGURE 14.12

Sectional view of double-suction, single-stage, horizontal-shaft split-casing pump. (Courtesy of Dresser Pump Division.)

FIGURE 14.13

Overall outside view of a double-suction, single-stage, horizontal-shaft split-casing pump. (Courtesy of Dresser Pump Division.)

example 14.4

A pump that has the characteristics given in Fig. 14.9 when operated at 2133.5 rpm is to be used to pump water at maximum efficiency under a head of 76 m. At what speed should the pump be operated, and what will the discharge be for these conditions?

Solution Since the diameter is fixed, the only change that will occur results from the change in speed (assuming negligible change due to viscous effects). The C_H, C_P, C_Q, and η for this pump operating at maximum efficiency against a head of 76 m are the same as for its operation at maximum efficiency with a speed of rotation of 2133.5 rpm, since both operating conditions correspond to the point of maximum efficiency in Fig. 14.10. Thus we can write

$$(C_H)_N = (C_H)_{2133.5\text{ rpm}}$$

Here N refers to the speed of rotation with $\Delta H = 76\text{ m}$. The graph of Fig. 14.9 indicates that $\Delta H = 90\text{ m}$ and $Q = 0.255\text{ m}^3/\text{s}$ at maximum efficiency for $N = 2133.5$ rpm. Thus

$$\frac{76\text{ m}}{N^2} = \frac{90\text{ m}}{(2133.5)^2}$$

$$N^2 = (2133.5)^2\left(\frac{76}{90}\right)$$

$$N = 2133.5\left(\frac{76}{90}\right)^{1/2} = 1960\text{ rpm} \quad \triangleleft$$

Using $(C_Q)_{1960} = (C_Q)_{2133.5\text{ rpm}}$ and solving for the ratio of discharge, we have

$$\frac{Q_{1960}}{Q_{2133.5}} = \frac{1960}{2133.5} = 0.919$$

$$Q_{1960} = 0.207\text{ m}^3/\text{s} \quad \triangleleft$$

example 14.5

The pump having the characteristics shown in Figs. 14.9 and 14.10 is a model of a pump that was actually used in one of the pumping plants of the Colorado River Aqueduct [see Daugherty and Franzini (4)]. For a prototype that is 5.33 times larger than the model and operates at a speed of 400 rpm, what head, discharge, and power are to be expected at maximum efficiency?

Solution From Fig. 14.10 we find values of 0.12, 5.2, and 0.69 for C_Q, C_H, and C_P, respectively, for the maximum-efficiency condition. Then for n = (400/60) rps and $D = 0.371 \times 5.33 = 1.98$ m, we solve for P, ΔH, and Q:

$$P = C_P \rho D^5 n^3 = 0.69(10^3 \text{ kg/m}^3)(1.98 \text{ m})^5 \left(\frac{400}{60}\text{s}^{-1}\right)^3 = 6200 \text{ kW} \quad \triangleleft$$

$$\Delta H = \frac{C_H D^2 n^2}{g} = \frac{5.2(1.98 \text{ m})^2 (400/60 \text{ s}^{-1})^2}{(9.81 \text{ m/s}^2)} = 92.4 \text{ m} \quad \triangleleft$$

$$Q = C_Q n D^3 = 0.12\left(\frac{400}{60}\text{ s}^{-1}\right)(1.98 \text{ m})^3 = 6.21 \text{ m}^3/\text{s} \quad \triangleleft$$

14.4

Specific Speed

From the discussion in preceding sections we have seen that a pump's performance is given by the values of its power and head coefficients (C_P and C_H) for a range of values of the discharge coefficient C_Q. It has been noted that certain types of machines are best suited for certain head and discharge ranges. For example, an axial-flow machine is best suited for low heads and high discharges, whereas a radial-flow machine is best suited for higher heads and lower discharges. The parameter used to pick the type of pump (or turbine) best suited for a given application is specific speed n_s. Specific speed is obtained by combining both C_H and C_Q in such a manner that the diameter D is eliminated:

$$n_s = \frac{C_Q^{1/2}}{C_H^{3/4}} = \frac{(Q/nD^3)^{1/2}}{[\Delta H/(D^2 n^2/g)]^{3/4}} = \frac{nQ^{1/2}}{g^{3/4}\Delta H^{3/4}}$$

Thus specific speed relates different types of pumps without reference to their sizes.

When efficiencies of different types of pumps are plotted against n_s, it is seen that certain types of pumps have higher efficiencies for certain ranges of n_s. In fact, in the range between the completely axial-flow machine and the completely radial-flow machine, there is a gradual change in impeller shape to accommodate the particular flow conditions with maximum efficiency (see Fig. 14.14).

It should be noted that the specific speed traditionally used for pumps in the United States is defined as $N_s = NQ^{1/2}/\Delta H^{3/4}$. Here the speed N is in revolutions per minute, Q is in gallons per minute, and ΔH is in feet. This form is not dimensionless. Therefore its values are much larger than those found for n_s (the conversion factor is 17,200). Most texts and references published before the introduction of the SI system of units use this traditional definition for specific speed.

FIGURE 14.14

Optimum efficiency and impeller design versus specific speed.

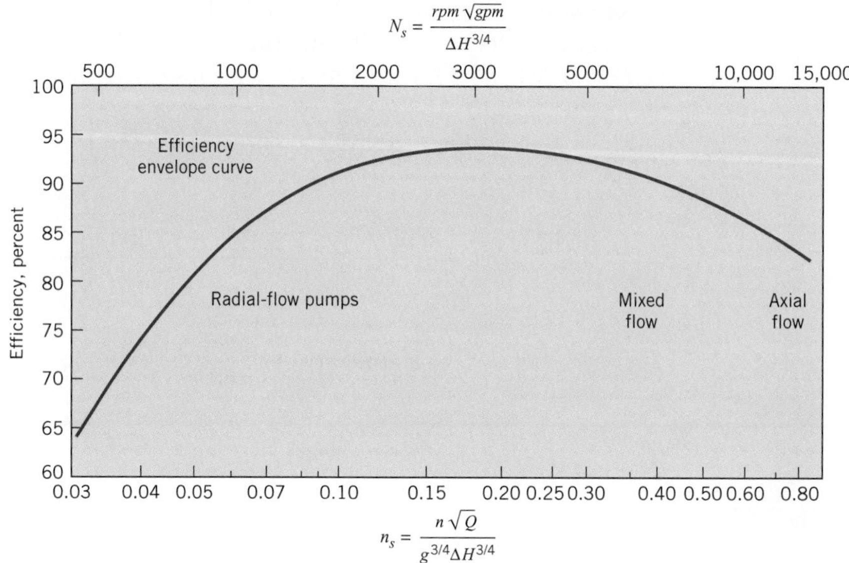

(*a*) Optimum efficiency and impeller designs versus specific speed n_s

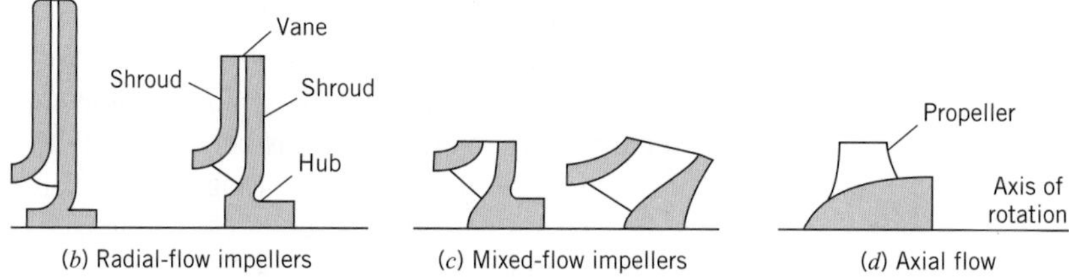

(*b*) Radial-flow impellers (*c*) Mixed-flow impellers (*d*) Axial flow

example 14.6

What type of pump should be used to pump water at the rate of 10 cfs and under a head of 600 ft? Assume $N = 1100$ rpm.

Solution In order to find the best configuration, we need to evaluate the specific speed. The rotational rate in rps is

$$n = \frac{1100}{60} = 18.33 \text{ rps}$$

Substituting the values for the specific speed gives

$$n_s = \frac{n\sqrt{Q}}{(gh)^{3/4}}$$

$$= \frac{18.33 \text{ rps} \times (10 \text{ cfs})^{1/2}}{(32.2 \text{ ft/s}^2 \times 600 \text{ ft})^{3/4}} = 0.035$$

From Fig. 14.14, we note that if the specific speed is less than 0.2 a radial flow pump should be used.

Suction Limitations of Pumps

The pressure at the suction side of a pump is most significant as to whether or not a pump will cavitate. As water flows past the impeller blades of a pump, local high-velocity flow zones produce low relative pressures (Bernoulli effect), and if these pressures reach the vapor pressure of the liquid, then cavitation will occur. For a given type of pump, operating at a given speed and a given discharge, there will be certain pressure at the suction side of the pump below which cavitation will occur. Pump manufacturers in their testing procedures always determine this limiting pressure and include it with their pump performance curves.

More specifically, the pressure that is significant is the difference in pressure between the suction side of the pump and the vapor pressure of the liquid being pumped. Actually, in practice, engineers express this difference in terms of pressure head, called the *net positive suction head,* which is abbreviated *NPSH*. To calculate NPSH for a pump that is delivering a given discharge, one first applies the energy equation from the reservoir from which water is being pumped to the section of the intake pipe at the suction side of the pump. Then subtract the vapor pressure head of the water to determine NPSH.

In Fig. 14.15, points 1 and 2 are the points between which the energy equation would be written to evaluate NPSH.

FIGURE 14.15

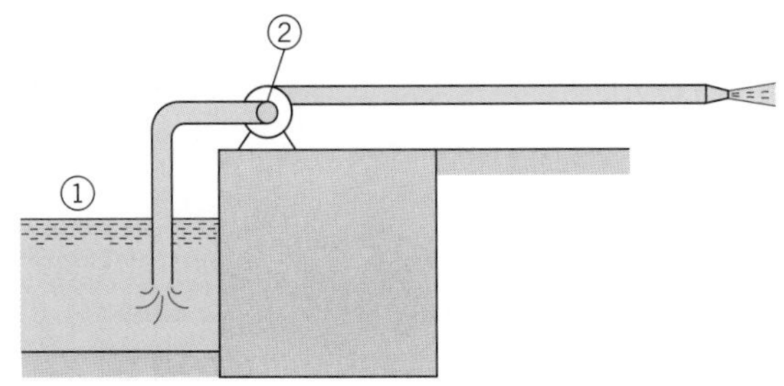

example 14.7

In Fig. 14.15 the pump delivers 2 cfs flow of 80°F water and the intake pipe diameter is 8 in. The pump intake is located 6 ft above the water surface level in the reservoir. What is the net positive suction head for these conditions?

Solution First write the energy equation from point 1 to point 2:

$$\frac{p_1}{\gamma} + \frac{V_1^2}{2g} + z_1 = \frac{p_2}{\gamma} + \frac{V_2^2}{2g} + z_2 + \sum h_L$$

where $p_1/\gamma = 34$ ft (assumed absolute atmospheric pressure head)

$$V_1 = 0$$

$$z_1 = 0$$

$$V_2 = Q/A = 2/0.349 = 5.73 \text{ ft/s}$$

$$V_2^2/2g = 0.510 \text{ ft}$$

$$z_2 = 6 \text{ ft}$$

Therefore $p_2/\gamma = 34 - 0.51 - 6.0 - \sum h_L$

Assume entrance loss coefficient = 0.10
Assume bend loss coefficient = 0.20
Neglect pipe friction loss

$$\sum h_L = \sum KV^2/2g = 0.3 \times 0.51 = 0.15 \text{ ft}$$

Thus $$p_2/\gamma = 34 - 0.51 - 6.0 - 0.15$$

$$p_2/\gamma = 27.3 \text{ ft}$$

The vapor pressure of 80°F water = 0.506 psia, so the vapor pressure head = $0.506 \times 144/62.4 = 1.17$ ft.

$$\text{NPSH} = \frac{p_2}{\gamma} - h_{\text{vap. press.}}$$

$$= 27.3 - 1.17$$

$$\approx 26.1 \text{ ft}$$

For a typical single-stage centrifugal pump with an intake diameter of 8 in. and pumping 2 cfs, the critical NPSH is normally about 10 ft; therefore, the pump of this example is operating well within the safe range with respect to cavitation susceptibility.

A more general parameter for indicating susceptibility to cavitation is specific speed. However, instead of using head produced (ΔH), one uses NPSH for the variable to the 3/4 power. This is

$$n_{ss} = \frac{nQ^{1/2}}{g^{3/4}(\text{NPSH})^{3/4}}$$

Here n_{ss} is called the suction specific speed. The more traditional suction specific speed used in the United States is $N_{ss} = NQ^{1/2}/(\text{NPSH})^{3/4}$, where N is in rpm, Q is in gpm, and NPSH is in feet. Analyses of data from pump tests show that the value of the suction specific speed is a good indicator of whether cavitation may be expected. For example, the Hydraulic Institute (5) indicates that the critical value of N_{ss} is 8500. The reader is directed to manufacturer's data or the Hydraulic Institute for more details about critical NPSH or N_{ss}.

example 14.8

For the pump of Example 14.7 what is the value of N_{ss}? Assume $N = 1750$ rpm.

$$Q = 2 \text{ cfs} = 898 \text{ gpm}$$

$$\text{NPSH} = 26.1 \text{ ft}$$

$$N_{ss} = (1750)(898)^{1/2}/(26.1)^{3/4}$$

$$N_{ss} = 4541$$

◁

This value of N_{ss} is much below the critical limit of 8500; therefore, it is in a safe operating range so far as cavitation is concerned.

A typical pump performance curve for a centrifugal pump that would be supplied by a pump manufacturer is shown in Fig. 14.16. The solid lines labeled from 5 in. to 7 in. represent different impeller sizes that can be accommodated by the pump housing. These curves give the head delivered as a function of discharge. The dashed lines represent the power required by the pump for a given head and discharge. Lines of constant efficiency are also shown. Obviously, when selecting an impeller you would like to have the operating point as close as possible to the point of maximum efficiency. The NPSH value gives the minimum head (absolute head) at the pump intake for which the pump will operate without cavitation.

Centrifugal Compressors

Centrifugal compressors are similar in design to centrifugal pumps. Because the density of the air or gases used is much less than the density of a liquid, the compressor must turn at much higher speeds than the pump does to effect a sizable pressure increase. If the

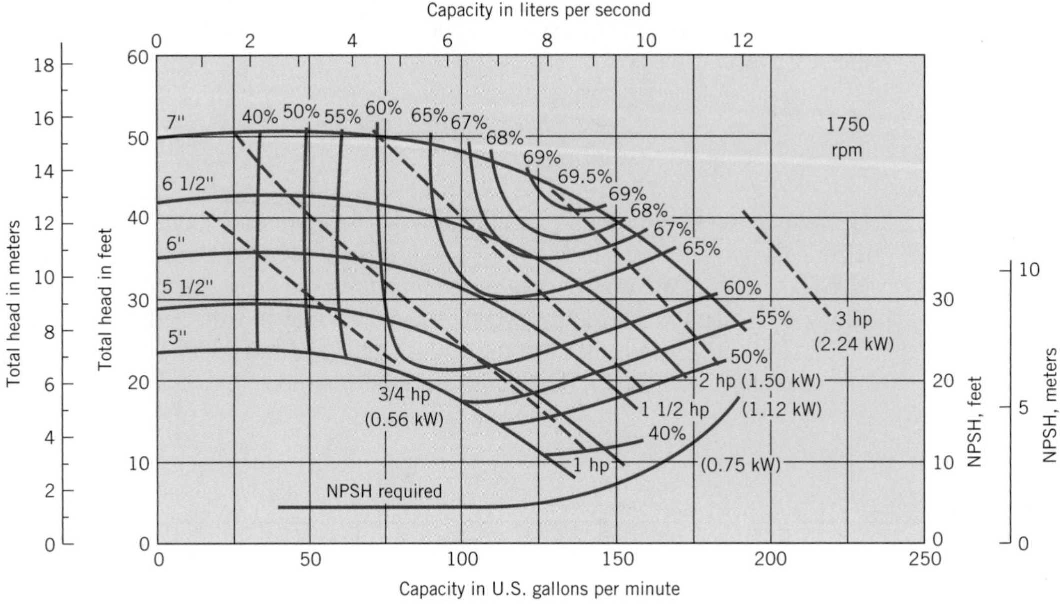

FIGURE 14.16

Centrifugal pump performance curve. [After McQuiston and Parker (6). Used with permission of John Wiley and Sons.]

compression process were isentropic and the gases ideal, the power necessary to compress the gas from p_1 to p_2 would be

$$P_{\text{theo}} = \frac{k}{k-1} Q_1 p_1 \left[\left(\frac{p_2}{p_1} \right)^{(k-1)/k} - 1 \right] \tag{14.17}$$

where Q_1 is the volume flow rate into the compressor. The power calculated using Eq. (14.17) is referred to as the *theoretical adiabatic power.* The efficiency of a compressor with no water cooling is defined as the ratio of the theoretical adiabatic power to the actual power required at the shaft. Ordinarily the efficiency improves with higher inlet-volume flow rates, increasing from a typical value of 0.60 at 0.6 m^3/s to 0.74 at 40 m^3/s. Higher efficiencies are obtainable with more expensive design refinements.

example 14.9

Determine the shaft power required to operate a compressor that compresses air at the rate of 1 m^3/s from 100 kPa to 200 kPa. The efficiency of the compressor is 65%.

Solution We first calculate the theoretical adiabatic power using Eq. (14.17) with $k = 1.4$.

$$P_{\text{theo}} = \frac{k}{k-1} Q_1 p_1 \left[\left(\frac{p_2}{p_1} \right)^{(k-1)/k} - 1 \right]$$

$$= (3.5)(1\ \text{m}^3/\text{s})(10^5\ \text{N/m}^2)[(2)^{0.286} - 1]$$

$$= 0.767 \times 10^5\ \text{N} \cdot \text{m/s} = 76.7\ \text{kW}$$

The shaft power required is

$$P_{\text{shaft}} = \frac{76.7}{0.65}\ \text{kW} = 118\ \text{kW}$$

◁

Cooling is necessary for high-pressure compressors because of the high gas temperatures resulting from the compression process. Cooling can be achieved through the use of water jackets or intercoolers that cool the gases between stages. The efficiency of water-cooled compressors is based on the power required to compress ideal gases isothermally, or

$$P_{\text{theo}} = p_1 Q_1 \ln \frac{p_2}{p_1} \qquad (14.18)$$

which is usually called the *theoretical isothermal power*. The efficiencies of water-cooled compressors are generally lower than those of noncooled compressors. If a compressor is cooled by water jackets, its efficiency characteristically ranges between 55% and 60%. The use of intercoolers results in efficiencies from 60% to 65%.

Application to Fluid Systems

The selection of a pump, fan, or compressor for a specific application depends on the desired flow rate. This process requires the acquisition or generation of a system curve for the flow system of interest and a performance curve for the fluid machine. The intersection of these two curves provides the operating point as discussed in Chapter 10.

As an example, consider using the centrifugal pump with the characteristics shown in Fig. 14.16 to pump water at 60°F from a wall into a tank as shown in Fig. 14.17. A pumping capacity of at least 80 gpm is required. Two hundred feet of standard schedule 40 2-in. galvanized iron pipe are to be used. There is a check valve in the system as well as an open gate valve. There is a 20-ft elevation between the well and the top of the fluid in the tank. Applying the energy equation developed in Chapter 7, the head required by the pump is

$$h_p = \Delta z + \frac{V^2}{2g}\left(\frac{fL}{D} + \sum K_L \right)$$

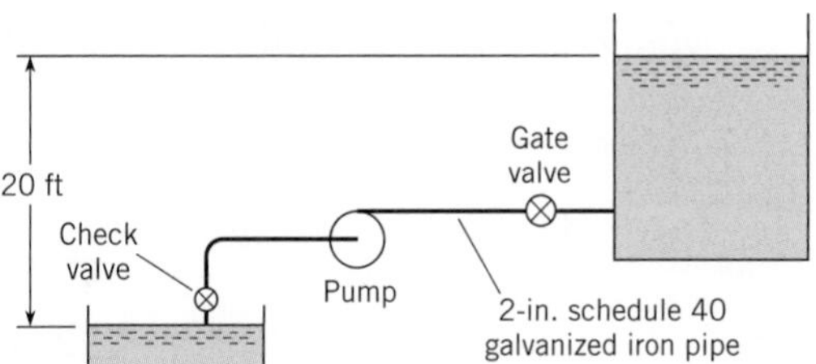

FIGURE 14.17

System for pumping water from a well into a tank.

where K_L are the coefficients of head loss for the entrance, check valve, gate valve, and sudden expansion loss entering the tank. Using representative values for the loss coefficients and evaluating the friction factor from the Moody diagram in Chapter 10, we find

$$h_p = 20 + 0.00305Q^2$$

where Q is the flow rate in gpm. This is the system curve.

The result of plotting the system curve on the pump performance curves is shown in Fig. 14.18. The locations where the lines cross are the operating points. We notice that a discharge of just over 80 gpm is achieved with the 6.5-in. impeller. Also, referring back to Fig. 14.16, the efficiency at this point is about 62%. To ensure that the design requirements are satisfied, the engineer may select the larger impeller, which has an operating point of 95 gpm. If the pump is to be used in continuous operation and the efficiency is important to operating costs, the engineer may choose to consider another pump that would have a higher efficiency at the operation point. An engineer experienced in the design of pump systems would be very familiar with the trade-offs for economy and performance and could make a design decision relatively quickly.

In some systems it may be advantageous to use two pumps in series or in parallel. If two pumps are used in series, the performance curve is the sum of the pump heads of the two machines at the same flow rate, as shown in Fig. 14.19*a*. This configuration would be desirable for a flow system with a steep system curve as shown in the figure. If two pumps are connected in parallel, the performance curve is obtained by adding the flow rates of the two pumps at the same pump heads, as shown in Fig. 14.19*b*. This configuration would be advisable for flow systems with shallow system curves as shown in the figure. The concepts presented here for pumps also apply to fans and compressors.

Turbines

A *turbine* is defined as a machine that extracts energy from a moving fluid. Much of the basic theory and most similarity parameters used for pumps also apply to turbines. However, there are some differences in physical features and terminology. Also, we have not yet considered details of the flow through the impellers of radial-flow machines. These topics will now be discussed.

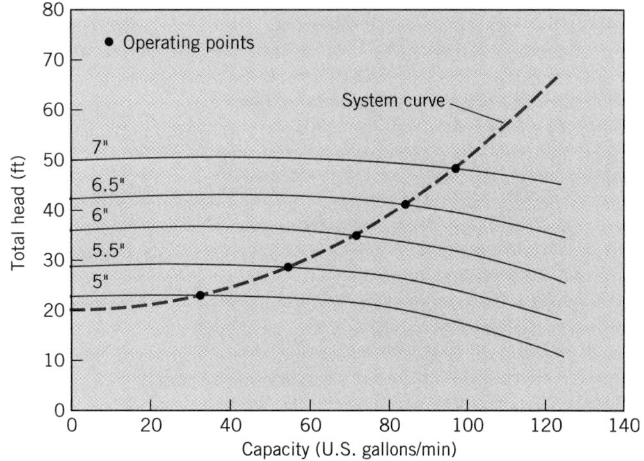

FIGURE 14.18

System and pump performance curves for pumping application.

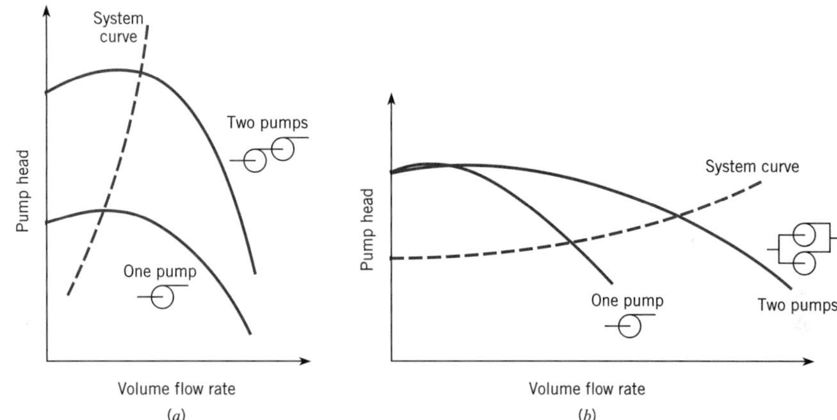

FIGURE 14.19

Pump performance curves for pumps connected in series (a) and in parallel (b).

The two main categories of hydraulic machines are the *impulse* and *reaction* turbines. In a reaction turbine, the water flow is used to rotate a turbine wheel or runner through the action of vanes or blades attached to the wheel. When the blades are oriented like a propeller, the flow is axial and the machine is called a Kaplan turbine. When the vanes are oriented like an impeller in a centrifugal pump, the flow is radial and the machine is called a Francis turbine. In an impulse turbine, the water accelerates through a nozzle and impinges on vanes attached to the rim of the wheel. This machine is called a Pelton wheel.

Impulse Turbine

In the impulse turbine a jet of fluid issuing from a nozzle impinges on vanes of the turbine wheel or *runner,* thus producing power as the runner rotates (see Fig. 14.20). Figure 14.21 shows a runner for the Henry Borden hydroelectric plant in Brazil. The

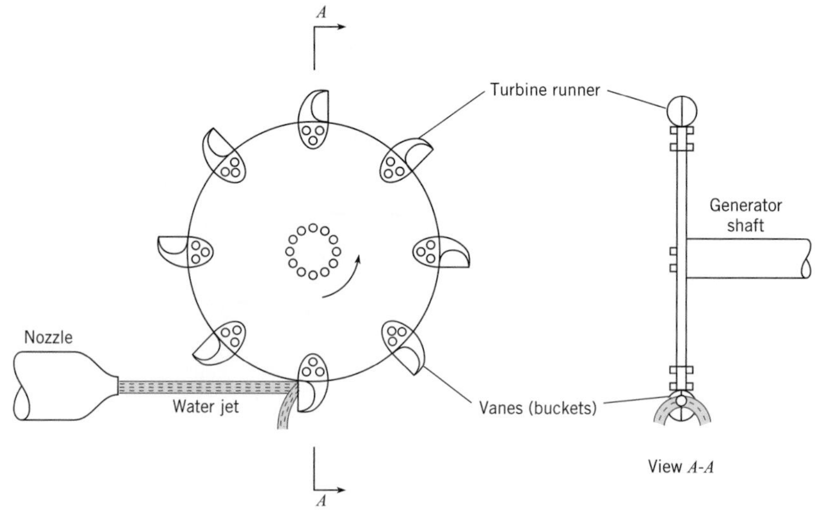

FIGURE 14.20

Impulse turbine.

FIGURE 14.21

Spare runner for the Henry Borden power plant in Brazil. (Courtesy of Voith Hydro Inc.)

primary feature of the impulse turbine with respect to fluid mechanics is the power production as the jet is deflected by the moving vanes. When the momentum equation is applied to this deflected jet, it can be shown [see Daugherty and Franzini (3)] for idealized conditions that the maximum power will be developed when the vane speed is one-half of the initial jet speed. With such conditions the exiting jet speed will be zero—all of the kinetic energy of the jet will have been expended in driving the vane. Thus if we apply the energy equation, Eq. (7.24), between the incoming jet and the exiting fluid (assuming negligible head loss and negligible kinetic energy at exit), we find that the head given up to the turbine is $h_t = V_j^2/2g$ and the power thus developed is

$$P = Q\gamma h_t \tag{14.19}$$

where Q is the discharge of the incoming jet, γ is the specific weight of jet fluid, and $h_t = V_j^2/2g$, or the velocity head of the jet. Thus Eq. (14.19) reduces to

$$P = \rho Q \frac{V_j^2}{2} \tag{14.20}$$

To obtain the torque on the turbine shaft, we apply the angular-momentum equation, Eq. (6.18), to a control volume, as shown in Fig. 14.22. Then for steady flow we have

$$\sum M = \sum_{cs} \mathbf{r}_o \times (\dot{m}_o \mathbf{v}_o) - \sum_{cs} \mathbf{r}_i \times (\dot{m}_i \mathbf{v}_i)$$

We assume that the exiting fluid has negligible angular momentum. The moment acting on the system is the torque T acting on the shaft. Thus the angular momentum equation reduces to

$$T = -\dot{m} r V_j \tag{14.21}$$

The mass flow rate across the control surface is ρQ, so the torque is

$$T = -\rho Q V_j r$$

The minus sign indicates that the torque applied to the system (to keep it rotating at constant angular velocity) is in the clockwise direction. However, the torque applied by the system to the shaft is in the counterclockwise direction, which is the direction of wheel rotation, so

$$T = \rho Q V_j r \tag{14.22}$$

The power developed by the turbine is $T\omega$, or

$$P = \rho Q V_j r \omega \tag{14.23}$$

Furthermore, if the velocity of the turbine vanes is $(1/2)V_j$ for maximum power, as noted earlier, we have $P = \rho Q V_j^2/2$, the same as Eq. (14.20).

FIGURE 14.22

Control-volume approach for the impulse turbine using the angular-momentum principle.

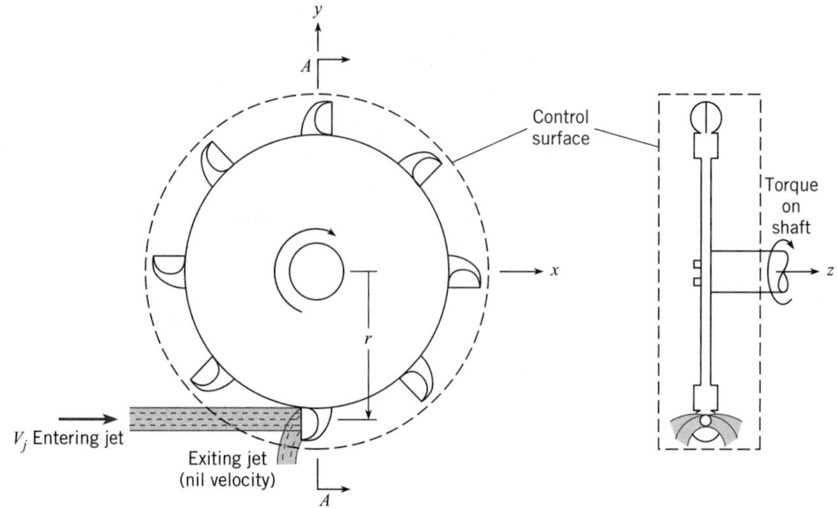

example 14.10

What power in kilowatts can be developed by the impulse turbine shown if the turbine efficiency is 85%? Assume that the resistance coefficient f of the penstock is 0.015 and the head loss in the nozzle itself is negligible. What will be the angular speed of the wheel, assuming ideal conditions ($V_j = 2V_{\text{bucket}}$), and what torque will be exerted on the turbine shaft?

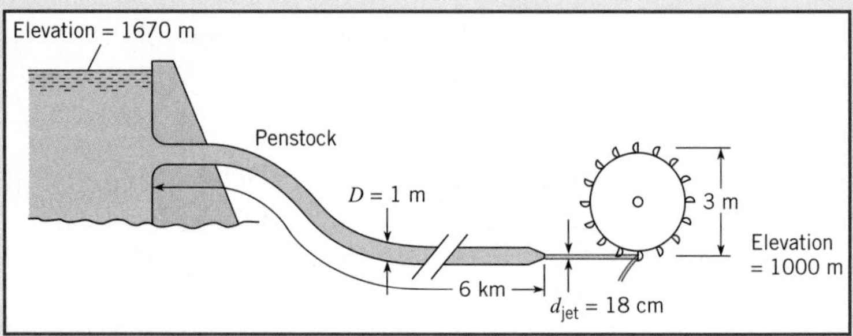

Solution First determine the jet velocity by applying the energy equation from the reservoir to the free jet before it strikes the turbine buckets.

$$\frac{p_1}{\gamma} + \frac{V_1^2}{2g} + z_1 = \frac{p_j}{\gamma} + \frac{V_j^2}{2g} + z_j + h_L$$

where $p_1 = 0$ $\quad p_j = 0$

$z_1 = 1670$ m $\quad z_j = 1000$ m

$V_1^2/2g = 0$ $\quad \gamma = 9810(\text{N/m}^3)$ at 10°C (assumed)

The penstock water velocity is

$$V_{\text{penstock}} = \frac{V_j A_j}{A_{\text{penstock}}} = 0.0324 V_j$$

Then
$$h_L = \frac{fL}{D}\frac{V^2}{2g} = \frac{0.015 \times 6000}{1}(0.0324)^2\frac{V_j^2}{2g} = 0.094\frac{V_j^2}{2g}$$

Now, solving the energy equation for V_j yields

$$V_j = \left(\frac{2g \times 670}{1.094}\right)^{1/2} = 109.6 \text{ m/s}$$

The gross power is

$$P = Q\gamma\frac{V_j^2}{2g} = \frac{\gamma A_j V_j^3}{2g}$$

$$= \frac{9810(\pi/4)(0.18)^2(109.6)^3}{2 \times 9.81} = 16{,}750 \text{ kW}$$

The power output of the turbine is

$$P = 16{,}750 \times \text{efficiency} = 14{,}238 \text{ kW} \quad \triangleleft$$

The tangential bucket speed will be $(1/2)V_j$. Therefore

$$V_{\text{bucket}} = \tfrac{1}{2}(109.6 \text{ m/s}) = 54.8/\text{s}$$

or
$$r\omega = 54.8 \text{ m/s}$$

Thus
$$\omega = \frac{54.8 \text{ m/s}}{1.5 \text{ m}} = 36.53 \text{ rad/s}$$

The wheel speed is

$$N = (36.53 \text{ rad/s})\frac{1 \text{ rev}}{2\pi \text{ rad}}(60 \text{ s/min}) = 349 \text{ rpm} \quad \triangleleft$$

Finally,
$$\text{Power} = T\omega$$

Thus
$$T = \frac{\text{power}}{\omega} = \frac{14{,}238 \text{ kW}}{36.53 \text{ rad/s}} = 390 \text{ kN} \cdot \text{m} \quad \triangleleft$$

Reaction Turbine

In contrast to the impulse turbine, where a jet under atmospheric pressure impinges on only one or two vanes at a time, flow in a reaction turbine is under pressure and reacts on all vanes of the impeller turbine simultaneously. Also, this flow completely fills the chamber in which the impeller is located (see Fig. 14.23). There is a drop in pressure from the outer radius of the impeller, r_1, to the inner radius, r_2. This is another point of difference with the impulse turbine, in which the pressure is the same for the entering and exiting flows. The original form of the reaction turbine, first extensively tested by J. B. Francis, had a completely radial-flow impeller (Fig. 14.24). That is, the flow passing through the impeller had velocity components only in a plane normal to the axis of the runner. However, more recent impeller designs, such as the mixed-flow and axial-flow types, are still called reaction turbines.

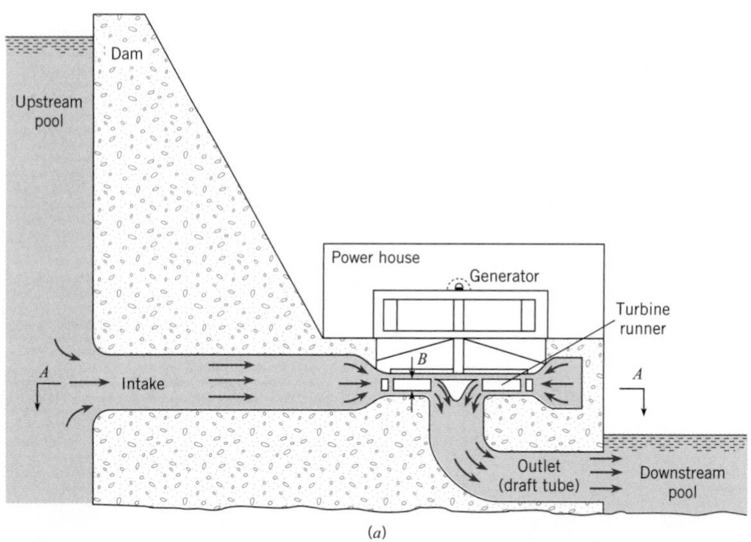

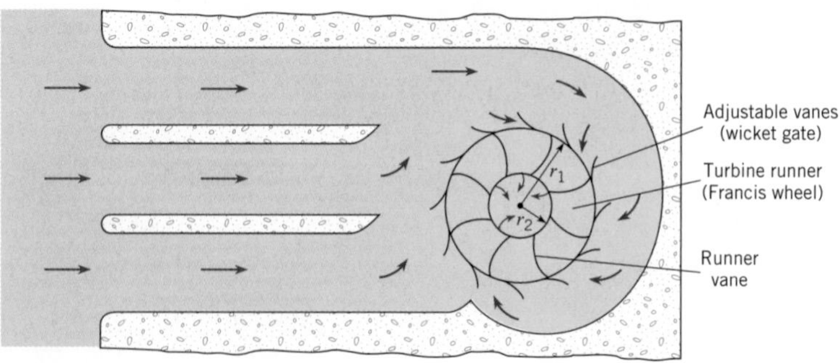

FIGURE 14.23

Schematic view of a reaction-turbine installation. (a) Elevation view. (b) Plan view, section A-A.

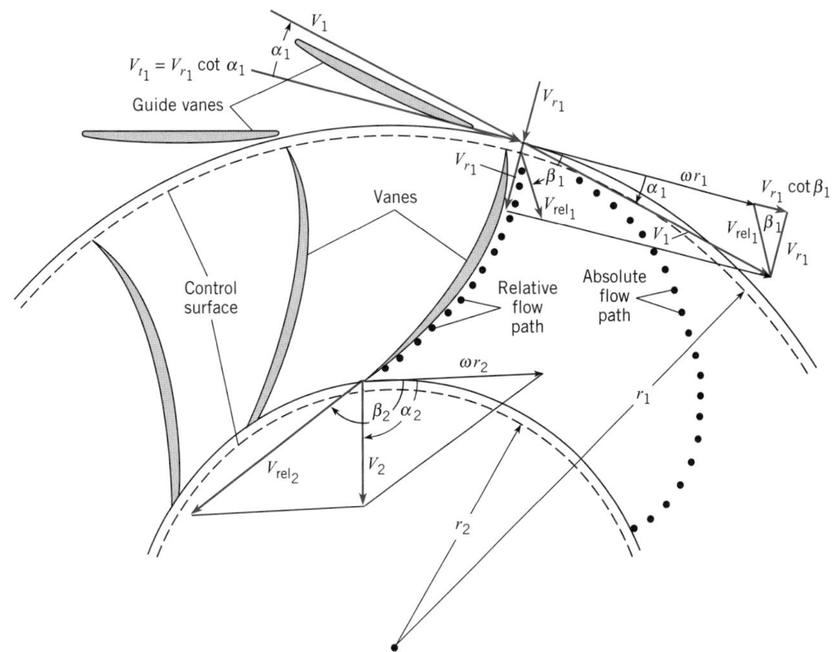

FIGURE 14.24

Velocity diagrams for the impeller for a Francis turbine.

Torque and Power Relations for the Reaction Turbine

As we did for the impulse turbine, we will use the angular-momentum equation to develop formulas for the torque and power for the reaction turbine. The segment of turbine runner shown in Fig. 14.24 depicts the flow conditions that occur for the entire runner. We can see that guide vanes outside the runner itself cause the fluid to have a tangential component of velocity around the entire circumference of the runner. Thus the fluid has an initial amount of angular momentum with respect to the turbine axis when it approaches the turbine runner. As the fluid passes through the passages of the runner, the runner vanes effect a change in the magnitude and direction of its velocity. Thus the angular momentum of the fluid is changed, which produces a torque on the runner. This torque drives the runner, which, in turn, generates power.

To quantify the above, we let V_1 and α_1 represent the incoming velocity and the angle of the velocity vector with respect to a tangent to the runner, respectively. Similar terms at the inner-runner radius are V_2 and α_2. Applying the angular-momentum equation for steady flow, Eq. (6.18), to the control volume shown in Fig. 14.24 yields

$$T = \dot{m}(-r_2V_2 \cos\alpha_2) - \dot{m}(-r_1V_1 \cos\alpha_1)$$

$$= \dot{m}(r_1V_1 \cos\alpha_1 - r_2V_2 \cos\alpha_2) \tag{14.24}$$

The power from this turbine will be $T\omega$, or

$$P = \rho Q\omega(r_1V_1 \cos\alpha_1 - r_2V_2 \cos\alpha_2) \tag{14.25}$$

Equation (14.25) shows that the power production is a function of the directions of the flow velocities entering and leaving the impeller—that is, α_1 and α_2.

It is interesting to note that even though the pressure varies within the flow in a reaction turbine, it does not enter into the expressions we have derived using the angular-momentum equation. The reason it does not appear is that the outer and inner control surfaces we chose are concentric with the axis about which we are evaluating the moments and angular momentum. The pressure forces acting on these surfaces all pass through the given axis; therefore they do not produce moments about the given axis.

Vane Angles

It should be apparent that the head loss in a turbine will be less if the flow enters the runner with a direction tangent to the runner vanes than if the flow approaches the vane with an angle of attack. In the latter case, separation will occur with consequent head loss. Thus vanes of an impeller designed for a given speed and discharge and with fixed guide vanes will have a particular optimum blade angle β_1. However, if the discharge is changed from the condition of the original design, the guide vanes and impeller vane angles will not "match" the new flow condition. Most turbines for hydroelectric installations are made with movable guide vanes on the inlet side to effect a better match at all flows. Thus α_1 is increased or decreased automatically through governor action to accommodate fluctuating power demands on the turbine.

To relate the incoming-flow angle α_1 and the vane angle β_1, we first assume that the flow entering the impeller is tangent to the blades at the periphery of the impeller. Likewise, the flow leaving the stationary guide vane is assumed to be tangent to the guide vane. In developing the desired equations, we will consider both the radial and the tangential components of velocity at the outer periphery of the wheel ($r = r_1$). We can easily compute the radial velocity, given Q and the geometry of the wheel, by the continuity equation:

$$V_{r_1} = \frac{Q}{2\pi r_1 B} \tag{14.26}$$

where B is the height of the turbine blades. The tangential (tangent to the outer surface of the runner) velocity of the incoming flow is

$$V_{t_1} = V_{r_1} \cot \alpha_1 \tag{14.27}$$

However, this tangential velocity is equal to the tangential component of the relative velocity in the runner, $V_{r_1} \cot \beta_1$, plus the velocity of the runner itself, ωr_1. Thus the tangential velocity, when viewed with respect to the runner motion, is

$$V_{t_1} = r_1\omega + V_{r_1} \cot \beta_1 \tag{14.28}$$

Now, by eliminating V_{t_1} between Eqs. (14.27) and (14.28), we have

$$V_{r_1} \cot \alpha_1 = r_1\omega + V_{r_1} \cot \beta_1 \tag{14.29}$$

Equation (14.29) can be rearranged to yield

$$\alpha_1 = \text{arccot}\left(\frac{r_1\omega}{V_{r_1}} + \cot\beta_1\right) \tag{14.30}$$

example 14.11

A Francis turbine is to be operated at a speed of 600 rpm and with a discharge of $4.0\ \text{m}^3/\text{s}$. If $r_1 = 0.60$ m, $\beta_1 = 110°$, and the blade height B is 10 cm, what should be the guide vane angle α_1 for a nonseparating flow condition at the runner entrance?

Solution

$$\alpha_1 = \text{arccot}\left(\frac{r_1\omega}{V_{r_1}} + \cot\beta_1\right)$$

where

$$r_1\omega = 0.6 \times 600\ \text{rpm} \times 2\pi\ \text{rad/rev} \times 1/60\ \text{min/s} = 37.7\ \text{m/s}$$

$$V_{r_1} = \frac{Q}{2\pi r_1 B} = \frac{4.00\ \text{m}^3/\text{s}}{2\pi \times 0.6\ \text{m} \times 0.10\ \text{m}} = 10.61\ \text{m/s}$$

$$\cot\beta_1 = -0.364$$

Then

$$\alpha_1 = \text{arccot}(3.55 - 0.364) = 17.4°$$

◁

Specific Speed for Turbines

Because of the attention focused on the production of power by turbines, the specific speed for turbines is defined in terms of power:

$$n_s = \frac{nP^{1/2}}{g^{3/4}\gamma^{1/2}h_t^{5/4}}$$

It should also be noted that large water turbines are innately more efficient than pumps. The reason for this is that as the fluid leaves the impeller of a pump, it decelerates appreciably over a relatively short distance. Also, because guide vanes are generally not used in the flow passages with pumps, large local velocity gradients develop, which in turn cause intense mixing and turbulence, thereby producing large head losses. In most turbine installations, the flow that exits the turbine runner is gradually reduced in velocity through a gradually expanding *draft tube,* thus producing a much smoother flow situation and less head loss than for the pump. For additional details of hydropower turbines, see Daugherty and Franzini (4).

Gas Turbines

The conventional gas turbine consists of a compressor that pressurizes the air entering the turbine and delivers it to a combustion chamber. The high-temperature, high-pressure gases resulting from combustion in the combustion chamber expand through a turbine, which both drives the compressor and delivers power. The theoretical efficiency (power delivered/rate of energy input) of a gas turbine depends on the pressure ratio between the combustion chamber and the intake; the higher the pressure ratio, the higher the efficiency. The reader is directed to Cohen et al. (7) for more detail.

Wind Turbines

Extraction of energy from the wind by a wind turbine is discussed frequently as an alternative energy source. In essence, the wind turbine is just a reverse application of the process of introducing energy into an airstream to derive a propulsive force. The wind turbine extracts energy from the wind to produce power. There is one significant difference, however. The theoretical upper limit of efficiency of a propeller supplying energy to an airstream is 100%; that is, it is theoretically possible, neglecting viscous and other effects, to convert all the energy supplied to a propeller into energy of the airstream. On the other hand, the theoretical upper limit of wind turbine efficiency as given by Glauert (8) is 16/27, or 59.3%. Thus the theoretical maximum power deliverable by a wind turbine with capture area A is

$$P_{\text{max}} = \frac{16}{27}\left(\frac{1}{2}\rho U^3 A\right)$$

where ρ is the air density and U is the wind speed. The capture area is the area swept by the wind turbine viewed from the wind direction. Other factors, such as swirl of the airstream and viscous effects, further reduce the wind turbine's efficiency.

The conventional wind turbine consists of a propeller mounted on a horizontal axis with a vane, or other device, to align the propeller shaft in the wind direction. In recent years considerable effort has been devoted to assessment of the Savonius rotor and the Darrieus turbine, both of which are vertical-axis turbines, as shown in Fig. 14.25. The Savonius rotor consists of two curved blades forming an S-shaped passage for the air flow. The Darrieus turbine consists of two or three airfoils attached to a vertical shaft; the unit resembles an egg beater. The advantage of vertical-axis turbines is that their operation is independent of wind direction. The Darrieus wind turbine is considered superior in performance but has a disadvantage in that it is not self-starting. Frequently, a Savonius rotor is mounted on the axis of a Darrieus turbine to provide the starting torque. For more information on wind turbines, refer to *Proceedings of the Cambridge Symposium on Wind Energy Systems* (9).

FIGURE 14.25

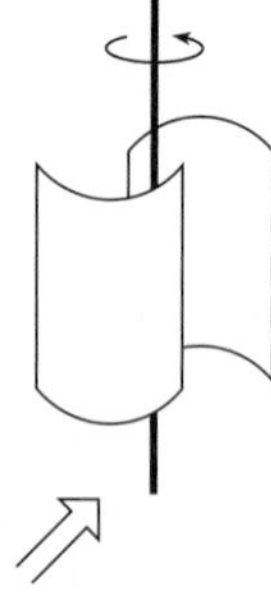

(*a*) Savonius rotor

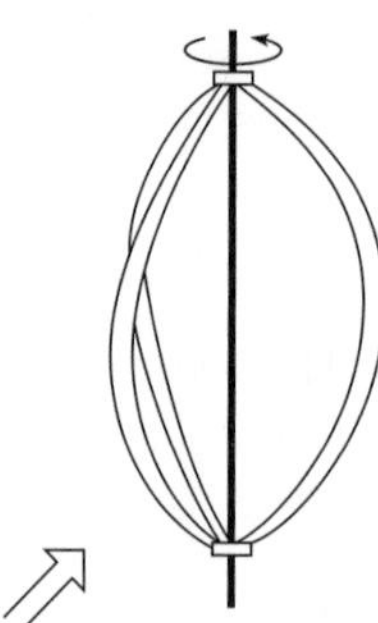

(*b*) Darrieus turbine

example 14.12

Calculate the minimum capture area necessary for a windmill that has to operate five 100-watt bulbs if the wind velocity is 20 km/h and the air density is 1.2 kg/m^3.

Solution The capture area is given by

$$A = P_{\max}\frac{54}{16}\frac{1}{\rho V^3}$$

A velocity of 20 km/h is equal to

$$20\ \text{km/h} = \frac{20 \times 1000}{3600} = 5.56\ \text{m/s}$$

The minimum capture area is

$$A = 500\ \text{W} \times \frac{54}{16} \times \frac{1}{1.2\ \text{kg/m}^3 \times (5.56\ \text{m/s})^3}$$

$$= 8.18\ \text{m}^2$$

14.8 Viscous Effects

In the foregoing sections, we developed similarity parameters to predict prototype results from model tests, neglecting viscous effects. The latter assumption is not necessarily valid, especially if the model is quite small. To minimize the viscous effects in modeling pumps, the Hydraulic Institute standards (5) recommend that the size of the model be such that the model impeller is not less than 30 cm in diameter. These same standards state that "the model should have complete geometric similarity with the prototype, not only in the pump proper, but also in the intake and discharge conduits."

Even with complete geometric similarity, one can expect the model to be less efficient than the prototype. An empirical formula proposed by Moody (10) is used for estimating prototype efficiencies of radial- and mixed-flow pumps and turbines from model efficiencies. That formula is

$$\frac{1-e_1}{1-e} = \left(\frac{D}{D_1}\right)^{1/5} \tag{14.31}$$

Here e_1 is the efficiency of the model and e is the efficiency of the prototype.

example 14.13

A model having an impeller diameter of 45 cm is tested and found to have an efficiency of 85%. If a geometrically similar prototype has an impeller diameter of 1.80 m, estimate its efficiency when it is operating under conditions that are dynamically similar to those in the model test ($C_{Q,\text{ model}} = C_{Q,\text{ prototype}}$).

Solution We apply Eq. (14.31) with the conditions $e = 0.85$ and $D/D_1 = 4$. Then

$$e = 1 - \frac{1 - e_1}{(D/D_1)^{1/5}} = 1 - \frac{0.15}{1.32} = 1 - 0.11 = 0.89$$

The efficiency of the prototype is estimated to be 89%. 

14.9 Summary

The thrust of a propeller is calculated using

$$F_T = C_T \rho n^2 D^4$$

where ρ is the fluid density, n is the rotational rate of the propeller, and D is the propeller diameter. The thrust coefficient C_T is a function of the advance ratio V_0/nD. The efficiency of a propeller is the ratio of the power delivered by the propeller to the power provided to the propeller.

$$\eta = \frac{F_T V_0}{P}$$

An axial-flow pump, or blower, consists of an impeller, much like a propeller, mounted in a housing. In a radial-flow, or centrifugal, pump, on the other hand, fluid enters near the eye of the impeller, passes through the vanes, and exits at the edge of the vanes. The head provided by a pump is quantified by the head coefficient, C_H, defined as

$$C_H = \frac{g\Delta H}{n^2 D^2}$$

where ΔH is the head across the pump. The head coefficient is a function of the discharge coefficient, which is

$$C_Q = \frac{Q}{nD^3}$$

where Q is the discharge. Pump performance curves show head delivered, power required, and efficiency as a function of discharge. The specific speed of a pump can be used to select an appropriate pump for a given application. Axial-flow pumps are best suited for high-discharge, low-head applications whereas radial-flow pumps are best suited for low-discharge, high-head applications.

Turbines convert the energy associated with a moving fluid to shaft work. The impulse turbine consists of a liquid jet impinging on vanes of a turbine wheel or runner. A reaction turbine consists of a series of rotating vanes where liquid enters from the outside and exits at the center. The pressure on the vanes provides the torque for the power. Wind turbines can be the conventional windmill design, the Darrieus turbine, or Savonius rotor. The maximum power derivable from a wind turbine is

$$P_{\max} = \frac{16}{27}\left(\frac{1}{2}\rho V_0^3 A\right)$$

where A is the capture area of the wind turbine (projected area from direction of wind) and V_0 is the wind speed.

References

1. Weick, F. E. *Aircraft Propeller Design*. McGraw-Hill, New York, 1930.

2. Weick, Fred E. "Full Scale Tests on a Thin Metal Propeller at Various Pit Speeds." *NACA Report*, 302 (January 1929).

3. Stepanoff, A. J. *Centrifugal and Axial Flow Pumps*, 2nd ed. John Wiley, New York, 1957.

4. Daugherty, Robert L., and Joseph B. Franzini. *Fluid Mechanics with Engineering Applications*. McGraw-Hill, New York, 1957.

5. Hydraulic Institute. *Centrifugal Pumps*. Hydraulic Institute, Parsippany, NJ, 1994.

6. McQuiston, F. C., and J. D. Parker. *Heating, Ventilating and Air Conditioning*. John Wiley, New York, 1994.

7. Cohen, H., G. F. C. Rogers, and H. I. H. Saravanamuttoo. *Gas Turbine Theory*. John Wiley, New York, 1972.

8. Glauert, H. "Airplane Propellers." *Aerodynamic Theory*, vol. IV, ed. W. F. Durand. Dover, New York, 1963.

9. *Proceedings of the Cambridge Symposium on Wind Energy Systems*. BHRA, Cranfield, England, 1976.

10. Moody, L. F. "Hydraulic Machinery." In *Handbook of Applied Hydraulics*, ed. C. V. Davis. McGraw-Hill, New York, 1942.

Problems

FLUID SOLUTIONS **14.1** What thrust is obtained from a propeller 3 m in diameter that has the characteristics given in Fig. 14.2 when the propeller is operated at an angular speed of 1400 rpm and an advance velocity of zero? Assume $\rho = 1.05 \text{ kg/m}^3$.

14.2 What thrust is obtained from a propeller 3 m in diameter that has the characteristics given in Fig. 14.2 when the propeller is operated at an angular speed of 1400 rpm and an advance velocity of 80 km/h? What power is required to operate the propeller under these conditions? Assume $\rho = 1.05 \text{ kg/m}^3$.

FLUID SOLUTIONS **14.3** A propeller 8 ft in diameter has the characteristics shown in Fig. 14.2. What thrust is produced by the propeller when it is operating at an angular speed of 1000 rpm and a forward speed of 25 mph? What power input is required under these operating conditions? If the forward speed is reduced to zero, what is the thrust? Assume $\rho = 0.0024 \text{ slugs/ft}^3$.

14.4 A propeller 8 ft in diameter, like the one for which characteristics are given in Fig. 14.2, is to be used on a swamp boat and is to operate at maximum efficiency when cruising. If the cruising speed is to be 30 mph, what should the angular speed of the propeller be?

14.5 For the propeller and conditions described in Prob. 14.4, determine the thrust and the power input.

FLUID SOLUTIONS **14.6** A propeller is being selected for an airplane that will cruise at 2000 m altitude, where the pressure is 60 kPa absolute and the temperature is 0°C. The mass of the airplane is 1200 kg, and the planform area of the wing is 10 m^2. The lift-to-drag ratio is 30:1. The lift coefficient is 0.4. The engine speed at cruise conditions is 3000 rpm. The propeller is to operate at maximum efficiency, which corresponds to a thrust coefficient of 0.025. Calculate the diameter of the propeller and the speed of the aircraft.

14.7 If the tip speed of a propeller is to be kept below $0.9c$, where c is the speed of sound, what is the maximum allowable angular speed of propellers having diameters of 2 m (6.56 ft), 3 m (9.84 ft), and 4 m (13.12 ft)? Take the speed of sound as 335 m/s (1099 ft/s).

14.8 A propeller 2 m in diameter, like the one for which characteristics are given in Fig. 14.2, is to be used on a swamp boat and is to operate at maximum efficiency when cruising. If the cruising speed is to be 40 km/h, what should the angular speed of the propeller be?

14.9 For the propeller and conditions described in Prob. 14.8, determine the thrust and the power input. Assume $\rho = 1.2 \text{ kg/m}^3$.

14.10 A propeller 2 m in diameter and like the one for which characteristics are given in Fig. 14.2 is used on a swamp boat. If the angular speed is 1000 rpm and if the boat and passengers have a combined mass of 300 kg, estimate the initial acceleration of the boat when starting from rest. Assume $\rho = 1.1 \text{ kg/m}^3$.

14.11 If a pump having the characteristics shown in Fig. 14.6 has a diameter of 40 cm and is operated at a speed of 1000 rpm, what will be the discharge when the head is 3 m?

14.12 If a pump that is geometrically similar to the one characterized in Fig. 14.7 is operated at the same speed (690 rpm) but is twice as large, $D = 71.2$ cm, what will the water discharge and the power demand be when the head is 10 m?

14.13 The pump used in the system shown has the characteristics given in Fig. 14.7. What discharge will occur under the conditions shown, and what power is required?

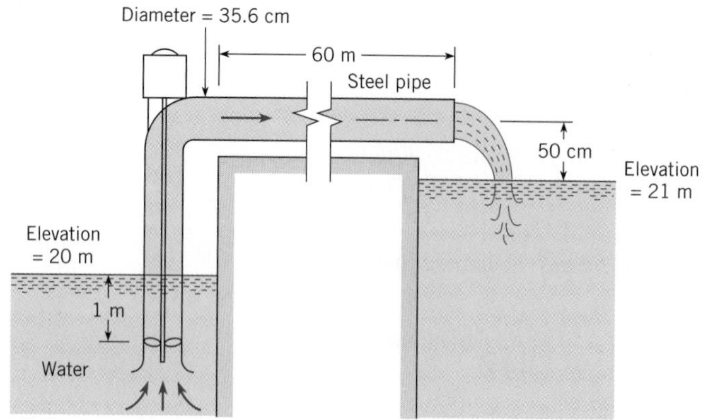

PROBLEMS 14.13, 14.14

14.14 If the conditions are the same as in Prob. 14.13 except that the speed is increased to 900 rpm, what discharge will occur, and what power is required for the operation?

14.15 For a pump having the characteristics given in Fig. 14.6 or 14.7, what water discharge and head will be produced at maximum efficiency if the pump diameter is 20 in. and the angular speed is 1100 rpm? What power is required under these conditions?

14.16 A pump has the characteristics given by Fig. 14.6. What discharge and head will be produced at maximum efficiency if the pump size is 50 cm and the angular speed is 45 rps? What power is required when pumping water under these conditions?

14.17 For a pump having the characteristics of Fig. 14.6, plot the head-discharge curve if the pump is 14 in. in diameter and is operated at a speed of 1000 rpm.

14.18 For a pump having the characteristics of Fig. 14.6, plot the head-discharge curve if the pump diameter is 60 cm and the speed is 690 rpm.

14.19 If a pump having the characteristics given in Fig. 14.9 is doubled in size but halved in speed, what will be the head and discharge at maximum efficiency?

14.20 An axial fan 2 m in diameter is used in a wind tunnel (test section 1.2 m in diameter; test section velocity of 60 m/s). The rotational speed of the fan is 1800 rpm. Assume the density of the air is constant at 1.2 kg/m^3. There are negligible losses in the tunnel. The performance curve of the fan is identical to that shown in Fig. 14.6. Calculate the power needed to operate the fan.

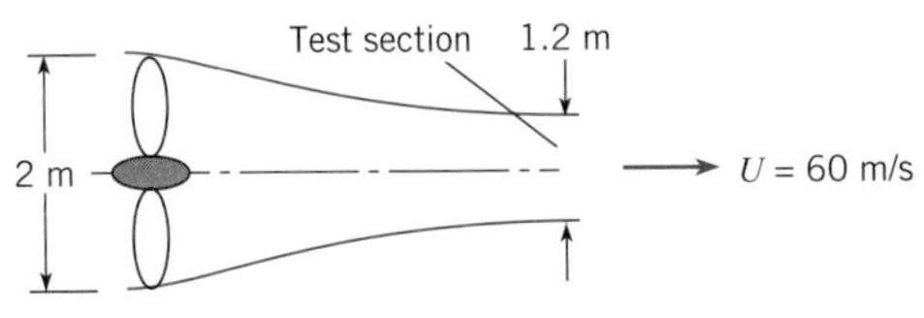

PROBLEM 14.20

14.21 A pump having the characteristics given in Fig. 14.9 pumps water from a reservoir at an elevation of 366 m to a reservoir at an elevation of 450 m through a 36-cm steel pipe. If the pipe is 610 m long, what will be the discharge through the pipe?

14.22 If a pump having the characteristics given in Fig. 14.9 or 14.10 is operated at a speed of 1600 rpm, what will be the discharge when the head is 150 ft?

14.23 If a pump having the performance curve shown is operated at a speed of 1600 rpm, what will be the maximum possible head developed?

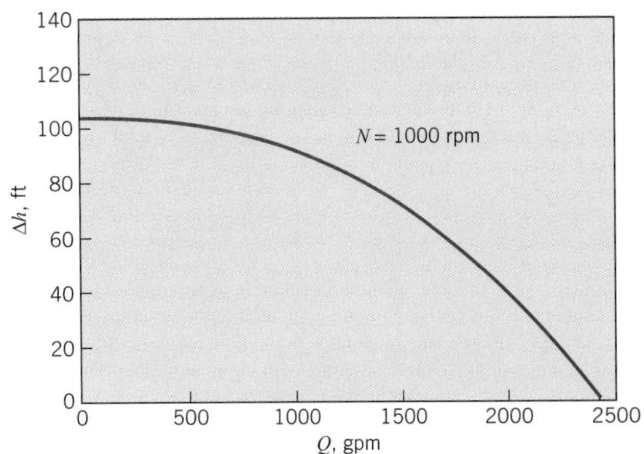

PROBLEM 14.23

14.24 If a pump having the characteristics given in Fig. 14.9 is operated at a speed of 30 rps, what will be the shutoff head?

14.25 If a pump having the characteristics given in Fig. 14.10 is 40 cm in diameter and is operated at a speed of 25 rps, what will be the discharge when the head is 50 m?

14.26 A centrifugal pump 20 cm in diameter is used to pump kerosene at a speed of 5000 rpm. Assume that the pump has the characteristics shown in Fig. 14.10. Calculate the flow rate, the pressure rise across the pump, and the power required if the pump operates at maximum efficiency. FLUID SOLUTIONS

14.27 Plot the five performance curves from Fig. 14.16 for the different impeller diameters in terms of the head and discharge coefficients. Use impeller diameter for D.

14.28 The pump curve for a given pump is represented by

$$h_{p,\text{pump}} = 20\left[1 - \left(\frac{Q}{100}\right)^2\right]$$

where $h_{p,\text{pump}}$ is the head provided by the pump in feet and Q is the discharge in gpm. The system curve for a pumping application is

$$h_{p,\text{sys}} = 5 + 0.002Q^2$$

where $h_{p,\text{sys}}$ is the head in feet required to operate the system and Q is the discharge in gpm. Find the operating point (Q) for (a) one pump, (b) two identical pumps connected in series, and (c) two identical pumps connected in parallel.

14.29 What is the suction-specific speed for the pump that is operating under the conditions given in Prob. 14.13? Is this a safe operation with respect to susceptibility to cavitation?

FLUID SOLUTIONS

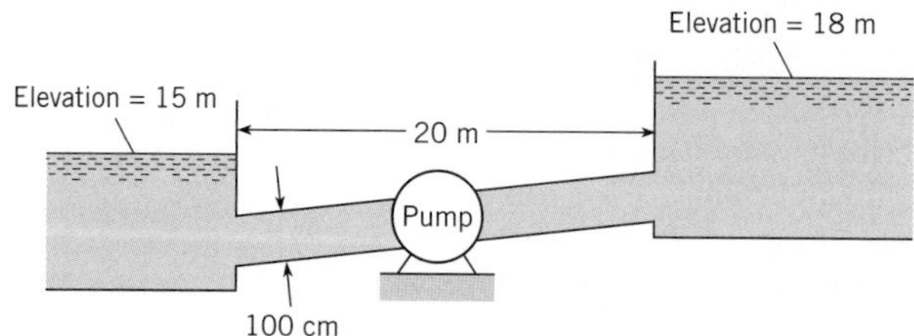

PROBLEM 14.34

14.30 What type of pump should be used to pump water at a rate of 10 cfs and under a head of 30 ft? Assume $N = 1500$ rpm.

14.31 For the most efficient operation, what type of pump should be used to pump water at a rate of 0.30 m^3/s and under a head of 8 m? Assume $n = 25$ rps.

14.32 What type of pump should be used to pump water at a rate of 0.40 m^3/s and under a head of 70 m? Assume $N = 1100$ rpm.

14.33 An axial-flow pump is to be used to lift water against a head (friction and static) of 15 ft. If the discharge is to be 5000 gpm, what maximum speed in revolutions per minute is allowed if the suction head is 5 ft?

FLUID SOLUTIONS **14.34** You want to pump water at a rate of 1.0 m^3/s from the lower to the upper reservoir shown in the figure. What type of pump would you use for this operation if the impeller speed is to be 600 rpm?

FLUID SOLUTIONS **14.35** An axial-flow blower is used for a wind tunnel that has a test section measuring 60 cm by 60 cm and is capable of airspeeds up to 40 m/s. If the blower is to operate at maximum efficiency at the highest speed and if the rotational speed of the blower is 2000 rpm at this condition, what are the diameter of the blower and the power required? Assume that the blower has the characteristics shown in Fig. 14.6.

14.36 An axial-flow blower is used to air-condition an office building that has a volume of 10^5 m^3. It is decided that the air in the building must be completely changed every 15 min. Assume that the blower operates at 600 rpm at maximum efficiency and has the characteristics shown in Fig. 14.6. Calculate the diameter and power requirements for two blowers operating in parallel.

14.37 Methane flowing at the rate of 1 kg/s is to be compressed by a noncooled centrifugal compressor from 100 kPa to 150 kPa. The temperature of the methane entering the compressor is 27°C. The efficiency of the compressor is 70%. Calculate the shaft power necessary to run the compressor.

14.38 A 12-kW (shaft output) motor is available to run a noncooled compressor for carbon dioxide. The pressure is to be increased from 90 kPa to 140 kPa. If the compressor is 60% efficient, calculate the volume flow rate into the compressor.

14.39 A water-cooled centrifugal compressor is used to compress air from 100 kPa to 400 kPa at the rate of 1 kg/s. The temperature of the inlet air is 15°C. The efficiency of the compressor is 50%. Calculate the necessary shaft power.

14.40 A penstock 1 m in diameter and 10 km long carries water from a reservoir to an impulse turbine. If the turbine is 85% efficient, what power can be produced by the system if the upstream reservoir elevation is 650 m above the turbine jet and the jet diameter is 16.0 cm? Assume that $f = 0.016$ and neglect head losses in the nozzle. What should the diameter of the turbine wheel be if it is to have an angular speed of 360 rpm? Assume ideal conditions for the bucket design $[V_{\text{bucket}} = (1/2)V_j]$.

14.41 Consider an idealized bucket on an impulse turbine that turns the water through 180°. Prove that the bucket speed should be one-half the incoming jet speed for a maximum power production. (*Hint:* Set up the momentum equation to solve for the force on the bucket in terms of V_j and V_{bucket}; then the power will be given by this force times V_{bucket}. You can use your mathematical talent to complete the problem.)

14.42 Consider a single jet of water striking the buckets of the impulse wheel as shown. Assume ideal conditions for power generation $[V_{\text{bucket}} = (1/2)V_j$ and the jet is turned through 180° of arc]. With the foregoing conditions, solve for the jet force on the bucket and then solve for the power developed. Note that this power is not the same as that given by Eq. (14.20)! Study the figure to resolve the discrepancy.

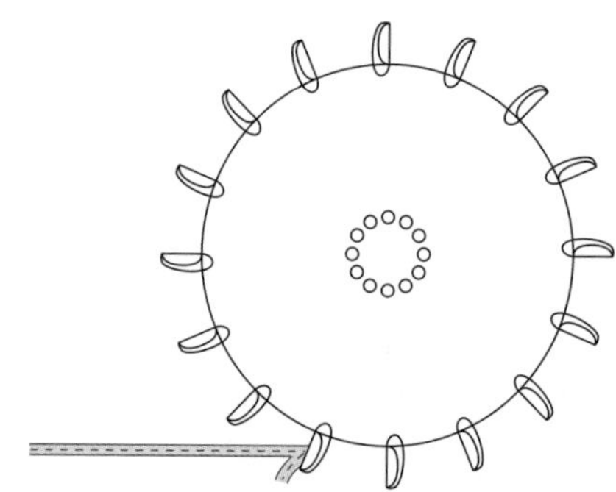

PROBLEM 14.42

14.43 a. For a given Francis turbine, $\beta_1 = 60°$, $\beta_2 = 90°$, $r_1 = 5$ m, $r_2 = 3$ m, and $B = 1$ m. What should α_1 be for a nonseparating flow condition at the entrance to the runner when the discharge rate is 126 m^3/s and $N = 60/\text{rpm}$?

b. What is the maximum attainable power with the conditions noted?

c. If you were to redesign the turbine blades of the runner, what changes would you suggest to increase the power production if the discharge and overall dimensions are to be kept the same?

14.44 A Francis turbine is to be operated at a speed of 60 rpm and with a discharge of 3.0 m^3/s. If $r_1 = 1.5$ m, $r_2 = 1.20$ m, $B = 30$ cm, $\beta_1 = 85°$, and $\beta_2 = 165°$, what should α_1 be for nonseparating flow to occur through the runner? What power and torque should result with this operation?

14.45 A Francis turbine is to be operated at a speed of 120 rpm and with a discharge of 113 m^3/s. If $r_1 = 2.5$ m, $B = 0.90$ m, and $\beta_1 = 45°$, what should α_1 be for nonseparating flow at the runner inlet?

14.46 Shown is a preliminary layout for a proposed small hydroelectric project. The initial design calls for a discharge of 8 cfs through the penstock and turbine. Assume 80% turbine efficiency. For this setup, what power output could be expected from the power plant? Draw the HGL and EGL for the system.

14.47 Calculate the maximum power derivable from a conventional horizontal-axis wind turbine with a propeller 2.5 m in diameter in a 50-km/h wind whose density is 1.2 kg/m^3.

14.48 A wind "farm" consists of 20 Darrieus turbines, each 15 m high. The total output from the turbines is to be 2 MW in a wind of 20 m/s and an air density of 1.2 kg/m^3. The Darrieus turbine has the shape of an arc of a circle. Find the minimum width, W, of the turbine needed to provide this power output.

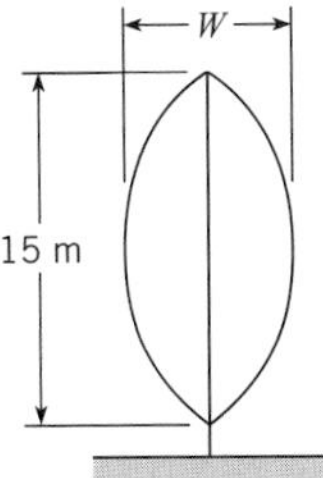

PROBLEM 14.48

14.49 A windmill is connected directly to a mechanical pump that is to pump water from a well 10 ft deep. The windmill is a conventional horizontal-axis type with a fan diameter of 10 ft. The efficiency of the mechanical pump is 80%. The density of the air is 0.07 lbm/ft^3. Assume the windmill delivers the maximum power available. What would the discharge of the pump be (in gallons per minute) for a 30-mph wind? (1 cfm = 7.48 gpm)

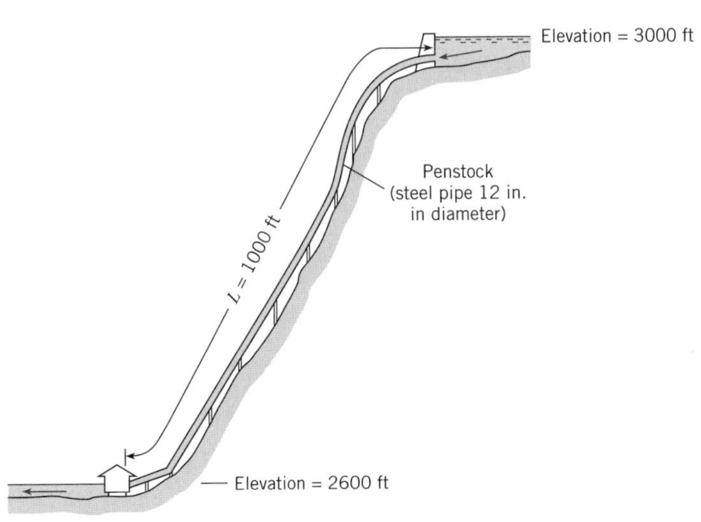

PROBLEM 14.46

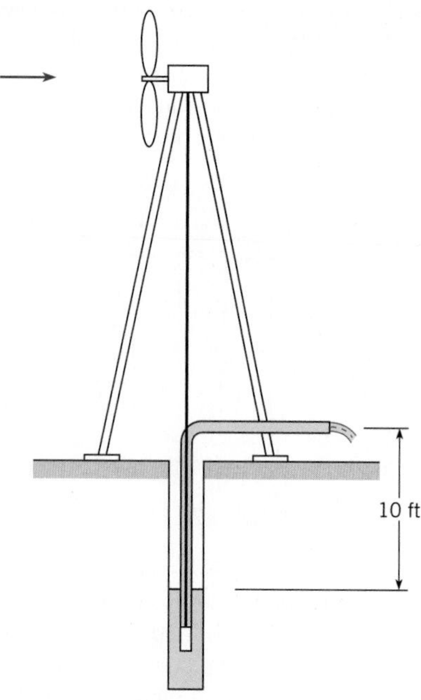

PROBLEM 14.49

14.50 Solve Prob. 10.102 but also include the choice of pump(s) in your design. Assume that the pump(s) will have the characteristics given in Fig. 14.10.

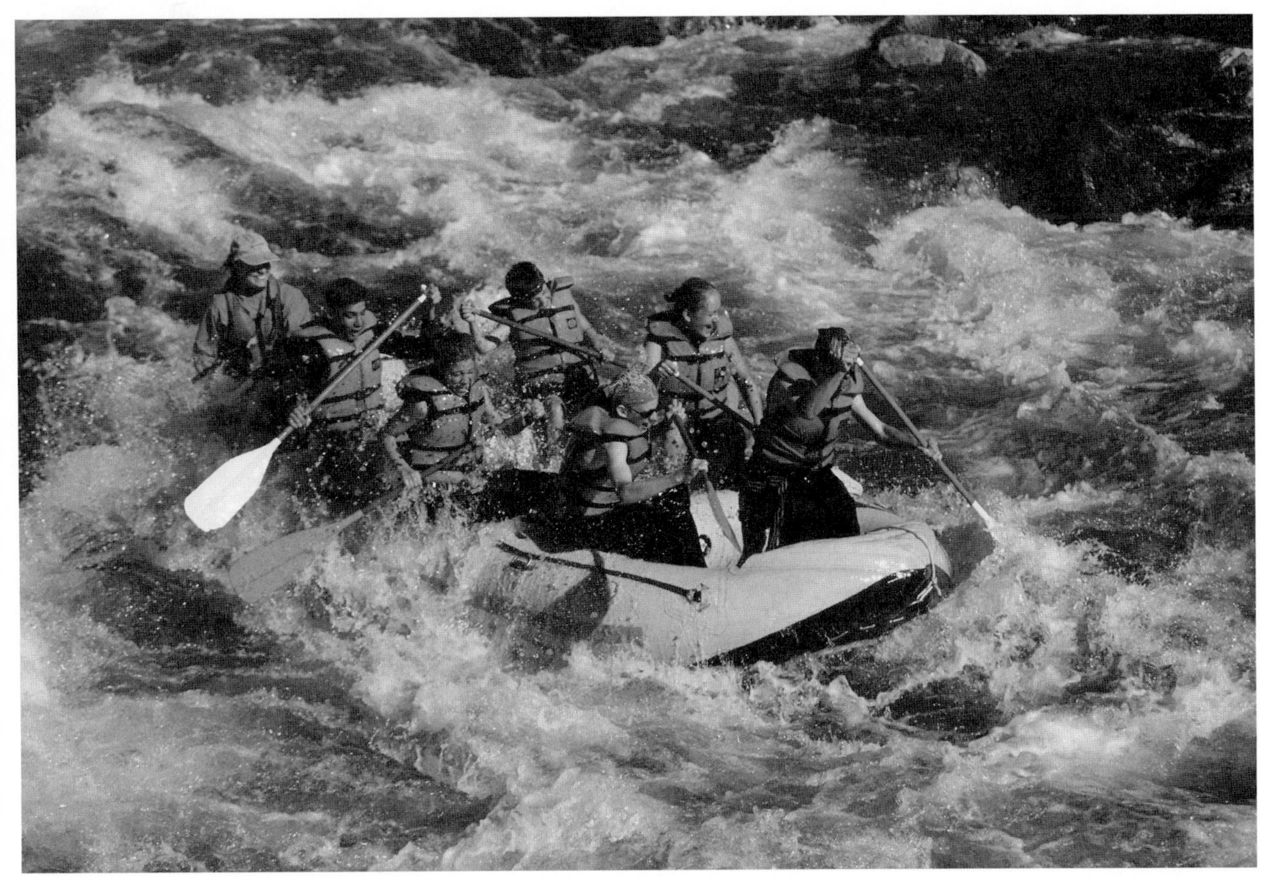

The river is a natural example of open channel flow. (Corbis Digital Stock)

C H A P T E R 15

Varied Flow in Open Channels

The primary difference between flow in closed conduits and flow in open channels is that in open channels there is a free surface (liquid surface is exposed to the atmosphere), and that surface is at atmospheric pressure, whereas in closed-conduit flow there is no free surface. As noted in Chapter 7, the hydraulic grade line (HGL) is a line showing where $p/\gamma = 0$ gage. Thus, for open-channel flow, the HGL is coincident with the free surface.

Almost all cases of open-channel flow are with water or waste-water as the flowing liquid. The Reynolds number for flow in open channels is given as $\text{Re} = VR_h/\nu$. If this Re is greater than 750, one can expect the flow to be turbulent. Because water has a kinematic viscosity of about 10^{-5} ft^2/s (in the traditional system of units), the flow of water in a channel will be turbulent if $VR_h \approx 750 \times 10^{-5}$ ft^2/s $= 0.0075$ ft^2/s. For most engineering problems involving the flow of water in channels, the velocity is rarely less than 1 ft/s and the hydraulic radius is seldom less than 0.1 ft. Thus the lower limit of VR_h for most engineering problems is 0.1 ft^2/s, which is an order of magnitude greater than the limit for onset of turbulence with water flow. Thus most open-channel flow with water as the medium will be turbulent.

In Chapter 10 we considered the case of steady uniform flow in open channels, that is, where the channel is straight and where the depth is uniform along the length of the channel. The depth for these uniform-flow conditions is called *normal* depth and is designated by y_n. The next more complicated classification of flow is steady nonuniform flow. By definition, this classification includes cases where the velocity is constant with respect to time but changes from section to section along the channel. The velocity change may be due simply to a change in depth along a straight prismatic channel, or it may also be a result of a change in channel configuration, such as a bend and/or change in cross-sectional shape and/or change in channel slope. This chapter focuses on the theory for and examples of steady nonuniform flow in open channels, usually referred to as varied flow.

Sections 15.1, 15.2, and 15.3 address cases of nonuniform flow in which the depth and velocity change markedly over a relatively short distance (this type of flow is often called *rapidly varied flow*). For these cases one can neglect the resistance of the channel walls and bottom. In Section 15.4 we consider cases where significant changes of depth and velocity occur over long reaches of the channel, and for this reason surface resistance *is* a significant variable in the flow process. This type of flow is called *gradually varied flow.*

The most complicated open-channel flow is unsteady nonuniform flow. An example of this is a breaking wave on a sloping beach. Theory and analysis of unsteady nonuniform flow are reserved for more advanced courses.

In Chapter 8 we noted that when gravity influences the flow pattern, the Froude number is a significant π-group. Up to this point in the text, except for problems involving dimensional analysis and flow over weirs, all of the problems considered have been ones for which either the Reynolds number or the Mach number was the significant π-group. In this chapter, we will focus on liquid flow in open channels, for which the force of gravity is a very significant variable. It will be shown that the Froude number is indeed a significant parameter for such flows.

15.1

Energy Relations in Open Channels

The Energy Equation Applied to Open-Channel Flow

The energy equation for open channels (see Fig. 15.1) is

$$\frac{p_1}{\gamma} + \alpha_1\frac{V_1^2}{2g} + z_1 = \frac{p_2}{\gamma} + \alpha_2\frac{V_2^2}{2g} + z_2 + h_L \tag{15.1}$$

We see from Fig. 15.1 that the following equalities hold:

$$\frac{p_1}{\gamma} + z_1 = y_1 + S_0\Delta x \qquad \text{and} \qquad \frac{p_2}{\gamma} + z_2 = y_2$$

Here S_0 is the slope of the channel bottom, and y is the depth of flow. Then if we assume $\alpha_1 = \alpha_2 = 1.0$, we can write Eq. (15.1) as

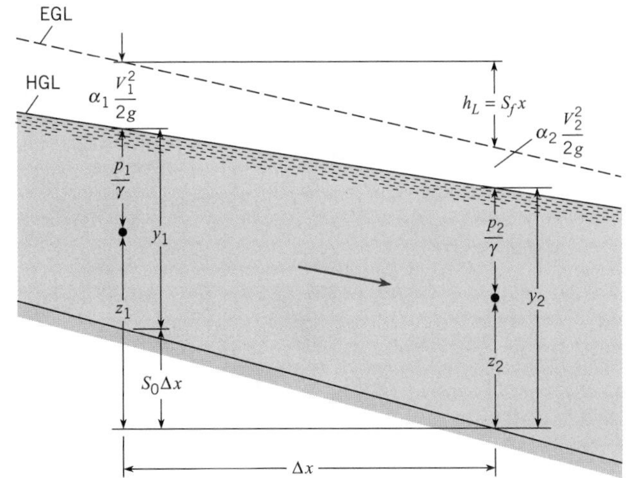

FIGURE 15.1

Definition sketch for flow in open channels.

$$y_1 + \frac{V_1^2}{2g} + S_0\Delta x = y_2 + \frac{V_2^2}{2g} + h_L \tag{15.2}$$

Now, if we consider the special case where the channel bottom is horizontal ($S_0 = 0$) and the head loss is zero ($h_L = 0$), Eq. (15.2) becomes

$$y_1 + \frac{V_1^2}{2g} = y_2 + \frac{V_2^2}{2g} \tag{15.3}$$

Specific Energy

The sum of the depth of flow and the velocity head is defined as the *specific energy*:

$$E = y + \frac{V^2}{2g} \tag{15.4}$$

Thus Eq. (15.3) states that the specific energy at section 1 is equal to the specific energy at section 2, or $E_1 = E_2$. The continuity equation between sections 1 and 2 is

$$A_1V_1 = A_2V_2 = Q \tag{15.5}$$

Therefore, Eq. (15.3) can be expressed as

$$y_1 + \frac{Q^2}{2gA_1^2} = y_2 + \frac{Q^2}{2gA_2^2} \tag{15.6}$$

Because A_1 and A_2 are both functions of the depth y, the magnitude of the specific energy at section 1 or 2 is solely a function of the depth at each section. If, for a given channel and given discharge, one plots depth versus specific energy, a relationship such as that shown in Fig. 15.2 is obtained. By studying Fig. 15.2 for a given value of specific energy,

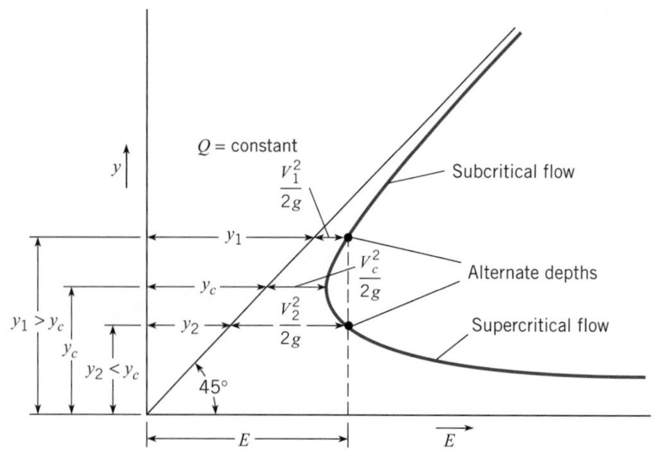

FIGURE 15.2

Relation between depth and specific energy.

we can see that the depth may be either large or small. In a physical sense, this means that for the small depth, the bulk of the energy of flow is in the form of kinetic energy ($Q^2/2gA^2$), whereas for a larger depth, most of the energy is in the form of potential energy. Flow under a *sluice gate* (Fig. 15.3) is an example of flow in which two depths occur for a given value of specific energy. The large depth and low kinetic energy occur upstream of the gate; the low depth and large kinetic energy occur downstream. The depths as used here are called *alternate depths*. That is, for a given value of E, the large depth is alternate to the low depth, or vice versa. Returning to the flow under the sluice gate, we find that if we maintain the same rate of flow but set the gate with a larger opening, as in Fig. 15.3*b*, the upstream depth will drop, and the downstream depth will rise. Thus we have different alternate depths and a smaller value of specific energy than before. This is consistent with the diagram in Fig. 15.2.

Finally, it can be seen in Fig. 15.2 that a point will be reached where the specific energy is minimum and only a single depth occurs. At this point, the flow is termed *critical*. Thus one definition of critical flow is the flow that occurs when the specific energy is minimum for a given discharge. The flow for which the depth is less than critical (velocity is greater than critical) is termed *supercritical flow,* and the flow for which the depth is greater than critical (velocity is less than critical) is termed *subcritical flow.* Using this terminology, we can see that subcritical flow occurs upstream and supercritical flow occurs downstream of the sluice gate in Fig. 15.3. It should be noted that some engineers refer to subcritical and supercritical flow as *tranquil* and *rapid* flow, respectively. We will consider other aspects of critical flow in the next section.

FIGURE 15.3

Flow under a sluice gate.

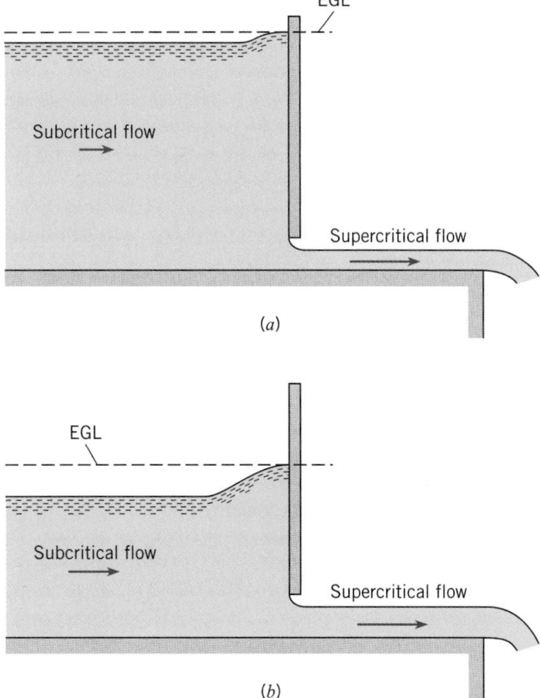

Characteristics of Critical Flow

We have already seen that critical flow occurs when the specific energy is minimum for a given discharge. The depth for this condition may be determined if we solve for dE/dy from $E = y + Q^2/2gA^2$ and set dE/dy equal to zero:

$$\frac{dE}{dy} = 1 - \frac{Q^2}{gA^3} \cdot \frac{dA}{dy} \tag{15.7}$$

However, $dA = T\,dy$, where T is the width of the channel at the water surface, as shown in Fig. 15.4. Then Eq. (15.7), with $dE/dy = 0$, will reduce to

$$\frac{Q^2 T_c}{gA_c^3} = 1 \tag{15.8}$$

or

$$\frac{A_c}{T_c} = \frac{Q^2}{gA_c^2} \tag{15.9}$$

If we define the *hydraulic depth D* as A/T, then Eq. (15.9) will become

$$D_c = \frac{Q^2}{gA_c^2} \tag{15.10}$$

$$D_c = \frac{V^2}{g} \tag{15.11}$$

Upon dividing Eq. (15.11) by D_c and taking the square root of it, we get

$$1 = \frac{V}{\sqrt{gD_c}} \tag{15.12}$$

Note: $V/\sqrt{gD_c}$ is the Froude number. Therefore, it has been shown that the Froude number is equal to unity when critical flow prevails.

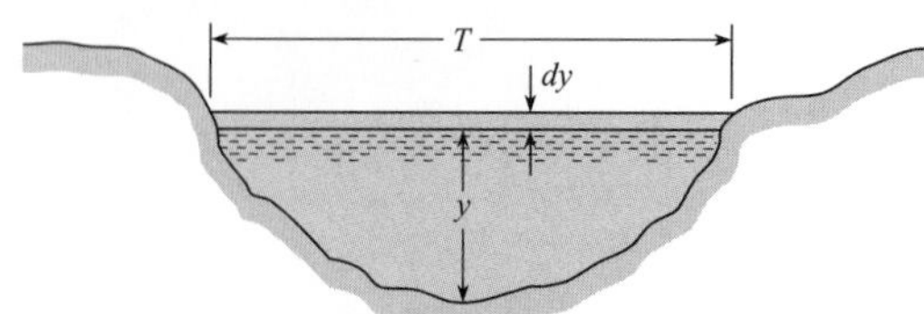

FIGURE 15.4

Open-channel relations.

example 15.1

Determine the critical depth in this trapezoidal channel for a discharge of 500 cfs. The width of the channel bottom is $B = 20$ ft, and the sides slope upward at an angle of 45°.

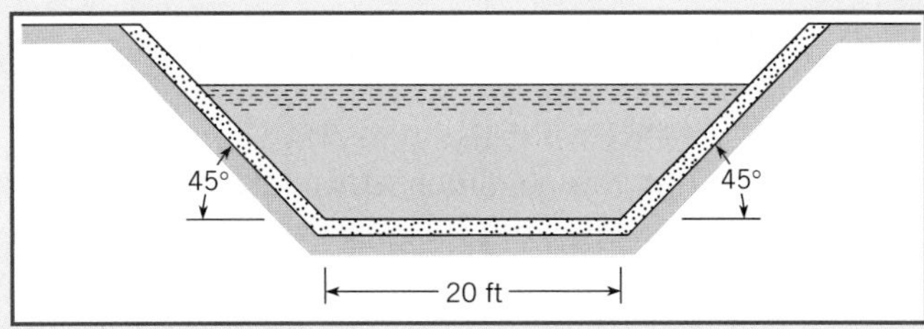

Solution Starting with Eq. (15.8),

$$\frac{Q^2 T_c}{g A_c^3} = 1$$

or

$$\frac{A_c^3}{T_c} = \frac{Q^2}{g}$$

Then, for $Q = 500$ cfs,

$$\frac{A_c^3}{T_c} = \frac{500^2}{32.2} = 7764 \text{ ft}^2$$

For this channel, $A = y(B + y)$ and $T = B + 2y$. Then by iteration (choose y and compute A^3/T), we can find y that will yield A^3/T equal to 7764 ft^2. Such a solution yields $y_c = 2.57$ ft. ◁

If a channel is of rectangular cross section, then A/T is the actual depth, and $Q^2/A^2 = q^2/y^2$, so the formula for critical depth [Eq. (15.9)] becomes

$$y_c = \left(\frac{q^2}{g}\right)^{1/3} \tag{15.13}$$

where q is the discharge per unit width of channel.

Critical flow may also be examined in terms of how the discharge in a channel varies with depth for a given specific energy. For example, consider flow in a rectangular channel where

$$E = y + \frac{Q^2}{2gA^2}$$

or

$$E = y + \frac{Q^2}{2gy^2B^2}$$

If we consider a unit width of the channel and let $q = Q/B$, then the above equation becomes

$$E = y + \frac{q^2}{2gy^2}$$

If we determine how q varies with y for a constant value of specific energy, we see that critical flow occurs when the discharge is maximum (Fig. 15.5).

Originally, the term *critical flow* probably related to the unstable character of the flow for this condition. If we refer to Fig. 15.2, we see that only a slight change in specific energy will cause the depth to increase or decrease a significant amount; this is a very unstable condition. In fact, observations of critical flow in open channels show that the water surface consists of a series of standing waves. Because of the unstable nature of the depth in critical flow, designing canals so that normal depth is either well above or well below critical depth is usually best. The flow in canals and rivers is usually subcritical; however, the flow in steep chutes or over spillways is supercritical.

In this section, various characteristics of critical flow have been explored. The main ones can be summarized as follows:

1. Critical flow occurs when

$$\frac{A^3}{T} = \frac{Q^2}{g}$$

2. Critical flow occurs when Fr $= 1$.
3. Critical flow occurs when the specific energy is minimum for a given discharge.
4. Critical flow occurs when the discharge is maximum for a given specific energy.
5. For rectangular channels, critical depth is given as $y_c = (q^2/g)^{1/3}$.

Occurrence of Critical Depth

Critical flow occurs when a liquid passes over a broad-crested weir (Fig. 15.6*a*). The principle of the broad-crested weir is illustrated by first considering a closed sluice gate that prevents water from being discharged from the reservoir (Fig. 15.6*b*). If the gate is

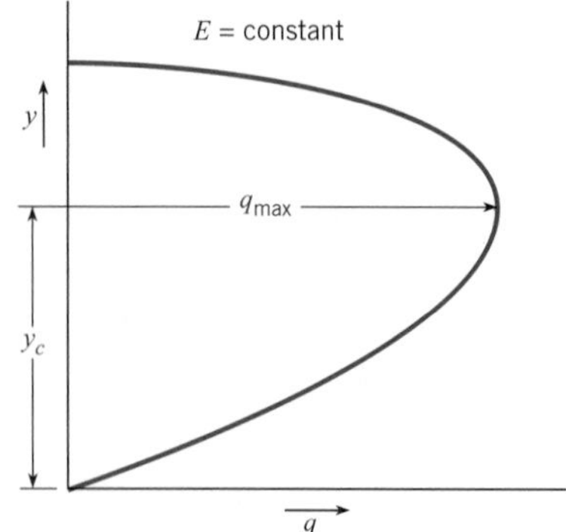

FIGURE 15.5

Variation of q and y with constant specific energy.

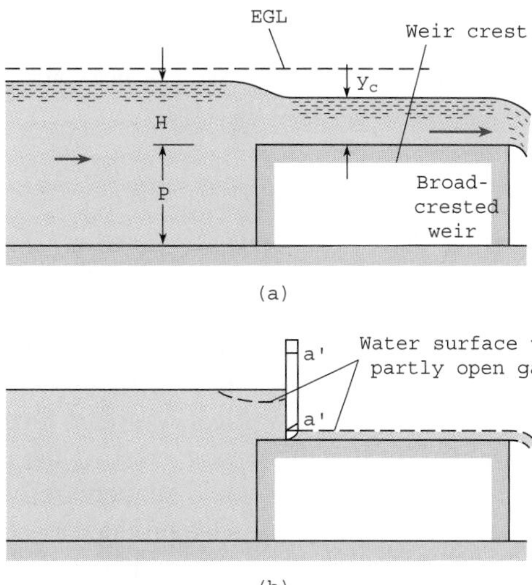

FIGURE 15.6

Flow over a broad-crested weir.

opened a small amount (gate position $a' - a'$), the flow upstream of the gate will be subcritical and the flow downstream will be supercritical (like the condition introduced in Fig. 15.3). As the gate is opened further, a point is finally reached where the depths immediately upstream and downstream of the gate are the same. This is the critical condition. At this gate opening and beyond, the gate has no influence on the flow; this is the condition shown in Fig. 15.6*a*, the broad-crested weir. If the depth of flow over the weir is measured, the rate of flow can easily be computed from Eq. (15.13):

$$q = \sqrt{gy_c^3}$$

or

$$Q = L\sqrt{gy_c^3} \tag{15.14}$$

where L is the length of the weir crest normal to the flow direction.

Because $y_c/2 = V_c^2/2g$ [from Eq. (15.11)], it is easily shown that $y_c = (2/3E)$, where E is the total head above the crest $(H + V_{\text{approach}}^2/2g)$; hence Eq. (15.14) can be rewritten as

$$Q = L\sqrt{g}\left(\frac{2}{3}\right)^{3/2} E^{3/2}$$

or

$$Q = 0.385L\sqrt{2g}E_c^{3/2} \tag{15.15}$$

For high weirs, the upstream velocity of approach is almost zero. Hence Eq. (15.15) can be expressed as

$$Q_{\text{theor}} = 0.385L\sqrt{2g}H^{3/2} \tag{15.16}$$

If the height P of the broad-crested weir is relatively small, then the velocity of approach may be significant, and the discharge produced will be greater than that given by Eq. (15.16). Also, head loss will have some effect. To account for these effects, a discharge coefficient C is defined as

$$C = Q/Q_{\text{theor}} \tag{15.17}$$

Then

$$Q = 0.385 CL\sqrt{2g}H^{3/2} \tag{15.18}$$

where Q is the actual discharge over the weir. An analysis of experimental data by Raju (8) shows that C varies with $H/(H + P)$ as shown in Fig. 15.7. The curve in Fig. 15.7 is for a weir with a vertical upstream face and a sharp corner at the intersection of the upstream face and the weir crest. If the upstream face is sloping at a 45° angle, the discharge coefficient should be increased 10% over that given in Fig. 15.7. Rounding of the upstream corner will also produce a coefficient of discharge as much as 3% greater.

Equation (15.18) reveals a definite relationship for Q as a function of the head, H. This type of discharge-measuring device is in the broad class of discharge meters called *critical-flow flumes*. Another very common critical-flow flume is the *Venturi flume,* which was developed and calibrated by Parshall (1). Figure 15.8 shows the essential features of the Venturi flume. The discharge equation for the Venturi flume is in the same form as Eq. (15.18), the only difference being that the experimentally determined coefficient C will have a different value from the C for the broad-crested weir. For more details on the Venturi flume, you may refer to Roberson et al. (2), Parshall (1), and Chow (3). The Venturi flume is especially useful for discharge measurement in irrigation systems because little head loss is required for its use and sediment is easily flushed through if the water happens to be silty.

The depth also passes through a critical stage in channel flow where the slope changes from a mild one to a steep one. A *mild slope* is defined as a slope for which the normal depth y_n is greater than y_c. Likewise, a *steep slope* is one for which $y_n < y_c$. This condition is shown in Fig. 15.9. Note that y_c is the same for both slopes in the figure because y_c is a function of the discharge only. However, normal depth (uniform flow depth) for the mild upstream channel is greater than critical, whereas the normal depth for the

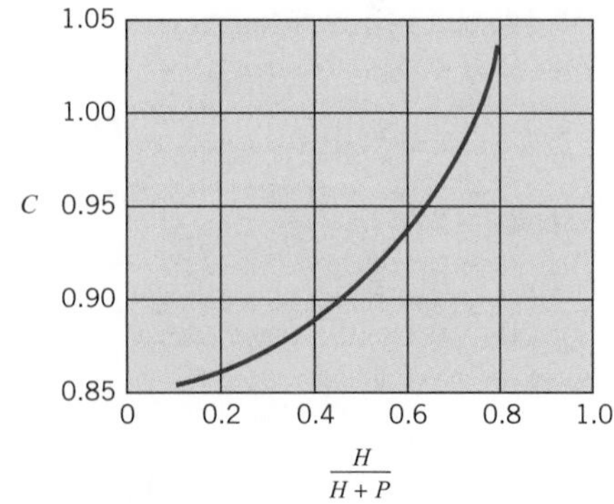

FIGURE 15.7

Discharge coefficient for a broad-crested weir for $0.1 < H/L < 0.8$.

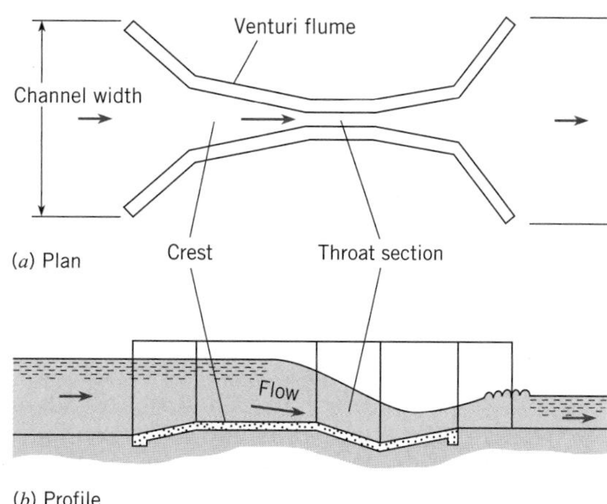

FIGURE 15.8

Flow through a Venturi flume.

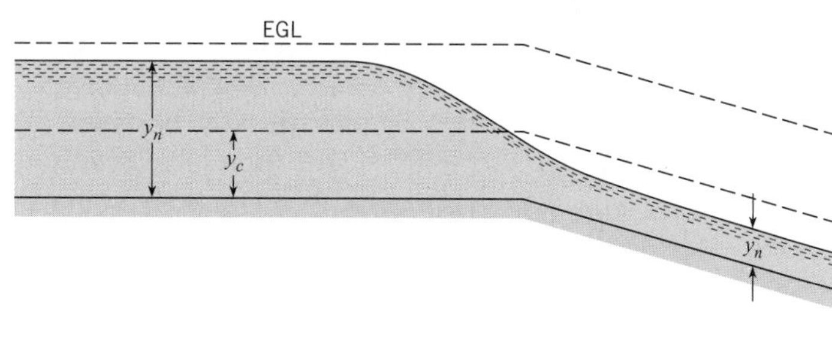

FIGURE 15.9

Critical depth at a break in grade.

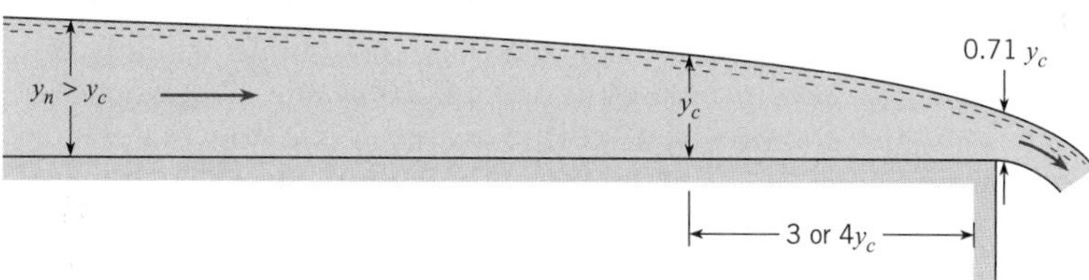

FIGURE 15.10

Critical depth at a free overfall.

steep downstream channel is less than critical; hence it is obvious that the depth must pass through a critical stage. Experiments show that critical depth occurs a very short distance upstream of the intersection of the two channels.

Another place where critical depth occurs is upstream of a free overfall at the end of a channel with a mild slope (Fig. 15.10). Critical depth will occur at a distance of 3 to $4y_c$ upstream of the brink. Such occurrences of critical depth (at a break in grade or at a brink) are useful in computing surface profiles because they provide a point for starting surface-profile calculations.*

* The procedure for making these computations starts on page 642.

Channel Transitions

Whenever a channel's cross-sectional configuration (shape or dimension) changes along its length, the change is termed a *transition*. We will use the basic concepts already presented in the first part of this chapter to show how the flow depth changes when the floor of a rectangular channel is increased in elevation or when the width of the channel is decreased. In these developments we assume negligible energy losses. We will look first at the case where the floor of the channel is raised (an upstep). Later in this section we will look at configurations of transitions used for subcritical flow from a rectangular to a trapezoidal channel.

Consider the rectangular channel shown in Fig. 15.11, where the floor rises an amount Δz. To help us in evaluating depth changes, we use a diagram of specific energy versus depth, which is similar to Fig. 15.2. This diagram is placed both at the section upstream of the transition and at the section just downstream of the transition. Because the discharge, Q, is the same at both sections, the given diagram is valid at both sections. As noted in Fig. 15.11, the depth of flow at section 1 can be either large (subcritical) or small (supercritical) if the specific energy E_1 is greater than that required for critical flow. It can also be seen in Fig. 15.11 that when the upstream flow is subcritical, a decrease in depth occurs in the region of the elevated channel bottom. This occurs because the specific energy at this section, E_2, is less than that at section 1 by the amount Δz. Therefore, the specific-energy diagram indicates that y_2 will be less than y_1. In a similar manner it can be seen that when the upstream flow is supercritical, the depth as well as the actual water-surface elevation increases from section 1 to section 2. A further note should be made about the effect on flow depth of a change in bottom-surface elevation. If the channel bottom at section 2 is at an elevation greater than that just sufficient to establish critical flow at section 2, then there is not enough head at section 1 to cause flow to occur over the rise under steady-flow conditions. Instead, the water level upstream will rise until it is just sufficient to reestablish steady flow.

When the channel bottom is kept at the same elevation but the channel is decreased in width, then the discharge per unit of width between sections 1 and 2 increases, but the specific energy E remains constant. Thus when we utilize the diagram of q versus depth for the given specific energy E, we note that the depth in the restricted section increases if the upstream flow is supercritical and decreases if it is subcritical (see Fig. 15.12).

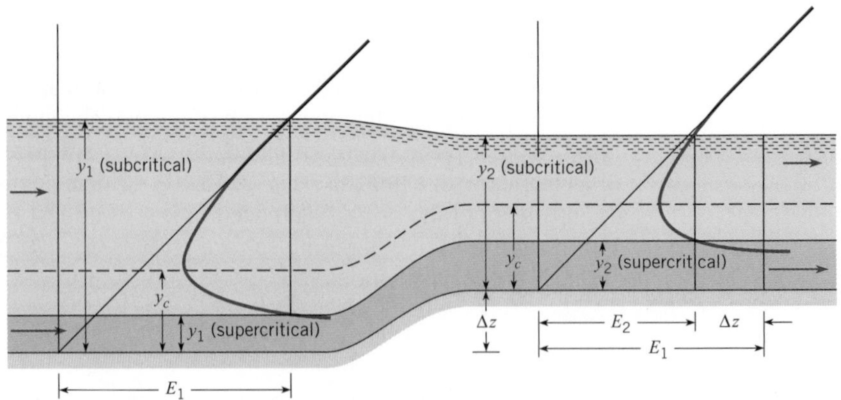

FIGURE 15.11

Change in depth with change in bottom elevation of a rectangular channel.

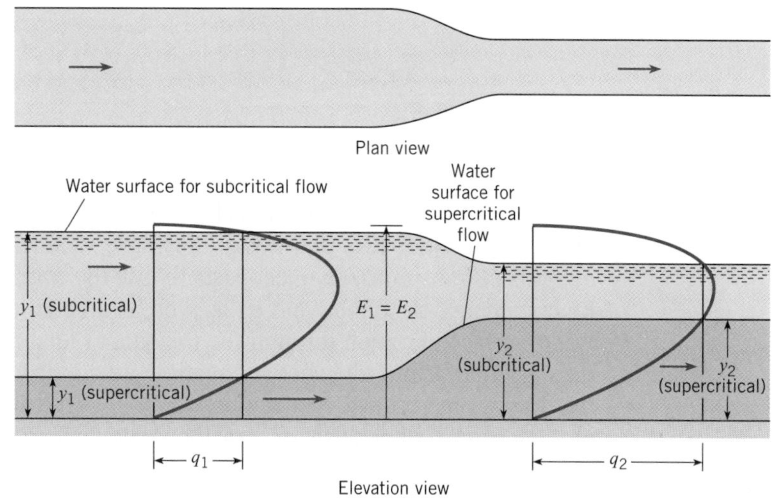

FIGURE 15.12

Change in depth with change in channel width.

The foregoing paragraphs describe gross effects for the simplest transitions. In practice, it is more common to find transitions between a channel of one shape (rectangular cross section, for example) and a channel having a different cross section (trapezoidal, for example). A very simple transition between two such channels consists of two straight vertical walls joining the two channels, as shown by half section in Fig. 15.13.

This type of transition can work, but it will produce excessive head loss because of the abrupt change in cross section and the ensuing separation that will occur. To reduce the head losses, a more gradual type of transition is used. Figure 15.14 is a half section of a transition similar to that of Fig. 15.13, but with the angle θ much greater than 90°. This is called a *wedge transition.*

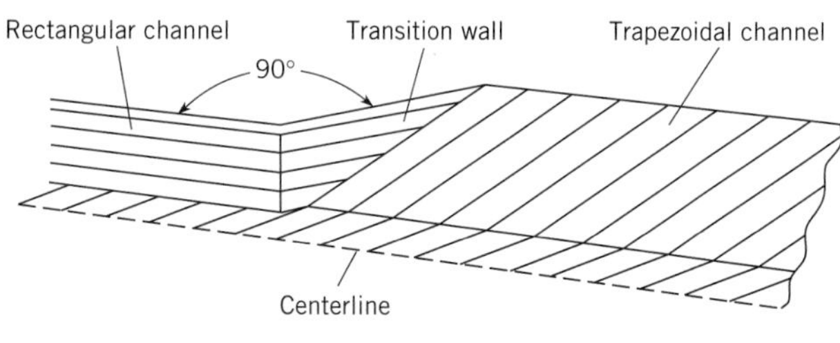

FIGURE 15.13

Simplest type of transition between a rectangular channel and a trapezoidal channel.

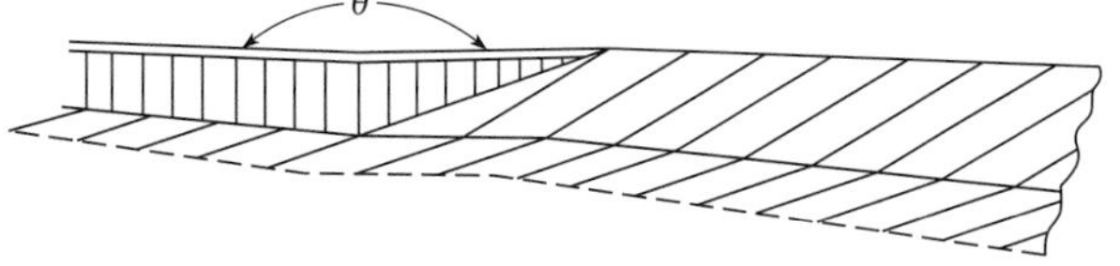

FIGURE 15.14

Half section of a wedge transition.

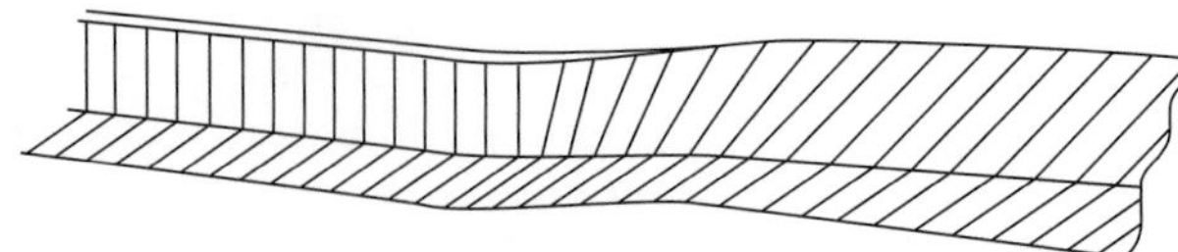

FIGURE 15.15
Half section of a warped-wall transition.

The *warped-wall* transition shown in Fig. 15.15 will yield even smoother flow than either of the other two, and it will thus have less head loss. In the practical design and analysis of transitions, engineers usually use the complete energy equation, including the kinetic energy factors α_1 and α_2 as well as a head loss term h_L, to define velocity and water-surface elevation through the transition. Analyses of transitions utilizing the one-dimensional form of the energy equation are applicable only if the flow is subcritical. If the flow is supercritical, then a much more involved analysis is required. For more details on the design and analysis of transitions, you are referred to Hinds (4), Chow (3), U.S. Bureau of Reclamation (5), and Rouse (6).

Wave Celerity

Wave celerity is the velocity at which an infinitesimally small wave travels relative to the velocity of the fluid. The following is a derivation of the wave celerity *c*.

Consider a small solitary wave moving with a velocity c in an otherwise calm body of liquid of small depth (Fig. 15.16*a*). Because the velocity in the liquid changes with time, this is a condition of unsteady flow. However, if we refer all velocities to a reference frame moving with the wave, the shape of the wave would be fixed and the flow would be steady. Then the flow is amenable to analysis with the Bernoulli equation. The steady-flow condition is shown in Fig. 15.16*b*. When the Bernoulli equation is written between a point on the surface of the undisturbed fluid and a point at the wave crest, we have

$$\frac{c^2}{2g} + y = \frac{V^2}{2g} + y + \Delta y \tag{15.19}$$

In Eq. (15.19), V is the velocity of the liquid in the section where the crest of the wave is located. From the continuity equation we have $cy = V(y + \Delta y)$. Hence

$$V = \frac{cy}{y + \Delta y}$$

and

$$V^2 = \frac{c^2 y^2}{(y + \Delta y)^2} \tag{15.20}$$

When Eq. (15.20) is substituted into Eq. (15.19), we obtain

$$\frac{c^2}{2g} + y = \frac{c^2 y^2}{2g[y^2 + 2y\Delta y + (\Delta y)^2]} + y + \Delta y \tag{15.21}$$

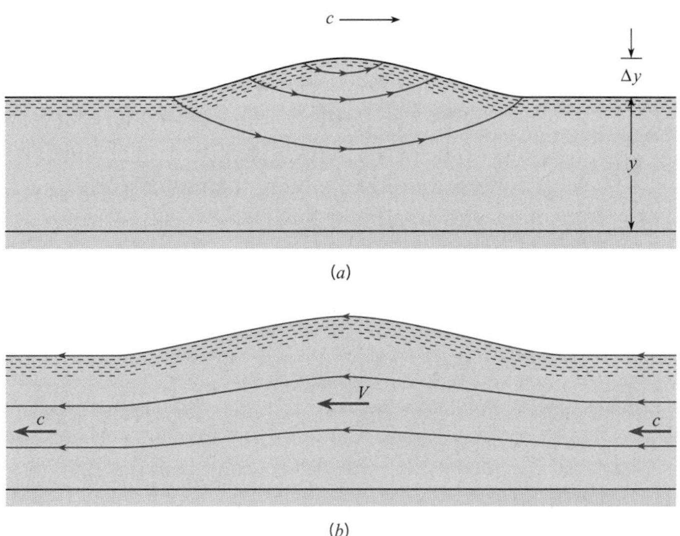

FIGURE 15.16

Solitary wave (exaggerated vertical scale). (a) Unsteady flow. (b) Steady flow.

Solving Eq. (15.21) for c after discarding terms with $(\Delta y)^2$, since we are assuming an infinitesimally small wave, we obtain the *wave celerity equation*

$$c = \sqrt{gy} \tag{15.22}$$

It has thus been shown that the speed of a small solitary wave is equal to the square root of the product of the depth and g.

The Hydraulic Jump

Occurrence of the Hydraulic Jump

When the flow is supercritical in an upstream section of a channel and is then forced to become subcritical in a downstream section (the change in depth can be forced by a sill in the downstream part of the channel or just by the prevailing depth in the stream further downstream), a rather abrupt change in depth usually occurs, and considerable energy loss accompanies the process. This flow phenomenon, called the *hydraulic jump* (Fig. 15.17), is often considered in the design of open channels and spillways of dams. For example, many spillways are designed so that a jump will occur on an *apron* of the spillway, thereby reducing the downstream velocity so that objectionable erosion of the river channel is prevented. If a channel is designed to carry water at supercritical velocities, the designer must be certain that the flow will not become subcritical prematurely. If it did, overtopping of the channel walls would undoubtedly occur, with consequent failure of the structure. Because the energy loss in the hydraulic jump is initially not known, the energy equation is not a suitable tool for analysis of the velocity–depth relationships. Therefore, the momentum equation is applied to the problem.

Derivation of Depth Relationships

Consider flow as shown in Fig. 15.17. Here it is assumed that uniform flow occurs both upstream and downstream of the jump and that the resistance of the channel bottom is negligible. The derivation is for a horizontal channel, but experiments show that the results of the derivation will apply to all channels of moderate slope ($S_0 < 0.02$). We start the derivation by applying the momentum equation in the x direction to the control volume shown in Fig. 15.18:

$$\sum F_x = \dot{m}_2 V_2 - \dot{m}_1 V_1$$

The forces are the hydrostatic forces on each end of the system; thus the following is obtained:

$$\bar{p}_1 A_1 - \bar{p}_2 A_2 = \rho V_2 A_2 V_2 - \rho V_1 A_1 V_1$$

or

$$\bar{p}_1 A_1 + \rho Q V_1 = \bar{p}_2 A_2 + \rho Q V_2 \tag{15.23}$$

In Eq. (15.23), $\bar{p}_1$ and $\bar{p}_2$ are the pressures at the centroids of the respective areas A_1 and A_2.

A representative problem might be to determine the downstream depth y_2 given the discharge and upstream depth. The left-hand side of Eq. (15.23) would be known because V, A, and p are all functions of y and Q, and the right-hand side is a function of y_2; therefore, y_2 can be solved for.

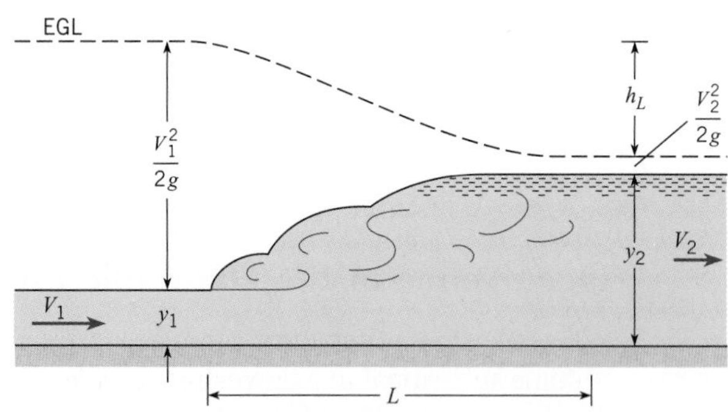

FIGURE 15.17

Definition sketch for the hydraulic jump.

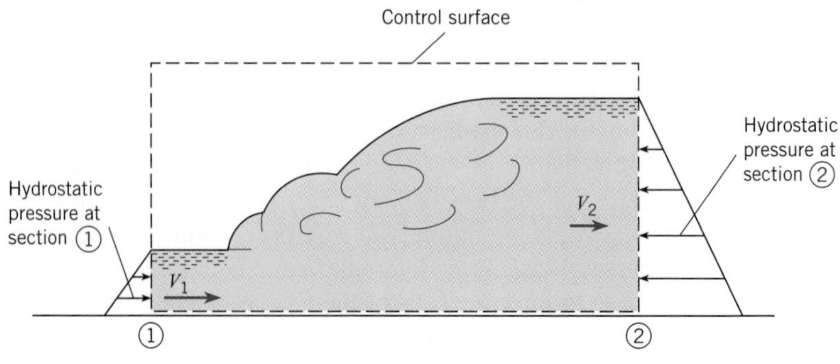

FIGURE 15.18

Control-volume analysis for the hydraulic jump.

example 15.2

Water flows in a trapezoidal channel at a rate of 300 cfs. The channel has a bottom width of 10 ft and side slopes of 1 vertical to 1 horizontal. If a hydraulic jump is forced to occur where the upstream depth is 1.0 ft, what will be the downstream depth and velocity? What are the values of Fr_1 and Fr_2?

Solution For the upstream section, the cross-sectional flow area is 11 ft². Therefore, the mean velocity $V_1 = Q/A_1 = 27.3\ \mathrm{ft/s}$. The hydraulic depth $D_1\ A_1/T_1 = 11\ \mathrm{ft}^2/12\ \mathrm{ft} = 0.9167\ \mathrm{ft}$.

Then
$$\mathrm{Fr}_1 = \frac{V_1}{\sqrt{gD_1}} = \frac{27.3\ \mathrm{ft/s}}{\sqrt{32.2\ \mathrm{ft/s}^2 \times 0.9167\ \mathrm{ft}}} = 5.02 \quad \triangleleft$$

The location of the centroid of the area A_1 can be obtained by taking moments of the subareas about the water surface:

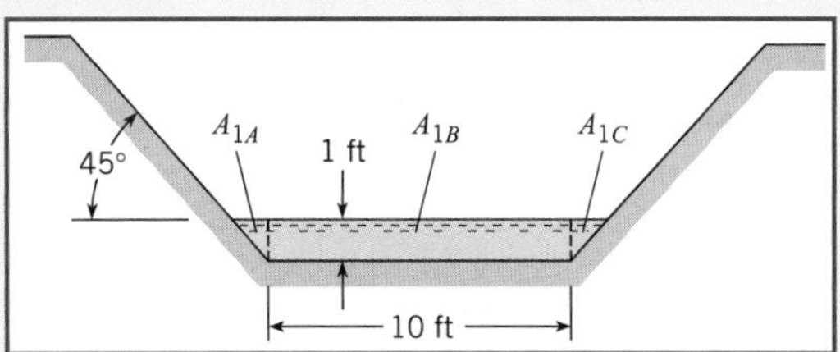

$$A_1\bar{y}_1 = A_{1A} \times 0.333\ \mathrm{ft} + A_{1B} \times 0.500\ \mathrm{ft} + A_{1C} \times 0.333\ \mathrm{ft}$$
$$(11\ \mathrm{ft}^2)\bar{y}_1 = (0.333\ \mathrm{ft})(0.500\ \mathrm{ft}^2 \times 2) + (0.50\ \mathrm{ft})(10.00\ \mathrm{ft}^2)$$
$$\bar{y} = 0.485\ \mathrm{ft}$$

Thus, the pressure at the centroid will be $62.4\ \mathrm{lbf/ft}^3 \times 0.485\ \mathrm{ft} = 30.26\ \mathrm{lbf/ft}^2$. Substituting the appropriate quantities into Eq. (15.23) yields

$$30.26 \times 11 + 1.94 \times 300 \times 27.3 = \bar{p}_2 A_2 + \rho Q V_2$$

or
$$\bar{p}_2 A_2 + \rho Q V_2 = 16{,}221\ \mathrm{lbf}$$

$$\gamma \bar{y}_2 A_2 + \frac{\rho Q^2}{A_2} = 16{,}221$$

where
$$\bar{y}_2 = \frac{\sum A_i y_i}{A_2} = \frac{By_2^2/2 + y_2^3/3}{A_2}$$

Then $$\gamma\left(\frac{By_2^2}{2}+\frac{y_2^3}{3}\right)+\frac{\rho Q^2}{(By+y^2)} = 16{,}221 \text{ lb}$$

Using $\gamma = 62.4 \text{ lbf/ft}^3$, $B = 10$ ft, $\rho = 1.94 \text{ slugs/ft}^3$, and $Q = 300 \text{ ft}^2\text{/s}$ and solving the above equation for y_2 yields

$$y_2 = 5.75 \text{ ft}$$

$$V_2 = \frac{Q}{A_2} = \frac{300}{57.5 + 5.75^2} = 3.31 \text{ ft/s}$$

Also $$D_2 = \frac{A}{T} = \frac{90.56}{21.5} = 4.21 \text{ ft}$$

$$\text{Fr}_2 = \frac{V}{\sqrt{gD}} = \frac{3.31}{\sqrt{32.2\times 4.21}} = 0.284$$

Hydraulic Jump in Rectangular Channels

If we write Eq. (15.23) for a unit width of a rectangular channel where $\bar{p}_1 = \gamma y_1/2$, $\bar{p}_2 = \gamma y_2/2$, $Q = q$, $A_1 = y_1$, and $A_2 = y_2$, we will have

$$\gamma\frac{y_1^2}{2} + \rho q V_1 = \gamma\frac{y_2^2}{2} + \rho q V_2 \tag{15.24a}$$

but $q = Vy$, so Eq. (15.24*a*) can be rewritten as

$$\frac{\gamma}{2}(y_1^2 - y_2^2) = \frac{\gamma}{g}(V_2^2 y_2 - V_1^2 y_1) \tag{15.24b}$$

The above equation can be further manipulated to yield

$$\frac{2V_1^2}{gy_1} = \left(\frac{y_2}{y_1}\right)^2 + \frac{y_2}{y_1} \tag{15.25}$$

The term on the left-hand side of Eq. (15.25) will be recognized as twice Fr_1^2. Hence Eq. (15.25) is written as

$$\left(\frac{y_2}{y_1}\right)^2 + \frac{y_2}{y_1} - 2\text{Fr}_1^2 = 0 \tag{15.26}$$

By use of the quadratic formula, it is easy to solve for y_2/y_1 in terms of the upstream Froude number. Thus we obtain the equation for *depth ratio* across a hydraulic jump:

$$\frac{y_2}{y_1} = \frac{1}{2}(\sqrt{1 + 8\text{Fr}_1^2} - 1) \tag{15.27}$$

or

$$y_2 = \frac{y_1}{2}(\sqrt{1 + 8\text{Fr}_1^2} - 1) \tag{15.28}$$

The other solution of Eq. (15.26) gives a negative downstream depth, which is not relevant here. Hence we have expressed the downstream depth in terms of the upstream depth and the upstream Froude number. In Eqs. (15.27) and (15.28), the depths y_1 and y_2 are said to be *conjugate* or *sequent* (both terms are in common use) to each other, in contrast to the alternate depths obtained from the energy equation. Numerous experiments show that the relation represented by Eqs. (15.27) and (15.28) is valid over a wide range of Froude numbers. Although no theory has been developed to predict the length of a hydraulic jump, experiments [see Chow (3)] show that the relative length of the jump, L/y_2, is approximately 6 for a range of Fr_1 from 4 to 18.

Head Loss in a Hydraulic Jump

In addition to determining the geometric characteristics of the hydraulic jump, it is often desirable to determine the head loss produced by it. This is obtained by comparing the specific energy before the jump to that after the jump, the head loss being the difference between the two specific energies. It can be shown that this *head loss* for a jump in a rectangular channel is

$$h_L = \frac{(y_2 - y_1)^3}{4y_1y_2} \tag{15.29}$$

For more information on the hydraulic jump, see Chow (2).

example 15.3

Water flows in a rectangular channel at a depth of 30 cm and with a velocity of 16 m/s, as shown in the figure. If a downstream sill (not shown in the figure) forces a hydraulic jump, what will be the depth and velocity downstream of the jump? What head loss is produced by the jump?

Solution To solve the problem we must know Fr_1, which is computed first:

$$\mathrm{Fr}_1 = \frac{V}{\sqrt{gy_1}} = \frac{16}{\sqrt{9.81(0.30)}} = 9.33$$

Next we compute y_2, using the depth ratio equation:

$$y_2 = \frac{0.30}{2}[\sqrt{1 + 8(9.33)^2} - 1] = 3.81 \text{ m} \quad \triangleleft$$

Then

$$V_2 = \frac{q}{y_2} = \frac{(16 \text{ m/s})(0.30 \text{ m})}{3.81\text{m}} = 1.26 \text{ m/s} \quad \triangleleft$$

From the equation for head loss,

$$h_L = \frac{(3.81 - 0.30)^3}{4(0.30)(3.81)} = 9.46 \text{ m} \quad \triangleleft$$

We can check the validity of the head loss equation since the head loss is equal to $E_1 - E_2$, or

$$h_L = \left(0.30 + \frac{16^2}{2 \times 9.81}\right) - \left(3.81 + \frac{1.26^2}{2 \times 9.81}\right) = 9.46 \text{ m}$$

The answers check.

Use of the Hydraulic Jump on Downstream End of Dam Spillway

We have already shown that the transition from supercritical to subcritical flow produces a hydraulic jump, and that the relative height of the jump (y_2/y_1) is a function of Fr_1. Because flow over the spillway of a dam invariably results in supercritical flow at the lower end of the spillway, and because flow in the channel downstream of a spillway is usually subcritical, it is obvious that a hydraulic jump must form near the base of the spillway (see Fig. 15.19). The downstream portion of the spillway, called the spillway *apron,* must be designed so that the hydraulic jump always forms on the concrete structure itself. If the hydraulic jump were allowed to form beyond the concrete structure, as in Fig. 15.20, severe erosion of the foundation material as a result of the high-velocity supercritical flow could undermine the dam and cause complete failure of it. One way to solve this problem might be to incorporate a long, sloping apron into the design of the

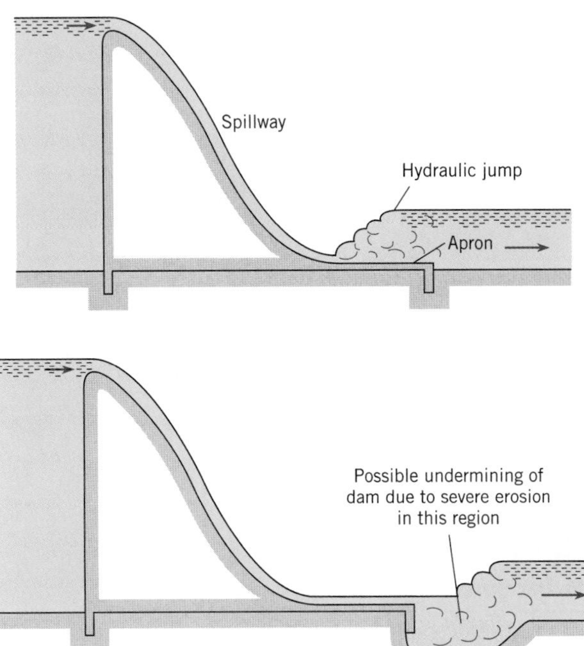

FIGURE 15.19

Spillway of dam and hydraulic jump.

FIGURE 15.20

Hydraulic jump occurring downstream of spillway apron.

spillway, as shown in Fig. 15.21. A design like this would work very satisfactorily from the hydraulics point of view. For all combinations of Fr_1 and water-surface elevation in the downstream channel, the jump would always form on the sloping apron. However, its main drawback is cost of construction. Construction costs will be reduced as the length, L, of the stilling basin is reduced. Much research has been devoted to the design of stilling basins that will operate properly for all upstream and downstream conditions and yet be relatively short to reduce construction cost. Research by the U.S. Bureau of Reclamation (7) has resulted in sets of standard designs that can be used. These designs include sills, baffle piers, and chute blocks, as shown in Fig. 15.22.

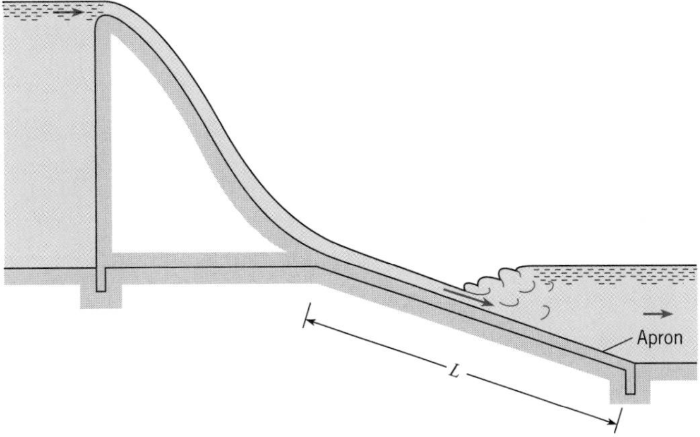

FIGURE 15.21

Long sloping apron.

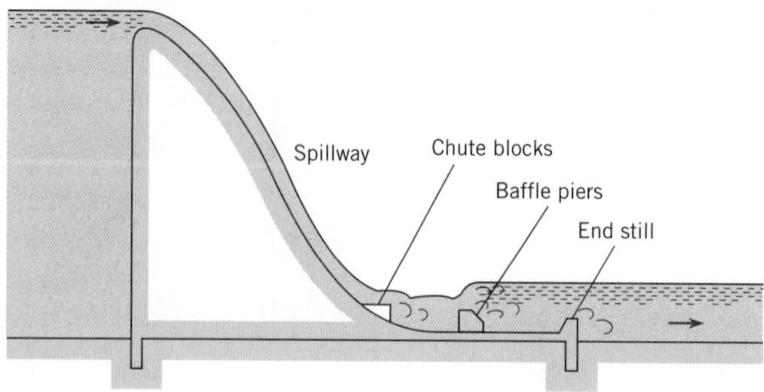

FIGURE 15.22

Spillway with stilling basin Type III as recommended by the U.S.B.R. (12).

Surge or Tidal Bore

Tides are generally low enough that the waves they produce are smooth and nondestructive. However, in some parts of the world the tides are so high that their entry into shallow bays or mouths of rivers causes surges to be formed that may be very hazardous to small boats. A *surge* is actually a moving hydraulic jump. Hence the same analytical methods that we used for the jump can be used to solve for the speed of the surge. In Fig. 15.23 a surge is shown coming into an otherwise still body of water. As indicated in Fig. 15.23, the flow is unsteady because throughout the surge itself the velocities are changing with time. However, if all velocities are measured in a coordinate system moving with the surge front, then a steady-flow pattern is obtained. Figure 15.24 shows the steady-flow pattern and control volume. Now the problem is directly analogous to the hydraulic-jump problem. We simply replace V_1 in Eq. (15.25) by V_s to yield

$$\frac{V_s}{\sqrt{gy_1}} = \left[\frac{y_2}{2y_1}\left(\frac{y_2}{y_1} + 1\right)\right]^{1/2} \tag{15.30}$$

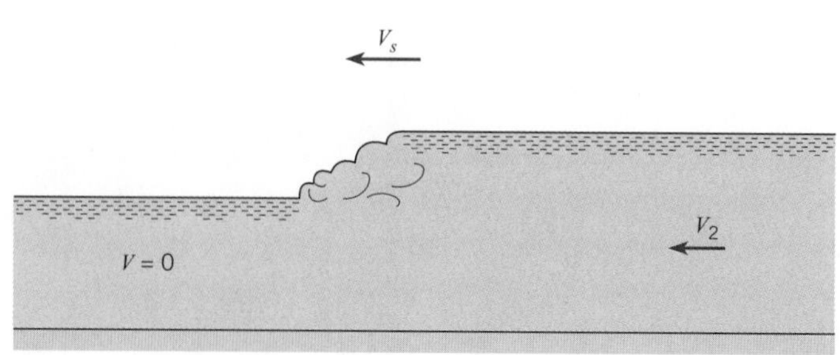

FIGURE 15.23

Surge moving into still water.

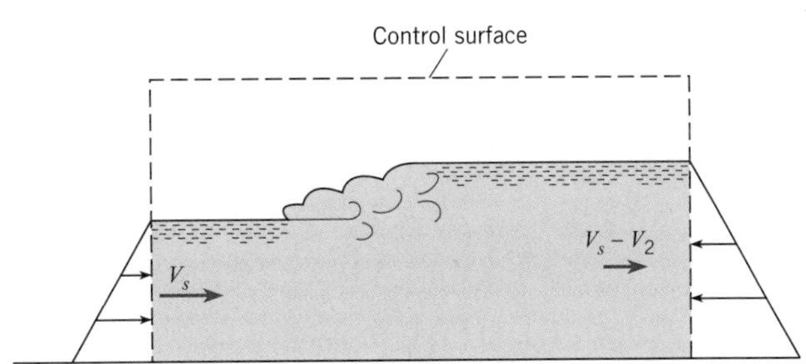

FIGURE 15.24

Control-volume analysis applied to the surge.

Gradually Varied Flow in Open Channels

In Sections 15.1, 15.2, and 15.3 we considered cases of rapidly varied flow, and in those examples the channel resistance was assumed to be negligible. For gradually varied flow, however, channel resistance is usually a significant factor in the flow process. We utilize the energy equation for the solution of this type of problem.

Basic Differential Equation for Gradually Varied Flow

There are a number of cases of open-channel flow in which the change in water-surface profile is so gradual that it is possible to integrate the relevant differential equation from one section to another to obtain the desired change in depth. This may be either an analytical integration or, more commonly, a numerical integration. In Section 15.1, the energy equation was written between two sections of a channel Δx distance apart. Because the only head loss here is the channel resistance, we can write h_L as Δh_f, and Eq. (15.2) becomes

$$y_1 + \frac{V_1^2}{2g} + S_0\Delta x = y_2 + \frac{V_2^2}{2g} + \Delta h_f \tag{15.31}$$

The friction slope S_f is defined as the slope of the EGL, or $\Delta h_f/\Delta x$. Then $\Delta h_f = S_f\Delta x$, and if we let $\Delta y = y_2 - y_1$ and

$$\frac{V_2^2}{2g} - \frac{V_1^2}{2g} = \frac{d}{dx}\left(\frac{V^2}{2g}\right)\Delta x \tag{15.32}$$

Eq. (15.31) becomes

$$\Delta y = S_0\Delta x - S_f\Delta x - \frac{d}{dx}\left(\frac{V^2}{2g}\right)\Delta x$$

Dividing through by Δx and taking the limit as Δx approaches zero gives us

$$\frac{dy}{dx} + \frac{d}{dx}\left(\frac{V^2}{2g}\right) = S_0 - S_f \tag{15.33}$$

The second term is rewritten as $[d(V^2/2g)/dy]\,dy/dx$, so that Eq. (15.33) simplifies to

$$\frac{dy}{dx} = \frac{S_0 - S_f}{1 + d(V^2/2g)/dy} \tag{15.34}$$

To put Eq. (15.34) in a more usable form, we express the denominator in terms of the Froude number. This is accomplished by observing that

$$\frac{d}{dy}\left(\frac{V^2}{2g}\right) = \frac{d}{dy}\left(\frac{Q^2}{2gA^2}\right) \tag{15.35}$$

After differentiating the right side of Eq. (15.35), the equation becomes

$$\frac{d}{dy}\left(\frac{V^2}{2g}\right) = \frac{-2Q^2}{2gA^3} \cdot \frac{dA}{dy}$$

But $dA/dy = T$ (top width), and $A/T = D$ (hydraulic depth); therefore,

$$\frac{d}{dy}\left(\frac{V^2}{2g}\right) = \frac{-Q^2}{gA^2D}$$

or

$$\frac{d}{dy}\left(\frac{V^2}{2g}\right) = -\mathrm{Fr}^2$$

Hence, when the expression for $d(V^2/2g)/dy$ is substituted into Eq. (15.34), we obtain

$$\frac{dy}{dx} = \frac{S_0 - S_f}{1 - \mathrm{Fr}^2} \tag{15.36}$$

This is the general differential equation for gradually varied flow. It is used to describe the various types of water-surface profiles that occur in open channels. Note that, in the derivation of the equation, S_0 and S_f were taken as positive when the channel and energy grade lines, respectively, were sloping downward in the direction of flow. Also note that y is measured from the bottom of the channel. Therefore, $dy/dx = 0$ if the slope of the water surface is equal to the slope of the channel bottom, and dy/dx is positive if the slope of the water surface is less than the channel slope.

Introduction to Water-Surface Profiles

In the design of projects involving the flow in channels (rivers or irrigation canals, for example), the engineer must often estimate the *water-surface profile* (elevation of the water surface along the channel) for a given discharge. For example, when a dam is being designed for a river project, the water-surface profile in the river upstream must be defined so that the project planners will know how much land to acquire to accommodate the upstream pool. The first step in defining a water-surface profile is to locate a point or points along the channel where the depth can be computed for a given discharge. For example, at a change in slope from mild to steep, critical depth will occur just upstream of the break in grade (see Section 15.1, page 621). At that point we can solve for y_c with Eq. (15.8) or Eq. (15.13). Also, for flow over the spillway of a dam, there will be a discharge equation for the spillway from which we can calculate the water-surface elevation in the reservoir at the face of the dam. Such points where there is a unique relationship between discharge and water-surface elevation are called *controls*. Once the water-surface elevations at these controls are determined, then the water-surface profile can be extended upstream or downstream from the control points to define the water-surface profile for the entire channel. The completion of the profile is done by numerical integration. However, before this integration is performed, it is usually helpful to the engineer to sketch in the profiles. To assist in the process of sketching the possible profiles, the engineer can refer to different categories of profiles (water-surface profiles have unique characteristics depending upon the relationship between normal depth, critical depth, and the actual depth of flow in the channel). This initial sketching of the profiles helps the engineer to scope the problem and to obtain a solution or solutions in a minimum amount of time. The next section describes the various types of water-surface profiles.

Types of Water-Surface Profiles

There are 12 different types of water-surface profiles for gradually varied flow in channels, and these are shown schematically in Fig. 15.25. Each profile is identified by a letter and number designator. For example, the first water-surface profile in column 1 of Fig. 15.25 is identified as an M1 profile. The letter indicates the type of slope of the channel—that is, whether the slope is mild (M), critical (C), steep (S), horizontal (H), or adverse (A). The slope is defined as mild if the uniform flow depth, y_n, is greater than the critical flow depth, y_c. Conversely, if y_n is less than y_c, the channel would be termed steep. Or if $y_n = y_c$, this would be a channel with critical slope. The designation M, S, or C is determined by computing y_n and y_c for the given channel for a given discharge. Equation (10.40) or (10.41) is used to compute y_n, and Eq. (15.8) or (15.13) is used to compute y_c. Figure 15.26 shows the relationship between y_n and y_c for the M, S, and C designations. As the name implies, a horizontal slope is one where the channel actually has a zero slope, and an adverse slope is one where the slope of the channel is upward in the direction of flow. Normal depth does not exist for these two cases (for example, water cannot flow at uniform depth in either a horizontal channel or one with adverse slope); therefore, they are given the special designations H and A, respectively.

The number designator for the type of profile relates to the position of the *actual* water surface in relation to the position of the water surface for uniform and critical flow in the channel. If the actual water surface is above that for uniform and critical flow ($y > y_n$; $y > y_c$), then that condition is given a 1 designation. If the actual water surface is between those for uniform and critical flow, then it is given a 2 designation; and if the actual water surface lies below those for uniform and critical flow, then it is given a 3 designation. Figure 15.27 depicts these conditions for mild and steep slopes.

Figure 15.28 shows how different water-surface profiles can develop in certain field situations. More specifically, if we consider in detail the flow downstream of the sluice gate (see Fig. 15.29), we see that the discharge and slope are such that the normal depth is greater than the critical depth; therefore the slope is termed mild. The actual depth of flow shown in Fig. 15.29 is less than both y_c and y_n. Hence a type 3 water-surface

FIGURE 15.25

Classification of water-surface profiles of gradually varied flow. [Adapted from Open Channel Hydraulics *by Chow (3). Copyright © 1959, McGraw-Hill Book Company, New York; used with permission of McGraw-Hill Book Company.]*

	Profiles in zone 1: $y > y_n$; $y > y_c$	Profiles in zone 2: $y_n \geq y \geq y_c$; $y_c \geq y \geq y_n$	Profiles in zone 3: $y < y_n$; $y < y_c$
Horizontal slope $y_n > y_c$	None (∞, y_n, y_c)	H2 (∞, y_n, y_c)	H3 (∞, y_n, y_c)
Mild slope $y_n > y_c$	M1 (y_n, y_c)	M2 (y_n, y_c)	M3 (y_n, y_c)
Critical slope $y_n = y_c$	C1 ($y_n = y_c$)	Unstable surface ($y_n = y_c$)	C3 ($y_n = y_c$)
Steep slope $y_n < y_c$	S1 (y_c, y_n)	S2 (y_c, y_n)	S3 (y_c, y_n)
Adverse slope	None (y_c)	A2 (y_c)	A3 (y_c)

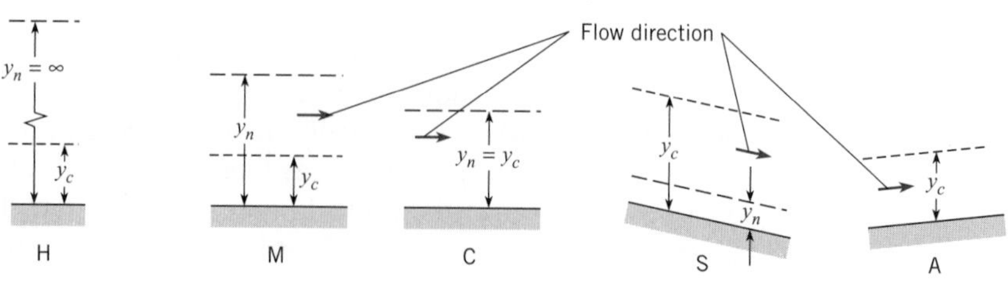

FIGURE 15.26

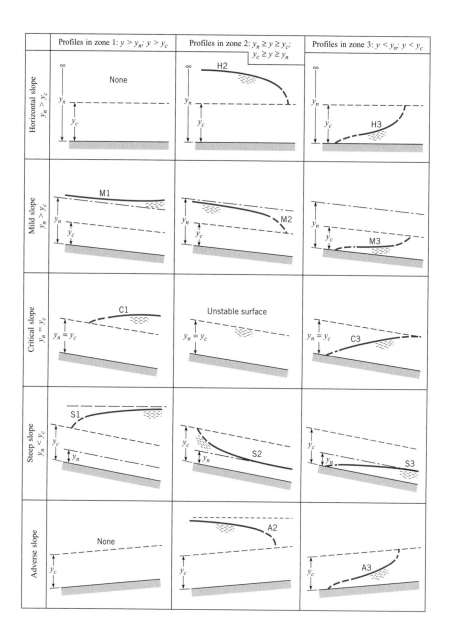

Letter designators as a function of the relationship between y_n and y_c.

profile exists. The complete classification of the profile in Fig. 15.29, therefore, is a mild type 3 profile, or simply an M3 profile. Using these designations, we would categorize the profile upstream of the sluice gate as type M1.

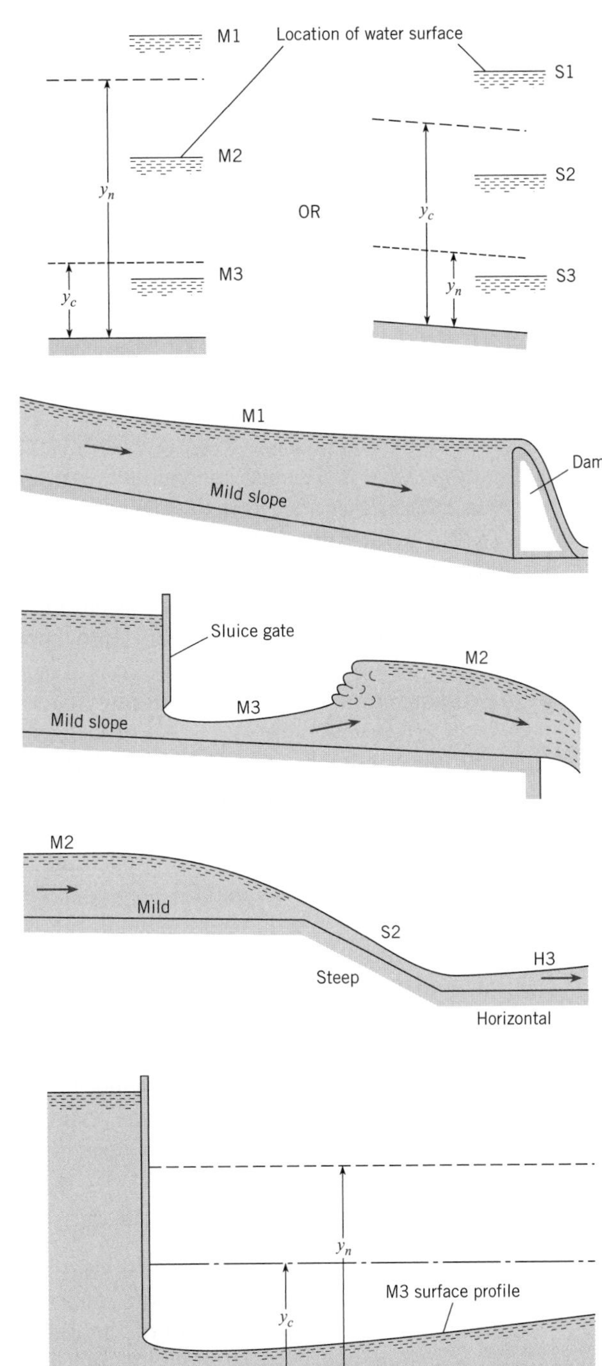

FIGURE 15.27

Number designator as a function of the location of the actual water surface in relation to y_n and y_c.

FIGURE 15.28

Water-surface profiles associated with flow behind a dam, flow under a sluice gate, and flow in a channel with a change in grade.

FIGURE 15.29

Water-surface profile, M3 type.

example 15.4

Classify the water-surface profile for the flow downstream of the sluice gate in Fig. 15.3 if the slope is horizontal, and that for the flow immediately downstream of the break in grade in Fig. 15.9.

Solution In Fig. 15.3 the actual depth is less than critical; thus the profile is type 3. The channel is horizontal; hence the profile is designated type H3. ◁

In Fig. 15.9 the actual depth is greater than normal but less than critical, so the profile is type 2. The uniform-flow depth (normal depth y_n) is less than the critical depth; hence the slope is steep. Therefore the water-surface profile is designated type S2. ◁

With the foregoing introduction to the classification of water-surface profiles, we can now refer to Eq. (15.36) to describe the shapes of the profiles. Again, for example, if we consider the M3 profile, we know that $\mathrm{Fr} > 1$ because the flow is supercritical ($y < y_c$), and that $S_f > S_0$ because the velocity is greater than normal velocity. Hence a head loss greater than that for normal flow must exist. Inserting these relative values into Eq. (15.36), we see that both the numerator and the denominator are negative. Thus dy/dx must be positive (the depth increases in the direction of flow), and as critical depth is approached, the Froude number approaches unity. Hence the denominator of Eq. (15.36) approaches zero. Therefore, as the depth approaches critical depth, $dy/dx \to \infty$. What actually occurs in cases where the critical depth is approached in supercritical flow is that a hydraulic jump forms and a discontinuity in profile is thereby produced.

Certain general features of profiles, as shown in Fig. 15.25, are evident. First, as the depth becomes very great, the velocity of flow approaches zero. Hence $\mathrm{Fr} \to 0$ and $S_f \to 0$ and dy/dx approaches S_0 because $dy/dx = (S_0 - S_f)(1 - \mathrm{Fr}^2)$. In other words, the depth increases at the same rate at which the channel bottom drops away from the horizontal. Thus the water surface approaches the horizontal. The profiles that show this tendency are types M1, S1, and C1. A physical example of the M1 type is the water-surface profile upstream of a dam, as shown in Fig. 15.28. The second general feature of several of the profiles is that those that approach normal depth do so asymptotically. This is shown in the S2, S3, M1, and M2 profiles. Also note in Fig. 15.25 that profiles that approach critical depth are shown by dashed lines. This is done because, near critical depth, either discontinuities develop (hydraulic jump) or the streamlines are very curved (such as near a brink). These profiles cannot be accurately predicted by Eq. (15.36) because this equation is based on one-dimensional flow, which, in these regions, is invalid.

Quantitative Evaluation of the Water-Surface Profile

In practice, most water-surface profiles are generated by numerical integration, that is, by dividing the channel into short reaches and carrying the computation for water-surface elevation from one end of the reach to the other. For one method, called the *direct step method,* the depth and velocity are known at a given section of the channel (one end of the reach), and one arbitrarily chooses the depth at the other end of the reach. Then the length of the reach is solved for. The applicable equation for quantitative evaluation of the water-surface profile is the energy equation written for a finite reach of channel, Δx:

$$y_1 + \frac{V_1^2}{2g} + S_0\Delta x = y_2 + \frac{V_2^2}{2g} + S_f\Delta x$$

From this equation we obtain

$$\Delta x(S_f - S_0) = \left(y_1 + \frac{V_1^2}{2g}\right) - \left(y_2 + \frac{V_2^2}{2g}\right)$$

or

$$\Delta x = \frac{(y_1 + V_1^2/2g) - (y_2 + V_2^2/2g)}{S_f - S_0} = \frac{(y_1 - y_2) + (V_1^2 - V_2^2)/2g}{S_f - S_0} \tag{15.37}$$

The procedure for evaluation of a profile starts by ascertaining which type applies to the given reach of channel (we use the methods of the preceding subsection). Then, starting from a known depth, we compute a finite value of Δx for an arbitrarily chosen change in depth. The process of computing Δx, step by step, up (negative Δx) or down (positive Δx) the channel is repeated until the full reach of channel has been covered. Usually small changes of y are taken, so that the friction slope is approximated by the following equation:

$$S_f = \frac{h_f}{\Delta x} = \frac{fV^2}{8gR} \tag{15.38}$$

Here V is the mean velocity in the reach and R is the mean hydraulic radius. That is, $V = (V_1 + V_2)/2$ and $R_h = (R_{h1} + R_{h2})/2$. It is obvious that a numerical approach of this type is ideally suited for solution by computer.

example 15.5

Water discharges from under a sluice gate into a horizontal rectangular channel at a rate of 1 m^3/s per meter of width as shown. What is the classification of the water-surface profile? Quantitatively evaluate the profile downstream of the gate and determine whether it will extend all the way to the abrupt drop 80 m downstream. Make the simplifying assumption that the resistance factor f is equal to 0.02 and that the hydraulic radius R_h is equal to the depth y.

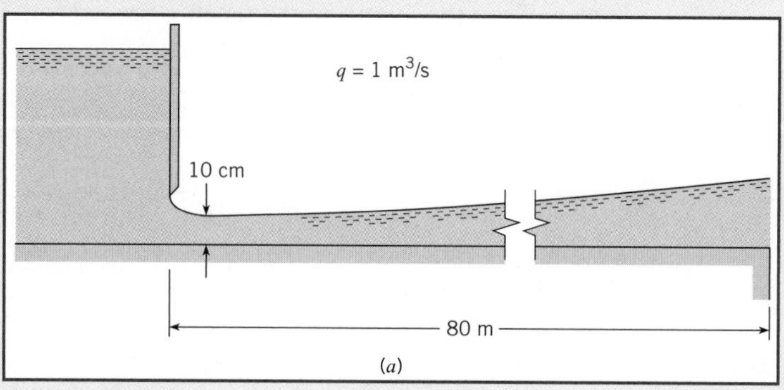

(a)

Solution First determine the critical depth y_c:

$$y_c = (q^2/g)^{1/3} = [(1^2 \text{ m}^4/\text{s}^2)/(9.81 \text{ m/s}^2)]^{1/3} = 0.467 \text{ m}$$

The depth of flow from the sluice gate is less than the critical depth. Hence the water-surface profile is classified as type H3. ◁

To solve for the depth versus distance along the channel, we apply Eqs. (15.37) and (15.38), using a numerical approach. The results of the computation are given in the table on page 649. From the numerical results we plot the profile shown in the accompanying figure, and we see that the *profile extends to the abrupt drop.* ◁

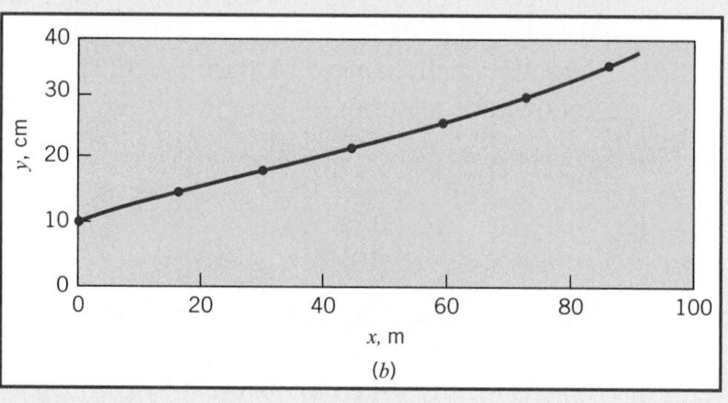

(b)

15.5 Summary

Nonuniform flow in open channels is characterized as either rapidly varied flow or gradually varied flow. In rapidly varied flow the channel resistance is negligible and flow changes (depth and velocity change) occur over relatively short distances.

The significant π-group is Froude number,

SOLUTION TO EXAMPLE 15.5

Section Number Downstream of Gate	Depth y, m	Velocity at Section V, m/s	Mean Velocity in Reach, $(V_1 + V_2)/2$	V^2	Mean Hydraulic Radius, $R_m = (y_1 + y_2)/2$	$S_f = \frac{f V_{\text{mean}}^2}{8gR_m}$	$\Delta x = \frac{(y_1 - y_2) + \frac{(V_1^2 - V_2^2)}{2g}}{(S_f - S_0)}$	Distance from Gate x, m
1 (at gate)	0.1	10	...	100	...	...	...	0
	...	...	8.57	73.4	0.12	0.156	15.7	
2	0.14	7.14	...	51.0	...	...	...	15.7
	...	...	6.35	40.3	0.16	0.064	15.3	
3	0.18	5.56	...	30.9	...	...	...	31.0
	...	...	5.05	25.5	0.20	0.032	15.1	
4	0.22	4.54	...	20.6	...	...	...	46.1
	...	...	4.19	17.6	0.24	0.019	13.4	
5	0.26	3.85	...	14.8	...	...	...	59.5
	...	...	3.59	12.9	0.28	0.012	12.4	
6	0.30	3.33	...	11.1	...	...	...	71.9
	...	...	3.13	9.8	0.32	0.008	10.9	
7	0.34	2.94	...	8.6	...	...	...	82.8

$$\frac{V}{\sqrt{gD_c}}$$

where D_c is the hydraulic depth, A/T. When the Froude number is equal to unity the flow is critical. Subcritical flow occurs when the Froude number is less than unity and supercritical when the Froude number is greater than unity.

A hydraulic jump usually occurs when the flow along the channel changes from supercritical to subcritical. The governing equation for hydraulic jump in a horizontal, rectangular channel is

$$y_2 = \frac{y_1}{2}\left(\sqrt{1 + 8\mathrm{Fr}_1^2} - 1\right)$$

The corresponding head loss in the hydraulic jump is

$$h_L = \frac{(y_2 - y_1)^3}{4y_1y_2}$$

When the flow along the channel changes from subcritical to supercritical flow, the head loss is assumed to be negligible and the depth and velocity relationship is governed by the change in elevation of the channel bottom and the specific energy, $y + V^2/2g$. Typical cases of this type of flow are

1. Flow under a sluice gate
2. An upstep in the channel bottom
3. Reduction in width of the channel

For gradually varied flow the governing differential equation is

$$\frac{dy}{dx} = \frac{S_0 - S_f}{1 - \mathrm{Fr}^2}$$

When this equation is integrated along the length of the channel the depth, y, is determined as a function of distance, x, along the channel. This yields the water surface profile for the reach of the channel.

References

1. Parshall, R. L. "The Improved Venturi Flume." *Trans. ASCE,* 89 (1926), 841–851.

2. Roberson, J. A., J. J. Cassidy, and M. H. Chaudhry. *Hydraulic Engineering*. John Wiley, New York, 1988.

3. Chow, Ven Te. *Open Channel Hydraulics*. McGraw-Hill, New York, 1959.

4. Hinds, J. "The Hydraulic Design of Flume and Siphon Transitions." *Trans. ASCE,* 92 (1928), pp. 1423–59.

5. U.S. Bureau of Reclamation. *Design of Small Canal Structures*. U.S. Dept. of Interior, U.S. Govt. Printing Office, Washington, DC, 1978.

6. Rouse, H. (ed.). *Engineering Hydraulics*. John Wiley, New York, 1950.

7. U.S. Bureau of Reclamation. *Hydraulic Design of Stilling Basin and Bucket Energy Dissipators*. Engr. Monograph no. 25, U.S. Supt. of Doc., 1958.

Problems

15.1 Water flows at a depth of 4 in. with a velocity of 28 ft/s in a rectangular channel. Is the flow subcritical or supercritical? What is the alternate depth?

15.2 The water discharge in a rectangular channel 16 ft wide is 900 cfs. If the depth of water is 3 ft, is the flow subcritical or supercritical?

15.3 The discharge in a rectangular channel 18 ft wide is 420 cfs. If the water velocity is 9 ft/s, is the flow subcritical or supercritical?

15.4 Water flows at a rate of 12 m^3/s in a rectangular channel 3 m wide. Determine the Froude number and the type of flow (subcritical, critical, or supercritical) for depths of 30 cm, 1.0 m, and 2.0 m. What is the critical depth?

15.5 For the discharge and channel of Prob. 15.4, what is the alternate depth to the 30-cm depth? What is the specific energy for these conditions?

15.6 Water flows at the critical depth with a velocity of 5 m/s. What is the depth of flow?

15.7 Water flows uniformly at a rate of 320 cfs in a rectangular channel that is 12 ft wide and has a bottom slope of 0.005. If n is 0.014, is the flow subcritical or supercritical?

15.8 The discharge in a trapezoidal channel is 10 m^3/s. The bottom width of the channel is 3.0 m, and the side slopes are 1 vertical to 1 horizontal. If the depth of flow is 1.0 m, is the flow supercritical or subcritical?

15.9 For the channel of Prob. 15.8, determine the critical depth for a discharge of 20 m^3/s.

15.10 A rectangular channel is 6 m wide, and the discharge of water in it is 18 m^3/s. Plot depth versus specific energy for these conditions. Let specific energy range from E_{min} to $E = 7$ m. What are the alternate and sequent depths to the 30-cm depth?

15.11 A long rectangular channel that is 4 m wide and has a mild slope ends in a free outfall. If the water depth at the brink is 0.35 m, what is the discharge in the channel?

15.12 A rectangular channel that is 15 ft wide and has a mild slope ends in a free outfall. If the water depth at the brink is 1.20 ft, what is the discharge in the channel?

15.13 A horizontal rectangular channel 14 ft wide carries a discharge of water of 500 cfs. If the channel ends with a free outfall, what is the depth at the brink?

15.14 What discharge of water will occur over a 2-ft-high, broad-crested weir that is 10 ft long if the head on the weir is 1.5 ft?

15.15 What discharge of water will occur over a 2-m-high, broad-crested weir that is 5 m long if the head on the weir is 60 cm?

15.16 The crest of a high, broad-crested weir has an elevation of 100 m. If the weir is 10 m long and the discharge of water over the weir is 25 m^3/s, what is the water-surface elevation in the reservoir upstream?

15.17 The crest of a high, broad-crested weir has an elevation of 300 ft. If the weir is 40 ft long and the discharge of water over the weir is 1200 cfs, what is the water-surface elevation in the reservoir upstream?

15.18 Water flows with a velocity of 3 m/s and at a depth of 3 m in a rectangular channel. What is the change in depth and in water-surface elevation produced by a gradual upward change in bottom elevation (upstep) of 30 cm? What would be the depth and elevation changes if there were a gradual downstep of 30 cm? What is the maximum size of upstep that could exist before upstream depth changes would result?

15.19 Water flows with a velocity of 2 m/s and at a depth of 3 m in a rectangular channel. What is the change in depth and in water-surface elevation produced by a gradual upward change in bottom elevation (upstep) of 60 cm? What would be the depth and elevation changes if there were a gradual downstep of 15 cm? What is the maximum size of upstep that could exist before upstream depth changes would result?

15.20 Assuming no energy loss, what is the maximum value of Δz that will permit the unit flow rate of 6 m^2/s to pass over the hump without increasing the upstream depth? Sketch carefully the water-surface shape from section 1 to section 2. On the sketch give values for Δz, the depth, and the amount of rise or fall in the water surface from section 1 to section 2.

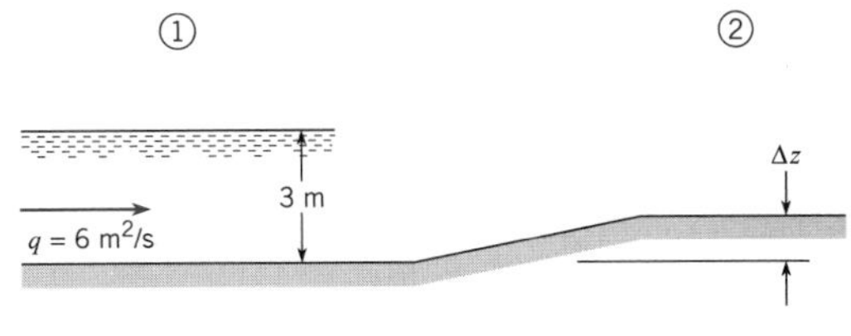

PROBLEM 15.20

15.21 Water flows with a velocity of 3 m/s in a rectangular channel 3 m wide at a depth of 3 m. What is the change in depth and in water-surface elevation produced when a gradual contraction in the channel to a width of 2.6 m takes place? Determine the greatest contraction allowable without altering the specified upstream conditions.

15.22 Because of the increased size of ships, the phenomenon called "ship squat" has produced serious problems in harbors where the draft of vessels approaches the depth of the ship channel. When a ship steams up a channel, the resulting flow situation is analogous to open-channel flow in which a constricting flow section exists (the ship reduces the cross-sectional area of the channel). The problem may be analyzed by referencing the water velocity to the ship and applying the energy equation. Thus, at the section of the channel where the ship is located, the relative water velocity in the channel will be greatest and the water level in the channel will be reduced as dictated by the energy equation. Consequently, the ship itself will be at a lower elevation than if it were stationary; this lowering is referred to as "ship squat." Estimate the squat of the fully loaded supertanker *Bellamya* when it is steaming at 5 kt (1 kt $= 0.515$ m/s) in a channel that is 35 m deep and 200 m wide. The draft of the *Bellamya* when fully loaded is 29 m. Its width and length are 63 m and 414 m, respectively.

FLUID SOLUTIONS **15.23** A rectangular channel that is 10 ft wide is very smooth except for a small reach that is roughened with angle irons attached to the bottom. Water flows in the channel at a rate of 200 cfs and at a depth of 1.0 ft upstream of the rough section. Assume frictionless flow except over the roughened part, where the total drag of all roughness (all of the angle irons) is assumed to be 2000 lbf. Determine the depth downstream of the roughness for the assumed conditions.

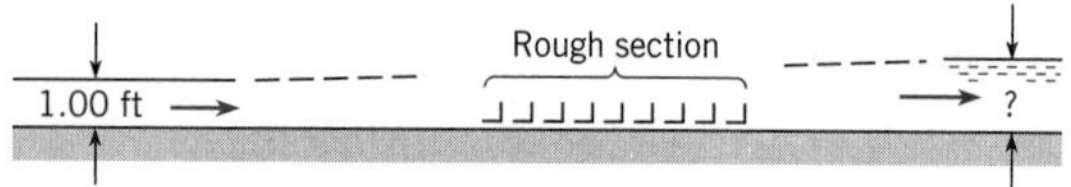

PROBLEM 15.23

15.24 Water flows from a reservoir into a steep rectangular channel that is 4 m wide. The reservoir water surface is 3 m above the channel bottom at the channel entrance. What discharge will occur in the channel?

15.25 A small wave is produced in a pond that is 8 in. deep. What is the speed of the wave in the pond?

15.26 A small wave in a pool of water having constant depth travels at a speed of 1.5 m/s. How deep is the water?

15.27 As waves in the ocean approach a sloping beach, they curve so that they are nearly parallel to the beach when they finally break (see the accompanying figure). Explain why the waves curve like this.

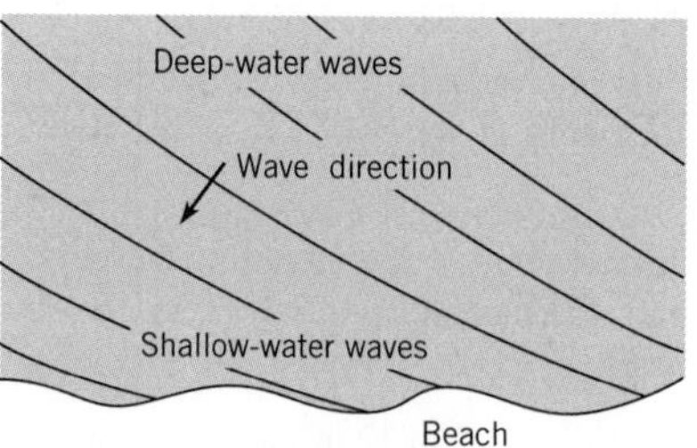

PROBLEM 15.27

15.28 The baffled ramp shown is used as an energy dissipator in a two-dimensional open channel. For a discharge of 18 cfs per foot of width, calculate the head lost, the power dissipated, and the horizontal component of force exerted by the ramp on the water.

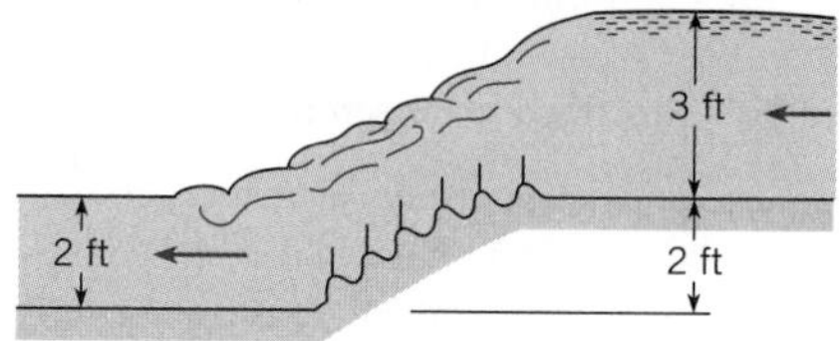

PROBLEM 15.28

15.29 The spillway shown has a discharge of 2.5 m^3/s per meter of width occurring over it. What depth y_2 will exist downstream of the hydraulic jump? Assume negligible energy loss over the spillway.

15.30 The flow of water downstream from a sluice gate in a horizontal channel has a depth of 30 cm and a flow rate of 3.60 m^3/m of width. The sluice gate is 2 m wide. Could a hydraulic jump be caused to form downstream of this section? If so, what would be the depth downstream of the jump?

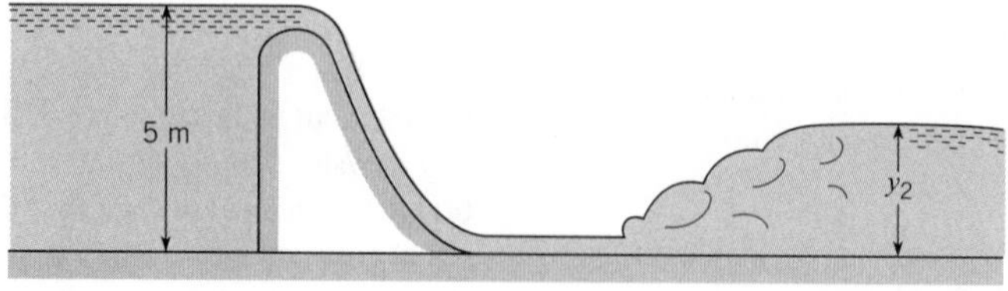

PROBLEM 15.29

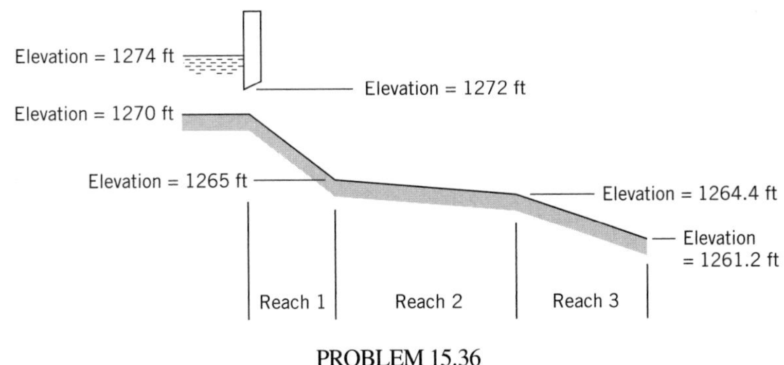

PROBLEM 15.36

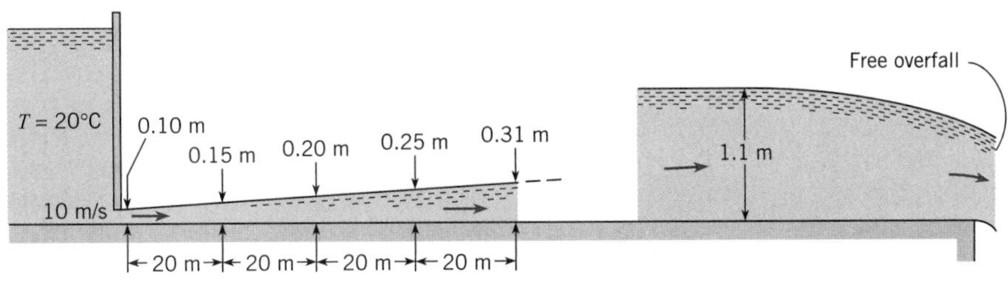

PROBLEM 15.37

15.31 Consider the dam and spillway shown in Fig. 15.19. Water is discharging over the spillway of the dam as shown. The elevation difference between the upstream pool level and the floor of the apron of the dam is 100 ft. If the head on the spillway is 5 ft and if a hydraulic jump forms on the horizontal apron, what is the depth of flow on the apron just downstream of the jump? Assume that the velocity just upstream of the jump is 95% of the maximum theoretical velocity. *Note:* The discharge over the spillway is given as $Q = KL\sqrt{2g}H^{3/2}$, where L is the length of the spillway, K is a coefficient (assume it has a value of 0.5), and H is the head on the spillway.

15.32 It is known that the discharge per unit width is 65 cfs/ft and that the height (H) of the hydraulic jump is 14 ft. What is the depth y_1?

PROBLEM 15.32

15.33 Water flows in a channel at a depth of 40 cm and with a velocity of 8 m/s. An obstruction causes a hydraulic jump to be formed. What is the depth of flow downstream of the jump? FLUID SOLUTIONS

15.34 Water flows in a trapezoidal channel at a depth of 40 cm and with a velocity of 10 m/s. An obstruction causes a hydraulic jump to be formed. What is the depth of flow downstream of the jump? The bottom width of the channel is 5 m, and the side slopes are 1 vertical to 1 horizontal.

15.35 A hydraulic jump occurs in a wide rectangular channel. If the depths upstream and downstream are 0.50 ft and 10 ft, respectively, what is the discharge per foot of width of channel?

15.36 The 20-ft-wide rectangular channel shown has three different reaches. $S_{0_1} = 0.01$; $S_{0_2} = 0.0004$; $S_{0_3} = 0.00317$; $Q = 500$ cfs; $n_1 = 0.015$; normal depth for reach 2 is 5.4 ft and that for reach 3 is 2.7 ft. Determine the critical depth and normal depth for reach 1 (use Manning's equation). Then classify the flow in each reach (supercritical, subcritical, critical), and determine whether a hydraulic jump could occur. In which reach(es) might it occur if it does occur?

15.37 Water flows from under the sluice gate as shown and continues on to a free overfall (also shown). Upstream from the overfall the flow soon reaches a normal depth of 1.1 m. The profile immediately downstream of the sluice gate is as it would be if there were no influence from the part nearer the overfall. Will a hydraulic jump form for these conditions? If so, locate its position. If not, sketch the full profile and label each part. Draw the energy grade line for the system.

FLUID SOLUTIONS **15.38** For the conditions of Prob. 15.37, estimate the shear stress on the smooth bottom 0.5 m downstream of the sluice gate. That shear stress τ falls within the range (a) $0 < \tau_0 \leq 1$ N/m², (b) $1 < \tau_0 \leq 10$ N/m², (c) $10 < \tau_0 \leq 40$ N/m², (d) 40 N/m$^2 < \tau_0$.

FLUID SOLUTIONS **15.39** Water is flowing as shown under the sluice gate in a horizontal rectangular channel that is 5 ft wide. The depths of y_0 and y_1 are 65 ft and 1 ft, respectively. What will be the horsepower lost in the hydraulic jump?

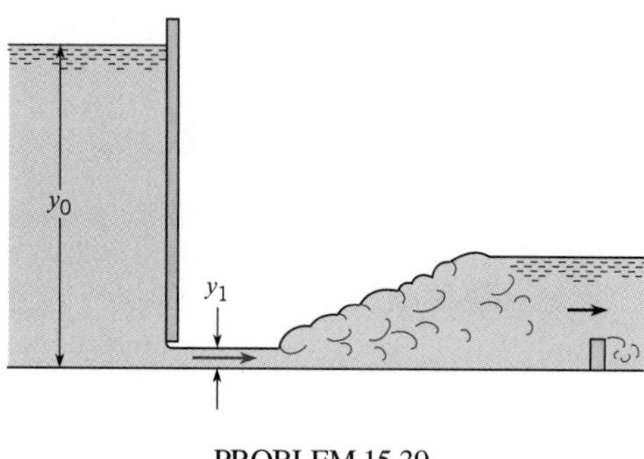

PROBLEM 15.39

15.40 Given the laboratory setup described in Problem 13.77 (p. 575), design a flume for the laboratory that will be used to verify the hydraulic jump relations given in Section 15.2. The design should include dimensions, valves, measuring devices, etc., but not stress or strain calculations on the structure itself.

15.41 Water flows uniformly at a depth $y_1 = 40$ cm in the concrete channel, which is 10 m wide. Estimate the height of the hydraulic jump that will form when a sill is installed to force it to form. Assume Manning's n value is $n = 0.012$.

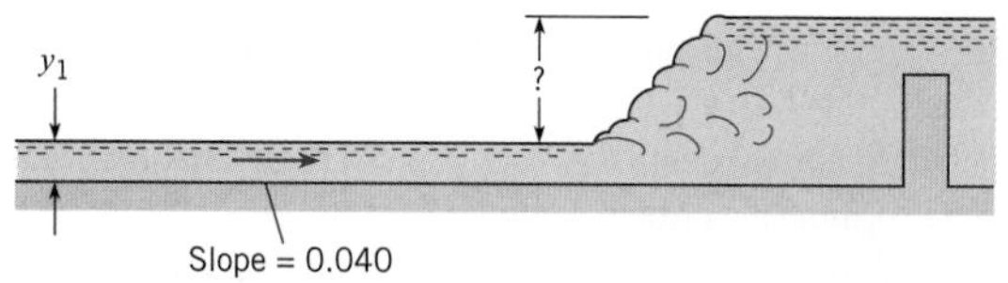

PROBLEMS 15.41, 15.42

15.42 For the derivation of Eq. (15.28) it is assumed that the bottom shearing force is negligible. For the conditions of Prob. 15.41, estimate the magnitude of the shearing force F_s associated with the hydraulic jump and then determine F_s/F_H, where F_H is the net hydrostatic force on the hydraulic jump.

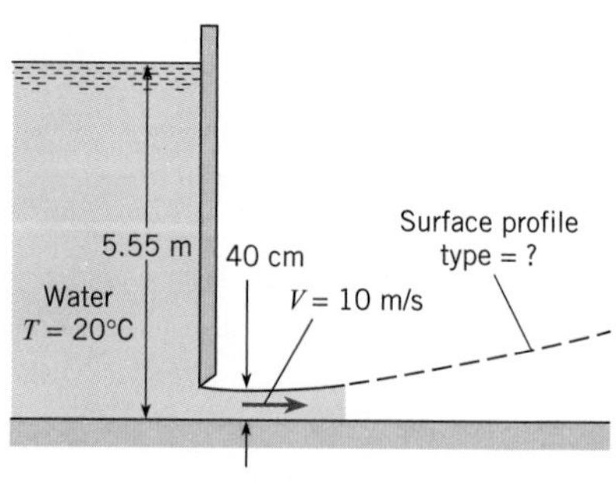

PROBLEM 15.43

15.43 The normal depth in the channel downstream of the sluice gate shown is 1 m. What type of water-surface profile occurs downstream of the sluice gate? Also, estimate the shear stress on the smooth bottom at a distance 0.5 m downstream of the sluice gate.

15.44 Water flows at a rate of 100 ft³/s in a rectangular channel 10 ft wide. The normal depth in that channel is 2 ft. The actual depth of flow in the channel is 4 ft. The water-surface profile in the channel for these conditions would be classified as (a) S1, (b) S2, (c) M1, (d) M2.

15.45 The water-surface profile labeled with a question mark is (a) M2, (b) S2, (c) H2, (d) A2.

15.46 The partial water-surface profile shown is for a rectangular channel that is 3 m wide and has water flowing in it at a rate of 5 m³/s. Sketch in the missing part of the water-surface profile and identify the type(s).

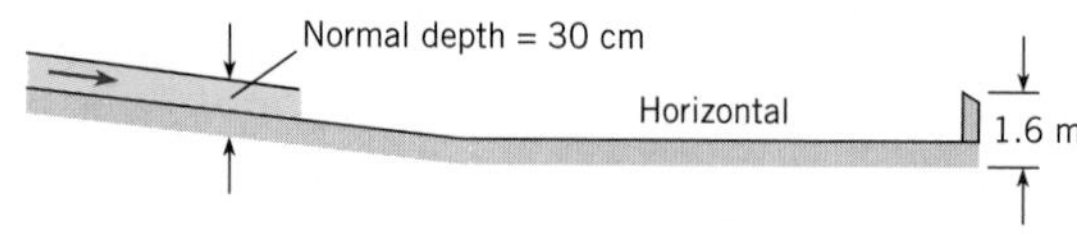

PROBLEM 15.46

15.47 A very long 10-ft-wide concrete rectangular channel with a slope of 0.0001 ends with a free overfall. The discharge in the channel is 120 cfs. One mile upstream the flow is uniform. What kind (classification) of water surface occurs upstream of the brink?

15.48 The horizontal rectangular channel downstream of the sluice gate is 10 ft wide, and the water discharge therein is 108 cfs. The water-surface profile was computed by the direct step method. If a 2-ft-high sharp-crested weir is installed at the end of the channel, do you think a hydraulic jump would develop in the channel? If so, approximately where would it be located? Justify your answers by appropriate calculations. Label any water-surface profiles that can be classified.

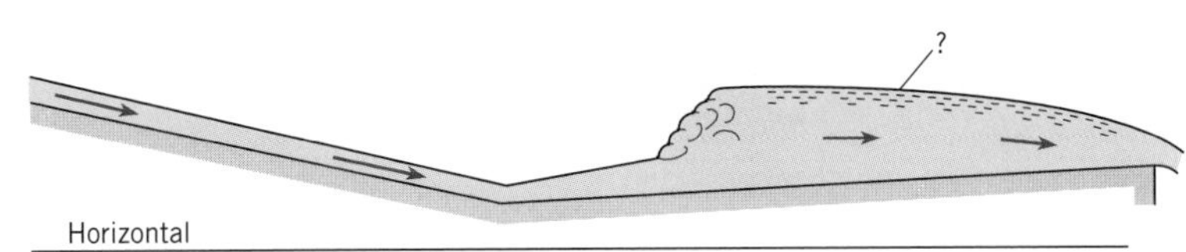

PROBLEM 15.45

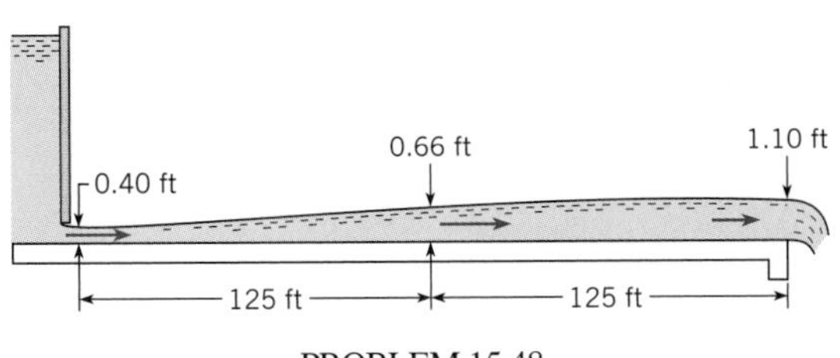

PROBLEM 15.48

15.49 The discharge per foot of width in this rectangular channel is 20 cfs. The normal depths for parts 1 and 3 are 0.5 ft and 1.00 ft, respectively. The slope for part 2 is 0.001 (sloping upward in the direction of flow). Sketch all possible water-surface profiles for flow in this channel, and label each part with its classification.

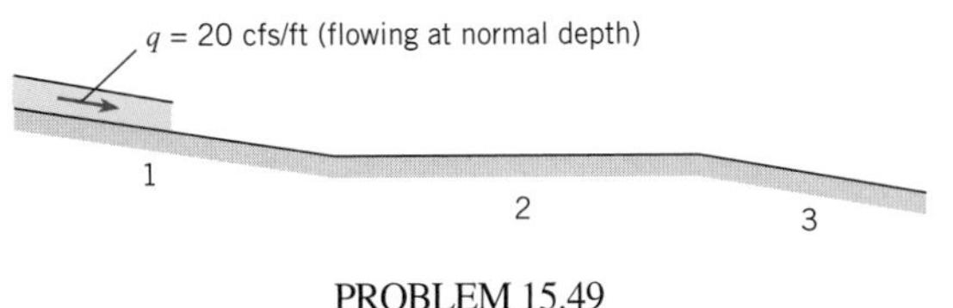

PROBLEM 15.49

15.50 Water flows from under a sluice gate into a horizontal rectangular channel at a rate of 3 m^3/s per meter of width. The channel is concrete, and the initial depth is 20 cm. Apply Eq. (15.37) to construct the water-surface profile up to a depth of 60 cm. In your solution, compute reaches for adjacent pairs of depths given in the following sequence: $d = 20$ cm, 30 cm, 40 cm, 50 cm, and 60 cm. Assume that f is constant with a value of 0.02. Plot your results.

15.51 A horizontal rectangular concrete channel terminates in a free outfall. The channel is 4 m wide and carries a discharge of water of 12 m^3/s. What is the water depth 300 m upstream from the outfall?

15.52 Consider the hydraulic jump shown for the long horizontal rectangular channel. What kind of water-surface profile (classification) is located upstream of the jump? What kind of water-surface profile is located downstream of the jump? If baffle blocks are put on the bottom of the channel in the vicinity of A to increase the bottom resistance, what changes are likely to occur given the same gate opening? Explain and/or sketch the changes.

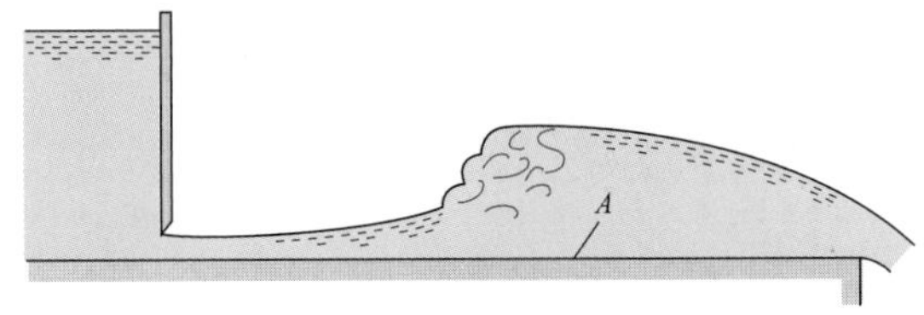

PROBLEM 15.52

15.53 The steep rectangular concrete spillway shown is 4 m wide and 500 m long. It conveys water from a reservoir and delivers it to a free outfall. The channel entrance is rounded and smooth (negligible head loss at the entrance). If the water-surface elevation in the reservoir is 2 m above the channel bottom, what will the discharge in the channel be? FLUID SOLUTIONS

15.54 The concrete rectangular channel shown is 3.5 m wide and has a bottom slope of 0.001. The channel entrance is rounded and smooth (negligible head loss at the entrance), and the reservoir water surface is 2.5 m above the bed of the channel at the entrance.

a. Estimate the discharge in it if the channel is 3000 m long.
b. Tell how you would solve for the discharge in it if the channel were only 100 m long.

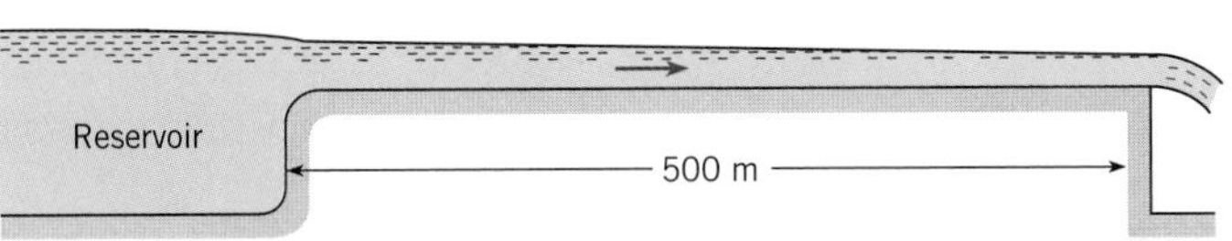

PROBLEM 15.53

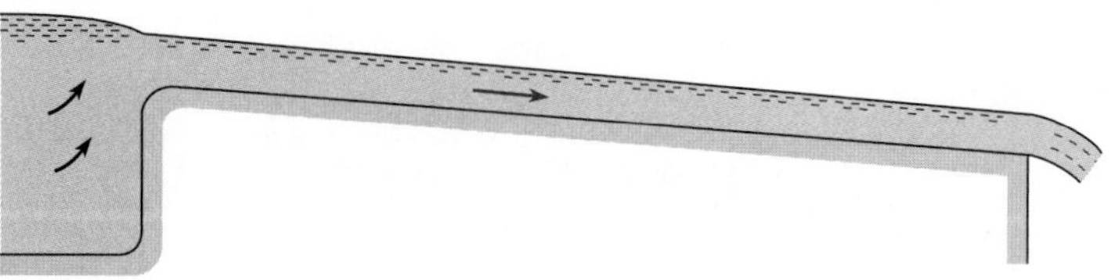

PROBLEM 15.54

15.55 A dam 50 m high backs up water in a river valley as shown. During flood flow, the discharge per meter of width, q, is equal to 10 m^3/s. Making the simplifying assumptions that $R = y$ and $f = 0.030$, determine the water-surface profile upstream from the dam to a depth of 6 m. In your numerical calculation, let the first increment of depth change be y_c; use increments of depth change of 10 m until a depth of 10 m is reached; and then use 2-m increments until the desired limit is reached.

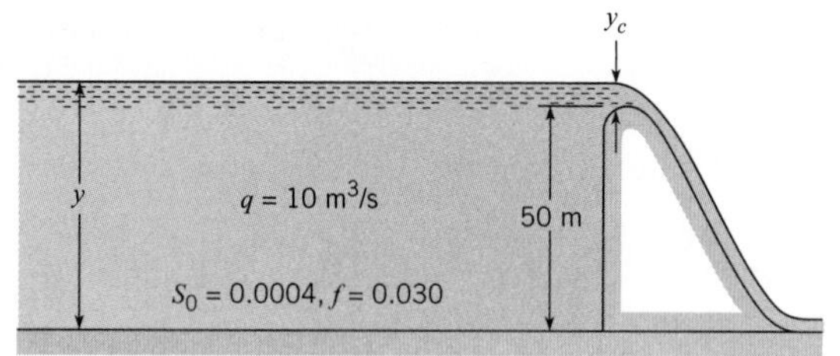

PROBLEM 15.55

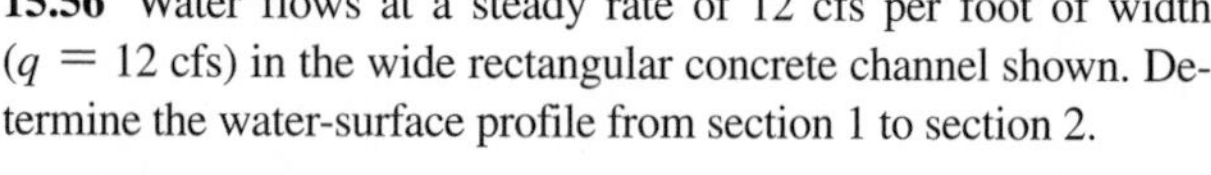

15.56 Water flows at a steady rate of 12 cfs per foot of width ($q = 12$ cfs) in the wide rectangular concrete channel shown. Determine the water-surface profile from section 1 to section 2.

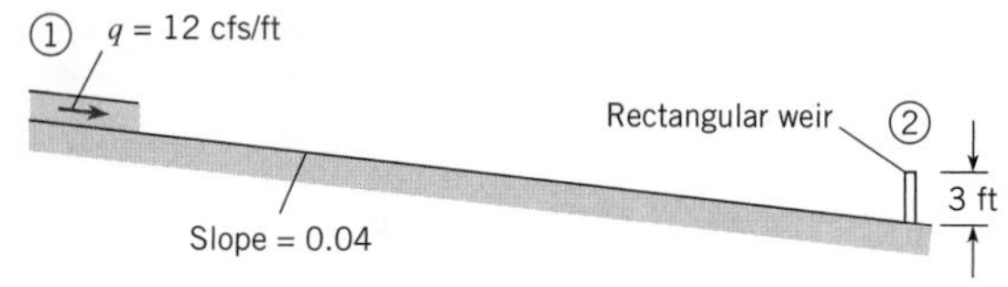

PROBLEM 15.56

Appendix

NOMENCLATURE AND DIMENSIONS		
Symbol	**Dimensions**	**Description**
A	L^2	Area
A_j	L^2	Jet area
A_0	L^2	Orifice area
A_*	L^2	Nozzle area at M = 1
a	L/T^2	Acceleration
b	...	Intensive property
B	L	Linear measure
B	...	Extensive property
Br		Brinell number
b	L	Linear measure
°C	θ	Temperature, Celsius
C_c	...	Coefficient of contraction
C_D	...	Coefficient of drag
C_d	...	Coefficient of discharge
C_f	...	Average shear stress coefficient
C_H	...	Head coefficient
C_L	...	Coefficient of lift
C_P	...	Power coefficient
C_p	...	Pressure coefficient
C_Q	...	Discharge coefficient
C_T	...	Thrust coefficient
C_v	...	Coefficient of velocity
c	...	Centi, multiple = 10^{-2}
c	L/T	Speed of sound
c_f	...	Local shear stress coefficient
c_p	$L^2/T^2\theta$	Specific heat at constant pressure
c_v	$L^2/T^2\theta$	Specific heat at constant volume
D	L	Diameter
D	L	Hydraulic depth
d	L	Diameter
d	L	Depth
E	ML^2/T^2	Energy
E	L	Specific energy
E_v	M/LT^2	Elasticity, bulk
e	L^2/T^2	Energy per unit mass
Fr	...	Froude number
F	ML/T^2	Force
°F	θ	Temperature, Fahrenheit
F_D	ML/T^2	Drag force
F_L	ML/T^2	Lift force
F_S	ML/T^2	Surface resistance
f	...	Resistance coefficient
G	...	Giga, multiple = 10^9
g	L/T^2	Acceleration due to gravity
g_c	...	Proportionality factor
H	L	Head

NOMENCLATURE AND DIMENSIONS (CONTINUED)		
Symbol	**Dimensions**	**Description**
h	L	Head
h	L	Piezometric head
h	L^2/T^2	Specific enthalpy
h_f	L	Friction head loss in pipe
h_L	L	Head loss
h_p	L	Head supplied by pump
h_t	L	Head given up to turbine
I	L^4	Area moment of inertia, centroidal
$\mathbf{i}$	...	Unit vector in x direction
J	ML^2/T^2	Joule, unit of work
$\mathbf{j}$	...	Unit vector in y direction
K	...	Flow coefficient
K	θ	Temperature, Kelvin
k	...	Ratio of specific heats
k	...	Kilo, multiple $= 10^3$
$\mathbf{k}$	...	Unit vector in z direction
k_s	L	Equivalent sand roughness size
L	L	Linear measure
l	L	Linear measure
ℓ	L	Linear measure
M	...	Mach number
$\mathbf{M}$	ML^2/T^2	Moment
M	M	Mass
M	...	Mega, multiple $= 10^6$
m	L	Meter
m	...	Milli, multiple $= 10^{-3}$
$\dot{m}$	M/T	Mass rate of flow
N	ML/T^2	Newton, unit of force
N	T^{-1}	Rotational speed
N_s	$L^{3/4}/T^{3/2}$	Specific speed
N_{ss}	$L^{3/4}/T^{3/2}$	Suction specific speed
n	T^{-1}	Frequency in Hertz
n	...	Manning's roughness coefficient
n	T^{-1}	Rotational speed
n_s	...	Specific speeds
n_{ss}	...	Suction specific speed
Pa	M/LT^2	Pascal, unit of pressure
P	ML^2/T^3	Power
p_*	M/LT^2	Pressure at M $=1$
p	M/LT^2	Pressure
p_t	M/LT^2	Total pressure
p_v	M/LT^2	Vapor pressure
Q	L^3/T	Discharge, volumetric flow rate
Q	ML^2/T^2	Heat transferred
q	L^2/T	Discharge per unit width
q	M/LT^2	Kinetic pressure
R_h	L	Hydraulic radius
R	ML/T^2	Reaction or resultant force
R	$L^2/\theta T^2$	Gas constant

NOMENCLATURE AND DIMENSIONS (CONTINUED)		
Symbol	**Dimensions**	**Description**
°R	θ	Temperature, Rankine
Re	. . .	Reynolds number
r	L	Linear measure in radial direction
S	L^2	Planform area
St	. . .	Strouhal number
S_0	. . .	Channel slope
s	$L^2/T^2\theta$	Specific entropy
S	. . .	Specific gravity
s	T	Time, second
s	L	Linear measure
T	M^2L^2/T^2	Torque
T	θ	Temperature
T_t	θ	Total temperature
T_*	θ	Temperature at M = 1
t	T	Time
U_0	L/T	Free-stream velocity
u	L/T	Velocity component, x direction
u	L^2/T^2	Internal energy per unit of mass
u_*	L/T	Shear velocity
u'	L/T	Velocity fluctuation in x direction
V	L/T	Velocity
V_0	L/T	Free-stream velocity
$\forall$	L^3	Volume
$\overline{V}$	L/T	Area-averaged velocity
v	L/T	Velocity component, y direction
v	L/T	Velocity relative to inertial reference frame
v'	L/T	Velocity fluctuation in y direction
W	ML^2/T^2	Work
W	ML/T^2	Weight
We	. . .	Weber number
W	ML^2/T^3	Watt, unit of power
w	L/T	Velocity component, z direction
x	L	Linear measure
y	L	Linear measure
y_c	L	Critical depth
y_n	L	Normal depth
z	L	Linear measure

NOMENCLATURE AND DIMENSIONS (CONTINUED)		
Symbol	**Dimensions**	**Description**
GREEK LETTERS		
α	. . .	Angular measure
α	. . .	Lapse rate
α	. . .	Kinetic energy coefficient
α	. . .	Angle of attack
β	. . .	Angular measure
Γ	L^2/T	Circulation
γ	M/L^2T^2	Specific weight
Δ	. . .	Increment
δ	L	Boundary layer thickness
δ'	L	Laminar sublayer thickness
δ'_N	L	Nom. laminar sublayer thickness
η	. . .	Efficiency
θ	. . .	Angular measure
κ	. . .	Turbulence constant
μ	M/LT	Viscosity, dynamic
μ	. . .	Micro, multiple $= 10^{-6}$
τ	M/LT^2	Shear stress
ν	L^2/T	Kinematic viscosity
π	. . .	3.1416
ρ	M/L^3	Mass density
ρ_*	M/L^3	Density at M = 1
ρ_t	M/L^3	Total density
Ω	T^{-1}	Vorticity
ω	T^{-1}	Angular speed
σ	M/T^2	Surface tension

FIGURE A.1

Centroids and moments of inertia of plane areas

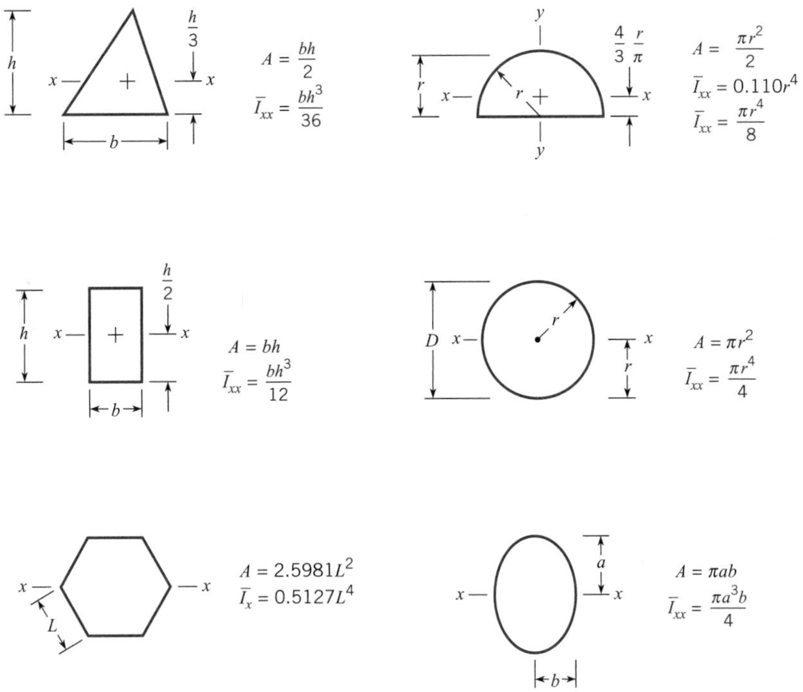

h, x, b, $\frac{h}{3}$

$$A = \frac{bh}{2}$$
$$\bar{I}_{xx} = \frac{bh^3}{36}$$

y, r, x, $\frac{4}{3}\frac{r}{\pi}$

$$A = \frac{\pi r^2}{2}$$
$$\bar{I}_{xx} = 0.110 r^4$$
$$\bar{I}_{xx} = \frac{\pi r^4}{8}$$

h, x, b, $\frac{h}{2}$

$$A = bh$$
$$\bar{I}_{xx} = \frac{bh^3}{12}$$

D, x, r

$$A = \pi r^2$$
$$\bar{I}_{xx} = \frac{\pi r^4}{4}$$

x, L

$$A = 2.5981 L^2$$
$$\bar{I}_x = 0.5127 L^4$$

x, a, b

$$A = \pi ab$$
$$\bar{I}_{xx} = \frac{\pi a^3 b}{4}$$

Volume and Area Formulas:

$$A_{\text{circle}} = \pi r^2 = \pi D^2/4$$
$$A_{\text{sphere surface}} = \pi D^2$$
$$\forall_{\text{sphere}} = \tfrac{1}{6}\pi D^3 = \tfrac{4}{3}\pi r^3$$

TABLE A.1 COMPRESSIBLE FLOW TABLES FOR AN IDEAL GAS WITH $k = 1.4$

M or M_1 = local number or Mach number upstream of a normal shock wave; p/p_t = ratio of static pressure to total pressure; ρ/ρ_t = ratio of static density to total density; T/T_t = ratio of static temperature to total temperature; A/A_* = ratio of local cross sectional area of an isentropic stream tube to cross-sectional area at the point where M = 1; M_2 = Mach number downstream of a normal shock wave; p_2/p_1 = static pressure ratio across a normal shock; T_2/T_1 = static pressure ratio across a normal shock wave; p_{t_2}/p_{t_1} = total pressure ratio across normal shock wave.

Subsonic Flow

M	p/p_t	ρ/ρ_t	T/T_t	A/A_*
0.00	1.0000	1.0000	1.0000	∞
0.05	0.9983	0.9988	0.9995	11.5914
0.10	0.9930	0.9950	0.9980	5.8218
0.15	0.9844	0.9888	0.9955	3.9103
0.20	0.9725	0.9803	0.9921	2.9630
0.25	0.9575	0.9694	0.9877	2.4027
0.30	0.9395	0.9564	0.9823	2.0351
0.35	0.9188	0.9413	0.9761	1.7780
0.40	0.8956	0.9243	0.9690	1.5901
0.45	0.8703	0.9055	0.9611	1.4487
0.50	0.8430	0.8852	0.9524	1.3398
0.52	0.8317	0.8766	0.9487	1.3034
0.54	0.8201	0.8679	0.9449	1.2703
0.56	0.8082	0.8589	0.9410	1.2403
0.58	0.7962	0.8498	0.9370	1.2130
0.60	0.7840	0.8405	0.9328	1.1882
0.62	0.7716	0.8310	0.9286	1.1657
0.64	0.7591	0.8213	0.9243	1.1452
0.66	0.7465	0.8115	0.9199	1.1265
0.68	0.7338	0.8016	0.9153	1.1097
0.70	0.7209	0.7916	0.9107	1.0944
0.72	0.7080	0.7814	0.9061	1.0806
0.74	0.6951	0.7712	0.9013	1.0681
0.76	0.6821	0.7609	0.8964	1.0570
0.78	0.6691	0.7505	0.8915	1.0471
0.80	0.6560	0.7400	0.8865	1.0382
0.82	0.6430	0.7295	0.8815	1.0305
0.84	0.6300	0.7189	0.8763	1.0237
0.86	0.6170	0.7083	0.8711	1.0179
0.88	0.6041	0.6977	0.8659	1.0129
0.90	0.5913	0.6870	0.8606	1.0089
0.92	0.5785	0.6764	0.8552	1.0056
0.94	0.5658	0.6658	0.8498	1.0031
0.96	0.5532	0.6551	0.8444	1.0014
0.98	0.5407	0.6445	0.8389	1.0003
1.00	0.5283	0.6339	0.8333	1.0000

Continued

TABLE A.1 COMPRESSIBLE FLOW TABLES FOR AN IDEAL GAS WITH $k = 1.4$ (CONTINUED)

Supersonic Flow					Normal Shock Wave			
M_1	p/p_t	ρ/ρ_t	T/T_t	A/A_*	M_2	p_2/p_1	T_2/T_1	p_{t_2}/p_{t_1}
1.00	0.5283	0.6339	0.8333	1.000	1.000	1.000	1.000	1.0000
1.01	0.5221	0.6287	0.8306	1.000	0.9901	1.023	1.007	0.9999
1.02	0.5160	0.6234	0.8278	1.000	0.9805	1.047	1.013	0.9999
1.03	0.5099	0.6181	0.8250	1.001	0.9712	1.071	1.020	0.9999
1.04	0.5039	0.6129	0.8222	1.001	0.9620	1.095	1.026	0.9999
1.05	0.4979	0.6077	0.8193	1.002	0.9531	1.120	1.033	0.9998
1.06	0.4919	0.6024	0.8165	1.003	0.9444	1.144	1.039	0.9997
1.07	0.4860	0.5972	0.8137	1.004	0.9360	1.169	1.046	0.9996
1.08	0.4800	0.5920	0.8108	1.005	0.9277	1.194	1.052	0.9994
1.09	0.4742	0.5869	0.8080	1.006	0.9196	1.219	1.059	0.9992
1.10	0.4684	0.5817	0.8052	1.008	0.9118	1.245	1.065	0.9989
1.11	0.4626	0.5766	0.8023	1.010	0.9041	1.271	1.071	0.9986
1.12	0.4568	0.5714	0.7994	1.011	0.8966	1.297	1.078	0.9982
1.13	0.4511	0.5663	0.7966	1.013	0.8892	1.323	1.084	0.9978
1.14	0.4455	0.5612	0.7937	1.015	0.8820	1.350	1.090	0.9973
1.15	0.4398	0.5562	0.7908	1.017	0.8750	1.376	1.097	0.9967
1.16	0.4343	0.5511	0.7879	1.020	0.8682	1.403	1.103	0.9961
1.17	0.4287	0.5461	0.7851	1.022	0.8615	1.430	1.109	0.9953
1.18	0.4232	0.5411	0.7822	1.025	0.8549	1.458	1.115	0.9946
1.19	0.4178	0.5361	0.7793	1.026	0.8485	1.485	1.122	0.9937
1.20	0.4124	0.5311	0.7764	1.030	0.8422	1.513	1.128	0.9928
1.21	0.4070	0.5262	0.7735	1.033	0.8360	1.541	1.134	0.9918
1.22	0.4017	0.5213	0.7706	1.037	0.8300	1.570	1.141	0.9907
1.23	0.3964	0.5164	0.7677	1.040	0.8241	1.598	1.147	0.9896
1.24	0.3912	0.5115	0.7648	1.043	0.8183	1.627	1.153	0.9884
1.25	0.3861	0.5067	0.7619	1.047	0.8126	1.656	1.159	0.9871
1.30	0.3609	0.4829	0.7474	1.066	0.7860	1.805	1.191	0.9794
1.35	0.3370	0.4598	0.7329	1.089	0.7618	1.960	1.223	0.9697
1.40	0.3142	0.4374	0.7184	1.115	0.7397	2.120	1.255	0.9582
1.45	0.2927	0.4158	0.7040	1.144	0.7196	2.286	1.287	0.9448
1.50	0.2724	0.3950	0.6897	1.176	0.7011	2.458	1.320	0.9278
1.55	0.2533	0.3750	0.6754	1.212	0.6841	2.636	1.354	0.9132
1.60	0.2353	0.3557	0.6614	1.250	0.6684	2.820	1.388	0.8952
1.65	0.2184	0.3373	0.6475	1.292	0.6540	3.010	1.423	0.8760
1.70	0.2026	0.3197	0.6337	1.338	0.6405	3.205	1.458	0.8557
1.75	0.1878	0.3029	0.6202	1.386	0.6281	3.406	1.495	0.8346
1.80	0.1740	0.2868	0.6068	1.439	0.6165	3.613	1.532	0.8127
1.85	0.1612	0.2715	0.5936	1.495	0.6057	3.826	1.569	0.7902
1.90	0.1492	0.2570	0.5807	1.555	0.5956	4.045	1.608	0.7674
1.95	0.1381	0.2432	0.5680	1.619	0.5862	4.270	1.647	0.7442
2.00	0.1278	0.2300	0.5556	1.688	0.5774	4.500	1.688	0.7209
2.10	0.1094	0.2058	0.5313	1.837	0.5613	4.978	1.770	0.6742

Continued

TABLE A.1 COMPRESSIBLE FLOW TABLES FOR AN IDEAL GAS WITH $k = 1.4$ (CONTINUED)

Supersonic Flow					Normal Shock Wave			
M_1	p/p_t	ρ/ρ_t	T/T_t	A/A_*	M_2	p_2/p_1	T_2/T_1	p_{t_2}/p_{t_1}
2.20	$0.9352^{-1\dagger}$	0.1841	0.5081	2.005	0.5471	5.480	1.857	0.6281
2.30	0.7997^{-1}	0.1646	0.4859	2.193	0.5344	6.005	1.947	0.5833
2.50	0.5853^{-1}	0.1317	0.4444	2.637	0.5130	7.125	2.138	0.4990
2.60	0.5012^{-1}	0.1179	0.4252	2.896	0.5039	7.720	2.238	0.4601
2.70	0.4295^{-1}	0.1056	0.4068	3.183	0.4956	8.338	2.343	0.4236
2.80	0.3685^{-1}	0.9463^{-1}	0.3894	3.500	0.4882	8.980	2.451	0.3895
2.90	0.3165^{-1}	0.8489^{-1}	0.3729	3.850	0.4814	9.645	2.563	0.3577
3.00	0.2722^{-1}	0.7623^{-1}	0.3571	4.235	0.4752	10.33	2.679	0.3283
3.50	0.1311^{-1}	0.4523^{-1}	0.2899	6.790	0.4512	14.13	3.315	0.2129
4.00	0.6586^{-2}	0.2766^{-1}	0.2381	10.72	0.4350	18.50	4.047	0.1388
4.50	0.3455^{-2}	0.1745^{-1}	0.1980	16.56	0.4236	23.46	4.875	0.9170^{-1}
5.00	0.1890^{-2}	0.1134^{-1}	0.1667	25.00	0.4152	29.00	5.800	0.6172^{-1}
5.50	0.1075^{-2}	0.7578^{-2}	0.1418	36.87	0.4090	35.13	6.822	0.4236^{-1}
6.00	0.6334^{-2}	0.5194^{-2}	0.1220	53.18	0.4042	41.83	7.941	0.2965^{-1}
6.50	0.3855^{-2}	0.3643^{-2}	0.1058	75.13	0.4004	49.13	9.156	0.2115^{-1}
7.00	0.2416^{-3}	0.2609^{-2}	0.9259^{-1}	104.1	0.3974	57.00	10.47	0.1535^{-1}
7.50	0.1554^{-3}	0.1904^{-2}	0.8163^{-1}	141.8	0.3949	65.46	11.88	0.1133^{-1}
8.00	0.1024^{-3}	0.1414^{-2}	0.7246^{-1}	190.1	0.3929	74.50	13.39	0.8488^{-2}
8.50	0.6898^{-4}	0.1066^{-2}	0.6472^{-1}	251.1	0.3912	84.13	14.99	0.6449^{-2}
9.00	0.4739^{-4}	0.8150^{-3}	0.5814^{-1}	327.2	0.3898	94.33	16.69	0.4964^{-2}
9.50	0.3314^{-4}	0.6313^{-3}	0.5249^{-1}	421.1	0.3886	105.1	18.49	0.3866^{-2}
10.00	0.2356^{-4}	0.4948^{-3}	0.4762^{-1}	535.9	0.3876	116.5	20.39	0.3045^{-2}

†x^{-n} means $x \cdot 10^{-n}$.

SOURCE: Abridged with permission from R. E. Bolz and G. L. Tuve, *The Handbook of Tables for Applied Engineering Sciences,* CRC Press, Inc., Cleveland, 1973.

FIGURE A.2

Absolute viscosities of certain gases and liquids [Adapted from Fluid Mechanics, *5th ed., by V. L. Streeter. Copyright © 1971, McGraw-Hill Book Company, New York. Used with permission of the McGraw-Hill Book Company.]*

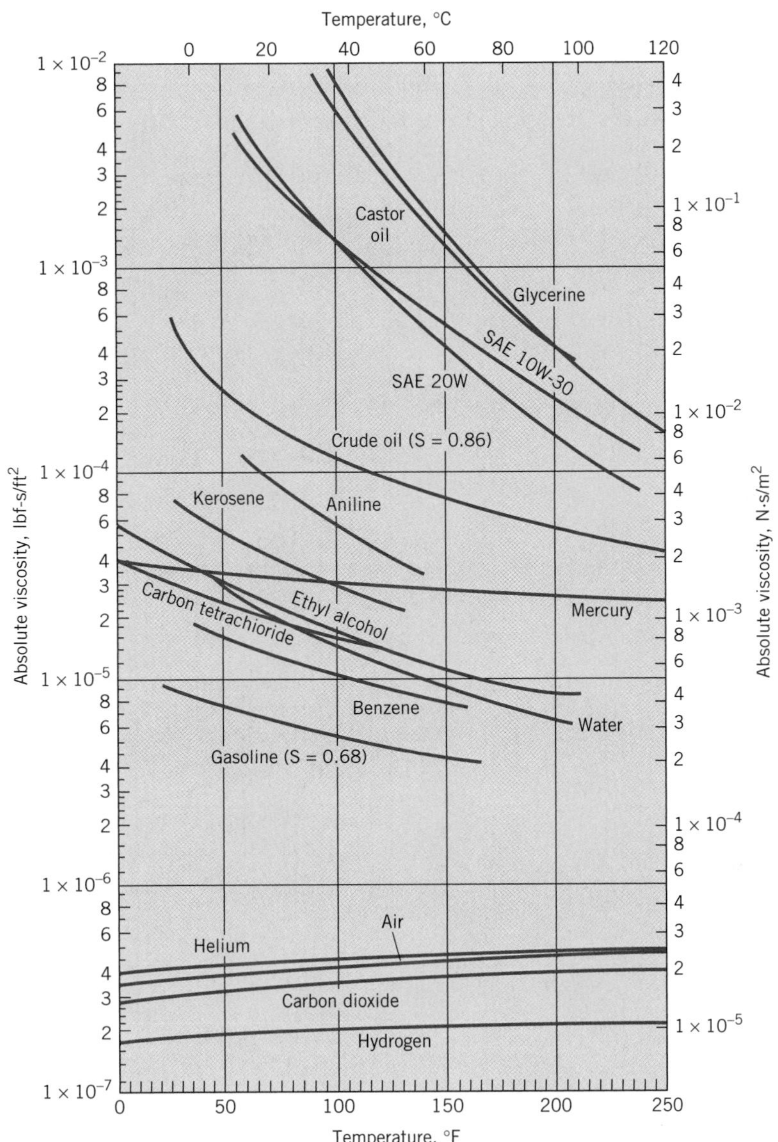

FIGURE A.3

Kinematic viscosities of certain gases and liquids. The gases are at standard pressure. [Adapted from Fluid Mechanics, *5th ed., by V. L. Streeter. Copyright © 1971, McGraw-Hill Book Company, New York. Used with permission of the McGraw-Hill Book Company.]*

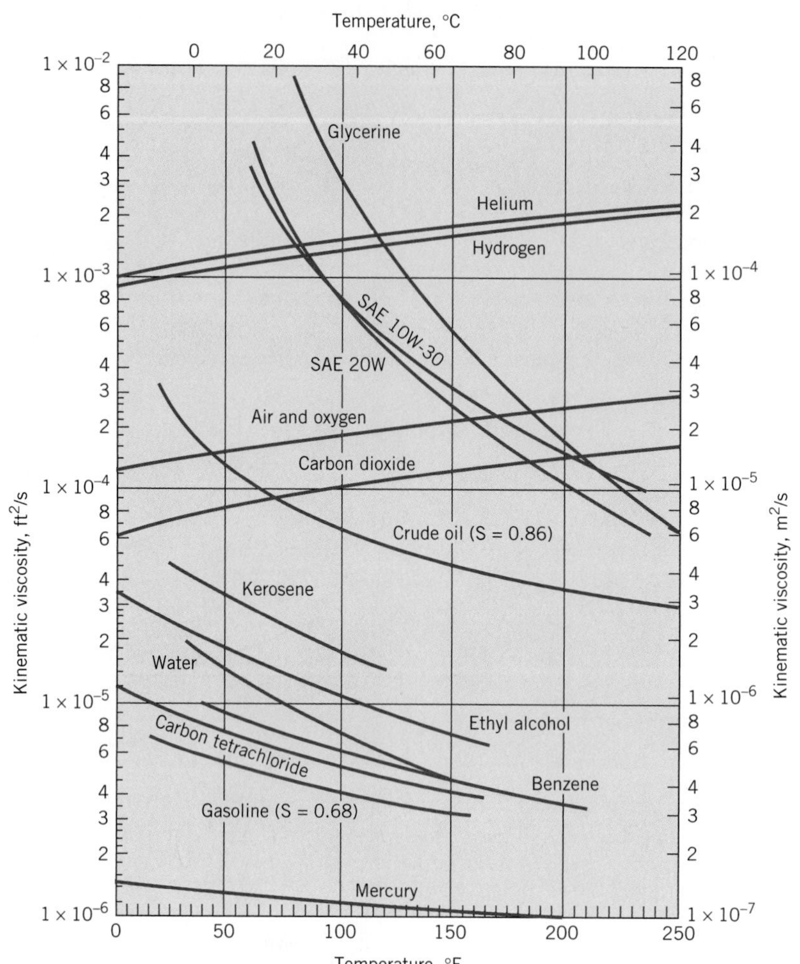

TABLE A.2 PHYSICAL PROPERTIES OF GASES AT STANDARD ATMOSPHERIC PRESSURE AND 15°C (59°F)

Gas	Density, kg/m^3 ($slugs/ft^3$)	Kinematic Viscosity, m^2/s (ft^2/s)	R, Gas Constant, J/kg K (ft-lbf/slug-°R)	c_p, $\frac{J}{kg\ K}$ $\left(\frac{Btu}{lbm\text{-}°R}\right)$	$k = \frac{c_p}{c_v}$	S, Sutherland's Constant, K (°R)
Air	1.22 (0.00237)	1.46×10^{-5} (1.58×10^{-4})	287 (1716)	1004 (0.240)	1.40	111 (199)
Carbon dioxide	1.85 (0.0036)	7.84×10^{-6} (8.48×10^{-5})	189 (1130)	841 (0.201)	1.30	222 (400)
Helium	0.169 (0.00033)	1.14×10^{-4} (1.22×10^{-3})	2077 (12,419)	5187 (1.24)	1.66	79.4 (143)
Hydrogen	0.0851 (0.00017)	1.01×10^{-4} (1.09×10^{-3})	4127 (24,677)	14,223 (3.40)	1.41	96.7 (174)
Methane (natural gas)	0.678 (0.0013)	1.59×10^{-5} (1.72×10^{-4})	518 (3098)	2208 (0.528)	1.31	198 (356)
Nitrogen	1.18 (0.0023)	1.45×10^{-5} (1.56×10^{-4})	297 (1776)	1041 (0.249)	1.40	107 (192)
Oxygen	1.35 (0.0026)	1.50×10^{-5} (1.61×10^{-4})	260 (1555)	916 (0.219)	1.40	

SOURCE: V. L. Streeter (ed.), *Handbook of Fluid Dynamics,* McGraw-Hill Book Company, New York, 1961; also R. E. Bolz and G. L. Tuve, *Handbook of Tables for Applied Engineering Science,* CRC Press, Inc. Cleveland, 1973; and *Handbook of Chemistry and Physics,* Chemical Rubber Company, 1951.

TABLE A.3 MECHANICAL PROPERTIES OF AIR AT STANDARD ATMOSPHERIC PRESSURE

Temperature	Density	Specific Weight	Dynamic Viscosity	Kinematic Viscosity
	kg/m^3	N/m^3	$N \cdot s/m^2$	m^2/s
–20°C	1.40	13.7	1.61×10^{-5}	1.16×10^{-5}
–10°C	1.34	13.2	1.67×10^{-5}	1.24×10^{-5}
0°C	1.29	12.7	1.72×10^{-5}	1.33×10^{-5}
10°C	1.25	12.2	1.76×10^{-5}	1.41×10^{-5}
20°C	1.20	11.8	1.81×10^{-5}	1.51×10^{-5}
30°C	1.17	11.4	1.86×10^{-5}	1.60×10^{-5}
40°C	1.13	11.1	1.91×10^{-5}	1.69×10^{-5}
50°C	1.09	10.7	1.95×10^{-5}	1.79×10^{-5}
60°C	1.06	10.4	2.00×10^{-5}	1.89×10^{-5}
70°C	1.03	10.1	2.04×10^{-5}	1.99×10^{-5}
80°C	1.00	9.81	2.09×10^{-5}	2.09×10^{-5}
90°C	0.97	9.54	2.13×10^{-5}	2.19×10^{-5}
100°C	0.95	9.28	2.17×10^{-5}	2.29×10^{-5}
120°C	0.90	8.82	2.26×10^{-5}	2.51×10^{-5}
140°C	0.85	8.38	2.34×10^{-5}	2.74×10^{-5}
160°C	0.81	7.99	2.42×10^{-5}	2.97×10^{-5}
180°C	0.78	7.65	2.50×10^{-5}	3.20×10^{-5}
200°C	0.75	7.32	2.57×10^{-5}	3.44×10^{-5}
	$slugs/ft^3$	lbf/ft^3	$lbf\text{-}s/ft^2$	ft^2/s
0°F	0.00269	0.0866	3.39×10^{-7}	1.26×10^{-4}
20°F	0.00257	0.0828	3.51×10^{-7}	1.37×10^{-4}
40°F	0.00247	0.0794	3.63×10^{-7}	1.47×10^{-4}
60°F	0.00237	0.0764	3.74×10^{-7}	1.58×10^{-4}
80°F	0.00228	0.0735	3.85×10^{-7}	1.69×10^{-4}
100°F	0.00220	0.0709	3.96×10^{-7}	1.80×10^{-4}
120°F	0.00213	0.0685	4.07×10^{-7}	1.91×10^{-4}
150°F	0.00202	0.0651	4.23×10^{-7}	2.09×10^{-4}
200°F	0.00187	0.0601	4.48×10^{-7}	2.40×10^{-4}
300°F	0.00162	0.0522	4.96×10^{-7}	3.05×10^{-4}
400°F	0.00143	0.0462	5.40×10^{-7}	3.77×10^{-4}

SOURCE: Reprinted with permission from R. E. Bolz and G. L. Tuve, *Handbook of Tables for Applied Engineering Science*, CRC Press, Inc., Cleveland, 1973.

TABLE A.4 APPROXIMATE PHYSICAL PROPERTIES OF COMMON LIQUIDS AT ATMOSPHERIC PRESSURE

Liquid and Temperature	Density, kg/m³ (slugs/ft³)	Specific Gravity (S), Water at 4°C is Ref.	Specific Weight, N/m³ (lbf/ft³)	Dynamic Viscosity, N · s/m² (lbf-s/ft²)	Kinematic Viscosity, m²/s (ft²/s)	Surface Tension, N/m* (lbf/ft)
Ethyl alcohol(3)(1)	799	0.79	7,850	1.2×10^{-3}	1.5×10^{-6}	2.2×10^{-2}
20°C (68°F)	(1.55)		(50.0)	(2.5×10^{-5})	(1.6×10^{-5})	(1.5×10^{-3})
Carbon tetrachloride(3)	1,590	1.59	15,600	9.6×10^{-4}	6.0×10^{-7}	2.6×10^{-2}
20°C (68°F)	(3.09)		(99.5)	(2.0×10^{-5})	(6.5×10^{-6})	(1.8×10^{-3})
Glycerine(3)	1,260	1.26	12,300	1.41	1.12×10^{-3}	6.3×10^{-2}
20°C (68°F)	(2.45)		(78.5)	(3×10^{-2})	(1.22×10^{-2})	(4.3×10^{-3})
Kerosene(2)(1)	814	0.81	8,010	1.9×10^{-3}	2.37×10^{-6}	2.9×10^{-2}
20°C (68°F)	(1.58)		(51)	(4×10^{-5})	(2.55×10^{-5})	(2.0×10^{-3})
Mercury(3)(1)	13,550	13.55	133,000	1.5×10^{-3}	1.2×10^{-7}	4.8×10^{-1}
20°C (68°F)	(26.3)		(847)	(3.2×10^{-5})	(1.3×10^{-6})	(3.3×10^{-2})
Sea water 10°C	1,026	1.03	10,070	1.4×10^{-3}	1.4×10^{-6}	
at 3.3% salinity	(1.99)		(64.1)	(3×10^{-5})	(1.5×10^{-5})	
Oils—38°C (100°F)						
SAE 10W(4)	870	0.87	8,530	3.6×10^{-2}	4.1×10^{-5}	
	(1.69)		(54.4)	(7.4×10^{-4})	(4.4×10^{-4})	
SAE 10W-30(4)	880	0.88	8,630	6.7×10^{-2}	7.6×10^{-5}	
	(1.71)		(55.1)	(1.4×10^{-3})	(8.2×10^{-4})	
SAE 30(4)	880	0.88	8,630	1.0×10^{-1}	1.1×10^{-4}	
	(1.71)		(55.1)	(2.0×10^{-3})	(1.2×10^{-3})	

*Liquid–air surface tension values.

SOURCES: (1) V. L. Streeter, *Handbook of Fluid Dynamics*, McGraw-Hill, New York, 1961; (2) V. L. Streeter, *Fluid Mechanics*, 4th ed., McGraw-Hill, New York, 1966; (3) A. A. Newman, *Glycerol*, CRC Press, Cleveland, 1968; (4) R. E. Bolz and G. L. Tuve, *Handbook of Tables for Applied Engineering Sciences*, CRC Press, Cleveland, 1973.

TABLE A.5 APPROXIMATE PHYSICAL PROPERTIES OF WATER* AT ATMOSPHERIC PRESSURE

Temperature	Density	Specific Weight	Dynamic Viscosity	Kinematic Viscosity	Vapor Pressure
	kg/m^3	N/m^3	$N \cdot s/m^2$	m^2/s	N/m^2 abs.
0°C	1000	9810	1.79×10^{-3}	1.79×10^{-6}	611
5°C	1000	9810	1.51×10^{-3}	1.51×10^{-6}	872
10°C	1000	9810	1.31×10^{-3}	1.31×10^{-6}	1230
15°C	999	9800	1.14×10^{-3}	1.14×10^{-6}	1700
20°C	998	9790	1.00×10^{-3}	1.00×10^{-6}	2340
25°C	997	9781	8.91×10^{-4}	8.94×10^{-7}	3170
30°C	996	9771	7.97×10^{-4}	8.00×10^{-7}	4250
35°C	994	9751	7.20×10^{-4}	7.24×10^{-7}	5630
40°C	992	9732	6.53×10^{-4}	6.58×10^{-7}	7380
50°C	988	9693	5.47×10^{-4}	5.53×10^{-7}	12,300
60°C	983	9643	4.66×10^{-4}	4.74×10^{-7}	20,000
70°C	978	9594	4.04×10^{-4}	4.13×10^{-7}	31,200
80°C	972	9535	3.54×10^{-4}	3.64×10^{-7}	47,400
90°C	965	9467	3.15×10^{-4}	3.26×10^{-7}	70,100
100°C	958	9398	2.82×10^{-4}	2.94×10^{-7}	101,300
	$slugs/ft^3$	lbf/ft^3	$lbf\text{-}s/ft^2$	ft^2/s	psia
40°F	1.94	62.43	3.23×10^{-5}	1.66×10^{-5}	0.122
50°F	1.94	62.40	2.73×10^{-5}	1.41×10^{-5}	0.178
60°F	1.94	62.37	2.36×10^{-5}	1.22×10^{-5}	0.256
70°F	1.94	62.30	2.05×10^{-5}	1.06×10^{-5}	0.363
80°F	1.93	62.22	1.80×10^{-5}	0.930×10^{-5}	0.506
100°F	1.93	62.00	1.42×10^{-5}	0.739×10^{-5}	0.949
120°F	1.92	61.72	1.17×10^{-5}	0.609×10^{-5}	1.69
140°F	1.91	61.38	0.981×10^{-5}	0.514×10^{-5}	2.89
160°F	1.90	61.00	0.838×10^{-5}	0.442×10^{-5}	4.74
180°F	1.88	60.58	0.726×10^{-5}	0.385×10^{-5}	7.51
200°F	1.87	60.12	0.637×10^{-5}	0.341×10^{-5}	11.53
212°F	1.86	59.83	0.593×10^{-5}	0.319×10^{-5}	14.70

*Notes: (1) Bulk modulus E_v of water is approximately 2.2 GPa (3.2×10^5 psi); (2) water–air surface tension is approximately 7.3×10^{-2} N/m (5×10^{-3} lbf/ft) from 10°C to 50°C.

SOURCE: Reprinted with permission from R. E. Bolz and G. L. Tuve, *Handbook of Tables for Applied Engineering Science*, CRC Press, Inc., Cleveland, 1973. Copyright © 1973 by The Chemical Rubber Co., CRC Press, Inc.

Answers to Even Problems

Chapter 2

2.2 $\rho_{CO_2} = 4.77\ \text{kg/m}^3$,

$\gamma_{CO_2} = 46.8\ \text{N/m}^3$

2.4 $M_2/M_1 = 1.5$

2.6 $W_{total} = 123$ lbf

2.8 $M = 3.49 \times 10^8$ slugs

$= 5.08 \times 10^9$ kg

2.10 $V\!\!\!\!- = 2.54\ \text{m}^3$,

$m = 5.66$ kg

2.12 $S_f = 0.972$, 13.7%

2.14 $\Delta v = 4.8 \times 10^{-6}\ \text{m}^2/\text{s}$

2.16 $\mu_{air}/\mu_{water} = 1.81 \times 10^{-2}$,

$v_{air}/v_{water} = 15.1$

2.18 $v/v_o = (p_o/p)(T/T_o)^{5/2}$

$(T_o + S)/(T + S)$

2.20 $v = 1.99 \times 10^{-5}\ \text{m}^2/\text{s}$

2.22 $v = 1.66 \times 10^{-3}\ \text{ft}^2/\text{s}$

2.24 $S = 903°$ R

2.26 $\mu = 1.32 \times 10^{-3}\ \text{lbf} \cdot \text{s/ft}^2$

2.28 $\tau = 1.25\ \text{lbf/ft}^2$

2.30 $\tau = 1.49\ \text{N/m}^2$

2.32 $\tau_{max} = 1.0\ \text{N/m}^2$, minimum on centerline

2.34 At the moving plate.

2.36 $u_t = (1/2\mu)(dp/ds)H^2$

2.38 (c)

2.40 $W = 18.1$ N

2.42 $T = 2.45 \times 10^{-4}\ \text{N} \cdot \text{m}$

2.46 $\Delta p = 22\ \text{MN/m}^2$

2.48 $\Delta p = 4\sigma/R$, $\Delta p = 73.0\ \text{N/m}^2$

2.50 $d(1/4\ \text{in.}) = 0.185$ in.

$d(1/8\ \text{in.}) = 0.369$ in.

$d(1/32\ \text{in.}) = 1.48$ in.

2.52 $\Delta p = 292\ \text{N/m}^2$

2.54 $\Delta h = 11.8$ mm

2.56 $D = 9.46$ mm

2.58 $T = 89.7°\text{C}$

Chapter 3

3.2 $F_{pull} = 1540$ lbf

3.4 $n = 5$

3.6 (b)

3.8 Higher in left tube

3.10 $p = 489.5$ kPa gage,

$p_{50}/p_{atm} = 5.83$

3.12 10 gpm pump with $3.0 \leq$ bore ≤ 3.6 inches or 15 gpm pump with $3.0 \leq$ bore ≤ 4.6 inches.

3.14 0.0826 m

3.16 $p_{max} = 127.5$ kPa, p_{max} at bottom of liquid with S = 3, $F_{CD} = 98.1$ kN

3.18 2850 lbf

3.20 288 kN (tension)

3.22 33.7 in.^{3}

3.24 103 kPa

3.26 1020 psfg

3.28 $p_A = +0.12$ psi

3.30 $p_B = -1.00$ kPa gage

3.32 27.5 psfg

3.34 $p_A = 5.72$ psig $= 39.5$ kPa gage

3.36 Water-mercury interface 0.13 ft above horizontal leg. Air-water interface 1.13 ft above horizontal leg. Air-mercury interface 0.203 ft above horizontal leg. $p_{max} = 172$ psfg

3.40 $p_A - p_B = 4.17$ kPa,

$h_A - h_B = -0.50$ m

3.42 $p_A - p_B = 108\ \text{lbf/ft}^2$,

$h_A - h_B = 3.32$ ft

3.44 $p_B - p_A = -1.57$ kPa,

$p_{zB} - p_{zA} = -0.589$ kPa

3.46 (c)

3.50 $z = 2.91$ km, $T = -1.21°\text{C}$

3.52 $T = 287\ \text{K} = 516°\text{R}$,

$p = 86.0\ \text{kPa} = 12.2$ psia,

$\rho = 1.04\ \text{kg/m}^3 = 0.00199\ \text{slugs/ft}^3$

3.54 $p = 1.32$ kPa, $\rho = 0.0197\ \text{kg/m}^3$

3.56 $F_{net} = 11{,}500$ lbf, $M = 0$

3.58 a, b, and e are valid statements

3.60 F (hydrostatic) $= 6075$ lbf/ft,

F (bottom tie) $= 8100$ lbf (tension)

3.62 $R_A = 759{,}000$ N at an angle of 30° below horizontal

3.64 $h = \ell/3$

3.66 Gate will stay in position

3.68 $F = 5\gamma W h^2/3\sqrt{3}$, $R_T/F = 3/10$

3.70 $P = 33{,}200$ lbf

372. Unstable

3.74 $V\!\!\!\!- = 3.04\ \text{ft}^3$

3.76 Wall A-A'

3.78 $F_V = 17.5$ kN, $F_H = 14.7$ kN,

$F_R = 22.9$ kN

3.80 $F_H = 375$ lbf/ft (to the right),

503.4 lbf/ft (downward),

$y_{cp} = 2.8$ ft (above the water surface)

3.82 Horizontal force $F_H = 61.1$ kN (to left). Line of action 0.125 m below the dome center of curvature. Vertical force $F_V =$ 20.6 kN (downward).

3.84 $V\!\!\!\!- = 0.0306\ \text{m}^3$,

$\gamma_{block} = 22.9\ \text{kN/m}^3$

3.86 $V\!\!\!\!- = 0.0315\ \text{m}^3$

3.88 Δlevel $= 0.0144$ in.

3.90 $h = 7.80$ m

3.92 $L = 2.24$ m

3.94 $\Delta h = 0.0255$ ft When ice melts, water surface level unchanged.

3.96 Pole will fall

3.98 $F = 3010$ N of scrap

3.100 $h = 0.456$ m

3.102 S = 0.938

3.104 S = 1.114 to 1.39

3.106 Stable

3.108 Block is unstable with the axis vertical.

3.110 Unstable

Chapter 4

4.2 (b) and (d), (a) and (b)

4.10 Unsteady flow

4.18 $\mathbf{a} = 28\mathbf{i} + 58\mathbf{j}\ \text{m/s}^2$

4.20 $a_{tot} = 14.14\ \text{m/s}^2$

4.22 $a_\ell = 5\ \text{ft/s}^2$, $\mathbf{a}_\ell = 0$

4.24 $a_\ell = 4q_0/(Bt_0)$
4.26 $a_\ell = 3.56\ \text{ft/s}^2$, $\mathbf{a}_c = 37.9\ \text{ft/s}^2$
4.28 $\partial p/\partial z = -59.9\ \text{lbf/ft}^3$
4.30 $p_2 = 187.2$ psfg
4.32 $\partial p/\partial z = -6000\ \text{N/m}^3$
4.34 $a_z = -141.3\ \text{ft/s}^2$
4.36 $dp/dx = -5335$ psf/ft
4.38 $p_B - p_A = -137.6$ psf,
$p_C - p_A = 47.9$ psf
4.40 $d_{max} = 4.758$ ft
4.42 (b)
4.44 $h = 6.21$ ft
4.46 $\ell = 0$
4.48 $h = 45.9$ cm
4.50 $V = 92.7$ fps
4.52 $V_o = 1.63$ m/s
4.54 $\theta = 41.8°$, $V_0 = (2\Delta p/\rho)^{1/2}$
4.56 (b)
4.58 $V = 70.7$ m/s
4.60 $V = 264$ ft/s
4.62 $p_B - p_C = 18$ kPa
4.64 $V_0 = 13.02$ m/s
4.66 $V_0 = 11.17$ m/s
4.68 $C_p = -0.687$,
$p_{gage} = -4.122$ kPa gage
4.70 $V_{true} = 69$ m/s
4.74 Irrotational
4.76 Irrotational
4.78 $\Delta\theta = 4y/(1-4y^2)$
4.80 $p = -12.56$ psig
4.82 $p_B - p_A = 48.14$ kPa
4.84 S = 3.68
4.86 $z_2 = 12.17$ cm
4.88 $a_n = 5g$
4.90 $\omega = (7.5g/\ell)^{1/2}$
4.92 $F = 23.2$ N
4.94 $z_1 = 13.5$ cm, $z_r = 21.5$ cm;
$r_1 = 19.15$ cm, $z_r = 61.15$ cm
4.96 $\omega = 17.7$ rad/s
4.98 $\partial p/\partial z = -34.8$ kPa/m (−1 m),
−9.810 kPa/m (0 m),
15.19 kPa/m (+1 m)
4.100 $z_{min} = 0.322$ ft above axis,
$\Delta p_{max} = 523$ psf
4.102 $p_A = 4.78$ psig
4.104 $\Delta p = 3.48$ kPa
4.106 $p_A - p_B = 263.2$ psf
4.110 $V_{10} = 100$ mph,
$a_c = 0.409\ \text{ft/s}^2$
4.114 Move toward center

Chapter 5

5.2 $Q = 4.19$ cfs, 1880 gpm
5.4 $\dot{m} = 0.239$ kg/s
5.6 $D = 1.25$ m
5.8 $\overline{V}/V_0 = 1/3$
5.10 $Q = 163.4$ cfs, 73,370 gpm
5.12 $Q = 5\ \text{m}^3/\text{s}$, $V = 5$ m/s,
$\dot{m} = 6.0$ kg/s
5.14 $Q = 0.93\ \text{m}^3/\text{s}$
5.16 $Q = 2.27 \times 10^{-3}\ \text{m}^3/\text{s}$
5.18 $q = 3.09\ \text{m}^2/\text{s}$, $V = 2.57$ m/s
5.20 $V = V_c/(1+n)$
5.22 $V = 1.74$ ft/s
5.24 $Q = 0.127$ cfs, 57 gpm
5.26 $a_c = V^2D/2h^2$
5.28
$V_{exit} = 2.8$ m/s, $a_{exit} = 3.6\ \text{m/s}^2$
5.30 $a_{2s} = -1258\ \text{m/s}^2$,
$a_{3s} = -2840\ \text{m/s}^2$
5.32 $t = 12.8$ s
5.34 (b)
5.36 All statements true
5.38 $y = 12\sqrt{2}d$
5.40 $V_R = 2/3$ ft/s
5.42 $V_1 = 14.4$ m/s, $V_2 = 36.0$ m/s
5.44 $V_{15} = 5.66$ m/s, $V_{20} = 6.37$ m/s
5.46 $V_B = 5$ m/s
5.48 $Q_B = +1.33$ cfm, leaving tank
5.50 Rising, $dh/dt = 1/8$ ft/s
5.52 $Q_p = 7.5$ cfs
5.54 $\dot{m} = 7.18$ slugs/s, $V_c = 20.4$ ft/s
S = 0.925
5.56 $\Delta t = 60$ s
5.58 $t = 5$ hr 50 min
5.60 $t = 55.5$ s
5.62 $\Delta t_{10\ \text{kPa}} = 5.48$ min,
$\Delta t_{0\ \text{kPa}} = 10.65$ min
5.64 $t = 43.5$ s
5.68 $\rho_e = 0.073\ \text{kg/m}^3$
5.70 $d\rho/dt = 250\ \text{kg/m}^3\text{s}$
5.72 $p_E = -40$ psf
5.74 $Q_f = 0.41\ \ell$pm, conc = 3.8%
5.76 $\Delta h = 0.25$ ft
5.78 $V_{e,max} = 3.62$ m/s,
$Q_{max} = 0.00362\ \text{m}^3/\text{s}$,
$L_{max} = 8850$ N
5.80 $F = 3.17$ lbf
5.82 $p_{r_0} = 540$ Pa gage,
$p_{1.1r_0} = 507$ Pa gage,
$p_{2r_0} = 127$ Pa gage
5.84 $V_0 = 21.65$ ft/s
5.86 $V_0 = 14.37$ m/s
5.88 $V_0 = 42.2$ ft/s
5.90 $V_0 = 9.54$ m/s
5.92 Cont. satisfied, rotational flow

Chapter 6

6.2 $F = 0.31$ N, $v = 57.6$ m/s
6.4 $\mu = 0.22$
6.6 $v_1 = 51.5$ ft/s
6.8 $F_1 = 182$ N, $F_2 = 169$ N
6.10 $\dot{m} = 200$ kg/s, $d = 7.14$ cm
6.12 $p_1 = 312$ kPa gage,
$F = 2.26$ kN (to left)
6.14 $p_{air} = 8.25$ atm
6.16 $T = 788$ lbf
6.18 $\mathbf{F} = (-571\mathbf{i} - 148\mathbf{j})$ lbf
6.20 $\mathbf{F} = (2.08\mathbf{i} + 7.76\mathbf{j})$ lbf

6.22 $v = 11.7$ m/s
6.24 $d_2 = (b_1/2)(1 - \cos\theta)$,
$d_3 = (b_1/2)(1 + \cos\theta)$
6.26 $T = 8.75$ kN (to left)
6.30 $F = -1840$ lbf (to left)
6.32 (d)
6.34 $\mathbf{F} = (-12.0\mathbf{i} + 0\mathbf{j} + 1.48\mathbf{k})$ kN
6.36 $F_x = -363$ kN
6.38 $\mathbf{F} = (-18.6\mathbf{i} - 6.12\mathbf{j} + 4.91\mathbf{k})$ kN
6.40 $\mathbf{F} = (-2400\mathbf{i} - 994\mathbf{j})$ lbf
6.42 $F_x = 1140$ N
6.44 $\Delta p = 125$ kPa
6.46 $F_x = 26.8$ lbf (to left)
6.48 $\mathbf{F} = (-2.62\mathbf{i} - 2.87\mathbf{j})$ kN
6.50 $F = -168$ lbf
6.52 $F_x = -3762$ lbf
6.54 $T = 1413$ N
6.56 $p = 13.3$ kPa gage, $F = 1.38$ kN/m
6.58 $\mathbf{F} = (-524\mathbf{i} - 58.9\mathbf{j})$ lbf
6.60 $\mathbf{F} = (36.8\mathbf{i} - 119\mathbf{j})$ N
6.64 $F_\tau = A(p_1 - p_2 - \rho U^2/3)$
6.66 $T = 472$ N
6.70 $\dot{m} = 2.205$ slugs/s, $v_{max} = 180$ ft/s,
$D = 109.3$ lbf
6.72 $\mathbf{F} = (816\mathbf{i} - 338\mathbf{j})$ N
6.74 $F_R = 29.8$ lbf
6.76 $P = 36.3$ hp/ft
6.78 $P = 32.3$ hp
6.80 $F_r = -60$ N (to left)
6.82 $F_y = 2\rho V^2 tr$
6.84 $T = 124$ kN
6.86 $v_{max} = 15.2$ m/s
6.88 $F = 1.02$ MN
6.90 $\Delta p = 6.97$ MPa
6.92 $L = 1.11$ km
6.94 $F = 44.4$ kN
6.98 $\mathbf{R} = (+465\mathbf{j} - 1530\mathbf{k})$ N,
$\mathbf{M} = (16.3\mathbf{j} - 413\mathbf{k})$ N · m
6.100 $\mathbf{R} = (+12.1\mathbf{i} - 3.1\mathbf{j})$ kN,
$\mathbf{M} = (-2.54\mathbf{k})$ kN · m

Chapter 7

7.2 $\dot{W}_s = 489$ kW
7.4 $\dot{W}_s = -330$ kW
7.6 $\bar{V} = (2/3)V_{max}$, $\alpha = 1.097$
7.8 (a) $\alpha = 1.0$, (b) $\alpha > 1.0$,
(c) $\alpha > 1.0$, (d) $\alpha > 1.0$
7.10 $\alpha = 27/20$
7.12 $\alpha = (1/4)[(n+2)(n+1)^3]/[(3n+2)(3n+1)]$,
$\alpha = 1.078$
7.14 $\alpha = 1.187$
7.16 $p_1 = 9.23$ psi
7.18 $p_A = -19.6$ kPa, $V_e = 14$ m/s
7.20 $Q = 5.40 \times 10^{-3}$ m^3/s
7.22 $p_2 = 8.23$ psig
7.24 $p_1 = 152.1$ Pa
7.26 $p_B = -2.38$ psig
7.28 $p_B = -45.3$ kPa-gage,
$Q = 0.266$ m^3/s
7.30 (b)
7.32 $p_2/\gamma = 13.16$ m
7.34 $h = 63.5$ m
7.36 $\dot{W} = 1.76$ MW
7.38 $\dot{W} = 41.2$ hp
7.40 $\dot{W} = 134.6$ kW
7.42 $\dot{W} = 14.89$ kW
7.44 $\dot{W} = 273$ hp
7.46 $t = 11.6$ min
7.48 $F_x = 11.0$ kN, $h_L = 0.996$ m
7.50 $h_L = 5.66$ ft
7.52 $F_{wall} = 197.5$ lbf
7.54 $h_L = 2.52$ ft
7.56 $Q = 0.302$ m^3/s
7.58 (a) right to left, (b) pump, (c) pipe *CA* smaller, (e) no vacuum
7.64 Possible if acceleration to left
7.68 $\dot{W} = 49.0$ hp
7.70 $z_L = 128.6$ ft
7.72 $Q = 0.320$ m^3/s, $p_m = -78.7$ kPa
7.76 $\dot{W} = 3.40$ kW
7.78 $F_d = (\pi D^2 \rho U^2/8)(D^2/d^2 - 1)^{-2}$,
$F_d = 0.372$ N

Chapter 8

8.2 (a) ML^2/T^2, (b) M/LT^2, (c) L/T, (d) dimensionless
8.4 $V^4\gamma/(g^2\sigma) = f(h^2\gamma/\sigma)$
8.6 $F_D/\mu\rho d = C$
8.8 $F/\rho V^2 D^2 = f(\rho VD/\mu, k_s/D, \omega D/V)$
8.10 $(\Delta p/\Delta\ell)(D^2/\mu V) = C$
8.12 $fR\sqrt{\rho/p} = f(k)$
8.14 $V = C\sqrt{\sigma/\rho\ell}$
8.16 $h/d = f(\sigma t^2/\rho d^3, \gamma t^2/\rho d, \mu t/\rho d^2)$
8.18 $eV/E = f(\sigma/E, Ed^2/V\dot{M}_p, d/D, Br)$
8.20 Viscous forces small
8.22 $V/\sqrt{gD} = f(\mu_l^2/\rho_l^2 D^3 g, \sigma/\rho_l D^2 g)$
8.24 Re and Fr
8.26 $F_D/\rho V^2 S = f(\omega^2 S/V^2)$
8.28 $V_m = 21.4$ m/s,
$F_{D,m}/F_{D,p} = 0.5$
8.30 $V_5 = 1.5$ m/s
8.32 $nd/V = f(\rho Vd/\mu)$
8.34 $F_p = 3.07$ lbf, 13.7 N
8.36 $\rho_m = 6.38$ kg/m^3
8.38 $V_m = 1.92$ ft/s
8.40 $V_{air} = 107.6$ m/s
8.42 $V_m = 4.5$ m/s
8.44 $V_a = 11.6$ m/s, $\Delta p_w = 7.33$ kPa
8.46 $F_p = 417$ lbf
8.48 $M_p = 141$ N · m,
$V_p = 0.179$ m/s
8.50 $p_m = 0.808$ MPa abs
8.52 Not rarefied

8.54 $D = 140\,\mu\text{m}$

8.56 $v_m/v_p = (L_m/L_p)^{3/2}$

8.58 $Q_p = 312.5\ \text{m}^3/\text{s}$, $V_p = 12.5\ \text{m/s}$

8.60 $V_m/V_p = 1/6$, $Q_m/Q_p = 1/7776$, $Q_m = 0.386\ \text{m}^3/\text{s}$

8.62 $V_p = 43.1\ \text{ft/s}$, $Q_p = 17{,}400\ \text{ft}^3/\text{s}$

8.64 $Q_p = 312.5\ \text{m}^3/\text{s}$, $t_p = 5$ min

8.66 $F_p = 3.83$ MN

8.68 $L_m/L_p = 0.0318$

8.70 $V_p = 25\ \text{ft/s}$, $F_p = 31{,}250$ lbf

8.72 (d)

Chapter 9

9.2 $\mu = 5.70 \times 10^{-3}\ \text{lbf-s/ft}^2$

9.6 (a) $u = 150y$ m/s, $v = 0$,
(b) rotational flow,
(c) continuity satisfied,
(d) $F_s = 180$ N

9.8 Greater shear stress on wire

9.10 $T = 222$ ft-lbf

9.12 $T = (2/3)\pi r_o^3 \mu\omega/\theta$

9.14 $T = 9.16 \times 10^{-4}\ \text{N} \cdot \text{m}$

9.16 $P = 0.0543$ W

9.18 $u_{max} = 9.18$ mm/s, $\overline{V} = 6.54$ mm/s

9.20 $d = 0.012$ in., $V = 0.137$ ft/s

9.22 $u_{max} = 0.150$ ft/s

9.24 $q = 4.65 \times 10^{-5}\ \text{m}^2/\text{s}$

9.26 $u_{max} = 8.31$ mm/s (upward)

9.28 $dp/ds = -464$ psf/ft

9.30 $u = \gamma y^2/2\mu + (U/L + \gamma L/2\mu)y$, $U = (\gamma/6\mu)L^2$

9.32 $Q = 6.62 \times 10^{-8}\ \text{m}^3/\text{s}$

9.34 $u = U(y/L)$

9.36 $F_s = 8.25$ N

9.38 $\delta/x = 0.0071$

9.40 (a)

9.42 $\delta = 4.46/\text{Re}_x^{1/2}$

9.44 $F_x = 5.15$ N

9.46 $u = U_o = 1$ m/s

9.48 $F_s = 1.385$ N, $F_{s,\text{ laminar}}/F_{s,\text{ total}} = 0.451$

9.50 $u = 0.171$ m/s

9.52 $F_s = 3.53$ N, $du/dy = 9.3 \times 10^4\ \text{s}^{-1}$

9.54 $U = 190$ ft/s, $F = 0.20$ lbf

9.58 $\delta^* = 1.5$ mm

9.60 $F_{s,30}/F_{s,10} = 2.59$

9.62 $T = 76.6$ N

9.64 $\delta' = 26.4 \times 10^{-6}$ m, $k_s > \delta'$

9.66 $F_s = 0.506$ N

9.70 $a = -0.410\ \text{m/s}^2$, $F_{s,\text{headwind}} = 0.455$ N, $F_{s,\text{tailwind}} = 0.232$ N, $x = 98.0$ m

9.72 $F_{s,100} = 1{,}353$ N, $P_{100} = 37.6$ kW, $F_{s,100} = 5{,}006$ N, $P_{100} = 278$ kW

9.74 $F_s = 68{,}967$ N

9.76 $F_s = 1.85$ MN, $P = 17.1$ MW, $\delta = 2.25$ m

9.78 (a) $V_m = 7.94$ m/s,
(b) $F_{sm} = 28.21$ N and $F_{wm} = 9.79$ N
(c) $F_p = 453{,}000$ N

9.80 $P = 1.62$ hp

Chapter 10

10.2 Upward, $V = 1.04$ m/s

10.4 $\Delta p = 293$ psi/100 ft

10.6 Parabolic, 0.75

10.8 $V = 0.81$ ft/s, $Q = 2.76 \times 10^{-4}$ cfs

10.10 $P = 1341$ W

10.12 (a) $V_{max} = 10$ cm/s,
(b) $f = 0.16$,
(c) $u_* = 0.00707$ m/s,
(d) $\tau = 0.020\ \text{N/m}^2$

10.14 (d)

10.16 $V_2 = 0.0409$ m/s

10.18 $D = 4.43 \times 10^{-4}$ m

10.22 $\nu = 8.91 \times 10^{-5}\ \text{m}^2/\text{s}$

10.24 Downward, $f = 0.076$, laminar, $\mu = 1.54 \times 10^{-3}\ \text{lbf-s/ft}^2$

10.26 Doubled

10.28 (a)

10.30 $\Delta p = 29.1$ kPa, $h_f = 2.97$ m, $p = 177$ W

10.32 $f = 0.015$

10.34 $\Delta h = 5.50$ m

10.36 $\nu = 2.0 \times 10^{-8}\ \text{m}^2/\text{s}$

10.38 $f = 0.0258$

10.40 $f = 0.0185$

10.42 $t = 29.7$ s

10.44 $P = 40.96$ hp

10.46 (a), (b), and (d)

10.48 (a) $\tau_o = 23.2\ \text{N/m}^2$,
(b) $\tau_1 = 20.0\ \text{N/m}^2$,
(c) $u_1 = 2.68$ m/s

10.50 $p_2 = 769$ kPa

10.52 (a) $V_2 = 1.81$ m/s,
(b) $h = 16.7$ cm

10.54 $t = 46.5$ min

10.56 $f = 0.038$

10.58 $p_A = 673$ kPa

10.60 $B \longrightarrow A$, $Q = 8.60$ cfs

10.62 $Q = 2.79\ \text{m}^3/\text{s}$, $Q_{rs} = 2.47\ \text{m}^3/\text{s}$, $P = 5.52$ MW

10.64 $P = 30.1$ kW

10.66 $Q = 20.4$ cfs

10.68 Use 90 in. pipe

10.70 26 in. steel pipe

10.72 $Q = 4.83$ cfs

10.74 (a) $P = 726$ W,
(b) $P = 3.03$ kW

10.76 $P = 7.49$ kW
10.78 $Q = 17.8$ cfs, $p_{min} = -3.03$ psig, $p_{max} = 12.1$ psig
10.80 $K_v = 18.5$
10.82 $Q = 3.11$ cfs
10.84 $\Delta p = 0.6$ in H_2O, $p = 38.9$ W
10.86 $P = 132$ kW
10.88 Will not act as pump
10.90 $Q = 4700$ gpm
10.92 $\Delta z = 12.8$ m
10.94 $Q = 10.8$ ft^3/s
10.96 $P = 386$ kW
10.100 $z_1 = 114$ m
10.106 $V_A/V_B = 1.26$
10.108 $Q_1 = 2$ cfs
10.110 $Q_{large}/Q_{small} = 3.86$
10.112 $Q_{12} = 6.45$ cfs, $Q_{14} = 7.75$ cfs, $Q_{16} = 10.8$ cfs, $h_{L,AB} = 107$ ft
10.114 $Q_v = 0.456$ m^3/s, $Q_p = 0.656$ m^3/s
10.116 $\Delta p_f = 1.77$ psf
10.118 $P_{loss} = 40.4$ kW
10.120 (b)
10.122 $Q = 347$ m^3/s
10.124 $Q = 243$ cfs
10.126 $Q = 10.6$ ft^3/s
10.128 $V = 5.74$ ft/s, $Q = 758$ cfs
10.130 $Q = 546$ cfs
10.132 $d = 4.92$ ft

Chapter 11

11.2 (d)
11.4 $C_D = 2.0$
11.6 $E = 77.2$ kJ = 18.4 Food calories
11.8 $M = 21.8$ kN · m
11.10 $F_D = 6.24 \times 10^6$ lbf
11.12 $V_{jet} = 9.45$ m/s
11.14 $F_D = 2250$ lbf
11.16 $F_{normal}/F_{edge} = 197$
11.18 $F_D = 18.6$ kN
11.20 $P = 9.47$ kW
11.22 Energy savings 14.7%
11.24 $P = 47.2$ kW
11.28 $V_{bicycle} = 20.0$ m/s
11.30 V(roof open) = 40.0 m/s, V(roof closed) = 44.3 m/s
11.32 $V = 117$ mph
11.34 $P = 12.1$ hp
11.38 $V_0 = 0.082$ ft/s (upward)
11.40 $V_0 = 1.55$ mm/s
11.44 $D = 0.0024$ in.
11.46 $V = 0.86$ m/s
11.48 $V_0 = 5.70$ m/s
11.50 $V_0 = 1.32$ m/s
11.52 $t = 51.3$ min
11.54 $T = 142$ N
11.56 $V = 2.28$ m/s (downward)
11.58 $D = 0.571$ m
11.60 $F_L = 0.265$ lbf
11.62 $F_L = 0.121$ lbf, $\delta = 1.43$ ft
11.64 $V_{max} = 17.7$ m/s, $V_{min} = 5.9$ m/s
11.66 $C_L = 1.30$ (based on planform area), $C_L = 2.60$ (based on projected area)
11.68 Dimensions: 1.14 ft wide by 4.56 ft long
11.70 (b)
11.74 $D = 458$ N, $P = 27.5$ kW
11.76 (b)
11.78 $t = 39.3$ min
11.80 $\alpha = 7°$, $F_D = 30$ lbf

Chapter 12

12.2 $c = 1055$ m/s
12.4 $\Delta c = 656$ m/s
12.6 $c = \sqrt{E_v/\rho}$, $c = 1483$ m/s
12.8 $T_t = 491$, $K = 218°C$
12.10 $V = 200$ m/s = 719 km/hr
12.12 $p_t = 285$ kPa, $T_t = 51°C$
12.14 $V = 346$ m/s, $p = 177$ kPa, $T = 407$ K
12.16 $T = 291$ K, $M = 0.192$, $p = 487$ kPa, $\dot{m} = 0.032$ kg/s
12.18 $C_p = (2/kM^2) \times \left\{[1 + (k-1)M^2/2]^{k/(k-1)} - 1\right\}$, $C_p(2) = 2.43$, $C_p(4) = 13.47$, $C_{p,\,inc} = 1.0$
12.20 $M_2 = 0.657$, $p_2 = 208$ kPa, $T_2 = 43°C$, $\Delta s = 35.6$ J/kg-K
12.22 M = 1.59
12.24 $V_1 = 1200$ m/s
12.28 $\dot{m} = 0.212$ kg/s
12.30 $p_t = 342$ kPa
12.32 $p_b = 87.2$ kPa
12.36 $A_T = 29.5$ cm^2
12.38 Underexpanded
12.40 $M = \sqrt{2}$, $A/A_* = 1.123$
12.42 (a) $A/A_* = 3.60$, $T = 2624$ N, (b) $T = 2627$ N
12.44 $A/A_* = 3.25$, distance from throat = 5.99 cm
12.46 $M_3 = 0.336$, $p_3 = 461$ kPa, $p_t = 499$ kPa
12.50 $L = 207$ ft, $p_* = 2.94$ psia
12.52 $\bar{f} = 0.0298$
12.54 $\dot{m} = 0.248$ kg/s
12.56 $p_1 = 105$ kPa, $V_1 = 47.6$ m/s, $\rho_1 = 1.27$ kg/m^3
12.58 $\Delta p = 45.5$ kPa

Chapter 13

13.2 $V \geq 0.06$ m/s

13.4 $\Delta h = 8.88$ mm

13.6 $V_0 = 18.1$ m/s

13.8 $Q = 0.267\ \text{m}^3/\text{s} = 9.42$ cfs $= 4{,}230$ gpm

13.10 $Q = 152\ \text{ft}^3/\text{s} = 9120$ cfm, $V_{max}/V_{min} = 1.24$, $\dot{m} = 11.1$ lbm/s

13.12 $r_m/D = 0.189$, $r_c/D = 0.351$, $\dot{m} = 16.4$ kg/s

13.14 $V = 0.623$ m/s

13.16 $C_c = 0.751$

13.20 $Q = 3.49$ cfs

13.22 $Q = 0.345$ cfs

13.24 $\Delta p_C = 225$ kPa, $\Delta p_F = 222$ kPa, $\Delta h\ (\text{both}) = \frac{1}{1.82}$ m

13.26 $Q = 0.0240\ \text{m}^3/\text{s}$

13.28 $\Delta p = 1610$ psf, $P = 32.0$ hp

13.30 $Q = 0.0226\ \text{m}^3/\text{s}$

13.32 $d = 0.754$ m

13.34 $\Delta h = 0.44$ m

13.36 $Q = 1.36$ cfs

13.38 $Q = 11.3\ \text{m}^3/\text{s}$

13.40 $Q = 6.08$ cfs

13.42 $Q = 0.00124\ \text{m}^3/\text{s}$

13.44 $h_L = 64V_0^2/2g$

13.46 $Q/Q_{std} = (\rho_{std}/\rho)^{1/2}$

13.48 (a) $V = (L/\Delta t)[\sqrt{1 + (c\Delta t/L)^2} - 1]$, (b) $V = c^2\Delta t/2L$, (c) $V = 22.5$ m/s

13.50 $Q = 0.0575\ \text{m}^3/\text{s}$

13.52 (b)

13.54 (c)

13.56 $H = 0.53$ ft, $Q = 2.54\ \text{ft}^3/\text{s}$

13.60 $H = 16.5$ in.

13.62 $H_R/H_T = 0.456$

13.64 $P = 1.00$ m

13.66 $H = 0.476$ ft

13.68 $M = 0.710$, $V = 235$ m/s

13.70 $M_1 = 1.79$, $V_1 = 521$ m/s

13.72 $\dot{m} = 0.00792$ kg/s

13.74 $\dot{m} = 6.35 \times 10^{-4}$ kg/s

13.80 $V = 3.96$ m/s, $U_V = 0.198$ m/s

13.82 $Q = 62.7$ cfs, $U_Q = 16.0$ cfs

Chapter 14

14.2 $F_T = 926$ N, $P = 35.7$ kW

14.4 $N = 1158$ rpm

14.6 $D = 1.69$ m, $V_o = 87.7$ m/s

14.8 $N = 1170$ rpm

14.10 $a = 0.782\ \text{m/s}^2$

14.12 $Q = 1.66\ \text{m}^3/\text{s}$, $P = 211$ kW

14.14 $Q = 0.32\ \text{m}^3/\text{s}$, $P = 13.5$ kW

14.16 $Q = 3.60\ \text{m}^3/\text{s}$, $\Delta h = 38.7$ m, $P = 1709$ MW

14.20 $P = 2.70$ MW

14.22 $Q = 5.50$ cfs

14.24 $H_{30} = 73.8$ m

14.26 $Q = 0.0833\ \text{m}^3/\text{s}$, $\Delta h = 146$ m, $P = 104.0$ kW

14.30 Mixed-flow pump

14.32 Radial-flow pump

14.34 Axial-flow pump

14.36 $D = 2.066$ m, $P = 27.6$ kW

14.38 $Q = 0.172\ \text{m}^3/\text{s}$

14.40 $P = 10.55$ MW, $D = 2.85$ m

14.42 $P = (1/2)\rho Q V_j^3/2$

14.44 $\alpha_1 = 8°25'$, $T = 44{,}671$ N · m, $P = 280.7$ kW

14.46 $P = 271$ hp

14.48 $W = 3.45$ m

Chapter 15

15.2 Supercritical (Fr = 4.04)

15.4 Fr(0.3) = 7.77, Fr(1.0) = 1.27, Fr(2.0) = 0.452, $y_c = 1.18$ m

15.6 $y_c = 2.55$ m

15.8 Subcritical

15.10 $y = 5.38$ m, $y_2 = 2.33$ m

15.12 $Q = 187$ cfs

15.14 $Q = 50.5$ cfs

15.16 $z_{water} = 101.4$ m

15.18 $\Delta y(+30) = -0.51$ m, w.s.(+30) = −0.21 m, $\Delta y(-30) = 0.40$ m, w.s. (−30) = +0.10 m, max $z_{step} = 0.43$ m

15.20 $\Delta z = 0.89$ m

15.22 Ship squat = 0.30 m

15.24 $Q = 35.5\ \text{m}^3/\text{s}$

15.26 $y = 0.23$ m

15.28 $h_L = 2.30$ ft, $P_{diss} = 4.70$ hp, $F_x = -51.2$ lbf

15.30 Yes, $y_2 = 2.82$ m

15.32 $y_1 = 1.08$ ft

15.34 $y_2 = 2.45$ m

15.36 $y_{c,1} = 2.69$ ft, $y_{n,1} = 1.86$ ft, reach 1, supercritical; reach 2, subcritical; reach 3, critical. $y_2 = 3.73$ m, between 1 and 2.

15.38 (d)

15.42 $F_s = 10{,}790$ N, $F_s/F_H = 0.044$

15.44 (c)

15.48 Yes, probably submerged against sluice gate

15.52 Upstream H3, downstream H2

15.54 $Q = 19.5\ \text{m}^3/\text{s}$

Index

C

D

E

J

K

L

M

R

S

T

U

V

W

Y

COMMONLY USED EQUIVALENT UNITS IN FLUID MECHANICS

Volume

Unit	Equivalent: cubic inches	liters	U.S. gallons	cubic feet	cubic yards	cubic meters
cubic inch	1	0.016 39	0.004 329	578.7×10^{-6}	21.43×10^{-6}	16.39×10^{-6}
liter	61.02	1	0.264 2	0.035 31	0.001 308	0.001
U.S. gallon	231.0	3.785	1	0.133 7	0.004 951	0.003 785
cubic foot	1728	28.32	7.481	1	0.037 04	0.028 32
cubic yard	46,660	764.6	202.0	27	1	0.764 6
meter^3	61,020	1000	264.2	35.31	1.308	1

Discharge (Flow Rate, Volume/Time)

Unit	gallons/minute	liters/second	feet^3/second	meters^3/second
gallons/minute	1	0.063 69	0.002 228	63.09×10^{-6}
liters/second	15.85	1	0.035 31	0.001
acre-feet/day	226.3	14.28	0.504 2	0.014 28
feet^3/second	448.8	28.32	1	0.028 32

Velocity

Unit	feet/day	kilometers/hour	feet/second	miles/hour	meters/second
feet/day	1	12.70×10^{-6}	11.57×10^{-6}	7.891×10^{-6}	3.528×10^{-6}
kilometers/hour	78,740	1	0.911 3	0.621 4	0.277 8
feet/second	86,400	1.097	1	0.681 8	0.304 8
miles/hour	126,700	1.609	1.467	1	0.447 0
meters/second	283,500	3.600	3.281	2.237	1

SOURCE: SI System of Units: Pamphlet prepared for the Universities Council on Water Resources by Peter C. Klingeman, 1976.